							17 7A	18 0
			13 3A	14 4A	15 5A	16 6A	1 **H** 1.0079	2 **He** 4.00260
			5 **B** 10.81	6 **C** 12.011	7 **N** 14.0067	8 **O** 15.9994	9 **F** 18.9984	10 **Ne** 20.179
	10	11 1B	12 2B	13 **Al** 26.9815	14 **Si** 28.0855	15 **P** 30.9738	16 **S** 32.06	17 **Cl** 35.453	18 **Ar** 39.948

28 **Ni** 58.69	29 **Cu** 63.546	30 **Zn** 65.39	31 **Ga** 69.72	32 **Ge** 72.59	33 **As** 74.9216	34 **Se** 78.96	35 **Br** 79.904	36 **Kr** 83.80
46 **Pd** 106.42	47 **Ag** 107.868	48 **Cd** 112.41	49 **In** 114.82	50 **Sn** 118.71	51 **Sb** 121.75	52 **Te** 127.60	53 **I** 126.905	54 **Xe** 131.29
78 **Pt** 195.08	79 **Au** 196.967	80 **Hg** 200.59	81 **Tl** 204.383	82 **Pb** 207.2	83 **Bi** 208.980	84 **Po** (209)	85 **At** (210)	86 **Rn** (222)
110 **Uun** (271)	111 **Uuu** (272)							

63 **Eu** 151.96	64 **Gd** 157.25	65 **Tb** 158.925	66 **Dy** 162.50	67 **Ho** 164.930	68 **Er** 167.26	69 **Tm** 168.934	70 **Yb** 173.04	71 **Lu** 174.967
95 **Am** (243)	96 **Cm** (247)	97 **Bk** (247)	98 **Cf** (251)	99 **Es** (254)	100 **Fm** (257)	101 **Md** (260)	102 **No** (259)	103 **Lr** (262)

See inside back cover for the names of the elements and their atomic masses.

Note: Atomic masses shown are the 1983 IUPAC values (maximum of six significant figures). Group numbers 1–18 in blue are the recommended notations.

Introduction to
Organic and
Biological Chemistry

Introduction to Organic and Biological Chemistry

Michael S. Matta
Antony C. Wilbraham
Dennis D. Staley

Southern Illinois University, Edwardsville

D. C. Heath and Company
Lexington, Massachusetts   Toronto

Address editorial correspondence to:
D. C. Heath and Company
125 Spring Street
Lexington, MA 02173

Acquisitions: Richard Stratton
Development: June Goldstein
Editorial Production: Ron Hampton
Design: Henry J. Rachlin
Photo Research: Toni Michaels
Art Editing: Penny Peters
Production Coordination: Charles Dutton

To Our Families
M. S. M.
A. C. W.
D. D. S.

Published simultaneously in Canada.

Printed in the United States of America.

International Standard Book Number: 0-669-39922-1

10 9 8 7 6 5 4 3 2 1

Preface

Introduction to Organic and Biological Chemistry is intended for students who have completed a one- or two-term course in the principles of chemistry and are completing a curriculum that requires an additional one- or two-term course in organic and biological chemistry. We assume a general knowledge of principles of atomic and molecular structure, chemical bonding, solutions, acid-base theory, and the gas laws. This text enables students to acquire a sound background in the principles and concepts of organic and biological chemistry. It also provides students with an understanding of how biological systems function at the molecular level. The scope and depth of coverage are appropriate for students pursuing majors in a wide variety of areas, including the allied health sciences, biology, nutrition, home economics, agriculture, forestry, and education.

Applications

Every chapter describes applications of chemical principles to the life sciences. These applications help sustain student interest and demonstrate that understanding chemistry is an absolute prerequisite to understanding how living organisms work. The connection between chemistry and the life sciences is highlighted in a chapter-opening feature called *Case in Point*. The *Case in Point* aims to provide students with an incentive to learn the chapter material and gain an understanding of the reasons that the chemistry covered in the chapter is important in the health-related sciences. Each *Case in Point* presents a vignette related to the chapter's content. The chemical background needed to understand the chemistry of the *Case in Point* is provided in the chapter, then applied in a *Follow-up to the Case in Point*. Additional health-related applications and other subjects of interest are highlighted in 46 brief essays entitled *A Closer Look*. Other health-related applications are embedded in the text.

Coherent Presentation of Principles

We have made no attempt to provide a comprehensive treatment of either organic or biological chemistry. Instead, we emphasize the principles of organic chemistry that are necessary for the understanding of biological molecules and systems. The student will learn that reactions in biochemical processes reflect the general reactions of organic chemistry. In industry and education, the fields of organic and biological chemistry have overlapped in recent years; the overlap is reflected in textbooks, and this book is no exception.

The material in Chapter 1, Chemical Bonding: Fundamental Concepts, can be presented in a lecture format or assigned as outside reading for students needing a brief review. The organic chemistry section, beginning with Chapter 2, describes structures of hydrocarbons; explains the reactions of important functional groups in simple molecules; then describes the complex, naturally occurring molecules that straddle organic and biological chemistry. The role played by weak forces in determining the physical properties of organic molecules is studied in some depth. Understanding this role gives insight into the biologically active structures of proteins and nucleic acids, as well as the organization of cell membranes.

The book is designed so that concepts particularly relevant to biochemistry recur at strategic places. Chapter 4, for example, provides a general introduction to oxidation reaction in organic chemistry that is reexamined and further developed in a biological context in Chapter 5. A section introducing esters and anhydrides of phosphoric acid in Chapter 13 helps demystify discussions of nucleic acid structure in Chapter 11, the biological importance of ATP in Chapter 14, and the energy yields of metabolic processes in subsequent chapters. Similarly, the importance of buffers in maintaining the body's acid-base balance is stressed in Chapter 13, and treated further in discussions of lactic acidosis in Chapter 15 and diabetic ketosis in Chapter 16.

Because many courses include large numbers of students of allied health sciences, Chapters 12 and 13 emphasize human nutrition, digestion, and body fluids. Instructors of courses with a different orientation may wish to proceed from Chapter 11 directly to a discussion of reducing power (Chapter 14) and then to metabolism (Chapter 15), reserving nutrition, digestion, and body fluids for the end if time permits.

Problem Solving

Many sections also contain worked *Examples,* that show detailed solutions to problems. The text features approximately 50 worked examples. The chapters also contain more than 150 *Practice Exercises,* which give students on-the-spot feedback so that they can see whether they have grasped an important concept, learned an important skill, or should review the material. The answers to all *Practice Exercises* are given at the end of the book. Approximately 1040 *Exercises, Additional Exercises,* and *Self-Test Questions* at the end of the chapters provide more practice in mastering the material. The *Exercises* are grouped and keyed to the sections of the text, again an opportunity for students to identify problem areas. The first exercise in each of these sections is similar to one presented as an *Example* or *Practice Exercise* in the chapter. The *Additional Exercises* that follow the *Exercises* take a global view of the chapter, incorporating material from more than one section. The answers to selected *Exercises* and *Additional Exercises* are provided at the end of the book. Finally, each chapter contains a *Self-Test* that students can use to review personal progress in meeting learning goals.

Study Aids

A *Chapter Outline* at the beginning of each chapter itemizes the topics to be covered in the chapter. Major topics in each chapter are introduced by *numbered section headings*. Each section contains one or more learning objectives labeled as *Aims*. A *Focus* statement in the margin summarizes the contents of the section. These statements as a group constitute both a good chapter outline and an excellent checklist for review. Throughout the text, *Key Terms* are printed in bold type and are defined where they first appear. Key terms are grouped at the end of the chapter and are included (with definitions) in a *Glossary* at the back of the text, for convenient reference. Each chapter also features a *Summary,* which briefly integrates the chapter's major ideas. In addition, the text contains margin notes that include interesting applications of environmental chemistry, biological chemistry, and health-related chemistry, as well as problem-solving tips supplementary to the many found in the main text.

Supplementary Materials

Supplementary materials for *Introduction to Organic and Biological Chemistry* include the following:

Study Guide   By Danny V. White of American River College and Joanne White. The chapters in the *Study Guide* provide additional explanations of important topics, abundant exercises for more practice in learning new concepts and skills, and self-tests.

Student Solutions Manual   By Michael S. Matta, Antony C. Wilbraham, and Dennis D. Staley. The *Student Solutions Manual* provides detailed solutions to all *Practice Exercises,* end-of-chapter odd-numbered *Exercises* and *Additional Exercises,* and annotated answers to the *Self-Tests.*

Experiments for Introduction to Organic and Biological Chemistry   By Michael S. Matta, Antony C. Wilbraham, and Dennis D. Staley. The laboratory manual provides sufficient organic and biological chemistry experiments that instructors can choose the most relevant for their course. The order of experimental topics follows the topics in the text. Each experiment has an introduction, hypotheses, objectives, a list of required materials and equipment, detailed descriptions of procedures, prelaboratory exercises, and laboratory report sheets with postlaboratory exercises. Suggestions for demonstrations and structural studies are also included.

Complete Solutions Manual   By Michael S. Matta, Antony C. Wilbraham, and Dennis D. Staley. The *Complete Solutions Manual* supplies detailed solutions to all *Practice Exercises,* all end-of-chapter *Exercises* and *Additional Exercises,* as well as annotated answers to the *Self-Tests.*

Test Item File   By Michael S. Matta, Antony C. Wilbraham, and Dennis D. Staley. The *Test Item File* includes multiple-choice questions arranged by chapter.

Computerized Test Item File   The *Test Item File* is available for IBM or Macintosh computers.

Instructor's Guide   By Michael S. Matta, Antony C. Wilbraham, and Dennis D. Staley. The *Instructor's Guide* contains suggested lecture schedules for one-semester and two-quarter courses. Sequences of laboratory experiments keyed to the text are also provided, as well as references to audiovisual and multimedia materials.

Instructor's Guide to Accompany Experiments for Introduction to Organic and Biological Chemistry. By Michael S. Matta, Antony C. Wilbraham, and Dennis D. Staley. The *Instructor's Laboratory Guide* provides procedures for the preparation of laboratory materials and solutions and answers to the prelaboratory and postlaboratory questions. Material disposal guidelines and time estimates for completing each experiment are also included.

Transparencies   A set of full-color transparencies for selected illustrations and tables in the text is available for classroom use.

Acknowledgments

We are grateful for the enthusiasm and unflagging energy of the staff at D. C. Heath and Company. We thank June Goldstein, Development Editor, for the unstinting gift of her editorial talents and especially for her kindness and patience toward sometimes irascible authors; Ron Hampton, Production Editor, for his careful supervision of this project; Henry Rachlin, Senior Designer, for his functional and attractive design; and Richard Stratton, Acquisitions Editor, for his support and faith in the project.

Others have made significant contributions to this book. Dick Morel made many suggestions that have been incorporated into the manuscript, figures, and photographs. Copy for several of the supplements was expertly prepared by John Matta of Moon Over Maine Graphics. Elizabeth and Leigh Anne Staley helped with many details of manuscript preparation.

We'd also like to thank the accuracy reviewers of the manuscript, galleys, pages, and supplements: Beverly Foote, Leland Harris, University of Arizona; Ruiess Van Fossen Bravo, Indiana University of Pennsylvania; John Goodenow, Lawrence Technological University; Robert Howell, University of Cincinnati; Catherine Keenan, Chaffey College; Thomas J. Nycz, Broward Community College; Roger Penn, Sinclair Community College; David B. Shaw, Madison Area Technical College; and Peggy Zitek.

Finally, we want to acknowledge the fine work of the reviewers of the manuscript: Beatrice Arnowich, Queensborough Community College; John Barbas, Valdosta State College; Kenneth F. Cerny, University of Massachusetts; Jack L. Dalton, Boise State University; John E. Davidson, Eastern Kentucky University; David V. Frank, Ferris State University; Leland Harris, University of Arizona; Robert G. Howell, University of Cincinnati– Raymond Walters College; David Hunter, Kaskaskia College; Philip M. Jaffe, Oakton Community College; Catherine A. Keenan, Chaffey College; Frank R. Milio, Towson State University; Kenneth E. Miller, Milwaukee Area Technical College; Frazier W. Nyasulu, University of Washington; Thomas J. Nycz, Broward Community College; Richard Peterson, Memphis State University; Edith M. Rand, East Carolina University; Theresa A. Salerno,

Mankato State University; David Saltzman, Santa Fe Community College; James Schooler, Jr., North Carolina Central University; David B. Shaw, Madison Area Technical College; Danny V. White, American River College; Karen Wiechelman, University of Southwestern Louisiana; and James E. Wiedman, Kaskaskia College.

M.S.M.
A.C.W.
D.D.S.

Chapter Opener

The connection between chemistry and the health sciences is highlighted in a chapter-opening feature called Case in Point. The chemical principles needed to understand the chemistry of the Case in Point are presented in the chapter, then applied to a Follow-up to the Case in Point.

CHAPTER OUTLINE

The maintenance of a consistent internal environment in the human body is an interactive and dynamic process. Organs and organ systems working together continuously adjust imbalances that occur. A breakdown in any part of the collaborating systems of the body can cause serious illness or even death. What role does chemistry play in these systems? Consider Larry's plight in the following Case in Point.

CASE IN POINT: A diabetic imbalance

 Larry, a 32-year-old teacher, collapses on the steps of the public library. The paramedics discover that he wears a medical alert tag identifying him as a diabetic. Upon his arrival at the hospital, the attending physician immediately orders several lab tests. One of these tests is measurement of the pH of Larry's blood. What imbalance does Larry's doctor suspect? And what does the chemical concept of pH have to do with Larry's illness? We will find out in Section 13.8.

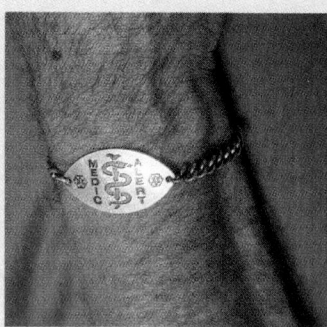

In case of an emergency, many people display their medical condition on a Medic Alert tag.

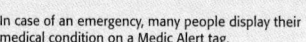

FOLLOW-UP TO THE CASE IN POINT: A diabetic imbalance

Metabolic acidosis is often a serious problem in uncontrolled diabetes, and it also occurs on a temporary basis after heavy exercise. Both conditions result in large influxes of protons from active tissues into the bloodstream. Larry, the teacher in the Case in Point earlier in this chapter, failed to keep his schedule of insulin injections. The acid his body produced as a result of a deficiency of insulin caused the pH of Larry's blood to drop to life-threatening levels. Larry's doctor suspected that diabetes-induced metabolic acidosis might be one of Larry's immediate problems. When the acidosis was confirmed by lab tests (see figure), Larry was given an intravenous drip of a sodium bicarbonate solution to restore his blood pH to normal. The sodium bicarbonate neutralized the excess acids in his blood, and his blood was literally titrated back to normal. With a proper schedule of insulin injections, Larry was able to leave the hospital in a few days.

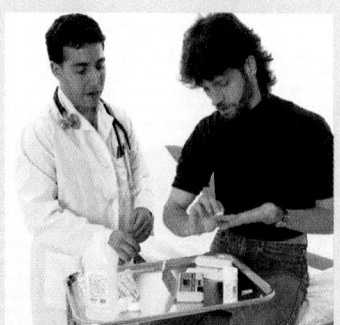

Testing for diabetes.

Closer Look Essays

Health-related and other interesting applications are explored in essays entitled A Closer Look. These special-interest boxes cover topics such as biomedical implants, cisplatin, food irradiation, the greenhouse effect, lactose intolerance, octane ratings of gasoline, DNA fingerprinting, and drug strategies for reducing serum cholesterol.

A Closer Look

Artificial Skin

Many burn victims lose large areas of skin. These people often die because vital fluids ooze out of their bodies, and they are too weakened to fight off bacterial infection. The essential requirement in treating burns is covering the damaged area quickly. When there is too little unburned skin on the patient's body for grafting, surgeons use skin from human cadavers or pigs. These grafts are usually rejected by the body after a few days and must be replaced.

An artificial skin that shows encouraging results was first used in the 1980s. The body does not recognize the artificial skin as foreign, and drugs that prevent rejection are not required. The wounds heal with little scarring.

Like natural skin, artificial skin has two layers (see figure). The bottom layer is a blend of a complex carbohydrate obtained from shark cartilage and collagen (a protein) extracted from cowhide. These components are added to water and acidified to produce short white fibers. The mixture is poured into a shallow pan and freeze-dried to remove the water. The fibers form a thin white sheet of material that is light and highly porous. The sheet is then baked in an oven to preserve its shape. A top layer is added. This layer, made from a thin sheet of a rubber-like plastic, is bonded with an adhesive to the fibrous sheet. The completed sheet of synthetic skin is soft and pliable as natural skin and about as thick as a paper towel. It is freeze-dried again and stored in a sterile container at room temperature until required. Ten square feet of material, enough to cover wounds over 50% of an adult body, can be made in a few days.

Artificial skin reacts with human flesh as does natural skin. The fibrous bottom layer, which is placed next to the burned area, is porous and provides scaffolding into which body cells migrate to make more collagen. Over a period of months, collagen in the artificial skin breaks down in the same way as collagen in healthy skin and sloughs off. Nerve fibers, which are still alive, and blood vessels in the flesh grow up into the new material. The upper plastic layer does not become part of the body but serves only as a protective flexible covering. Within a week or so after grafting, small areas of this layer are removed and replaced with thin patches of the patient's own skin. The need to cover the artificial graft with natural skin, although inconvenient, is not a serious drawback because it can be done later when the patient's condition is improved.

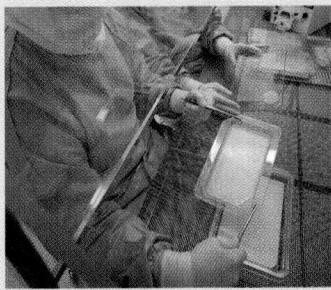
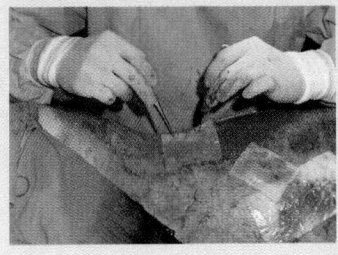

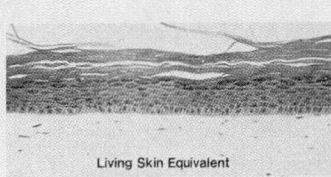
Living Skin Equivalent

A skin graft using artificial skin can be a lifesaving procedure.

(page 390)

A Closer Look

Dietary Fiber

Have you had your fiber today? Many experts recommend that people eat high-fiber diets in order to prevent colon cancer. The nondigestible carbohydrates that we eat constitute dietary fibers. Dietary fibers consist of polysaccharides that cannot be hydrolyzed to monosaccharides and therefore cannot be absorbed into the bloodstream. Cellulose is insoluble fiber. Cellulose from vegetable leaves and stalks provides you with dietary bulk and helps to prevent constipation. There are also soluble fibers. The soluble dietary fibers are noncellulosic. Pectins from fruits, which are used to thicken jams and jellies, are examples of soluble fibers. The pectins are polyhydroxy compounds that have a carboxylic acid group at one end of the molecule and an aldehyde group at the other. Vegetable "gums" are also soluble fibers.

Fruits and vegetables are high in dietary fiber.

Which foods can we eat to ensure that we get enough dietary fiber? Wheat, brown rice, and bran cereals are high in insoluble fiber. Oats, barley, carrots, and fruits are high in soluble fiber. Peas and beans are a good source of both soluble and insoluble fiber.

(page 222)

Worked Examples

The text contains approximately 150 worked examples, each titled for easy reference. The examples include detailed solutions.

Practice Exercises

Nearly 300 practice exercises within the chapter provide on-the-spot feedback so students can check their understanding of an important concept or skill. Answers are provided at the back of the book.

2.4 Straight-chain alkanes

AIM: To name and recognize structural, condensed, and molecular formulas of the straight-chain hydrocarbons containing up to 10 carbon atoms.

Focus

Straight-chain alkanes can be constructed using more than two carbon atoms.

We can build alkanes with more than two carbons by the same method used to build ethane. *Alkanes with more than two carbons strung together like the links of a chain are called* **straight-chain alkanes.** There is no need to resort to electron dot structures for each chain. We can just write the symbol for carbon as many times as we need in order to get the chain length and then fill in with hydrogens and lines representing covalent bonds. Remember that each carbon has four covalent bonds.

EXAMPLE 2.1   Writing structural formulas for straight-chain alkanes

Draw complete structural formulas for the straight-chain alkanes that have three and four carbons.

SOLUTION

Draw three carbons and four carbons in a row, and then put in lines to show the carbon-carbon bonds.

$$C—C—C \qquad C—C—C—C$$

Next, add enough covalent bonds to give each carbon a total of four bonds.

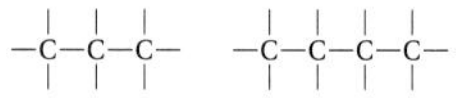

$$
\begin{array}{ccc}
\vert & \vert & \vert \\
-C-C-C- \\
\vert & \vert & \vert
\end{array}
\qquad
\begin{array}{cccc}
\vert & \vert & \vert & \vert \\
-C-C-C-C- \\
\vert & \vert & \vert & \vert
\end{array}
$$

Finally, complete the structures by adding hydrogens to fill out the carbon-hydrogen bonds.

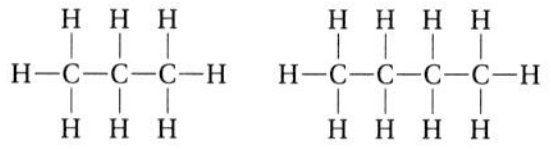

$$
\begin{array}{ccc}
H & H & H \\
\vert & \vert & \vert \\
H-C-C-C-H \\
\vert & \vert & \vert \\
H & H & H
\end{array}
\qquad
\begin{array}{cccc}
H & H & H & H \\
\vert & \vert & \vert & \vert \\
H-C-C-C-C-H \\
\vert & \vert & \vert & \vert \\
H & H & H & H
\end{array}
$$

PRACTICE EXERCISE 2.1
Draw complete structural formulas for the straight-chain alkanes with five and six carbons.

Drawing structural formulas

Sometimes we will find it convenient to draw complete structural formulas—that is, to show *all* the atoms and bonds in a molecule. There are, however, many ways to draw shorthand or *condensed structural formulas*.

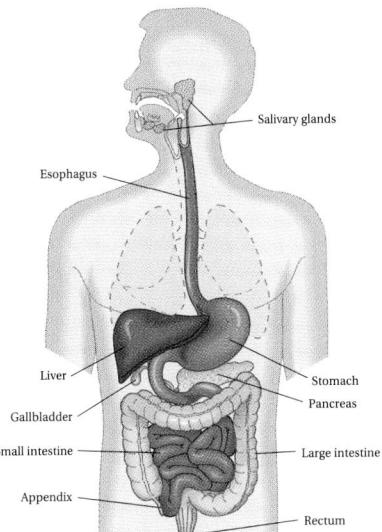

Figure 12.1
Organs of the human digestive system.

glycogen to the disaccharide maltose. No digestion of fats, proteins, or nucleic acids takes place in the mouth.

12.3 The stomach

AIM: To name the enzyme that hydrolyzes protein molecules in the stomach and characterize the environment in which it works.

Focus

Protein digestion begins in the stomach.

The mass of food mixed with saliva is swallowed and enters the stomach. The environment of the stomach is quite different from that of the mouth. The fluid of the stomach is very acidic owing to the presence of **gastric juice,** *a fluid secreted by cells that line the stomach.* The gastric juice is about 0.1 *M* hydrochloric acid. (A 0.1 *M* solution of a strong acid has a pH of 1.)

Most enzymes have a pH optimum in solutions near pH 7. Pepsin, however, the protease of the stomach, is an exception; it is most active at the pH of gastric juice. Pepsin, formed from pepsinogen by autoactivation or by the action of hydrochloric acid, begins to catalyze the hydrolysis of large pro-

Aim Statements
Aim statements accompany each numbered section and provide learning objectives for the student.

Focus Statements
Focus statements summarize the content of each numbered section.

Illustrations and Photographs

The illustrations and photographs provide visual representations to help students understand chemistry and make it more inviting to them.

Alpha helix

In some proteins, regions of the backbone of the peptide chain are coiled into a spiral shape called an **alpha helix,** *similar to a corkscrew.* As Figure 9.3 shows, a corkscrew must be turned in a right-handed, or clockwise, direction to penetrate a cork. The alpha helixes of proteins are always right-handed. The helixes are held together by hydrogen bonds, shown in Figure 9.4, formed between the hydrogen of an N—H of a peptide bond and the carbonyl oxygen of another peptide bond group four residues away in the same peptide chain.

The tightness of coiling is such that 3.6 amino acid residues of the peptide backbone make each full turn of the alpha helix. There are no amino acid residue side chains inside the alpha helix; they are located on the outside. The cyclic amino acid proline does not fit well into the peptide backbone of alpha helixes. Alpha helixes in long protein chains often end at a place where proline residues occur in the primary structure. Proline is

Figure 9.3
A corkscrew must be turned in a right-handed, or clockwise, direction to penetrate a cork.

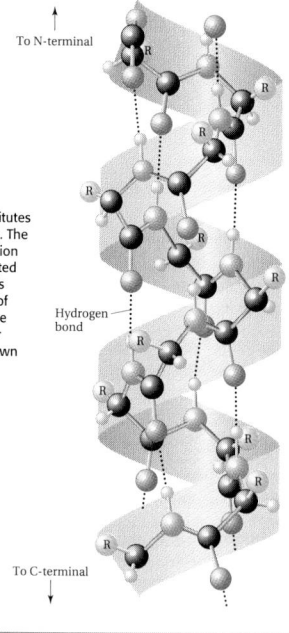

Figure 9.4
A peptide chain twisted into a right-handed alpha helix constitutes a protein's secondary structure. The N-terminal to C-terminal direction is from top to bottom. The dotted lines show the hydrogen bonds between the carbonyl oxygen of one amino acid residue and the N—H hydrogen of another, four amino acid residues further down the chain.

To N-terminal

Hydrogen bond

To C-terminal

(a)   **(b)**   **(c)**

Figure 2.13
Shown is (a) fossil tree-fern imprint in coal, (b) an exposed coal seam, and (c) a sample of coal tar.

bon, is also used as a fuel in many industrial processes. Ammonia is often converted to ammonium sulfate for use as a fertilizer. Coal tar (Fig. 2.13c) can be further distilled to yield many useful chemicals, including benzene, toluene, naphthalene, phenol, and pitch.

Over the past few years, growth in the world's demand for petroleum has outstripped the discovery of new deposits. Since petroleum is a nonrenewable resource, there eventually will be insufficient quantities to support the world's energy and other demands. Scientists and engineers are working to develop replacement sources of energy such as the sun, winds, tides, and geothermal springs.

Some of the clothes you wear and the music you play (CDs and tapes) are made from the same finite supply of hydrocarbons that you burn in your car and use to heat your home. Does this give you cause for concern?

Marginal Notes

Marginal notes provide interesting applications of environmental chemistry, biological chemistry, health-related chemistry, and problem-solving tips.

Chapter Summaries

Each chapter has a summary section to reinforce and integrate key concepts in the chapter.

Methyl ethyl ketone is used industrially to remove wax from lubricating oils when they are refined. It is also a common solvent in nail polish remover.

SUMMARY

Aldehydes and ketones contain the carbonyl group ($-C=O$). The carbonyl carbon in an aldehyde has at least one hydrogen attached (R—CHO), but the carbonyl carbon in a ketone has no hydrogens (R—CO—R). Formaldehyde (H_2CO) is the simplest aldehyde; acetone (CH_3COCH_3) is the simplest ketone. The physical and chemical properties of aldehydes and ketones are influenced by the very polar carbonyl group. Molecules of aldehydes and ketones can attract each other through polar-polar interactions. These compounds have higher boiling points than the corresponding alkanes but lower boiling points than the corresponding alcohols. Aldehydes and ketones can accept hydrogen bonds, and those with low molar mass are completely soluble in water.

Aldehydes and ketones are produced by the oxidation of primary alcohols and secondary alcohols, respectively. Aldehydes are usually more reactive than ketones and are good reducing agents. An aldehyde can be oxidized to the corresponding carboxylic acid, but ketones resist further oxidation. Addition reactions are characteristic of both aldehydes and ketones. Addition of water to the carbon-oxygen bond of the carbonyl group forms hydrates. Addition of an alcohol produces hemiacetals and hemiketals. The reaction of alcohols with hemiketals and hemiacetals produces acetals and ketals, respectively.

The most important industrial aldehydes and ketones are formaldehyde, acetaldehyde, acetone, and methyl ethyl ketone. A 40% aqueous solution of formaldehyde, called formalin, is commonly used to preserve biological specimens. Many aldehydes and ketones have fragrant aromas.

Summary of Reactions **127**

SUMMARY OF REACTIONS

Here are the reactions of aldehydes and ketones presented in this chapter.

Aldehydes

1. Preparation of aldehydes:

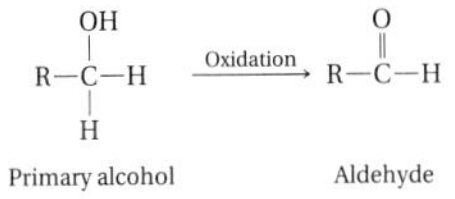

Primary alcohol → Aldehyde (Oxidation)

2. Oxidation of aldehydes:

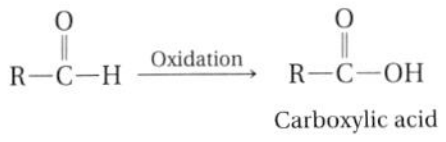

Aldehyde → Carboxylic acid (Oxidation)

3. Reaction of aldehydes with Tollens' reagent:

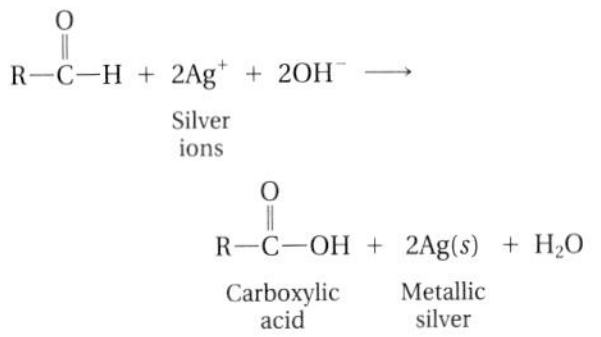

$R-C(=O)-H + 2Ag^+ + 2OH^- \longrightarrow R-C(=O)-OH + 2Ag(s) + H_2O$
(Silver ions) (Carboxylic acid + Metallic silver)

4. Reaction of aldehydes with Benedict's reagent:

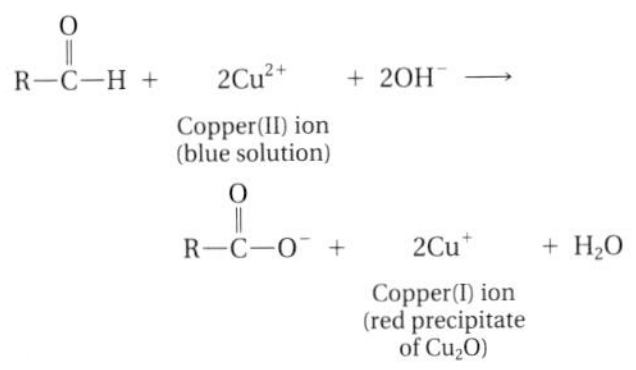

$R-C(=O)-H + 2Cu^{2+} + 2OH^- \longrightarrow R-C(=O)-O^- + 2Cu^+ + H_2O$
(Copper(II) ion, blue solution) (Copper(I) ion, red precipitate of Cu_2O)

5. Addition of water to aldehydes:

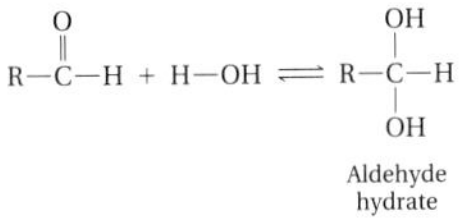

$R-C(=O)-H + H-OH \rightleftharpoons R-C(OH)(H)-OH$ (Aldehyde hydrate)

6. Addition of alcohol to an aldehyde followed by a second reaction with an alcohol:

Aldehyde + RO—H ⇌ Hemiacetal + RO—H → Acetal + H_2O

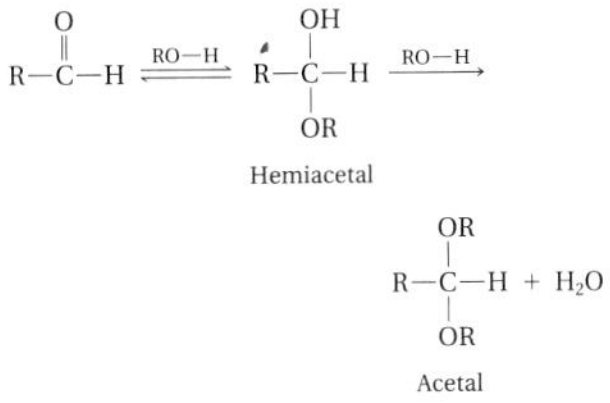

Ketones

1. Preparation of ketones:

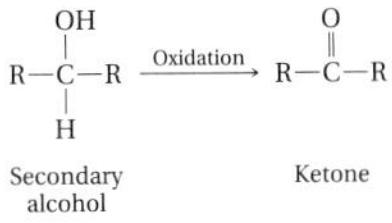

Secondary alcohol → Ketone (Oxidation)

2. Addition of alcohol to a ketone followed by a second reaction with an alcohol:

Ketone + RO—H ⇌ Hemiketal + RO—H → Ketal + H_2O

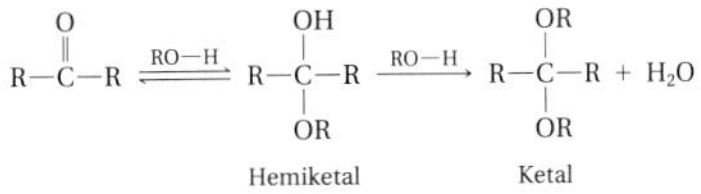

Key Terms

Key terms are printed in bold type and are defined where they first appear. They are also grouped at the end of the chapter and in a glossary at the back of the text, for convenient reference.

End-of-Chapter Exercises

The end-of-chapter exercises are keyed to chapter sections. Additional exercises are not keyed to any section or topic and incorporate material from more than one section. The odd-numbered exercises are answers in the back of the text.

---

4. Hydrolysis of amides. (The amide may be simple, monosubstituted, or disubstituted.)

$$R-\overset{\overset{O}{\|}}{C}-\overset{\cdot\cdot}{N}H_2 + HO-H \xrightarrow[\text{H}^+ \text{or OH}^-]{\text{Heat}} R-\overset{\overset{O}{\|}}{C}-OH + H-\overset{\cdot\cdot}{N}H_2$$

KEY TERMS

Alkaloid (6.8)	Anilide (6.4)	Free amine (6.2)	Polyamide (6.5)
Amide (6.4)	Antihistamine (6.7)	Hallucinogen (6.7)	Protonated amine (6.2)
Amine (6.1)	Arylammonium ion (6.2)	Hypnotic (6.9)	Quaternary ammonium
Alkylammonium ion (6.2)	Barbiturate (6.9)	Neurotransmitter (6.7)	salt (6.3)
Ammonium salt (6.2)	Decongestant (6.7)	Opiate (6.8)	Sedative (6.9)

EXERCISES

Amines (Sections 6.1, 6.2, 6.3)

6.15 Name or write a structural formula for the following amines. Classify them as primary, secondary, or tertiary amines.
(a) $(CH_3)_2NH$ (b) *p*-chloro-*N*-methylaniline
(c) (structure of aniline with NH_2 and Cl substituent on benzene ring)
(d) diethylmethylamine

6.16 Write structural formulas and name the following amines. Classify each amine as primary, secondary, or tertiary.
(a) diethylamine   (b) $(CH_3CH_2)_2NCH_3$
(c) butylamine   (d) (structure: CH_3N with H bonded to benzene ring)

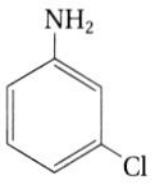

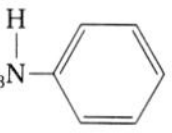

6.17 What is the meaning of the term *heterocyclic amine*?

6.18 Draw the structure and give the name of (a) an aromatic heterocyclic amine and (b) an aliphatic heterocyclic amine.

6.19 Draw structural formulas for (a) pyrimidine and (b) purine. Derivatives of these two compounds are found in what biologically important molecules?

6.20 Draw the structure of pyrrole. List some of the naturally occurring molecules that contain the pyrrole ring system.

6.21 Why are amines weak bases?

6.22 Draw the general formulas for (a) an unprotonated (free) amine and (b) a protonated amine.

6.23 Write the structure and name the expected products of each of the following reactions.
(a) $CH_3CH_2NH_2 + HCl$ (b) $(CH_3)_2NH + HNO_3$
(c) $CH_3NH_2 + H_2SO_4$ (d) $(CH_3)_3CNH_2 + HCl$

6.24 Draw the structure and name the organic product for each of the following reactions.
(a) $CH_3NH_3{}^+I^- + NaOH$
(b) $(CH_3)_2NH + CH_3Cl$
(c) $CH_3I + NH_3$
(d) (structure of benzene ring with $NH_3{}^+Cl^-$) $+ NaOH$ 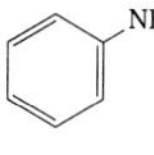
(e) $(CH_3CH_2)_3N + CH_3CH_2Cl$

6.25 Write an equation for the dissociation of the dimethylammonium ion. Why does an aqueous solution of dimethylammonium chloride test acidic?

6.26 Draw the structure of tetramethylammonium iodide. What happens if this compound is treated with sodium hydroxide?

Amides (Sections 6.4, 6.5, 6.6)

6.27 Name or write structural formulas for the following amides.
(a) $CH_3\overset{\overset{O}{\|}}{C}NH_2$ (b) $CH_3CH_2\overset{\overset{O}{\|}}{C}NHCH_3$
(c) (benzene ring with $\overset{\overset{O}{\|}}{C}NHCH_3$) 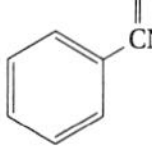
(d) *N*-ethyl-*N*-methylpropanamide
(e) acetanilide

Self-Tests

Each chapter's review material ends with a Self-Test that students can use to gauge their progress in meeting learning goals.

SELF-TEST (REVIEW)

True/False

1. A dehydrogenation reaction is a reduction reaction.
2. Hydrogen bonding accounts for the relatively high boiling point of acetaldehyde.
3. The reaction of equal moles of an alcohol and an aldehyde gives an acetal.
4. Oxidation of a tertiary alcohol gives a ketone.
5. You would expect propanal to have a higher boiling point than propanol.
6. Propanal should have higher water solubility than hexanal.
7. One mole of methanol would release more energy upon complete oxidation than one mole of methane.
8. Both aldehydes and ketones have a carbonyl group.
9. All aldehydes and ketones give a positive Tollens' test.
10. Four pairs of electrons are shared in the carbonyl bond formed between an oxygen and a carbon atom.

Multiple Choice

11. Which of the following statements about the carbon-oxygen double bond of the carbonyl group is *false*?
    (a) The bond is polar.
    (b) The carbon has a partial negative charge.
    (c) The bonding electrons are unequally shared between the carbon and oxygen.
    (d) The oxygen has two unshared pairs of electrons.
12. Acetaldehyde would be likely to form hydrogen bonds with
    (a) formaldehyde.   (b) octane.   (c) water.
    (d) acetone.
13. On the basis of your knowledge of intermolecular forces, which of the following would you expect to have the highest boiling point?
    (a) propanal   (b) propane   (c) acetone
    (d) 1-propanol

14. Which of the following compounds contains a di-ether linkage?
    (a) a hemiacetal   (b) chloral hydrate
    (c) a ketal   (d) camphor
15. Which of the following substances is used as a preservative of biological specimens?
    (a) paraldehyde   (b) formalin
    (c) methyl ethyl ketone   (d) cinnamaldehyde
16. Which of the following substances can undergo an addition reaction with methanol?
    (a) propane   (b) methyl ethyl ether
    (c) propanol   (d) propanal
17. A structural isomer of 2-butanone is
    (a) diethyl ether.   (b) *tert*-butyl alcohol.
    (c) diethyl ketone.   (d) butanal.
18. Which of the following compounds would release the most energy upon oxidation to carbon dioxide?
    (a) ethanol   (b) acetic acid, CH_3COOH
    (c) ethane   (d) acetaldehyde
19. The oxidation of 2-methyl-2-butanol with $K_2Cr_2O_7$ and H_2SO_4 would give
    (a) 2-methyl-2-butanone.
    (b) isopropyl alcohol and ethane.
    (c) 2-methyl-2-butanal.
    (d) none of the above.
20. In a positive Tollens' test,
    (a) silver ions are oxidized to silver atoms.
    (b) the aldehyde is an oxidizing agent.
    (c) a silver mirror is formed.
    (d) more than one are correct.
21. In the reaction of substance A with substance B, substance A loses oxygen. Which of the following is true?
    (a) Substance B is an oxidizing agent.
    (b) Substance A is reduced.
    (c) Substance B is reduced.
    (d) Substance A is a reducing agent.

Brief Contents

Contents

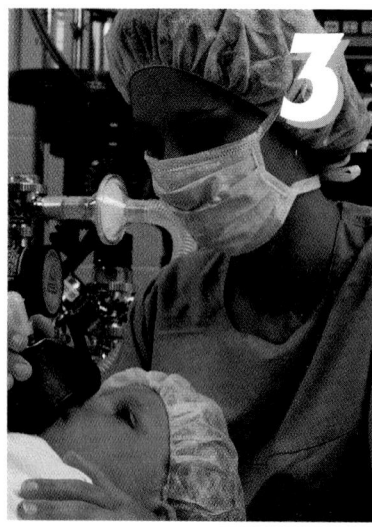

Acids and Their Derivatives: 132
Reactions of the Carboxyl Group

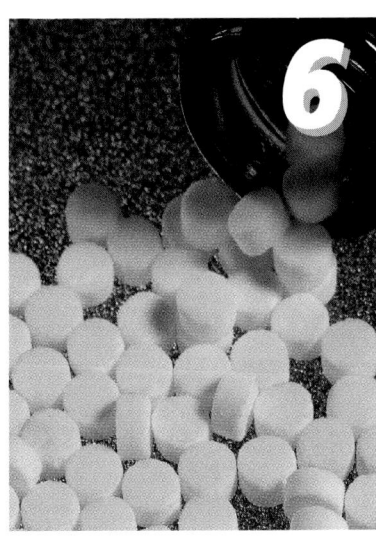

Amines and Amides: 171
Organic Nitrogen Compounds

Nucleic Acids: 319

The Molecular Basis of Heredity

Digestion and Nutrition: 355

Materials for Living

Body Fluids: 387

Maintaining the Body's Internal Environment

Energy and Life: 414

Sources and Uses of Energy in Living Organisms

Carbohydrates in Living Organisms: 441

At the Core of Metabolism

Case in Point: **Carbohydrate loading** 442

Lipid Metabolism: 474

Fat Chemistry in Cells

Case in Point: **Carnitine deficiency** 475

Metabolism of Nitrogen Compounds: 497

Nitrogen and Life

Case in Point: **A painful episode** 498

Health-Related Topics

Chemical Bonding

Fundamental Concepts

The precise shapes of crystals result from the organization of their atomic building blocks in regular repeating patterns. The photograph is a kidney stone magnified 240 times.

Scientists of 150 years ago believed the ability to produce carbon compounds rested exclusively with living organisms. The production of carbon compounds in living organisms—organic chemistry—was thought to be directed by a mysterious "vital force." Vitalism was rudely shattered in 1818 when the German chemist Friedrich Woehler (1800–1882) made urea, a carbon-containing compound found in urine, in the absence of any living agent. *Organic chemistry now includes the study of the structures, properties, and reactions of all carbon compounds, regardless of their origins.* Organic chemists soon discovered that carbon compounds could be synthesized—built from simpler materials—by ordinary chemical reactions. Since this discovery, millions of organic compounds have been synthesized or isolated from natural sources. Since most biological molecules contain carbon, organic chemistry is entwined with *biological chemistry* or *biochemistry, the chemistry conducted by living organisms.*

In order to prepare for organic chemistry, we will briefly review chemical bonding in this chapter. In subsequent chapters we will learn about the structure and reactivity of organic compounds and take a biochemical journey through the living cell.

Thousands of organic compounds are used for health-related applications. How have these compounds been identified? In the pharmaceutical industry, for example, how does a medical researcher know what compound may have a medicinal use? This is the subject of the Case in Point.

CASE IN POINT: A new era of medicinal drug discovery

Traditionally, promising compounds for use as medicinal drugs have been isolated from plants or microorganisms, or prepared by chemists in the laboratory, then screened for medical effectiveness by tests in animals and finally in humans. All of these procedures are costly and time-consuming. History shows that for every 10,000 substances tested, about 20 prove promising enough to enter animal trials, about 10 proceed to human trials, and only one gains approval as a new drug (see figure). Can this process be streamlined so that new pharmaceuticals for the treatment of deadly diseases, such as AIDS, can be brought to the marketplace more quickly and at a more reasonable cost? We will examine the possibilities in Section 1.4.

Taxol, a promising antitumor agent, is obtained from the bark of the Pacific yew tree. Taxol has also been synthesized in the laboratory.

1.1 Atomic structure and electron configuration

AIMS: To show how electrons are configured around the nuclei of atoms. To use the periodic table to find the number of valence electrons in an atom. To draw electron dot structures for representative elements.

Focus

Energies of electrons in atoms are confined to certain values.

After the discovery of electrons and atomic nuclei, scientists were eager to learn how these particles are put together to form atoms. An early model of the structure of the atom envisioned tiny electrons embedded in a large nucleus, rather like blueberries in a muffin. In 1913 Niels Bohr (1885–1962), a young Danish physicist, made a revolutionary proposal.

Energies of electrons

Bohr suggested that the electrons in atoms have certain fixed energies. Bohr's idea can be illustrated by a ladder. The rungs of a ladder are analogous to the permissible energies of electrons. The lowest rung corresponds to the lowest energy level (Fig. 1.1). Just as a person climbs up or down a ladder by going from rung to rung, an electron can jump from one energy level to another. But the regions between the energy levels are forbidden. A person on a ladder cannot stand between the rungs, and electrons in an atom cannot have an energy between the energy levels. To jump from one energy level to another, an electron must gain or lose just the right amount of energy. *A* **quantum of energy** *is the amount of energy required to move an electron from its present energy level to the next higher one.* The energies of electrons are said to be *quantized.* The term *quantum leap,* used to describe an abrupt change, comes from this concept. The amount of energy gained or lost by an electron as it jumps between energy levels is not the same. Unlike the evenly spaced rungs of a ladder, the energy levels in an atom are *not equally spaced.* The energy levels of electrons become closer as the electrons get farther from the nucleus. The outermost electrons are the easiest to move from one energy level to a higher level.

Location of electrons

Where are an atom's electrons in relation to the nucleus? Surprisingly, there is no exact answer to this question. Consider a thrown baseball, which has a fairly large mass and moves, at most, at 100 miles per hour. The position of the baseball at any instant can be determined with great accuracy. But an electron has a very low mass and moves much faster than a baseball. It is impossible to pinpoint the location of such an object at any instant. All anyone can do is describe the chances that an electron will be in a certain place at any instant. As a result of the uncertainty inherent in electron movements, the locations of electrons in atoms are generally described by *atomic orbitals. An* **atomic orbital** *is a region in space where there is a good chance of finding an electron.* An atomic orbital is usually portrayed as a blurry cloud that is most dense where there is a good chance of finding the electron, and less dense where the chance of finding the electron is slight.

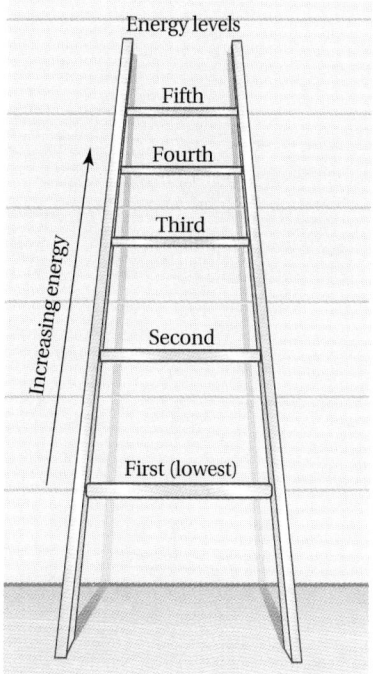

Energy levels

Fifth

Fourth

Third

Second

First (lowest)

Increasing energy

Figure 1.1
The energy levels in an atom are analogous to the rungs on a ladder. The higher the energy level occupied by the electron, the more energetic the electron.

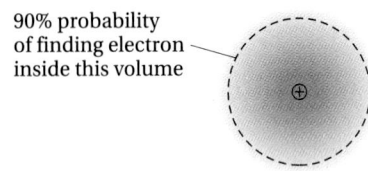

90% probability
of finding electron
inside this volume

Figure 1.2
An atomic orbital is shown as a blurry cloud of negative charge. The cloud is most dense where the probability of finding the electron is high and is least dense where the probability is slight.

We can imagine the negative charge spread out over the entire atomic orbital. Figure 1.2 illustrates a cloudy volume of the kind typically used to show an atomic orbital. There is at least a small chance of finding the electron a considerable distance from the nucleus. Atomic orbitals are usually drawn so that there is a 90 percent chance of finding the electron somewhere within its boundaries. Different shapes for atomic orbitals are denoted by letters: *s orbitals* are spherical, and *p orbitals* are dumbbell-shaped; the shapes of *d orbitals* and *f orbitals* are far more complex. Figure 1.3 shows the shapes of *s*, *p*, and *d* orbitals. The shapes of the *f* orbitals are not shown.

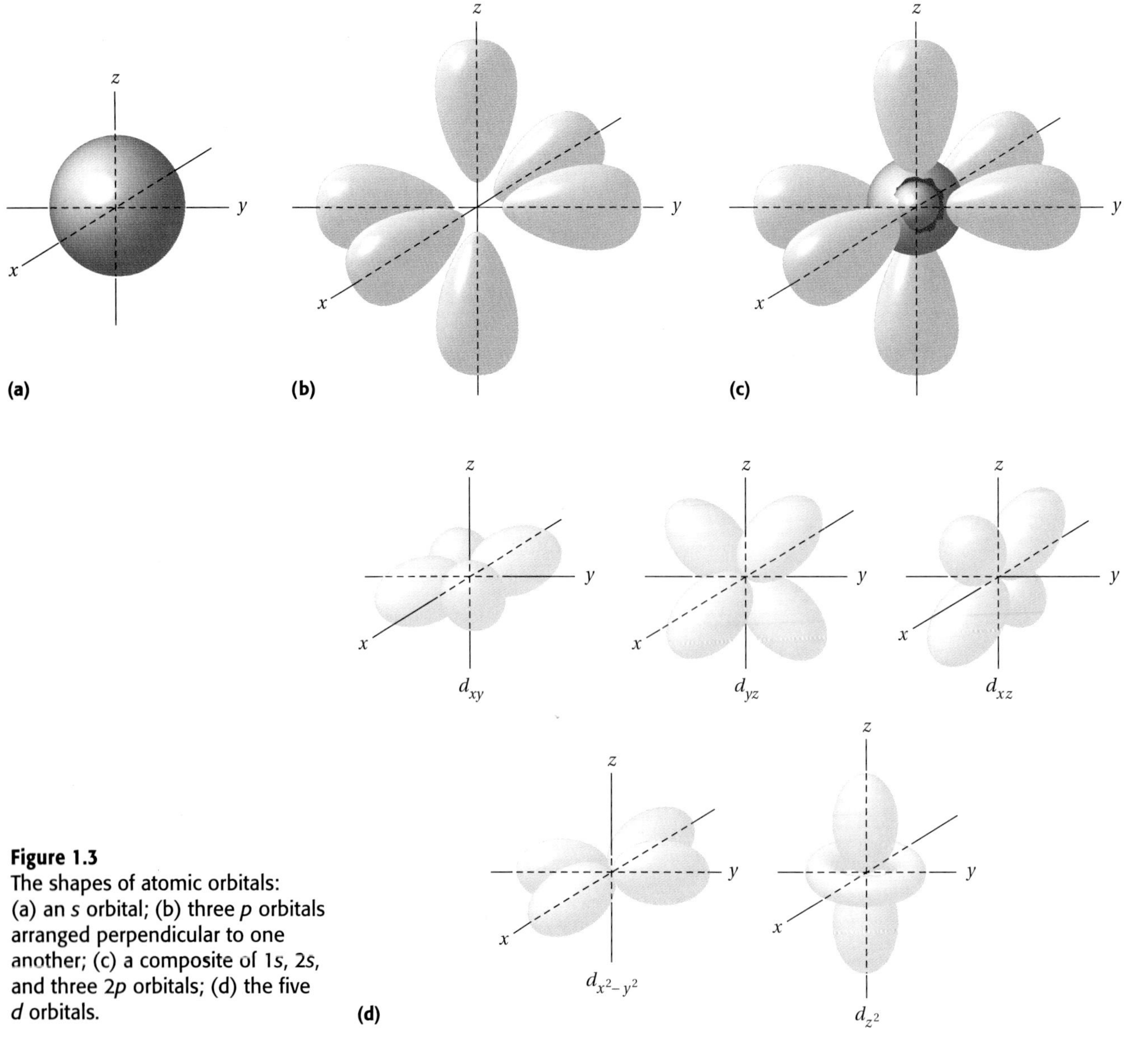

(a) **(b)** **(c)**

d_{xy} d_{yz} d_{xz}

$d_{x^2-y^2}$ d_{z^2}

Figure 1.3
The shapes of atomic orbitals: (a) an *s* orbital; (b) three *p* orbitals arranged perpendicular to one another; (c) a composite of 1*s*, 2*s*, and three 2*p* orbitals; (d) the five *d* orbitals.

(d)

Table 1.1 Summary of Principal Energy Levels, Sublevels, and Orbitals

Principal energy level	Number of sublevels	Type of sublevel
1	1	$1s$ (1 orbital)
2	2	$2s$ (1 orbital), $2p$ (3 orbitals)
3	3	$3s$ (1 orbital), $3p$ (3 orbitals), $3d$ (5 orbitals)
4	4	$4s$ (1 orbital), $4p$ (3 orbitals), $4d$ (5 orbitals), $4f$ (7 orbitals)

The energies of electrons in atoms and the atomic orbitals that they occupy are related. We can think of the energies of the electrons that occupy atomic orbitals in terms of *principal energy levels* and *sublevels*. The principle energy levels are designated in order of increasing energy as 1, 2, 3, and so forth. Within each *principal energy level,* electrons have slightly different energies corresponding to *energy sublevels.* The number of energy sublevels within a principal energy level is equal to the number of the principal level. The first principal energy level has only one sublevel, called $1s$. The second principal energy level has two sublevels, called $2s$ and $2p$. The $2p$ sublevel is of higher energy than the $2s$ and consists of three p orbitals of equal energy, with the long axis of each p orbital perpendicular to the other two. It is convenient to label the axes $2p_x$, $2p_y$, and $2p_z$. The second principal energy level has a total of four orbitals: $2s$, $2p_x$, $2p_y$, and $2p_z$. The third principal energy level has three sublevels, called $3s$, $3p$, and $3d$. The $3d$ sublevel consists of five d orbitals of equal energy. Thus, the third principal energy level has nine orbitals. The fourth principal energy level has four sublevels, called $4s$, $4p$, $4d$, and $4f$. The $4f$ sublevel consists of seven f orbitals of equal energy. The fourth principal energy level, then, has 16 orbitals. The sublevels and orbitals for each principal energy level are summarized in Table 1.1.

PRACTICE EXERCISE 1.1

How many electrons are in the second principal energy level of an atom of (a) oxygen and (b) phosphorus?

PRACTICE EXERCISE 1.2

How many orbitals are in each of the following?
(a) $3p$ sublevel   (b) $2s$ sublevel   (c) $4f$ sublevel
(d) $4p$ sublevel   (e) $3d$ sublevel   (f) third principal energy level

Electron configurations

It is the nature of things to seek the lowest possible energy. High-energy systems are unstable, and unstable systems undergo changes that lose energy and become more stable. In the world of the atom, electrons and the nucleus interact to make the most stable arrangement, or configuration, possible. *The arrangements of electrons around the nuclei of atoms are* called **electron configurations.**

Three rules govern the electron configurations of atoms: the *Aufbau principle*, the *Pauli exclusion principle*, and *Hund's rule*.

1. **The Aufbau principle:** *Electrons enter orbitals of lowest energy first.* We will use a box (□) to represent an atomic orbital.

2. **The Pauli exclusion principle:** *An atomic orbital may describe, at most, two electrons.* In order to occupy the same orbital, two electrons must have opposite spins; that is, the electron spins must be paired. (Spin is a quantum property of electrons and may be clockwise or counterclockwise.) We use a vertical arrow (↑) to indicate an electron and its direction of spin (↑ or ↓), and we write an orbital containing paired electrons as [↑↓].

3. **Hund's rule:** *When electrons occupy orbitals of equal energy, one electron enters each orbital until all the orbitals contain one electron with spins parallel (in the same direction, as in* [↑] [↑] [↑]*).* Second electrons then add to each orbital so that their spins are paired with the first electron in the orbital. Table 1.2 shows the electron configurations of atoms of nine selected elements. An oxygen atom, for example, contains eight electrons. The orbital of lowest energy, 1*s*, gets one electron, then a second of opposite spin. The next orbital to fill is 2*s*. Three electrons then go, one each, into the three 2*p* orbitals, which have equal energy. The remaining electron now pairs with an electron occupying one of the 2*p* orbitals.

A convenient shorthand method for showing the electron configuration of an atom involves writing the energy level and the symbol for every sublevel occupied by an electron. A superscript indicates the number of electrons occupying that sublevel. As shown in Table 1.3, for hydrogen with one electron in a 1*s* orbital, the electron configuration is written $1s^1$; for helium with two electrons in a 1*s* orbital it is $1s^2$; for oxygen with two electrons in a 1*s* orbital, two electrons in a 2*s* orbital, and four electrons in 2*p* orbitals, it is $1s^2, 2s^2, 2p^4$.

Table 1.2 Electron Configurations for Some Selected Elements

	Energy							
Element	**1s**	**2s**	**$2p_x$**	**$2p_y$**	**$2p_z$**	**3s**	**Electron configuration**	**Number of valence electrons**
H	[↑]	[ ]	[ ]	[ ]	[ ]	[ ]	$1s^1$	1
He	[↑↓]	[ ]	[ ]	[ ]	[ ]	[ ]	$1s^2$	2
Li	[↑↓]	[↑]	[ ]	[ ]	[ ]	[ ]	$1s^2 2s^1$	1
C	[↑↓]	[↑↓]	[↑]	[↑]	[ ]	[ ]	$1s^2 2s^2 2p^2$	4
N	[↑↓]	[↑↓]	[↑]	[↑]	[↑]	[ ]	$1s^2 2s^2 2p^3$	5
O	[↑↓]	[↑↓]	[↑↓]	[↑]	[↑]	[ ]	$1s^2 2s^2 2p^4$	6
F	[↑↓]	[↑↓]	[↑↓]	[↑↓]	[↑]	[ ]	$1s^2 2s^2 2p^5$	7
Ne	[↑↓]	[↑↓]	[↑↓]	[↑↓]	[↑↓]	[ ]	$1s^2 2s^2 2p^6$	8
Na	[↑↓]	[↑↓]	[↑↓]	[↑↓]	[↑↓]	[↑]	$1s^2 2s^2 2p^6 3s^1$	1

Table 1.3 The Electron Configurations of Some of the Known Elements

Element	Atomic number	Electron structure*	Element	Atomic number	Electron structure*
H	1	$1s^1$	K	19	[Ar] $4s^1$
He	2	$1s^2$	Ca	20	[Ar] $4s^2$
Li	3	$1s^2\,2s^1$	Sc	21	[Ar] $3d^1\,4s^2$
Be	4	$1s^2\,2s^2$	Ti	22	[Ar] $3d^2\,4s^2$
B	5	$1s^2\,2s^2\,2p^1$	V	23	[Ar] $3d^3\,4s^2$
C	6	$1s^2\,2s^2\,2p^2$	Cr	24	[Ar] $3d^5\,4s^1$
N	7	$1s^2\,2s^2\,2p^3$	Mn	25	[Ar] $3d^5\,4s^2$
O	8	$1s^2\,2s^2\,2p^4$	Fe	26	[Ar] $3d^6\,4s^2$
F	9	$1s^2\,2s^2\,2p^5$	Co	27	[Ar] $3d^7\,4s^2$
Ne	10	$1s^2\,2s^2\,2p^6$	Ni	28	[Ar] $3d^8\,4s^2$
Na	11	[Ne] $3s^1$	Cu	29	[Ar] $3d^{10}\,4s^1$
Mg	12	[Ne] $3s^2$	Zn	30	[Ar] $3d^{10}\,4s^2$
Al	13	[Ne] $3s^2\,3p^1$	Ga	31	[Ar] $3d^{10}\,4s^2\,4p^1$
Si	14	[Ne] $3s^2\,3p^2$	Ge	32	[Ar] $3d^{10}\,4s^2\,4p^2$
P	15	[Ne] $3s^2\,3p^3$	As	33	[Ar] $3d^{10}\,4s^2\,4p^3$
S	16	[Ne] $3s^2\,3p^4$	Se	34	[Ar] $3d^{10}\,4s^2\,4p^4$
Cl	17	[Ne] $3s^2\,3p^5$	Br	35	[Ar] $3d^{10}\,4s^2\,4p^5$
Ar	18	[Ne] $3s^2\,3p^6$	Kr	36	[Ar] $3d^{10}\,4s^2\,4p^6$

*For elements beyond neon, the configurations are simplified by using the symbol for the preceding noble gas rather than its complete electron configuration.

EXAMPLE 1.1

Writing electron configurations of atoms

Using Figure 1.4, write the electron configuration of (a) silicon and (b) cobalt.

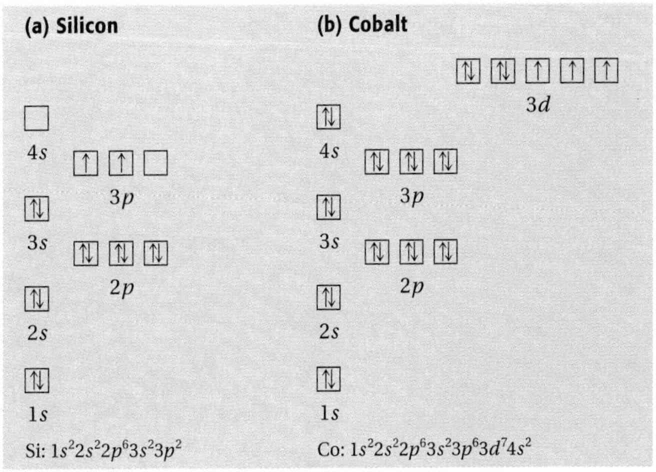

Figure 1.4
Aufbau diagram. To determine the order of filling of energy levels in a many-electron atom, begin with the 1s orbital, follow the arrows and fill in order of increasing energy.

SOLUTION

Silicon has 14 electrons; cobalt has 27 electrons. Using Figure 1.4, start placing electrons in the orbitals with the lowest energy (1s). Remember that there are a maximum of 2 electrons in each orbital and that electrons do not pair up in orbitals of equal energy until necessary.

$B = 5e^- = 1s^2 2s^2 2p^1 = 1 \text{ unpaired } e^-$

$F = 9e^- = 1s^2 2s^2 2p^5 = 1 \text{ "}$

PRACTICE EXERCISE 1.3

Write electron configurations for atoms of the elements boron and fluorine. How many unpaired electrons do each of the following have?
(a) an atom of boron
(b) an atom of fluorine.

Valence Electrons

Valence
valere: (Latin)   to be strong

Knowing electron configurations is important because the number of **valence electrons**—*the electrons in the highest occupied principal energy level of an element's atoms*—largely determines the chemical properties of the element. You can find the number of valence electrons in an atom of a Group A element by looking up the group number of that element in the periodic table. For example, hydrogen, in Group 1A, has one valence electron; carbon, in Group 4A, has four; nitrogen and phosphorus, in Group 5A, have five; and oxygen and sulfur, in Group 6A, have six. The exceptions to this rule are the noble gases that make up Group 0. Of these, helium has two valence electrons and all the others have eight.

Electron dot structures

Since valence electrons are usually the only electrons involved in chemical reactions, it is customary to show only the valence electrons in *electron dot structures.* **Electron dot structures,** *also called Lewis dot structures, depict valence electrons as dots;* the inner electrons and the atomic nuclei are represented by the symbol for the element being considered. Table 1.4 shows the electron dot structures for the atoms of some selected elements.

Table 1.4 Electron Dot Structures of Selected Elements

Period	Group							
	1A	**2A**	**3A**	**4A**	**5A**	**6A**	**7A**	**0**
1	H·							He:
2	Li·	·Be·	·Ḃ·	·Ċ·	·N̈·	:Ö·	:F̈·	:N̈e:
3	Na·	·Mg·	·Äl·	·Ṡi·	·P̈·	:S̈·	:C̈l·	:Är:
4	K·	·Ca·	·Ġa·	·Ġe·	·Äs·	:S̈e·	:B̈r·	:K̈r:

PRACTICE EXERCISE 1.4

How many valence electrons do each of the following atoms have?
(a) potassium   (b) carbon   (c) magnesium   (d) oxygen

1.2 Ionization and ionic bonding

AIMS: To show how ions are formed. To define an ionic bond and an ionic compound.

Focus

Ionic compounds are formed by electrostatic attractions between oppositely charged ions.

Ion
ienai (Greek):   to go or move

The principal ions of the blood are the sodium ion (Na^+), chloride ion (Cl^-), and hydrogen carbonate ion (HCO_3^-).

There are many ionic compounds that have medicinal uses. For example: ammonium carbonate, $(NH_4)_2CO_3$, is an expectorant; potassium permanganate, $KMnO_4$, is a topical anti-infective agent; magnesium sulfate (Epsom salts), $MgSO_4$, is a cathartic; barium sulfate, $BaSO_4$, is a radio-opaque substance used in X-ray work.

Except for the noble gases, the atoms of all the elements are reactive. Atoms lose energy and gain stability when they form compounds. In forming compounds, atoms lose, gain, or share valence electrons and attain noble gas electron configurations.

When atoms lose or gain valence electrons to attain noble gas electron configurations, the atoms are said to be *ionized. Ionized atoms have positive or negative charges and are called* **ions.** *The loss of one or more valence electrons from atoms produces positively charged ions called* **cations.** *The gain of one or more valence electrons produces negatively charged ions called* **anions.** For example, a sodium atom, with one valence electron, and a chlorine atom with seven valence electrons, both achieve greater stability by the transfer of one electron from the sodium to the chlorine.

$$Na \cdot \: + \: \cdot \ddot{\underset{..}{Cl}} : \: \longrightarrow \: Na^+ \: + \: \ddot{\underset{..}{:Cl}} :^-$$

By losing one electron, sodium achieves the electron configuration of neon; by gaining one electron, chlorine achieves the electron configuration of argon. As a consequence of this electron transfer, a sodium atom has become a sodium cation (Na^+) and a chlorine atom has become a chloride anion (Cl^-).

Compounds that contain ions are called **ionic compounds.** Ionic compounds often consist of positively charged metal ions and negatively charged nonmetal ions. The anions and cations in ionic compounds are strongly attracted to one another by electrostatic forces. *The forces of attraction between oppositely charged ions constitute* **ionic bonds.** Ionic compounds are electrically neutral because the number of negative and positive charges are equal. Table salt (sodium chloride, NaCl) is an ionic compound. Sodium chloride is composed of equal numbers of positively charged sodium ions (Na^+) and negatively charged chloride ions (Cl^-). The superscript plus sign indicates one unit of positive charge on the sodium ion; the superscript minus sign indicates one unit of negative charge on the chloride ion. Sodium ions and chloride ions play an important role in biological systems. Not only must these ions be present, but they must also exist in proportions that maintain healthy systems in the organism.

EXAMPLE 1.2   Illustrating the formation of an ion

Write an equation for the formation for each of the following:
(a) selenide ion   (b) strontium ion

SOLUTION

(a) A selenium atom gains two electrons to form a selenide ion, attaining an electron configuration like that of the noble gas krypton.

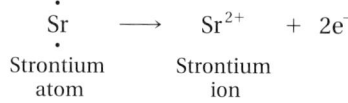

$$: \ddot{Se} \cdot \ + \ 2e^- \ \longrightarrow \ : \ddot{Se} :^{2-}$$

| Selenium atom | Selenide ion |

(b) A strontium atom loses two electrons to form a strontium ion, attaining an electron configuration like that of the noble gas krypton.

$$\dot{\underset{\cdot}{Sr}} \ \longrightarrow \ Sr^{2+} \ + \ 2e^-$$

| Strontium atom | Strontium ion |

PRACTICE EXERCISE 1.5

What charge will the cation have when the following elements lose their valence electrons?
(a) calcium (Ca)   (b) potassium (K)   (c) aluminum (Al)

Ca^{+2} K^{+1} Al^{+3}

PRACTICE EXERCISE 1.6

What charge will the anion have when the following elements gain valence electrons and attain noble gas configurations?
(a) sulfur (S)   (b) nitrogen (N)   (c) fluorine (F)

S^{-2} N^{-3} F^{-1}

1.3 Covalent bonding using electron dot structures

AIM: To define a covalent bond and give examples of covalent compounds.

Focus

The stability of molecules is conferred by electron sharing.

Some atoms—particularly those of hydrogen and the nonmetallic elements in Groups 4A, 5A, 6A, and 7A—share electrons to attain stable electron configurations. **Covalent bonds** *are the result of electron-sharing between atoms. A* **molecule** *is an electrically neutral species consisting of two or more atoms of the same or different elements joined by covalent bonds.*

Single covalent bonds

Hydrogen (H_2) is the simplest molecule. A hydrogen atom (H) has a single valence electron and is quite unstable. By sharing electrons to form a hydrogen molecule, each hydrogen atom attains the stable electron configuration of helium, which has two valence electrons. *A* **single covalent bond** *is the result of sharing a pair of valence electrons:*

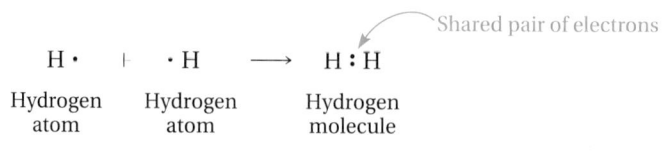

Shared pair of electrons

$$H \cdot \ + \ \cdot H \ \longrightarrow \ H : H$$

| Hydrogen atom | Hydrogen atom | Hydrogen molecule |

Sometimes it is helpful to show the pair of electrons in a covalent bond as a dash, as in H—H for hydrogen. Notations of this type are **structural formulas**—*chemical formulas that show the arrangements of atoms in molecules.* There are two things to remember about the dashes between the atoms in structural formulas: They always indicate a pair of shared electrons, and they are never used to show ionic bonds.

Water and ammonia provide two more examples of chemical bond formation through electron pair sharing. In the water molecule, two hydrogen atoms share electrons with one oxygen atom:

$$2H\cdot \ + \ \cdot \overset{\displaystyle ..}{\underset{\displaystyle ..}{O}}\cdot \ \longrightarrow \ H:\overset{\displaystyle ..}{\underset{\displaystyle ..}{O}}:H \ \text{ or } \ H-\overset{\displaystyle ..}{\underset{\displaystyle ..}{O}}-H \quad \text{Unshared pairs}$$

Hydrogen atoms	Oxygen atom	Water molecule

The hydrogen and oxygen atoms attain stable, noble gas configurations by electron-sharing. The oxygen in water also has two pairs of valence electrons that do not participate in covalent bonding to hydrogen. *Pairs of valence electrons that do not participate in covalent bonding are called* **unshared pairs.**

Ammonia, a suffocating gas, is formed in a similar way:

$$3H\cdot \ + \ \cdot \overset{\displaystyle .}{\underset{\displaystyle ..}{N}}\cdot \ \longrightarrow \ H:\overset{\displaystyle ..}{N}:H \ \text{ or } \ \overset{\displaystyle H}{\underset{\displaystyle H}{H-N-}} \quad \text{Unshared pair}$$

Hydrogen atoms	Nitrogen atom	Ammonia molecule

The ammonia molecule has one unshared pair.

EXAMPLE 1.3	**Drawing electron dot structures**

Hydrogen chloride (HCl) is a diatomic molecule with a single covalent bond. Draw the electron dot structure for HCl.

SOLUTION $H\cdot \ + \ \cdot \overset{..}{\underset{..}{C}}l: \ \longrightarrow \ H:\overset{..}{\underset{..}{C}}l:$

To form a single covalent bond, hydrogen and chlorine atoms must share a pair of electrons. First write electron dot structures for the two atoms; then show the electron-sharing:

$$H\cdot \ + \ \cdot \overset{\displaystyle ..}{\underset{\displaystyle ..}{C}}l: \ \longrightarrow \ H:\overset{\displaystyle ..}{\underset{\displaystyle ..}{C}}l: \quad \text{Shared electron pair}$$

Through electron sharing, hydrogen and chlorine atoms attain the electron configurations of the noble gases helium and argon, respectively.

PRACTICE EXERCISE 1.7

Draw electron dot structures and write structural formulas for the following covalent molecules, which have only single covalent bonds:
(a) H_2S (b) PH_3 (c) ClF   (d) H_2O_2

a) $2H\cdot + \cdot \overset{..}{\underset{.}{S}}: \ = \ H-\overset{..}{\underset{.}{S}}:$ b) $3H\cdot + \overset{\cdot}{P}\cdot \ = \ \overset{..}{H-P-H}$ c) $\cdot \overset{..}{\underset{..}{F}}: + \cdot \overset{..}{\underset{..}{C}}l: = \ :\overset{..}{\underset{..}{F}}-\overset{..}{\underset{..}{C}}l:$
                                                                    |
                                                                    H
                                          $\overset{\displaystyle |}{H}$ d) $2H + 2O = H-\overset{..}{O}-\overset{..}{O}-H$

An increased amount of carbon dioxide in the atmosphere is a major contributor to an increase in the greenhouse effect. This change would increase the Earth's average temperature and result in significant climatic changes.

Double and triple covalent bonds

Atoms can sometimes share more than one pair of electrons to attain stable, noble gas electron configurations. **Double covalent bonds** *contain two shared pairs of electrons and* **triple covalent bonds** *contain three shared pairs.*

A carbon atom has four valence electrons and needs four more to attain a noble gas configuration. The carbon dioxide molecule contains two oxygens that each share two electrons with carbon to form two carbon-oxygen double bonds:

$$:\overset{..}{\underset{..}{O}}: \quad + \quad \cdot\overset{}{\underset{}{C}}\cdot \quad + \quad :\overset{..}{\underset{..}{O}}: \quad \longrightarrow \quad :\overset{..}{\underset{..}{O}}::C::\overset{..}{\underset{..}{O}}: \quad \text{or} \quad O{=}C{=}O$$

| Oxygen atom | Carbon atom | Oxygen atom | Carbon dioxide molecule |

The nitrogen molecule contains a triple covalent bond. Each nitrogen atom has five valence electrons and needs three more to attain the electron configuration of neon:

$$:\overset{\cdot}{N}\cdot \quad + \quad \cdot\overset{\cdot}{N}: \quad \longrightarrow \quad :N::N: \quad \text{or} \quad N{\equiv}N$$

| Nitrogen atom | Nitrogen atom | Nitrogen molecule |

Coordinate covalent bonds

Sometimes one atom provides both bonding electrons in a covalent bond. This is a **coordinate covalent bond.** The polyatomic ammonium ion (NH_4^+) has a coordinate covalent bond. It is formed when a proton (hydrogen ion) is attrcted to the unshared electron pair of an ammonia molecule. The nitrogen of NH_4^+ carries a positive charge.

Unshared pair

$$H^+ \quad + \quad H:\overset{\overset{\displaystyle H}{..}}{\underset{\displaystyle H}{N}}:H \quad \longrightarrow \quad \left[H:\overset{\overset{\displaystyle H}{..}}{\underset{\displaystyle H}{N}}:H \right]^+ \quad \text{or} \quad H{-}\overset{\overset{\displaystyle H}{|}}{\underset{\displaystyle H}{N}}{\overset{+}{-}}H$$

| Proton | Ammonia molecule | Ammonium ion |

Once formed, a coordinate covalent bond is no different from any other covalent bond because electrons are indistinguishable from each other. The only difference between a coordinate covalent bond and other covalent bonds is the source of the electrons. An important anticancer agent, cisplatin, contains coordinate covalent bonds, as described in A Closer Look: Cisplatin: Coordinate Covalent Bonds and Chemotherapy.

PRACTICE EXERCISE 1.8

The hydronium ion (H_3O^+) contains a coordinate covalent bond. It forms when a proton is attracted to an unshared electron pair of a water molecule. Write the electron dot structure and structural formula for the hydronium ion, and indicate the location of the positive charge.

A Closer Look

Cisplatin: Coordinate Covalent Bonds and Chemotherapy

Cisplatin is an important cancer chemotherapy drug used in the treatment of cancer. Cisplatin has the formula $Pt(NH_3)_2Cl_2$. The cisplatin molecule has two coordinate covalent bonds. Each coordinate covalent bond is formed between the platinum ion, Pt^{2+}, and an unshared pair of electrons of each ammonia molecule.

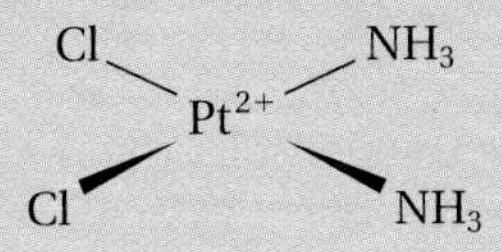

$$Cl \quad NH_3$$
$$Pt^{2+}$$
$$Cl \quad NH_3$$

Cancers of the ovaries, testes, lungs, and others are treated using cisplatin. The anticancer activity of the compound arises from its ability to bind to the genetic material (DNA) to prevent the division of tumor cells (see figure). Although this compound was first isolated in the mid-1800s, its use as a potential cancer chemotherapy agent was not recognized until the mid-1960s. The ability of cisplatin to prevent the division of cells was discovered accidently during an experiment studying

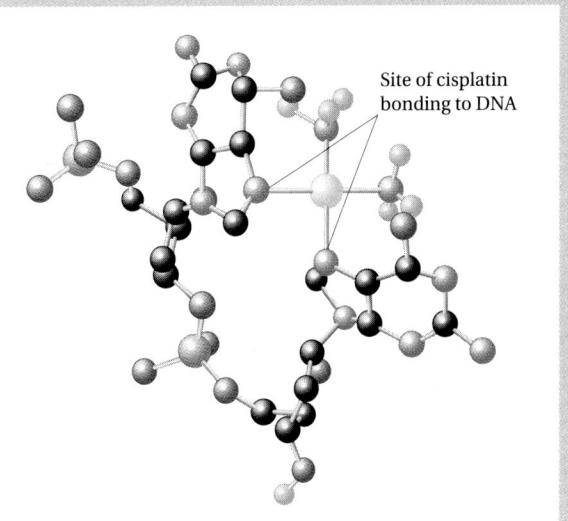

Site of cisplatin bonding to DNA

Cisplatin binds to two nitrogen atoms in DNA thereby preventing the division of tumor cells.

the effect of electricity on bacterial growth. The compound was formed during the experiment by a reaction between the platinum ions from platinum electrodes and ammonia and chloride ions in the bacteria culture medium.

1.4 Shapes of molecules

AIM: To explain the shapes of simple covalent molecules by using electron-pair repulsion theory.

Focus

Molecules are three-dimensional.

Just as a photograph or sketch may fail to do justice to a person's appearance, electron dot structures and structural formulas fail to reflect the three-dimensional shape of molecules. We are interested in knowing something about the shapes of molecules, since complex molecules in biological systems must fit together correctly for important chemical reactions to occur.

Predicting molecular shapes by electron-pair repulsion theory

We can get some basic ideas about molecular shapes by studying simple molecules such as methane. A carbon atom shares its four valence elec-

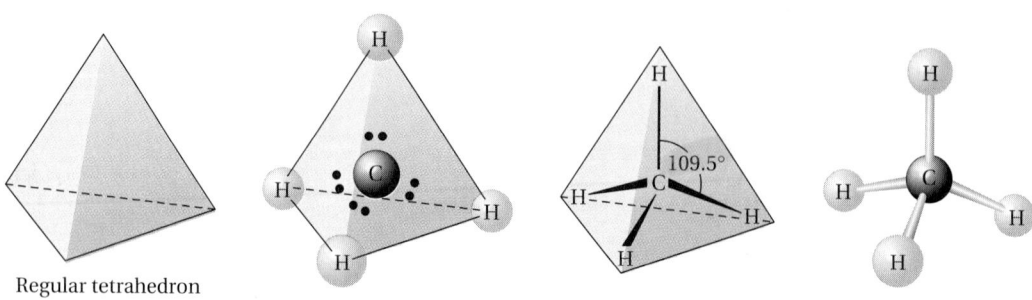

Regular tetrahedron

Figure 1.5
The hydrogens in methane are at the corners of a regular tetrahedron. This orientation places the bonding pairs as far apart as possible.

Tetrahedron
tetra: (Greek)   four
hedra: (Greek)   face

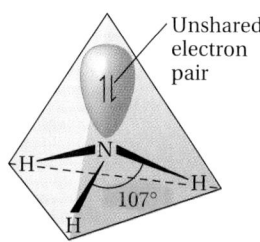

(a) Ammonia

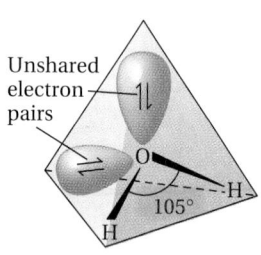

(b) Water

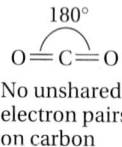

No unshared
electron pairs
on carbon

(c) Carbon dioxide

Figure 1.6
The unshared pairs and the hydrogens in ammonia (a) and water (b) are at the corners of a tetrahedron; the carbon dioxide molecule (c) is linear.

trons with four hydrogen atoms in the methane molecule (CH_4), which may be written as either its electron dot structure or its structural formula:

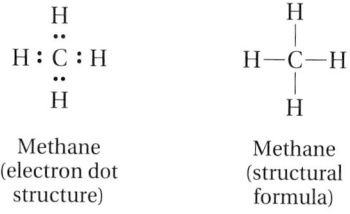

Methane
(electron dot
structure)

Methane
(structural
formula)

But neither formula really describes the shape of the methane molecule. Experimental measurements show that the hydrogens in the methane molecule are at the four corners of a geometric solid, the regular tetrahedron, as shown in Figure 1.5. In this arrangement all of the H—C—H angles are 109.5°, the tetrahedral angle. *One concept that explains this shape says that since electron pairs repel, molecules adjust their shapes so that the valence-electron pairs are as far apart as possible.* In a molecule such as methane, with four bonding electron pairs and no unshared pairs, the bond pairs are farthest apart when the angle between the central carbon and its attached hydrogens is 109.5°, the H—C—H bond angle found by experiment. Any other arrangement tends to bring two bonding pairs of electrons closer together.

Unshared pairs of electrons are important when we are trying to predict the shapes of molecules using the electron-pair repulsion theory. Like the carbon in methane (CH_4), the nitrogen in ammonia (NH_3) is surrounded by four pairs of valence electrons that take a tetrahedral orientation (Fig. 1.6a). But note that one of the pairs in ammonia is an unshared pair. The unshared pair takes up more space than the bonding pairs and strongly repels them, pushing them closer together so that the H—N—H bond angle is only 107°.

In a water molecule, oxygen forms single covalent bonds with two hydrogen atoms and there are two unshared pairs. The bonding and unshared pairs of electrons form a tetrahedral arrangement around the central oxygen (Fig. 1.6b). The water molecule is flat but bent. With two

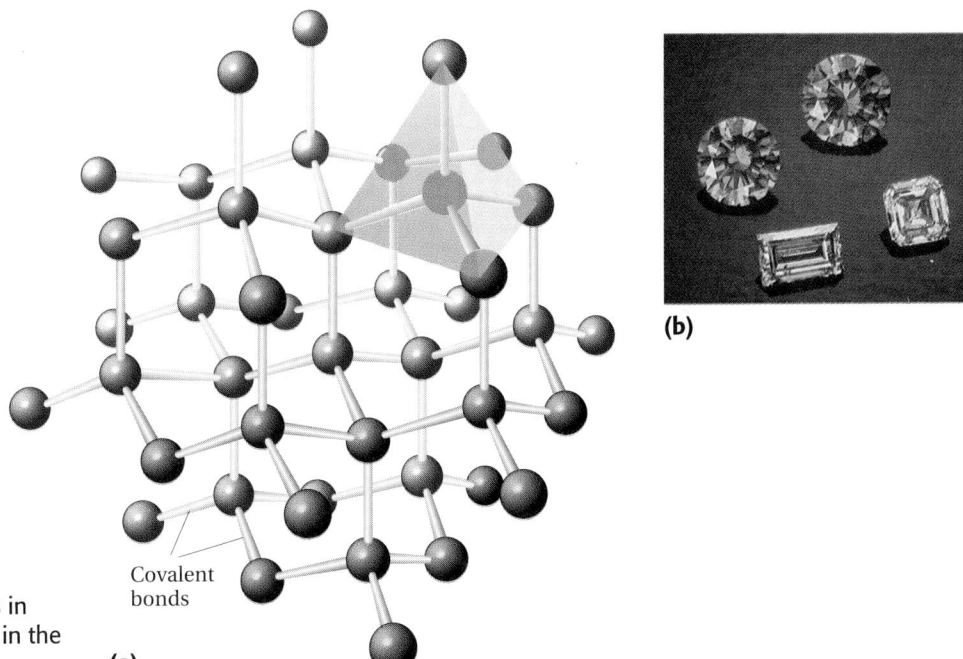

Figure 1.7
The arrangement of carbons in diamond, (a), is manifested in the shape of diamonds, (b).

Covalent bonds

(a)

(b)

unshared pairs repelling the bonding pairs, the H—O—H bond angle is compressed to about 105°.

A tetrahedral arrangement does not always minimize repulsions between unshared pairs. Carbon dioxide, for example, is a linear molecule (Fig. 1.6c). The carbon in CO_2 has no unshared pairs, and the double bonds joining the oxygens to the carbon are farthest apart when the O—C—O bond angle is 180° (when the atoms are in a straight line).

The structure of a diamond is a three-dimensional framework in which all the atoms are carbons joined by single covalent bonds (Fig. 1.7). The incredible symmetry inherent in the diamond structure is a result of bonding electron-pair repulsions. As in methane, all the bond angles (in this instance C—C—C) are 109.5°. Diamond-cutting requires breaking a multitude of these bonds. A Closer Look: Buckyball: A Third Form of Carbon, describes another form of carbon that has recently been discovered.

Molecular models

Stereochemistry
stereos: (Greek)   solid

Stereochemistry *is the study of the spatial arrangements of the atoms in molecules.* Chemists and chemistry students often find molecular models very useful in stereochemistry. Two popular models are the ball-and-stick and space-filling types.

Ball-and-stick models resemble a child's tinker toy; bonds are represented by sticks or springs that fit into holes drilled into colored balls representing atoms. There is a different color for each kind of atom—hydrogens may be white, oxygens, red; nitrogens, blue; carbons, black; and so forth.

A Closer Look

Buckyball: A Third Form of Carbon

For many years scientists believed that carbon came in two basic forms: hard, shiny diamond and soft, dull graphite. Both of these forms of carbon have their uses. Diamond, besides being used for jewelry, is important in making metal-cutting tools because it is so hard. Graphite, in which carbon atoms are bonded together in flat plates that slide easily over each other, has many applications as a lubricant.

Recently, chemists have discovered a fascinating third form of carbon. This substance has the molecular formula C_{60}. Although C_{60} had been previously detected in interstellar space, it was unknown on Earth. The carbons of the C_{60} molecule are arranged to form a spherical molecule (see figure, part a). Because of its structure, chemists have nicknamed the C_{60} compound "buckyball." The name comes from R. Buckminster Fuller (1895–1983), inventor of the geodesic dome, which the buckyball closely resembles (see figure, part b). You might also recognize that the positions of the carbon atoms in buckyball are fixed at the corners of pentagons and hexagons, placed exactly in the arrangement of these shapes that is sewn together to make a soccer ball. Despite its recent discovery on Earth, buckyball is a minor component in soot obtained by burning such ordinary objects as a plastic milk container.

Scientists are now working to discover practical applications for buckyball. Buckyball molecules would seem to be naturals as lubricants; because they are spherical, they could act as molecule-sized ball bearings. Buckyball with metal atoms trapped in the carbon cage could prove useful as electrical semiconductors or superconductors.

There are also exciting potential medical uses for buckyball. Because of its ability to "tie-up" energetic molecules in the body that are linked to the formation of cancer cells, buckyball may someday be part of a preventative medicine for cancer. It has been proposed that radioactive atoms enclosed inside buckyball molecules could be used for antitumor therapy in cancer patients. In a test-tube experiment, a compound made by modifying buckyball has been shown to block the action of an enzyme needed for the reproduction of the human immunodeficiency virus (HIV), the causative agent in AIDS. Only time will tell whether some day buckyball, or similar molecules, will be used to deliver medicines attached to, or enclosed in the carbon cage, to specific cells of the body.

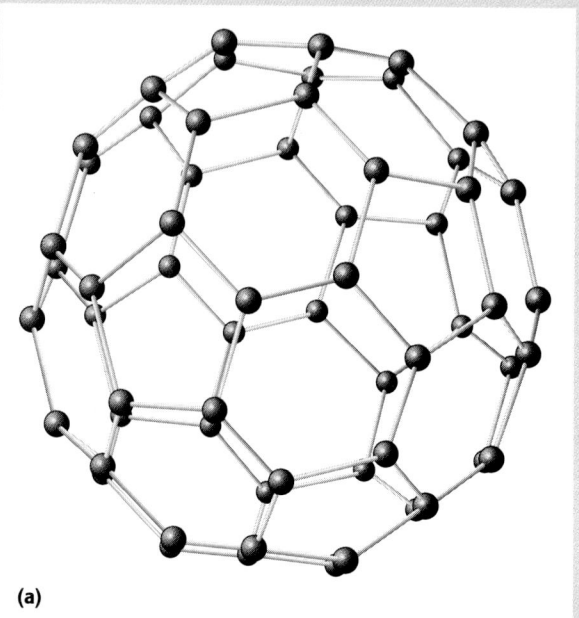

(a)

(b)

(a) Buckyball, the latest form of carbon and (b) the geodesic dome of the religious center at Southern Illinois University at Edwardsville.

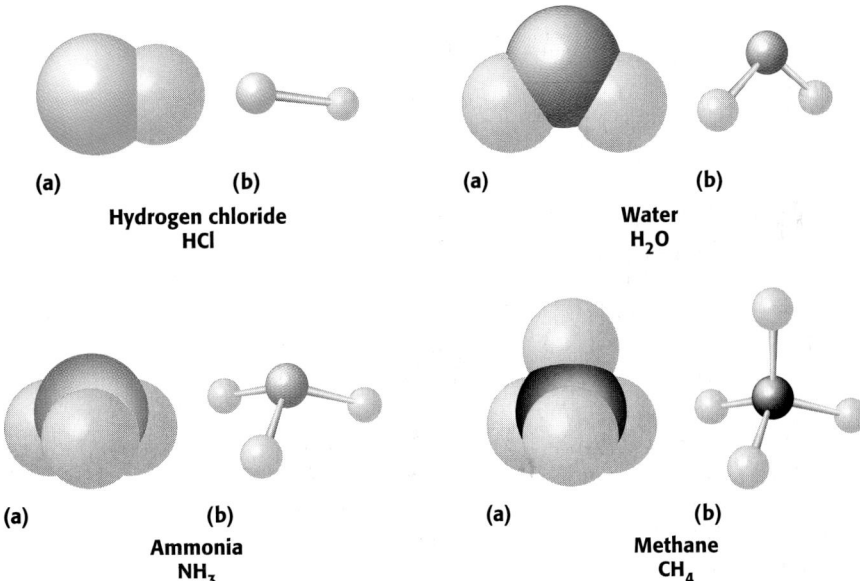

Figure 1.8
Space-filling and ball-and-stick
molecular models.

The holes are drilled at the tetrahedral angle, and the number of holes corresponds to the number of bonds usually formed by the atom—oxygen gets two holes; nitrogen, three; carbon, four. Hydrogens get only one hole, and unshared pairs are ignored. Ball-and-stick models accurately show spatial relationships between atoms in a molecule, but they do not show accurate relationships between atomic sizes and bond lengths.

Space-filling models are more accurate; the atomic sizes are to scale, and molecules more closely resemble the true shapes of the molecules they depict. A disadvantage of space-filling models is that the atoms in a molecule are so tightly packed that stereochemical relationships are often obscured.

Figure 1.8 shows ball-and-stick and space-filling models of some of the molecules we have been discussing. You may wish to borrow some molecular models or make your own from toothpicks and modeling clay to represent the molecular structures shown in the figure.

PRACTICE EXERCISE 1.9

Just as carbon and hydrogen form methane (CH_4), carbon and chlorine form carbon tetrachloride (CCl_4). Draw the electron dot structure for CCl_4, and discuss the shape of the molecule.

The shapes of molecules are of interest to researchers in the allied health fields because complex molecules in biological systems must fit together correctly for important chemical reactions to occur. The fit of biological molecules is also important in the discovery of medicinal drugs. In introducing the Case in Point for this chapter, we said that historically, the discovery of therapeutic drugs has been an expensive, time-consuming process. In the Follow-up to the Case in Point, we will learn about another method that could facilitate the discovery of new pharmaceuticals.

FOLLOW-UP TO CASE IN POINT: A new era of medicinal drug discovery

Pharmaceutical chemists hope to eliminate much of the tedious, but necessary, screening and preliminary testing of potential pharmaceutical drugs by using a method called molecular modeling. Molecular modeling blends experimental facts and theories about molecules. Using molecular modeling, molecular properties such as shapes and charges can be predicted. The method can also provide information about the ways that molecules interact with each other in chemical reactions. Molecular modeling has been embraced by many branches of chemistry, but it has been greeted with special enthusiasm by the pharmaceutical industry. Because a drug requires 10 to 12 years of testing to gain approval for therapeutic use, no currently available drug was discovered by molecular modeling. However, pharmaceutical chemists are using the method to identify promising variations of compounds already in use as pharmaceuticals. They are also searching for new candidate compounds for medicinal use. To see how this might be done, suppose we want to find an ion that will prevent the passage of calcium ions into heart muscle through the passages called ion channels. Such substances, called calcium ion-channel blockers, can relieve angina, or heart pain. If we know the molecular structure of the ion channel, we can use computerized molecular modeling to test a large number of ions as calcium ion-channel blockers. Our modeling could show which ions will enter the ion channel and make a tight fit, much as a cork blocks the opening of a bottle. If our modeling studies could help us identify the top 20 potential channel blockers out of, say, 10,000 compounds, we could prepare these compounds for experimental testing.

The software programs for molecular modeling studies contain fundamental information about

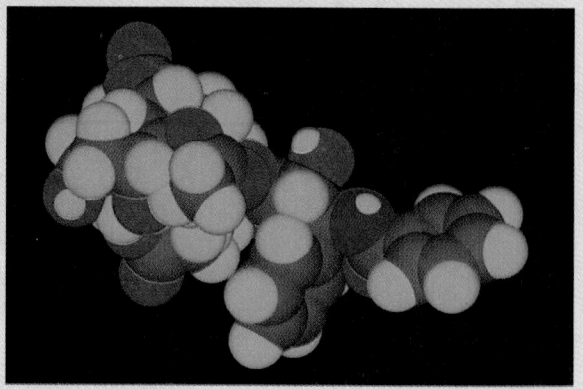

An example of a computer-generated molecular model.

chemical bonding and the shapes of ions and molecules. When an ionic or molecular structure is introduced into the program, essentially as a rough drawing, the computer calculates such properties as the bond lengths and the overall three-dimensional shape of the ion or molecule (see figure). In most instances, the three-dimensional structures produced by the computer are in near-perfect agreement with experimentally determined structures. Two or more structures can then be brought together to study how they interact, as in the case of a calcium-ion channel and an ion that could be used as a blocker.

Today, thanks to relatively inexpensive, powerful computers, the average chemist can explore many detailed aspects of molecular structure. The savings in time and money gained through the use of molecular modeling can be enormous. Today's pharmaceutical chemists have only begun to tap the potential of molecular modeling, but this powerful tool for the study of molecular structures and interactions has a bright future.

1.5 Polarity of covalent bonds

AIMS: To distinguish between nonpolar and polar covalent bonds. To use electronegativity values to determine whether a bond is ionic, polar covalent, or nonpolar covalent.

Focus

The properties of molecules are determined by the character of their covalent bonds.

We have seen that covalent bonds are formed by electron-sharing between atoms. Not all covalent bonds are the same, however. The character of these bonds in a given molecule depends on the kind and number of atoms that are joined together. These features in turn determine the properties of the molecules.

Polar bonds

The bonding pairs of electrons in covalent bonds are pulled, as in a tug-of-war, between the nuclei of the atoms sharing the electrons. When the atoms involved are the same, as in H_2, N_2, and the halogens (F_2, Cl_2, Br_2, I_2), the bonding electrons are shared equally and the contest is a standoff. The covalent bonds in these molecules are said to be *nonpolar. Covalent bonds in which the bonding electrons are shared equally by both atoms are* **nonpolar bonds.**

The situation changes when two different atoms are joined by a covalent bond. Since one atom is always able to attract electrons more strongly than the other, it gains a greater share of the bonding electrons. *When the bonding electron pair is unequally distributed between the atoms, the bond becomes a* **polar bond.** It follows that the atom with stronger electron attraction in a polar bond acquires a slight negative charge. The weaker atom acquires a slight positive charge. *The ability of an atom to attract electrons to itself in a covalent bond is called the atom's* **electronegativity.** Table 1.5 presents the electronegativities of some common elements. The higher the electronegativity value, the greater is the ability of an atom to attract electrons to itself. Electronegativities also tell us whether two atoms will form an ionic bond, a polar bond, or a nonpolar bond. If the electronegativity difference between two atoms is greater than about 2.0, a bond between the two atoms is usualy ionic or very polar; if it less than about 2.0 but greater than 0.0, the bond is usually polar and covalent; if it is 0.0, the bond is nonpolar.

Consider the hydrogen chloride molecule (HCl). Hydrogen has an electronegativity value of 2.1, and chlorine has an electronegativity value of 3.0. So the covalent bond in hydrogen chloride is polar with chlorine acquiring a slight negative charge ($\delta-$) and hydrogen acquiring a slight positive charge ($\delta+$). The polarity of the bond can be represented as

$$\overset{\delta+ \quad \delta-}{H-Cl} \quad or \quad H-\overset{\longleftarrow}{Cl}$$

In the latter representation the arrowhead always points to the more electronegative atom. The O—H bonds in the water molecule are also polar because the very electronegative oxygen pulls the bonding electrons away

Table 1.5 Electronegativity Values for Some Elements

H 2.1							He 0.0
Li 1.0	Be 1.5	B 2.0	C 2.5	N 3.0	O 3.5	F 4.0	Ne 0.0
Na 0.9	Mg 1.2	Al 1.5	Si 1.8	P 2.1	S 2.5	Cl 3.0	Ar 0.0
K 0.8	Ca 1.0	Ga 1.6	Ge 1.8	As 2.0	Se 2.4	Br 2.8	Kr 0.0

from hydrogen. This results in the oxygen acquiring a slight negative charge and the hydrogens acquiring a slight positive charge. The bond polarities are shown as

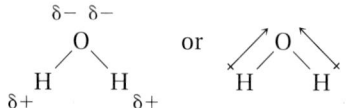

EXAMPLE 1.4   **Predicting the polarity of chemical bonds**

What type of bond—polar covalent, nonpolar covalent, or ionic—will form between atoms of the following pairs of elements?
(a) Na and Cl   (b) N and H   (c) F and F   (d) C and Cl.

SOLUTION

(a) The electronegativity difference between Na (0.9) and Cl (3.0) is 2.1. The difference is greater than 2; the bond is ionic.

(b) The electronegativity difference between N (3.0) and H (2.1) is 0.9. The difference is less than 2 but greater than zero; the bond is polar covalent.

(c) When identical atoms share electrons the electronegativity difference is zero and the bond is nonpolar covalent.

(d) The electronegativity difference between C (2.5) and Cl (3.0) is 0.5; the bond is polar covalent.

PRACTICE EXERCISE 1.10

The bonds between the following pairs of elements are covalent. Arrange them according to polarity, naming the most polar first:
(a) H—Cl   (b) H—C   (c) H—F   (d) H—O   (e) H—H
(f) S—Cl

Polar molecules

The presence of a polar bond in a molecule often makes the entire molecule polar. In the hydrogen chloride molecule the partial charges on the hydrogen and chlorine produce electrically charged regions, or poles, one slightly positive and one slightly negative. *A molecule that has oppositely charged electrical poles is a* **polar molecule.**

The effect of individual polar bonds on the polarity of an entire molecule depends upon the shape of the molecule and the orientation of the polar bonds. A carbon dioxide molecule, for example, has two polar bonds and is linear:

$$O=C=O$$

The carbon and oxygens lie along the same axis, therefore the bond polari-

[handwritten: 94 / .10 / 9.40]

ties cancel because they are in opposite directions. Carbon dioxide is a non-polar molecule.

The water molecule also has two polar bonds, but the situation is different from that of carbon dioxide because the molecule is bent. The bond polarities do not cancel, so a water molecule is polar.

Bond dissociation energies of covalent bonds

A large quantity of heat is liberated when hydrogen atoms combine to form hydrogen molecules. This is evidence that the product is more stable than the reactants. Indeed, the covalent bond in the hydrogen molecule is so strong that it requires 104 kcal of energy to dissociate 1 mol of hydrogen molecules to hydrogen atoms. We say it has a **bond dissociation energy**—*the energy required to break a single covalent bond*—of 104 kcal/mol.

$$H-H + 104 \text{ kcal} \longrightarrow H\cdot + \cdot H$$

The carbon-carbon single covalent bond found in numerous organic molecules is also a strong bond. It has a bond dissociation energy of about 83 kcal/mol. The ability of carbon to form strong carbon-carbon bonds helps to account for the stability of carbon compounds. Table 1.6 gives bond dissociation energies for several representative covalent bonds. Compounds with only C—C and C—H single covalent bonds are quite unreactive chemically, in part because of the high dissociation energies of these bonds.

Table 1.6 Bond dissociation energies for covalent bonds

Bond	Bond energy (kcal/mol)
H—H	104
C—H	94
C—O	85
C—C	83
C=C	157
C≡C	217
C—N	73
S—S	62
Cl—Cl	58
N—N	50
Br—Br	46
I—I	36
O—O	34

[handwritten: 9.4 / 4 / 37.6]

PRACTICE EXERCISE 1.11

Assuming the bond dissociation energy is the same for each bond, how many kilocalories would be required to dissociate all the C—H single bonds in 0.10 mol of methane?

[handwritten: 94]

[handwritten: ? kcal = 0.10 mol · 94 kcal / mol]

1.6 Attractions between molecules

[handwritten: 9.4 kcal · 4 bonds = 38 kcal/mol]

AIM: To name and describe the weak attractive forces between molecules.

Focus

Molecules are held to each other by weak attractive forces.

Weak attractive forces contribute to the maintenance of the three-dimensional structures of proteins, nucleic acids, and other important biological molecules.

The type of bonding in a compound dramatically affects its physical properties. We have already touched on this point by noting that most ionic compounds are crystalline solids. However, some covalent compounds are gases, some are liquids, and some are solids at room temperature. The greater variety of physical properties among covalent compounds occur because many types and strengths of weak attractions between covalent molecules are possible. There are several types of attractions between molecules, each of which is much weaker than either an ionic or covalent bond. Do not, however, underestimate the power of these forces simply because they are weak. Among other things, they are responsible for the organization of molecules in living systems. Although covalent bonds hold the atoms of biological molecules together, the weak forces hold the molecules to each other. Without the weak forces, there would be no life.

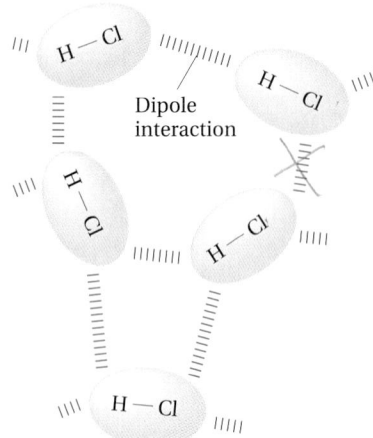

Figure 1.9
Polar molecules are attracted to each other by van der Waals forces called dipole interactions.

van der Waals forces

The weakest attractions between molecules are collectively called **van der Waals forces** after the nineteenth century Dutch chemist Johannes van der Waals. Two major van der Waals forces are dispersion forces and dipole interactions.

Dispersion forces, *the weakest of all molecular interactions, vary in strength depending on a molecule's size and shape.* They are thought to be caused by the motion of electrons and are sufficiently strong enough to cause nonpolar bromine to exist as a liquid and iodine to exist as a solid.

Dipole interactions *result from attractions between polar molecules.* Considerably stronger than dispersion forces, dipole interactions occur when polar molecules are attracted to one another (Fig. 1.9). Dipole interactions between polar molecules and attractions between oppositely charged ions both result from electrostatic forces, but the charges on dipolar molecules are only a fraction of those on ions. Dipole interactions are therefore much weaker than ionic bonds. Dipolar interactions in water, for example, result in a weak attraction of water molecules for each other. Each O—H bond in the water molecule is highly polar; the oxygen, because of its greater electronegativity, acquires a slight negative charge; the hydrogens acquire a slight positive charge. Polar molecules attract one another, so the positive region of one water molecule attracts the negative region of another. This dipolar attraction is very weak, however, when compared with the strength of another weak force—*hydrogen-bonding*—between water molecules.

Hydrogen bonding

Hydrogen bonds *are attractive forces in which hydrogen covalently bonded to a very electronegative atom is also weakly bonded to an unshared electron pair of an electronegative atom in the same molecule or in a nearby molecule.* Hydrogen-bonding always involves hydrogen because it is the only chemically reactive element whose valence electrons are not shielded from the nucleus by a layer of underlying valence electrons. The unequal sharing of electrons in a very polar covalent bond between hydrogen and an electronegative atom like oxygen, nitrogen, or fluorine leaves the hydrogen nucleus quite electron deficient. The hydrogen makes up for its deficiency by sharing a nonbonding electron pair on a nearby electronegative atom. The resulting hydrogen bond, the strongest of the weak forces, has about 5% of the strength of an average covalent bond. Hydrogen bonds are extremely important in determining the properties of water and biological molecules like proteins.

Figure 1.10 shows hydrogen-bonding between water molecules. In water the individual molecules change positions by sliding over one another; but because of hydrogen-bonding, most do not have enough kinetic energy to escape at room temperature. Hydrogen bonds are the reason for water's high boiling point and high surface tension, to name but two of its many unusual properties.

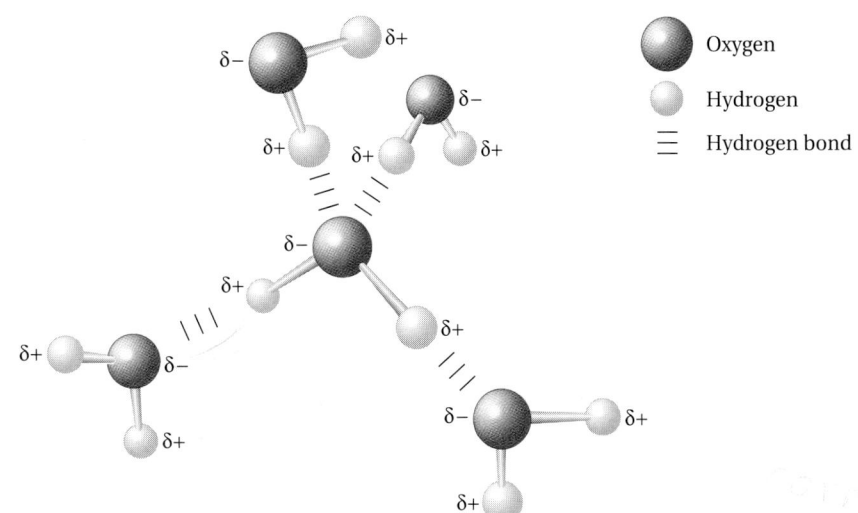

Figure 1.10
Hydrogen-bonding between water molecules. The hydrogen bond has about 5% of the strength of an average covalent bond.

PRACTICE EXERCISE 1.12

Draw structural formulas and show hydrogen bonding between one ammonia molecule and one water molecule.

SUMMARY

The ways in which electrons are arranged around the nucleus of atoms are called electron configurations. The modern model of an atom depicts the probability of finding electrons in certain positions around the nucleus as electron clouds. The Aufbau principle, the Pauli exclusion principle, and Hund's rule guide the filling of atomic orbitals by electrons.

Atoms are held together in compounds by chemical bonds. Chemical bonds result from the sharing or transfer of valence electrons. The transfer of one or more valence electrons between atoms produces positively and negatively charged ions—cations and anions. The attraction between an anion and a cation is an ionic bond. A substance with ionic bonds constitutes an ionic compound.

When atoms share electrons to achieve the stable electron configuration of a noble gas, they form covalent bonds. A shared pair of valence electrons constitutes a single covalent bond. Sometimes two or three pairs of electrons may be shared to give double or triple covalent bonds. As a general rule, molecules adjust their three-dimensional shapes so that the valence-electron pairs around a central atom are as far apart as possible.

When like atoms are joined by a covalent bond, the bonding electrons are shared equally and the bond is nonpolar. When the atoms in a bond are not the same, the bonding electrons are shared unequally and the bond is polar. The degree of polarity of a bond between any two atoms is determined by consulting a table of electronegativities. Some molecules are polar because they contain polar covalent bonds. The attractions between opposite poles of polar molecules constitute dipole interactions. The dipole interaction is one of several weak attractions between molecules that determine whether a covalent compound will be a solid, a liquid, or a gas.

KEY TERMS

Anion (1.2)
Atomic orbital (1.1)
Aufbau principle (1.1)
Bond dissociation energy
  (1.5)
Cation (1.2)
Coordinate covalent bond
  (1.3)
Covalent bond (1.3)

Dipole interaction (1.6)
Dispersion force (1.6)
Double covalent bond
  (1.3)
Electron configuration
  (1.1)
Electron dot structure (1.1)
Electronegativity (1.5)
Hund's rule (1.1)

Hydrogen bond (1.6)
Ion (1.2)
Ionic bond (1.2)
Ionic compound (1.2)
Molecule (1.3)
Nonpolar bond (1.5)
Pauli exclusion principle
  (1.1)
Polar bond (1.5)

Polar molecule (1.6)
Quantum of energy (1.1)
Single covalent bond (1.3)
Stereochemistry (1.4)
Structural formula (1.3)
Triple covalent bond (1.3)
Unshared pair (1.3)
Valence electron (1.1)
van der Waals force (1.6)

EXERCISES

Atomic structure and electron configuration
(Section 1.1)

1.13 How many electrons are in the third energy level of an atom of chlorine?

1.14 An atom of an element has two electrons in the first energy level and five electrons in the second energy level. What is the element?

1.15 Give the symbols of the elements whose atoms have the following electron configurations:
(a) $1s^2 2s^2 2p^4$ (b) $1s^2 2s^2 2p^6 3s^2$
(c) $1s^2 2s^2 2p^1$ (d) $1s^2 2s^2 2p^6 3s^2 3p^4$

1.16 Write the symbols of the elements that are represented by these energy level diagrams.

1.17 Define the term *valence electron*. Write the number of valence electrons for each of the following atoms:
(a) nitrogen (b) sulfur (c) fluorine
(d) lithium

1.18 In which of the following sets do all the elements have the same number of valence electrons?
(a) O, Fe, He (b) O, S, Cl (c) Mg, Ca, Ba
(d) F, I, Br (e) Na, Mg, Al

Ionization and ionic bonding
(Section 1.2)

1.19 Why has an anion a negative charge and a cation a positive charge?

1.20 Complete the table:

	Atomic number	Number of electrons	Symbol of the element	Charge on the ion
(a)	11	10	Na	1+
(b)	16	18	S	2–
(c)	8	10	O	2–
(d)	20	18	Ca	+2
(e)	9	10	___	1–

Covalent bonding
(Section 1.3)

1.21 Define each of the following terms and give an example of each:
(a) single covalent bond
(b) unshared pair of electrons

1.22 Draw the electron dot structures of these molecules:
(a) F_2 (b) HCl (c) HCCH (d) HCN

1.23 Define the following terms:
(a) coordinate covalent bond
(b) double covalent bond

1.24 Why can compounds containing C—N and C—O single bonds form coordinate covalent bonds with H^+, but compounds containing only C—H and C—C single bonds cannot bond with H^+?

Shapes of molecules
(Section 1.4)

1.25 Explain how the electron pair repulsion theory can be used to predict bond angles in the following:
(a) methane (b) ammonia (c) water

1.26 If one of the hydrogens in methane is replaced by chlorine, explain how the shape of the molecule will change (if at all).

Polarity of covalent bonds (Section 1.5)

1.27 With use of the table of electronegativities on page 19, determine which member of each of the following pairs has the higher electronegativity:
(a) S and O   (b) C and S   (c) H and C
(d) C and N

1.28 Draw the electron dot structure for each of the following molecules and identify polar covalent bonds by assigning $\delta+$ and $\delta-$ to the appropriate atoms:
(a) HOOH   (b) BrCl   (c) HBr

1.29 Define the term *bond dissociation energy.*

1.30 What explanation can you offer to account for the fact that the N—N and O—O single covalent bonds are much weaker than the C—C single bond?

Attractions between molecules (Section 1.6)

1.31 Briefly discuss the nature and significance of van der Waals forces.

1.32 What is a hydrogen bond? Which compound in each pair exhibits hydrogen bonding?
(a) H_2S and H_2O (b) HCl and HF

Additional Exercises

1.33 Sketch the shapes of the *s* and *p* orbitals. How are the *p* orbitals oriented in relation to one another?

1.34 What is the Pauli exclusion principle and how does it affect the filling of atomic orbitals by electrons?

1.35 In your own words state Hund's rule. Draw electron configurations (see Table 1.3 on page 7) for the elements Si, P, and S and use them as examples to explain how Hund's rule is applied.

1.36 Explain why each of these electron configurations is incorrect.
(a) $1s^2 2s^1 2p^1$
(b) $1s^2 2s^2 2p^7$
(c) $1s^2 2s^2 2p^6 3s^2 3p^6 3d^{10}$
(d) $1s^2 2s^2 2p^6 3s^2 3p^5 4s^2 4s^6$

1.37 How many electrons can occupy each of the following?
(a) an *s* orbital   (b) a *p* orbital   (c) a *d* orbital?

1.38 Explain why these electron configurations are not possible for atoms of elements in their most stable state.
(a) $1s^2 2s^2 2p^6 3s^3$ (b) $1s^2 2s^2 2p^4 3s^1$
(c) $1s^2 2s^3 2p^1$ (d) $1s^2 2s^2 2p^6 3s^2 3p^1 4s^2$

1.39 What effect does the unshared pair of electrons have on the shape of an ammonia molecule?

1.40 In which group on the periodic table would an element with these characteristics be found?
(a) The element gains two electrons to form an anion.
(b) The element has four valence electrons.
(c) The element loses one electron to form a cation.
(d) The element has a stable electron configuration.

1.41 Draw the electron dot structure for each of these molecules.
(a) BrCl (one single bond).
(b) HCN (one single and one triple bond).
(c) SO_2 (one single and one double bond).
(d) HCCH (two single and one triple bond).

1.42 What relationship exists between the electron dot structure of an element and the location of the element in the periodic table?

SELF-TEST (REVIEW)

True/False

1. A triple covalent bond consists of a total of 3 shared electrons.

2. Every element in group 2A of the periodic table has two valence electrons.

3. Ionic compounds are typically formed between two nonmetals.

4. When atoms bond together, each atom attains a noble gas electron configuration.

5. It is possible to have polar bonds in a nonpolar molecule.

6. Based on Table 1.6 on page 21, it would be easier to break a carbon-hydrogen (C—H) single bond than a carbon-nitrogen (C—N) bond.

7. An aluminum atom has three valence electrons.

8. The ionic charge of an oxide ion is −2.

9. An oxygen atom has two unpaired valence electrons.

10. In a coordinate covalent bond, both bonding electrons are supplied by one of the atoms.

Multiple Choice

11. In an ionic compound, a sulfur atom would generally
(a) lose two electrons and form an anion
(b) lose two electrons and form a cation
(c) gain two electrons and form an anion
(d) gain two electrons and form a cation

12. The name of the bond between a carbon and an oxygen atom in carbon dioxide, CO_2, is
(a) double covalent   (b) coordinate covalent
(c) single covalent   (d) triple chorale

13. Which of these bonds would be the most polar? (Use Table 1.5 on page 19)
(a) C—Cl   (b) C—S   (c) C—O   (d) C—N

14. The correct electron dot formula of phosphorus, P, is
(a) · P ·   (b) · P :   (c) · P ·   (d) : P :

15. Which of these is the strongest attractive force?
(a) hydrogen bond   (b) dipole interaction
(c) dispersion forces   (d) single covalent bond

16. The correct electron configuration of oxygen is
(a) $1s^2, 2s^2, 2p^4$ (b) $1s^2, 2s^1, 2p^2, 3s^1, 3p^2, 4s^1$
(c) $1s^2, 2s^2, 2p^2, 3s^2$ (d) $1s^2, 2s^6$

17. The number of protons, neutrons, and electrons in an atom of argon-40 is respectively
(a) 18, 18, 22   (b) 22, 18, 18
(c) 18, 22, 18   (d) 18, 40, 18

18. Which of these molecules would not be polar?
(a) HBr   (b) CS_2 (linear)
(c) CH_3Cl (tetrahedral)   (d) H_2S (bent)

19. Which of these atoms has the same number of valence electrons as an atom of magnesium?
(a) Na   (b) Ne
(c) Ar   (d) Ca

20. How many electrons are shared in a single covalent bond?
(a) one electron   (b) eight electrons
(c) four electrons   (d) two electrons

21. Which of these compounds would not have covalent bonds?
(a) NO_2 (b) KF
(c) CO   (d) CBr_4

22. What kinds of bonds are found in a molecule of silicon dioxide?
(a) single covalent   (b) hydrogen
(c) double covalent   (d) triple covalent

23. How many electrons are there in the third energy level of a silicon atom?
(a) 14   (b) 4
(c) 12   (d) 8

24. The shape attained when four pairs of bonding electrons orient themselves according to electron-pair repulsion theory is
(a) square   (b) triangle
(c) quartet   (d) tetrahedral

25. The total number of shared electrons in a triple covalent bond is
(a) 2   (b) 3
(c) 6   (d) 9

26. An ion with an atomic number of 12 and ten electrons has a charge of
(a) 2+   (b) 1−
(c) 1+   (d) 2−

27. A water molecule can be best described as
(a) linear   (b) tetrahedral
(c) nonpolar   (d) bent

28. In which of the following bonds would oxygen have a partial positive charge?
(a) H—O   (b) N—O
(c) S—O   (d) F—O

29. Water molecules would hydrogen-bond with all of the following molecules except:
(a) CH_4 (b) H_2O
(c) NH_3 (d) HCN

30. Which of the following is not used to determine the order of filling of atomic orbitals?
(a) Pauli exclusion principle
(b) van der Waal's principle
(c) Aufbau principle
(d) Hund's rule

Carbon Chains and Rings

The Foundations of Organic Chemistry

Organic chemistry is the chemistry of the compounds of carbon. Daily, we encounter numerous carbon-based compounds. Paraffin wax, one of the many organic products obtained from petroleum, is used to make candles.

The element carbon constitutes only two-tenths of 1% of the mass of the elements found in the Earth's crust. Oxygen, silicon, aluminum, and iron are much more abundant, but carbon is the basis of life on Earth. Nearly one out of every ten atoms in our bodies is a carbon atom. Most of the rest are hydrogen and oxygen, mainly combined as water. Since *all* organic material is carbon based, so is the food our bodies burn as fuel.

Over 10 million organic compounds have been isolated or synthesized. Many have proved useful as medicines, plastics, fibers, fertilizers, insecticides, and a host of other products. In this and most of the following chapters we will explore the immense structural and chemical diversity of carbon compounds.

The combustion of carbon compounds is an important source of energy. Carbon compounds in coal and natural gas are burned for heat throughout the world. Like other chemical reactions, this common and widespread chemical reaction of carbon compounds must be handled with respect, as illustrated in the Case in Point.

CASE IN POINT: **Carbon monoxide poisoning**

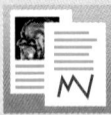

Claudine lives in the upper midwestern United States. A friend arrived early one morning to take Claudine on a shopping trip. Her friend rang Claudine's doorbell, but oddly, no one answered. The friend looked through a garage window and saw Claudine's car which suggested that she was at home. Concerned, the friend went to the nearest public phone and called the police. When they entered the house, the police found Claudine near the front door. She was conscious but disoriented. Her breathing was shallow, and her pulse was weak. Her face appeared unusually flushed. There was no strong odor in the house, and there was no evidence of foul play. Claudine was immediately taken to the hospital, where she was treated for carbon monoxide poisoning. We will find out how Claudine was poisoned and how she was treated in Section 2.11.

Carbon monoxide is a colorless, odorless, tasteless, and nonirritating gas. Although it is difficult to detect with the senses, people can be warned of carbon monoxide's presence by installing a carbon monoxide detector.

2.1 Carbon compounds

AIM: To describe the ways that carbon atoms are bonded in organic compounds.

Organic compounds exhibit dramatically different chemical and physical properties because they have different structures. At standard conditions, some are solids, some are liquids, and some are gases. Some taste sweet, and others taste sour. Some are poisons, but others are essential for life. In order to understand these properties of organic molecules, it is necessary to explore and understand their structures. Three simple principles can provide a basic understanding of the structure and chemistry of organic molecules:

1. Carbon atoms can form covalent bonds with hydrogen atoms.
2. Carbon atoms can form covalent bonds with other carbon atoms to build carbon chains.
3. Carbon atoms can form covalent bonds with other elements, especially oxygen, nitrogen, phosphorus, sulfur, and the halogens.

   We begin our study of organic chemistry with bonding between carbon and hydrogen atoms, and then we will examine bonding between carbon atoms. We will discuss carbon's bonding to other elements in later chapters.

2.2 Hydrocarbons

AIMS: To draw the electron dot and a structural formula of methane. To use orbital hybridization to describe the shape of the methane molecule.

Hydrocarbons *are compounds whose molecular structures contain only hydrogen and carbon.* The simplest hydrocarbons are the **alkanes**—*hydrocarbons that contain only single covalent bonds.* The smallest organic molecule is the alkane called *methane* (CH_4). Methane, a gas at standard temperature and pressure, is the major component of natural gas. It is sometimes called "marsh gas" because it is formed by the action of bacteria on decaying vegetation in swamps and other marshy areas.

The carbon-hydrogen bond in methane

A carbon atom contains four valence electrons. Four hydrogen atoms, each with one valence electron, can form four covalent carbon-hydrogen bonds. This combination is a molecule of methane:

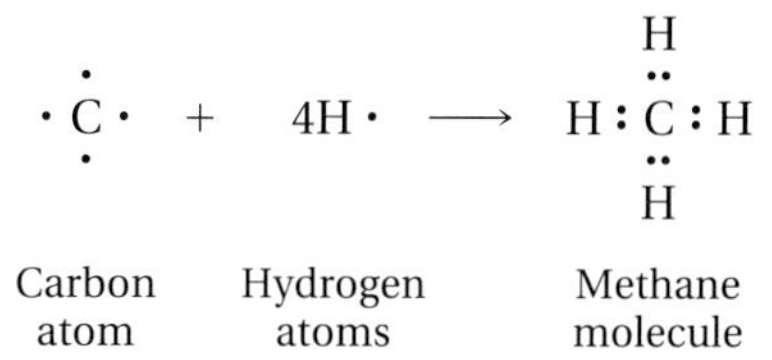

$$\cdot \overset{\displaystyle\cdot}{\underset{\displaystyle\cdot}{C}} \cdot \;+\; 4H\cdot \;\longrightarrow\; H\!:\!\overset{\displaystyle H}{\underset{\displaystyle H}{\overset{\cdot\cdot}{\underset{\cdot\cdot}{C}}}}\!:\!H$$

| Carbon atom | Hydrogen atoms | Methane molecule |

There is an important principle here: *Because a carbon atom contains four valence electrons, in making organic compounds it nearly always forms four covalent bonds.* Remembering this principle will help you to write complete and correct structures for organic molecules.

As you can see, the molecular formula of methane is CH_4. However, molecular formulas are not very useful in organic chemistry. Molecular formulas cannot, by themselves, contain enough information about a molecule, such as the arrangement of the atoms and the bonding between the atoms. For this reason, organic chemists prefer to write **structural formulas**—*formulas that show covalent bonds and the arrangement of atoms in each molecule.* Organic chemists usually abbreviate the bonding electron pairs in carbon-containing molecules as short lines. It will help to remember that the line between the atomic symbols represents *two electrons.* For example:

$$H-\overset{\displaystyle H}{\underset{\displaystyle H}{C}}-H$$

Line represents shared electron pair

Methane molecule

The shape of methane

Electron-pair repulsion theory predicts a tetrahedral shape for the methane molecule. This prediction agrees with experimental work that shows that methane is a completely symmetrical tetrahedral molecule; all the C—H bonds are identical, and all the H—C—H bond angles are 109.5 degrees (Fig. 2.1).

Electron configurations in atoms can be described by atomic orbitals. The bonding in methane and other molecules also can be described by atomic orbitals. When one atomic orbital overlaps with another atomic orbital, each with one electron, the pair of electrons is shared, and a covalent bond is formed. We can illustrate this concept with the hydrogen molecule, H_2. The 1s atomic orbitals of two hydrogen atoms overlap in the formation of a hydrogen molecule. Two electrons, one from each hydrogen atom, are available for sharing and a covalent bond is formed (Fig. 2.2).

However, if we attempt to generate a picture of methane by simple atomic orbital overlap, the method fails. A carbon atom has only two unpaired electrons, one in each of two 2p orbitals:

$$\boxed{\uparrow\downarrow} \quad \boxed{\uparrow\downarrow} \quad \boxed{\uparrow} \quad \boxed{\uparrow} \quad \boxed{\ }$$
$$\;\; 1s \qquad 2s \qquad\quad 2p$$

If the two electrons of the 2p orbitals were each shared with an electron of the 1s orbital of hydrogen atoms, we would get a molecule with the formula

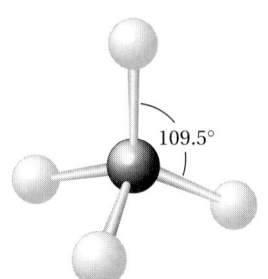

Figure 2.1
Ball-and-stick model of the methane molecule.

109.5°

Figure 2.2
The overlap of the 1s orbitals of two hydrogen atoms, each with one electron, produces a covalent electron pair bond. The product is a hydrogen molecule, H_2.

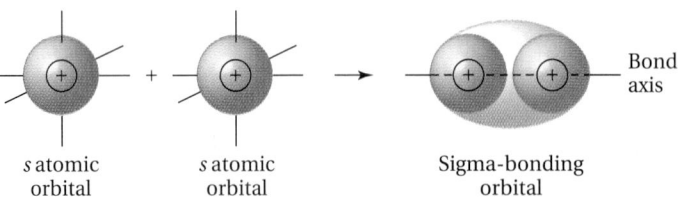

s atomic orbital   +   s atomic orbital   →   Sigma-bonding orbital   Bond axis

(handwritten margin notes: "where does from?", "3 comes from", "↑ I thought an orbital only existed if an e⁻ was in it.")

CH$_2$ instead of CH$_4$. One solution to this dilemma is **orbital hybridization**—*the mixing of two or more different atomic orbitals.* In methane, a carbon atom's 2s orbital and three 2p orbitals mix to form four identical sp^3 (read "s-p-three") hybrid atomic orbitals, each with one electron. Note that the number of hybrid orbitals that are formed is equal to the number of orbitals that are mixed. Because each sp^3 hybrid orbital is composed of 75% of dumbbell-shaped p orbitals and only 25% of a spherical s orbital, its appearance is similar to a fattened p atomic orbital. Figure 2.3 shows how the balloon-shaped sp^3 orbitals point toward the corners of a

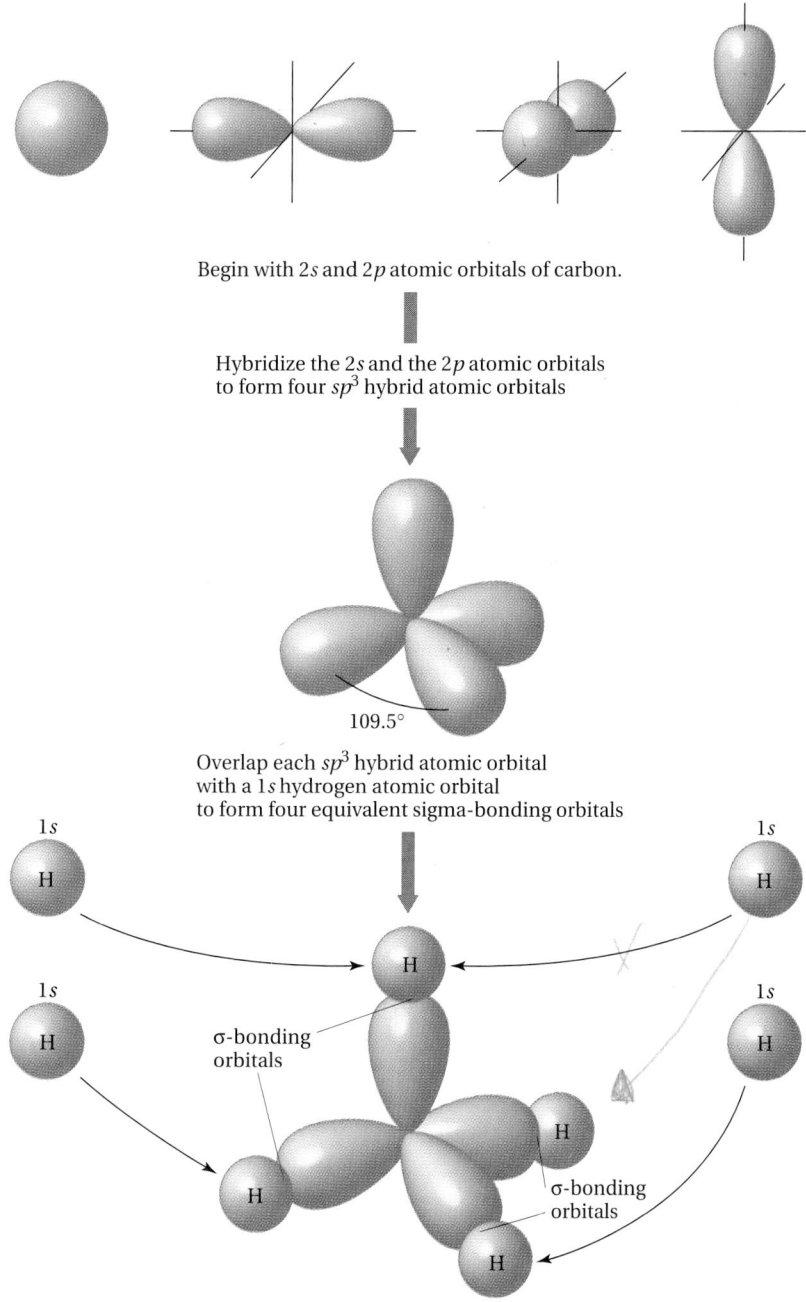

Begin with 2s and 2p atomic orbitals of carbon.

Hybridize the 2s and the 2p atomic orbitals to form four sp^3 hybrid atomic orbitals

109.5°

Overlap each sp^3 hybrid atomic orbital with a 1s hydrogen atomic orbital to form four equivalent sigma-bonding orbitals

1s   1s   1s   1s

σ-bonding orbitals

σ-bonding orbitals

Methane

Figure 2.3
Steps leading to an orbital description of the methane molecule. Hybridization of the 2s and 2p orbitals of a carbon atom produces four sp^3 hybrid atomic orbitals that point to the corners of a tetrahedron. Each of these hybrid orbitals overlaps with a 1s orbital of hydrogen to produce a methane molecule.

tetrahedron. The tetrahedral angle of 109.5 degrees between any two hybrid orbitals minimizes unfavorable interactions among the orbitals. Now the four sp^3 orbitals of carbon, each of which has one electron, can overlap with the $1s$ orbitals of four hydrogen atoms, each with one electron. The product is methane, CH_4. The carbon-hydrogen bonds in methane are all of a type called *sigma bonds.* Imagine slowly rotating the bond between a carbon nucleus and a hydrogen nucleus. The bond looks exactly the same throughout the rotation because it is symmetrical along the bond axis. *A covalent bond that is completely symmetrical along the axis connecting two atomic nuclei is a* **sigma bond** (the Greek letter sigma is σ). The covalent bond in the hydrogen molecule is a sigma bond. Carbon-hydrogen single bonds, carbon-carbon single bonds, and single bonds between carbon and other elements are also sigma bonds.

The structure of the resulting methane molecule is in agreement with experiment; it has four identical C—H bonds and is tetrahedral. Since sp^3 orbitals extend farther in space than either s or p orbitals, the overlap of a carbon sp^3 orbital with a hydrogen $1s$ orbital is greater than is possible with either a $2s$ or a $2p$ orbital of carbon. This greater overlap results in an unusually strong covalent bond.

2.3 Carbon-carbon bonds

AIM: To show with an electron dot and a structural formula how a carbon-carbon bond is formed in ethane, C_2H_6.

Focus

Ethane is the smallest alkane containing a carbon-carbon bond.

The unique ability of carbon to make stable carbon-carbon bonds and form chains is the major reason for the vast number of organic molecules. A few other elements, notably silicon, form short chains, but most silicon chains are unstable in an oxygen environment.

Like methane, ethane is a gas at standard temperature and pressure. Let us see how ethane could be formed from carbon and hydrogen. Two carbon atoms could share a pair of electrons to form a carbon-carbon covalent bond:

$$\cdot \overset{\cdot}{\underset{\cdot}{C}} \cdot + \cdot \overset{\cdot}{\underset{\cdot}{C}} \cdot \longrightarrow \cdot \overset{\cdot}{\underset{\cdot}{C}} - \overset{\cdot}{\underset{\cdot}{C}} \cdot$$

Carbon atoms   Electron pair bond

Now the remainder of the valence shell electrons of each carbon could be paired with the electrons from six hydrogen atoms to make a molecule of ethane:

$$\cdot \overset{\cdot}{\underset{\cdot}{C}} - \overset{\cdot}{\underset{\cdot}{C}} \cdot + 6H \cdot \longrightarrow H - \overset{\overset{\displaystyle H}{|}}{\underset{\underset{\displaystyle H}{|}}{C}} - \overset{\overset{\displaystyle H}{|}}{\underset{\underset{\displaystyle H}{|}}{C}} - H$$

Ethane

2.4 Straight-chain alkanes

AIM: To name and recognize structural, condensed, and molecular formulas of the straight-chain hydrocarbons containing up to 10 carbon atoms.

Focus

Straight-chain alkanes can be constructed using more than two carbon atoms.

We can build alkanes with more than two carbons by the same method used to build ethane. *Alkanes with more than two carbons strung together like the links of a chain are called* **straight-chain alkanes.** There is no need to resort to electron dot structures for each chain. We can just write the symbol for carbon as many times as we need in order to get the chain length and then fill in with hydrogens and lines representing covalent bonds. Remember that each carbon has four covalent bonds.

EXAMPLE 2.1

Writing structural formulas for straight-chain alkanes

Draw complete structural formulas for the straight-chain alkanes that have three and four carbons.

SOLUTION

Draw three carbons and four carbons in a row, and then put in lines to show the carbon-carbon bonds.

$$C-C-C \qquad C-C-C-C$$

Next, add enough covalent bonds to give each carbon a total of four bonds.

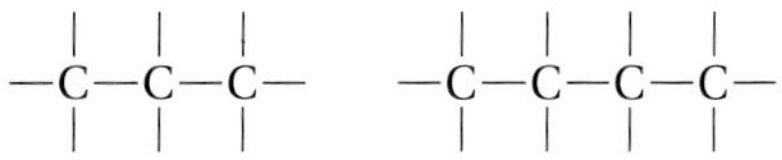

$$
\begin{array}{ccc}
| & | & | \\
-C-C-C- \\
| & | & |
\end{array}
\qquad
\begin{array}{cccc}
| & | & | & | \\
-C-C-C-C- \\
| & | & | & |
\end{array}
$$

Finally, complete the structures by adding hydrogens to fill out the carbon-hydrogen bonds.

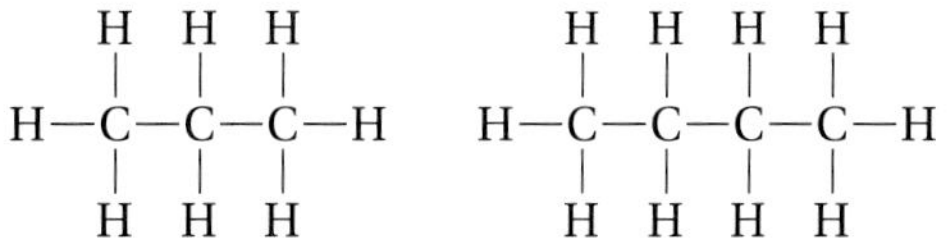

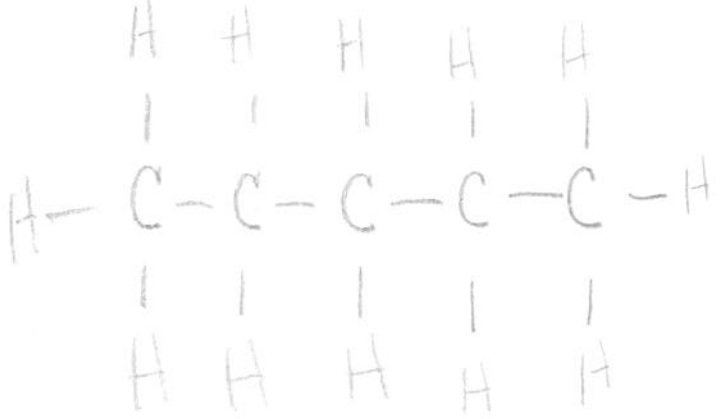

$$
\begin{array}{ccc}
H & H & H \\
| & | & | \\
H-C-C-C-H \\
| & | & | \\
H & H & H
\end{array}
\qquad
\begin{array}{cccc}
H & H & H & H \\
| & | & | & | \\
H-C-C-C-C-H \\
| & | & | & | \\
H & H & H & H
\end{array}
$$

PRACTICE EXERCISE 2.1

Draw complete structural formulas for the straight-chain alkanes with five and six carbons.

Drawing structural formulas

Sometimes we will find it convenient to draw complete structural formulas—that is, to show *all* the atoms and bonds in a molecule. There are, however, many ways to draw shorthand or *condensed structural formulas.*

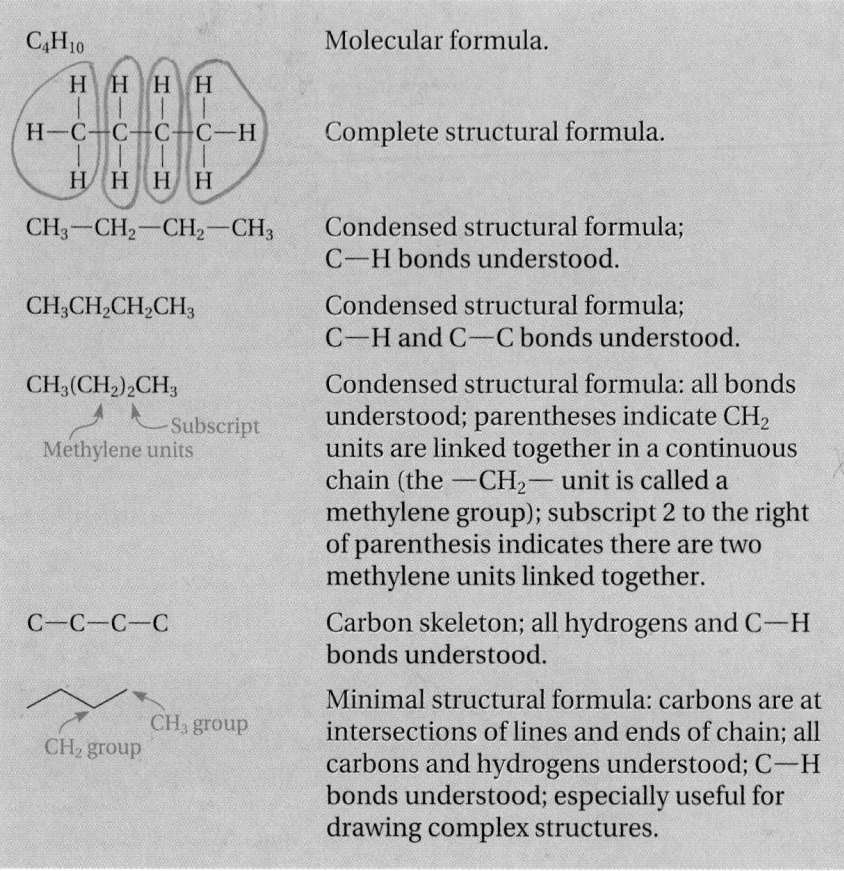

C_4H_{10} — Molecular formula.

Complete structural formula.

$CH_3—CH_2—CH_2—CH_3$ — Condensed structural formula; C—H bonds understood.

$CH_3CH_2CH_2CH_3$ — Condensed structural formula; C—H and C—C bonds understood.

$CH_3(CH_2)_2CH_3$
Methylene units —Subscript

Condensed structural formula: all bonds understood; parentheses indicate CH_2 units are linked together in a continuous chain (the —CH_2— unit is called a methylene group); subscript 2 to the right of parenthesis indicates there are two methylene units linked together.

C—C—C—C — Carbon skeleton; all hydrogens and C—H bonds understood.

CH_2 group CH_3 group

Minimal structural formula: carbons are at intersections of lines and ends of chain; all carbons and hydrogens understood; C—H bonds understood; especially useful for drawing complex structures.

Figure 2.4
Ways to represent structural formulas—in this case for a straight-chain alkane containing four carbons.

Condensed structural formulas *leave out part of the structure of a molecule.* What has been left out has to be understood by the person looking at the structure. Figure 2.4 shows several ways to draw condensed structural formulas for a straight-chain alkane containing four carbons.

PRACTICE EXERCISE 2.2

Draw all versions of the condensed structural formulas for straight-chain alkanes containing five and six carbons.

Origins of names of organic compounds

Table 2.1 shows the names and condensed structures for straight-chain alkanes containing up to 10 carbons. These names are recommended by the International Union of Pure and Applied Chemistry (IUPAC). *The* **IUPAC system** *is the basis for a precise, internationally accepted system of naming compounds.* We will use the IUPAC system to name many compounds.

Table 2.1 Structural Formulas of the First 10 Straight-chain Alkanes

Name	Molecular formula	Structural formula *unbranched*	Boiling point (°C)
methane	CH_4	CH_4	−161.0
ethane	C_2H_6	CH_3CH_3	−88.5
propane	C_3H_8	$CH_3CH_2CH_3$	−42.0
butane	C_4H_{10}	$CH_3CH_2CH_2CH_3$	0.5
pentane	C_5H_{12}	$CH_3CH_2CH_2CH_2CH_3$	36.0
hexane	C_6H_{14}	$CH_3CH_2CH_2CH_2CH_2CH_3$	68.7
heptane	C_7H_{16}	$CH_3CH_2CH_2CH_2CH_2CH_2CH_3$	98.5
octane	C_8H_{18}	$CH_3CH_2CH_2CH_2CH_2CH_2CH_2CH_3$	125.6
nonane	C_9H_{20}	$CH_3CH_2CH_2CH_2CH_2CH_2CH_2CH_2CH_3$	150.7
decane	$C_{10}H_{22}$	$CH_3CH_2CH_2CH_2CH_2CH_2CH_2CH_2CH_2CH_3$	174.1

Propane, often used for heating in rural areas where natural gas is not available, is more dense than air. If a propane leak develops in a building, propane can collect in a basement and is an explosion hazard. Because of its density, it is difficult to air out a propane-filled room.

All organic compounds can be named by the IUPAC system, but many are still called by their *common names,* which bear no relation to the IUPAC name or to the molecular structures of the compounds themselves. Common names for organic compounds isolated from nature often reflect the source of the compounds—for example, penicillin from the mold *Penicillium notatum.* Like the IUPAC names for many complex organic compounds, the IUPAC name for penicillin is too long and unwieldy for everyday use. In such instances, we will use the common names exclusively. The common names for even some simple organic compounds are so deeply ingrained in organic chemistry and biochemistry that we will mention them along with the IUPAC names.

PRACTICE EXERCISE 2.3

Refer to Table 2.1 to name the following alkanes:

(a) 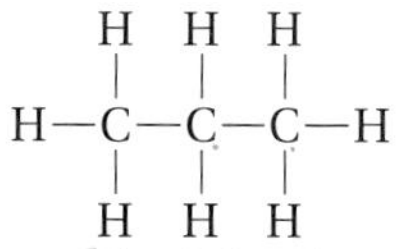 propane
(b) 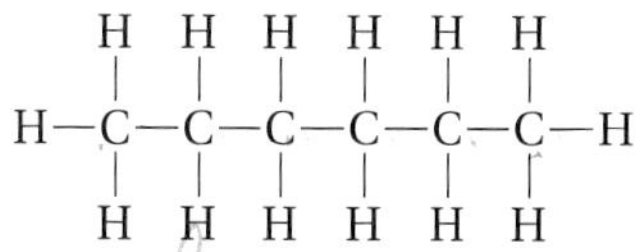 hexane
(c) 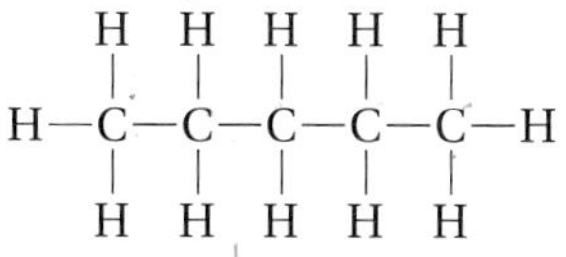 pentane

True shapes of straight-chain alkanes

Although they are usually drawn in a straight line, the carbon chains of straight-chain alkanes are actually zigzag shaped, as shown for hexane in Figure 2.5(a). The zigzag shape occurs because the bonds extending from

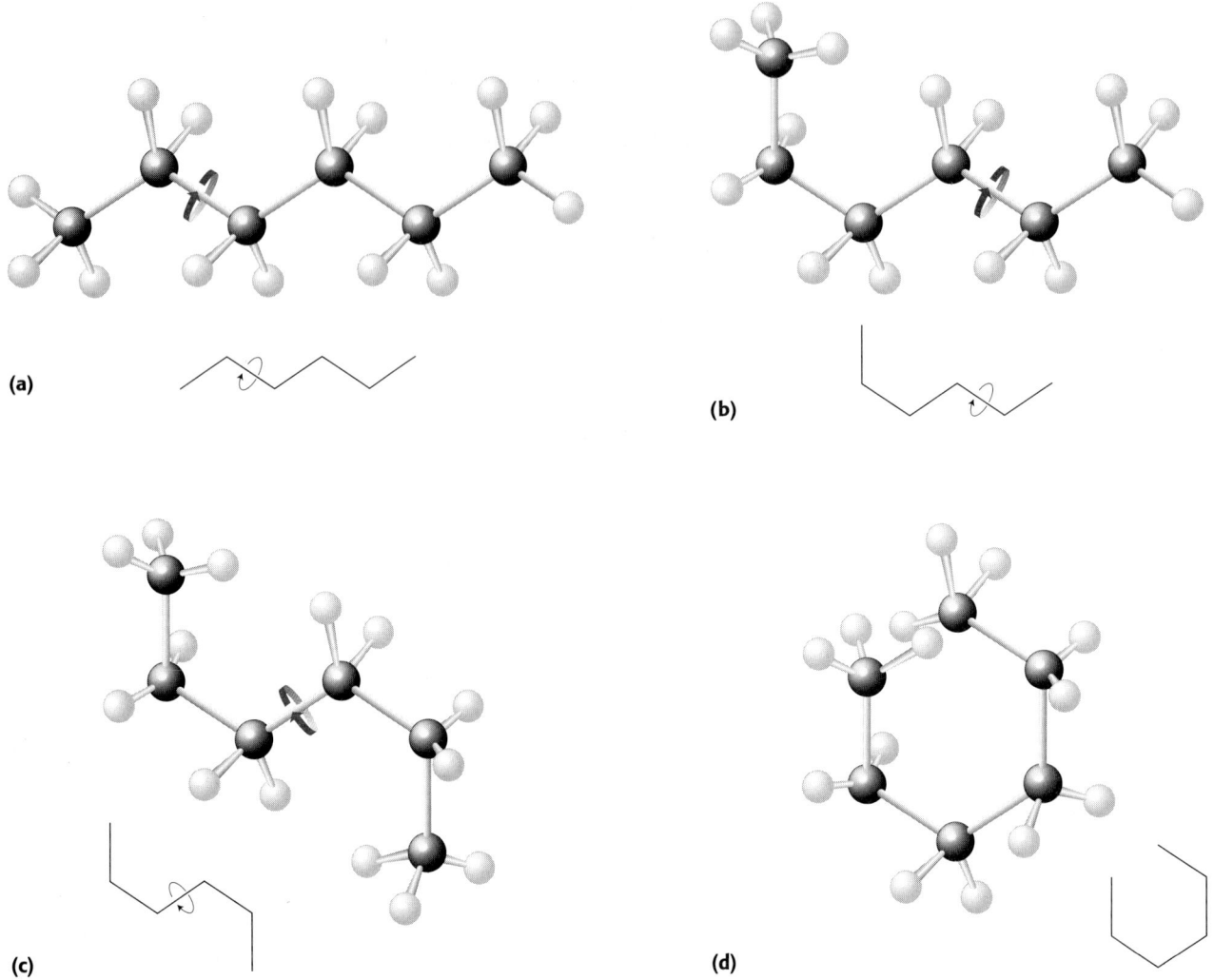

(a)

(b)

(c)

(d)

Figure 2.5
Ball-and-stick model of the extended form of hexane (a). Structures (b), (c), and (d) show a few of the possible coilings of the chain by rotations about carbon-carbon bonds.

The female mushroom fly produces a mixture of straight-chain alkanes from C_{15} to C_{30} that attracts the male fly. Such insect attractants are called *pheromones.*

the carbons in the chain are in a tetrahedral arrangement, and the C—C—C bond angles are 109.5 degrees, the normal tetrahedral bond angle. The zigzag form of a straight-chain alkane is the most stable shape the molecule can achieve because it puts the constituent atoms in the chain as far apart as they can be. However, the carbons in alkanes can rotate about carbon-carbon single bonds, as shown in Figure 2.5(b–d). At room temperature the freedom of rotation is nearly complete. Thus a beaker of hexane molecules is dynamic—it resembles the activity of a can of live worms much more than the inactivity on a plate of spaghetti.

2.5 Branched-chain alkanes

AIMS: To name an alkane according to IUPAC rules, given the structural formula of the alkane. To draw the structural formula of an alkane, given its IUPAC name.

Hydrocarbon groups can substitute for hydrogens on the straight-chain alkanes. For example, the middle hydrogen of propane could be replaced by the hydrocarbon group CH_3-, as shown in the following structures:

$$
\begin{array}{ccc}
& H & \\
& | & \\
CH_3 & -C- & CH_3 \\
& | & \\
& H &
\end{array}
\qquad
\begin{array}{ccc}
& CH_3 & \longleftarrow \text{Substituent} \\
& | & \\
CH_3 & -C- & CH_3 \\
& | & \\
& H &
\end{array}
$$

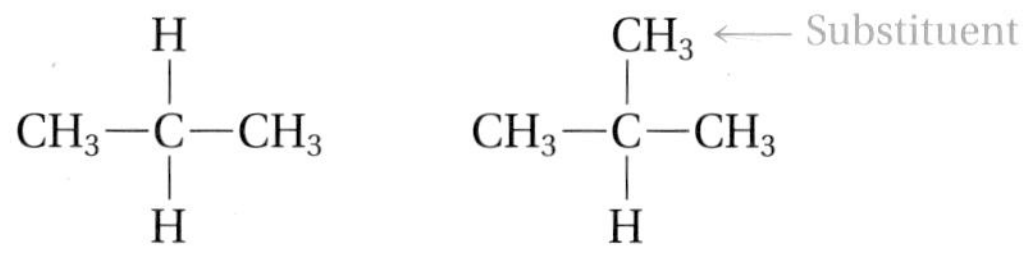

Propane

Groups that replace hydrogen on carbon are called **substituents.** *When the substituents are hydrocarbon groups, they are called* **alkyl groups.**

Naming branched-chain alkanes

Alkyl groups are named by removing the *-ane* ending from the alkane name and adding *-yl.* For example, the group CH_3- is derived from the hydrocarbon methane (the parent hydrocarbon) and is called a *methyl* group. The group CH_3CH_2-, derived from the parent hydrocarbon ethane, is called an *ethyl* group, and the group $CH_3CH_2CH_2-$, derived from the parent hydrocarbon propane, is a *propyl* group. The straight-chain alkyl group derived from butane is called a *butyl* group ($CH_3CH_2CH_2CH_2-$). The presence of alkyl groups as substituents in a hydrocarbon molecule turns straight-chain alkanes into *branched-chain alkanes. In a* **branched-chain alkane,** *hydrocarbon substituents (alkyl groups) are attached to a parent straight-chain alkane.* The following structures are examples of branched-chain alkanes. Hydrogens have been omitted in these structures.

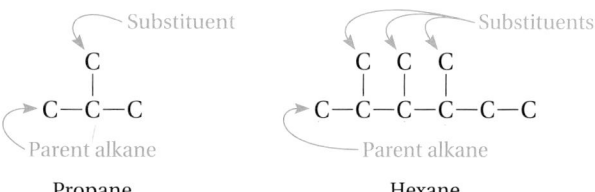

Propane   Hexane

We will use five rules to help us determine the IUPAC names of branched-chain hydrocarbons:

1. Find the longest chain of carbons in the molecule. This chain is used as the parent name of the compound.

2. Number the carbons in the main chain in sequence, starting at the end that will give the alkyl substituents attached to the chain the *smallest numbers.*

3. Give the appropriate position number and name for each alkyl group.

4. Attach an appropriate prefix to groups that appear more than once in the structure: *di-* (twice), *tri-* (three times), *tetra-* (four), and *penta-* (five).

5. List the names of alkyl substituents in alphabetical order. For purposes of alphabetizing, the prefixes *di-*, *tri-*, and so on are ignored.

Examples 2.2 to 2.4 show how to apply these rules to naming branched-chain hydrocarbons. If we know the IUPAC name of a compound, the structure is easy to reconstruct, as shown in Examples 2.5 and 2.6.

EXAMPLE 2.2   **Naming a branched-chain alkane**

What is the IUPAC name for the following branched-chain alkane?

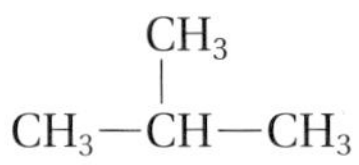

$$CH_3$$
$$|$$
$$CH_3-CH-CH_3$$ *2-methylpropane*

SOLUTION

Number the longest chain of carbons in the molecule (*rule 1*). Since the longest chain contains three carbons, the parent name of the compound will be *propane*.

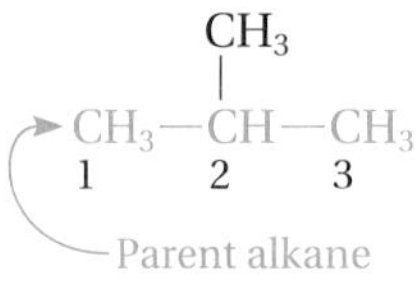

$$CH_3$$
$$|$$
$$CH_3-CH-CH_3$$
$$1 \quad 2 \quad 3$$
— Parent alkane

Propane

The chain could be numbered in other ways as follows, but notice that in this molecule, a methyl substituent always ends up in the 2 position.

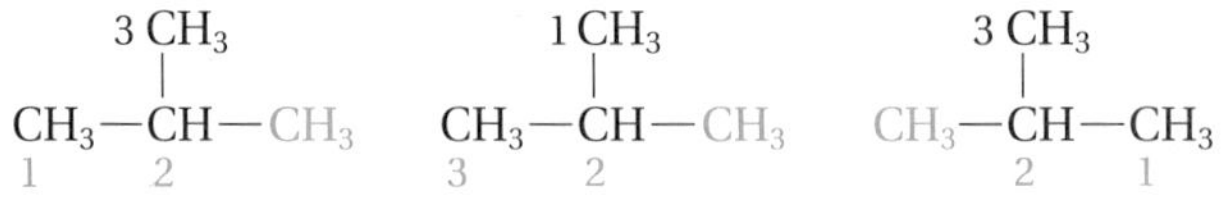

$3\ CH_3$	$1\ CH_3$	$3\ CH_3$
$\|$	$\|$	$\|$
$CH_3-CH-CH_3$	$CH_3-CH-CH_3$	$CH_3-CH-CH_3$
$1 \quad 2$	$3 \quad 2$	$2 \quad 1$

The 2 position is the lowest number the methyl substituent can have (*rule 2*). Hyphens are used to separate position numbers and substituent names. The substituent is a 2-methyl group, no matter how the numbering is done (*rule 3*).

Rules 4 and 5 do not apply, since we have only one substituent.

Name the compound by combining the substituent number and name, *2-methyl*, with the name of the parent chain, *propane*. The name of this compound appears to be *2-methylpropane*. The number 2 is not necessary here, however, because there is no other place a methyl substituent can be attached to one propane. The compound is *methylpropane*. Notice that we write the name of the alkane as one word, *not* as two words, as in methyl propane.

EXAMPLE 2.3 **Naming a branched-chain hydrocarbon**

Use the IUPAC rules to name this branched-chain alkane:

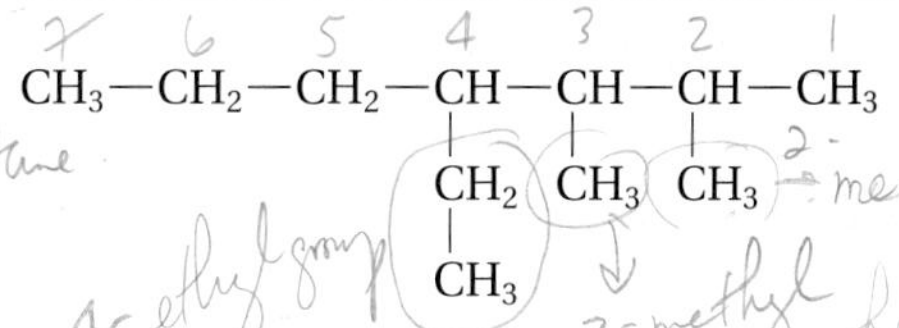

$$CH_3—CH_2—CH_2—CH—CH—CH—CH_3$$

(handwritten annotations: 4-ethyl-2,3-dimethylheptane, 4-ethyl group, 3-methylheptane, 2-methyl group)

SOLUTION

The longest chain in our example contains seven carbons, so the parent straight-chain alkane is heptane (*rule 1*).

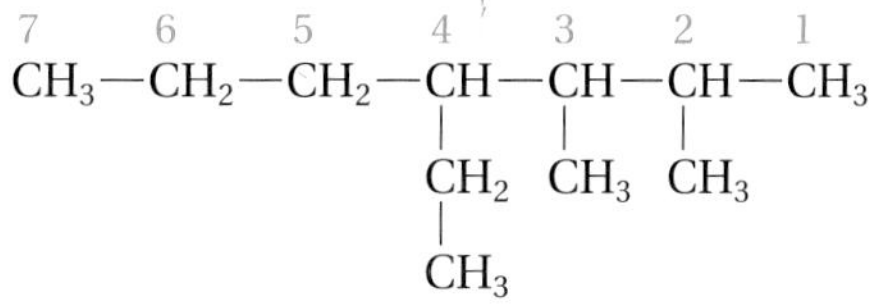

$$\overset{7}{CH_3}—\overset{6}{CH_2}—\overset{5}{CH_2}—\overset{4}{CH}—\overset{3}{CH}—\overset{2}{CH}—\overset{1}{CH_3}$$
$$\quad\quad\quad\quad\;\; CH_2 \quad CH_3 \quad CH_3$$
$$\quad\quad\quad\quad\;\; CH_3$$

Number the chain carbons to give the lowest possible numbers to the substituents (*rule 2*). This has already been done in the preceding structure. In this instance, the numbers go from right to left, which places the substituents at carbon atoms 2, 3, and 4. If the chain were numbered from left to right, the substituents would be at positions 4, 5, and 6—these are higher numbers and would violate the rule.

The example contains two methyl groups and one ethyl group. In this example the substituents and positions are 2-methyl, 3-methyl, and 4-ethyl (*rule 3*).

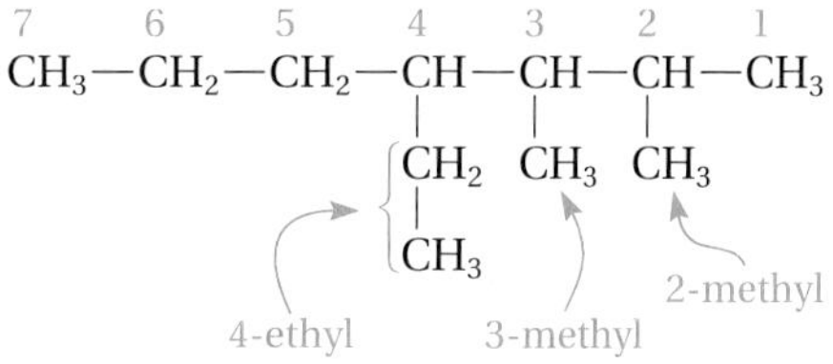

$$\overset{7}{CH_3}—\overset{6}{CH_2}—\overset{5}{CH_2}—\overset{4}{CH}—\overset{3}{CH}—\overset{2}{CH}—\overset{1}{CH_3}$$
$$\quad\quad\quad\quad CH_2 \quad CH_3 \quad CH_3$$
$$\quad\quad\quad\quad CH_3$$

4-ethyl   3-methyl   2-methyl

This example has two methyl substituents. To name these, group them together by using the prefix *di-* to make the word *dimethyl* (*rule 4*). Also designate the positions of the methyl groups. Indicate the name and numbers as *2,3-dimethyl*. Commas are used to separate numbers in naming organic compounds.

Now list the substituent names in alphabetical order, ignoring the prefixes in alphabetizing (*rule 5*). The substituents in the example are written *4-ethyl-2,3-dimethyl*.

Recalling that the parent name is heptane, combine the substituent names with the name of the parent to obtain the name of the compound. According to the IUPAC rules, the name of the compound is *4-ethyl-2,3-dimethylheptane*.

EXAMPLE 2.4   **Naming alkanes by the IUPAC system**

Name the following compound according to the IUPAC system:

$$CH_3-CH_2-\underset{\underset{\overset{|}{CH_2}}{\underset{\overset{|}{CH_2}}{\overset{|}{\underset{\overset{|}{CH_3}}{C}}}}-CH_3$$

3,3-dimethylhexane

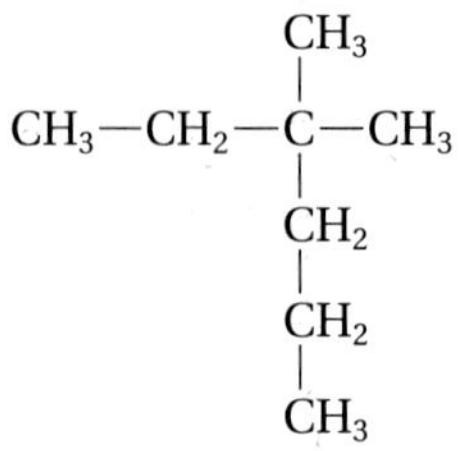

SOLUTION

Notice that the longest chain of carbons is *not* written in a straight line in the molecule.

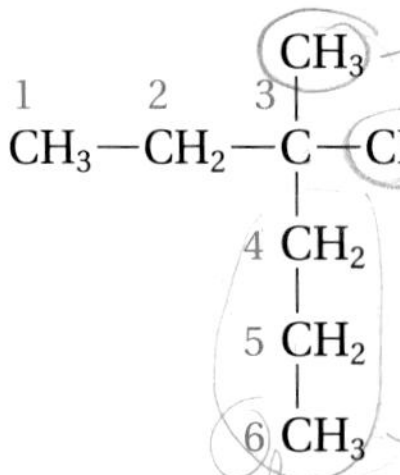

methyl group
methyl group
3,3-dimethylhexane
hexane

The longest chain has six carbon atoms, so the parent straight-chain alkane name is hexane. The chain is numbered so that the methyl substituents have the lowest possible numbers. When a carbon atom carries two identical substituents, the number is repeated. The compound is 3,3-dimethylhexane.

a) 2-methylbutane
b) 3-methylpentane
c) 3-ethylhexane

PRACTICE EXERCISE 2.4

Name the following compounds according to the IUPAC system.

(a) $\underset{4}{CH_3}-\underset{3}{CH_2}-\underset{2}{\underset{\overset{|}{CH_3}}{CH}}-\underset{1}{CH_3}$ *2-methyl*

(b) $\underset{1}{CH_3}-\underset{2}{\underset{\overset{|}{CH_2}}{\underset{\overset{|}{CH_3}}{CH}}}-\underset{3}{CH_2}-\underset{4}{CH_3}$ *3-methylp*

(c) $\underset{5}{CH_2}-\underset{4}{CH_2}-\underset{3}{\underset{\overset{|}{CH_2}}{\underset{\overset{|}{CH_3}}{CH}}}-\underset{2}{CH_2}-\underset{1}{CH_3}$

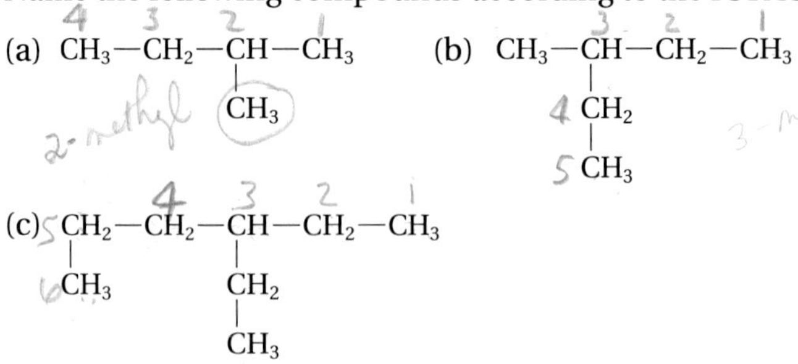

EXAMPLE 2.5   **Constructing a structural formula from a name**

Draw a condensed structural formula for 3,3-dimethylpentane.

$$\underset{1}{CH_3}-\underset{2}{CH_2}-\underset{\underset{3}{\underset{\overset{|}{CH_3}}{C}}}{\overset{\overset{CH_3}{|}}{C}}-\underset{4}{CH_2}-\underset{5}{CH_3}$$

SOLUTION

Draw five carbons and connect them with a line to show the single carbon-carbon bonds of pentane, the parent hydrocarbon. Number the carbons.

$$C-C-C-C-C$$
$$1 \quad 2 \quad 3 \quad 4 \quad 5$$

Attach two methyl groups to the 3 position of the structure.

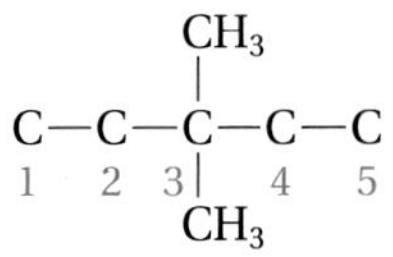

$$
\begin{array}{c}
\text{CH}_3 \\
| \\
C-C-C-C-C \\
1 \quad 2 \quad 3| \quad 4 \quad 5 \\
\text{CH}_3
\end{array}
$$

Complete the structure by filling in hydrogens so that each carbon has four single covalent bonds. This gives the structure of 3,3-dimethylpentane.

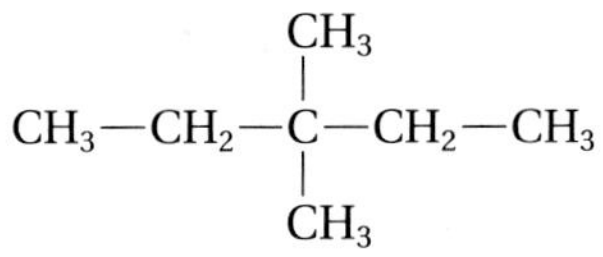

$$
\begin{array}{c}
\text{CH}_3 \\
| \\
\text{CH}_3-\text{CH}_2-\text{C}-\text{CH}_2-\text{CH}_3 \\
| \\
\text{CH}_3
\end{array}
$$

3,3-Dimethylpentane

EXAMPLE 2.6 — **Constructing a structural formula from a name**

Draw a condensed structural formula for 2,2,4-trimethylpentane.

SOLUTION

As in the preceding example, draw and number the parent chain, which is pentane.

$$C-C-C-C-C$$
$$1 \quad 2 \quad 3 \quad 4 \quad 5$$

According to the name, the structure has two methyl groups in the 2 position and one methyl group in the 4 position. It will be convenient to add only the carbons of the methyl groups and fill out the structure with hydrogens later.

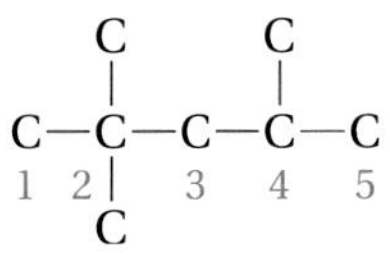

$$
\begin{array}{c}
C \quad\quad C \\
| \quad\quad | \\
C-C-C-C-C \\
1 \quad 2| \quad 3 \quad 4 \quad 5 \\
C
\end{array}
$$

Now fill in the hydrogens to produce the completed structure.

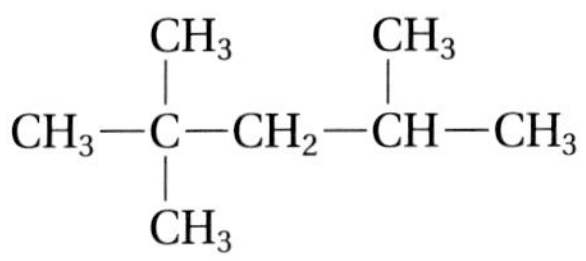

$$
\begin{array}{c}
\text{CH}_3 \quad\quad \text{CH}_3 \\
| \quad\quad\quad | \\
\text{CH}_3-\text{C}-\text{CH}_2-\text{CH}-\text{CH}_3 \\
| \\
\text{CH}_3
\end{array}
$$

2,2,4-Trimethylpentane

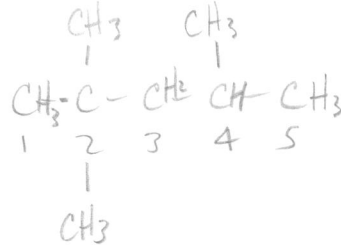

(handwritten notes in margins)

a) CH₃ CH₃ → does it matter whether top or bottom NO

CH₃ CH CH CH₃ CH₂ CH₃
1   2   3   4   5   6

b) CH₃ CH₂ CH CH₂ CH CH₂ CH₃
          | |
          CH₂
          |
          CH₃

PRACTICE EXERCISE 2.5

Draw a condensed structural formula for each of the following compounds:

(a) 2,3-dimethylhexane

(b) 3-ethylheptane

(c) 3-ethyl-2,4-dimethyloctane

c) CH₃ – CH – CH – CH – CH₂ – CH₂ – CH₂ – CH₃
          | | |
          CH₃   CH₂   CH₃

(handwritten: book answer CH₂CH₃ ok.)

2.6 Structural isomers

AIM: To name and draw structural isomers of hydrocarbons.

Structural isomers have the same molecular formula but different molecular structures and properties.

We can draw the structures of two or more alkanes that have the same molecular formula but different molecular structures. *Compounds that have the same molecular formula but different molecular structures are called* **structural isomers.** For example, there are two different molecules with the formula C_4H_{10}: butane and methylpropane. They are structural isomers of each other.

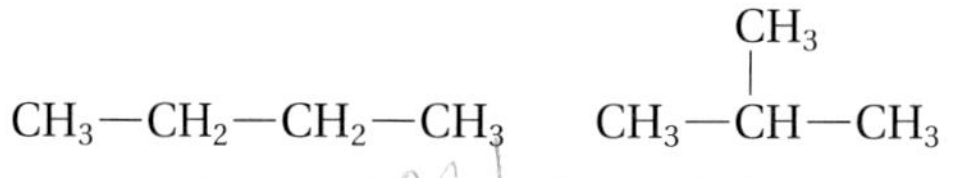

$$CH_3-CH_2-CH_2-CH_3 \qquad CH_3-\underset{\overset{|}{CH_3}}{CH}-CH_3$$

Butane
(C_4H_{10}) *(handwritten: 0°C bp)*

Methylpropane
(C_4H_{10}) *(handwritten: –10.2°C bp)*

The existence of structural isomers is the reason organic chemists do not often use molecular formulas. To be clear, they need to use IUPAC names or structural formulas. Structural isomers have different physical properties, such as boiling points or melting points, and different chemical reactivities. For example, butane boils at 0 °C, and its structural isomer, methylpropane, boils at −10.2 °C. In general, the more highly branched the hydrocarbon structure, the lower is its boiling point compared with its other structural isomers. Pentane, C_5H_{12}, provides another example.

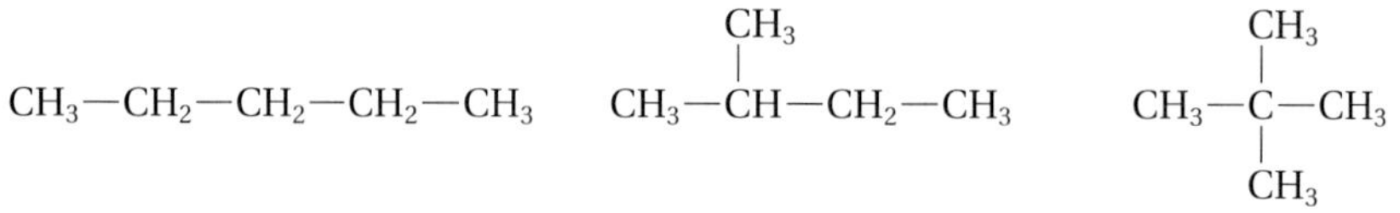

Structural isomerism possible with carbon compounds helps explain the immense number of organic compounds. Octane, C_8H_{18}, has 18 structural isomers. Decane, $C_{10}H_{22}$, has 75 isomers. The alkane with 20 carbon atoms, $C_{20}H_{42}$, has 336,319 isomers. All these isomers do not occur in nature, but any one of them could theoretically be made in the laboratory.

$$CH_3-CH_2-CH_2-CH_2-CH_3 \qquad CH_3-\underset{\overset{|}{CH_3}}{CH}-CH_2-CH_3 \qquad CH_3-\underset{\overset{|}{CH_3}}{\overset{\overset{CH_3}{|}}{C}}-CH_3$$

Pentane
(C_5H_{12})
(bp 36 °C)

Methylbutane
(C_5H_{12})
(bp 28 °C)

Dimethylpropane
(C_5H_{12})
(bp 9 °C)

PRACTICE EXERCISE 2.6

There are five structural isomers with the molecular formula C_6H_{14}. Draw and name each one. (For convenience, you may wish to draw only the carbon skeleton for each structure.)

(handwritten answers:)

① C–C–C–C–C–C ✓ hexane

② C–C–C–C–C ✓
        |
        C
2-methylpentane

③ C–C–C–C ✓
      | |
      C C
2,3-dimethylbutane

④ C–C–C–C ?
        |
        C
        |
        C 3-methylpentane

⑤ C–C–C–C
      |
      C
2,2-dimethylbutane

2.7 Cycloalkanes

AIM: To name and draw structural isomers for cycloalkanes.

Focus

Ends of carbon chains can join
to form rings.

The ends of the carbon chains of some alkanes are joined. *Alkanes with joined ends are called* **cycloalkanes,** and the structures formed are called *rings* (Table 2.2). Many biologically important molecules such as cholesterol and the steroid hormones contain carbon rings. The parent cyclo-

Table 2.2 Names and Structures for Some Cycloalkanes

Name	Structure	Shorthand structure
cyclopropane	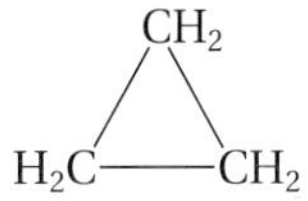 CH_2 ⟋△⟍ H_2C——CH_2	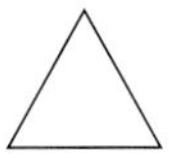 △
cyclobutane	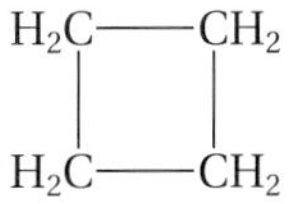 H_2C——CH_2 H_2C——CH_2	▢
cyclopentane	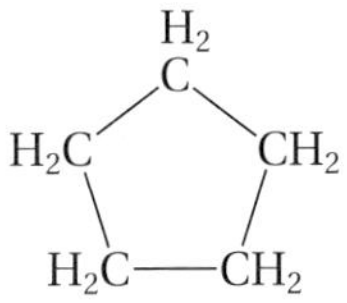 H_2 C H_2C CH_2 H_2C——CH_2	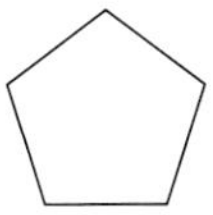 ⬠
cyclohexane	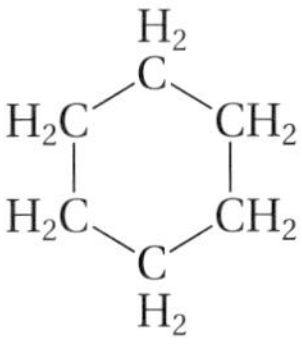 H_2 C H_2C CH_2 H_2C CH_2 C H_2	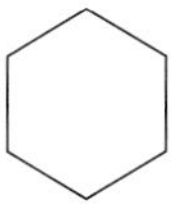 ⬡
cycloheptane	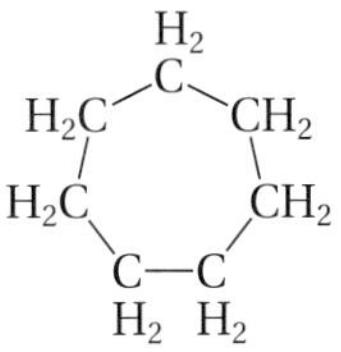 H_2 C H_2C CH_2 H_2C CH_2 C—C H_2 H_2	 ⬡
cyclooctane	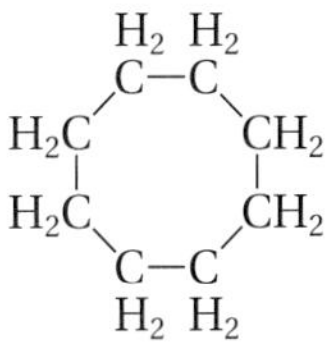 H_2 H_2 C—C H_2C CH_2 H_2C CH_2 C—C H_2 H_2	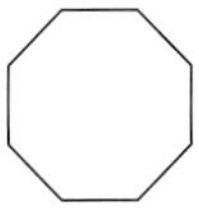 ⯃

alkanes are named somewhat like the parent straight-chain alkanes. In this case, however, the prefix *cyclo-* is attached to the name of the parent hydrocarbon.

Substituted cycloalkanes

Substituted cycloalkanes are numbered and named according to the same IUPAC rules as straight-chain alkanes. Here are a few examples:

Ethylcyclohexane
(a number is not
needed if there
is only one
substituent)

1,3-Dimethylcyclobutane

1,3,6-Trimethylcyclooctane

The simplest cycloalkane, cyclopropane, C_3H_6, is used as an inhalation anesthetic. Since cyclopropane is nonpolar, it is not very soluble in blood. However, sufficient cyclopropane dissolves and is transported by the blood to the brain, where it is transferred to brain cells. Cyclopropane is both flammable and explosive and must be handled carefully.

Rings containing from 3 to more than 30 carbons are found in nature. Five- and six-membered rings, the cyclopentanes and cyclohexanes, are the most abundant. Of the common unsubstituted cycloalkanes, cyclopropane and cyclobutane are gases at standard temperature and pressure. Cyclopropane is sometimes used as a surgical anesthetic. The remainder of the cycloalkanes are oily liquids.

True shapes of cycloalkanes

Carbon-carbon bonds are easier to break in smaller rings than in larger ones. In cyclopropane and cyclobutane rings, the C—C—C bond angles are 60 and 90 degrees, respectively. These rings are severely strained by this compression of the normal tetrahedral bond angle, which is 109.5 degrees. As a result, cyclopropane and cyclobutane rings are much less stable than cyclopentane or cyclohexane rings.

Placing carbon in a cycloalkane ring locks the carbon so that its freedom to rotate is severely restricted. The highly strained cyclopropane ring is locked into a flat, rigid shape (Fig. 2.6). Cyclobutane rings are also highly strained. In fact, it is impossible to make ball-and-stick models of cyclopropane and cyclobutane without using flexible springs instead of rigid sticks to join the ring carbons. The C—C—C bond angles in cyclopentane rings are nearly normal—about 108 degrees—and cyclopentane rings are not very strained. Cyclopentane and larger rings are also rather flexible; they readily "pucker" or bend. The puckered form of cyclopentane resembles an open envelope. If cyclohexane rings were flat, their C—C—C bond angles would be 120 degrees, and the ring would be strained, since 120 degrees is much larger than 109.5 degrees. In cyclohexane and larger rings, however, puckering reduces ring strain and produces normal tetrahedral bond angles of 109.5 degrees. One puckered shape that a cyclohexane molecule takes resembles a reclining chair. The cyclohexane ring can flip from one chair form to another through an intermediate shaped like a boat. The

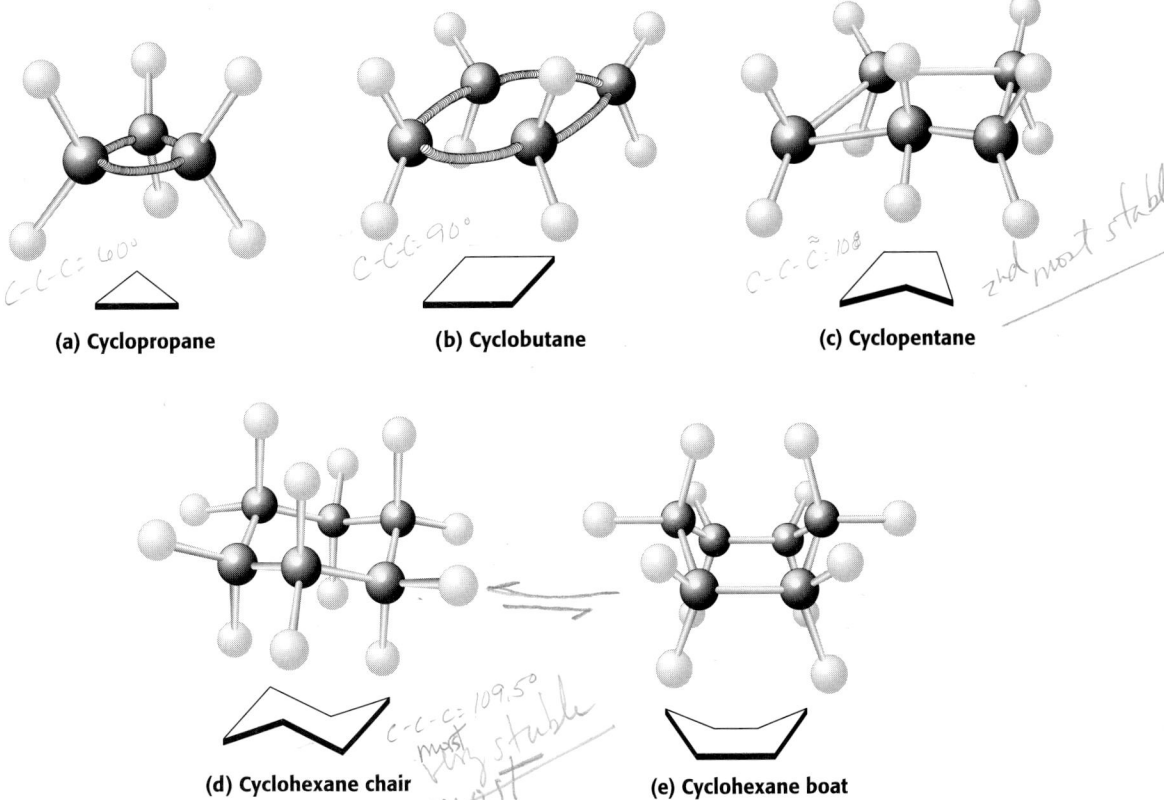

(a) Cyclopropane   C-C-C: 60°

(b) Cyclobutane   C-C-C: 90°

(c) Cyclopentane   C-C-C̃: 108°   2nd most stable

(d) Cyclohexane chair   C-C-C: 109.5°   most stable   most stable   most

(e) Cyclohexane boat

Figure 2.6
Ball-and-stick models showing the true shapes of cycloalkane rings. Flexible connectors (springs) must be used to connect the ring carbon atoms in strained rings such as cyclopropane (a) and cyclobutane (b), since the holes in the carbon atoms of the models are drilled at angles of 109.5 degrees and the angles in the three-membered and four-membered rings are only 60 and 90 degrees, respectively. The cyclopentane ring (c) is puckered and shaped like an open envelope. Cyclohexane rings are also puckered and can take the shape of a chair (d) or a boat (e).

chair and boat forms of cyclohexane are in dynamic equilibrium, but the chair forms are more stable and are therefore highly favored over the boat.

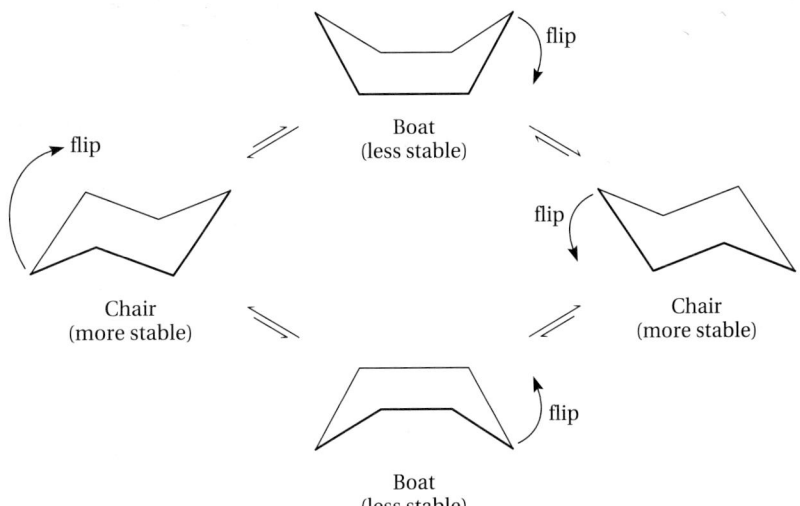

The reason the boat form of cyclohexane is less stable than the chair form is that ring hydrogens in the boat form infringe on each other's space, which is an unfavorable situation. These interfering hydrogens are called *flagpole hydrogens* because they resemble flagpoles on the bow and stern of the cyclohexane boat. In the chair form of cyclohexane, the ring hydrogens are in positions where they do not interfere with each other.

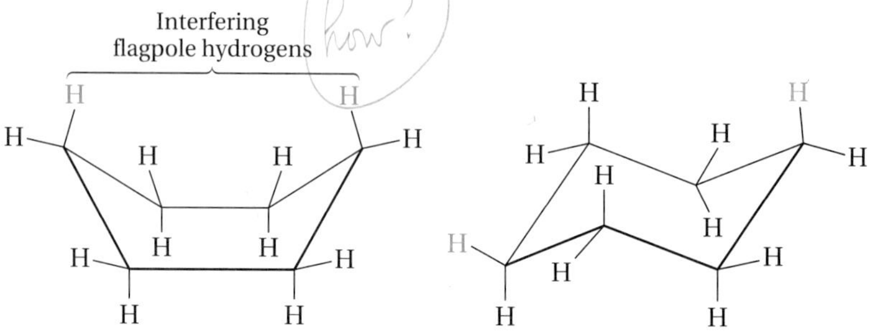

Interfering
flagpole hydrogens   *how?*

(a) Boat   **(b) Chair**

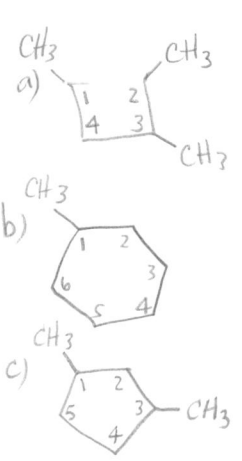

PRACTICE EXERCISE 2.7

Draw a carbon skeleton for each of the following:

(a) 1,2,3-trimethylcyclobutane

(b) methylcyclohexane   (c) 1,3-dimethylcyclopentane

2.8 Multiple bonds

AIM: To name and draw structural isomers of alkenes, cycloalkenes, and alkynes, identifying cis and trans geometric isomers.

Focus

Carbon atoms can combine to make double or triple bonds.

The straight-chain alkanes, branched-chain alkanes, and cycloalkanes have—are *saturated* with—the largest number of hydrogens that making four bonds to carbon will permit. *For this reason, the straight-chain alkanes, branched-chain alkanes, and cycloalkanes are called* **saturated hydrocarbons.** In saturated hydrocarbons, all the carbon-carbon bonds in alkanes and cycloalkanes are single bonds. But multiple bonds between carbons also exist. *Organic compounds containing carbon-carbon double bonds are called* **alkenes.** *Alkenes contain fewer than the largest possible number of hydrogens in their molecular structures; for this reason, they are sometimes called* **unsaturated compounds.** This is the carbon-carbon double bond found in alkenes:

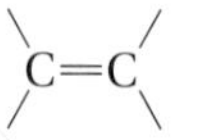

$$\diagup C=C \diagdown$$ *Carbon*

Naming alkenes

To name an alkene by the IUPAC system, we find the longest straight chain in the *molecule that contains the double bond.* This is the parent alkene. It gets the root name of the alkane with the same number of carbons but the ending -*ene* rather than -*ane*. For example, the alkene with two carbons is called *ethene;* the alkene with three carbons is called propene.

$$
\begin{array}{c}
\text{H} \qquad\qquad \text{H}\\
\diagdown \qquad\qquad \diagup\\
\text{C}=\text{C}\\
\diagup \qquad\qquad \diagdown\\
\text{H} \qquad\qquad \text{H}
\end{array}
\qquad\qquad
\begin{array}{c}
\text{H} \;\; \text{H}\\
| \;\; |\\
\text{CH}_3-\text{C}=\text{C}-\text{H}
\end{array}
$$

Ethene   Propene
(ethylene)   (propylene)

Ethene and propene, the smallest alkenes, are often called by the common names *ethylene* and *propylene.*

In naming alkenes, we number the longest chain containing the double bond so that the positions of the carbon atoms of the double bond get the lowest possible numbers. Substituents on the chain are named and numbered the same way as for the alkanes. Here are some examples of the structures and IUPAC names of simple alkenes.

$$
\begin{array}{c}
\text{H} \;\; \text{H}\\
| \;\; |\\
\text{H}-\text{C}=\text{C}-\text{CH}_2-\text{CH}_3
\end{array}
\qquad
\begin{array}{c}
\text{H} \;\; \text{H}\\
| \;\; |\\
\text{CH}_3-\text{C}=\text{C}-\text{CH}_3
\end{array}
\qquad
\begin{array}{c}
\text{CH}_3 \;\; \text{H} \;\; \text{H}\\
| \;\; | \;\; |\\
\text{CH}_3-\text{CH}-\text{C}=\text{C}-\text{CH}_3
\end{array}
$$

1-Butene   2-Butene   4-Methyl-2-pentene

(1-butene and 2-butene are structural isomers of C_4H_8)

Double bonds are also found in ring compounds. The same change in the parent cycloalkane name from -*ane* to -*ene* applies in naming the cycloalkenes. Some examples of the cycloalkenes are given here.

Cyclobutene   Cyclopentene   Cyclohexene

Shape of the double bond

There is no rotation about carbon-carbon double bonds. Experimental evidence shows that the four bonds that project from a pair of double-bonded carbons lie in a plane and are 120 degrees apart (Fig. 2.7). This is the maximum separation that can be obtained without breaking bonds, and it is what would be predicted by the electron pair repulsion theory.

Recall that we used orbital hybridization to obtain a description of the shape of the methane molecule. The orbital description of the carbon-carbon double bond in ethene relies on hybridization as well, but the hybridization of carbon in ethene is different from that in methane. Use Figure 2.8 to visualize the steps leading to the structure of ethene. Remember that each orbital represents one electron.

In ethene, the 2*s* orbital of carbon mixes with two 2*p* orbitals to give three hybrid atomic orbitals called sp^2. Note that the number of hybrid orbitals is always the same as the number of orbitals mixed. Each hybrid orbital is separated from the other two by 120 degrees. Overlap of one of the sp^2 hybrid orbitals of one carbon with that of another results in formation

Ethylene, C_2H_4, is used as a ripening agent for tomatoes and citrus fruits.

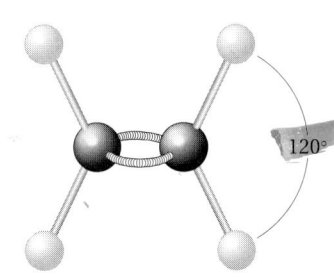

Figure 2.7
Ball-and-stick model of ethene (common name ethylene). Rotation about the double bond is prohibited. All the atoms are in the plane of the page.

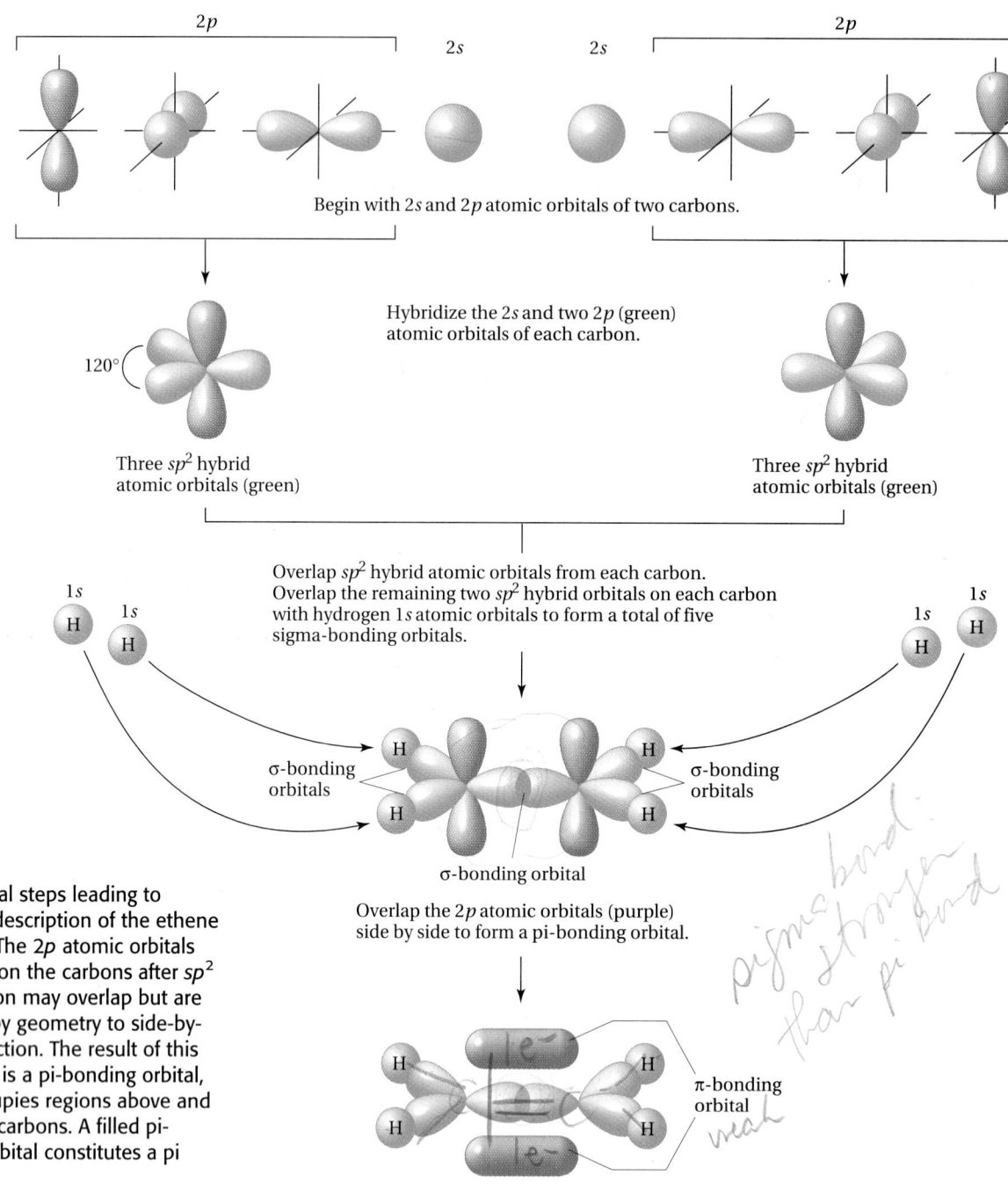

Begin with 2s and 2p atomic orbitals of two carbons.

Hybridize the 2s and two 2p (green)
atomic orbitals of each carbon.

120°

Three sp^2 hybrid
atomic orbitals (green)

Three sp^2 hybrid
atomic orbitals (green)

Overlap sp^2 hybrid atomic orbitals from each carbon.
Overlap the remaining two sp^2 hybrid orbitals on each carbon
with hydrogen 1s atomic orbitals to form a total of five
sigma-bonding orbitals.

1s   1s   1s   1s
H   H   H   H

σ-bonding   σ-bonding
orbitals   orbitals

σ-bonding orbital

Figure 2.8
Hypothetical steps leading to
an orbital description of the ethene
molecule. The 2p atomic orbitals
remaining on the carbons after sp^2
hybridization may overlap but are
restricted by geometry to side-by-
side interaction. The result of this
interaction is a pi-bonding orbital,
which occupies regions above and
below the carbons. A filled pi-
bonding orbital constitutes a pi
bond.

Overlap the 2p atomic orbitals (purple)
side by side to form a pi-bonding orbital.

π-bonding
orbital

Ethene

of a carbon-carbon sigma bond. Similar overlap of 1s orbitals of hydrogen
with the remaining sp^2 orbitals on the carbons gives four carbon-hydrogen
sigma bonds. Each carbon still has a 2p orbital. These 2p orbitals overlap
side by side to form a type of bond that we have not seen before. *A covalent
bond formed by side-by-side overlap of p atomic orbitals is called a* **pi bond**
(the Greek letter pi is π).

The completed bonds of the ethene molecule consist of five sigma bonds and one pi bond. Ethene is a molecule held together by five sigma bonds and one pi bond.

Like a sigma bond, the pi bond is a two-electron covalent bond. In writing structural formulas, we draw the pi bond as a line like any other covalent bond. Pi bonds are weaker than sigma bonds, however; so in chemical reactions of carbon-carbon double bonds the pi bond is broken in preference to the sigma bond.

Geometric isomerism

Geometric isomerism arises from a difference in orientation of atoms that leads to a difference in molecular shapes. Many biological reactions are shape-sensitive. For example, the odor of a compound depends on both the size and the shape of its molecule. A change from a *cis* to a *trans* form changes the shape of a molecule and may very well change its odor.

Trans
trans (Latin): across
cis
cis (Latin): on this side

The lack of rotation around carbon-carbon double bonds has an important structural implication. Looking at the structure of 2-butene in Figure 2.9, we see that there are two possible arrangements of the methyl groups with respect to the rigid double bond. *One arrangement has the substituents on opposite sides of the double bond; this is called the* **trans** *configuration. The other arrangement has the methyl groups on the same side of the double bond; this is called the* **cis** *configuration.* Trans-*2-butene and* cis-*2-butene are* **geometric isomers**—*they have the same molecular formulas but differ in the geometry of their substituents.* Like other structural isomers, isomeric 2-butenes are distinguishable by their different physical and chemical properties. The groups on the carbons of the double bond need not be the same. Geometric isomerism is possible whenever there is one hydrogen and one substituent on each carbon of the double bond. For example:

don't have to be identical to be trans

$$CH_3\quad H$$
$$C=C$$
$$H\quad CH_2CH_3$$
trans-2-Pentene

$$CH_3\quad CH_2CH_3$$
$$C=C$$
$$H\quad H$$
cis-2-Pentene

$$CH_3\quad H$$
$$C=C$$
$$CH_3CH_2\quad H$$
2-Methyl-1-butene
(no *cis*, *trans* isomers)

Meth
Eth
Prop
But
Pent

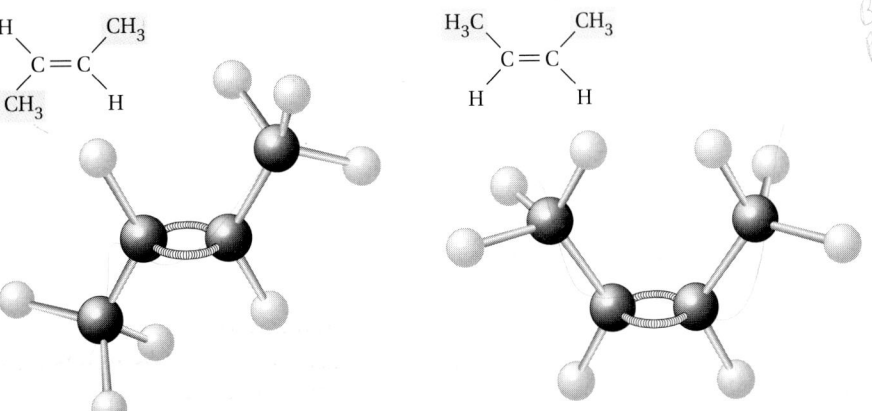

Figure 2.9
Structural formulas and ball-and-stick models of the geometric isomers of 2-butene.

(a) *trans*-2-Butene (bp 1°C)

(b) *cis*-2-Butene (bp 4°C)

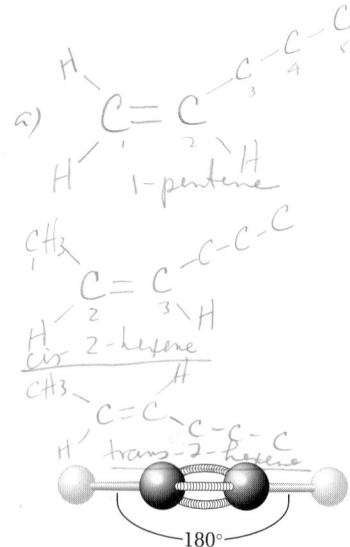

a) (handwritten)
$$C=C$$
with H, H, and C-C-C-C chain
1-pentene

b) (handwritten)
CH_3
$$C=C$$
H, 2, 3, H, C-C-C
cis 2-hexene

CH_3
$$C=C$$
H, trans-2-hexene, C-C-C

Figure 2.10
Ball-and-stick model of ethyne
(common name acetylene).
180°

d) (handwritten)
CH_3, CH_3
$$C=C$$
CH_3, 2, 3, CH_3

PRACTICE EXERCISE 2.8

Draw structural formulas for the following alkenes. If a compound has geometric isomers, draw both the *cis* and the *trans* forms.

(a) 1-pentene (b) 2-hexene (c) 2-methyl-2-hexene
(d) 2,3-dimethyl-2-butene

c) (handwritten)
CH_3
$$C=C$$
CH_3, 2, 3, H, C-C-C

Alkynes

Organic compounds containing carbon-carbon triple bonds are called **alkynes.** Like alkenes, alkynes are unsaturated compounds. This is the carbon-carbon triple bond found in alkynes:

$$—C≡C—$$

Alkynes are quite rare in nature. The simplest alkyne is the gas ethyne, C_2H_2. The common name for ethyne is *acetylene*—the fuel burned in oxyacetylene torches used in welding. The single bonds that extend from the carbons involved in the carbon-carbon triple bond of ethyne are separated by the maximum bond angle, 180 degrees (Fig. 2.10). Ethyne is a linear molecule.

The development of an orbital picture for ethyne requires us to go through the same steps that we used for methane and ethene: hybridization of carbon atomic orbitals and orbital overlap. As with methane and ethene, remember that each orbital represents one electron. The description of ethyne that best fits the experimental data is obtained if the 2*s* atomic orbital of carbon is mixed with *only one* of the three available 2*p* atomic orbitals (Fig. 2.11). The result of this mixing is two *sp* atomic orbitals; two 2*p* orbitals remain unused. Overlap of one of the *sp* hybrid orbitals of one carbon with that of another carbon produces a carbon-carbon sigma bond. The remaining *sp* orbital on each carbon overlaps with the 1*s* orbital of a hydrogen to form two more sigma-bonding orbitals. The remaining pair of *p* atomic orbitals on each carbon now overlaps side by side to form *two* pi bonds. The bonding in ethyne consists of three sigma bonds and two pi bonds. Each pi bond has two regions in which the probability of finding the bonding electrons is high, so the pi bonds in ethyne resemble four sausages tightly packed around the carbon-carbon sigma bond.

2.9 Aromatic compounds

AIM: To describe the bonding, structure, and chemical properties of benzene and other simple aromatic and fused-ring aromatic compounds.

Focus

Benzene is the simplest arene.

The organic compounds we have seen thus far are all *aliphatic.* **Aliphatic** *compounds contain only carbon-carbon single, double, or triple bonds.* Now we will turn to the group of carbon compounds known as *arenes.* **Arenes** *are a class of carbon compounds that have structures based on the parent molecule benzene,* C_6H_6. All arenes contain a benzene ring or a

$C_n H_n$ (handwritten)

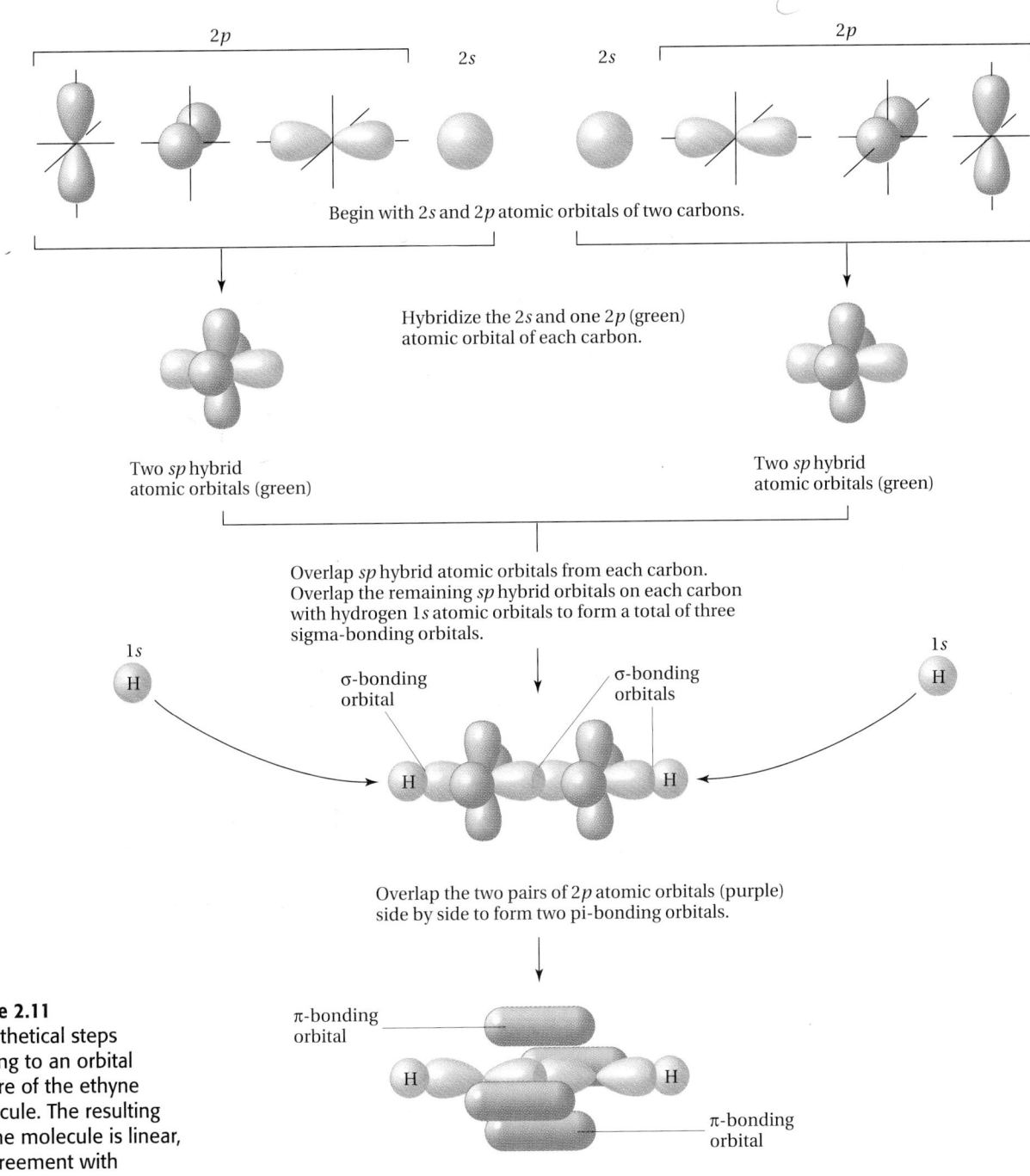

2p   2s   2s   2p

Begin with 2s and 2p atomic orbitals of two carbons.

Hybridize the 2s and one 2p (green)
atomic orbital of each carbon.

Two *sp* hybrid
atomic orbitals (green)

Two *sp* hybrid
atomic orbitals (green)

Overlap *sp* hybrid atomic orbitals from each carbon.
Overlap the remaining *sp* hybrid orbitals on each carbon
with hydrogen 1s atomic orbitals to form a total of three
sigma-bonding orbitals.

1s

1s

σ-bonding
orbital

σ-bonding
orbitals

Overlap the two pairs of 2p atomic orbitals (purple)
side by side to form two pi-bonding orbitals.

Figure 2.11
Hypothetical steps
leading to an orbital
picture of the ethyne
molecule. The resulting
ethyne molecule is linear,
in agreement with
experimental evidence.

π-bonding
orbital

π-bonding
orbital

Ethyne

related system of rings. **Arenes** *are also called* **aromatic compounds**
*because the earliest known examples of these compounds had pleasant
odors.* Benzene, C_6H_6, is the simplest arene. The benzene molecule has a
six-membered carbon ring with a hydrogen attached to each carbon. This
leaves one electron on each carbon free to participate in a double bond.

"Connecting the dots" to complete the structural formula of benzene, we see that there are *two different ways* to form three double bonds:

These two structural formulas only describe the extremes of electron sharing between any two adjacent carbons in benzene. One extreme is a normal single bond; the other is a normal double bond. *These two extremes of bonding—single and double bonds—are called* **resonance structures.** When resonance structures can be drawn for a certain molecule or ion, that molecule or ion is more stable than others for which only one structural formula can be drawn. As a result, benzene resists chemical reactions to which any ordinary alkene easily succumbs.

A more accurate description of bonding in benzene is an average picture or **resonance hybrid** *of these forms.* (The offspring of a jackass and a mare, two extreme forms of the horse family, is a hybrid, a neuter mule. The hybrid resembles each of its parents but also has its own unique characteristics.) When organic chemists talk about benzene and related arenes, they usually call this hybrid bonding *aromatic character* or *aromaticity.* The benzene ring is a perfectly flat molecule; ring puckering or bending would make it less stable.

The orbital description of benzene is helpful in furthering our understanding of the low chemical reactivity of benzene and related arenes compared with alkenes such as ethene. As mentioned previously, all orbitals represent one electron. The carbons of benzene are sp^2 hybridized, and the formation of the carbon-hydrogen and carbon-carbon sigma bonds of benzene is similar to their formation in ethene. Once again, one $2p$ atomic orbital is available on each carbon. In benzene, however, each of these p orbitals overlaps side by side with *both* of its neighbors (Fig. 2.12). The

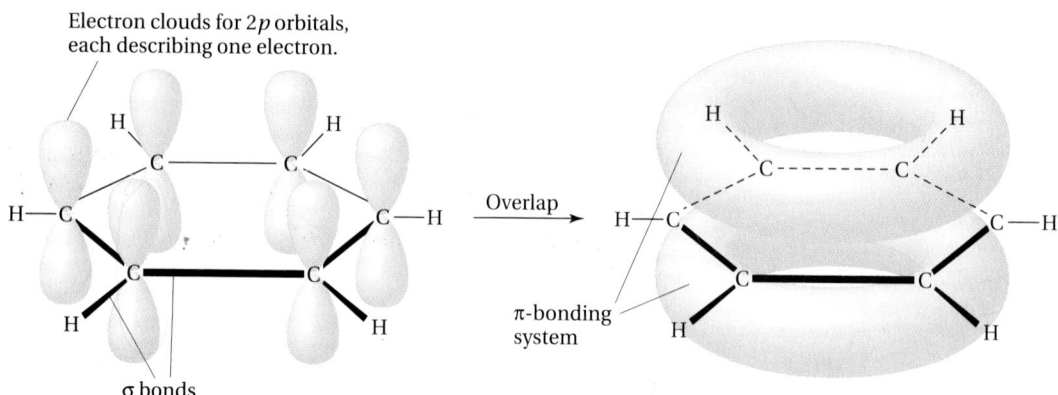

Figure 2.12

Side-by-side overlap of $2p$ atomic orbitals results in an extensive pi-bonding orbital in the benzene molecule.

result is a pi bond in which there is a high probability of finding the electrons in doughnut-shaped regions above and below the carbon ring. The orbital picture shows that the pi electrons in benzene are no longer identifiable with any particular carbon. Thus the electrons in the pi-bonding orbitals of benzene are said to be *delocalized*. This bonding is quite different from that in the carbon-carbon double bond of an alkene. In the double bond of an alkene, the pi electrons are *localized* above and below the two carbons in the bonding orbitals.

The low chemical reactivity of benzene and related compounds compared with that of alkenes results from their extensive pi-bonding systems. The more the electrons in pi bonds are smeared over a molecule, the lower the energy of the bonding electrons and the more stable the molecule. Therefore, the pi bonds of benzene are less chemically reactive than those of ordinary alkenes.

When drawing an abbreviated structure of benzene, chemists often inscribe either a complete or dashed circle in the ring to indicate the delocalized pi-bonding electrons. The inscribed circle is useful for showing the aromatic character of benzene and related compounds, but it does not give any indication of *how many* electrons are involved. For this reason, we will stick with the time-honored benzene structure shown at the right.

The benzene ring is found in many compounds used medicinally: the analgesic aspirin, the decongestant ephedrine, the painkiller and cough suppressant codeine, and the antibiotic sulfa drugs.

Alternative representations of benzene

Naming arenes

Most compounds containing alkyl substituents attached to the benzene ring are named as derivatives of benzene. Here are some examples of benzene derivatives with one substituent:

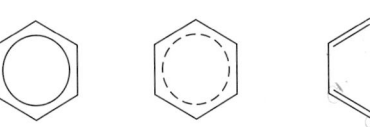

Methylbenzene (toluene)

Ethylbenzene

Another name for methylbenzene is *toluene.* Sometimes the benzene ring is named as a substituent on an alkane or cycloalkane. In such instances, the benzene ring is called a *phenyl group,* as in 3-phenylhexane or phenylcyclohexane.

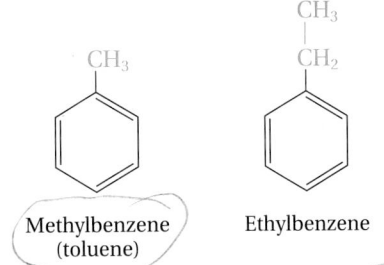

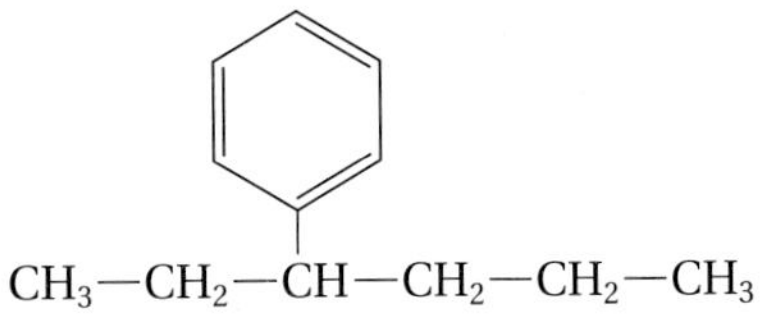

$CH_3-CH_2-CH-CH_2-CH_2-CH_3$

3-Phenylhexane

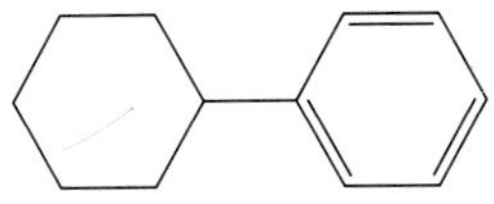

Phenylcyclohexane

PRACTICE EXERCISE 2.9

Name the following compounds:

(a) 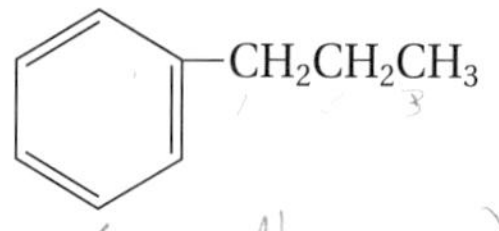 [benzene ring]—$CH_2CH_2CH_3$

(propylbenzene)
or 1-phenylpropane

(b) 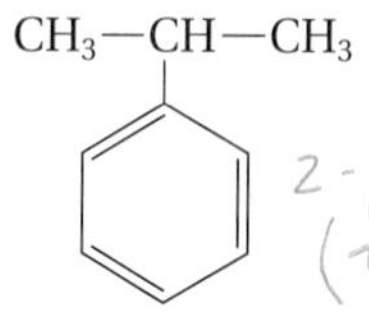 $\overset{1}{CH_3}-\overset{2}{CH}-\overset{3}{CH_3}$ with benzene ring attached to C2

2-phenylpropane
(isopropylbenzene)

(c) 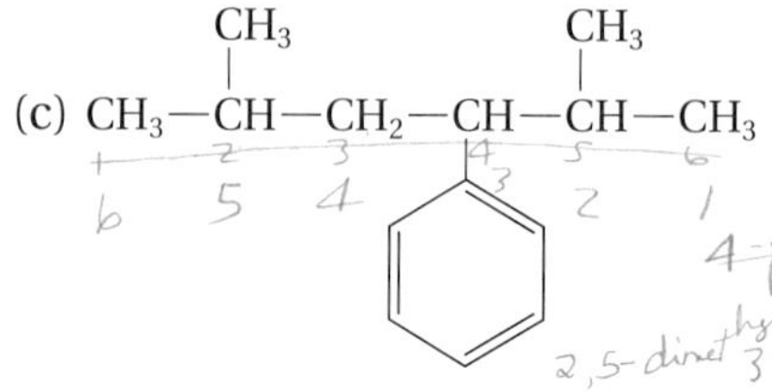 $CH_3-\overset{CH_3}{\underset{|}{CH}}-CH_2-\overset{}{CH}-\overset{CH_3}{\underset{|}{CH}}-CH_3$ with benzene ring attached

4-phenyl-2,5-dimethylhexane
2,5-dimethyl-3-phenylhexane

2,3-dimethyl-hexane 3-phenyl

Structural isomers of benzene derivatives

There are derivatives of benzene with two substituents of the same kind. Such derivatives are called *disubstituted benzenes*. There are three different structural isomers of the liquid aromatic compound dimethylbenzene, $C_6H_4(CH_3)_2$. The dimethylbenzenes are also called *xylenes*. The difference between the structural isomers of dimethylbenzene is the relative positions of the methyl groups on the benzene ring. We can distinguish among the possible positions of the methyl groups by assigning the methyl groups the numbers 1,2; 1,3; or 1,4. It is also permissible in naming disubstituted benzenes to use the prefixes *ortho-, meta-,* and *para-* (abbreviated *o, m,* and *p*) in place of numbers. *Ortho-* means that the substituents are on adjacent carbons on the benzene ring. *Meta-* substituents are separated by one ring carbon, and *para-* substituents are separated by two ring carbons. The physical properties of the structural isomers of dimethylbenzene are different, as indicated by their boiling points.

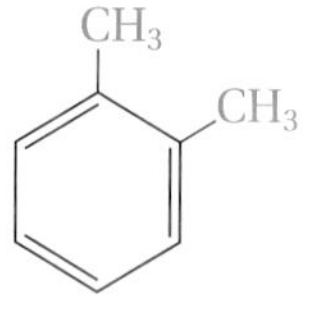

1,2-Dimethylbenzene
(*o*-dimethylbenzene,
o-xylene)
(bp 144 °C)

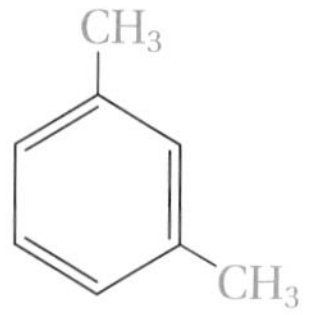

1,3-Dimethylbenzene
(*m*-dimethylbenzene,
m-xylene)
(bp 139 °C)

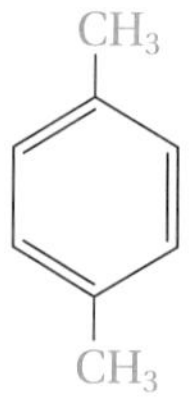

1,4-Dimethylbenzene
(*p*-dimethylbenzene,
p-xylene)
(bp 138 °C)

Fused-ring aromatics

Fused-ring (polycyclic) aromatic compounds *are derivatives of benzene in which carbons are shared between benzene rings.* Naphthalene, used in mothballs, is the simplest fused-ring aromatic. Anthracene is found in anthracite coal, and the carbon skeleton of phenanthrene forms the basic structure of the steroids, among which are the sex hormones. Benzpyrene is

A Closer Look

Hydrocarbons and Health

Some hydrocarbons have medical applications. Cyclopropane is an anesthetic. Mineral oil is a mixture of hydrocarbons of high molar mass. Petroleum jelly (Vaseline is one brand name) is a semisolid mixture of solid and liquid aliphatic hydrocarbons of high molar mass. Mineral oil and petroleum jelly (see figure) are used to soften skin and protect it from exposure to water. The hydrocarbon components of mineral oil and petroleum jelly have low vapor pressures, and these products are essentially odorless. Mineral oil is indigestible and is sometimes used as a mild laxative because of its lubricating properties.

Petroleum jelly, gasoline, paint thinners, and mineral oil are all examples of hydrocarbons found in the home.

Many polycyclic aromatic compounds are carcinogenic (cancer-causing). Smoking is a chief source of these carcinogenic chemicals. More than 300 carcinogens have been identified in cigarette smoke. One such compound is the hydrocarbon benzpyrene. This carcinogen is converted by the body into highly reactive compounds that react with DNA molecules in cell nuclei to cause the transformation of normal cells into cancerous ones. Benzene is implicated as a causative agent in leukopenia (low white cell count) and in certain types of leukemia. Toluene poses less of a hazard than benzene, although prolonged exposure to its vapors can cause headaches, nausea, and vomiting. Breathing toluene vapors also can produce a narcotic effect. The permissible exposure of chemical workers to benzene and other aromatic hydrocarbons is regulated by law.

Every year hospitals admit patients who have swallowed gasoline. The saturated aliphatic hydrocarbons in gasoline are not particularly toxic, since they are chemically unreactive in our bodies. The main threat to health from the ingestion of gasoline is from toxic additives. Gasoline in the lungs can damage lung tissue, causing the symptoms of pneumonia. Anyone who swallows gasoline or other liquid hydrocarbons should not be made to vomit. That would increase the chance of the hydrocarbon getting from the stomach into the lungs. The greatest threat to health from gasoline is its high flammability. Gasoline should never be used near a source of flames or sparks.

an aromatic hydrocarbon that poses a threat to health. A Closer Look: Hydrocarbons and Health describes some applications and hazards of aliphatic and aromatic hydrocarbons to health.

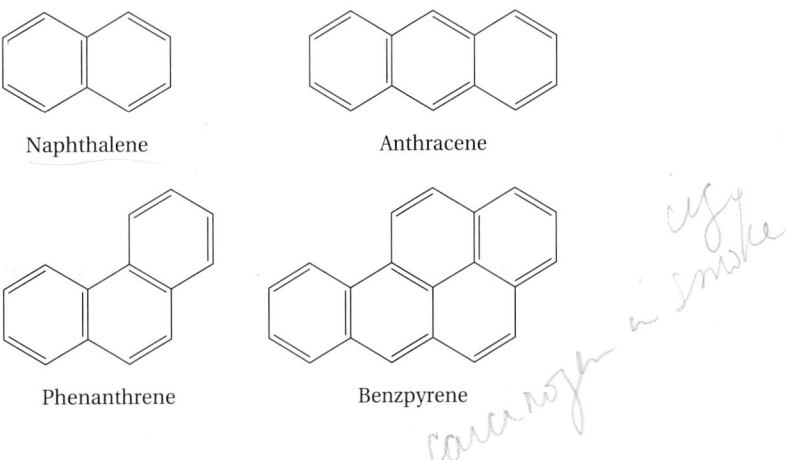

Naphthalene

Anthracene

Phenanthrene

Benzpyrene

2.10 Sources of hydrocarbons

AIM: To characterize hydrocarbons in terms of their source and use.

Petroleum and coal are the sources of most hydrocarbons used in the home, in industry, and in the laboratory. Major carbon compounds for organic chemicals and energy needs are supplied by petroleum, natural gas, and coal. Petroleum, or crude oil, originated in marine life buried in the sediments of the oceans millions of years ago. A combination of heat and pressure, and perhaps the action of bacteria, changed this residue of marine life into petroleum and gas. Vast deposits of crude oil were discovered in the United States in 1859 and in the Middle East in 1908. Crude oil is a complex mixture of hydrocarbons, mostly alkanes and cycloalkanes; it also includes small amounts of other organic compounds that contain sulfur, oxygen, and nitrogen. The components of crude oil can be separated by *refining.* **Petroleum refining** *consists of distilling crude oil to divide it into fractions of hydrocarbons according to boiling point.* Fractional distillation separates petroleum into "fractions" of hydrocarbons: butane (gas), octane (gasoline), dodecane (kerosene), fuel oil, lubricating oil, paraffin wax, and tar—a thick black residue. The distillation fractions, each of which contains several different hydrocarbons, are shown in Table 2.3. Modern petroleum refining also employs **cracking**—*a controlled process by which catalysts are used to break down or rearrange natural hydrocarbons into more useful molecules.* The hydrocarbon composition of gasoline affects its performance as a fuel, as discussed in A Closer Look: Octane Ratings of Gasoline.

Natural gas is found overlying deposits of oil or in separate pockets in rock. Natural gas is an important source of alkanes of low molecular mass. A typical natural gas composition is methane 80%, ethane 10%, propane 4%, and butane 2%. The balance is nitrogen and higher-molecular-mass hydrocarbons. Most natural gas contains a small percentage of helium and is an important source of this noble gas. Natural gas is used as a fuel and is distributed through pipes; the individual gases may be liquefied and sold as fuel for heating, portable cookers, and welding equipment.

Table 2.3 Fractions Obtained from Crude Oils*

Fraction	Composition of carbon chains	Boiling range (°C)
natural gas	C_1 to C_4	below 20
petroleum ether (solvent)	C_5 to C_6	30 to 60
naphtha (solvent)	C_7	60 to 90
gasoline	C_6 to C_{12}	75 to 200
kerosene	C_{12} to C_{15}	200 to 300
fuel oils, mineral oil	C_{15} to C_{18}	300 to 400
lubricating oil, petroleum jelly, greases, paraffin wax	C_{16} to C_{24}	over 400

*The undistilled residue is asphalt.

A Closer Look

Octane Ratings of Gasoline

When we fill the tank of an automobile, we must decide what octane of gasoline to buy. Today's automobile manufacturers generally recommend the octane rating that will give good engine performance. If our engine is out of tune, however, or if we choose an octane rating that is lower than recommended, we may experience engine knock. The engine may ping if we accelerate too rapidly or knock and lose power when the automobile goes up a steep hill. The knock characteristics of a gasoline depend on the structure of the hydrocarbons in the fuel. Straight-chain hydrocarbons cause engine knock and are relatively poor fuels for today's automobiles. Branched-chain hydrocarbons have better antiknock characteristics. The octane or antiknock rating is an arbitrary scale defined by the petroleum industry. The knock created by burning heptane in a standard engine is assigned a value of zero octane; the knock from 2,2,4-trimethylpentane (isooctane) is assigned a value of 100 octane. The octane rating of a gasoline is equal to the percentage of isooctane in a heptane-isooctane mixture that performs the same way in the standard engine (see figure). For example, a gasoline giving the same knock as a mixture of 90% isooctane and 10% heptane has an octane

Gasoline is a mixture of many different hydrocarbons. The octane rating of a gasoline, shown on the delivery pump, gives an indication of the engine performance you may expect using that grade of gasoline.

rating of 90. The percentage of branched-chain hydrocarbons in gasoline can be increased by catalytic cracking in petroleum refining (see Sec. 2.11). High-octane gasoline is produced by reforming the structure of the C_4–C_{18} straight-chain hydrocarbons to give a highly branched C_6–C_{10} fraction with octane ratings of 90 to 110. Gasoline available today is usually 87 octane, 89 octane, and 92 octane.

Geologists think that coal originated about 300 million years ago. When huge tree ferns and mosses growing in swampy tropical regions died, they formed thick layers of decaying vegetation, which were eventually covered by soil and rock. The pressure of soil, rocks, and earth with the heat from the Earth's interior eventually turned these remains into coal (Fig. 2.13a). Coal consists largely of condensed ring compounds of very high molecular mass. These compounds have a high proportion of carbon compared with hydrogen. Coal is obtained from underground and surface mines, where it is found in seams (Fig. 2.13b) from 1 to 3 m thick. In the United States, coal mines are usually less than 100 m deep, and much of the coal is so close to the surface that it is strip-mined. Many coal mines in Europe and other parts of the world go down 1000 to 1500 m. The hydrocarbons in coal are mostly aromatic—benzene, naphthalene, anthracene, and a host of others. Distillation of coal yields coal gas, coke, ammonia, and coal tar. Coal gas consists mainly of hydrogen, methane, and carbon monoxide, all of which are flammable and good fuels. Coke, the solid residue of almost pure car-

(a)

(b)

(c)

Figure 2.13
Shown is (a) fossil tree-fern leaf imprint in coal, (b) an exposed coal seam, and (c) a sample of coal tar.

bon, is also used as a fuel in many industrial processes. Ammonia is often converted to ammonium sulfate for use as a fertilizer. Coal tar (Fig. 2.13c) can be further distilled to yield many useful chemicals, including benzene, toluene, naphthalene, phenol, and pitch.

Over the past few years, growth in the world's demand for petroleum has outstripped the discovery of new deposits. Since petroleum is a nonrenewable resource, there eventually will be insufficient quantities to support the world's energy and other demands. Scientists and engineers are working to develop replacement sources of energy such as the sun, winds, tides, and geothermal springs.

Some of the clothes you wear and the music you play (CDs and tapes) are made from the same finite supply of hydrocarbons that you burn in your car and use to heat your home. Does this give you cause for concern?

2.11 Properties of hydrocarbons

AIM: To characterize the physical and chemical properties of hydrocarbons.

Focus

Hydrocarbons float on water and burn in air.

The public transit system of the city of Santa Fe, New Mexico, operates entirely on compressed natural gas.

Anyone who has seen footage of the oil slick from a wrecked oil tanker can observe two physical properties common to all hydrocarbons: They are insoluble in water and less dense than water (they float). If the tanker has been involved in a collision, a chemical property common to all hydrocarbons often is evident: combustion. Under controlled conditions, however, the burning of hydrocarbons enormously benefits people. Of the more than 500,000 tons of crude oil and 560 million tons of coal produced yearly in the United States, most is burned to produce energy. Methane, the major constituent of natural gas, burns with a hot, clean flame:

$$CH_4(g) + 2O_2(g) \rightarrow CO_2(g) + 2H_2O(g) + heat$$

Propane and butane are also good heating fuels and are often sold in liquid

Figure 2.14
Propane is the fuel that is also known as *bottled gas.*

form under pressure in tanks as liquid petroleum gas (LPG), as shown in Figure 2.14.

A sufficient supply of oxygen is necessary to oxidize a hydrocarbon fuel completely to carbon dioxide and obtain the greatest amount of heat. Complete combustion of a hydrocarbon gives a blue flame. Incomplete combustion gives a yellow flame, due to the formation of small, glowing carbon particles that are deposited as soot when they cool. Carbon monoxide, an odorless toxic gas, is also formed along with carbon dioxide and water during incomplete combustion. Generally, unsaturated compounds, such as the aromatic compounds found in coal, are sootier fuels than the alkanes.

The burning of hydrocarbons found in coal and petroleum are not without environmental hazards, as described in A Closer Look: The Greenhouse Effect, on the following page. For the effect of the products of incomplete combustion of hydrocarbons on people's health, recall Claudine from the Case in Point for this chapter. Claudine was the victim of carbon monoxide poisoning. In the Follow-up, below, we will see how she was poisoned and how she was treated for her condition.

FOLLOW-UP TO THE CASE IN POINT: Carbon monoxide poisoning

Claudine's house has a furnace that burns natural gas piped in from the local gas company. The incomplete combustion of the gas produces highly toxic but odorless carbon monoxide. Ordinarily, the carbon monoxide and other combustion products are carried out of the house by a flue pipe. Unknown to Claudine, the flue pipe worked loose from the furnace, permitting carbon monoxide to escape into the house. Before the incident in which she was hospitalized, Claudine had experienced increasingly severe headaches caused by the increasing level of carbon monoxide in her house. The paramedics who transported Claudine to the hospital recognized the characteristic flushed skin of carbon monoxide poisoning, a result of interference of the gas with the body's use of oxygen. Carbon monoxide combines with the hemoglobin of the blood, making it useless as an oxygen carrier. Fortunately, the carbon monoxide can be replaced by oxygen. The paramedics administered oxygen, and by the time Claudine arrived at the hospital, her condition had significantly improved. Claudine suffered no permanent effects from her ordeal. Fortunately, she was discovered before she suffered permanent damage to her nervous system or brain or

even death. Others are not so lucky. Every year hundreds of people lose their lives as a result of carbon monoxide poisoning from faulty heating devices.

Less dramatic incidents of carbon monoxide poisoning are experienced by millions of people every day. Cigarette smoke is a source of low-level carbon monoxide poisoning. A burning cigarette produces a great deal of carbon monoxide. It can take several hours to replace the carbon monoxide in a smoker's blood after only one cigarette. Low levels of carbon monoxide poisoning can cause headaches, dizziness, nausea, and sluggishness. The reduced amount of oxygen in the blood of smokers makes the heart pump harder to supply oxygen to tissues. As a result, smoking contributes to the incidence of heart disease and heart attacks. Even nonsmokers can suffer from low-level carbon monoxide poisoning. The air around traffic jams often reaches dangerously high concentrations of carbon monoxide. Automobiles recycle gases from the combustion of gasoline in order to burn more of the carbon monoxide to carbon dioxide. This has helped to lower the level of carbon monoxide in automobile exhaust emissions.

A Closer Look

The Greenhouse Effect

Since the onset of the industrial revolution in the 19th century, human activities have been adding vast amounts of carbon dioxide into the Earth's atmosphere. People burn fossil fuels, petroleum and coal, for energy. Grasslands and forests are being burned to clear land for crops. Each year, more than 8 billion (8×10^9) metric *tons* of carbon dioxide pour into the Earth's atmosphere (1 metric ton is 1000 kg). The amount of atmospheric carbon dioxide is increasing at about 0.5% per year and is already 25% higher than preindustrial levels.

The increased concentration of atmospheric carbon dioxide due to burning of fossil fuels and plants has led to a problem called the *greenhouse effect*. Ordinarily, much of the Sun's energy would be lost by the radiation of heat from the Earth's surface back into space. However, high concentrations of carbon dioxide in the atmosphere reflect the heat back to Earth, making the surface warmer (see figure). Other gases such as methane are also thought to contribute to the warming effect. The warming of the Earth's surface by reflected heat is called the *greenhouse effect* because it operates in much the same way as the panes of glass in a greenhouse. The glass in a greenhouse is transparent to visible light but absorbs infrared radiation, which is absorbed as heat. The heat reflects from the glass, raising the temperature inside the building. Scientists have predicted that a rise of only a few degrees in the average temperature of the Earth is likely to disrupt weather patterns. These weather changes would affect the growing of food crops. Polar ice caps also could melt, inundating coastal cities with water. Because of concerns about the greenhouse effect, scientists are investigating the possibility of converting carbon dioxide into solid compounds that could be readily disposed of without adding to the atmosphere's carbon dioxide load.

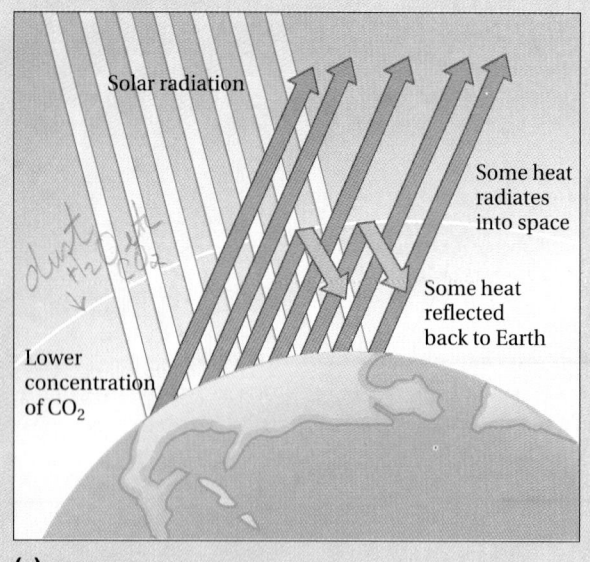

(a)

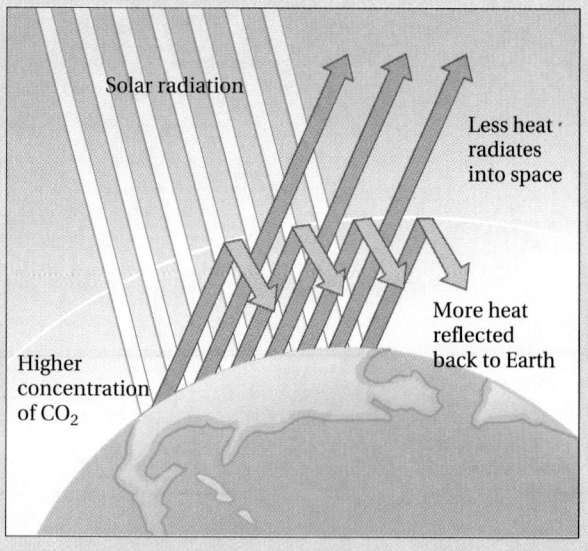

(b)

The Earth is warmed by the Sun's radiant energy. (a) The warming effect is controlled because some heat radiates from the Earth's surface back into space. (b) The *greenhouse effect* results when higher concentrations of atmospheric carbon dioxide reflect more heat back to Earth, making its surface warmer.

SUMMARY

Organic chemistry deals with carbon compounds. Carbon makes covalent bonds with other carbons to form chain and ring compounds. Hydrocarbons are compounds containing only carbon and hydrogen. Many hydrocarbons exhibit structural isomerism; structural isomers have the same molecular formula but different molecular structures. Alkanes, alkenes, and alkynes are the principal classes of aliphatic hydrocarbons.

Alkanes contain only carbon-carbon single bonds. The four bonds extending from each carbon in an alkane are arranged in a regular tetrahedron; the angle between any two bonds is 109.5 degrees. The groups attached to single bonds in straight-chain alkanes rotate freely about the bonds at room temperature.

Alkenes contain one or more carbon-carbon double bonds. The two bonds that extend from each carbon of a carbon-carbon double bond are 120 degrees apart. All four bonds extending from the pair of carbons in the carbon-carbon double bond lie in the same plane. The carbons in the carbon-carbon double bond do not rotate. Alkenes that have one group besides hydrogen attached to each carbon of the double bond exhibit geometric isomerism, a special case of structural isomerism. Geometric isomers are *cis* or *trans* according to whether substituent groups—one on each carbon of the double bond—are on the same side (*cis*) or on opposite sides (*trans*) of the double bond. Like structural isomers, geometric isomers differ in their physical and chemical properties.

Alkynes contain one or more carbon-carbon triple bonds. The bonds that extend from the triple bond are 180 degrees apart. Ethyne (C_2H_2), the simplest alkyne, is a linear molecule.

Aromatic hydrocarbons are related to the hydrocarbon benzene. Benzene is rather unusual among hydrocarbons. It appears to be a six-membered carbon ring containing three alternating double bonds. However, the interior bonds of the benzene ring are hybrid or aromatic bonds—somewhere in between double and single bonds. Benzene is less reactive than alkenes. Many aromatic compounds, such as naphthalene, are fused-ring compounds.

Aliphatic hydrocarbons come from petroleum. Aromatic hydrocarbons come from coal. All hydrocarbons, aliphatic and aromatic, have three properties in common: They are insoluble in water, they are less dense than water, and they burn in air.

KEY TERMS

Aliphatic compound (2.9)
Alkane (2.2)
Alkene (2.8)
Alkyl group (2.5)
Alkyne (2.8)
Arene (2.9)
Aromatic compound (2.9)
Branched-chain alkane (2.5)

Cis configuration (2.8)
Condensed structural formula (2.4)
Cracking (2.10)
Cycloalkane (2.7)
Fused-ring aromatic compound (2.9)
Geometric isomer (2.8)
Hydrocarbon (2.2)

IUPAC system (2.4)
Orbital hybridization (2.2)
Petroleum refining (2.10)
Pi bond (2.8)
Polycyclic aromatic compound (2.9)
Resonance hybrid (2.9)
Resonance structure (2.9)
Saturated compound (2.8)

Sigma bond (2.2)
Straight-chain alkane (2.4)
Structural formula (2.2)
Structural isomer (2.6)
Substituent (2.5)
Trans configuration (2.8)
Unsaturated compound (2.8)

EXERCISES

Saturated Hydrocarbons (Sections 2.2–2.7)

2.10 Draw complete structural formulas and condensed structural formulas and give the IUPAC names for the first five straight-chain alkanes with an odd number of carbon atoms.

2.11 Name the alkanes or alkyl groups that have the following formulas.

(a) CH_3CH_2- (b) $CH_3(CH_2)_6CH_3$ (c) C_3H_8

(d) 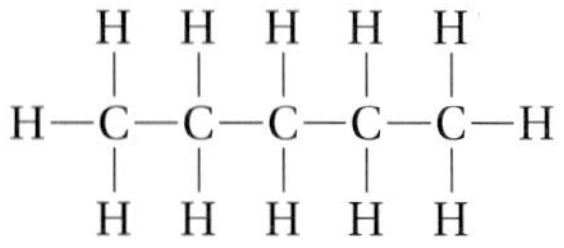
$$H-\overset{\overset{H}{|}}{C}-\overset{\overset{H}{|}}{\underset{\underset{H}{|}}{C}}-\overset{\overset{H}{|}}{\underset{\underset{H}{|}}{C}}-\overset{\overset{H}{|}}{\underset{\underset{H}{|}}{C}}-\overset{\overset{H}{|}}{\underset{\underset{H}{|}}{C}}-H$$

(e) $CH_3CH_2CH_2CH_2CH_3$ (f) $CH_3CH_2CH_2-$

2.12 What system is used for naming branched-chain alkanes? Briefly state the rules.

2.13 What is a substituent? What is the name of a branched-chain alkane substituent?

2.14 Draw structural formulas for the following branched-chain alkanes.
(a) 2,3-dimethylbutane
(b) 2,2-dimethylbutane
(c) 2,2,3-trimethylbutane
(d) 3,4-diethyl-4,5-dimethyloctane
(e) 4-ethyl-2,3,4-trimethylnonane
(f) 3,5-diethyl-2,3-dimethyl-5-propyldecane

2.15 Give the IUPAC name for each of the following structural formulas.

(a) 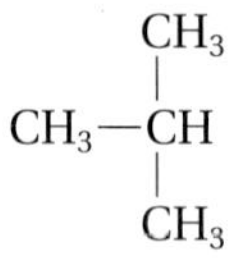
$$CH_3-\overset{\overset{CH_3}{|}}{\underset{\underset{CH_3}{|}}{CH}}$$

(b) 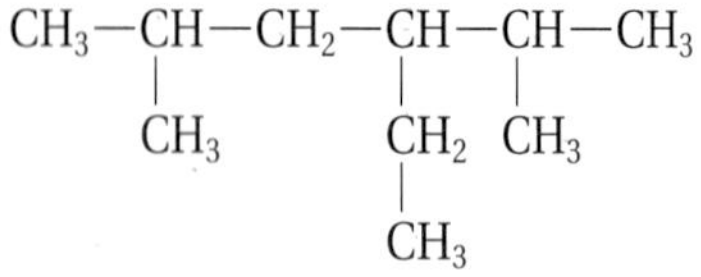
$$CH_3-CH-CH_2-CH-CH-CH_3$$
with CH_3 below first CH, and CH_2, CH_3 below the last CHs, and CH_3 below CH_2

(c) 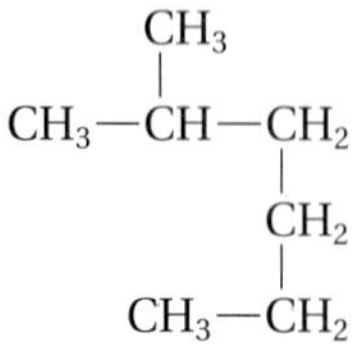
$$CH_3-\overset{\overset{CH_3}{|}}{CH}-CH_2$$
$$CH_2$$
$$CH_3-CH_2$$

(d) 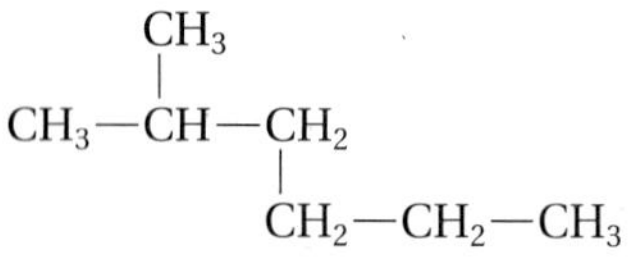
$$CH_3-\overset{\overset{CH_3}{|}}{CH}-CH_2$$
$$CH_2-CH_2-CH_3$$

(e) 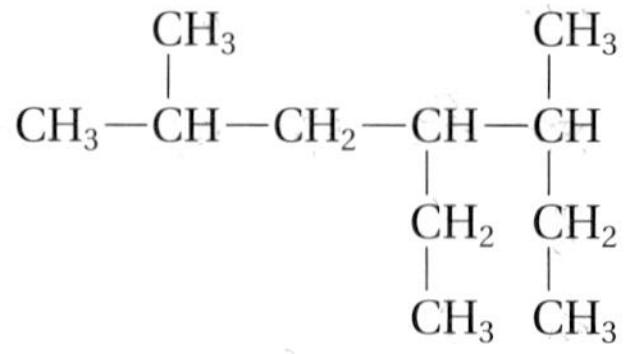
$$CH_3-\overset{\overset{CH_3}{|}}{CH}-CH_2-\overset{\overset{CH_3}{|}}{CH}-CH$$
$$\overset{\overset{}{}}{CH_2} \quad CH_2$$
$$CH_3 \quad CH_3$$

(f) 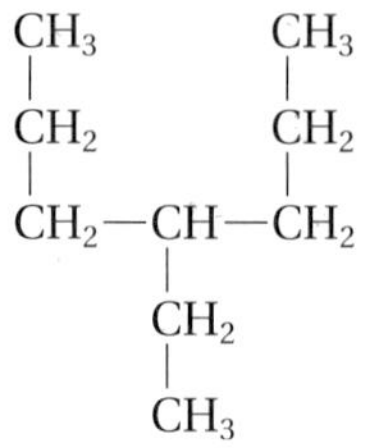
$$\overset{CH_3}{|} \quad \overset{CH_3}{|}$$
$$\overset{CH_2}{|} \quad \overset{CH_2}{|}$$
$$CH_2-CH-CH_2$$
$$CH_2$$
$$CH_3$$

2.16 What are structural isomers?

2.17 Draw one structural isomer of each of the following.

(a) 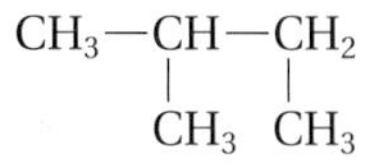
$$CH_3-\overset{\overset{}{}}{CH}-CH_2$$
$$CH_3 \quad CH_3$$

(b) 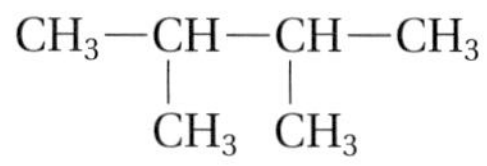
$$CH_3-CH-CH-CH_3$$
$$CH_3 \quad CH_3$$

(c) 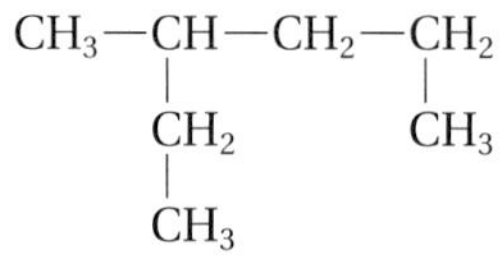
$$CH_3-CH-CH_2-CH_2$$
$$CH_2 \quad CH_3$$
$$CH_3$$

(d) 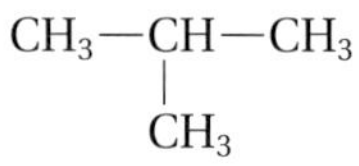
$$CH_3-CH-CH_3$$
$$CH_3$$

2.18 Which of the following are structural isomers?

(a) 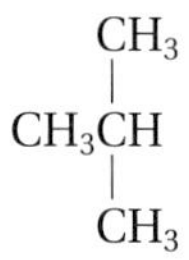
$$CH_3CH$$
$$\overset{\overset{CH_3}{|}}{} \quad CH_3$$

(b) 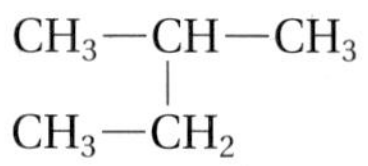
$$CH_3-CH-CH_3$$
$$CH_3-CH_2$$

(c) 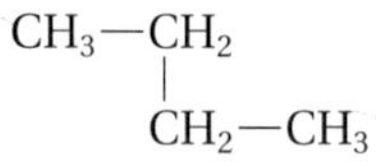
$$CH_3-CH_2$$
$$CH_2-CH_3$$

(d) 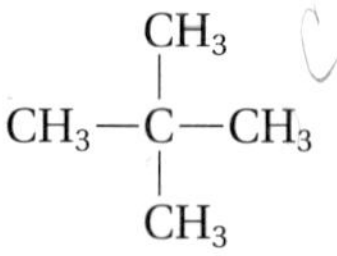
$$CH_3-\overset{\overset{CH_3}{|}}{\underset{\underset{CH_3}{|}}{C}}-CH_3$$

(e) 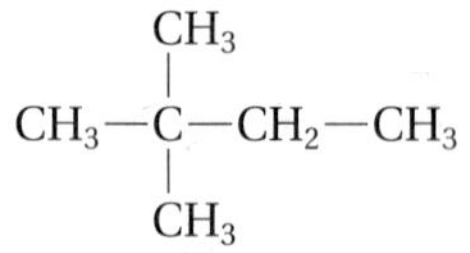
$$CH_3-\overset{\overset{CH_3}{|}}{\underset{\underset{CH_3}{|}}{C}}-CH_2-CH_3$$

(f) 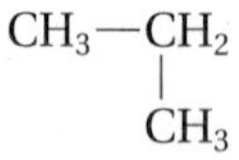
$$CH_3-CH_2$$
$$CH_3$$

2.19 A student incorrectly names a compound 1,3-dimethylpropane. Draw a structural formula and correctly name this compound.

2.20 Explain why cyclopropane and cyclobutane are strained molecules but cyclopentane and cyclohexane are not.

2.21 Name the following substituted cycloalkanes.

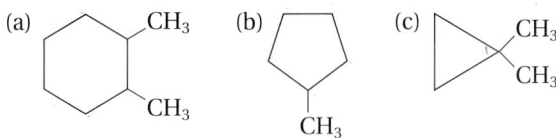

(a)   (b)   (c)

Unsaturated Hydrocarbons (Section 2.8)

2.22. Explain the difference between saturated and unsaturated hydrocarbons.

2.23 How are alkenes named by the IUPAC system?

2.24 Show how the lack of rotation about a carbon-carbon double bond leads to geometric isomerism. Use the geometric isomers of 2-pentene to illustrate your answer.

2.25 Does 2,4-dimethyl-2-hexene have geometric isomerism? Explain.

2.26 Draw a structural formula for each of the following alkenes. Include both *cis* and *trans* forms if the compound has geometric isomerism.
(a) 2-pentene   (b) 2-methyl-2-pentene

2.27 Draw a carbon skeleton for each of the following alkenes. Include both *cis* and *trans* forms if the compound has geometric isomerism.
(a) 2-pentene   (b) 3-hexene

2.28 Give a systematic name for each of the following alkenes.

(a) $CH_3CHCH_2CH=CH_2$ (b)
$\quad\quad\ |$
$\quad\quad CH_3$

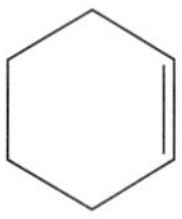

(c) $CH_3CH=CH_2$ (d) 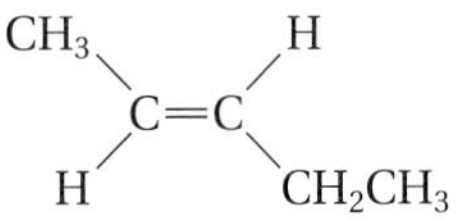 CH_3 $\quad\quad$ H
$\quad\quad\quad\quad\quad\quad\quad\quad\quad$ C=C
$\quad\quad\quad\quad\quad\quad\quad$ H $\quad\quad$ CH_2CH_3

2.29 Name each of these unsaturated compounds.

(a) $CH_3CH_2CH=CH_2$ (b) $CH_3(CH_2)_5CH=CCH_3$
$\quad\quad\quad\quad\quad\quad\quad\quad\quad\quad\quad\quad\quad\quad\quad\quad\quad |$
$\quad\quad\quad\quad\quad\quad\quad\quad\quad\quad\quad\quad\quad\quad\quad\quad\quad CH_3$

(c) 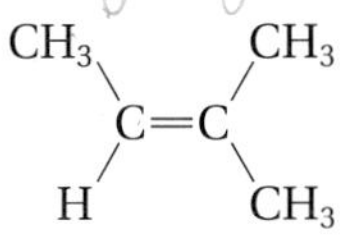 CH_3 $\quad\quad$ CH_3 (d) $CH_2=CHCH_2CH_2$
$\quad\quad\quad$ C=C $\quad\quad\quad\quad\quad\quad\quad\quad\quad\quad |$
$\quad$ H $\quad\quad$ CH_3 $\quad\quad\quad\quad\quad\quad\quad CH_2CH_3$

2.30 Draw a structural formula for each alkene with the molecular formula C_5H_{10}. Name each compound.

Aromatic Compounds (Section 2.9)

2.31 What is meant by the term *aromatic character*?

2.32 Describe the shape of a benzene molecule.

2.33 Explain why both the following structures represent 1,2-diethylbenzene.

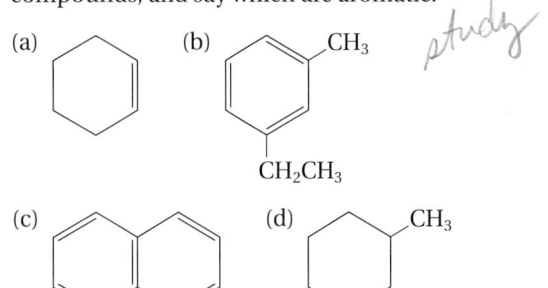

(a) 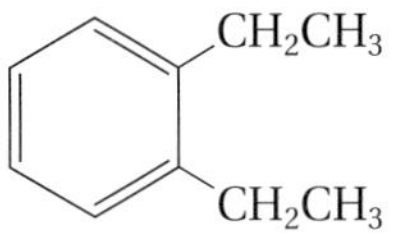   (b)

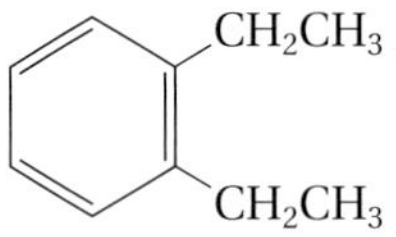

2.34 How many different compounds can be formed when two ethyl groups are attached to a benzene ring?

2.35 Name or draw a structural formula for the following compounds, and say which are aromatic.

(a)   (b)   CH₃

(c)   (d)   CH₃

(e) toluene   (f) ethylbenzene

2.36 Draw a structural formula or name the following compounds. Classify each compound as an aliphatic or aromatic compound.
(a) *p*-diethylbenzene
(b) anthracene
(c) 2-methyl-3-phenylpentane
(d) *p*-xylene

(e) 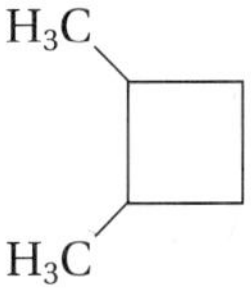 H_3C

$\quad$ H_3C

(f)

$\quad\quad\quad\quad CH_2-CH_2-CH_3$

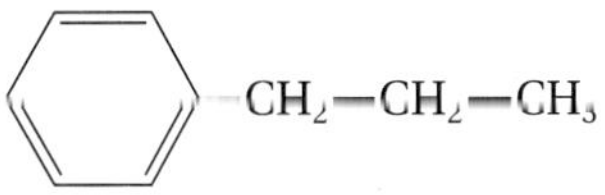

Petroleum and Coal (Sections 2.10, 2.11)

2.37 What is the major difference between crude oil and coal as a source of hydrocarbons?

2.38 Define the terms (a) *petroleum refining* and (b) *cracking*.

Additional Exercises

2.39 Briefly describe (a) hydrocarbon, (b) alkane, (c) alkene, (d) alkyne, (e) cycloalkane, and (f) arene.

2.40 Write a balanced equation for the complete combustion of pentane.

2.41 Match the following:

(a) no carbon-carbon bond rotation 3
(b) structural isomer 10
(c) cyclopropane 6
(d) cracking 7
(e) ethene 8
(f) contains a carbon-carbon triple bond 1
(g) methane 9
(h) butane 2
(i) benzene 5
(j) *trans*-2-butene 4

(1) ethyne
(2) straight-chain alkane
(3) carbon-carbon double and triple bonds
(4) geometric isomer
(5) simplest aromatic triple bond compound
(6) strained ring
(7) petroleum refining
(8) contains carbon-carbon double bond
(9) marsh gas
(10) same formula, different structure

2.42 Draw the structural formula for each of the following:
(a) 3,3-dimethylheptane   (b) 4,4-dipropylnonane
(c) methylpropane   (d) 3-ethylhexane

2.43 Give the IUPAC name for

(a)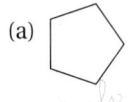
(b) $CH_3-CH=C-CH_3$
          |
          CH_3

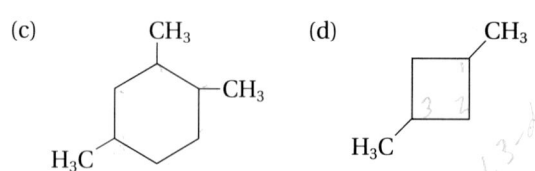

2.44 The following names are incorrect. Explain. (*Hint:* Write a structural formula for the incorrectly named compound and then rename it correctly.)
(a) 2-ethylbutane
(b) 3-methylcyclohexane
(c) 1,2-dimethylpentane
(d) 4-ethylhexane
(e) 3-dimethylheptane
(f) 2-ethyl-3,3-dimethylbutane

2.45 How many *cycloalkanes* have the molecular formula C_5H_{10}? Draw each of their structures and give the correct IUPAC names.

2.46 Draw condensed structural formulas for the following compounds; show all carbon-carbon bonds.
(a) 4-methyl-2-pentene
(b) 2-phenylheptane
(c) *trans*-2-heptene
(d) 3-methylcyclohexene
(e) 3,3-dimethyl-1-butene
(f) 4,5-diethyl-3,6,7-trimethyldecane
(g) 5,5-diethyl-2,6,7-trimethyl-3-octene

SELF-TEST (REVIEW)

True/False

1. All unsaturated compounds have double bonds. False they can have 3 bonds, be benzene
2. Hydrocarbons readily dissolve in water. F
3. The designations *cis* and *trans* are used to distinguish structural isomers. F geometric
4. An alkyne has a triple covalent bond between two carbon atoms. T
5. Any hydrocarbon molecule containing seven carbon atoms will be called a *heptene*. F
6. Most alkenes tend to be more chemically reactive than benzene. F
7. The hybridization of the bonding orbitals in the carbon-carbon double bond of an alkene is sp^3. F sp^2 → alkene
8. The boiling points of hydrocarbons tend to increase with increasing chain length. T

9. Structural isomers and geometric isomers have identical molecular formulas. T
10. Cycloalkanes with more than four carbon atoms in the ring are planar (flat) molecules. F

Multiple Choice

11. All the following are unsaturated hydrocarbons except
(a) 3-octene.   (b) benzene.   (c) cyclobutane.
(d) ethyne.

12. *Cis-trans* geometric isomerism is possible in
(a) 1-butane.   (b) 1-butene.   (c) 2-pentene.
(d) pentane.

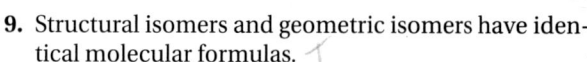

13. Hydrocarbons in general are
(a) soluble in water.   (b) chemically very reactive.
(c) less dense than water.   (d) all of the above.

14. Rotation about a carbon-carbon bond is possible in
(a) benzene.   (b) ethyne.   (c) propane.
(d) ethene.

15. The name of an alkyl group that contains two carbon atoms is
(a) diphenyl.   (b) ethyl.   (c) dimethyl.
(d) propyl.

16. Name this compound: $CH_3CH(CH_3)C(CH_3)_3$.
(a) 2,2,3-trimethylbutane
(b) tetramethylpropane
(c) 1,1,1,2-tetramethylpropane
(d) 2,3,3,3-tetramethylpropane

17. In the electron dot structure of butane, C_4H_{10}, what is the total number of electrons shared between all carbon atoms?
(a) 4   (b) 6   (c) 8   (d) 10

18. A hydrocarbon with nine carbon atoms joined by single bonds in a straight chain is called
(a) heptane.   (b) nonane.   (c) hexene.
(d) trinane.

19. Structural isomers have
(a) the same melting points.
(b) the same molecular formulas.
(c) the same densities.
(d) all of the above.

20. A structural isomer of hexane is
(a) 2,2-dimethylbutane.   (b) cyclohexane.
(c) benzene.   (d) 2-methylpentene.

21. Which of the following words or phrases do not describe benzene?
(a) aromatic   (b) puckered ring
(c) hybrid bonding   (d) arene

22. A correct name for benzene with one CH_3— substituent attached to the ring is
(a) benzylmethane.   (b) methene benzene.
(c) methylbenzene.   (d) methylene benzene.

23. The number of fused rings in naphthalene is
(a) 2.   (b) 3.   (c) 4.   (d) 5.

24. The major products of combustion of an alkane are
(a) carbon and water.
(b) water and carbon monoxide.
(c) carbon dioxide and oxygen.
(d) water and carbon dioxide.

25. The name of $(CH_3)_2$—C=CH_2 is
(a) methylpropene.   (b) dimethylethene.
(c) trimethylmethene.   (d) 1-butene.

26. The most stable form of cyclohexane
(a) is planar.
(b) is called the chair form.
(c) is called the inert form.
(d) is called the boat form.

27. How many isomers, including geometric isomers, are there for the molecular formula C_4H_8?
(a) three   (b) four   (c) five   (d) six

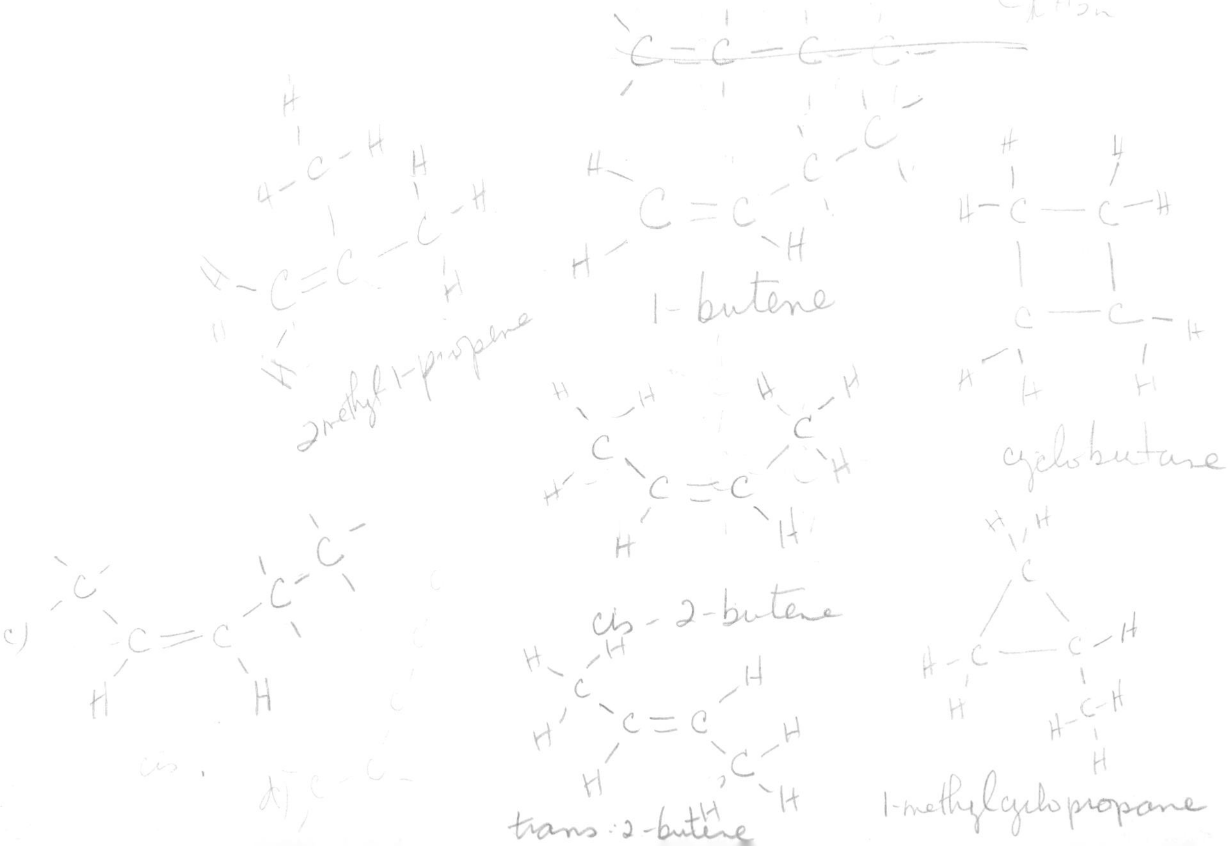

3

Halocarbons, Alcohols, and Ethers

The Polar Bond in Organic Molecules

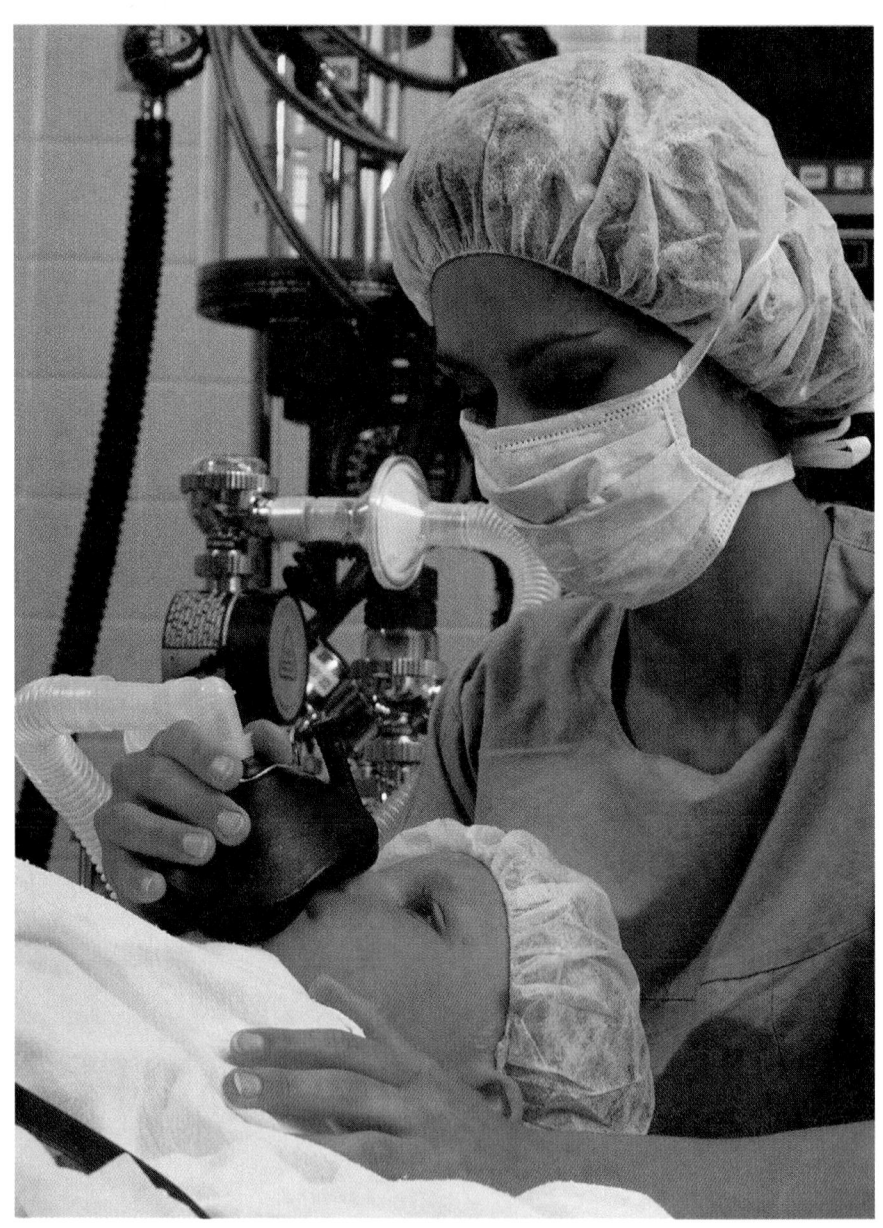

A general anesthetic produces unconsciousness and insensitivity to pain.

In this chapter we will begin our study of organic compounds containing carbon bonded to other elements. Some of these compounds are important biological molecules. Others are used in the health sciences as antiseptics and anesthetics. The compounds considered here are those with carbon-halogen bonds, carbon-oxygen single bonds, and carbon-sulfur single bonds.

Among the compounds we will study in this chapter are alcohols, which contain carbon-oxygen bonds. Ethylene glycol, an alcohol, played a prominent role in a guilty verdict in a murder trial in St. Louis, Missouri. The case is worth noting not only because it supposedly involved an alcohol but also because it shows how scientists subsequently established the innocence of the accused murderer.

Case in Point: A mistaken conviction

In 1989, a woman brought her ill 3-month-old son to the emergency room of a St. Louis hospital, where the attending physician diagnosed the symptoms of ethylene glycol poisoning. After the child recovered, he was taken from his parents and placed in a foster home. The unfortunate child died soon after a visit from his mother, who was charged with first-degree murder for poisoning her son. A commercial laboratory and a hospital laboratory claimed to have found ethylene glycol in the boy's blood and in a bottle of milk he was fed by his mother shortly before he died. The mother was convicted and sentenced to life in prison for the murder of her oldest son by ethylene glycol poisoning. However, this murder case was not as open-and-shut as it seemed, as we will learn in Section 3.7.

Ethylene glycol is the principal ingredient in antifreeze solutions. Although it has a sweet taste, ethylene glycol is quite toxic.

3.1 Halocarbons

AIMS: To name and draw structures of simple halocarbons. To name and draw the structures of the common alkyl groups containing up to four carbon atoms.

Focus

Compounds containing carbon-halogen bonds are called halocarbons.

Halocarbons *are a class of organic compounds containing covalently bonded halogens: fluorine, chlorine, bromine, or iodine.* For example, the compounds CH_3Cl and CH_3Br are halocarbons. Although very few halocarbons are found in nature, they are nevertheless readily prepared and used for many purposes, such as anesthetics and insecticides. *Halocarbons in which a halogen is attached to a carbon of an aliphatic chain are* **alkyl halides;** *halocarbons in which a halogen is attached to a carbon of an arene ring are* **aryl halides.**

The IUPAC rules for naming halocarbons are very similar to those we cited in Chapter 2 for naming substituted alkanes, except that halogen groups must now be added to our repertoire of substituents. Table 3.1 shows the names used for halogen groups when they are substituents on carbon. Common names of a few simple halocarbons are still used. These common names consist of two parts. The first part names the hydrocarbon entity of the molecule as an alkyl group, such as methyl or ethyl. The second part names the halogen as if it were an ion. Methyl chloride, CH_3Cl, is an example. Remember, however, that the bonding in a halocarbon is covalent, not ionic. Alkyl groups besides those of methyl, ethyl, and propyl covered in Chapter 11 have been given names. Table 3.2 shows some of these alkyl groups and the vinyl group (CH_2=CH—), which is derived from ethene. The table also describes the use of the prefix *iso-* and the terms *secondary* and *tertiary* in alkyl group names. Some structures and IUPAC and common names (in parentheses) for some simple halocarbons follow. Studying them and Examples 3.1 and 3.2 will help you understand the naming of halocarbons. Example 3.3 shows how to write a structure when the compound name is given.

Table 3.1 Names of Halogens as Substituent Groups

Halogen	Substituent name
fluorine	fluoro-
chlorine	chloro-
bromine	bromo-
iodine	iodo-

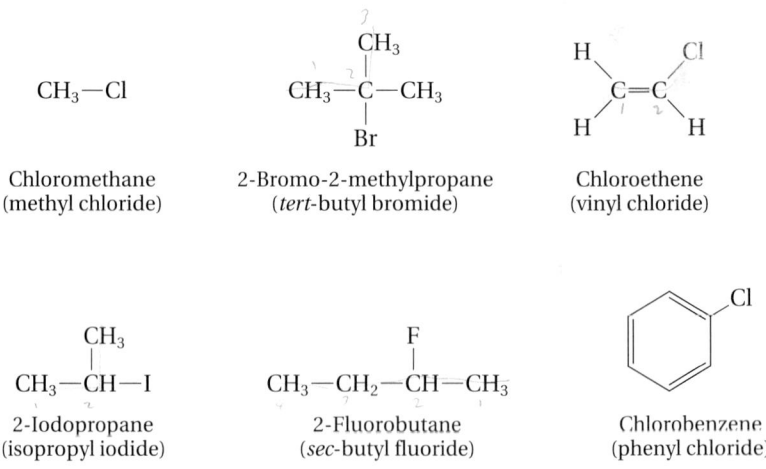

CH₃—Cl

Chloromethane
(methyl chloride)

2-Bromo-2-methylpropane
(*tert*-butyl bromide)

Chloroethene
(vinyl chloride)

2-Iodopropane
(isopropyl iodide)

2-Fluorobutane
(*sec*-butyl fluoride)

Chlorobenzene
(phenyl chloride)

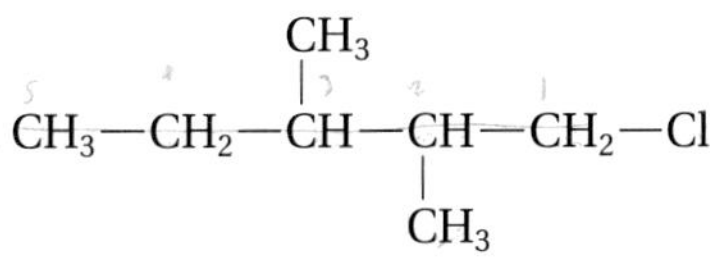

$$CH_3-CH_2-CH-CH-CH_2-Cl$$

with CH_3 above the third carbon and CH_3 below the fourth carbon

1-Chloro-2,3-dimethylpentane
(no common name)

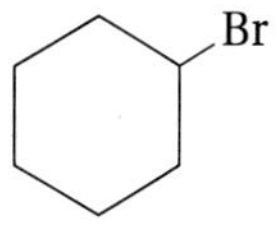

Br on cyclohexane ring

Bromocyclohexane
(cyclohexyl bromide)

Table 3.2 Names of Some Common Alkyl Groups

Name	Alkyl group	Ball-and-stick models	Remarks
isopropyl	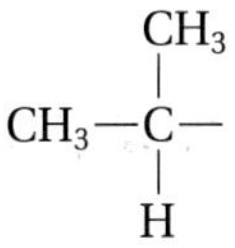 CH_3-C- with CH_3 above and H below $CH_3 CH-$ with CH_3 above	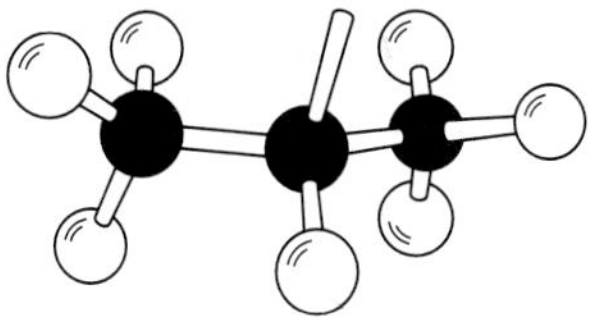	The prefix *iso-* is reserved for carbon chains that are straight except for the presence of a methyl group on the carbon second from the unsubstituted end of the longest chain.
isobutyl	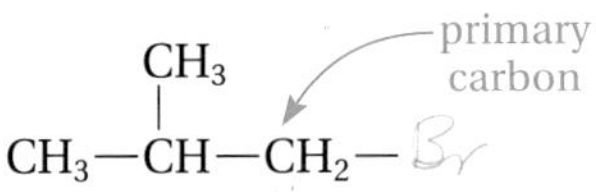 $CH_3-CH-CH_2-Br$ with CH_3 above — primary carbon	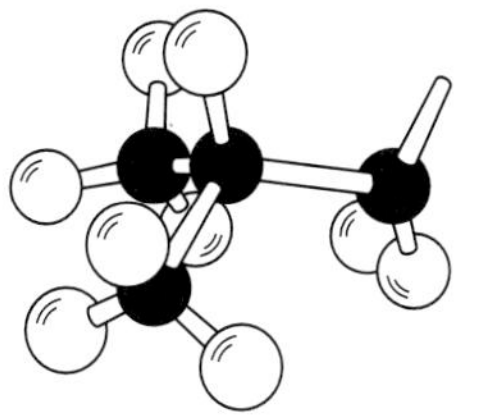	Note the use of the prefix *iso-*. The carbon joining this alkyl group to another group is bonded to one other carbon; it is a *primary carbon*.
secondary butyl (*sec*-butyl)	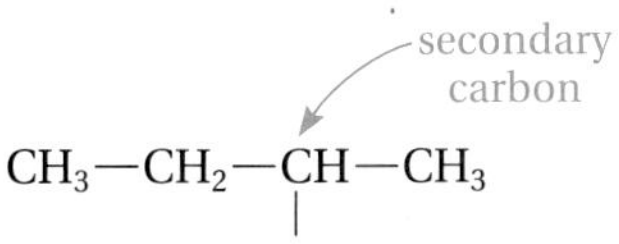 $CH_3-CH_2-CH-CH_3$ — secondary carbon M C R B	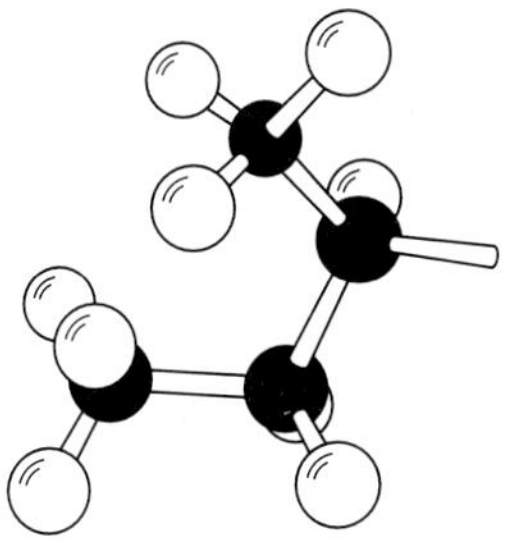	The carbon joining this alkyl group to another group is bonded to two other carbons; it is a *secondary carbon*.
tertiary butyl (*tert*-butyl)	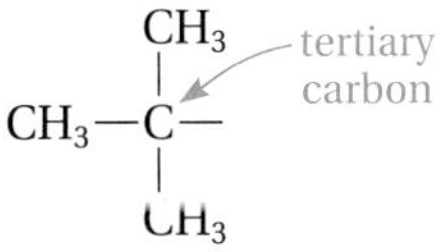 CH_3-C- with CH_3 above (tertiary carbon) and CH_3 below	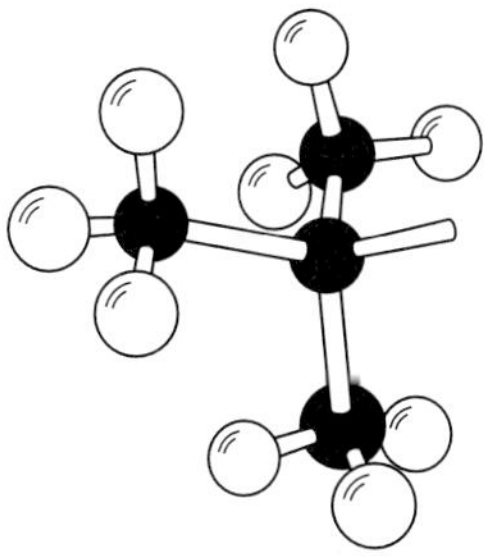	The carbon joining this alkyl group to another group is bonded to three carbons; it is a *tertiary carbon*.
vinyl	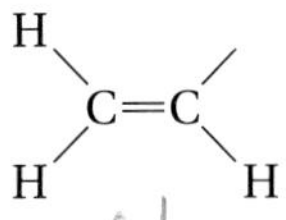 $C=C$ with H, H on left and H, H on right alkene substituent	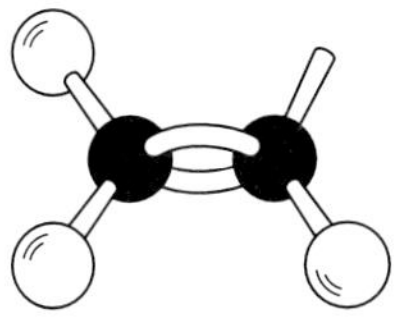	When used as an alkyl group in giving compounds common names, this group, derived from ethene, is called *vinyl*.

EXAMPLE 3.1

Naming an alkyl halide

What are the IUPAC and common names for $CH_3CH_2CH_2Cl$?

SOLUTION

You may wish to review the rules given in Chapter 2 for the IUPAC naming of alkanes and alkyl substituents. The rules will be the same for halocarbons, except that the halogen must be named as a substituent. The longest carbon chain in our molecule contains three carbons, making propane the parent alkane.

$$\underset{3\quad\;\;2\quad\;\;1}{CH_3CH_2CH_2Cl}$$

1-chloropropane or propyl chloride

A chlorine atom substituent is called *chloro-;* this substituent is given the lowest possible number on the parent chain—in this case position 1. The IUPAC name for the compound is *1-chloropropane.* We write the common name of the compound by naming the $CH_3CH_2CH_2$— portion of the molecule as the alkyl group, *propyl,* and adding the word *chloride.* The common name for this compound is *propyl chloride.*

EXAMPLE 3.2

Naming an alkyl fluoride

Give the IUPAC and common names for the following compound.

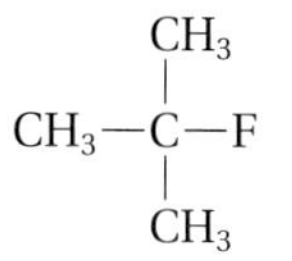

$$\begin{array}{c} CH_3 \\ | \\ CH_3-C-F \\ | \\ CH_3 \end{array}$$

2-fluoro-2-methylpropane

2-fluorobutyl

2-fluoro-2-methylpropane

SOLUTION

The longest chain is three carbons long, making propane the parent alkane.

$$\begin{array}{c} CH_3 \\ | \\ \underset{3}{CH_3}-\underset{2}{C}-F \\ | \\ \underset{1}{CH_3} \end{array}$$

2-fluoro-2-methylpropane

tertiary butyl fluoride

tert-butyl fluoride

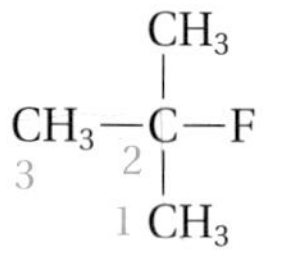

Fluoro and *methyl* groups are substituents on the propane chain, both in the 2 position. Listing the substituent names in alphabetical order and attaching the name of the parent alkane gives the IUPAC name for the compound: *2-fluoro-2-methylpropane.* The common name for the alkyl group with three methyl groups bonded to the same carbon is *tertiary butyl,* which we abbreviate as *tert*-butyl in writing the name (see Table 3.2). Since the halogen substituent is fluorine, the common name for the compound is *tertiary butyl fluoride,* written as *tert*-butyl fluoride.

PRACTICE EXERCISE 3.1

Write the IUPAC name for each of the following halocarbons.

(a) 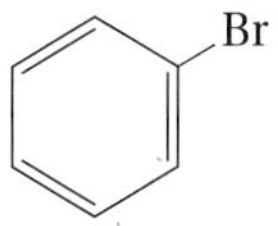 [benzene ring with Br]   *C — C — Cl*   (b) CH_3CH_2Cl (c) [Cl above] $CH_3CHCH{=}CH_2$

bromobenzene   *1-chloroethane*   *3-chloro-1-butene*
ethyl chloride

EXAMPLE 3.3 | **Writing a condensed structural formula for a halocarbon**

Write a condensed structural formula for 2-bromo-3,3-dimethylpentane.

[handwritten: Cl CH₃ above, CH₃ — C — C — CH₂ CH₃, with H and CH₃ below]

SOLUTION

Use the information in the name to construct this halocarbon. The parent alkane is pentane, so write a chain of five carbons.

$$C{-}C{-}C{-}C{-}C$$
$$1 \quad 2 \quad 3 \quad 4 \quad 5$$

Add the substituents at the proper positions. Substituents are *bromo-* at the 2 position and two *methyl* groups on the 3 position of the parent chain.

$$
\begin{array}{cccccc}
 & & Br & CH_3 & & \\
 & & | & | & & \\
C & - & C & - C & - C & - C \\
 & & & | & & \\
 & & & CH_3 & &
\end{array}
$$

Fill in the structure with the number of hydrogens needed to give each carbon four covalent bonds. The structure of 2-bromo-3,3-dimethylpentane must be

$$
\begin{array}{c}
\quad\quad\quad Br \quad CH_3 \\
\quad\quad\quad | \quad\quad | \\
CH_3{-}CH{-}C{-}CH_2{-}CH_3 \\
\quad\quad\quad\quad\quad | \\
\quad\quad\quad\quad\quad CH_3
\end{array}
$$

2-Bromo-3,3-dimethylpentane

PRACTICE EXERCISE 3.2

Give the structural formula for each of the following.

[handwritten: a) CH₃—C—Cl with CH₃ above and H below]

(a) isopropyl chloride

(b) 1-iodo-2,2-dimethylpentane   *[handwritten b) I CH₃ above, CH₂ C - CH₂ CH₂ CH₃, CH₃ below]*

(c) 1-bromo-4-ethylbenzene

[handwritten c) benzene ring with Br at top (position 1), 4 below, CH₂CH₃]

Some alkyl and aryl halocarbons currently in use for a variety of purposes are shown in Table 3.3. Some halocarbons that were widely used in the 1980s, however, are no longer made or used, for reasons discussed in A Closer Look: Chlorofluorocarbons and the Ozone Layer.

[handwritten: CH₃ above, CH₃ — C — Cl, H below]

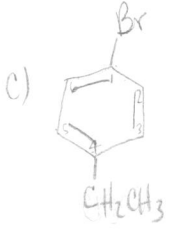

Table 3.3 Some Halocarbons and Their Uses

Halocarbon	Use
$CH_3—CH_2—Cl$ Chloroethane (ethyl chloride)	A local anesthetic, its rapid evaporation on the skin (bp 13 °C) cools nerve endings and cuts down transmission of pain.
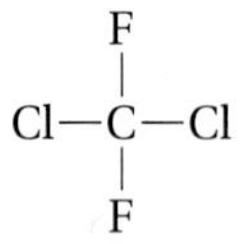 Dichlorodifluoromethane (Freon 12) and 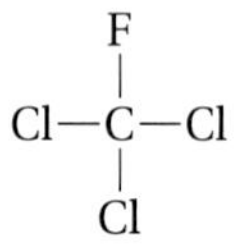 Trichloromonofluoromethane (Freon 11)	Although they were once widely used as refrigerants, Freons are now banned. They are permitted, however, as propellants to deliver inhalation aerosols of medication to asthmatics. (Freons are also known as *chlorofluorocarbons,* or CFCs. CFCs cause depletion of ozone in the stratosphere.)
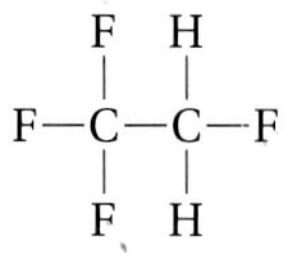 1,1,1,2-tetrafluoroethane (HFC-134a)	Hydrofluorocarbons (HFCs) have been developed as replacements for CFCs (Freons) in car air-conditioning units. Most cars are equipped with systems that use HFC-134a. HFCs do not cause ozone depletion.
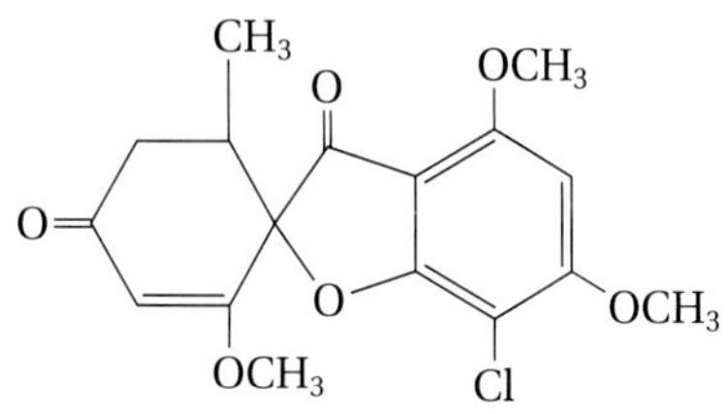 Griseofulvin	Used in treatment of fungal infections. Obtained from *Penicillium.*
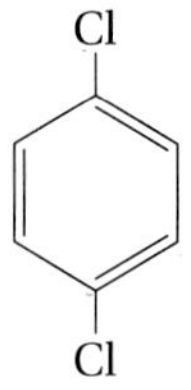 *p*-Dichlorobenzene (1,4-dichlorobenzene)	Used as a moth repellent.
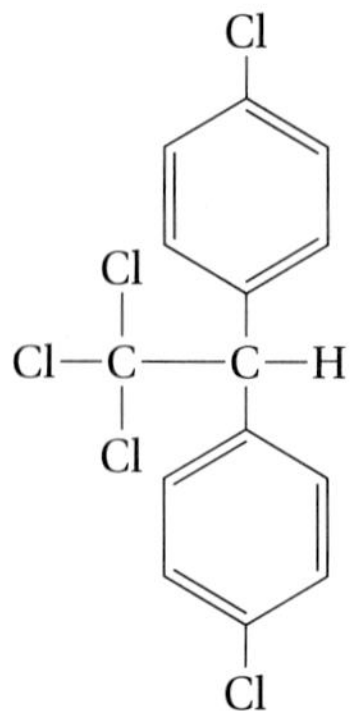 Dichlorodiphenyltrichloroethane (DDT)	Synthesized in 1874, the insecticidal properties of DDT were not recognized until the 1940s. Its use has been banned in the United States because of its persistence in the environment; it is nonbiodegradable.

Chlorofluorocarbons and the Ozone Layer

Halocarbon molecules that contain chlorine as well as fluorine are known as *chlorofluorocarbons,* or *CFCs,* or *Freons.*

$$Cl-\underset{\underset{Cl}{|}}{\overset{\overset{Cl}{|}}{C}}-F \qquad Cl-\underset{\underset{F}{|}}{\overset{\overset{Cl}{|}}{C}}-F \qquad F-\underset{\underset{F}{|}}{\overset{\overset{Cl}{|}}{C}}-\underset{\underset{F}{|}}{\overset{\overset{Cl}{|}}{C}}-F$$

Freon 11   Freon 12   Freon 13

CFCs are gases or low boiling liquids that are chemically inert, nontoxic, nonflammable, and insoluble in water. These properties made them good candidates for refrigerants in air conditioners and as propellants for aerosol cans of hair sprays, deodorants, and inhalation medications (see figure, part a).

Ozone is an important natural component of the stratosphere, the layer of the atmosphere that ranges from 11 to 48 km above the Earth. The ozone molecules shield the Earth's plants and animals from life-destroying ultraviolet radiation. When an ozone molecule in the stratosphere absorbs ultraviolet radiation, it is converted to an oxygen molecule (O_2) and an oxygen atom ($O\cdot$).

$$O_3(g) \xrightarrow[\text{radiation}]{\text{Ultraviolet}} O_2(g) + O\cdot(g)$$

The chemical inertness of CFCs allows them to remain in the environment for a long time. Eventually, they find their way into the stratosphere, where the carbon-chlorine bond in CFCs is broken by energy from ultraviolet light. The chlorine atom ($Cl\cdot$) that is produced combines with an ozone molecule in the stratosphere, to give a chlorine oxide radical ($ClO\cdot$) and an oxygen molecule.

$$Cl\cdot(g) + O_3(g) \longrightarrow ClO\cdot(g) + O_2(g)$$

The $ClO\cdot$ radical then reacts with an oxygen atom ($O\cdot$), formed when ozone absorbs ultraviolet light, to produce another chlorine atom and an oxygen molecule.

$$ClO\cdot(g) + O\cdot(g) \longrightarrow Cl\cdot(g) + O_2(g)$$

This process is repeated many times. It has been estimated that the breaking of a single C—Cl bond of a CFC molecule results in the destruction of 4000 or more ozone molecules in the stratosphere. The destruction of the ozone layer permits larger amounts of harmful ultraviolet radiation to reach the Earth. The effect of ozone depletion is manifested in an increased incidence of skin cancers and crop damage. In 1985, scientists discovered a "hole" in the ozone layer over Antarctica (see figure, part b). Following the concern that this discovery generated, the Montreal Protocol on Substances that Deplete the Ozone Layer went into effect in January 1989. It was signed by 24 nations. The protocol calls for CFCs to be phased out by the year 1996. In the meantime, alternatives are being sought.

One class of compounds that show promise as alternatives are the hydrofluorocarbons, or HFCs. They are possible substitutes for ozone-depleting CFCs because they contain no chlorine and therefore cannot catalyze ozone destruction. However, an air-conditioning system designed for CFC refrigerant will not operate on HFC refrigerants. Most new cars and trucks sold in the United States are now equipped with air-conditioning systems that use HFCs.

$CCl_2F_2 + light \rightarrow CClF_2 + Cl\cdot$

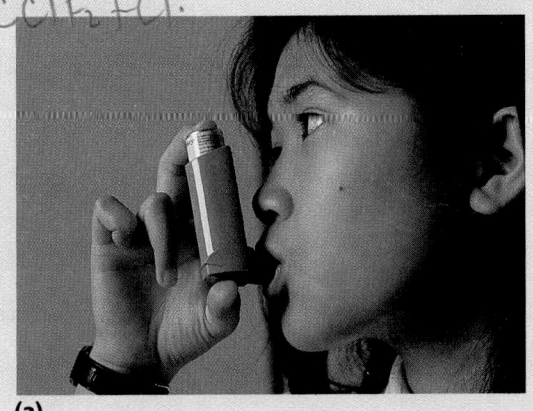

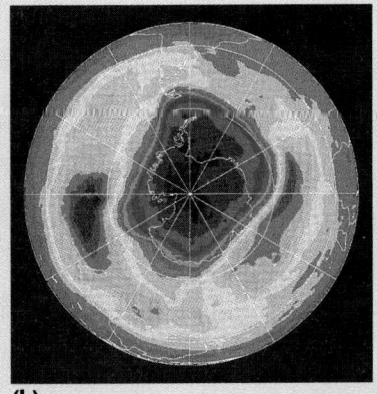

CFCs are used to deliver precise doses of medication directly into the lungs of asthmatics (a). The hole in the ozone layer over Antarctica is clearly visible in this NASA photograph (b).

(a)   (b)

3.2 Halocarbons from alkenes

AIMS: *To distinguish among halogenation, hydrohalogenation, and hydrogenation reactions. To contrast an addition reaction of an alkene with a substitution reaction with benzene. To use Markovnikov's rule to predict the major products in an addition reaction. To describe a polymerization reaction.*

Focus

Halogens and hydrogen halides add to multiple carbon-carbon bonds.

Up to this point we have focused on the structures of hydrocarbon chains and rings, which are essential components of every organic compound. Yet in most chemical reactions involving organic molecules, the hydrocarbon skeletons of the molecules are chemically inert. The chemistry of the alkanes is relatively limited. Most organic chemistry involves substituents attached to hydrocarbon chains. These groups act as a unit and often contain oxygen, nitrogen, sulfur, phosphorus, or halogens. Such groups are called **functional groups**—*the chemically functional parts of the molecule.*

The double and triple bonds of alkenes and alkynes are chemically reactive and are considered functional groups. So are the carbon-halogen groups of halocarbons. Carbon-carbon single bonds are not easy to break. However, recall from Section 2.8 that the pi bond of the double bond in alkenes is somewhat weaker than a carbon-carbon single bond. It is possible for a compound of general structure X—Y to react with a double bond under appropriate conditions. *A reaction in which two molecules of reactant combine to form a single product is called an* **addition reaction.** Here is the general equation for addition reactions of alkenes:

$$\ce{\overset{\diagdown}{\diagup}C=C\overset{\diagup}{\diagdown} + X-Y \longrightarrow -\overset{\overset{\displaystyle X}{|}}{C}-\overset{\overset{\displaystyle Y}{|}}{C}-}$$

Addition reactions are an important method of introducing new functional groups into organic molecules. Halocarbons can be produced by addition reactions in which *the adding reactant or reagent X—Y is a halogen molecule, such as Cl—Cl* (**halogenation**) *or a hydrogen halide, such as H—Cl* (**hydrohalogenation**). Although *hydrogenation* is a method for producing alkanes rather than halocarbons, we include the *addition of hydrogen, H—H, to double bonds* (**hydrogenation**) in this section on additions to alkenes.

Halogenation

When the reagent X—Y is a halogen molecule such as chlorine or bromine, the product of the reaction is a disubstituted halocarbon:

$$\ce{\underset{H}{\overset{H}{\diagdown}}C=C\underset{H}{\overset{H}{\diagup}} + Br-Br \longrightarrow H-\overset{\overset{\displaystyle Br}{|}}{\underset{\underset{\displaystyle H}{|}}{C}}-\overset{\overset{\displaystyle Br}{|}}{\underset{\underset{\displaystyle H}{|}}{C}}-H}$$

Ethene (colorless)   Bromine (brownish orange)   1,2-Dibromoethane (colorless)

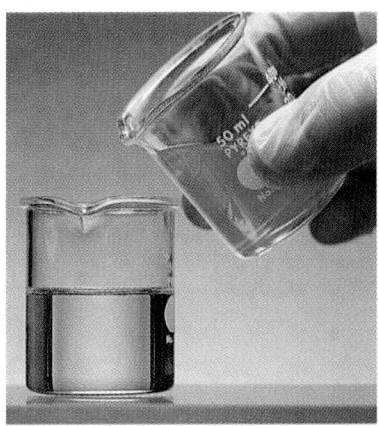

(a)

(b)

Figure 3.1
(a) The beaker on the left contains an alkene; the beaker on the right is a solution of bromine in dichloromethane.
(b) The bromine solution loses its brownish orange color when it is added to the alkene.

The addition of bromine to carbon-carbon multiple bonds is often used as a chemical test for unsaturation in an organic molecule. Bromine has a brownish orange color, but most organic compounds of bromine are colorless. The test for unsaturation is done by adding a few drops of a 1% solution of bromine in dichloromethane (CH_2Cl_2) to the suspected alkene. Loss of the brownish orange color is a positive test for unsaturation (Fig. 3.1).

PRACTICE EXERCISE 3.3

Write the structure for the expected product from each of the following reactions.

(a) $CH_2{=}CHCH_2CH_3 + Br_2 \longrightarrow$

(b) [hexene structure] $+ Cl_2 \longrightarrow$ [dichlorocyclohexane structure]

(c) $CH_3CH{=}CHCH_3 + I_2 \longrightarrow$

a) $CH_2 = CH CH_2 CH_3 + Br_2 \longrightarrow CH_2 - CH CH_2 CH_3$ *b)* *c) $CH_3 CH - CH CH_3$*

Hydrohalogenation

Hydrogen halides such as HBr or HCl also can add to the double bond. Since the product contains only one substituent, it is called a *monosubstituted* halocarbon. The addition of hydrogen chloride to ethene is an example of hydrohalogenation.

[Structure: ethene + H—Cl → chloroethane]

Ethene | Hydrogen | Chloroethane
(ethylene) | chloride | (ethyl chloride)

alkene + hydrogen halides → monosubstituted halocarbon

The addition of hydrogen halides to alkenes more complex than ethene can, in principle, give equal amounts of two different structural isomers, as shown for propene:

[Structure: propene + H—Cl → 2-chloropropane + 1-chloropropane]

Propene | 2-Chloropropane | 1-Chloropropane
(propylene) | (isopropyl chloride) | (propyl chloride)
 | (major product) | (minor product)

alkene + hydrogen halides →

In reality, more of one structural isomer is formed. This follows **Markovnikov's rule:** *The major product of hydrohalogenation is the one in which the hydrogen of the hydrogen halide ends up on the carbon of the double bond that already has more hydrogens.* This general rule was devised by Vladimir Markovnikov (1838–1904), a Russian chemist. You can usually be certain of writing the correct major product in reactions of this type by thinking of where the hydrogen of the adding reagent is going and remembering that "birds of a feather flock together."

The benzene ring resists addition reactions, since addition would destroy the stable aromatic electron system. If benzene is treated with a halogen in the presence of a catalyst, however, *substitution* of a ring hydrogen by a halogen group occurs. *In* **aromatic substitution,** *the hydrogen on a* *benzene ring is replaced by another group.* Iron compounds are often used as catalysts for aromatic substitution reactions—a rusty nail dropped in the reaction flask works fine.

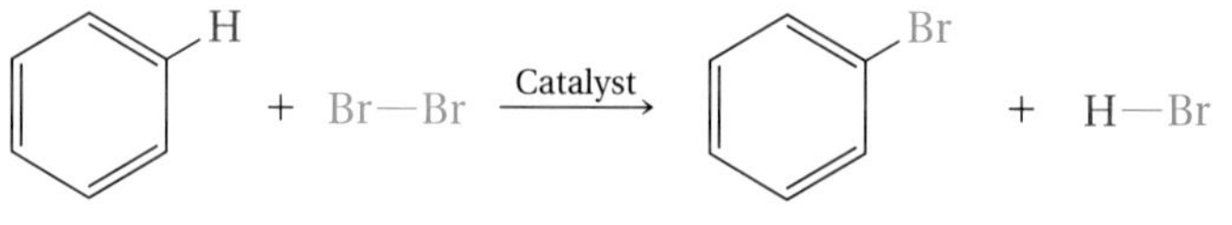

| Benzene | Bromine | Bromobenzene (phenyl bromide) | Hydrogen bromide |

benzene + halogen ——cata——>

EXAMPLE 3.4

Predicting the major product of hydrohalogenation

Which is the major addition product in the following reaction?

major product

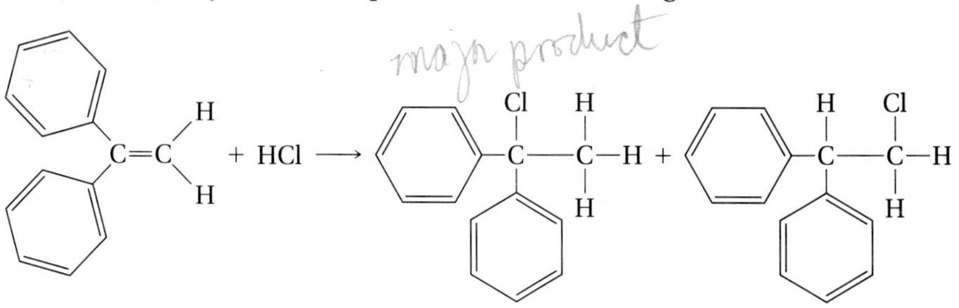

SOLUTION

The reaction is a hydrohalogenation in which hydrogen chloride adds to the double bond of the alkene. Markovnikov's rule predicts that the hydrogen of HCl will end up on the carbon of the reacting double bond that has more hydrogens; the chlorine of HCl will end up on the carbon that has fewer attached hydrogens. In the reactant, one of the carbons of the double bond has two hydrogens; the other carbon of the double bond has none. The major product of the reaction should be

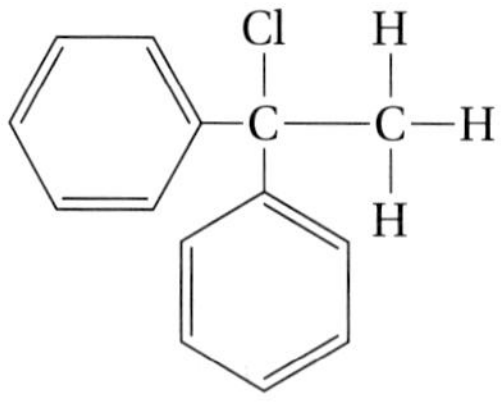

Major product

PRACTICE EXERCISE 3.4

Give the structure(s) for the expected product(s) from each of the following reactions.

(a)

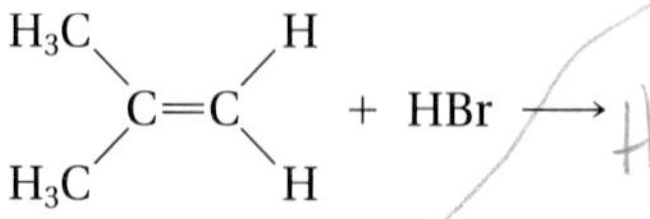

does it matter where Br or H is attached no different in both if both

major product   minor product

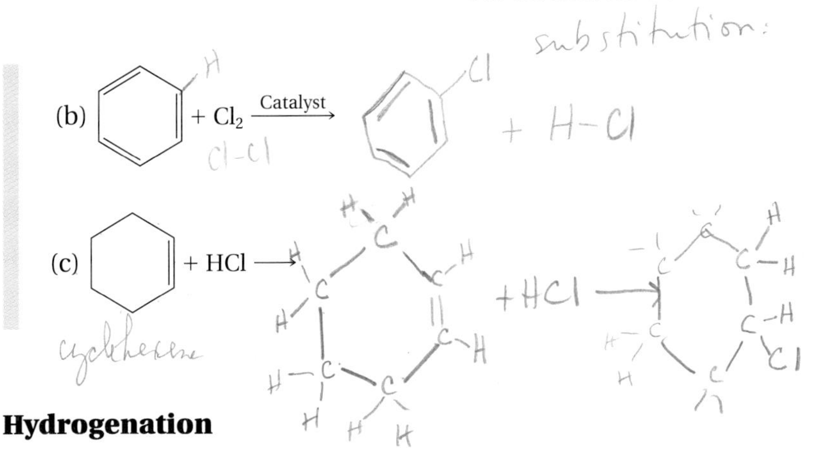

(handwritten annotations)

substitution:

(b) [benzene ring] + Cl₂ $\xrightarrow{\text{Catalyst}}$ [chlorobenzene ring] + H—Cl

Cl—Cl

(c) [cyclohexene ring] + HCl ⟶ [structural drawing] + HCl ⟶ [structural drawing with Cl]

cyclohexene

Hydrogenation

Besides halogens and hydrogen halides, hydrogen also adds to carbon-carbon double bonds to give alkanes. These hydrogenation reactions usually require a catalyst. Finely divided platinum (Pt) or palladium (Pd) are often used. Figure 3.2 shows an industrial chemist working with an assortment of catalysts. Here are two examples:

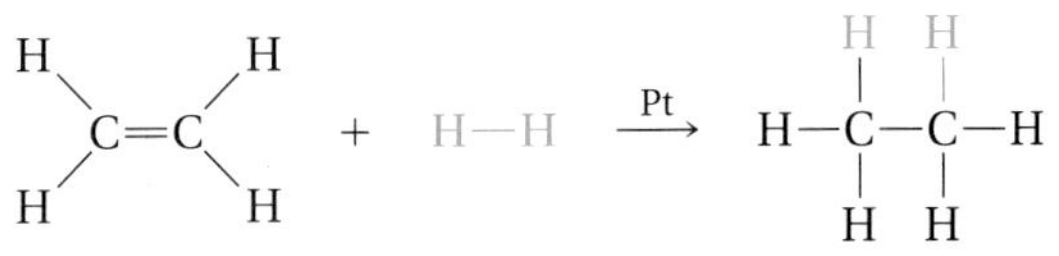

Figure 3.2
A collection of catalysts. Catalysts are used to speed up a wide variety of chemical reactions.

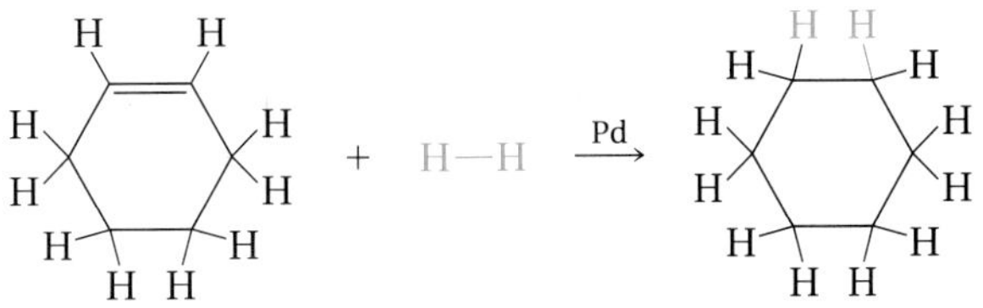

[structure] Ethene + H—H $\xrightarrow{\text{Pt}}$ Ethane

[structure] Cyclohexene + H—H $\xrightarrow{\text{Pd}}$ Cyclohexane

add H

Hydrogenation of a double bond is a *reduction reaction*—ethene is reduced to ethane, for example, and cyclohexene is reduced to cyclohexane. We will examine reduction reactions of organic compounds in more detail in Chapter 4.

Benzene resists hydrogenation, just as it resists halogenation and hydrohalogenation. At high temperatures and high pressures of hydrogen and in the presence of a catalyst, however, three molecules of hydrogen reduce one molecule of benzene to cyclohexane:

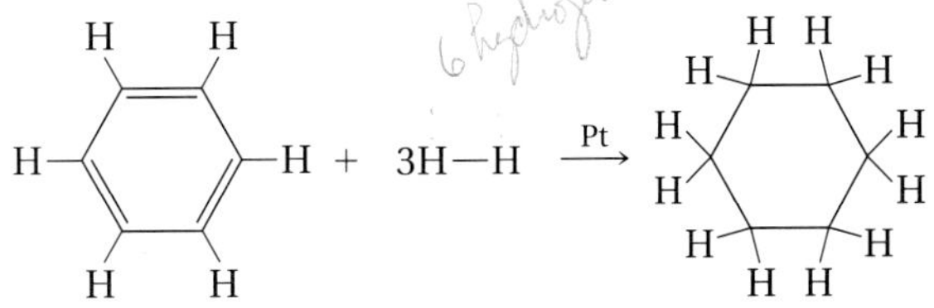

6 hydrogen

Benzene + 3H—H $\xrightarrow{\text{Pt}}$ Cyclohexane

Benzene   Hydrogen   Cyclohexane

Polymerization reactions

The flammability of hydrocarbons decreases as the halogen content increases. Because they cannot be burned at ordinary combustion temperatures, the destruction of polyhalogenated hydrocarbons (such as PVC) is a serious waste-disposal problem.

Many useful **polymers**—*long molecules formed from smaller molecular units*—are made from alkene *monomers*. **Monomers** *are molecules that add to each other to form the repeating units of the polymer.* Many halocarbon polymers have useful properties. Two examples are polyvinyl chloride and polytetrafluoroethene. Vinyl chloride (CH_2=CHCl) is the monomer of polyvinyl chloride.

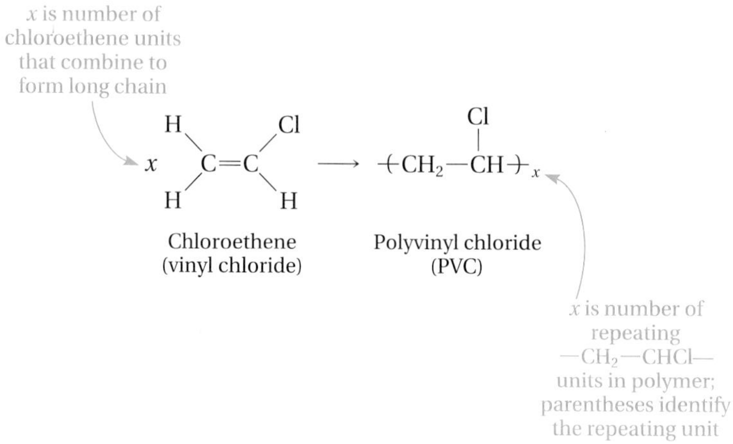

x is number of chloroethene units that combine to form long chain

Chloroethene
(vinyl chloride)

Polyvinyl chloride
(PVC)

x is number of repeating —CH_2—CHCl— units in polymer; parentheses identify the repeating unit

Polyvinyl chloride (PVC) is used for pipes in plumbing and is also produced in sheets for use as tough plastic covering on upholstery. Polytetrafluoroethene (Teflon or PTFE) is the product of the polymerization of tetrafluoroethene (F_2C=CF_2) monomers.

$$x CF_2=CF_2 \longrightarrow \quad +CF_2-CF_2+_x$$

Tetrafluoroethene   Teflon (PTFE)

Teflon is resistant to most chemicals. A coating of Teflon on the inside surfaces of pots and pans makes them nonstick. Teflon is used in the manufacture of medical instruments and is also produced as a tape for sealing pipe fittings.

Other alkene molecules can react to form polymers, as shown here for ethene:

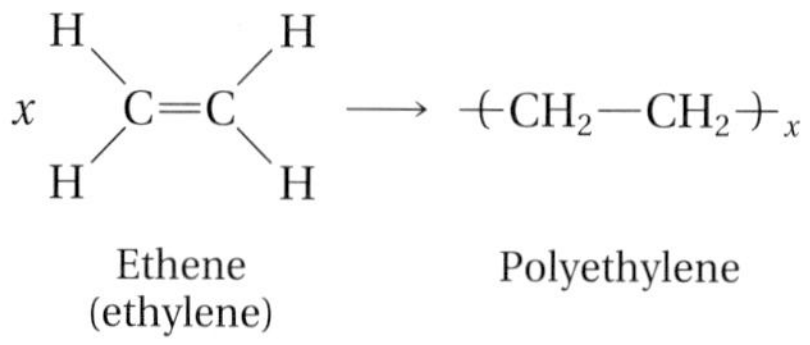

Ethene
(ethylene)

Polyethylene

Ethene is the monomer of polyethylene. Polyethylene is an important industrial product. Chemically resistant and easy to clean, it is used to make refrigerator dishes, plastic milk bottles, laboratory wash bottles, and many other familiar items. By shortening or lengthening the carbon chains, chemists can control the physical properties of polyethylene. Polyethylene containing relatively short chains ($x = 100$) has the consistency of paraffin wax; that with long chains ($x = 1000$) is harder and more rigid, like plastic milk containers. Polypropylene is prepared by polymerization of propene.

$$x\text{CH}{=}\text{CH}_2 \overset{\text{CH}_3}{\longrightarrow} {+}\overset{\text{CH}_3}{\text{CH}{-}\text{CH}_2}{+}_x$$

Propene
(propylene)

Polypropylene

Polypropylene makes a stiffer polymer than polyethylene, and it is used extensively in utensils and containers.

Polystyrene is prepared by the polymerization of styrene (vinyl benzene):

$$x\text{CH}_2{=}\text{CH} \longrightarrow {+}\text{CH}_2{-}\text{CH}{+}_x$$

Styrene
(vinyl benzene)

Polystyrene

Polystyrene is a poor heat conductor when produced as a foam. It is used to insulate homes and in the manufacture of molded items such as coffee cups and coolers.

Polyisoprene is the polymer that constitutes natural rubber. It is used in the manufacture of boots, tires, and rubber tubing. The polyisoprene molecule contains *cis* double bonds:

$$x\text{CH}_2{=}\text{CCH}{=}\text{CH}_2 \longrightarrow$$

$$\overset{\text{CH}_3}{|}$$

Isoprene

cis-Polyisoprene

Figure 3.3 shows a small selection of the wide variety of articles made from polymers.

Figure 3.3
Synthetic polymers play an important role in our daily lives.

3.3 Alcohols

AIMS: *To name and draw structures of alcohols, glycols, and phenols. To identify an alcohol as being primary, secondary, or tertiary. To identify the uses of some common alcohols.*

Focus

Alcohols are organic derivatives of water.

Alcohols *are compounds in which one hydrogen of the water molecule is replaced by a hydrocarbon chain or ring. The functional group* —OH *in alcohols is called a* **hydroxy function** *or* **hydroxyl group.** In organic chemistry, the symbol **R** is often used to represent any carbon chains or carbon rings. Therefore, the general formula of an alcohol can be written as ROH.

hydroxyl group

Water molecule

Alcohol molecule

Chemists often arrange aliphatic alcohols into structural categories according to the number of R groups attached to the *carbon* that is bonded to the hydroxyl group. The R groups in a molecule may be the same or may be different. Only one R group and two hydrogens are attached to the C—OH of a primary (abbreviated 1°) alcohol. Two R groups and one hydrogen are attached to the C—OH of a secondary (2°) alcohol. A tertiary (3°) alcohol has three R groups and no hydrogens attached to the C—OH.

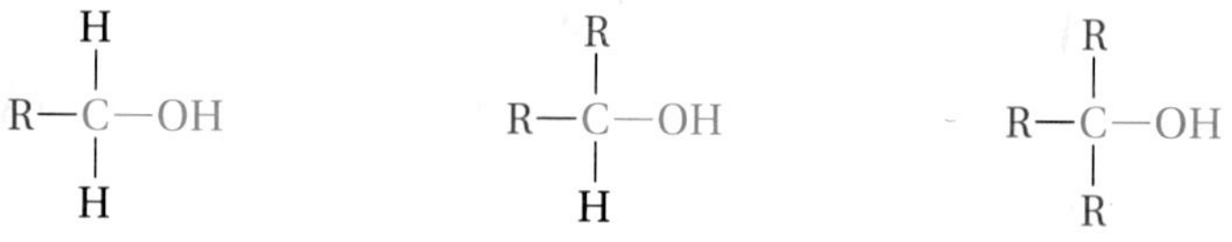

$$
\begin{array}{ccc}
\overset{\displaystyle H}{\underset{\displaystyle H}{R-C-OH}} & \overset{\displaystyle R}{\underset{\displaystyle H}{R-C-OH}} & \overset{\displaystyle R}{\underset{\displaystyle R}{R-C-OH}} \\
\text{Primary (1°) alcohol} & \text{Secondary (2°) alcohol} & \text{Tertiary (3°) alcohol}
\end{array}
$$

EXAMPLE 3.5   **Determining whether an alcohol is primary, secondary, or tertiary**

Identify each of the following compounds as a primary, secondary, or tertiary alcohol.

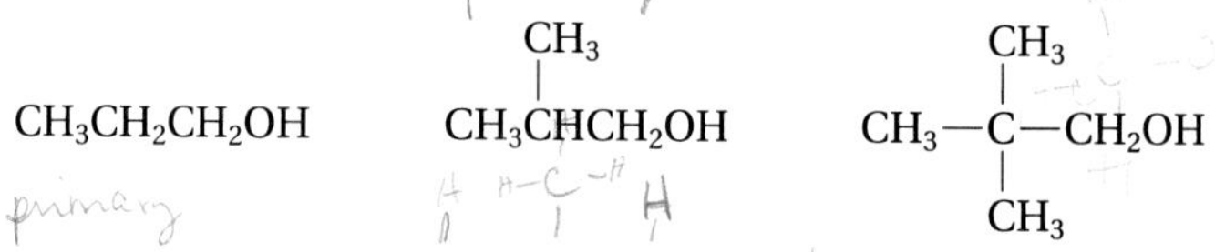

$$CH_3CH_2CH_2OH \qquad \underset{\underset{\displaystyle CH_3}{|}}{CH_3CHCH_2OH} \qquad \underset{\underset{\displaystyle CH_3}{|}}{\overset{\overset{\displaystyle CH_3}{|}}{CH_3-C-CH_2OH}}$$

SOLUTION

The alcohols in this example have increasingly complex hydrocarbon chains. However, it is the number of alkyl groups bonded to the carbon in C—OH that determines whether an alcohol is primary, secondary, or tertiary: A primary alcohol has one R group and two hydrogens attached, a secondary alcohol has two R groups and one hydrogen attached, and a tertiary alcohol has three R groups and no hydrogens attached. In each of the examples, the carbon bonded to the hydroxyl group has only one R group and two hydrogens attached. Therefore, each of these alcohols is a primary alcohol.

Naming alcohols

To name straight-chain and substituted alcohols by the IUPAC system, we drop the *-e* ending of the name of the parent alkane and add the ending *-ol*. The parent alkane is the longest continuous chain of carbons that includes the carbon attached to the hydroxyl group. In numbering the longest continuous chain, we give the position of the hydroxyl group the lowest possible number. Alcohols containing two, three, and four —OH substituents are named *diols*, *triols*, and *tetrols*, respectively. Common names of aliphatic alcohols are written in the same way as those for the halocarbons. The alkyl group—methyl, for example—is named and followed by the word *alcohol*, as in methyl alcohol. Compounds with more than one —OH substituent are called glycols. Here are some simple aliphatic alcohols along with their IUPAC and common names:

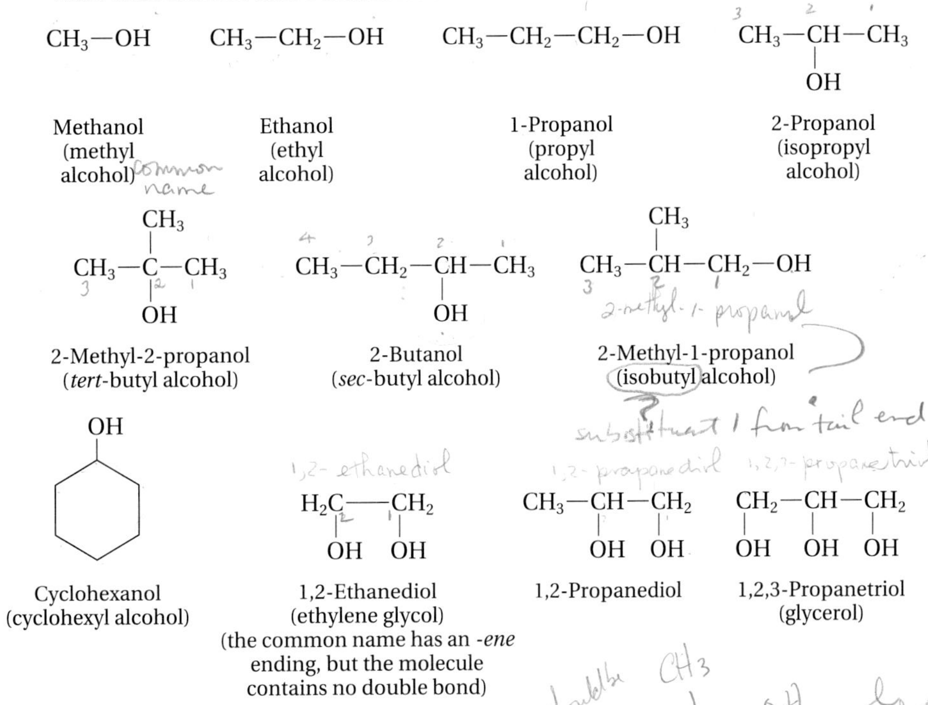

CH_3-OH CH_3-CH_2-OH $CH_3-CH_2-CH_2-OH$ $CH_3-CH-CH_3$
                                                                            |
                                                                            OH

Methanol (methyl alcohol) *common name*   Ethanol (ethyl alcohol)   1-Propanol (propyl alcohol)   2-Propanol (isopropyl alcohol)

CH_3
|
CH_3-C-CH_3
|
OH

$CH_3-CH_2-CH-CH_3$
|
OH

CH_3
|
$CH_3-CH-CH_2-OH$

2-Methyl-2-propanol (*tert*-butyl alcohol)   2-Butanol (*sec*-butyl alcohol)   2-Methyl-1-propanol (isobutyl alcohol)

Cyclohexanol (cyclohexyl alcohol)

H_2C-CH_2
| |
OH OH

$CH_3-CH-CH_2$
| |
OH OH

$CH_2-CH-CH_2$
| | |
OH OH OH

Cyclohexanol (cyclohexyl alcohol)   1,2-Ethanediol (ethylene glycol) (the common name has an *-ene* ending, but the molecule contains no double bond)   1,2-Propanediol   1,2,3-Propanetriol (glycerol)

PRACTICE EXERCISE 3.5

Name the following alcohols.

(a) $CH_3CH_2CH_2CH_2OH$ (b) CH_3 (c) $CH_3CH_2CHCH_2OH$
                                         | |
                                   CH_3CHOH CH_3
                                         |
                                      CH_3

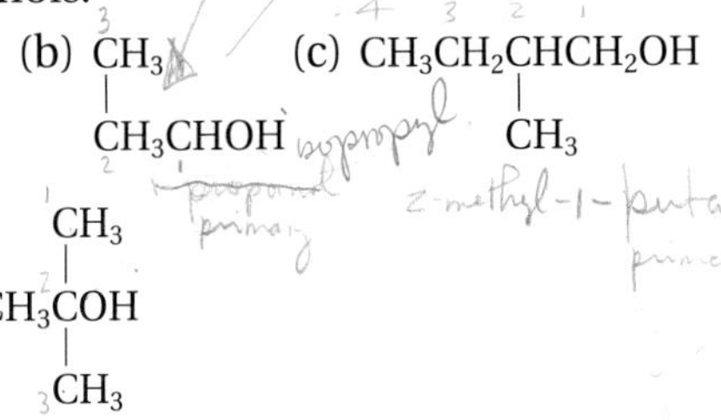

(d) [cyclohexane ring]—OH (e) CH_3COH
                                        |
                                      CH_3

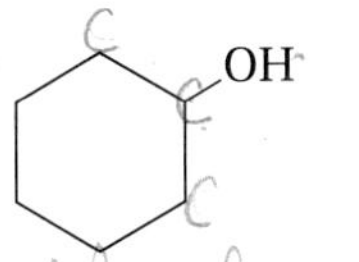

PRACTICE EXERCISE 3.6

Classify each of the alcohols in Practice Exercise 3.5 as primary, secondary, or tertiary.

Uses of alcohols

Many aliphatic alcohols are used in the laboratory, in the health sciences, and in industry. Isopropyl alcohol, a colorless, nearly odorless liquid (bp 82 °C), is the rubbing alcohol used for massages. It is also used as a base for perfumes, creams, lotions, and other cosmetics. Ethylene glycol (bp 197 °C) is the principal ingredient of antifreeze. Its advantages over other high-boiling liquids are its solubility in water and a freezing point of −17.4 °C. If

H—C—C—C—H
  | | |
  OH OH OH

1,2,3-propanetriol

Drinking during pregnancy increases the chance of miscarriage, may cause low birth weight, and is linked to fetal alcohol syndrome. Characteristics of this disorder include abnormal limb development, facial abnormalities such as cleft palate, heart defects, and lower-than-average intelligence.

—C—C—OH
  | |

ethanol

water is added to ethylene glycol, the mixture freezes at an even lower temperature—a 50% (v/v) aqueous solution of ethylene glycol freezes at −36 °C. Glycerol is a viscous, sweet-tasting, water-soluble liquid. It is used as a lubricant in suppositories and as a moistening agent in cosmetics, foods, and drugs. Glycerol is also an important component of fats and oils.

Ethanol (bp 78.5 °C) is the most important alcohol. Also called *grain alcohol*, much ethanol is produced by yeast fermentation of sugar. Archaeological evidence indicates that cave dwellers had produced alcoholic beverages by using fermentation. Since those early times, people have pursued and perfected that art. Virtually every grain, fruit, or vegetable has been fermented: corn, wheat, rye, rice, grapes, tomatoes, dandelions, elderberries, cherries, potatoes, and even cactus pulp. The *proof* of an alcoholic beverage is twice the alcohol content. For example, 100 proof spirits contain 50% ethanol. Ethanol itself is tasteless and odorless. The differences in taste among alcoholic beverages are the result of other products formed during the fermentation process. Among these products are the fusel oils, which include propyl, isopropyl, and other straight- and branched-chain aliphatic alcohols. Fusel oils are often toxic and are one cause of the hangover effects of excessive drinking.

Most ethanol is produced by the fermentation of corn starch and cane sugar. Ethanol is an alternative fuel for use in automobiles. It is mixed with gasoline to form *gasohol.* At a concentration of 10% ethanol, gasohol can be used in a standard automobile engine.

Ethanol is a hypnotic or sleep-inducing drug. Its abuse is a serious public health problem. Some experts estimate that one in nine adult Americans have a drinking problem of some kind. There is no evidence that moderate drinking is harmful.

Ethanol for industrial use is often *denatured.* **Denatured alcohol** *is ethanol that contains an added substance to make it toxic.* Methanol, sometimes called *wood alcohol* because before 1925 it was prepared by the distillation of wood, is often the denaturant. The inability to remember the difference between grain alcohol and wood alcohol has spelled tragedy for many people. Wood alcohol is extremely toxic: 10 mL has been reported to cause permanent blindness—and as little as 30 mL, death. Some undenatured ethanol is used in laboratories under careful supervision. Pure ethanol is labeled *absolute alcohol.* A preparation containing 5% water is also available. Nobody should ever attempt to drink laboratory alcohol.

Phenols

Phenols *are compounds in which a hydroxyl group is attached directly to an aromatic ring.* Many phenolic compounds are found in nature. Shown below are phenol, the parent compound, and some natural and synthetic derivatives.

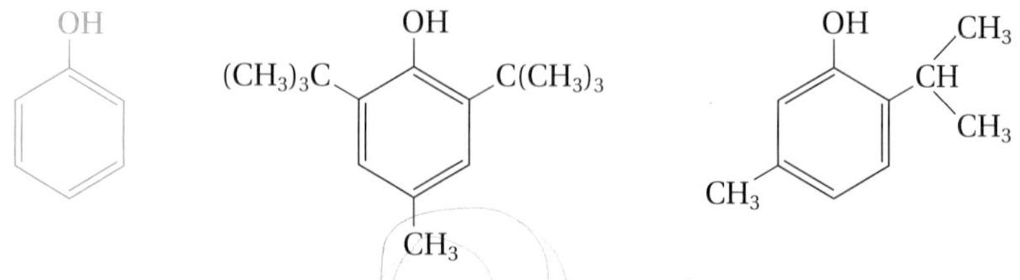

Phenol   BHT   Thymol

Phenol was formerly used as an antiseptic in hospitals (see A Closer Look: Phenolic Antiseptics and Disinfectants). A synthetic derivative of phenol, butylated hydroxy toluene (BHT), is widely used as a food preservative. BHT is often put in food wrappers rather than in the food itself to keep it from imparting a rather antiseptic flavor to the product.

A Closer Look

Phenolic Antiseptics and Disinfectants

Phenol, also called *carbolic acid,* was first used as a medical antiseptic in 1867 by Joseph Lister. He demonstrated that the incidence of infections after surgical procedures was dramatically reduced when solutions of phenol were used to clean the surgical instruments, the operating room, and the patient's skin. Today's use of phenol for medicinal purposes is more limited. When phenol is absorbed through the skin, ingested, or inhaled, it is toxic. It also can cause severe chemical burns. However, it is available as an antiseptic in aqueous or alcoholic solutions that contain a maximum of 1.5% phenol. Phenol is an ingredient in a variety of throat lozenges that can contain up to 50 mg phenol per lozenge. It is a useful topical drug for treating sore throats because it numbs the inflamed area.

A number of derivatives of the parent phenol are used today as antiseptics. Many mouthwashes and throat lozenges include alkyl-substituted phenols as their active ingredients for pain relief (see figure). The compound 4-hexylresorcinol has a superior antibacterial action to that of phenol. It is also much less toxic. Therefore, it is an ingredient of choice in mouthwashes and throat lozenges.

OH

OH

CH₂CH₂CH₂CH₂CH₂CH₃

4-hexylresorcinol

Phenols and methyl phenols (cresols) are commonly used as disinfectants in hospitals and around the home. Disinfectants are formulated for use on inanimate objects, not living tissues. Phenol's germicidal properties result because these chemicals have the ability to disrupt the microorganisms' cell wall permeability.

The naturally occurring phenol eugenol is found in cloves and is used in dentistry to relieve toothaches. Methyl salicylate, another phenol derivative commonly known as oil of wintergreen, is a flavoring agent and is used in liniments to relieve sore muscles. Another derivative of phenol, thymol, is used as an antiseptic in mouthwash preparations.

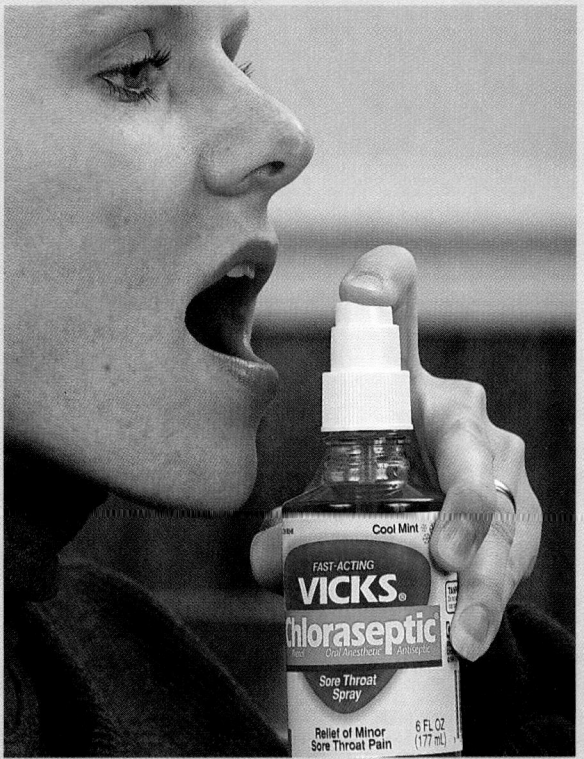

Phenol and its derivatives are the active ingredients in many of the preparations used for oral hygiene.

3.4 Making alcohols

AIM: To illustrate the synthesis of alcohols by addition and displacement reactions.

Focus

Alcohols can be synthesized by addition and displacement reactions.

An organic chemist who wants to make an alcohol may find over a dozen ways to do the job. We will discuss two ways: addition of water to carbon-carbon double bonds of alkenes and reactions of halocarbons with hydroxide ions.

Addition of water to alkenes

The general reaction for adding water to the double bond in an alkene is

$$\underset{\text{Alkene}}{\overset{}{\diagdown}C=C\overset{}{\diagup}} + \underset{\text{Water}}{H-OH} \xrightarrow{H^+} \underset{\text{Alcohol}}{-\overset{|}{\underset{|}{C}}-\overset{|}{\underset{|}{C}}-}$$

The addition of water to an alkene is called a **hydration reaction.** Hydration reactions usually require heating of the alkene and water at about 100 °C in the presence of a trace of strong acid. The acid serves as a catalyst for the reaction; hydrochloric or sulfuric acid is generally used. For alkenes with alkyl substituents on the double bond, Markovnikov's rule applies to the preparation of alcohols, just as it does to the preparation of halocarbons. Hydrogen goes to the carbon of the double bond with more hydrogens; —OH goes to the other carbon. This is shown for the addition of water to propene:

$$\underset{\substack{\text{Propene}\\\text{(propylene)}}}{\overset{H}{\underset{H_3C}{\diagdown}}C=C\overset{H}{\underset{H}{\diagup}}} + H-OH \xrightarrow[100\,°C]{HCl} \underset{\substack{\text{2-Propanol}\\\text{(isopropyl alcohol)}\\\text{(major product)}}}{CH_3-\overset{H}{\underset{OH}{C}}-\overset{H}{\underset{H}{C}}-H} + \underset{\substack{\text{1-Propanol}\\\text{(minor product)}}}{CH_3-\overset{H}{\underset{H}{C}}-\overset{H}{\underset{OH}{C}}-H}$$

Reactions of halocarbons with hydroxide ions

The reactions of halocarbons with hydroxide ions to give alcohols are examples of *displacement reactions. In a* **displacement reaction,** *a reactant replaces a substituent group (—X) on another reactant.* The substituent group —X that is displaced is usually a halogen, most often a chloro-, bromo-, or iodo- substituent. The general displacement reaction for the preparation of alcohols from halocarbons is

$$\underset{\text{Halocarbon}}{R-X} + \underset{\substack{\text{Hydroxide}\\\text{ion}}}{OH^-} \xrightarrow[100\,°C]{H_2O} \underset{\text{Alcohol}}{R-OH} + \underset{\substack{\text{Halide}\\\text{ion}}}{X^-}$$

Chemists usually use aqueous solutions of sodium or potassium hydroxide as the source of hydroxide ions. Fluoro- groups are not easily displaced, and fluorocarbons are seldom, if ever, used to make alcohols. Here are a few specific examples of displacement reactions:

$$CH_3{-}I \quad + \quad KOH \xrightarrow[100\,°C]{H_2O} CH_3{-}OH + \quad KI$$

Iodomethane   Potassium   Methanol   Potassium
(methyl iodide)   hydroxide   iodide

$$
\begin{array}{ccc}
& CH_3 & \\
& | & \\
H-C-Cl \\
& | & \\
& CH_3 &
\end{array}
+ \quad NaOH \xrightarrow[100\,°C]{H_2O}
\begin{array}{ccc}
& CH_3 & \\
& | & \\
H-C-OH \\
& | & \\
& CH_3 &
\end{array}
+ \quad NaCl
$$

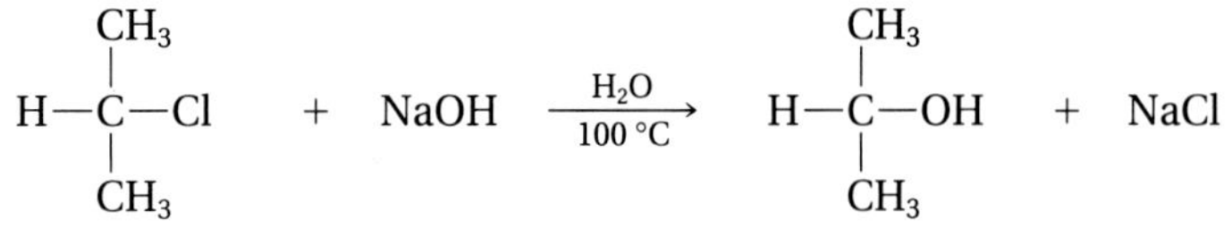

2-Chloropropane   Sodium   2-Propanol   Sodium
(isopropyl chloride)   hydroxide   (isopropyl alcohol)   chloride

$$\bigcirc{-}Br \quad + \quad NaOH \xrightarrow[100\,°C]{H_2O} \bigcirc{-}OH \quad + \quad NaBr$$

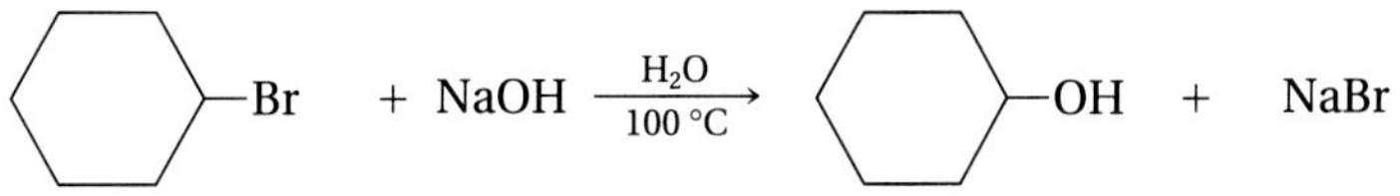

Bromocyclohexane   Cyclohexanol   Sodium
(cyclohexyl bromide)   (cyclohexyl alcohol)   bromide

Phenols can be prepared from aromatic halides. Because of the stability of the benzene ring, the reaction must be performed at high pressure and a temperature of 350 °C.

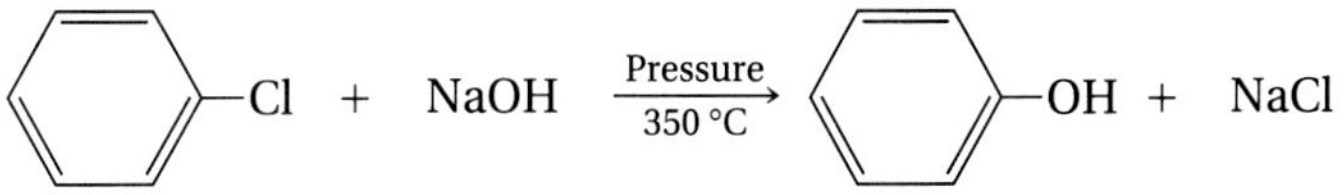

$$\bigcirc{-}Cl \quad + \quad NaOH \xrightarrow[350\,°C]{Pressure} \bigcirc{-}OH \quad + \quad NaCl$$

Chlorobenzene   Sodium   Phenol   Sodium
(phenyl chloride)   hydroxide   chloride

3.5 Elimination reactions

AIM: To use Saytzeff's rule to predict the major product of an elimination reaction.

Focus

Halocarbons and alcohols undergo elimination to produce alkenes.

You may recall that hydrogen halides and water add to double bonds to yield halocarbons and alcohols, respectively. However, the addition process also can be reversed—water can be removed from alcohols and hydrogen halides or halogens can be removed from halocarbons to produce alkenes. *Reversals of addition reactions are* **elimination reactions.** The general reaction for elimination is

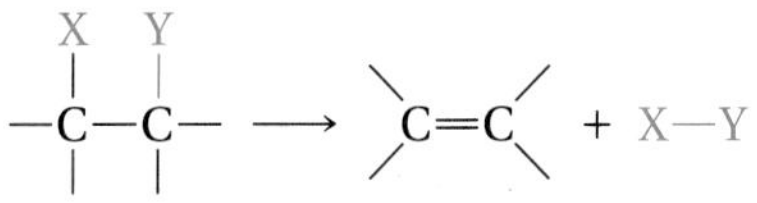

$$
\begin{array}{cc}
X & Y \\
| & | \\
-C-C- \\
| & |
\end{array}
\longrightarrow \quad {>}C{=}C{<} \quad + \quad X{-}Y
$$

Dehydrohalogenation

Dehydrohalogenation *is the elimination of hydrogen halides such as HCl or HBr from a carbon-carbon bond.* The elimination of a hydrogen halide usually requires heating the halocarbon in a concentrated solution of a strong base such as sodium hydroxide or potassium hydroxide. As the acidic hydrogen halide is formed, it is immediately neutralized by the strong base to form a salt:

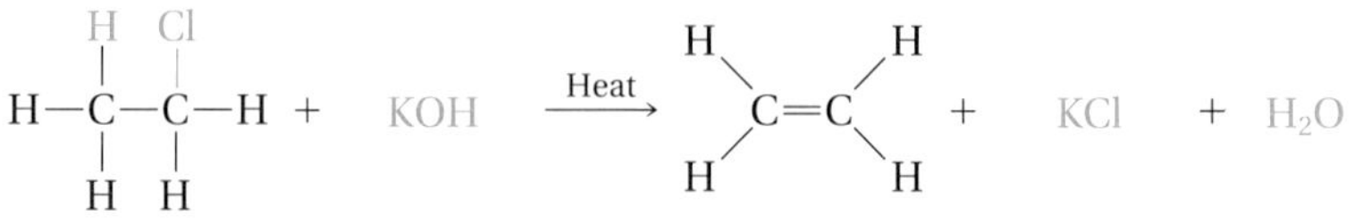

Chloroethane (ethyl chloride) Potassium hydroxide → Ethene (ethylene) + Potassium chloride + Water

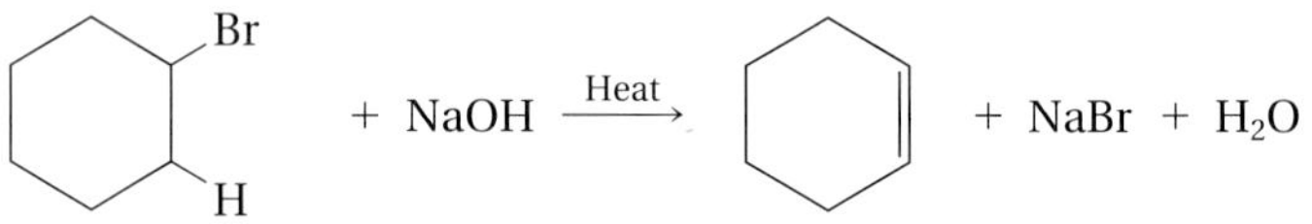

Bromocyclohexane (cyclohexyl bromide) Sodium hydroxide → Cyclohexene + Sodium bromide + Water

Dehydration

Concentrated sulfuric acid is such a strong dehydrating agent that it can remove the elements of water from organic compounds that themselves contain no free water. For example, concentrated sulfuric acid removes water from table sugar, $C_{12}H_{22}O_{11}$, leaving a residue of carbon. In a similar reaction, concentrated sulfuric acid can char paper, cotton, and wool and destroy skin tissue.

Dehydration *is the elimination of water.* To prepare an alkene from an alcohol, a solution of the alcohol in concentrated sulfuric acid is usually heated to 180 °C. The alkene is collected by distillation as it is formed.

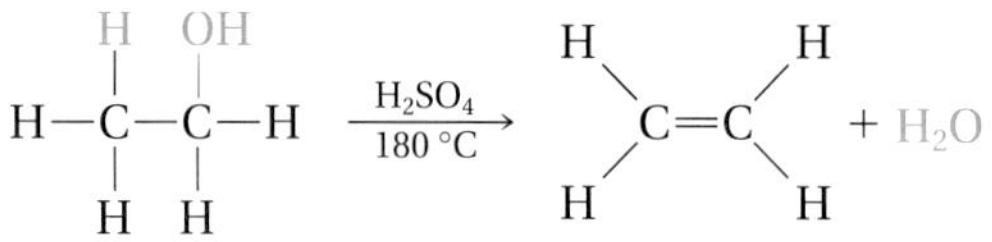

bp 78.5 °C → bp −104 °C

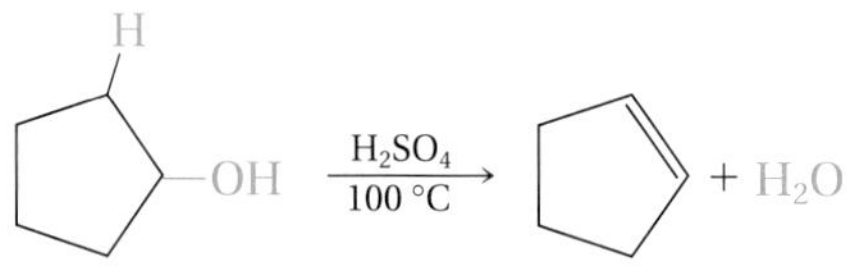

bp 140 °C → bp 44 °C

Dehalogenation

Dehalogenation *is the elimination of halogen from a carbon-carbon bond of a halocarbon.* Alkenes also can be prepared by elimination of halogen from an alkyl dichloride or dibromide using finely divided zinc metal as a reagent.

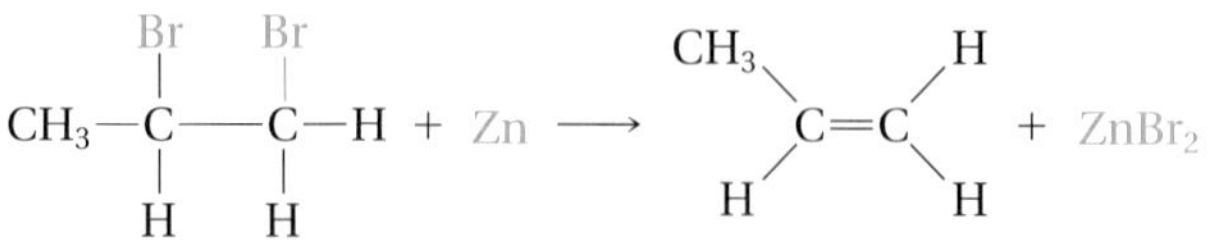

Direction of elimination

Sometimes more than one alkene can be produced during an elimination reaction. In such instances, we can predict which alkene will be the major product by **Saytzeff's rule:** *The major product of elimination is the alkene with the largest number of carbon groups on the carbons of the double bond.* For example:

$$\underset{\substack{\text{2-Butanol}\\(\textit{sec}\text{-butyl alcohol})}}{CH_3{-}\overset{\overset{\displaystyle OH}{|}}{C}H{-}CH_2{-}CH_3} \xrightarrow[180\,^\circ C]{H_2SO_4} \underset{\substack{\text{2-Butene}\\\text{Major product (90\%)}\\\text{(2 carbon groups on}\\\text{the carbons of the}\\\text{double bond)}}}{CH_3{-}CH{=}CH{-}CH_3} + \underset{\substack{\text{1-Butene}\\\text{Minor product (10\%)}\\\text{(1 carbon group on}\\\text{the carbons of the}\\\text{double bond)}}}{H_2C{=}CH{-}CH_2{-}CH_3}$$

PRACTICE EXERCISE 3.7

Write the product or products you would expect to get from each of the following elimination reactions.

(a) $CH_3\overset{\overset{\displaystyle OH}{|}}{C}HCH_3 \xrightarrow[\text{Heat}]{H_2SO_4}$ __________ $+ H_2O$

(b) $CH_3\overset{\overset{\displaystyle Cl}{|}}{C}H{-}\overset{\overset{\displaystyle Cl}{|}}{C}HCH_3 + Zn \longrightarrow$ __________ $+ ZnCl_2$

(c) $CH_3\overset{\overset{\displaystyle Br}{|}}{C}HCH_2CH_3 + KOH \xrightarrow{\text{Heat}}$ __________ + __________ $KBr + H_2O$

3.6 Alkoxides

AIM: To explain the behavior of alcohols and phenols as weak acids.

Focus

Alcohols are very weak acids.

Water dissociates to a slight extent to hydronium ions and hydroxide ions. Alcohols are very similar to water in this respect:

$$2H{-}\ddot{\underset{\cdot\cdot}{O}}{-}H \longrightarrow H{-}\overset{\overset{\displaystyle H^+}{}}{\ddot{\underset{\cdot\cdot}{O}}}{-}H + H{-}\ddot{\underset{\cdot\cdot}{O}}{:}^-$$

$$2R{-}\ddot{\underset{\cdot\cdot}{O}}{-}H \longrightarrow R{-}\overset{\overset{\displaystyle H^+}{}}{\ddot{\underset{\cdot\cdot}{O}}}{-}H + R{-}\ddot{\underset{\cdot\cdot}{O}}{:}^-$$

And just as reactive metals such as sodium react vigorously with water to form the strong bases called *hydroxides, reactive metals react vigorously with alcohols to form strong bases called* **alkoxides:**

$$2H{-}OH + 2Na \longrightarrow 2H{-}O^- + 2Na^+ + H_2$$

Hydroxide
ion

$$2R{-}OH + 2Na \longrightarrow 2R{-}O^- + 2Na^+ + H_2$$

Alkoxide
ion

Because of their weak acidic character, aqueous solutions of alcohols are essentially neutral. Phenols are much stronger acids than alcohols, but they are still very weak acids. Phenols produce slightly acidic solutions.

PRACTICE EXERCISE 3.8

Write the equation for the reaction of sodium with each of the following:
(a) methanol   (b) phenol.

3.7 Ethers

AIMS: To name and draw structures of ethers. To illustrate the synthesis of an ether from a halocarbon and an alkoxide ion.

Focus

Ethers are disubstituted derivatives of water.

Ethers *are compounds in which both hydrogens of water are replaced by carbon chains or rings.* The general formula for ethers is R—O—R. The R stands for any alkyl or aryl group.

O
H   H

Water molecule

O
R   R

Ether molecule

The alkyl or aryl groups joined by the ether linkage are named in alphabetical order and are followed by the word *ether.* For example:

$$CH_3CH_2{-}O{-}CH_3$$

Ethylmethyl ether

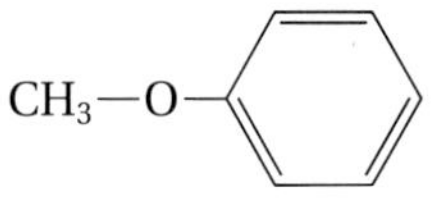

$$CH_3{-}O{-}\text{⬡}$$

Methylphenyl ether
(anisole)

Ethylmethyl ether and methylphenyl ether are asymmetrical ethers because the R groups attached to the oxygen are different. When both R groups are the same, the ether is symmetrical. Symmetrical ethers are named by using the prefix *di-.* For example:

$$CH_3CH_2{-}O{-}CH_2CH_3$$

Diethyl ether

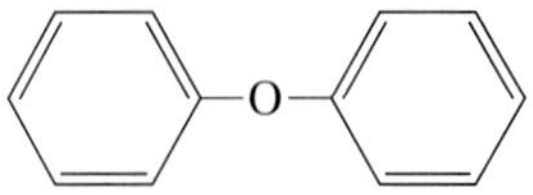

⬡—O—⬡

Diphenyl ether
(phenyl ether)

Many modern anesthetics contain halogen and ether functional groups, as described in A Closer Look: Halocarbon and Ether Anesthetics.

Halocarbon and Ether Anesthetics

Anesthetics have alleviated a great deal of pain and suffering during surgery. *Local anesthetics* make one part of the body insensitive to pain but leave the patient conscious. *General anesthetics* act on the brain to produce unconsciousness and insensitivity to pain. Many general anesthetics are halocarbons or ethers or contain both kinds of functional groups in their molecular structures.

Diethyl ether (C_2H_5—O—C_2H_5), the first general anesthetic, was introduced into surgery in 1846 by William Morton, a Boston dentist (see figure). In 1847, chloroform ($CHCl_3$) was introduced as a general anesthetic. Both diethyl ether and chloroform can cause undesirable side effects. Today, neither compound is used as an anesthetic in the Western Hemisphere.

Why are compounds such as diethyl ether and chloroform general anesthetics? The potency of an anesthetic is related to its solubility in fatty tissue. One theory for the action of general anesthetics is that they dissolve in the fatlike membranes of nerve cells of the brain (neurons). This changes the properties of the membranes. As a consequence, the activity of the neurons is depressed, leading to anesthesia.

The fat-solubility theory carries over into modern anesthetics. Today's anesthetics include relatively nonpolar fluorine-containing organic compounds such as halothane (Fluothane, $CF_3CHBrCl$), enflurane (Ethrane, Efrane, $CHFClCF_2$—O—CHF_2) and isoflurane (Forane, CF_3CHCl—O—CHF_2) are also used. All these compounds are inhalant anesthetics. These compounds are nonflammable and relatively safe, and the patient recovers rapidly from their effects.

Halothane is nonexplosive. The start of anesthesia is rapid, but slower than for anesthetics of greater solubility in membranes such as enflurane and isoflurane.

Enflurane is a stable liquid that is somewhat less volatile than halothane. Enflurane provides rapid anesthesia and rapid recovery for the patient. Although enflurane is broken down in the liver to produce fluoride ions, elevated fluoride levels in the blood are not considered a problem.

Isoflurane has physical, pharmacologic, and clinical properties that are similar to those of halothane and enflurane. This anesthetic is a more potent muscle relaxant than halothane, and induction of anesthesia is relatively rapid.

General anesthesia is usually induced with the administration of an intravenous anesthetic, regardless of the inhalant anesthetic subsequently used for anesthesia maintenance. The most commonly used induction agent is the barbiturate thiopental, also known as Pentothal. In many instances, low concentrations of halothane, enflurane, or isoflurane are used in conjunction with nitrous oxide (N_2O).

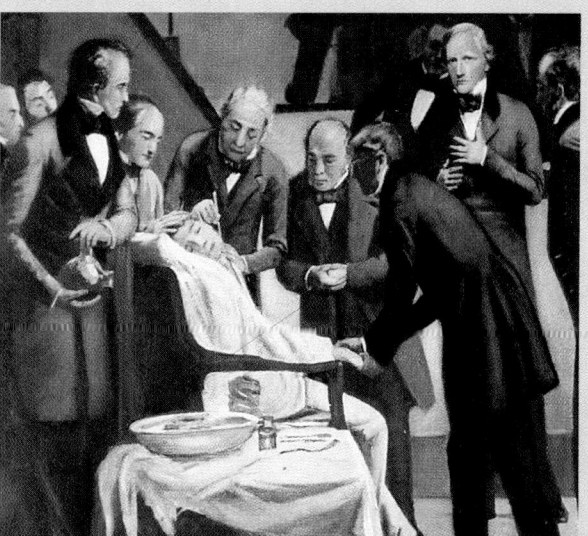

The first use of ether as an anesthetic in 1846 is depicted in this painting.

Just as alcohols can be prepared by using hydroxide ions to displace halogen from a halocarbon, ethers can be prepared by using alkoxide ions. For example:

$$\underset{\text{Iodomethane}}{CH_3{-}I} + \underset{\text{Sodium ethoxide}}{CH_3CH_2O^-Na^+} \longrightarrow \underset{\text{Ethylmethyl ether}}{CH_3CH_2{-}O{-}CH_3} + \underset{\text{Sodium iodide}}{NaI}$$

The ether linkage is often found in rings. *Rings that contain elements other than carbon are called* **heterocyclic rings** *or heterocycles.* Here are some common oxygen-containing heterocycles:

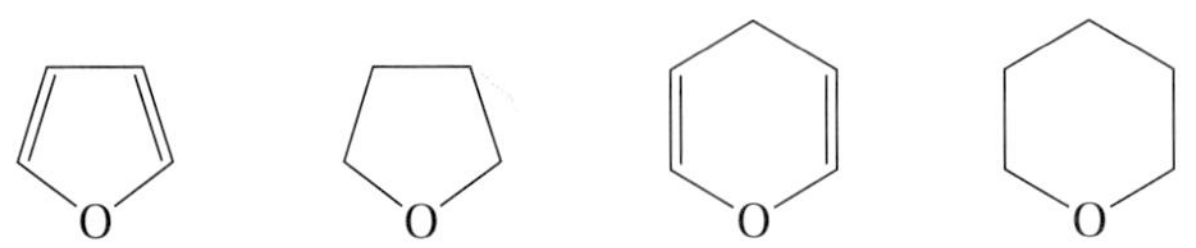

Furan Tetrahydrofuran Pyran Tetrahydropyran

The sex attractant of the female gypsy moth is the following epoxide. A very small quantity of the synthetic material is placed in traps to attract the male moth. Trapping helps to control the pest, which damages deciduous forests and fruit trees in eastern North America.

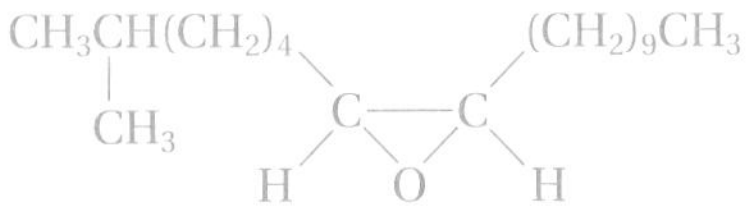

The fundamental ring structures of furan and pyran are found in many natural sugars. Compounds containing an oxygen atom in a three-membered ring are called *epoxides.* Epoxyethane is the simplest example.

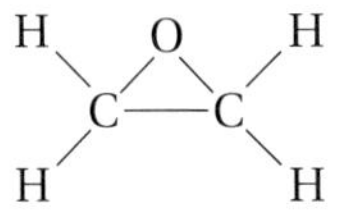

Epoxyethane
(ethylene oxide)

Other epoxides are used to make cements and adhesives. With the exception of the epoxides, the ether linkage is very resistant to chemical modification.

Since they are three-membered rings, epoxide rings are highly strained. Therefore, epoxides are much more reactive than other ethers. For example,

Follow-up to the Case in Point: A mistaken conviction

You may recall from the Case in Point that a mother was accused in 1989 of murdering her oldest son by ethylene glycol poisoning. She was convicted and sentenced to life in prison. The case might have ended there, except for a quirk of fate and the efforts of Dr. William Sly and Dr. James Shoemaker of the St. Louis University School of Medicine. The quirk of fate was that while the accused mother was in custody in 1990, she gave birth to a second son, who soon began to exhibit the same symptoms as his late older brother. It was impossible that she had poisoned her youngest son. The St. Louis University scientists, who had followed the case on television, recognized that a rare inherited disease, methylmalonic acidemia, has symptoms very similar to those of ethylene glycol poisoning. Contrary to findings reported in previous blood tests, new tests undertaken by Dr. Shoemaker revealed no evidence of ethylene glycol in the blood of either child. Dr. Piero Rinaldo of the Yale University School of Medicine verified from blood samples that both sons had been born with methylmalonic acidemia. In September 1991, the mother's conviction was reversed and all charges were dismissed. Thanks to science and caring scientists, she has resumed her life with her husband and remaining son.

the epoxide ring of epoxyethane is easily opened. In aqueous solution containing a trace of strong acid, the product is ethylene glycol:

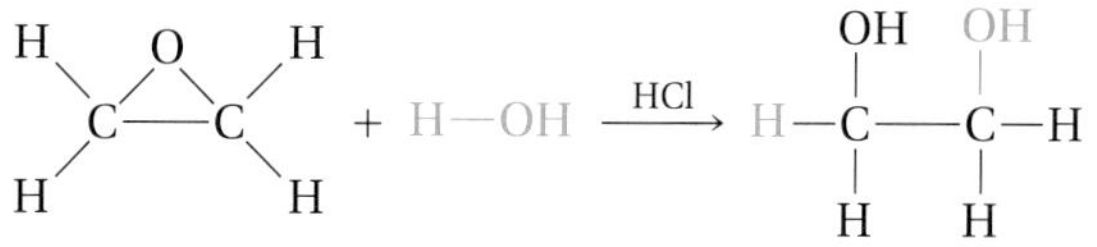

Although very useful as antifreeze, ethylene glycol is toxic and can be fatal if ingested, a fact that led to the accusation of murder described in the Case in Point early in this chapter.

3.8 Physical properties

AIM: To relate differences in boiling point and solubility to the molecular structures of hydrocarbons, halocarbons, alcohols, and ethers.

Focus

The physical properties of organic molecules depend on their molecular structure.

Thus far we have seen the aliphatic and aromatic hydrocarbons, halocarbons, alcohols, and ethers. Except for an occasional comment, not much has been said about their physical properties. This is not an oversight. Discussion of the physical properties of all four classes of compounds at one time will help us understand why the properties are what they are.

Boiling points

Hydrocarbons and halocarbons of low molar mass tend to be gases or low-boiling liquids. Hydrocarbon molecules such as the alkanes are nonpolar. The electron pair in a carbon-hydrogen or carbon-carbon bond is about equally shared by the nuclei of the elements involved. The carbon-halogen bond is only slightly polar. Attractions between molecules, because of hydrogen bonding, require that the molecules contain hydrogen attached to very electronegative atoms such as oxygen. There is no hydrogen bonding in hydrocarbons and halocarbons. Consequently, the forces that hold hydrocarbon or halocarbon molecules together in the liquid state are very weak. Table 2.1 shows that all alkanes containing fewer than five carbons are gases at room temperature.

Boiling points of closely related organic compounds usually increase as molar mass increases. The data in Table 3.4 show this principle. Remember that a pure liquid boils when enough heat energy has been supplied to let molecules in the liquid escape. The sum of weak forces holding heavy nonpolar molecules together in a liquid is greater than the sum of weak forces holding light nonpolar molecules together.

Like water, alcohols are capable of intermolecular hydrogen bonding. Alcohols therefore boil at higher temperatures than alkanes and halocarbons containing comparable numbers of atoms (Table 3.5).

Ethers usually have lower boiling points than alcohols of comparable molar mass, but they have higher boiling points than comparable hydrocarbons and halocarbons.

Table 3.4 Molar Masses and Boiling Points of the Chloromethanes Compared with Those of Methane

Molecular structure	Name	Molar mass (g)	Boiling point (°C)
CH_4	methane	16	−161
CH_3Cl	chloromethane (methyl chloride)	50.5	−24
CH_2Cl_2	dichloromethane (methylene chloride)	85.0	40
$CHCl_3$	trichloromethane (chloroform)	129.5	61
CCl_4	tetrachloromethane (carbon tetrachloride)	154.0	74

Table 3.5 Boiling Points of Alcohols and Comparable Alkyl and Aryl Chlorides

Alcohol	Boiling point (°C)	Alkyl chloride	Boiling point (°C)
CH_3OH	65	CH_3Cl	−64
CH_3CH_2OH	78	CH_3CH_2Cl	13
$CH_3CH_2CH_2OH$	97	$CH_3CH_2CH_2Cl$	47
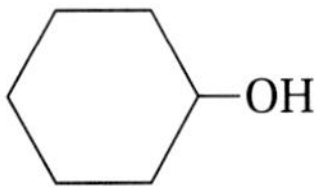 ⬡—OH	162	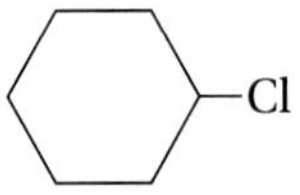 ⬡—Cl	143
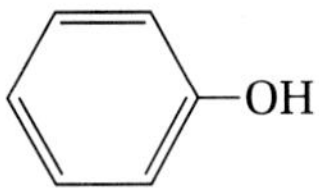 ⬡—OH	182	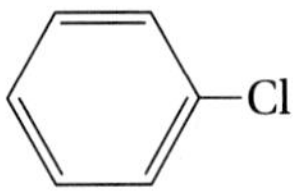 ⬡—Cl	132

Solubility in water

Hydrophobic
hudor (Greek): water
phobos (Greek): fear

The hydrocarbon parts of chains and rings of organic molecules are **hydrophobic** *("water-hating")—repelled by water.* Oil and water don't mix. If we mix two nonpolar liquids, however, they form a solution. A good rule of thumb is that "like dissolves like."

With the principle that "like dissolves like" in mind, how would we expect alcohols to behave with respect to their solubilities in water? Since alcohols are derivatives of water, we might expect them to have similar properties. And to a point, this is correct. Alcohols of up to four carbons are soluble in water in all proportions. The solubility of alcohols with four or more carbons in the chain is usually much less. For example, the solubility of 1-butanol is only 7.9 g/100 mL of water. The reason is that alcohols consist of two parts: the carbon chain and the hydroxyl group. These parts are in opposition to each other. The carbon chain is nonpolar and hydrophobic, but the hydroxyl group forms hydrogen bonds with water. *Groups such* as —*OH that interact strongly with water, usually by hydrogen bonding, are called* **hydrophilic** *("water-loving") groups.* Alcohols with short carbon chains are soluble in water. Those with longer carbon chains will not dissolve. And some alcohols whose carbon chains are not too long are only slightly soluble.

Ethers are more soluble in water than hydrocarbons and halocarbons but less soluble than alcohols of approximately the same molar mass. The reason is that the oxygens in ethers are hydrogen-bond acceptors, but ethers have no hydroxyl hydrogens to donate in hydrogen bonding. This lower solubility compared with alcohols is overcome in molecules with more than one ether linkage. Dioxane, a cyclic compound with two ether linkages, is soluble in water in all proportions; diethyl ether, with the same number of carbons but only one ether linkage, is not.

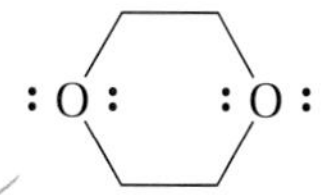

Hydrophilic
hudor (Greek):   water
philos (Greek):   loving

$: O :$ $: O :$ $CH_3CH_2—O—CH_2CH_3$

Dioxane
(soluble in water
in all proportions)

Diethyl ether
(solubility in water:
8 g/100 mL)

PRACTICE EXERCISE 3.9

Name and classify by functional groups the following compounds, and identify the one that is most polar.
(a) CH_3OCH_3 (b) CH_3CH_2Cl (c) CH_3CH_2OH
(d) $CH_3CH_2CH_3$

PRACTICE EXERCISE 3.10

Arrange the compounds in Practice Exercise 3.9 in order of increasing polarity then comment on their relative boiling points and water solubilities.

3.9 Sulfur compounds

AIM: To identify names, structures, and uses of some common thiols, thioethers, and disulfides.

Just as alcohols and ethers are organic derivatives of water, thioalcohols and thioethers are organic derivatives of hydrogen sulfide, H_2S.

Thioalcohols and thioethers

Thioalcohols, *compounds with the general formula R—S—H, are also called* **thiols.** *Thiols are also called* **mercaptans**—*a name coined because of their ready ability to "capture" or react with the element mercury.* The odor of

Focus

Thiols and thioethers are sulfur analogues of alcohols and ethers.

the fluid that skunks eject to protect themselves from predators comes partially from two thiols.

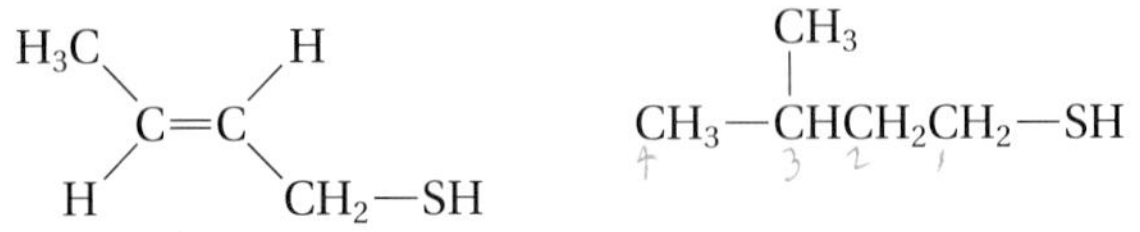

trans-2-Butenethiol

3-Methyl-1-butanethiol

Thiol

theion (Greek):   sulfur

The odor from a skunk can be removed from a contaminated object by treatment with a solution of the following composition: 1 quart 3% H_2O_2, 1/4 cup baking soda, and 1 teaspoon liquid soap. After treatment, the object should be rinsed with tap water.

Sulfur compounds are well known among chemists as the most foul smelling of all organic compounds. Sometimes, however, even an odor can be put to good use. Natural gas is odorless, but the gas piped to a gas range has a characteristic smell. The gas company has added a trace of methanethiol or ethanethiol to the natural gas so that leaks can be detected.

$$CH_3-SH \qquad CH_3CH_2-SH$$

Methanethiol
(methyl mercaptan)

Ethanethiol
(ethyl mercaptan)

In very low concentrations, the odors of some thiols and sulfides are even desirable in cooking. The pungent smell of onions comes from propanethiol.

$$CH_3CH_2CH_2-SH$$

Propanethiol
(propyl mercaptan)

Thioethers, *compounds with the general formula R—S—R, are also called sulfides.* The characteristic aroma and flavor of garlic comes from divinyl sulfide.

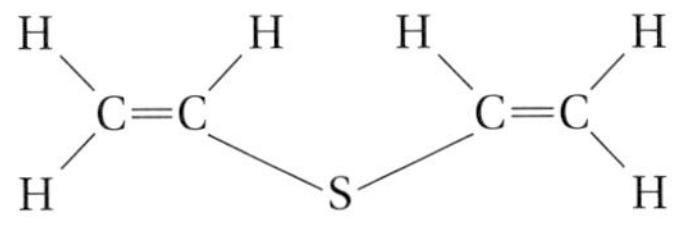

Divinyl sulfide

Disulfides

Sulfur atoms can form relatively stable bonds with other sulfur atoms. The sulfur-sulfur bond is not as stable as the carbon-carbon bond, and sulfur does not form long chains. But **disulfides,** *organic compounds of the general structural formula R—S—S—R,* are found in many proteins, especially those of hair, hooves, and nails (see Sec. 9.6). A third compound in the defense fluid of skunks is a disulfide:

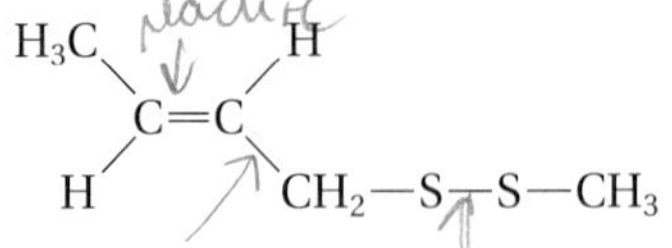

Disulfides are prepared by the mild oxidation of thiols. Thiols permitted to stand in air spontaneously oxidize to disulfides:

$$2CH_3S-H \xrightarrow{O_2} CH_3S-SCH_3$$

Methanethiol   Methyl disulfide

Once formed, disulfides can be further oxidized to sulfonic acids by hydrogen peroxide:

$$CH_3S-SCH_3 \xrightarrow{H_2O_2} 2CH_3SO_3H$$

Methyl disulfide   Methyl sulfonic acid

Disulfides are easily reduced to thiols. Although many reducing agents are suitable, hydrogen works well:

$$CH_3S-SCH_3 \xrightarrow{H_2} 2CH_3-SH$$

Methyl disulfide   Methanethiol

3.10 Polyfunctional compounds

AIM: To recognize the functional groups of a given polyfunctional molecule.

Focus

Many organic molecules contain more than one functional group.

So far our introduction to functional group chemistry has been focused on molecules with only one kind of functional group. **Polyfunctional** *organic compounds contain two or more functional groups.* Tetrahydrocannabinol is an example of a polyfunctional molecule, as shown in the following example.

EXAMPLE 3.6   **Identifying functional groups**

Identify the functional groups in the tetrahydrocannabinol molecule.

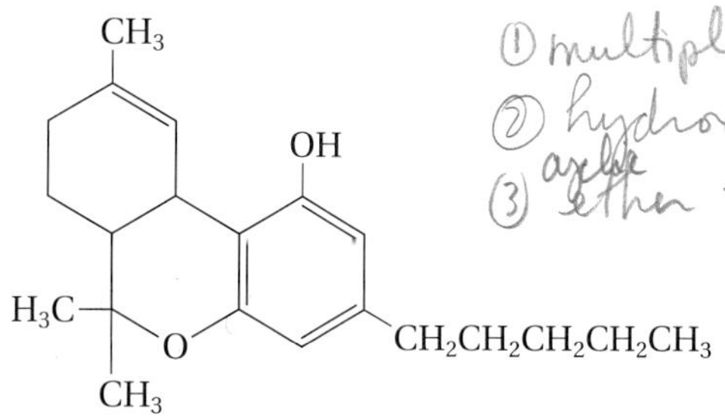

Tetrahydrocannabinol

SOLUTION

Inspect the molecule to find the functional groups, such as multiple carbon-carbon bonds, hydroxyl groups, ether linkages, and so forth. The tetrahydrocannabinol molecule contains three functional groups: a cyclic ether linkage, a phenolic hydroxyl group, and a carbon-carbon double bond. These functional groups are in color in the following structure.

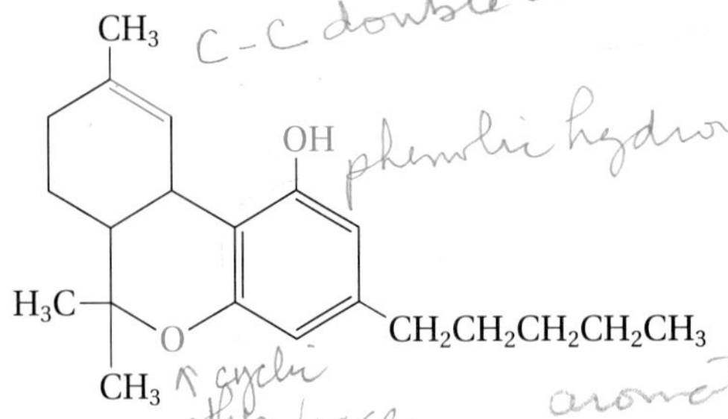

(handwritten annotations: "why is aromatic ring not included in functional group like", "C–C double bond", "phenolic hydroxyl group", "a cyclic ether linkage", "aromatic ring")

Tetrahydrocannabinol is the active ingredient of the *Cannabis sativa* (marijuana) plant. The effects of marijuana use on health and society are still being debated. One interesting finding is that smoking marijuana reduces the pressure of the optic fluid and may therefore be helpful in relieving the symptoms of glaucoma, a serious eye disease.

Another polyfunctional compound is hexachlorophene, which is both an aromatic halocarbon and a phenol: *(handwritten: "hydroxyl groups attached directly to an aromatic ring")*

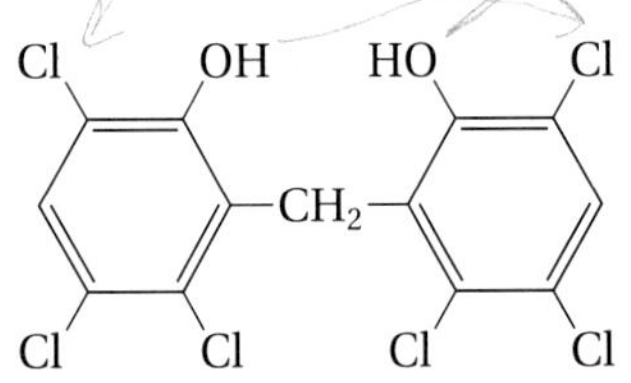

Hexachlorophene

Hexachlorophene is an antiseptic. Until recently, it was an important ingredient in germicides and soaps used in hospitals. Its use was curbed, however, when it was found that the babies of female hospital workers who frequently washed their hands with hexachlorophene soap had a higher incidence of birth defects than babies born to women in the general population.

PRACTICE EXERCISE 3.11

The structure that follows is a urushiol, a family of compounds that are the irritants in poison ivy. It is a polyfunctional compound with the following structure. How many functional groups can you identify?

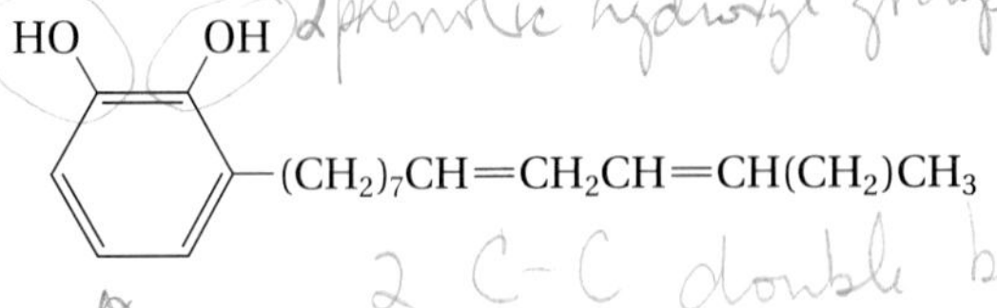

HO   OH $-(CH_2)_7CH=CH_2CH=CH(CH_2)CH_3$

(handwritten: "2 phenolic hydroxyl group", "2 C–C double bonds", "1 aromatic ring")

PRACTICE EXERCISE 3.12

Estradiol is an important female sex hormone. Describe the structural features and identify the functional groups in the estradiol molecule.

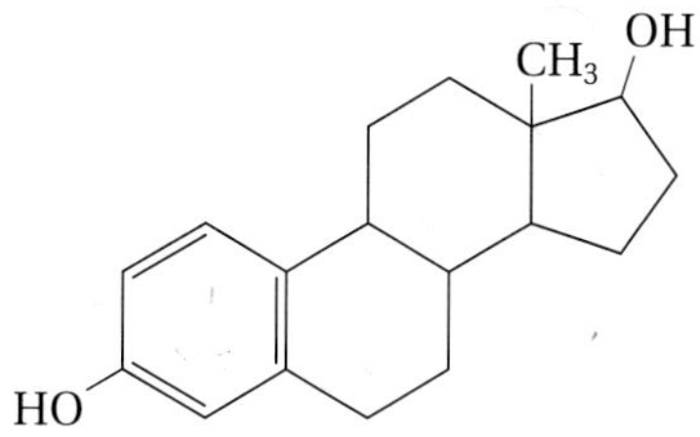

[handwritten notes:]
structural features
① aromatic ring fused
② c cyclo alkanes
③ methyl substituents.
Functional groups
① 1 phenolic hydroxyl group.
② aromatic ring
③ hydroxyl group of an aliphatic
   alkyl cyclic alcohol

SUMMARY

Halocarbons are like hydrocarbons except that one or more hydrogens is replaced by a halogen—fluorine, chlorine, bromine, or iodine. Halocarbons may be saturated or unsaturated aliphatic compounds. They also may be aromatic. Many useful polymers are made from haloalkenes.

Replacement of a hydrocarbon hydrogen by a hydroxyl group, —OH, gives alcohols. Alcohols are organic derivatives of water, R—OH. They may be primary, secondary, or tertiary. Methanol, ethanol, isopropyl alcohol, ethylene glycol, and glycerol are alcohols of commercial importance. Glycerol is also a major constituent of plant and animal fats and oils. Aromatic alcohols are called phenols. Ethers have the general formula R—O—R.

The chemistry of the alkenes, halocarbons, alcohols, and ethers is related. Halocarbons are synthesized from alkenes by halogenation and hydrohalogenation. Alcohols are prepared by hydration of alkenes or from halocarbons by displacement of a halide ion by a hydroxide ion. In additions of hydrogen halides to double bonds, Markovnikov's rule tells us: If more than one halo-carbon can be formed, then the hydrogen of the adding reagent usually ends up on the carbon of the double bond that has the most hydrogens. Ethers are made by displacement reactions, except that halocarbons react with alkoxide ions rather than hydroxide ions.

Halocarbons and alcohols undergo elimination reactions to form alkenes. When two or more alkenes can be formed by elimination of a hydrogen halide, Saytzeff's rule tells us: The alkene with the most carbon groups attached to the double bond is the major product.

Alcohol molecules hydrogen bond to each other and to water. Alcohols therefore have higher boiling points and greater water solubility than hydrocarbons, halocarbons, or ethers.

Thioalcohols (thiols) and thioethers (sulfides) are sulfur derivatives of alcohols and ethers; they have the general structures R—SH and R—S—R, respectively. These sulfur-containing groups and disulfide groups (R—S—S—R) are found in many biological molecules.

REACTION SUMMARY

Here is a summary of the reactions covered in this chapter.

Halogenation (the halogen is usually Cl_2 or Br_2):

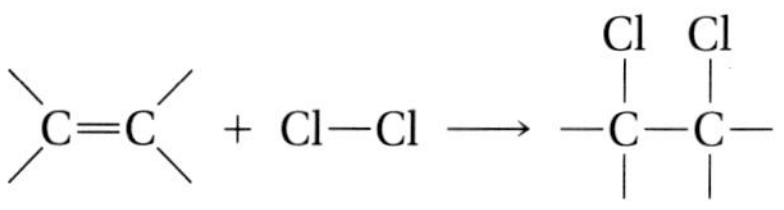

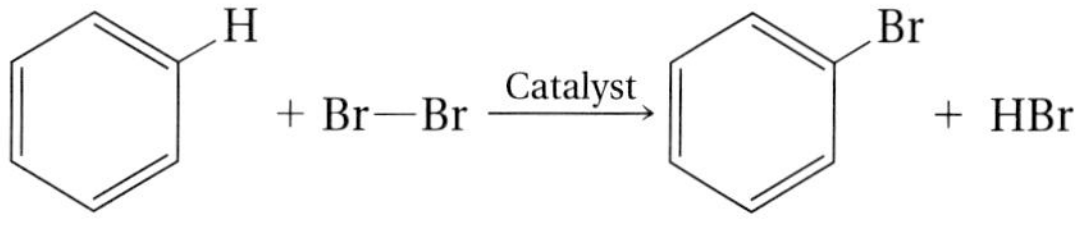

Hydrohalogenation (the hydrohalogen is usually HCl, HBr, or HI):

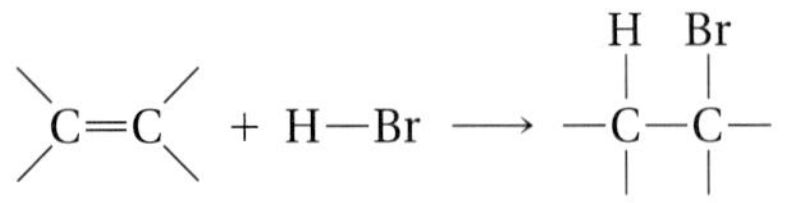

$$\diagdown C=C \diagup + H{-}Br \longrightarrow {-}\overset{\displaystyle H}{\underset{\displaystyle |}{C}}{-}\overset{\displaystyle Br}{\underset{\displaystyle |}{C}}{-}$$

Hydration:

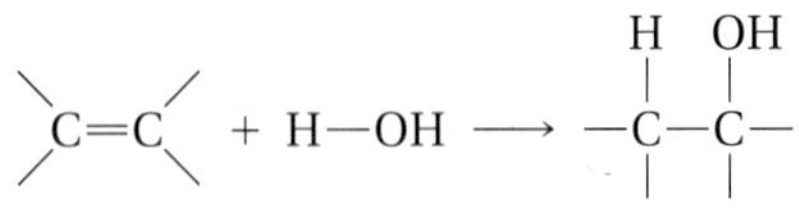

$$\diagdown C=C \diagup + H{-}OH \longrightarrow {-}\overset{\displaystyle H}{\underset{\displaystyle |}{C}}{-}\overset{\displaystyle OH}{\underset{\displaystyle |}{C}}{-}$$

Polymerization:

$$x\,C=C \xrightarrow{\text{Catalyst}} {+}\overset{|}{C}{-}\overset{|}{C}{+}_x$$

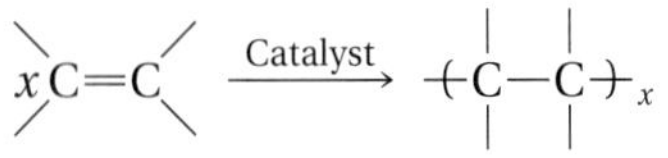

Displacement (Br⁻, Cl⁻, or I⁻ is usually displaced):

$$R{-}Cl + {}^-OH \longrightarrow R{-}OH + Cl^-$$
$$R{-}I + {}^-OR \longrightarrow R{-}OR + I^-$$

Dehydrohalogenation (Br⁻, Cl⁻, or I⁻ is usually displaced):

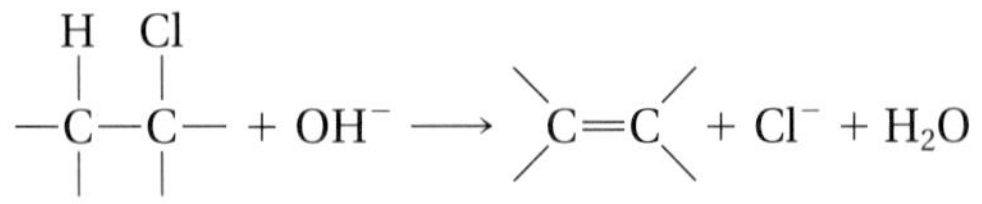

$$-\overset{H}{\underset{|}{C}}{-}\overset{Cl}{\underset{|}{C}}{-} + OH^- \longrightarrow \diagdown C=C \diagup + Cl^- + H_2O$$

Dehydration:

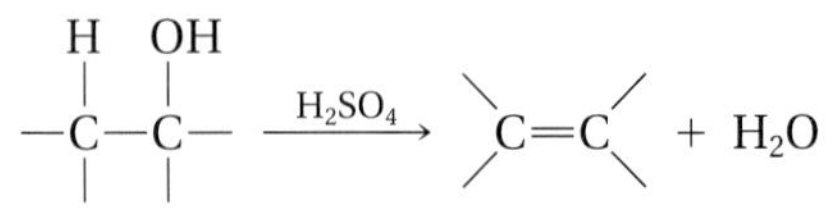

$$-\overset{H}{\underset{|}{C}}{-}\overset{OH}{\underset{|}{C}}{-} \xrightarrow{H_2SO_4} \diagdown C=C \diagup + H_2O$$

Dehalogenation:

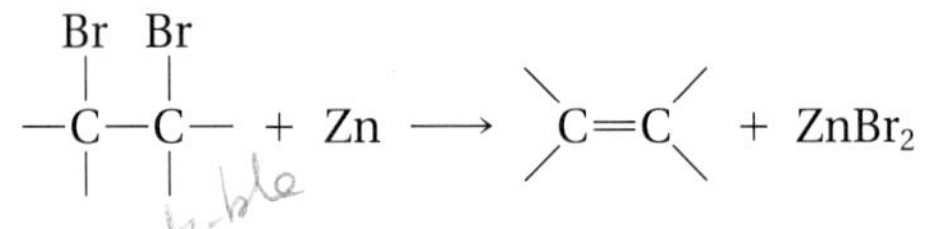

$$-\overset{Br}{\underset{|}{C}}{-}\overset{Br}{\underset{|}{C}}{-} + Zn \longrightarrow \diagdown C=C \diagup + ZnBr_2$$

KEY TERMS

Addition reaction (3.2)
Alcohol (3.3)
Alkoxide (3.6)
Alkyl halide (3.1)
Aromatic substitution (3.2)
Aryl halide (3.1)
Dehalogenation (3.5)
Dehydration (3.5)
Dehydrohalogenation (3.5)

Denatured alcohol (3.3)
Displacement reaction (3.4)
Disulfide (3.9)
Elimination reaction (3.5)
Ether (3.7)
Functional group (3.2)
Halocarbon (3.1)
Halogenation (3.2)
Heterocyclic ring (3.7)

Hydration reaction (3.4)
Hydrogenation (3.2)
Hydrohalogenation (3.2)
Hydrophilic (3.8)
Hydrophobic (3.8)
Hydroxy function (3.3)
Hydroxyl group (3.3)
Markovnikov's rule (3.2)
Mercaptan (3.9)
Monomers (3.2)

Phenol (3.3)
Polyfunctional molecule (3.10)
Polymer (3.2)
Saytzeff's rule (3.5)
Thioalcohol (3.9)
Thioether (3.9)
Thiol (3.9)

EXERCISES

Halocarbons (Sections 3.1, 3.2)

3.13 Name or write a structural formula for the following halocarbons.
(a) *m*-dichlorobenzene   (b) $CH_2{=}CHCH_2Cl$ 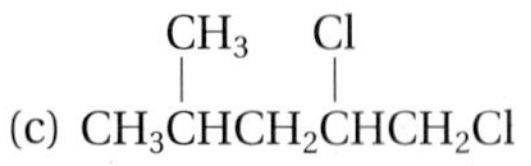
(c) $CH_3\overset{CH_3}{\underset{|}{C}}HCH_2\overset{Cl}{\underset{|}{C}}HCH_2Cl$
(d) 1,2-dichlorocyclohexane

3.14 Write a structural formula for or name each of the following.
(a) 1,2,2-trichlorobutane   (b) 1,3,5-tribromobenzene 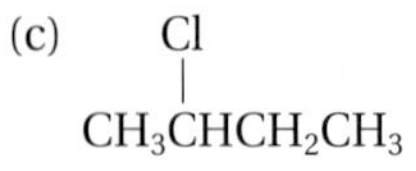
(c) $CH_3\overset{Cl}{\underset{|}{C}}HCH_2CH_3$
(d) $CH_2{=}CHBr$

3.15 Write the structural formula(s) for the product(s) from the following reactions.
(a) $CH_3CH_2CH{=}CH_2 + Cl_2 \longrightarrow$
(b) $CH_3CH_2CH{=}CH_2 + HBr \longrightarrow$
(c) 3-methyl-2-pentene + hydrogen (with platinum catalyst)

3.16 Write structural formula(s) and name(s) of product(s) of each reaction.
(a) $CH_3CH_2CH{=}CH_2 + H_2 \xrightarrow{Pt}$
(b) cyclohexene + bromine $\longrightarrow$
(c) $CH_3CH_2CH{=}C(CH_3)_2 + HBr \longrightarrow$

3.17 Write structural formulas and give IUPAC names for all the isomers of $C_3H_6Cl_2$.

3.18 Write structural formulas and give IUPAC names for all the isomers of C_4H_9Br.

Polymers and Polymerization (Section 3.2)

3.19 What is the structure of the repeating units in a polymer in which the monomer is (a) 1-butene and (b) 1,2-dichloroethene?

3.20 The *trans* isomer of polyisoprene is a hard material known as *gutta percha*. What is the structure of the repeating unit in this polymer? see p.95

Alcohols (Sections 3.3, 3.4, 3.5, 3.6)

3.21 Write the IUPAC name or give the structural formula for each of the following alcohols.

(a) $CH_3CH_2CHCH_3$
        |
        OH

(b) CH_3CHCH_2OH
        |
        CH_3

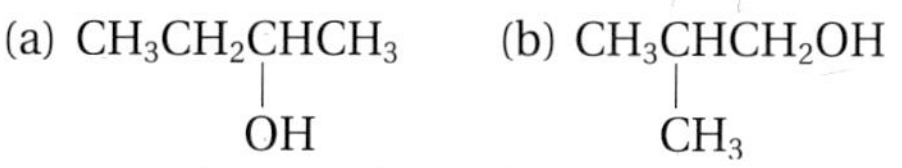

(c) 1,2-ethanediol   (d) *p*-bromophenol

3.22 Give the structural formula or IUPAC name for each of the following alcohols.

(a) 3-methyl-2-butanol   (b) cyclopentanol

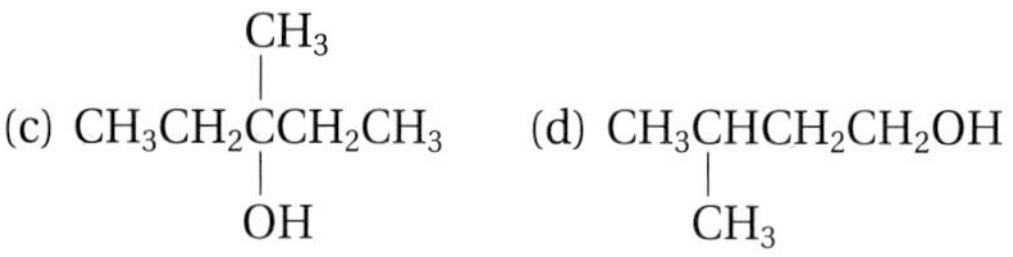

(c) $CH_3CH_2CCH_2CH_3$
          |
          OH
       with CH_3 above C

(d) $CH_3CHCH_2CH_2OH$
        |
        CH_3

3.23 Identify each of the alcohols in Exercise 3.22 as primary, secondary, or tertiary.

3.24 Which of the alcohols in Exercise 3.22 are not secondary alcohols?

3.25 Describe two ways of synthesizing alcohols. Use specific examples to illustrate the reactions involved.

3.26 Name three types of elimination reactions that produce alkenes from halocarbons and alcohols.

3.27 Write the structures and give the names of the alcohols produced in the following reactions.

(a) $CH_2{=}CH_2 + H_2O \xrightarrow[100\ °C]{H^+}$

(b) $CH_3CH{=}CHCH_3 + H_2O \xrightarrow[100\ °C]{H^+}$

(c) 2-methyl-1-butene + $H_2O \xrightarrow[100\ °C]{H^+}$

3.28 What are the structures and IUPAC names of the alcohols formed in these reactions?

(a) 3-hexene + $H_2O \xrightarrow[100\ °C]{H^+}$

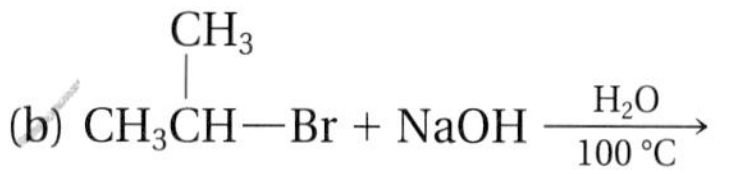

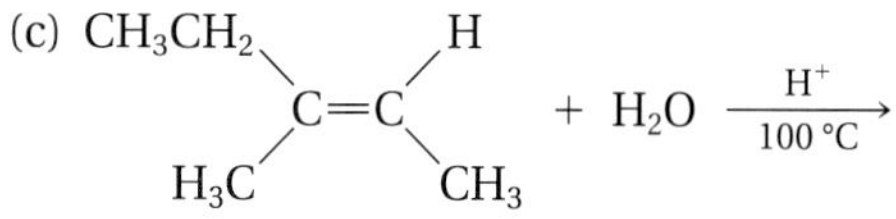

(b) $CH_3CH{-}Br + NaOH \xrightarrow[100\ °C]{H_2O}$
        with CH_3 above CH

(c) CH_3CH_2 and H, H_3C and CH_3 on C=C $+ H_2O \xrightarrow[100\ °C]{H^+}$

3.29 What are the structures and names of the alkenes that result when the following compounds undergo elimination?

(a) cyclohexanol   (b) 1,2-dibromopropane

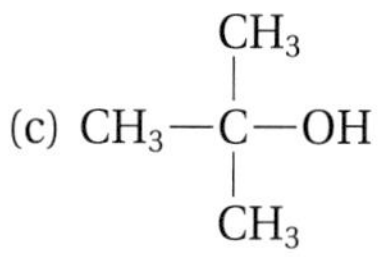

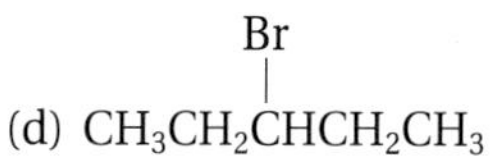

(c) $CH_3{-}C{-}OH$ with CH_3 above and CH_3 below

(d) $CH_3CH_2CHCH_2CH_3$ with Br above

3.30 Write the structures and give the names of the alkenes produced in the following elimination reactions.

(a) $CH_3CH{-}CCH_2CH_3$ with Br, Br above and CH_3 below second C

(b) $CH_3CH_2C{-}OH$ with CH_3 above and CH_3 below

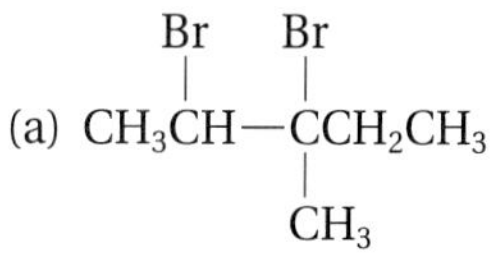

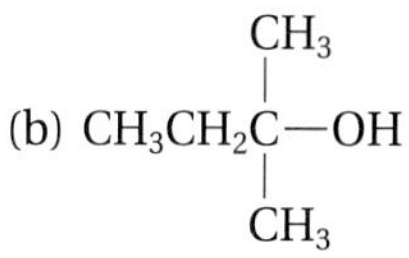

(c) 2-methyl-2-butanol   (d) bromocyclohexane

Ethers (Section 3.7)

3.31 Name or write structural formulas for the following ethers.
(a) ethylphenyl ether
(b) furan
(c) $CH_3CH_2OCH_2CH_2CH_2CH_3$
(d) $CH_3CH_2CH_2OCH_2CH_2CH_3$

3.32 Write structural formulas or name each of the following ethers.
(a) $CH_3OCH_2CH_3$ (b) $CH_2{=}CHOCH{=}CH_2$
(c) tetrahydropyran   (d) ethylpropyl ether

3.33 What are the structures of the ethers produced by the following reactions:

(a) (benzene ring)$-ONa + CH_3Br \longrightarrow$ ____ $+ NaBr$

(b) $CH_3OK +$ (cyclohexane ring)$-Cl \longrightarrow$ ____ $+ KCl$

(c) $CH_3CHONa + CH_3CH_2Br \longrightarrow$ ____ $+ NaBr$
        |
        CH_3

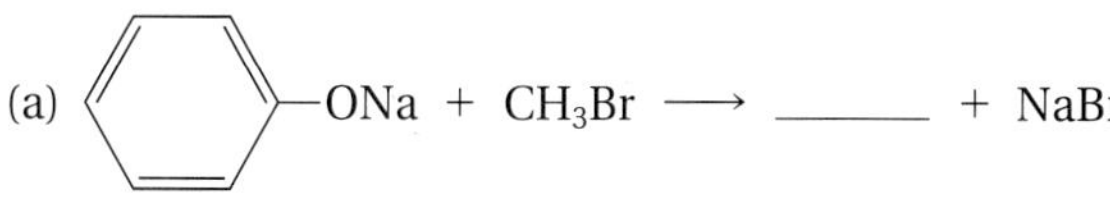

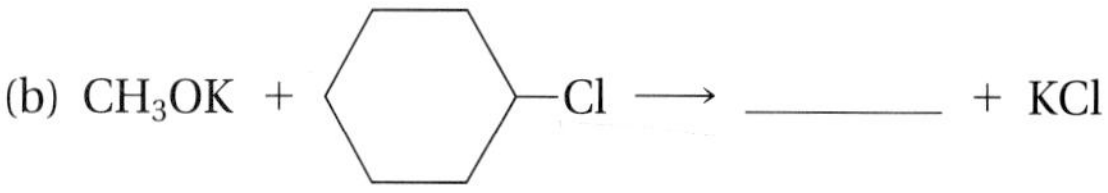

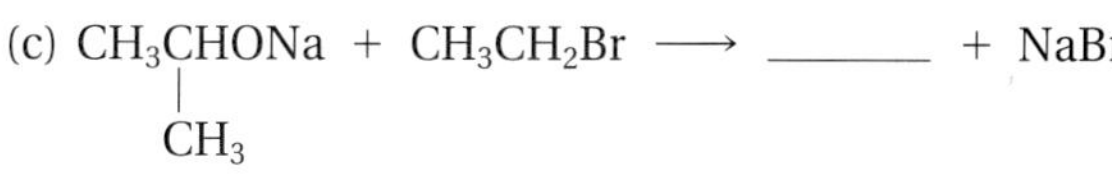

3.34 Classify each of these compounds as an alcohol, a phenol, or an ether.

(a)

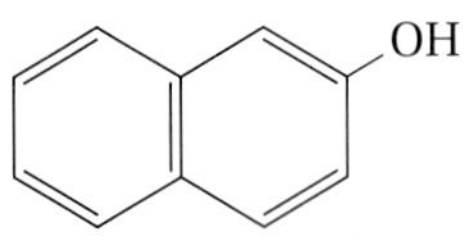

(b)

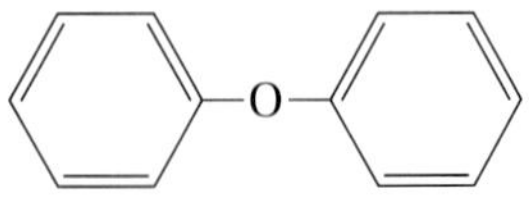

(c)

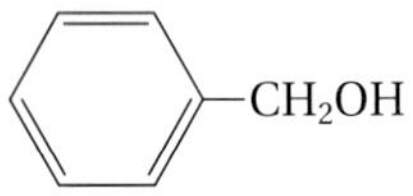

(d)

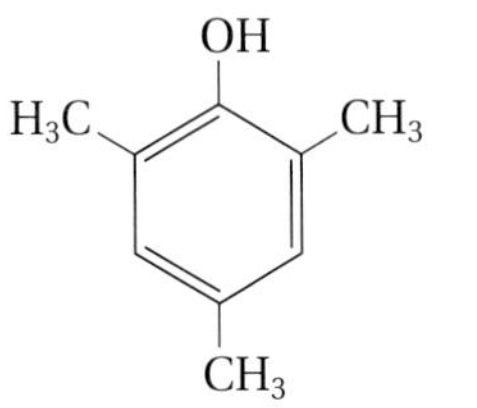

(e) 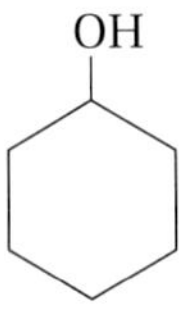 OH

(f) 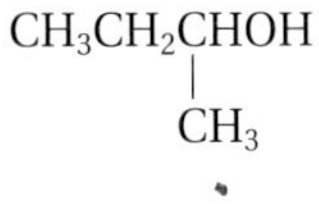 CH_3CH_2CHOH
$\qquad CH_3$

Physical Properties of Organic Molecules (Section 3.8)

3.35 Explain why ethanol, CH_3CH_2OH, is soluble in water in all proportions but decanol, $CH_3(CH_2)_9OH$, is almost insoluble in water.

3.36 Explain why diethyl ether is more soluble in water than dihexyl ether. Would you expect propane to be more soluble than diethyl ether in water? Why?

3.37 Show how hydrogen bonds form between molecules of the following pairs of compounds: (a) water-water, (b) water-methanol, and (c) methanol-methanol.

3.38 Describe why hydrogen bonds can form between molecules of water and diethyl ether but not between molecules of diethyl ether.

Polyfunctional Molecules (Section 3.10)

3.39 The compound whose structure is drawn here has no common name, but it has a number of functional

groups and other structural features. See how many you can identify.

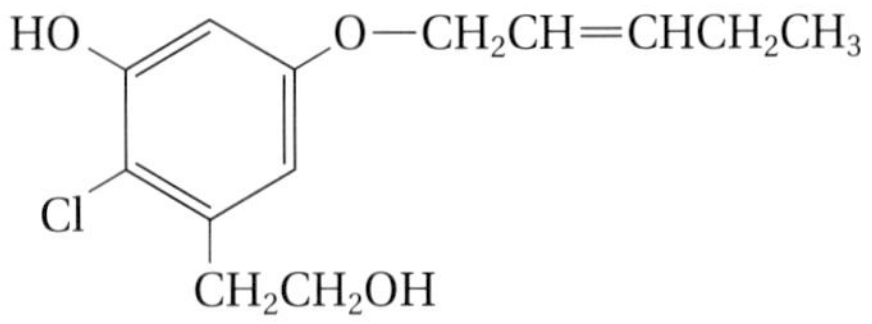

3.40 Cholesterol is a compound that is in our diet and also synthesized in the liver. Sometimes it is deposited on the inner walls of blood vessels, causing hardening of the arteries. Describe the structural features and functional groups of this important molecule.

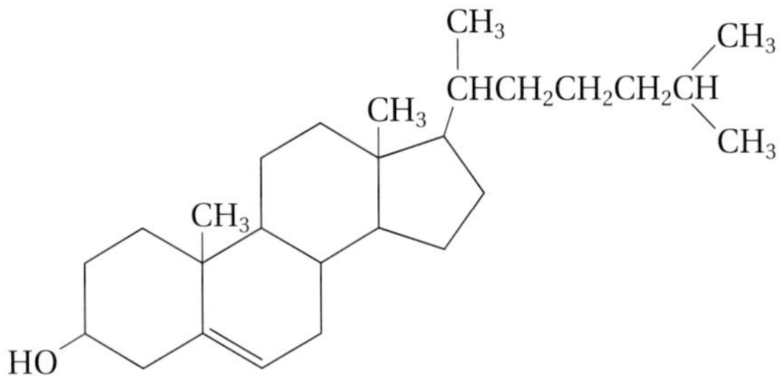

Additional Exercises

3.41 Which member of each of the following pairs of compounds would you predict to have the higher boiling point? Explain your answers.

(a)

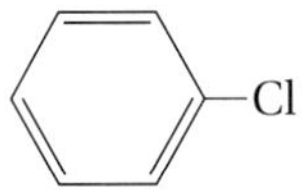

—Cl or —OH

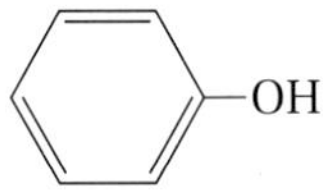

(b) $CH_3OCH_2CH_3$ or $CH_3CH_2CH_2OH$

(c) 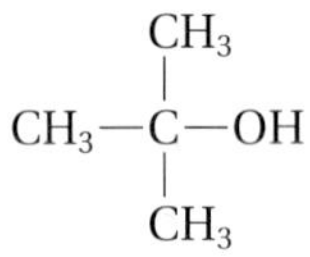 $CH_3 - \underset{\underset{CH_3}{|}}{\overset{\overset{CH_3}{|}}{C}} - OH$ or $CH_3CH_2CH_2CH_2OH$

(d) CH_3Cl or CH_3OH

(e)

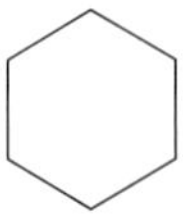

or —Cl

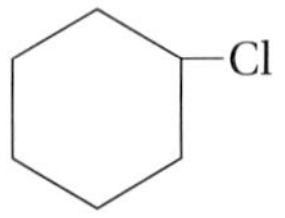

3.42 Both compounds A and B have the molecular formula C_2H_6O. When compound A reacts with sodium metal, hydrogen gas is produced. Compound B does not react with sodium metal. The boiling points of compounds A and B are 78.5 and $-23.7\,°C$, respectively. Write the structural formulas and names of the two compounds.

3.43 Write the structural formulas for the major and minor products from the following reactions.

(a)

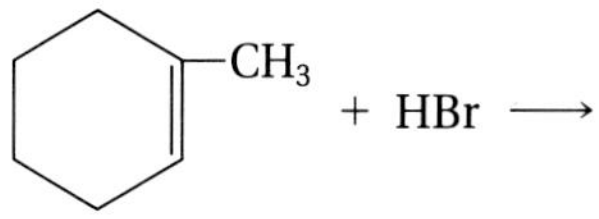

(b)

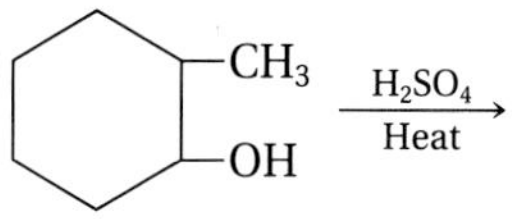

(c)

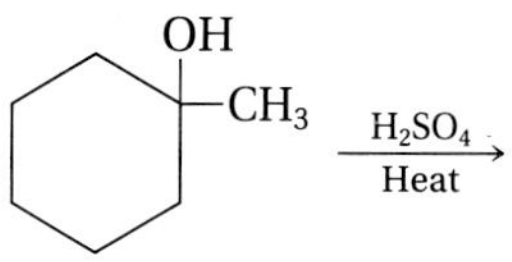

(d)

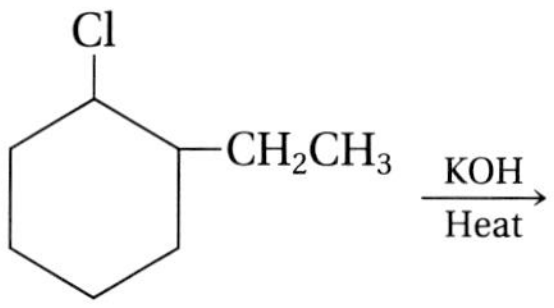

3.44 Complete the following reactions by writing the structural formulas of the products.

(a)

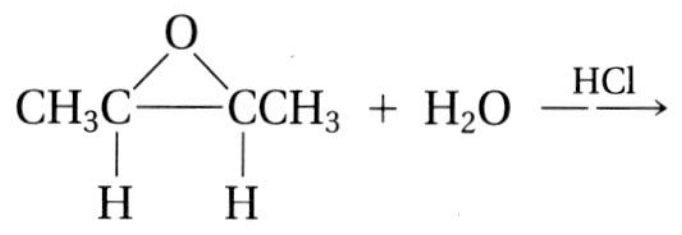

(b)

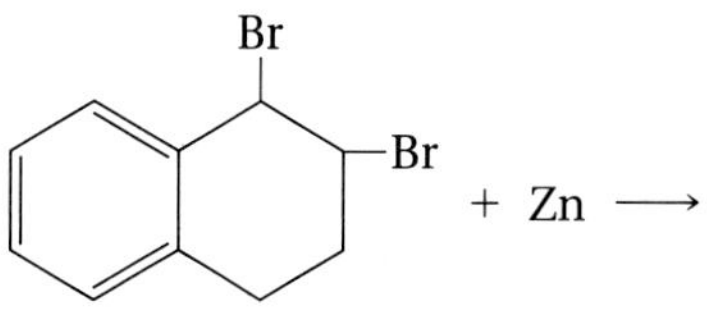

(c)

(d)

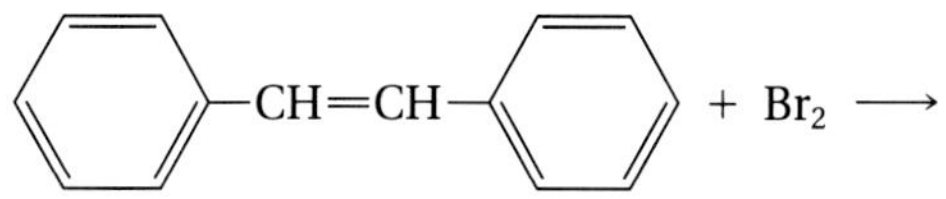

3.45 List the following compounds in the expected order of increasing solubility in water.

(a)

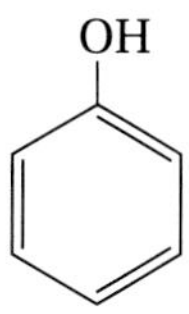

(b)

(c) $CH_3CH_2OCH_2CH_3$ (d)

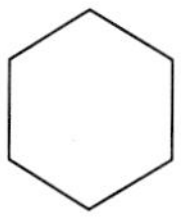

3.46 Give the reagents necessary to make the following transformations.

(a) $CH_3CH{=}CH_2 \longrightarrow CH_3\overset{\text{OH}}{\underset{\,}{C}}HCH_3$

(b) 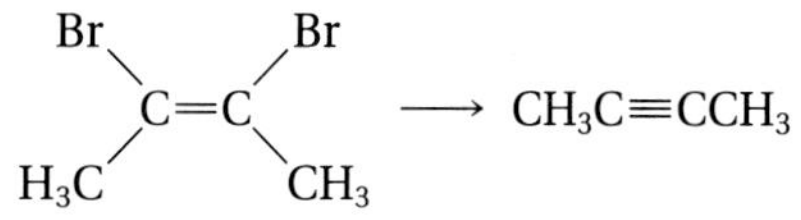
$\longrightarrow CH_3C{\equiv}CCH_3$

(c)

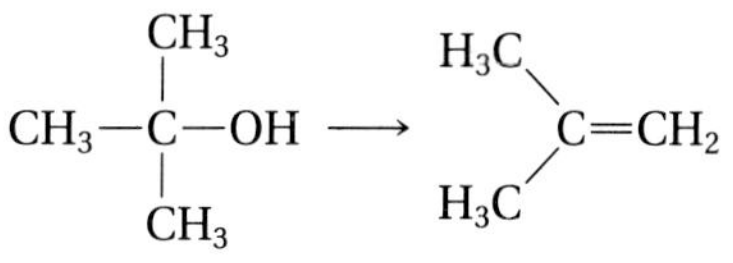

(d) $CH_3CH_2OH \longrightarrow CH_3CH_2O^-Na^+$

SELF-TEST (REVIEW)

True/False

1. As a general rule, the hydrocarbon portion of halocarbons is chemically inert. T
2. A disulfide molecule contains a single covalent bond between two sulfur atoms. T
3. Although similar in molar mass, ethanol has a higher boiling point than dimethyl ether. T
4. The hydrocarbon chain end of a decanol molecule is hydrophilic. F

C − C − C − C − C − C − C − C − C − OH

5. The reaction of an alcohol and sodium hydroxide gives an alkoxide.

6. Both hydrogenation and halogenation are examples of addition reactions.

7. Pyran is an example of a cyclic ether.

8. Although similar in molar mass, chloroethane is more soluble in water than 1-propanol.

9. A propylene glycol molecule has two hydroxyl groups.

10. Both aromatic and aliphatic alcohols are weakly acidic.

Multiple Choice

11. A hydrocarbon added to an orangish solution of bromine in carbon tetrachloride turns the solution colorless. The hydrocarbon is probably
    (a) 2-butene. (b) isopropyl alcohol.
    (c) ethyl ether. (d) benzene.

12. A low-molar-mass alcohol dissolved in water gives
    (a) an acidic solution. (b) a neutral solution.
    (c) a basic solution. (d) none of the above.

13. An ether can be formed by the reaction of a halocarbon with
    (a) an alkene. (b) an alkoxide ion.
    (c) an alcohol. (d) water.

14. Which of the following reactions is most typical of alkenes?
    (a) substitution (b) addition
    (c) replacement (d) elimination

15. Which of the following compounds, all of similar molar mass, would you expect to be most soluble in water?
    (a) CH_3CH_2-Cl (b) $CH_3CH_2CH_2-F$
    (c) $CH_3CH_2CH_2CH_3$ (d) $CH_3CH_2CH_2-OH$

16. The reaction of water with ethene in the presence of an acid catalyst
    (a) results in the formation of ethanol.
    (b) is an example of a replacement reaction.
    (c) produces a polymer.
    (d) More than one answer is correct.

17. A compound once widely used as an anesthetic is
    (a) $CH_3CH_2CH_2OH$. (b) $CH_3CH_2-O-CH_2CH_3$.
    (c) 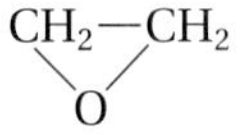 CH_2-CH_2 (d) $CH_3CH_2CH_2SH$
         \ /
          O

18. Which of the following compounds, all of similar molar mass, has the highest boiling point?
    (a) CH_3CH_2-F (b) $CH_3CH_2CH_3$
    (c) CH_3-O-CH_3 (d) CH_3CH_2-OH

19. Which of the following compounds is found in antifreeze?
    (a) glycerol (b) ethylene glycol (c) propanol
    (d) propanethiol

20. Another name for isobutyl chloride is
    (a) 2-chloro-2-methylpropane.
    (b) chlorobutane.
    (c) 2-chloro-3-methylbutane.
    (d) 1-chloro-2-methylpropane.

21. Which of the following could be used as starting materials to produce 1-butene as the major product? (*Hint:* Use Saytzeff's rule.)
    (a) 2-butanol (with H_2SO_4)
    (b) 1,2-dibromobutane (with zinc)
    (c) 2-chlorobutane (with base)
    (d) all of the above

22. Using Markovnikov's rule, what would you predict the major organic product of the following reaction to be?

$$CH_3-CH_2-C=CH-CH_3 + HBr \longrightarrow$$
$$|$$
$$CH_3$$

    (a) 3-methylpentane
    (b) 3-bromo-3-methylpentane
    (c) 2-bromo-3-methylpentane
    (d) 2,3-dibromo-3-methylpentane

23. Phenols are characterized by
    (a) their behavior as bases. (b) ether linkages.
    (c) an —OH group attached to a benzene ring.
    (d) their use as flavoring agents.

24. Which of the following compounds *does not* contain the element sulfur?
    (a) methyl ethyl sulfide (b) propyl mercaptan
    (c) 2-propanethiol (d) epoxyethane

25. A correct name for the secondary alcohol containing four carbon atoms is
    (a) 2-butanol. (b) 3-butanol.
    (c) 2-methyl-2-propanol. (d) ethanediol.

26. Benzene typically undergoes which of the following types of reactions?
    (a) elimination (b) addition (c) substitution
    (d) polymerization

C
|
·C — C — C — Cl
1 chloro-2methylpropane

Aldehydes and Ketones

Introduction to the Carbonyl Group

Br   H
| |
$CH_3-CH_2-C-C-CH_3$
| |
CH_3 H

3-bromo-2-methyl pentane

The aroma and flavor of vanilla, nutmeg, cloves, and many other spices are due to aldehydes and ketones.

OH
|
C — C — C — C

The carbonyl group consists of a carbon atom and an oxygen atom joined by a double bond ($\mathrm{C{=}O}$). Present in carbohydrates, fats, proteins, and steroids, the carbonyl group is one of the most common functional groups in nature. In this chapter we will learn about the chemistry of the carbonyl group in two classes of compounds, aldehydes and ketones. This is only a brief introduction to an important functional group that will appear again and again in following chapters.

Although aldehydes and ketones are important components of cells, some of the harmful effects of drinking ethanol (CH_3CH_2OH) result from the effect of an aldehyde, acetaldehyde (CH_3CHO), on a variety of bodily functions. These effects are magnified in a developing fetus, as described in the Case in Point.

CASE IN POINT: Fetal alcohol syndrome

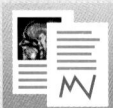
Marla is an alcoholic and has been in and out of treatment for her drinking problems since she was a teenager. Her physician is aware of her alcoholism and at one point in the past helped place Marla in an alcohol abuse center. Her treatment had gone well, but when Marla became pregnant, her physician became concerned. The doctor told Marla that it was especially important that she abstain from alcohol during her pregnancy. The physician was concerned that the child might be born with *fetal alcohol syndrome*, a condition that can result when pregnant women consume alcohol. We will learn more about the relationships among ethanol consumption, acetaldehyde, and fetal alcohol syndrome in Section 4.4.

The surgeon general warns that pregnant women should not drink alcoholic beverages because of the risk of birth defects.

4.1 Aldehydes and ketones

AIM: To describe the carbon-oxygen bond of the carbonyl group of aldehydes and ketones.

Focus

The general structures of aldehydes and ketones are similar.

The functional group known as the **carbonyl group** ($>C{=}O$)*—a carbon atom and an oxygen atom joined by a double bond*—is found in compounds called *aldehydes* and *ketones.*

Structures of aldehydes and ketones

Aldehydes *are organic compounds in which the carbonyl carbon—the carbon to which the oxygen is bonded—is always joined to at least one hydrogen.* The general formula for an aldehyde is

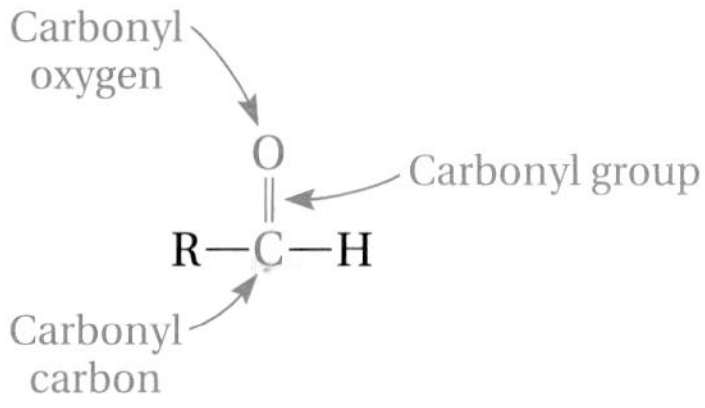

This structural formula is often abbreviated to RCHO.

Ketones *are organic compounds in which the carbonyl carbon is joined to two other carbons:*

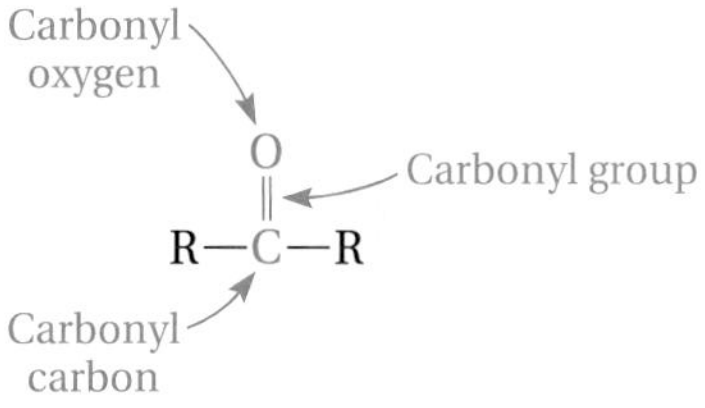

The abbreviated form for a ketone is RCOR.

Note the similarity in structure of aldehydes and ketones. Because they both contain the carbonyl group, the chemistry of aldehydes and ketones is similar. Both aldehydes and ketones are highly reactive, but aldehydes are generally the more reactive of the two classes.

Naming aldehydes and ketones

The IUPAC system may be used for naming aldehydes. We must first identify the longest hydrocarbon chain that contains the carbonyl carbon. The *-e* ending of the hydrocarbon is replaced by *-al* to designate an aldehyde. Using the IUPAC system, we name the aldehydes methanal, ethanal, propanal, butanal, and so forth. In naming substituted aldehydes, the longest chain is counted starting from the carbon of the aldehyde group.

EXAMPLE 4.1   **Naming a substituted aldehyde by the IUPAC system**

What is the IUPAC name for the following compound?

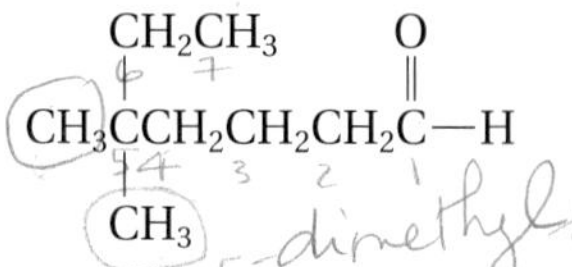

$$CH_3CCH_2CH_2CH_2C-H$$

(with annotations: CH₂CH₃ group, CH₃, "5,5-dimethylheptanal")

SOLUTION

The longest continuous chain contains seven carbons, making heptane the parent alkane. In counting the carbons in the chain, we number the chain so that the carbon of the aldehyde group is at position 1. The carbonyl carbon is understood to be at position 1 and will not appear in the final name of the aldehyde.

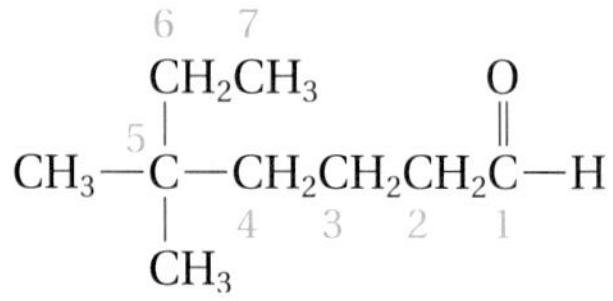

$$\overset{6}{C}H_2\overset{7}{C}H_3$$
$$CH_3-\overset{5}{C}-\overset{4}{C}H_2\overset{3}{C}H_2\overset{2}{C}H_2\overset{1}{C}-H$$
$$CH_3$$

Obtain the aldehyde name by dropping the -e ending from the alkane name and adding the ending -al; our aldehyde is a *heptanal.* The substituents are named in the same way as for hydrocarbons. The molecule has two methyl groups in the 5 position of the parent chain, which are designated *5,5-dimethyl.* Combine this with the parent name to obtain the complete name of the aldehyde: *5,5-dimethylheptanal.*

PRACTICE EXERCISE 4.1

Give the IUPAC names for each of the following aldehydes.
(a) CH_3CH_2CHO
(b) 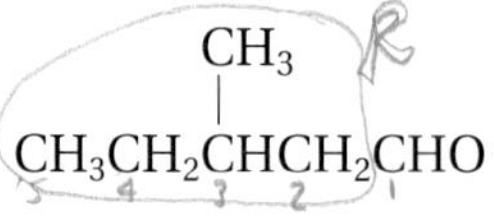
$$CH_3$$
$$CH_3CH_2CHCH_2CHO$$
(c) $CH_3CH_2CH_2CH_2CHO$
(d)
$$Cl$$
$$CH_3CHCH_2CHO$$

(handwritten notes:)
a) $CH_3 CH_2 - C - H$ propanal
b) 3-methylpentanal
c) pentanal
d) 3-chlorobutanal
(b) $CH_3CH_2CHCH_2 - C-H$ with CH₃

Ketones also can be named by the IUPAC system. We show that a compound is a ketone by changing the ending of the longest carbon chain that contains the carbonyl group from -e to -one, as demonstrated in Table 4.1. If there are several locations in the chain where the carbonyl group could be placed, its position is designated by the lower number.

EXAMPLE 4.2   **Naming ketones by the IUPAC system**

Using the IUPAC system, name the following two structural isomers of $C_5H_{10}O$:

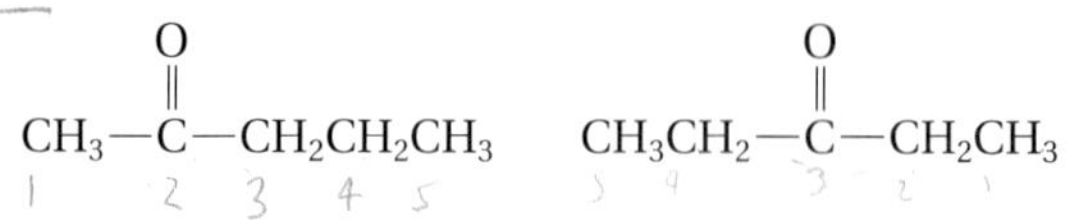

$$CH_3-\overset{O}{C}-CH_2CH_2CH_3 \qquad CH_3CH_2-\overset{O}{C}-CH_2CH_3$$

(handwritten: 2-pentanone   3-pentanone)

Table 4.1 Some Common Aldehydes and Ketones

Condensed formula	Structural formula	IUPAC name	Common name
Aldehydes			
HCHO	$H-\overset{\overset{\displaystyle O}{\|\|}}{C}-H$	methanal	formaldehyde
CH_3CHO	$CH_3-\overset{\overset{\displaystyle O}{\|\|}}{C}-H$	ethanal	acetaldehyde
CH_3CH_2CHO	$CH_3-CH_2-\overset{\overset{\displaystyle O}{\|\|}}{C}-H$	propanal	propionaldehyde
$CH_3CH_2CH_2CHO$	$CH_3-CH_2-CH_2-\overset{\overset{\displaystyle O}{\|\|}}{C}-H$	butanal	butyraldehyde
C_6H_5CHO	benzene ring $-\overset{\overset{\displaystyle O}{\|\|}}{C}-H$	benzaldehyde	benzaldehyde
$C_6H_5CH{=}CHCHO$	benzene ring $-CH{=}CH-\overset{\overset{\displaystyle O}{\|\|}}{C}-H$	3-phenyl-2-propenal	cinnamaldehyde
Ketones			
CH_3COCH_3	$CH_3-\overset{\overset{\displaystyle O}{\|\|}}{C}-CH_3$	propanone	acetone (dimethyl ketone)
$CH_3COC_2H_5$	$CH_3-\overset{\overset{\displaystyle O}{\|\|}}{C}-CH_2-CH_3$	butanone	methyl ethyl ketone
$C_6H_5COC_6H_5$	benzene rings with $\overset{\overset{\displaystyle O}{\|\|}}{C}$	diphenylmethanone	benzophenone (diphenyl ketone)
$C_6H_{10}O$	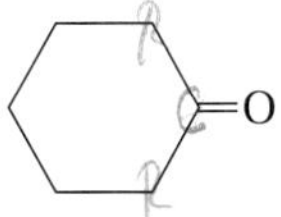 cyclohexane ring ${=}O$	cyclohexanone	cyclohexanone

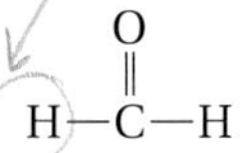

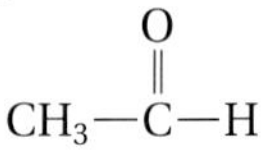

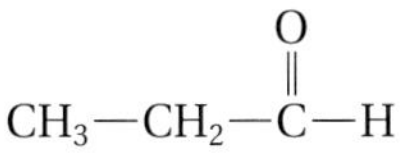

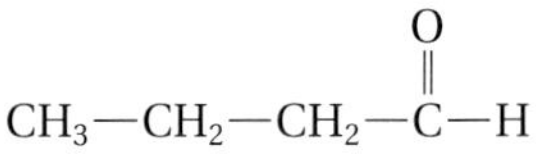

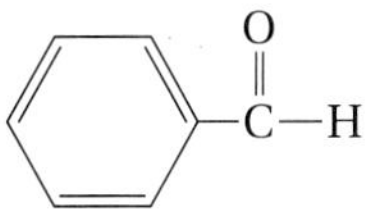

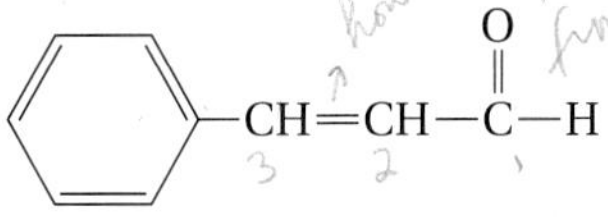

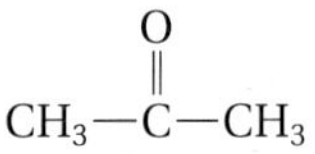

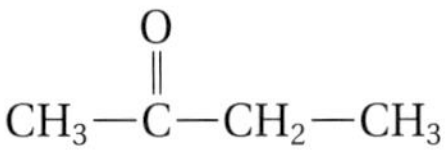

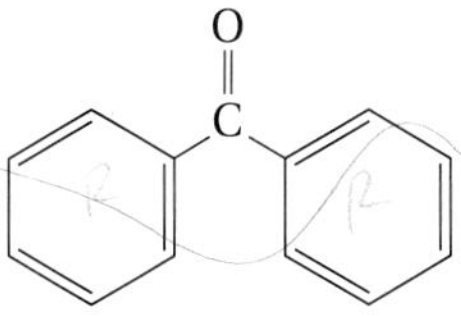

Handwritten annotations in margins:
- (B)CHO
- can H be R group or If R why H₃C group? I thought R was H₃C group yes
- know all these both names
- f / a / p / b / b
- how would you know there is a double here from the name
- benzene ring $-CH_2-CH_2-\overset{O}{C}-H$?
- 3 2
- why not 2-butanone since it could be 1-butanal?
- don't learn
- R (various)

SOLUTION

In both compounds the longest chains that contain the carbonyl group have five carbons, making pentane the parent alkane. Indicate that the compounds are ketones by dropping the *-e* ending from pentane and adding the ending *-one*. The compounds are *pentanones*. Give the position of the carbonyl groups in the carbon chain. The numbering that gives the lowest position number to the isomer on the left is 2 and on the right, 3.

These compounds are 2-pentanone and 3-pentanone, respectively.

$$CH_3-\overset{\overset{\displaystyle O}{\|}}{C}-CH_2CH_2CH_3 \qquad CH_3CH_2-\overset{\overset{\displaystyle O}{\|}}{C}-CH_2CH_3$$

2-Pentanone   3-Pentanone

PRACTICE EXERCISE 4.2

Give the IUPAC name for each of the following ketones.

(a)
$$\underset{\substack{6\ \ \ 5\ \ \ 4\ \ \ 3\ 2\ \ \ 1}}{CH_3CH_2CH_2\overset{\overset{\displaystyle O}{\|}}{C}CH_2CH_3}$$

3- hexanone

(b)
$$\underset{\substack{1\ \ \ 2\ \ 3\ \ 4}}{CH_3\overset{\overset{\displaystyle O}{\|}}{C}CH_2CH_3}$$

why not 2? 2- butanone *both answer: butanone only.*

(c)
$$\underset{\substack{5\ \ \ 4\ \ \ 3\ \ 2\ \ \ 1}}{CH_3\overset{\overset{\displaystyle CH_3}{|}}{CH}CH_2\overset{\overset{\displaystyle O}{\|}}{C}CH_3}$$

4 -methyl -2 pentanone

Common names for aldehydes and ketones are frequently used. The common names for methanal and ethanal are *formaldehyde* and *acetaldehyde*, respectively. The common names of the ketones are obtained by naming each of the alkyl groups attached to the carbonyl carbon and adding the word *ketone*. The sole exception is dimethyl ketone, which almost everyone calls *acetone*, a versatile solvent.

EXAMPLE 4.3

Writing a structure for a ketone

Write the structure for ethylisopropyl ketone.

SOLUTION

The common name of a ketone indicates the alkyl or aryl groups bonded to either side of the carbonyl group. Write a carbonyl group and attach ethyl and isopropyl groups to obtain the structure of the ketone.

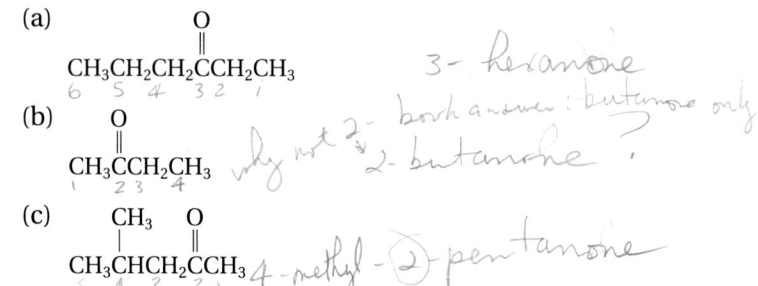

$$CH_3CH_2\overset{\overset{\displaystyle O}{\|}}{C}\underset{\underset{\displaystyle CH_3}{|}}{C}HCH_3$$

PRACTICE EXERCISE 4.3

Write the structure of each of the following compounds.

(a) acetaldehyde   (b) acetone   (c) diethyl ketone
(d) 3-methylbutanal

4.2 The carbonyl group

AIMS: To name and draw structures of simple aldehydes and ketones. To explain how intermolecular interactions of the carbonyl group affect the boiling point and water solubility of aldehydes and ketones.

Focus

The polar carbonyl group influences the physical properties of aldehydes and ketones.

Electron sharing in the carbon-oxygen double bond of the carbonyl group is similar to that of the carbon-carbon double bond in alkenes. Compare ethylene, $H_2C{=}CH_2$, with formaldehyde, $H_2C{=}O$ (Fig. 4.1). A carbon atom has only four unshared electrons in its valence shell, whereas oxygen has six. This means that carbon can make four covalent bonds, as in ethylene, but oxygen can make only two, as in the carbonyl group. The oxygen in a carbonyl group has two unshared pairs of electrons, but the carbon in ethylene has none.

In $C{=}C$ bonds, the bonding electrons are shared equally. But oxygen is much more electronegative than carbon. The $C{=}O$ bond is very polar because oxygen draws the bonding electrons away from carbon. The oxygen in the carbonyl group carries a partial negative charge, and the carbon carries a partial positive charge:

$$\overset{\delta+}{\underset{}{}}C{=}\overset{\delta-}{O}$$

Hydrogen bonding cannot take place between molecules of aldehydes or ketones because they lack O—H bonds. However, polar-polar interactions are possible:

$$\overset{\delta+}{}C{=}\overset{\delta-}{O}{-}{-}{-}\overset{\overset{\delta-}{O}}{\underset{}{C}}{-}{-}{-}\overset{\delta-}{O}{=}\overset{\delta+}{C}$$

Polar-polar interaction

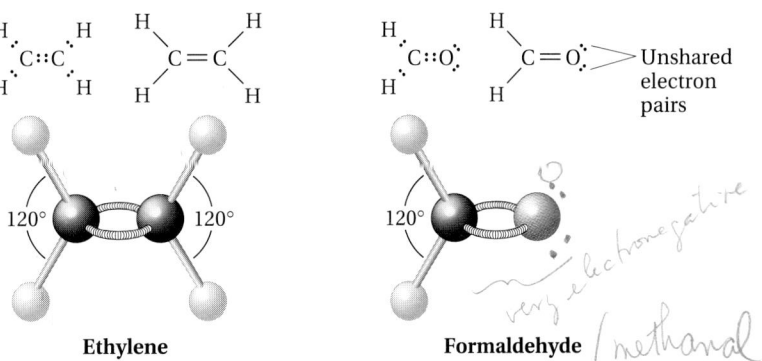

Ethylene   **Formaldehyde** *(methanal)*

120°   120°   120°

Unshared electron pairs

Figure 4.1
Comparison of the structural formulas and ball-and-stick models of ethylene and formaldehyde. The oxygen of the carbonyl group of formaldehyde has two unshared pairs of electrons.

Boiling points

Aldehydes and ketones with low molar masses are very volatile and highly flammable.

Aldehydes and ketones cannot form intermolecular hydrogen bonds because they lack hydroxyl (—OH) groups. Consequently, they have boiling points lower than those of the corresponding alcohols. The aldehydes and ketones can attract one another through polar-polar interactions of their carbonyl groups, however, and their boiling points are higher than those of the corresponding alkanes. Table 4.2 shows how the boiling points increase in the sequence alkane, aldehyde, and alcohol for compounds containing one and two carbons. Except for formaldehyde, which is an irritating, pungent gas, all aldehydes and ketones are either liquids or solids at 20 °C (Table 4.3). The vapors of aldehydes and ketones contribute to the pleasant odors of many natural products, as described in A Closer Look: Flavors and Fragrances.

Table 4.2 Some Characteristics of One- and Two-Carbon Alkanes, Aldehydes, and Alcohols

Compound	Formula	Boiling point (°C)	Comments
One Carbon			
methane	CH_4	−161	no hydrogen bonding or polar-polar interactions
formaldehyde	HCHO	−21	polar-polar interactions
methanol	CH_3OH	65	hydrogen bonding
Two Carbons			
ethane	C_2H_6	−89	no hydrogen bonding or polar-polar interactions
acetaldehyde	CH_3CHO	20	polar-polar interactions
ethanol	CH_3CH_2OH	78	hydrogen bonding

Table 4.3 Physical Constants of Some Aldehydes and Ketones

Compound	Melting point (°C)	Boiling point (°C)	Solubility in water (g/100 mL)
Aldehydes			
formaldehyde	−92	−21	completely miscible
acetaldehyde	−123	20	completely miscible
butyraldehyde	−99	76	4
benzaldehyde	−26	179	0.3
Ketones			
acetone	−95	56	completely miscible
methyl ethyl ketone	−86	80	25
diethyl ketone	−42	101	5
benzophenone	48	306	insoluble

Flavors and Fragrances

All pleasant odors and tastes come from chemicals. Aldehydes are responsible for the delightful odors of vanilla, cinnamon, and almonds. Other odors such as that of camphor, which has been used in medicine for thousands of years, are due to ketones.

Perfumes are made by blending certain odoriferous substances. Fine perfumes may contain more than 100 ingredients. Perfumes are classified by the dominant odor. Odors of rose, gardenia, lily of the valley, and jasmine are the floral group. Aromas of clove, nutmeg, cinnamon, and carnation provide the spicy blends. Cedar and sandalwood make up the woody group. The Orientals group combines the spicy and woody odors with the sweet odors of balsam and vanilla and accentuates the mixture with the odors of musk and civet. The odors of aldehydes dominate the aldehydic group, which have fruity character.

Natural products of vegetable and animal origin traditionally have been used as starting materials for perfume manufacture. Mixtures of essential oils (which give perfumes their odor or *essence*) and waxes are obtained from these natural products by extraction into organic solvents. Removal of the solvent gives a residue called a *concrete*. If the concrete is treated with ethanol, the wax remains behind and the essential oil dissolves in the alcohol. Essential oils are removed from citrus peel by pressing. Individual compounds used in perfumes may be isolated from the essential oils by distillation.

Some animal secretions contain odorous compounds that improve the lasting quality of perfumes. These substances and their components act as fixatives, preventing the more volatile ingredients in the perfume from evaporating too rapidly. Such materials are usually used as solutions in ethanol and give high-quality perfume its strength, character, and tenacity. The traditional animal products include castor or castoreum from the beaver, civet from the civet cat (see figure), ambergris from the sperm whale, and musk from the musk deer. Today, because of ecological concerns, chemists synthesize the compounds that produce the musk odor. Musk, the most commonly known musky substance, has a penetrating, persistent odor. Natural musk is produced by a gland under the skin of the abdomen of the male musk deer. The odorous substance in musk is a ketone, muscone, which has the chemical name 3-methylcyclopentadecanone. The civet cat, a native of Africa, southern Europe, and Asia, marks its territory with another ketone, civetone. Civetone, like the musk of the deer, has a musky odor that is excellent for use in perfumes.

The civet cat and musk deer produce secretions that are used as fixatives in perfumes—they give the perfume a lasting odor.

Solubility in water

Aldehydes and ketones can form hydrogen bonds with polar water molecules (Fig. 4.2). Formaldehyde, acetaldehyde, and acetone are soluble in water in all proportions. As the length of the hydrocarbon chain increases, water solubility decreases; when the carbon chain exceeds five

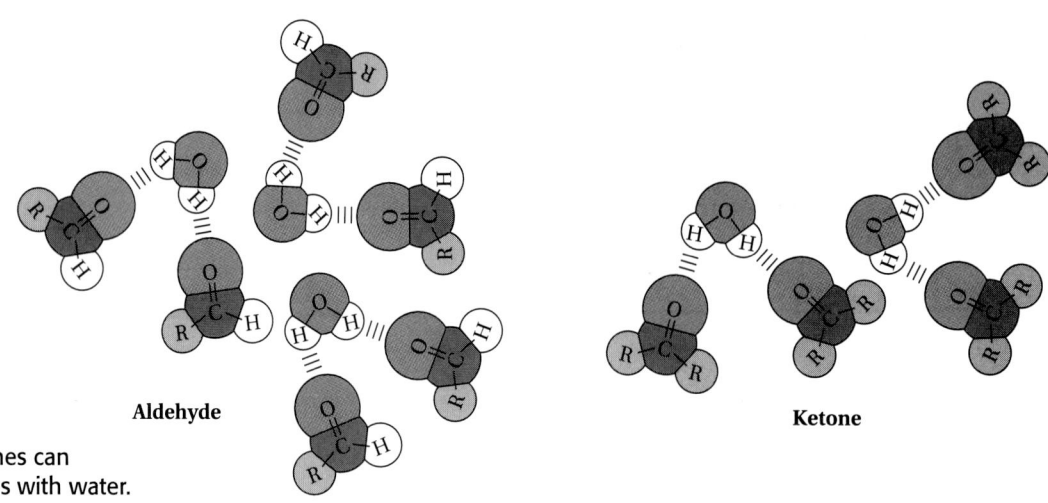

Figure 4.2
Aldehydes and ketones can
form hydrogen bonds with water.

Aldehyde

Ketone

or six carbons, solubility of both aldehydes and ketones is very low. All aldehydes and ketones are soluble in nonpolar solvents.

> **PRACTICE EXERCISE 4.4**
> Arrange the following substances in order of increasing solubility in water:
> (a) acetone   (b) butanal   (c) pentanol
> (d) benzophenone   (e) benzaldehyde

4.3 Redox reactions of organic compounds

AIMS: To describe the processes of oxidation and reduction in organic chemistry in terms of the loss or gain of oxygen, hydrogen, or electrons. To relate the energy content of a molecule to its degree of oxidation or reduction.

Focus

Many reactions of organic compounds involve oxidation or reduction.

Now that we are familiar with functional groups in organic chemistry, we can examine oxidation-reduction reactions of organic molecules. We are interested in redox reactions of organic molecules because they are important in energy production in living organisms. Oxidation reactions are energy-releasing, and the more reduced a carbon compound is, the more energy the compound can release upon its complete oxidation. You may recall the principles of redox reactions: *Oxidation reactions* involve a gain of oxygen, a loss of hydrogen, or a loss of electrons; *reduction reactions* involve a loss of oxygen, a gain of hydrogen, or a gain of electrons. Oxidation and reduction reactions must be coupled; if a compound is oxidized in a reaction, some other compound in the reaction must be reduced. We will discuss electron transfers in redox reactions in living organisms in Chapter 14. Our concern with redox reactions in this chapter will focus on those reactions that involve oxygen and hydrogen. For example, methane, a saturated

[handwritten top margin: oxidation of alkane to CO2]

[handwritten top margin: metabolism of fat n oil in body. goes thru same re-dox]

hydrocarbon, can be oxidized in steps to carbon dioxide by alternately gaining oxygens and losing hydrogens. Methane can be oxidized to methanol, then to formaldehyde, then to formic acid, and finally to carbon dioxide.

[handwritten: alcohol | dehydrogenation | aldehyde/ketone | functional group | a a c bonded to 2 O]

[handwritten labels near structures: alkane; gain 1 O; lose 2H need heat; gain 1 O; lose 2H need 1 t]

| Methane | Methanol (methyl alcohol) | Methanal (formaldehyde) | Methanoic acid (formic acid) | Carbon dioxide |

Methane — Most energetic molecule

Carbon dioxide — Least energetic molecule

[handwritten: c reduced has no oxygen]

[handwritten: most oxidized]

In any series consisting of an alkane, alcohol, aldehyde (or ketone), carboxylic acid, and carbon dioxide, the alkane is the least oxidized (most reduced) compound and the carbon dioxide is the most oxidized (least reduced) compound. In biological systems, fats are similar in molecular structure to alkanes, and sugars are similar to alcohols. The more reduced a compound is, the more energy it can release upon its complete oxidation to carbon dioxide. Therefore, the complete oxidation of a carbon atom in a fat can produce more energy than the complete oxidation of a carbon atom in a sugar. The oxidation of carbon compounds to carbon dioxide can be reversed. For example, in photosynthesis, green plants can extract carbon dioxide from the atmosphere and produce sugars by a series of reduction reactions (see Sec. 14.2).

Many fat molecules have long hydrocarbon chains as part of their structure. As such, they are highly reduced molecules and have a relatively high energy content compared with other nutrient molecules.

We can use the loss and gain of hydrogen to find the relative degree of oxidation of organic molecules that contain carbon-carbon double bonds and carbon-carbon triple bonds. For example, ethane (an alkane) can lose hydrogen and go to ethene (an alkene) and then to ethyne (an alkyne).

| Ethane | Ethene | Ethyne |

Least oxidized (most reduced) — Ethane

Most oxidized (least reduced) — Ethyne

[handwritten: Successive oxidation cause loosing H (or gaining O)]

Each of these losses of hydrogen represents an oxidation of the compound that loses the hydrogen. Ethane is the least oxidized compound, and ethyne is the most oxidized. The fewer the number of hydrogens on a carbon-carbon bond, the more oxidized is the bond. *A redox reaction involving the loss of hydrogen from an organic molecule is called a* **dehydrogenation reaction.** The loss of hydrogen from ethane to give ethene and the loss of hydrogen from methanol to give methanal are examples of dehydrogenation reactions. Strong heating and a catalyst are usually necessary to make dehydrogenation reactions occur in the laboratory. In living organisms, dehydrogenation reactions are catalyzed by enzymes called *dehydrogenases* and occur at very mild conditions. Like other oxidation reactions, dehydrogenation reactions can be reversed. Alkynes can be reduced to alkenes and alkenes can be reduced to alkanes by addition of hydrogen to the double bond.

EXAMPLE 4.4   **Identifying the relative degree of oxidation**

List the compounds 2-propanol, propane, and propanone in order from most reduced to most oxidized.

SOLUTION

Write the structure of each compound:

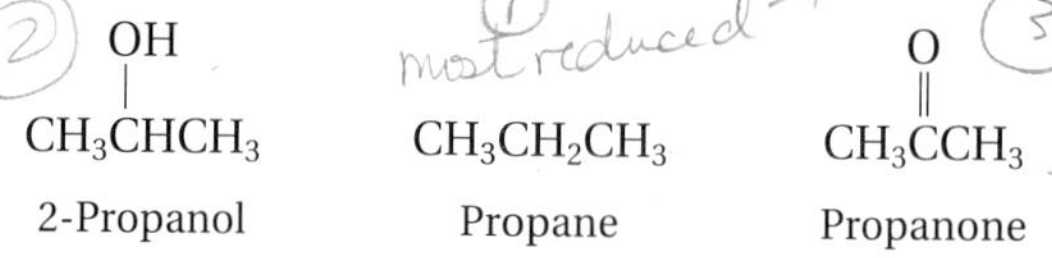

OH		O
CH_3CHCH_3	$CH_3CH_2CH_3$	CH_3CCH_3
2-Propanol	Propane	Propanone

Propane is most reduced; it has the maximum number of hydrogens. Propanone is the most oxidized. It has the same number of oxygens as 2-propanol but fewer hydrogens. The order is propane, 2-propanol, and propanone.

PRACTICE EXERCISE 4.5

Indicate the most oxidized compound in each pair.
(a) 1-butyne and 1-butene
(b) propanal and propane
(c) cyclohexane and cyclohexanol
(d) 3-pentanol and 3-pentanone

4.4 Redox reactions involving aldehydes and ketones

AIMS: *To write structures for the aldehyde and ketone products (if any) of the oxidation of primary, secondary, and tertiary alcohols. To write structures for the products of the reduction of aldehydes and ketones.*

The transformation of an alcohol to an aldehyde or ketone can be accomplished by an oxidation reaction. Conversely, aldehydes and ketones can be reduced to alcohols.

Focus

Redox reactions are used to interconvert aldehydes or ketones and alcohols.

Oxidation of alcohols

Primary alcohols can be oxidized to aldehydes, and secondary alcohols can be oxidized to ketones. These oxidations are represented as follows:

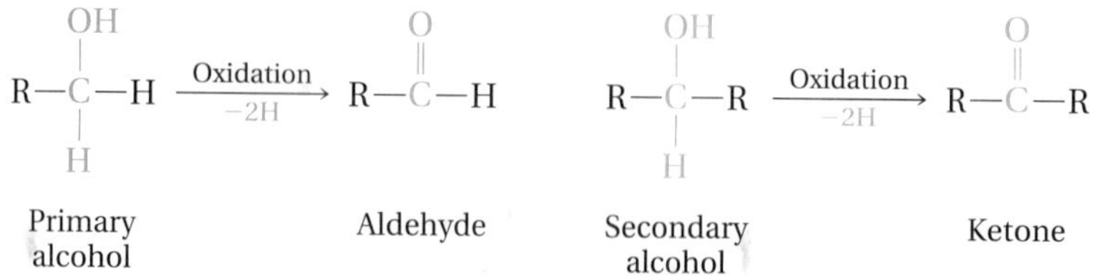

| Primary alcohol | Aldehyde | Secondary alcohol | Ketone |

Tertiary alcohols are not oxidized because there is no hydrogen to remove from the carbon bearing the hydroxyl group.

Oxidation of the primary alcohols methanol and ethanol by warming them at about 50 °C with acidified potassium dichromate ($K_2Cr_2O_7$) produces formaldehyde and acetaldehyde, respectively:

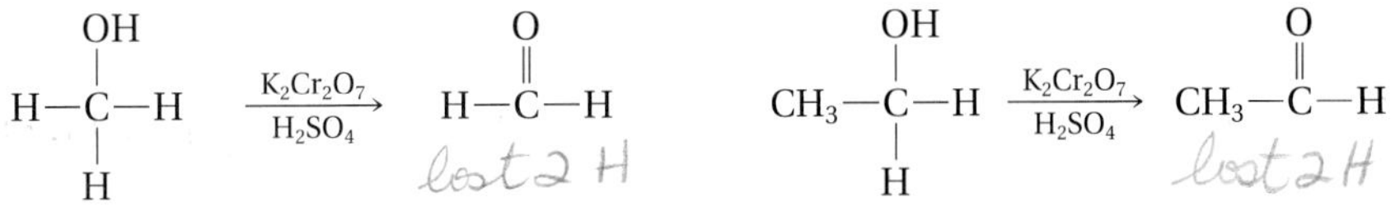

$$H-\underset{\underset{H}{|}}{\overset{\overset{OH}{|}}{C}}-H \xrightarrow[H_2SO_4]{K_2Cr_2O_7} H-\overset{\overset{O}{\|}}{C}-H \qquad CH_3-\underset{\underset{H}{|}}{\overset{\overset{OH}{|}}{C}}-H \xrightarrow[H_2SO_4]{K_2Cr_2O_7} CH_3-\overset{\overset{O}{\|}}{C}-H$$

lost 2 H   *lost 2 H*

Methanol	Methanal	Ethanol	Ethanal
(methyl alcohol)	(formaldehyde)	(ethyl alcohol)	(acetaldehyde)
(bp 65 °C)	(bp −21 °C)	(bp 78 °C)	(bp 21 °C)

The preparation of an aldehyde by this method is often a problem because aldehydes are easily oxidized to carboxylic acids:

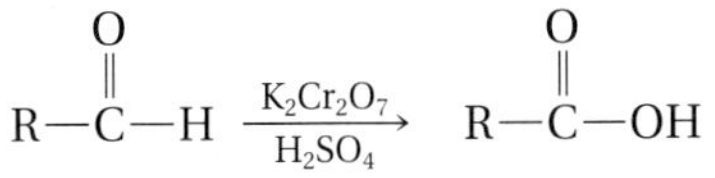

$$R-\overset{\overset{O}{\|}}{C}-H \xrightarrow[H_2SO_4]{K_2Cr_2O_7} R-\overset{\overset{O}{\|}}{C}-OH$$

Aldehyde   Carboxylic acid

Further oxidation is not a problem with aldehydes that have low boiling points, such as acetaldehyde, because the product can be distilled from the reaction mixture as it is formed.

Oxidation of the secondary alcohol 2-propanol by warming with acidified potassium dichromate produces acetone:

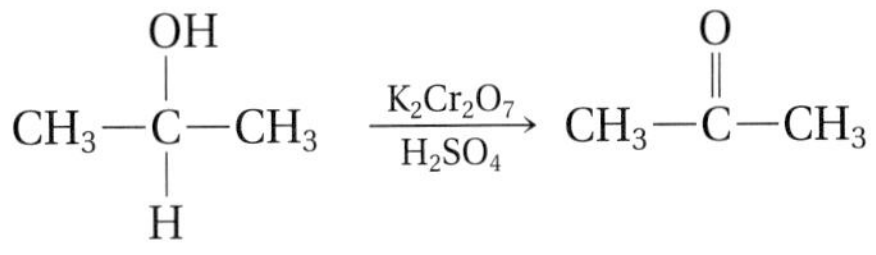

$$CH_3-\underset{\underset{H}{|}}{\overset{\overset{OH}{|}}{C}}-CH_3 \xrightarrow[H_2SO_4]{K_2Cr_2O_7} CH_3-\overset{\overset{O}{\|}}{C}-CH_3$$

2-Propanol   Propanone
(isopropyl alcohol)   (acetone)

Ketones are resistant to further oxidation. There is no need to remove them from the reaction mixture during the course of the reaction.

EXAMPLE 4.5   **Writing an equation for oxidation of an alcohol**

Write an equation for the oxidation of 4-methyl-2-hexanol.

SOLUTION

A secondary alcohol is oxidized to a ketone. The equation is

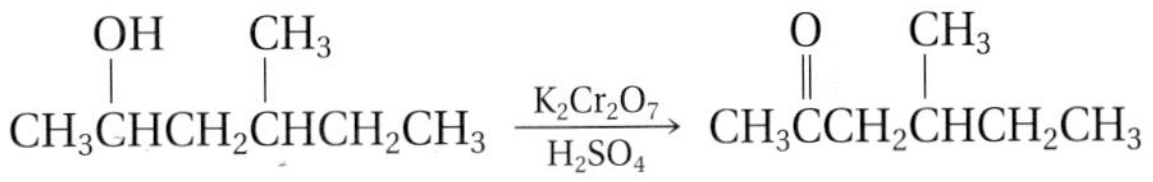

$$CH_3\underset{\underset{OH}{|}}{C}HCH_2\underset{\underset{CH_3}{|}}{C}HCH_2CH_3 \xrightarrow[H_2SO_4]{K_2Cr_2O_7} CH_3\overset{\overset{O}{\|}}{C}CH_2\underset{\underset{CH_3}{|}}{C}HCH_2CH_3$$

Since the reactant is a secondary alcohol, oxidation to a carboxylic acid will not occur.

(handwritten margin notes, top left) both answer only butanone? can only 2 oxygen no? why

(handwritten, top right) $CH_3CH_2CH_2\overset{\displaystyle O}{\overset{\|}{C}}-H$

PRACTICE EXERCISE 4.6

What products are expected when the following compounds are oxidized? *(handwritten)* butanol → butanal

(a) $CH_3CH_2CH_2CH_2OH$ *(handwritten)* primary alcohol → acid n aldehyde

(b)
$$\underset{\displaystyle CH_3CH_2CHCH_3}{\overset{\displaystyle OH}{|}}$$
(handwritten) 2-butanol → 2-butanone $CH_3CH_2\overset{\displaystyle O}{\overset{\|}{C}}CH_3$ secondary alcohol → ketone

(c)
$$\underset{\displaystyle \underset{\displaystyle CH_3}{|}}{\overset{\displaystyle OH}{\overset{\displaystyle |}{CH_3CH_2CCH_3}}}$$
(handwritten) tertiary alcohol → NR   2 methyl-2-butanol

(d)
(structure of cyclobutanol with OH)
(handwritten) secondary alcohol → ketone   cyclobutanol → cyclobutanone *(structure of cyclobutanone)*

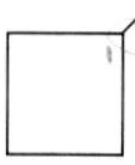

PRACTICE EXERCISE 4.7

Give the name and structure of the alcohol you must oxidize to make the following compounds.

(a) CH_3CH_2CHO *(handwritten)* $CH_3CH_2CH_2OH$ 1-propanol

(b)
$$CH_3CH_2\overset{\displaystyle O}{\overset{\|}{C}}CH_3$$
(handwritten) $CH_3CH_2\overset{\displaystyle OH}{\overset{\displaystyle |}{C}}HCH_3$ 2-butanol (sec-alcohol)

(c)
$$\underset{\displaystyle \underset{\displaystyle CH_3}{|}}{CH_3CH_2CHCHO}$$
(handwritten) $CH_3CH_2\overset{\displaystyle }{\underset{\displaystyle \underset{\displaystyle CH_3}{|}}{C}}H-CH_2-OH$ 2-methylbutanol

(handwritten margin notes, left side)
a) propanal (propionaldehyde)
b) 2-butanone
c) 2-methyl butanal   name   2-methyl butyraldehyde

Oxidation of alcohols in the liver

Ethanol is oxidized in the liver to acetaldehyde. Acetaldehyde is then oxidized to acetic acid and finally to carbon dioxide and water:

$$CH_3CH_2-OH \longrightarrow CH_3-\overset{\displaystyle O}{\overset{\|}{C}}-H \longrightarrow CH_3-\overset{\displaystyle O}{\overset{\|}{C}}-OH \longrightarrow CO_2 + H_2O$$

Ethanol   Acetaldehyde   Acetic acid   Carbon dioxide   Water

Consumption of large quantities of ethanol causes the buildup of high concentrations of acetaldehyde in the blood. This can lead to a sharp decrease in blood pressure, a more rapid heartbeat, and a generally uncomfortable feeling—a hangover. Continued overindulgence in ethanol eventually leads to yellowing and hardening of the liver, called *cirrhosis* (Fig. 4.3), because of the sustained high levels of acetaldehyde. Cirrhosis is an irreversible degeneration of functioning liver cells.

Methanol, sometimes called *wood alcohol,* is extremely toxic. When methanol enters the body, it is quickly absorbed into the bloodstream and passes to the liver, where it is oxidized to formaldehyde. Formaldehyde is a very reactive compound. It destroys the catalytic power of enzymes and causes liver tissue to become hard. This is why formaldehyde solutions are

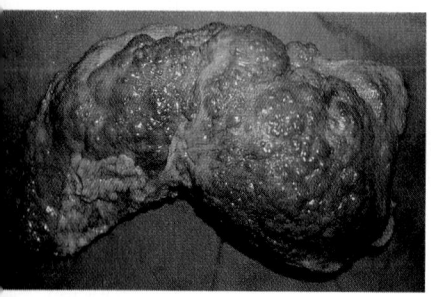

Figure 4.3
Chronic alcohol consumption can cause cirrhosis of the liver.

FOLLOW-UP TO THE CASE IN POINT: Fetal alcohol syndrome

Fortunately, Marla was able to follow her physician's advice, and at the end of a full-term pregnancy, she gave birth to a healthy baby girl. The doctor's concern about Marla's history of alcoholism resulted from the knowledge that alcohol consumption can cause great damage to the fetus. The fetus processes ethanol more slowly than the mother, so the injurious effects of alcohol and its oxidation product, acetaldehyde, affect the fetus for a longer time. A pregnant woman who drinks alcohol passes these effects in a magnified form to her unborn child. The damaging effects of alcohol on the unborn are called *fetal alcohol effects* (FAE). The range of FAE can be mild to severe. Some FAE are so broad and long-lasting that they can cripple a child's physical, social, and academic development. *Fetal alcohol syndrome* (FAS) is the name given to extremely severe FAE. FAS children can be underweight or below normal height and have abnormally small heads or other abnormal facial features, crossed eyes, underdeveloped jawbones,

cleft palates, and dysfunction of the central nervous system. FAS is the leading cause of mental retardation in children, and the retardation may be profound. The symptoms of FAE/FAS last a lifetime, and new symptoms often appear as the child grows up. Some children who do not have visible signs of FAE/FAS at birth develop them over time. For example, an infant who appears normal at birth may display slowed development in walking or talking as a toddler.

When a woman drinks heavily while she's pregnant, the chances are 30% to 40% that her baby will have FAE/FAS. If the mother also smokes, the chances of FAE/FAS increase, and the severity is intensified. FAE/FAS researchers recommend that women abstain from alcohol consumption and smoking during pregnancy. Fathers are not relieved of responsibility in preventing FAE/FAS. One study shows the birth weight of babies born to fathers who drink regularly averages 181 g less than children born to fathers who drink only occasionally.

used to preserve biological specimens. When methanol is ingested, temporary or permanent blindness may occur.

The consumption of even moderate amounts of ethanol by pregnant women can have devastating effects on the unborn fetus. In the Case in Point for this chapter, you may recall that Marla, a pregnant woman with a history of alcoholism, was strongly advised by her doctor to avoid alcohol during her pregnancy. In the Follow-up to the Case in Point, above, we will learn more about the consequences of using alcohol during pregnancy.

Reduction of aldehydes and ketones

Aldehydes and ketones can be reduced to alcohols by the addition of hydrogen, H—H, to the —C=O double bond. Reductions of an aldehyde and a ketone can be represented as follows:

$$
\underset{\text{Aldehyde}}{R-\overset{\displaystyle O}{\overset{\|}{C}}-H} \xrightarrow[\text{+2H}]{\text{Reduction}} \underset{\text{Primary alcohol}}{R-\overset{\displaystyle OH}{\underset{\displaystyle H}{\overset{|}{\underset{|}{C}}}}-H} \qquad \underset{\text{Ketone}}{R-\overset{\displaystyle O}{\overset{\|}{C}}-R} \xrightarrow[\text{+2H}]{\text{Reduction}} \underset{\text{Secondary alcohol}}{R-\overset{\displaystyle OH}{\underset{\displaystyle H}{\overset{|}{\underset{|}{C}}}}-R}
$$

As we can see from these reactions, the reduction of aldehydes produces primary alcohols, and the reduction of ketones produces secondary alcohols. A variety of reagents are available for the reduction of —C=O groups

An important biological example of reduction of an aldehyde occurs in fermentation. At the last step in the metabolism of glucose, yeast and other organisms reduce acetaldehyde to ethanol.

to —CH—OH groups. For example, hydrogen gas can be used with a platinum or palladium catalyst.

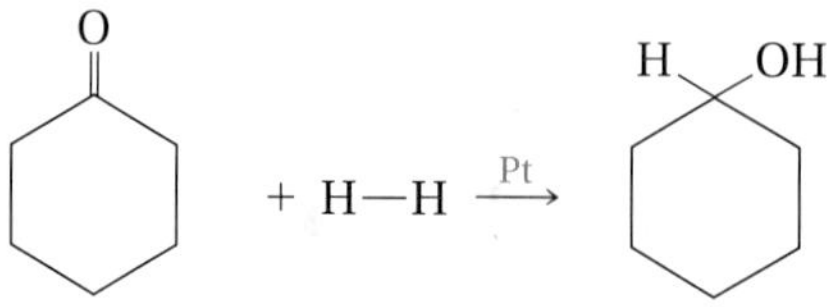

Cyclohexanone + H—H $\xrightarrow{Pt}$ Cyclohexanol

The hydrogenation of C=O bonds is similar to the hydrogenation of C=C bonds discussed previously. Other reducing reagents for C=O bonds include compounds called *hydrides*. Lithium aluminum hydride (LiAlH$_4$) and sodium tetrahydroborate (NaBH$_4$) are often used.

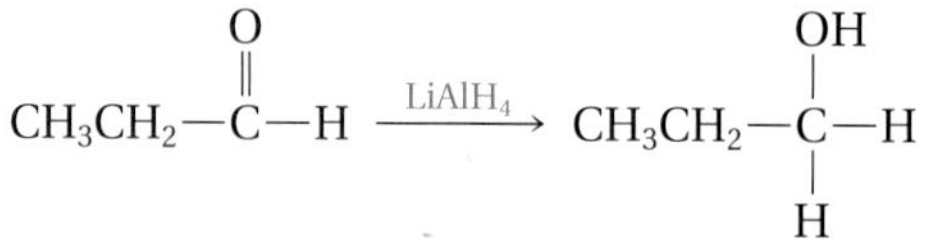

$$CH_3CH_2-\overset{\overset{\displaystyle O}{\|}}{C}-H \xrightarrow{LiAlH_4} CH_3CH_2-\overset{\overset{\displaystyle OH}{|}}{\underset{\underset{\displaystyle H}{|}}{C}}-H$$

Propanal — 1-Propanol (propyl alcohol)

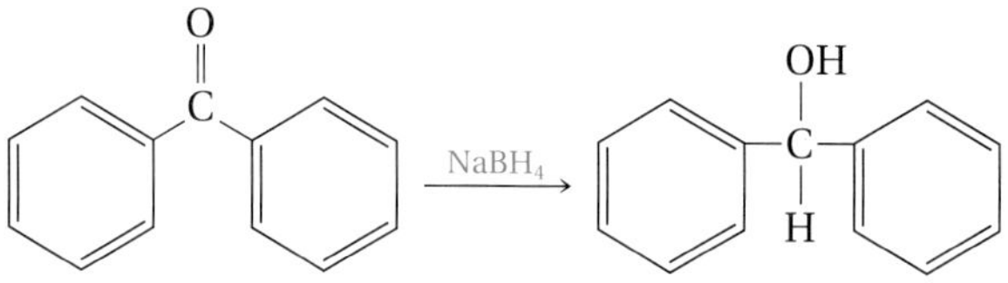

Diphenylmethanone (benzophenone) $\xrightarrow{NaBH_4}$ Diphenylmethanol

Reductions of aldehydes and ketones in living organisms are catalyzed by dehydrogenase enzymes. Enzymes work reversibly, and dehydrogenases catalyze both the oxidation of alcohols and the reduction of aldehydes and ketones.

4.5 Aldehyde detection

AIM: To describe the results of a Tollens', a Benedict's, or a Fehling's test on an aldehyde, a ketone, and an alpha-hydroxy ketone.

Focus

Aldehydes can be distinguished from ketones by using mild oxidizing agents.

Aldehydes and ketones react with a wide variety of compounds. In general, however, aldehydes are more reactive than ketones. Chemists have taken advantage of the ease with which an aldehyde can be oxidized to develop several visual tests for their detection. The most widely used tests for aldehyde detection are Tollens', Benedict's, and Fehling's.

Tollens' test

The mild oxidizing agent used in this test, **Tollens' reagent,** *is an alkaline solution of silver nitrate.* It is clear and colorless. To prevent the silver ions from precipitating as silver oxide (Ag$_2$O) at the high pH, a few drops of an ammonia solution are added. The ammonia forms a water-soluble complex with silver ions:

$$Ag^+(aq) + 2NH_3(aq) \rightarrow [Ag(NH_3)_2]^+(aq)$$

When an aldehyde is oxidized with Tollens' reagent, the corresponding carboxylic acid is formed, and simultaneously, silver ions are reduced to metallic silver. For example, acetaldehyde goes to acetic acid. The silver is usually deposited as a mirror on the inside surface of the reaction vessel. The appearance of a silver mirror is a positive test for an aldehyde. If acetaldehyde is treated with Tollens' reagent, the reaction is

$$\underset{\text{Acetaldehyde}}{CH_3-\overset{\displaystyle O}{\overset{\|}{C}}-H} + \underset{\substack{\text{Tollens'}\\ \text{reagent}}}{2[Ag(NH_3)_2]^+} + 2OH^- \longrightarrow \underset{\substack{\text{Acetic acid}\\ \text{(as ammonium salt)}}}{CH_3-\overset{\displaystyle O}{\overset{\|}{C}}-O^-NH_4^+} + \underset{\substack{\text{Silver}\\ \text{(mirror)}}}{2Ag(s)} + 3NH_3 + H_2O$$

The aldehyde, acetaldehyde, is oxidized to a carboxylic acid, acetic acid; it is a reducing agent. The silver ions are reduced to metallic silver; they are oxidizing agents. Mirrors (Fig. 4.4) are often silvered by using Tollens' reagent. The commercial process uses glucose or formaldehyde as the reducing agent.

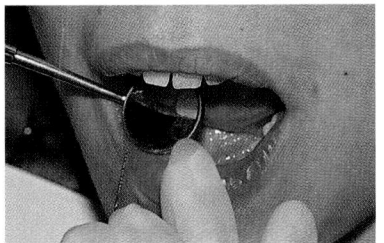

Figure 4.4
The reflective coating on a dentist's mirror is produced on the back of a sheet of glass when formaldehyde reduces silver ions in solution to silver metal. The formaldehyde is oxidized to formic acid in the process.

PRACTICE EXERCISE 4.8

You have two unlabeled test tubes, one containing pentanal and the other containing 2-pentanone. What simple test could you do to find out which tube contains pentanal and which tube contains 2-pentanone?

Benedict's and Fehling's tests

Benedict's *and* **Fehling's reagents** *are deep blue alkaline solutions of copper sulfate* of slightly differing compositions. When an aldehyde is oxidized with Benedict's or Fehling's reagents, a brick red precipitate of copper(I) oxide (Cu_2O) is obtained. The reaction with acetaldehyde is

$$\underset{\text{Acetaldehyde}}{CH_3-\overset{\displaystyle O}{\overset{\|}{C}}-H} + \underset{\substack{\text{Copper(II)}\\ \text{ion complex}\\ \text{(blue solution)}}}{2Cu^{+2}} + 5OH^- \longrightarrow \underset{\substack{\text{Acetic acid}\\ \text{(as acetate}\\ \text{ion)}}}{CH_3-\overset{\displaystyle O}{\overset{\|}{C}}-O^-} + \underset{\substack{\text{Copper(I)}\\ \text{oxide}\\ \text{(red precipitate)}}}{Cu_2O(s)} + 3H_2O$$

The acetaldehyde is oxidized to acetic acid; copper(II) ions (Cu^{2+}) are reduced to copper(I) ions (Cu^+).

Alpha-hydroxy ketones

Ketones are not usually oxidized by mild oxidizing agents such as Tollens' and Benedict's solutions. However, ketones that contain a carbonyl group attached to a carbon that bears a hydroxyl group give positive tests with Tol-

Persons in good health do not have sugar in their urine. Clinitest tablets are used to screen for sugar in urine. The reaction of the tablet with the urine gives colored products corresponding to differing levels of sugar. The reaction chemistry is that of a Benedict's test.

aldehyde

lens', Benedict's, and Fehling's reagents. *These compounds are called* **alpha-hydroxy ketones,** *which have this general formula:*

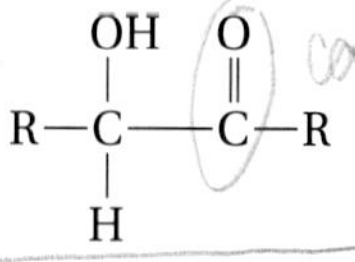

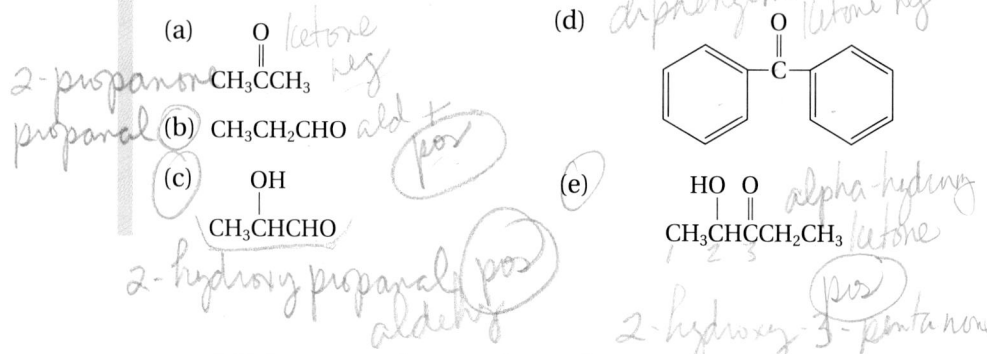

carbonyl group attached to the C that has hydroxyl gr.

Alpha-hydroxy ketone

PRACTICE EXERCISE 4.9

Determine which of the following substances give a positive test (red precipitate) with Benedict's reagent.

Cu+

(a)

2-propanone   *ketone neg*

$$CH_3CCH_3$$ with O double bond

(b) CH_3CH_2CHO *ald.* *pos*

propanal

(c)   OH

$$CH_3CHCHO$$

2-hydroxy propanal (pos) aldehyde

(d) *diphenylmethanone ketone neg*

structure of diphenyl ketone (benzophenone)

(e)   HO O

$$CH_3CHCCH_2CH_3$$ *alpha-hydroxy ketone*

2-hydroxy-3-pentanone (pos)

4.6 Additions to the carbonyl group

AIM: To illustrate with equations the formation of a hydrate, a hemiacetal and an acetal, and a hemiketal and a ketal.

The most characteristic reactions of both aldehydes and ketones are *addition* reactions. A wide variety of compounds add to the carbon-oxygen double bond of the carbonyl group. We have already seen additions to carbon-carbon double bonds in Sections 3.2 and 3.4, and additions to carbon-oxygen double bonds are quite similar. Our main concern here is with the addition of water and alcohols to the carbonyl group.

Addition of water

Water will add to most aldehydes and ketones to form **hydrates**—*compounds that have two hydroxyl groups on the same carbon.* The reaction is reversible. With few exceptions, the equilibrium lies to the left, in favor of the starting materials.

$$R-\overset{\overset{\displaystyle O}{\|}}{C}-H(R) \;+\; H-OH \;\rightleftharpoons\; R-\overset{\overset{\displaystyle OH}{|}}{\underset{\underset{\displaystyle OH}{|}}{C}}-H(R)$$

Carbonyl group   Water   Hydrate
of aldehyde
or ketone

One carbonyl compound that forms a stable hydrate is chloral. The reaction is

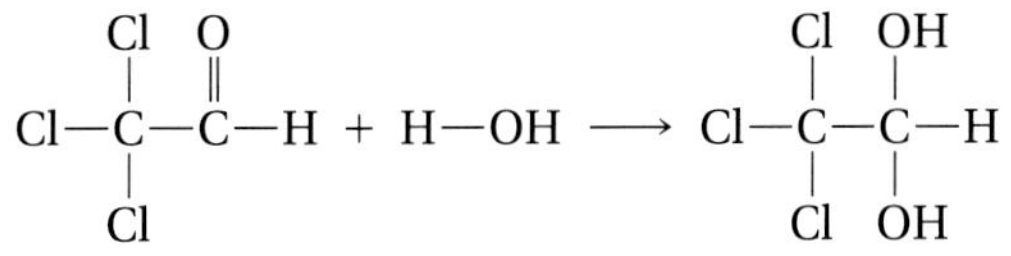

$$
\begin{array}{ccc}
\underset{\substack{|\\ Cl}}{\overset{\substack{Cl \quad O\\ |\quad\; \|}}{Cl-C-C-H}} + H-OH & \longrightarrow & \underset{\substack{|\quad\; |\\ Cl \;\; OH}}{\overset{\substack{Cl \;\; OH\\ |\quad\; |}}{Cl-C-C-H}} \\
\text{Chloral} \qquad \text{Water} & & \text{Chloral} \\
& & \text{hydrate}
\end{array}
$$

trichloroethanal

Chloral hydrate is a colorless crystalline solid that is very soluble in water. It is one of the few stable organic molecules that has two hydroxyl groups on the same carbon. Chloral hydrate is a powerful sedative (relaxant) and hypnotic (sleep inducer).

Addition of alcohols

The addition of alcohols to aldehydes and ketones is very similar to the addition of water. *The product of the reaction between an aldehyde and an alcohol is called a* **hemiacetal.** A hemiacetal is a compound with the general structure

$$
\underset{\substack{|\\ OR}}{\overset{\substack{OH\\ |}}{R-C-H}} \quad \text{Hemiacetal carbon}
$$

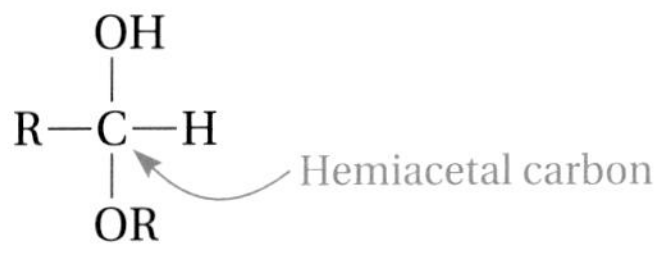

Hemiacetal

With acetaldehyde and methanol, the reaction for formation of a hemiacetal is

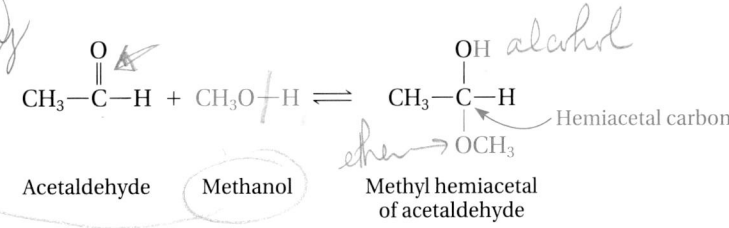

$$
\underset{\text{Acetaldehyde}}{\overset{\substack{O\\ \|}}{CH_3-C-H}} + \underset{\text{Methanol}}{CH_3O{-}H} \rightleftharpoons \underset{\substack{\text{Methyl hemiacetal}\\ \text{of acetaldehyde}}}{\overset{\substack{OH\\ |}}{\underset{\substack{|\\ OCH_3}}{CH_3-C-H}}} \quad \text{Hemiacetal carbon}
$$

alcohol *ether*

The addition of the alcohol is readily reversible, so many hemiacetals are quite unstable. In the presence of an excess of alcohol, a hemiacetal rapidly reacts with another molecule of alcohol to produce a relatively stable *acetal. An* **acetal** *is a compound with the general structure*

$$
\underset{\substack{|\\ OR}}{\overset{\substack{OR\\ |}}{R-C-H}} \quad \text{Acetal carbon}
$$

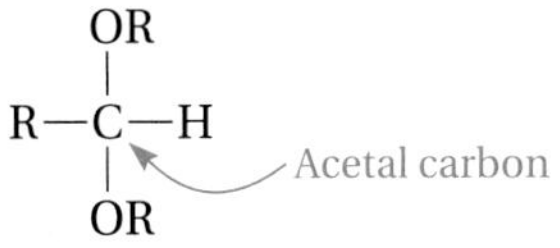

Acetal

Acetals contain a carbon that has a hydrogen and two OR groups attached and therefore contain a diether linkage. The following reaction shows the formation of an acetal from a hemiacetal:

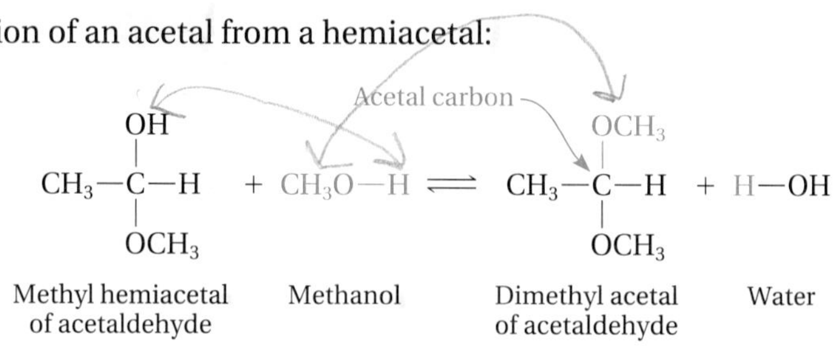

$$CH_3\!-\!\underset{\underset{\textstyle OCH_3}{|}}{\overset{\overset{\textstyle OH}{|}}{C}}\!-\!H \;+\; CH_3O\!-\!H \;\rightleftharpoons\; CH_3\!-\!\underset{\underset{\textstyle OCH_3}{|}}{\overset{\overset{\textstyle OCH_3}{|}}{C}}\!-\!H \;+\; H\!-\!OH$$

Acetal carbon

Methyl hemiacetal   Methanol   Dimethyl acetal   Water
of acetaldehyde   of acetaldehyde

Ketones undergo similar additions with alcohols to form *hemiketals* and *ketals*. **Hemiketals** *are compounds with the general structure*

$$R\!-\!\underset{\underset{\textstyle OR}{|}}{\overset{\overset{\textstyle OH}{|}}{C}}\!-\!R$$

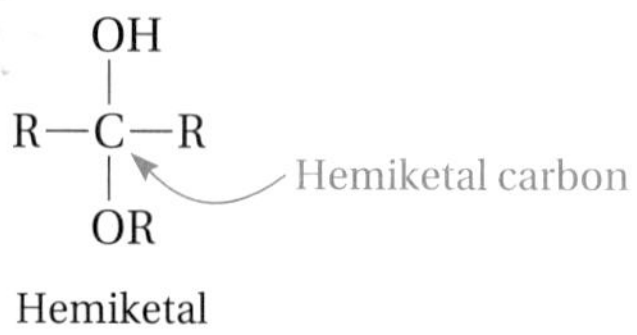

Hemiketal carbon

Hemiketal

Ketals *are compounds with the general structure*

$$R\!-\!\underset{\underset{\textstyle OR}{|}}{\overset{\overset{\textstyle OR}{|}}{C}}\!-\!R$$

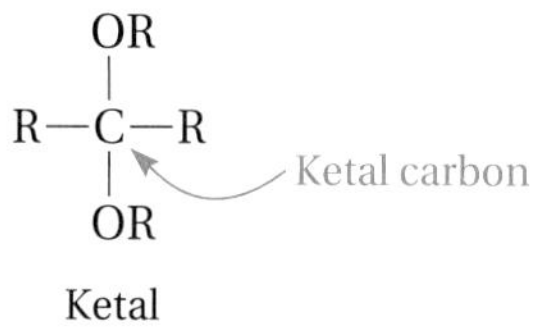

Ketal carbon

Ketal

Hemiketals and ketals have two carbons attached to their central carbons; hemiacetals and acetals have only one carbon attached. Hemiketals are often unstable, but the ketals are relatively stable. The following equation shows the formation of a hemiketal and a ketal from a ketone and an alcohol.

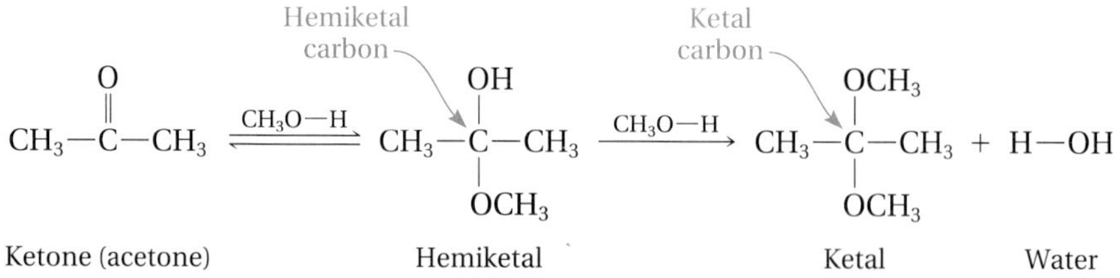

$$CH_3\!-\!\overset{\overset{\textstyle O}{||}}{C}\!-\!CH_3 \;\underset{}{\overset{CH_3O-H}{\rightleftharpoons}}\; CH_3\!-\!\underset{\underset{\textstyle OCH_3}{|}}{\overset{\overset{\textstyle OH}{|}}{C}}\!-\!CH_3 \;\overset{CH_3O-H}{\longrightarrow}\; CH_3\!-\!\underset{\underset{\textstyle OCH_3}{|}}{\overset{\overset{\textstyle OCH_3}{|}}{C}}\!-\!CH_3 \;+\; H\!-\!OH$$

Hemiketal carbon   Ketal carbon

Ketone (acetone)   Hemiketal   Ketal   Water

The hemiacetal, hemiketal, acetal, and ketal groups are important in the chemistry of sugars.

4.7 Uses of aldehydes and ketones

AIM: To state the names and uses of some important aldehydes and ketones.

A wide variety of aldehydes and ketones have been isolated from plants and animals. Many of them, particularly those with high molar masses, have fragrant or penetrating odors. They are usually known by their common names, which indicate their natural source or perhaps a characteristic property. Aromatic aldehydes are often used as flavoring agents. Benzaldehyde, also known as *oil of bitter almonds,* is a constituent of the almond. It is a colorless liquid with a pleasant almond odor. Cinnamaldehyde imparts the characteristic odor to oil of cinnamon. Vanillin, which is responsible for the popular vanilla flavor, was once obtainable only from the podlike capsules of certain climbing orchids. Today most vanillin is synthetically produced.

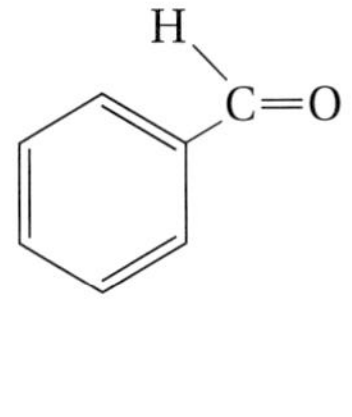

Benzaldehyde 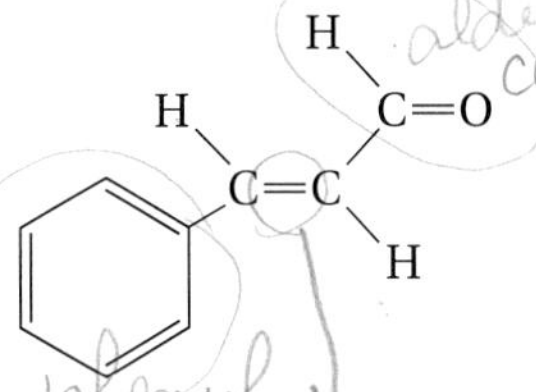   Cinnamaldehyde 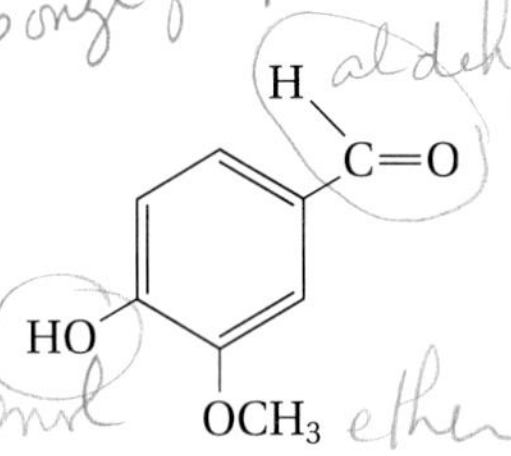   Vanillin

Vanillin is an interesting molecule because it has a number of different functional groups. It possesses an aldehyde group and an aromatic ring, making it an aromatic aldehyde. It also contains an ether linkage and a phenol functional group.

Camphor is a naturally occurring ketone obtained from the bark of the camphor tree. It has a penetrating and fragrant odor. Long known for its medicinal properties, it is an analgesic often found in liniments. Two other natural ketones, beta-ionine and muscone, are both used in perfumes. Beta-ionine is the scent of violets. Muscone, obtained from the scent gland of the male musk deer, has a ring structure containing 15 carbons.

The sex hormones are included in the long list of naturally occurring aldehydes and ketones. The molecule of the male sex hormone testosterone has a cyclic ketone group. One of the primary female sex hormones, progesterone, has two ketone functional groups.

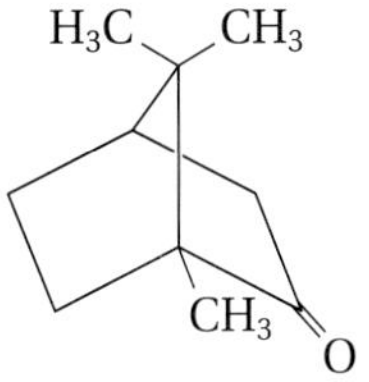

Camphor 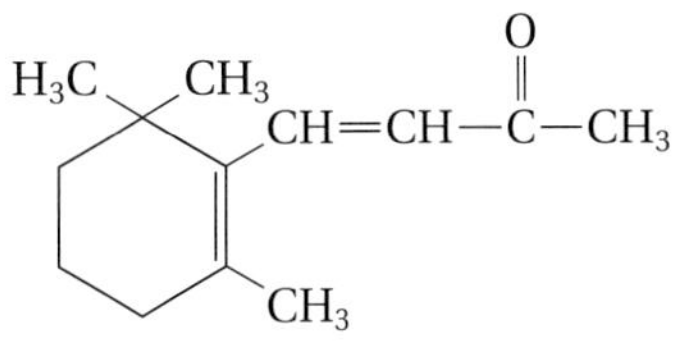   Beta-ionine 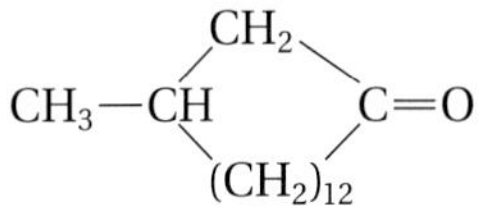   Muscone

---

EXAMPLE 4.6

Recognizing the structural features of an organic molecule

What structural features can you identify in the cinnamaldehyde molecule?

SOLUTION

Structural features include functional groups, carbon rings, and multiple bonds. The cinnamaldehyde molecule has a carbonyl (aldehyde) group and a phenyl group arranged in a *trans* configuration about a carbon-carbon double bond.

PRACTICE EXERCISE 4.10

Cortisone is secreted by the adrenal gland and has been used to treat arthritic conditions. How many functional groups can you identify?

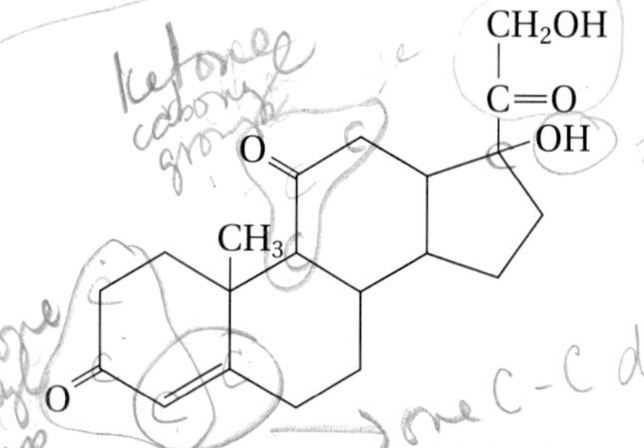

[handwritten annotations:]
C=O
book does not mention but "2-aliphatic —OH group"
ketone carbonyl group
ketone carbonyl
alpha hydroxy ketone
hydroxyl group.
one C–C double bond

The naturally occurring aldehyde 11-*cis*-retinal is essential to the production of light-sensitive molecules within the retina of the eye. How this aldehyde permits us to see is described in A Closer Look: The Chemistry of Vision.

A Closer Look

The Chemistry of Vision

The chemistry of vision is very complex. How does the action of light on the retina of the eye produce a visual image in the brain? Not all the chemical mechanisms involved in the visual cycle are known. However, it is known that a *cis-trans* isomerization plays a central role. This isomerization involves the change from a *cis* double bond in the aldehyde 11-*cis*-retinal to a *trans* double bond in its isomer all-*trans*-retinal, as seen below.

Cones and rods, so named because of their distinctive shapes, are the two kinds of photoreceptor cells in the retina (see figure). Cones function in bright light and are responsible for color perception; rods function in dim light and are responsible for black and white vision. Rods and

Scanning electron micrograph (×1755) of the retina showing photoreceptors (rods and cones).

cones contain *chromophores*, molecules that absorb light. The chromophore in the rod cells is 11-*cis*-retinal. When the 11-*cis*-retinal is bound to a protein called *opsin*, it forms the light-sensi-

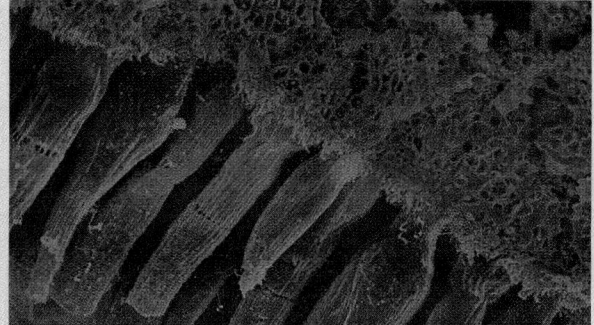

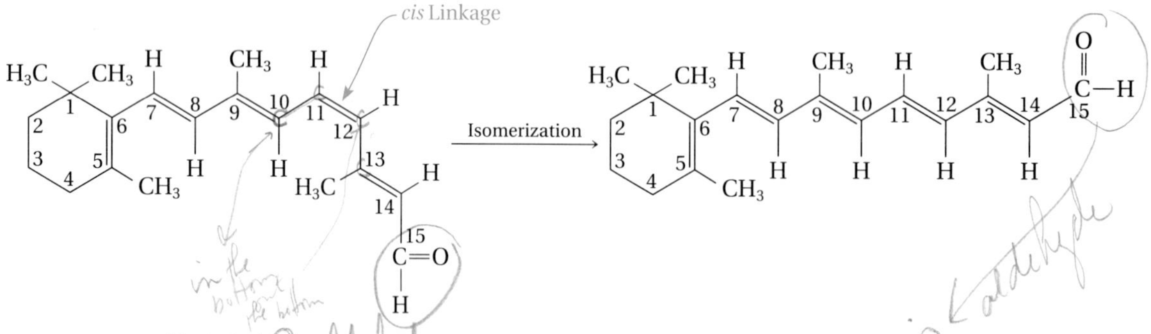

11-*cis*-Retinal — *aldehyde*   all-*trans*-Retinal — *aldehyde*

cis Linkage
Isomerization
in the bottom the bottom

Phenol   Formaldehyde

Bakelite

Figure 4.5

The simplest aldehyde, formaldehyde, is a colorless gas with an irritating odor. It is very important industrially but inconvenient to handle in the gaseous state. Formaldehyde is usually available as a 40% aqueous solution, known as *formalin,* or as a white, solid polymer, known as *paraformaldehyde.* When paraformaldehyde is gently heated, it decomposes and gives off formaldehyde:

$$HO\text{---}(CH_2O)_x\text{---}H \xrightarrow{\text{Heat}} xH\overset{\overset{\displaystyle O}{\|}}{-}C\text{---}H$$

Paraformaldehyde   Formaldehyde

Formalin is used to preserve biological specimens. The formaldehyde in solution combines with protein in tissues to make them hard and insoluble in water. This prevents the specimen from decaying. Formalin also can be used as a general antiseptic. The greatest use of formaldehyde is in the manufacture of synthetic resins. When it is polymerized with phenol, a phenolformaldehyde resin known as *Bakelite* is formed (Fig. 4.5). Bakelite is an

tive molecule *rhodopsin.* As you can see, 11-*cis*-retinal has a molecular shape that permits it to fit exactly on the surface of the opsin molecule:

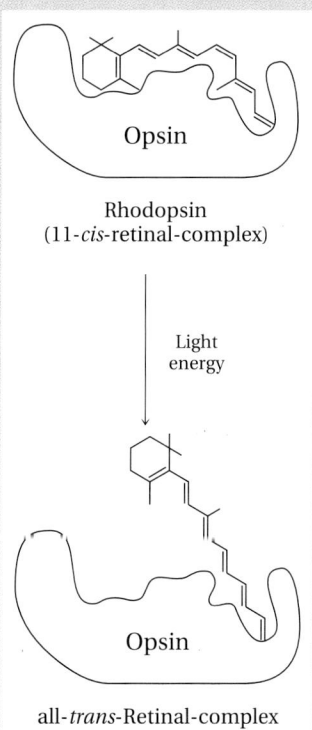

Rhodopsin
(11-*cis*-retinal-complex)

Light energy

Opsin

all-*trans*-Retinal-complex

When rhodopsin absorbs light energy, the *cis* double bond of the chromophore is broken, and the molecule re-forms as the *trans* isomer, all-*trans*-retinal. This change in shape of the chro-

mophore makes it impossible for the *trans* isomer to remain bound to opsin in the same way as the *cis* isomer. As a consequence, the shape of the entire rhodopsin molecule changes. The altered rhodopsin then undergoes further structural changes during which a nerve impulse is generated at the optic nerve and transmitted to the brain. It is this event that permits sight. Subsequently, the rhodopsin complex breaks apart, all-*trans*-retinal is liberated from opsin, and the cycle repeats.

The photosensitivity of rod cells depends on a continuous supply of the chromophore 11-*cis*-retinal. Where does this chromophore come from? It turns out that the precursor for 11-*cis*-retinal is all-*trans*-retinol, an alcohol, better known as *vitamin A.* Liver, eggs, butter, fish, and carrots are rich dietary sources of vitamin A. In carrots it is obtained indirectly from beta-carotene, the pigment that makes carrots orange. When one molecule of beta-carotene is cleaved during digestion, it produces two molecules of vitamin A (see Sec. 12.7):

all-*trans*-Retinol
(Vitamin A)

excellent electrical insulator. At one time it also was used for making billiard balls.

Acetaldehyde is a colorless, volatile liquid with an irritating odor. It is a versatile starting material and is used for the manufacture of many compounds. When acetaldehyde is heated with an acid catalyst, it polymerizes to a liquid called *paraldehyde:*

Acetaldehyde   Paraldehyde

Paraldehyde was once used as a sedative and hypnotic. Its use has diminished because of its unpleasant odor and the discovery of more effective substitutes.

The most important industrial ketone is acetone, a colorless, volatile liquid that boils at 56 °C. It is used as a solvent for resins, plastics, and varnishes. Moreover, it is miscible with water in all proportions. Acetone is produced in the body as a by-product of fat metabolism. Its concentration is usually less than 1 mg/100 mL of blood. However, in uncontrolled diabetes mellitus, acetone is produced in larger quantities, causing its level in the body to increase dramatically. It is expelled by the body as a waste product in urine and in severe cases may even be detected in the breath.

Methyl ethyl ketone is used industrially to remove wax from lubricating oils when they are refined. It is also a common solvent in nail polish remover.

SUMMARY

Aldehydes and ketones contain the carbonyl group (—C=O). The carbonyl carbon in an aldehyde has at least one hydrogen attached (R—CHO), but the carbonyl carbon in a ketone has no hydrogens (R—CO—R). Formaldehyde (H_2CO) is the simplest aldehyde; acetone (CH_3COCH_3) is the simplest ketone. The physical and chemical properties of aldehydes and ketones are influenced by the very polar carbonyl group. Molecules of aldehydes and ketones can attract each other through polar-polar interactions. These compounds have higher boiling points than the corresponding alkanes but lower boiling points than the corresponding alcohols. Aldehydes and ketones can accept hydrogen bonds, and those with low molar mass are completely soluble in water.

Aldehydes and ketones are produced by the oxidation of primary alcohols and secondary alcohols, respectively. Aldehydes are usually more reactive than ketones and are good reducing agents. An aldehyde can be oxidized to the corresponding carboxylic acid, but ketones resist further oxidation. Addition reactions are characteristic of both aldehydes and ketones. Addition of water to the carbon-oxygen bond of the carbonyl group forms hydrates. Addition of an alcohol produces hemiacetals and hemiketals. The reaction of alcohols with hemiketals and hemiacetals produces acetals and ketals, respectively.

The most important industrial aldehydes and ketones are formaldehyde, acetaldehyde, acetone, and methyl ethyl ketone. A 40% aqueous solution of formaldehyde, called formalin, is commonly used to preserve biological specimens. Many aldehydes and ketones have fragrant aromas.

SUMMARY OF REACTIONS

Here are the reactions of aldehydes and ketones presented in this chapter.

Aldehydes

1. Preparation of aldehydes:

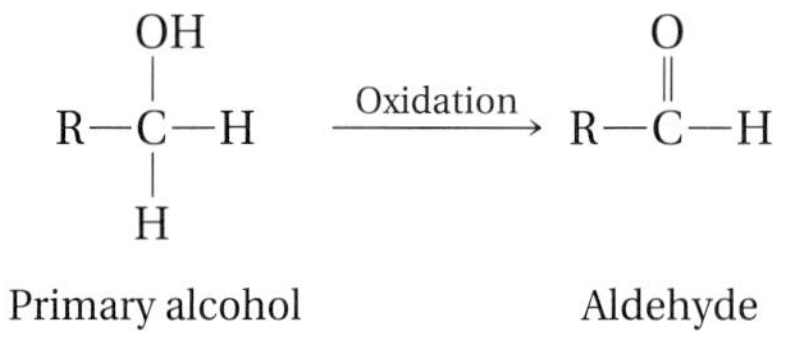

$$R-\underset{\underset{H}{|}}{\overset{\overset{OH}{|}}{C}}-H \xrightarrow{\text{Oxidation}} R-\overset{\overset{O}{\|}}{C}-H$$

Primary alcohol   Aldehyde

2. Oxidation of aldehydes:

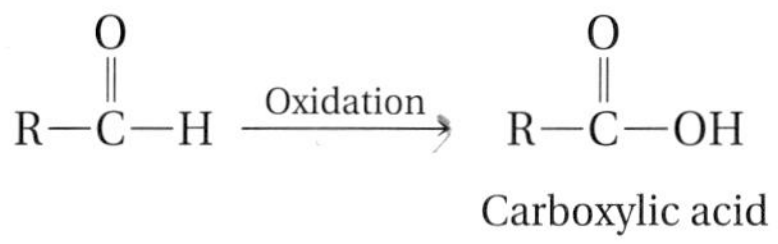

$$R-\overset{\overset{O}{\|}}{C}-H \xrightarrow{\text{Oxidation}} R-\overset{\overset{O}{\|}}{C}-OH$$

Carboxylic acid

3. Reaction of aldehydes with Tollens' reagent:

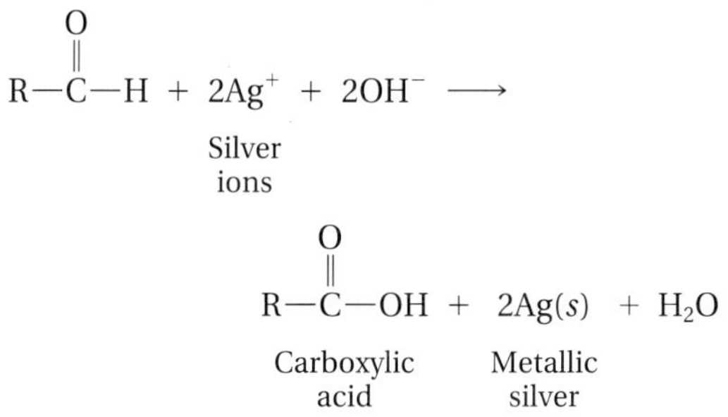

$$R-\overset{\overset{O}{\|}}{C}-H + 2Ag^+ + 2OH^- \longrightarrow$$

Silver ions

$$R-\overset{\overset{O}{\|}}{C}-OH + 2Ag(s) + H_2O$$

Carboxylic acid   Metallic silver

4. Reaction of aldehydes with Benedict's reagent:

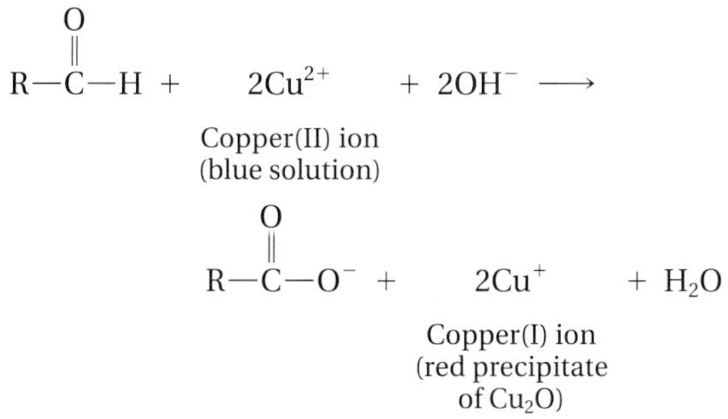

$$R-\overset{\overset{O}{\|}}{C}-H + \quad 2Cu^{2+} \quad + 2OH^- \longrightarrow$$

Copper(II) ion (blue solution)

$$R-\overset{\overset{O}{\|}}{C}-O^- + \quad 2Cu^+ \quad + H_2O$$

Copper(I) ion (red precipitate of Cu_2O)

5. Addition of water to aldehydes:

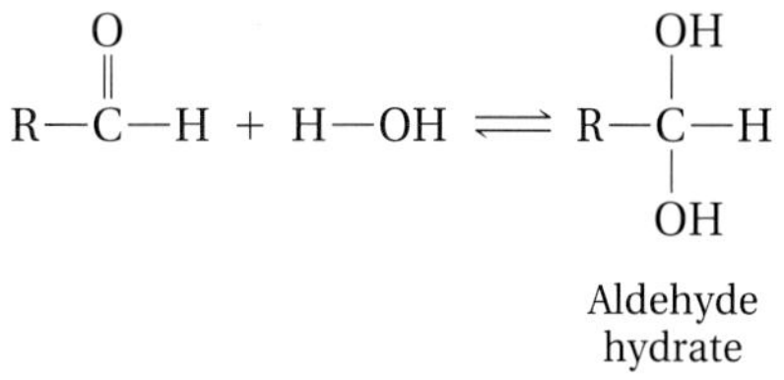

$$R-\overset{\overset{O}{\|}}{C}-H + H-OH \rightleftharpoons R-\underset{\underset{OH}{|}}{\overset{\overset{OH}{|}}{C}}-H$$

Aldehyde hydrate

6. Addition of alcohol to an aldehyde followed by a second reaction with an alcohol:

$$R-\overset{\overset{O}{\|}}{C}-H \overset{RO-H}{\rightleftharpoons} R-\underset{\underset{OR}{|}}{\overset{\overset{OH}{|}}{C}}-H \xrightarrow{RO-H}$$

Hemiacetal

$$R-\underset{\underset{OR}{|}}{\overset{\overset{OR}{|}}{C}}-H + H_2O$$

Acetal

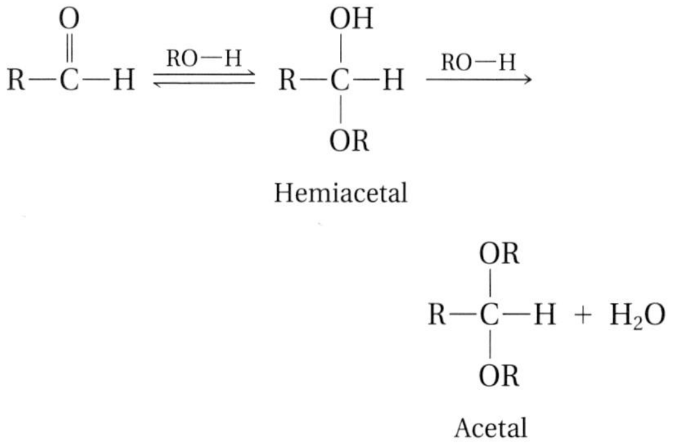

Ketones

1. Preparation of ketones:

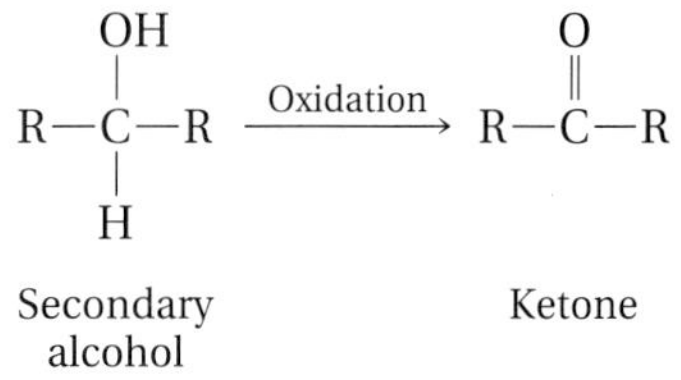

$$R-\underset{\underset{H}{|}}{\overset{\overset{OH}{|}}{C}}-R \xrightarrow{\text{Oxidation}} R-\overset{\overset{O}{\|}}{C}-R$$

Secondary alcohol   Ketone

2. Addition of alcohol to a ketone followed by a second reaction with an alcohol:

$$R-\overset{\overset{O}{\|}}{C}-R \overset{RO-H}{\rightleftharpoons} R-\underset{\underset{OR}{|}}{\overset{\overset{OH}{|}}{C}}-R \xrightarrow{RO-H} R-\underset{\underset{OR}{|}}{\overset{\overset{OR}{|}}{C}}-R + H_2O$$

Hemiketal   Ketal

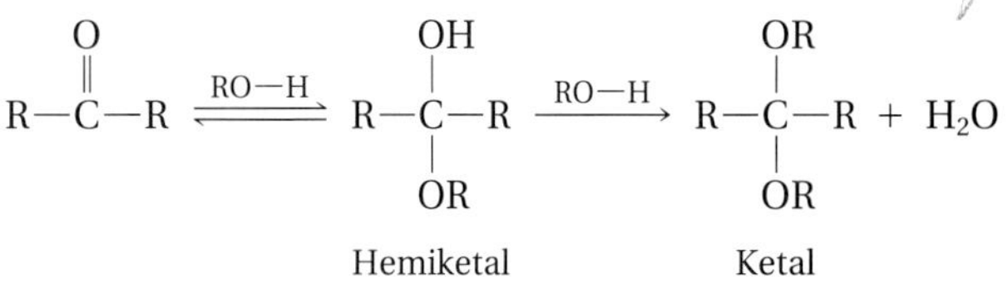

KEY TERMS

Acetal (4.6)
Aldehyde (4.1)
Alpha-hydroxy ketone (4.5)

Benedict's reagent (4.5)
Carbonyl group (4.1)
Dehydrogenation reaction (4.3)

Fehling's reagent (4.5)
Hemiacetal (4.6)
Hemiketal (4.6)
Hydrate (4.6)

Ketal (4.6)
Ketone (4.1)
Tollens' reagent (4.5)

EXERCISES

The Carbonyl Group (Sections 4.1, 4.2)

4.11 What is the carbonyl group? Explain why a carbon-carbon double bond is nonpolar, but a carbon-oxygen double bond is very polar.

4.12 Aldehydes and ketones are carbonyl compounds. Draw their general formulas.

4.13 Name the following aldehydes and ketones.
(a) CH_3CHO *ethanal a acetaldehyde*
(b)

 —CHO *benzaldehyde*

(c) CH_3
   |
CH_3CHCH_2CHO *3-methyl butanal*

(d)

 *diphenyl methanone*

(e) CH_3 O
   | ||
$CH_3CHCH_2CCH_2CH_3$ *5-methyl-3-hexanone*

4.14 Name the following aldehydes and ketones.
(a) O   (b)
    ||
 CH_3CCH_3

(c)   O   (d)
   ||
$CH_3CH_2CCH_2CH_2CH_3$ —CH_2CHO *2-phenylethanal*

(e) CH_3 Cl
    | |
$CH_3CHCH_2CHCH_2CHO$ *3-chloro-5-methylhexanal*

4.15 Propane ($CH_3CH_2CH_3$) and acetaldehyde (CH_3CHO) have the same molar mass, but propane boils at −42 °C, whereas acetaldehyde boils at 20 °C. Why are the boiling points so different? *H bonding & dipole-dipole*

4.16 Explain why aldehydes and ketones cannot form intermolecular hydrogen bonds but can form hydrogen bonds with water molecules.

Redox Reactions of Organic Compounds (Section 4.3)

4.17 Classify each reaction as an oxidation or a reduction.
(a) ethene ⟶ ethyne *O −2H*
(b) decanol ⟶ decanal *O −2H*
(c) cyclopentanone ⟶ cyclopentane *R loss gain of H*
(d) 3-hexanol ⟶ 3-hexanone *O loss of H*

4.18 Justify each classification in Exercise 4.17 in terms of loss or gain of oxygen and/or hydrogen.

Preparation of Aldehydes and Ketones (Section 4.4)

4.19 Give the names and structures of the expected oxidation products from
(a) 1-propanol   (b)   OH
                    |
             $CH_3CH_2CCH_2CH_3$ *NR*
                   |
                 CH_3

(c)   OH   (d) cyclohexanol
      |
$CH_3CH_2CHCH_2CH_3$

4.20 Write the structure of the product of the oxidation of each compound. Name each product.
(a) CH_3 (b) $CH_3(CH_2)_6CH_2OH$
     |
CH_3CHOH

(c) 2-methyl-1-butanol   (d) acetaldehyde

4.21 Write the name and structure for the aldehyde or ketone that must be reduced to make each of the following alcohols.
(a) methanol   (b)   OH
                   |
               CH_3CHCH_3

(c) CH_3
    |
CH_3CHCH_2OH

4.22 Write the name and structure for the aldehyde or ketone that must be reduced to make each of the following alcohols.
(a) 2-methylcyclopentanol   (b) CH_3CH_2OH
(c) 2-hexanol

Detection of Aldehydes (Section 4.5)

4.23 Which of the oxidation products in Exercise 4.20 will give a positive Benedict's test?

4.24 Which of the carbonyl compounds that were reduced in Exercise 4.22 will give a positive Tollens' test?

Addition Reactions (Section 4.6)

4.25 Write the structure for the hemiacetal that is formed when ethanol is added to propanal.

4.26 Write the structure for the acetal that is formed when an excess of methanol is added to ethanal.

4.27 Which of the following structures are acetals, hemiacetals, or neither?

(a) 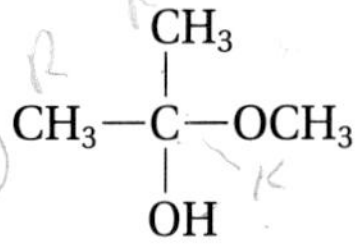
$$CH_3-\underset{\underset{OH}{|}}{\overset{\overset{CH_3}{|}}{C}}-OCH_3$$

(b)
$$CH_3-\underset{\underset{OH}{|}}{\overset{\overset{OH}{|}}{C}}-CH_3$$

(c)
$$CH_3CH_2\underset{\underset{}{|}}{\overset{\overset{OCH_2CH_3}{|}}{C}HOCH_2CH_3}$$

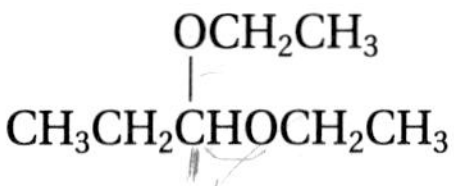

(d) 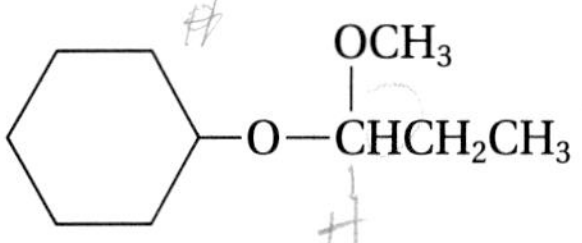
a cyclohexane ring with $-O-\underset{\underset{}{}}{\overset{\overset{OCH_3}{|}}{C}HCH_2CH_3}$

4.28. Which of the following are acetals, hemiacetals, or neither?

(a) 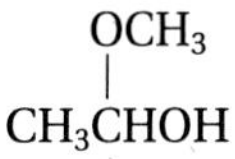
$$CH_3\underset{\underset{H}{|}}{\overset{\overset{OCH_3}{|}}{C}HOH}$$

(b)
a benzene ring with $\underset{\underset{H}{|}}{\overset{\overset{OCH_3}{|}}{C}-OCH_3}$

(c)
$$CH_3CH_2-\underset{\underset{H}{|}}{\overset{\overset{OCH_3}{|}}{C}HOH}$$

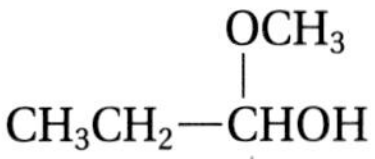

(d) 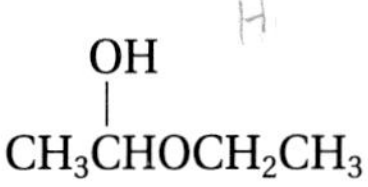
$$CH_3\underset{\underset{}{}}{\overset{\overset{OH}{|}}{C}HOCH_2CH_3}$$

Additional Exercises

4.29 Give the product of each of the following reactions. If there is none, write "no reaction."

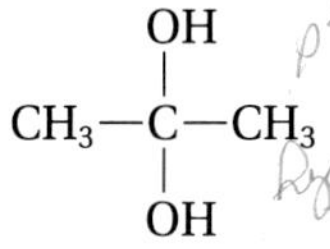

(a) 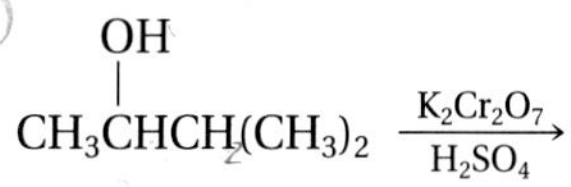
$$CH_3\underset{\underset{}{}}{\overset{\overset{OH}{|}}{C}HCH(CH_3)_2} \xrightarrow[H_2SO_4]{K_2Cr_2O_7}$$

(b)
$$CH_3CH_2\underset{\underset{}{}}{\overset{\overset{CH_3}{|}}{C}HCH_2OH} \xrightarrow[H_2SO_4]{K_2Cr_2O_7}$$

(c)
$$CH_3CH_2\underset{\underset{CH_3}{|}}{\overset{\overset{OH}{|}}{C}CH_2CH_3} \xrightarrow[H_2SO_4]{K_2Cr_2O_7}$$

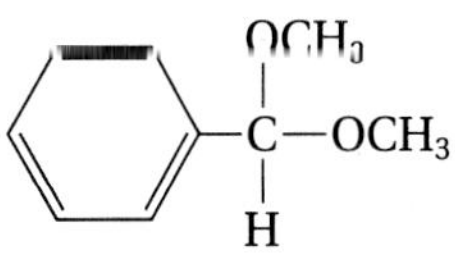

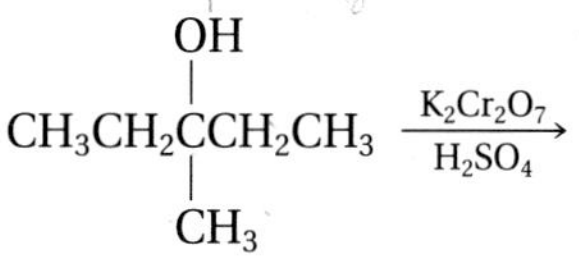

(d) 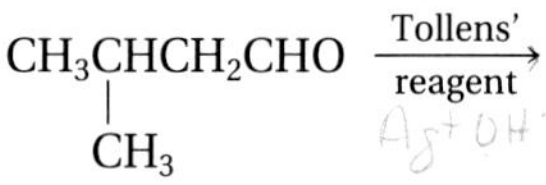
$$CH_3\underset{\underset{CH_3}{|}}{\overset{}{C}HCH_2CHO} \xrightarrow[\text{reagent}]{\text{Tollens'}}$$

(e) 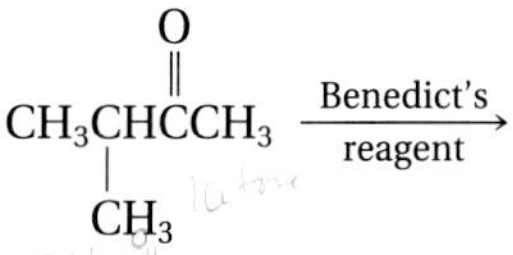
$$CH_3\underset{\underset{CH_3}{|}}{\overset{\overset{O}{\|}}{C}HCCH_3} \xrightarrow[\text{reagent}]{\text{Benedict's}}$$

(f) $CH_3CHO + H_2 \xrightarrow[\text{Pressure}]{Ni + Heat}$

(g) $CH_3CHO + CH_3OH \longrightarrow$

(h) 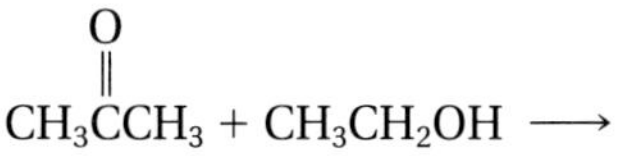
$$CH_3\overset{\overset{O}{\|}}{C}CH_3 + CH_3CH_2OH \longrightarrow$$

4.30 Give the major organic product of each of the following reactions.

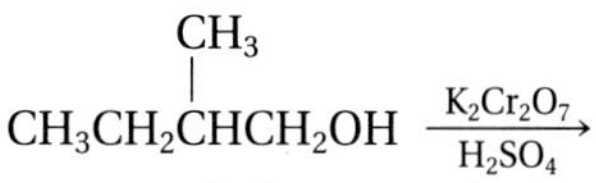

(a)
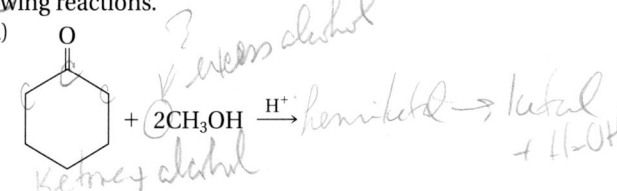
$+ 2CH_3OH \xrightarrow{H^+}$

(b) 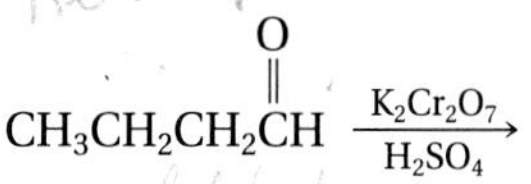
$$CH_3CH_2CH_2\overset{\overset{O}{\|}}{C}H \xrightarrow[H_2SO_4]{K_2Cr_2O_7}$$

(c)
a cyclohexane ring with $\overset{\overset{O}{\|}}{C}H \xrightarrow[\text{reagent}]{\text{Benedict's}}$

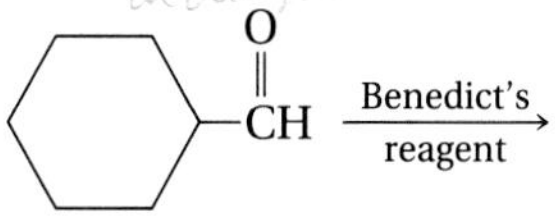

(d) 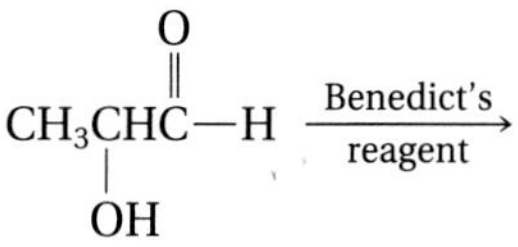
$$CH_3\underset{\underset{OH}{|}}{\overset{\overset{O}{\|}}{C}HC}-H \xrightarrow[\text{reagent}]{\text{Benedict's}}$$

4.31 Identify the following structures as acetals, hemiacetals, ketals, or hemiketals.

(a)
$$CH_3-\underset{\underset{OCH_2CH_3}{|}}{\overset{\overset{CH_3}{|}}{C}}-OCH_3$$
ketal *ether* *ether*

(b)
$$CH_3-\underset{\underset{OCH_3}{|}}{CH}-OH$$
hemiacetal

(c)
2 ether linkages *acetal*

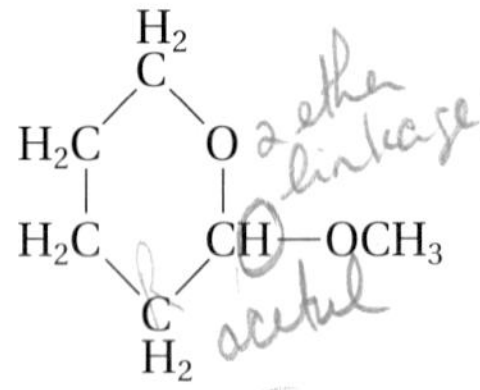

(d)
hemiketal

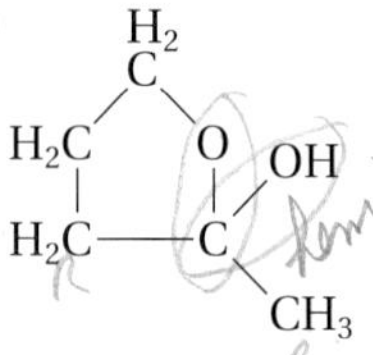

(e) $CH_3CH_2-\underset{\underset{OCH_2CH_3}{|}}{CH}-OCH_3$
2 ether link + H *hemiacetal*

(f)
ether *acetal*

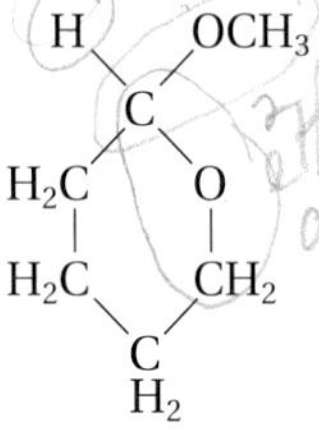

4.32 Name the following compounds by the IUPAC system.
(a) $CH_3CH_2CH_2CHO$
butanal

(b) $CH_3CH_2CCH_3$ with $\overset{\parallel}{O}$
2-butanone

(c) $ClCH_2CH_2CHO$
3-chloro propanal

(d) $CH_3CHBrCCH_2CH_3$ with $\overset{\parallel}{O}$
2-bromo 3-pentanone

(e) $FCH_2CHCH_2CCH_2CH_2CH_3$ with $\underset{OH}{|}$ and $\overset{\parallel}{O}$
4 heptanone

(f) $CH_3CCH_2CBr_3$ with $\overset{\parallel}{O}$
4-tribromo-2-butanone

4.33 The boiling points of ethyl methyl ether and propanone are 11 and 56 °C, respectively, but their molecular weights are very similar. Explain.

4.34 Which pairs of the following molecules form hydrogen bonds? Draw structures to show the hydrogen bond, if formed.
(a) acetone and diethyl ether *no*
(b) acetone and acetone *no*
(c) acetone and water *yes*
(d) acetone and acetaldehyde *no*
(e) acetaldehyde and water *yes*

4.35 You are given a compound and told that it is either $CH_3CH_2COCH_3$ or $CH_3CH_2CH_2CHO$. What laboratory test(s) would you perform to identify the compound?

4.36 Draw structural formulas for the following.
(a) 3-hexanone (b) butyraldehyde (c) propanal
(d) diisobutyl ketone (e) methanal
(f) methyl phenyl ketone

4.37 Name the following structures. (Where appropriate, give both the common name and the IUPAC name.)
(a) HCHO *methanal formaldehyde*
(b) CH_3CCH_3 with $\overset{\parallel}{O}$ *acetone propanone*
(c) $CH_3CH_2CH_2CHO$ *butanal butyr aldeh.*
(d) (benzene ring)—CHO *phenyl methanal benzaldehyde*
(e) (cyclohexanone) *cyclohexanone*
(f) (benzene ring)—C(=O)—CH₃ *phenyl ethanone*
(g) $CH_3CH_2-C-CH_2CH_3$ with $\overset{\parallel}{O}$

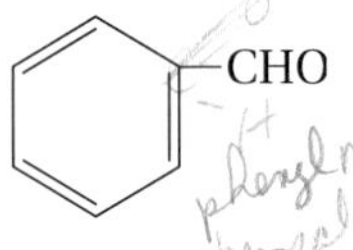

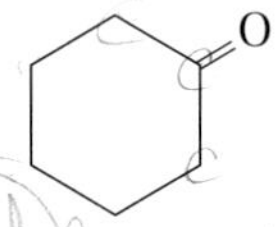

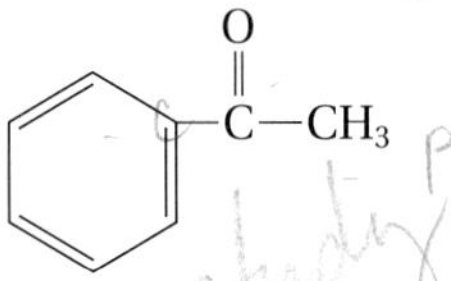

4.38 Write equations for the following reactions. Identify the class of compound(s) formed.
(a) the oxidation of 2-propanol
(b) the addition of water to acetone
(c) the reduction of formaldehyde
(d) the stepwise addition of two moles of ethanol to one mole of acetaldehyde

4.39 Which member of each of these pairs of compounds would release more energy upon complete oxidation?
(a) butanol, butanal
(b) ethene, ethane
(c) carbon dioxide, formaldehyde
(d) isopropyl alcohol, acetone

most reduced releases more energy upon oxidation.

SELF-TEST (REVIEW)

True/False

1. A dehydrogenation reaction is a reduction reaction.
2. Hydrogen bonding accounts for the relatively high boiling point of acetaldehyde.
3. The reaction of equal moles of an alcohol and an aldehyde gives an acetal.
4. Oxidation of a tertiary alcohol gives a ketone.
5. You would expect propanal to have a higher boiling point than propanol.
6. Propanal should have higher water solubility than hexanal.
7. One mole of methanol would release more energy upon complete oxidation than one mole of methane.
8. Both aldehydes and ketones have a carbonyl group.
9. All aldehydes and ketones give a positive Tollens' test.
10. Four pairs of electrons are shared in the carbonyl bond formed between an oxygen and a carbon atom.

Multiple Choice

11. Which of the following statements about the carbon-oxygen double bond of the carbonyl group is *false*?
    (a) The bond is polar.
    (b) The carbon has a partial negative charge.
    (c) The bonding electrons are unequally shared between the carbon and oxygen.
    (d) The oxygen has two unshared pairs of electrons.
12. Acetaldehyde would be likely to form hydrogen bonds with
    (a) formaldehyde.   (b) octane.   (c) water.
    (d) acetone.
13. On the basis of your knowledge of intermolecular forces, which of the following would you expect to have the highest boiling point?
    (a) propanal   (b) propane   (c) acetone
    (d) 1-propanol

14. Which of the following compounds contains a di-ether linkage?
    (a) a hemiacetal   (b) chloral hydrate
    (c) a ketal   (d) camphor
15. Which of the following substances is used as a preservative of biological specimens?
    (a) paraldehyde   (b) formalin
    (c) methyl ethyl ketone   (d) cinnamaldehyde
16. Which of the following substances can undergo an addition reaction with methanol?
    (a) propane   (b) methyl ethyl ether
    (c) propanol   (d) propanal
17. A structural isomer of 2-butanone is
    (a) diethyl ether.   (b) *tert*-butyl alcohol.
    (c) diethyl ketone.   (d) butanal.
18. Which of the following compounds would release the most energy upon oxidation to carbon dioxide?
    (a) ethanol   (b) acetic acid, CH_3COOH
    (c) ethane   (d) acetaldehyde
19. The oxidation of 2-methyl-2-butanol with $K_2Cr_2O_7$ and H_2SO_4 would give
    (a) 2-methyl-2-butanone.
    (b) isopropyl alcohol and ethane.
    (c) 2-methyl-2-butanal.
    (d) none of the above.
20. In a positive Tollens' test,
    (a) silver ions are oxidized to silver atoms.
    (b) the aldehyde is an oxidizing agent.
    (c) a silver mirror is formed.
    (d) more than one are correct.
21. In the reaction of substance *A* with substance *B*, substance *A* loses oxygen. Which of the following is true?
    (a) Substance *B* is an oxidizing agent.
    (b) Substance *A* is reduced.
    (c) Substance *B* is reduced.
    (d) Substance *A* is a reducing agent.

5

Acids and Their Derivatives

Reactions of the Carboxyl Group

Citric acid, a carboxylic acid, gives
oranges and other citrus fruits
their sour taste and low pH.

CHAPTER OUTLINE

What causes the sting of ant venom, the sour taste of unripe fruit, and the tart taste of vinegar? All are caused by carboxylic acids (RCO_2H), which along with their derivatives are the major topic of this chapter. We also will examine some organic derivatives of inorganic acids that are important to biology and health. One carboxylic acid that has proven to be valuable in the treatment of *acne vulgaris*, or common acne, is the subject of the following Case in Point.

CASE IN POINT: Acne and tretinoin

Lana passed her first 12 years of life without a blemish on her skin. When she was 13 years old, Lana noticed her first pimple. Soon her face was covered with red blotches, some of which formed white caps of pus. She also began to find blackheads around her nose and on her cheeks. Although most adolescents experience common acne, Lana's acne seemed worse than most. When Lana developed a large boil on her nose, her parents decided to seek medical advice. The doctor agreed that Lana's acne was worse than usual and gave Lana a cream containing a carboxylic acid called *tretinoin* to apply to her face (see figure). Within a few weeks, Lana's skin had cleared up considerably. What is the chemical structure of tretinoin? How does this carboxylic acid work to relieve acne? We will learn about this remarkable compound in Section 5.1.

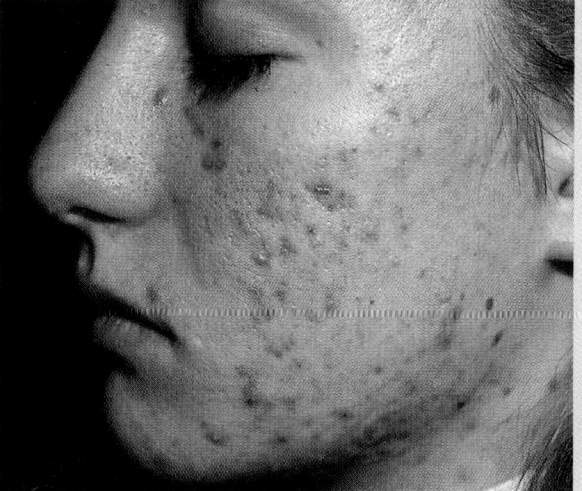

Preparations containing tretinoin are effective treatments for acne.

5.1 Carboxylic acids

AIMS: To name and draw structures of common carboxylic acids and organic acid salts. To distinguish among dicarboxylic acids, polyfunctional carboxylic acids, and fatty acids.

Carboxylic acids *are a class of organic compounds characterized by the* **carboxyl group,** *whose name comes from the words* carbonyl *and* hydroxyl, *its two functional groups.*

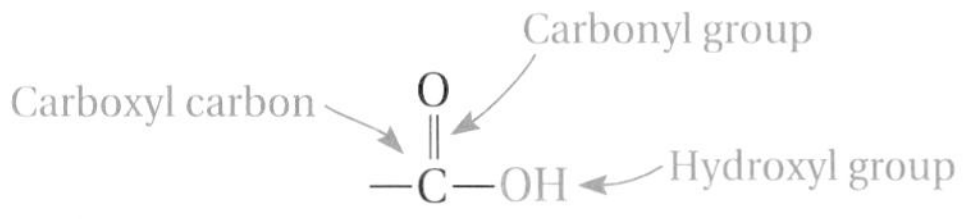

Carbonyl group

Carboxyl carbon

$$O$$
$$\parallel$$
$$-C-OH$$

Hydroxyl group

Carboxyl group
(also written —CO_2H or —COOH)

The general formula of a carboxylic acid is RCOOH. Carboxylic acids are weak acids because they dissociate to a slight extent in aqueous solution to give a carboxylate ion and a hydrogen ion:

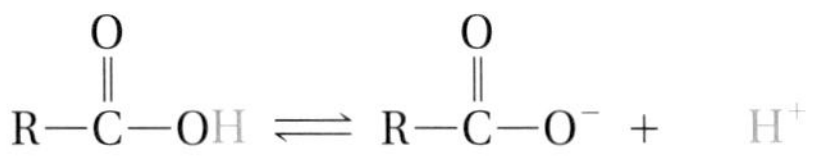

$$
\underset{}{R-\overset{\overset{\textstyle O}{\parallel}}{C}-OH} \rightleftharpoons \underset{\text{Carboxylate ion}}{R-\overset{\overset{\textstyle O}{\parallel}}{C}-O^-} + \underset{\text{Hydrogen ion}}{H^+}
$$

Naming carboxylic acids

In the IUPAC system for naming carboxylic acids, the -*e* ending of the parent alkane is replaced by the ending -*oic acid.* The parent alkane is the hydrocarbon with the longest continuous carbon chain containing the carboxyl group. The first four straight-chain aliphatic acids (common names in parenthesis) are

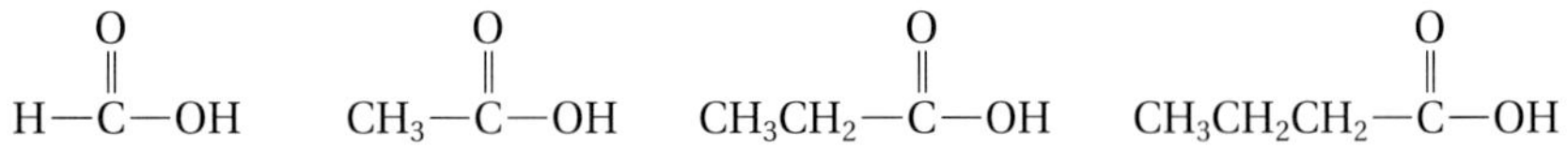

$$
\underset{\substack{\text{Methanoic acid}\\\text{(formic acid)}}}{H-\overset{\overset{\textstyle O}{\parallel}}{C}-OH} \qquad \underset{\substack{\text{Ethanoic acid}\\\text{(acetic acid)}}}{CH_3-\overset{\overset{\textstyle O}{\parallel}}{C}-OH} \qquad \underset{\substack{\text{Propanoic acid}\\\text{(propionic acid)}}}{CH_3CH_2-\overset{\overset{\textstyle O}{\parallel}}{C}-OH} \qquad \underset{\substack{\text{Butanoic acid}\\\text{(butyric acid)}}}{CH_3CH_2CH_2-\overset{\overset{\textstyle O}{\parallel}}{C}-OH}
$$

In naming carboxylic acids by the IUPAC system, we number the carboxyl carbon as carbon 1. Each substituent on the chain is identified by its name and a number indicating its position on the chain. Here are the structures and IUPAC names of two substituted carboxylic acids:

$$
\overset{3}{C}l-\overset{}{C}H_2-\overset{2}{C}H_2-\overset{1}{C}OOH \qquad\qquad \overset{4}{C}H_3-\overset{3}{C}H-\overset{2}{C}H-\overset{1}{C}OOH
$$

$$
\underset{}{}\overset{}{\underset{OH\ \ Br}{|\ \ \ \ |}}
$$

3-Chloropropanoic acid   2-Bromo-3-hydroxybutanoic acid

The simplest aromatic acid is benzoic acid, a name derived from the parent hydrocarbon benzene:

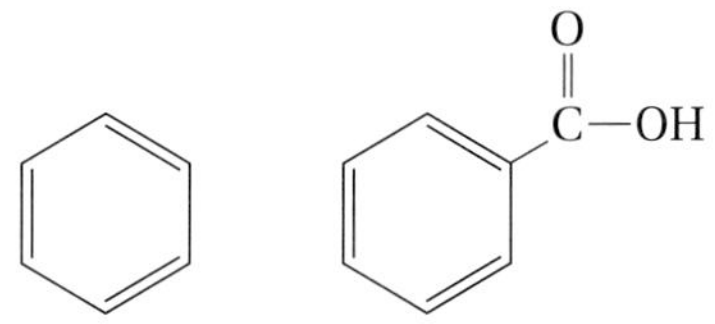

Benzene   Benzoic acid

EXAMPLE 5.1   **Naming a carboxylic acid**

Give the IUPAC name for the following carboxylic acid.

$$\underset{5\quad4\quad3\quad2\quad1}{HOCH_2CH_2CH_2CHCO_2H} \overset{Br}{|}$$

2-bromo-5-hydroxypentanoic acid

SOLUTION

We number the longest carbon chain so that the carboxyl group is in position 1.

$$\underset{5\quad4\quad3\quad2\quad1}{HOCH_2CH_2CH_2CHCO_2H} \overset{Br}{|}$$

The longest carbon chain contains five carbons, making *pentane* the parent alkane. We remove the *-e* ending from the word *pentane* and add *-oic acid* to the root name *pentan-*. Our compound is a *pentanoic acid*. Now we name and number the substituents in the carbon chain; the name for a —Br substituent is *bromo,* and the name for an —OH substituent is *hydroxy.* The substituents in our compound are *2-bromo* and *5-hydroxy.* We combine the substituent names and numbers in alphabetical order with the parent acid name to obtain the complete name of the carboxylic acid: *2-bromo-5-hydroxypentanoic acid.*

Concentrated acetic acid is known as *glacial* acetic acid because it freezes to an icelike solid in cold stockrooms or laboratories that are below 17 °C.

Carboxylic acids are abundant and widely distributed in nature. Many have common names derived from a Greek or Latin word describing their natural sources. For example, methanoic acid is usually called by its common name, *formic acid.* Formic acid is produced by ants. The common name for ethanoic acid is *acetic acid.* Acetic acid is produced when wine turns sour and becomes vinegar. The pungent aroma of vinegar comes from its acetic acid. Common names for many carboxylic acids are used more often than IUPAC names.

Fatty acids

The straight-chain carboxylic acids were first isolated from fats and are called **fatty acids.** Propionic acid, the three-carbon acid, literally means

Lactobacillic acid (shown below), a fatty acid from a bacterial source, contains the highly strained cyclopropyl ring.

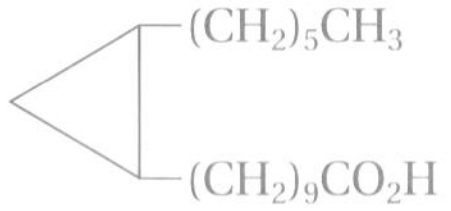

$-(CH_2)_5CH_3$

$-(CH_2)_9CO_2H$

"first fatty acid." The four-carbon acid, butyric acid, can be obtained from butterfat.

The low-molar-mass members of the aliphatic carboxylic acid series are colorless, volatile liquids. They have sharp odors. The smells of rancid butter and dirty feet are due in part to butyric acid. The smell of goats is caused by the 6-, 8-, and 10-carbon (C-6, C-8, C-10) straight-chain acids. The longer members of the series are nonvolatile, waxy solids with low melting points. Stearic acid (C-18) is obtained from beef fat. It is used to make wax candles. Stearic acid and other long-chain fatty acids have very little odor. Table 5.1 lists the names and formulas for some common saturated aliphatic carboxylic acids.

Many fatty acids contain carbon-carbon double bonds and are referred to as *unsaturated fatty acids.* Table 5.2 lists some names and formulas for some unsaturated aliphatic carboxylic acids. Recall from the Case in Point in the introduction to this chapter that Lana was afflicted with chronic acne. In our follow-up we will see how she was helped by a rather amazing unsaturated carboxylic acid.

Dicarboxylic acids

Dicarboxylic acids *are compounds that have two carboxyl groups.* Here are the first five members of the series; if you want to memorize their names, the expression "**O**h **M**y, **S**uch **G**ood **A**pples" is helpful.

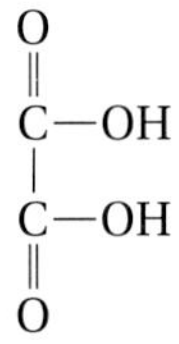

| Oxalic acid | 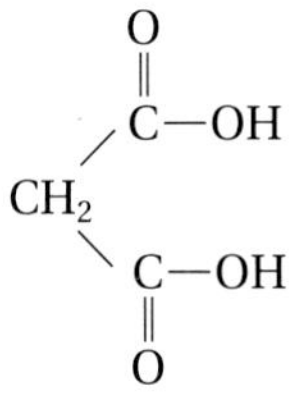 Malonic acid | 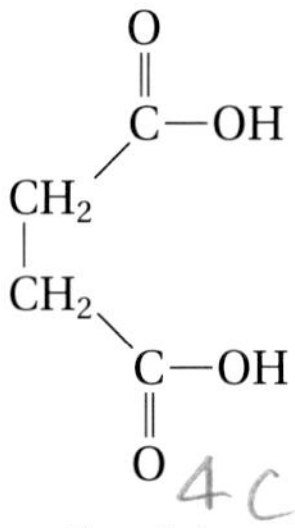 Succinic acid | 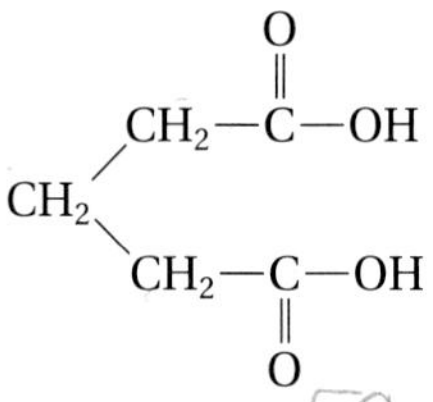 Glutaric acid | 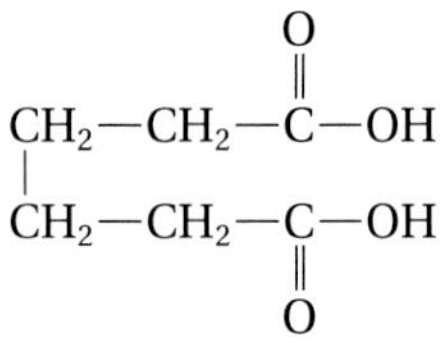 Adipic acid |

Table 5.1 Saturated Aliphatic Carboxylic Acids

Formula	Carbon atoms	Common name	Melting point (°C)	Derivation of name
HCOOH	1	formic acid	8	Lat. *formica,* ant
CH₃COOH	2	acetic acid	17	Lat. *acetum,* vinegar
CH₃CH₂COOH	3	propionic acid	−22	Gk. *protos,* first; *pion,* fat
CH₃(CH₂)₂COOH	4	butyric acid	−6	Lat. *butyrum,* butter
CH₃(CH₂)₄COOH	6	caproic acid	−3	Lat. *caper,* goat
CH₃(CH₂)₆COOH	8	caprylic acid	16	Lat. *caper,* goat
CH₃(CH₂)₈COOH	10	capric acid	31	Lat. *caper,* goat
CH₃(CH₂)₁₀COOH	12	lauric acid	44	laurel seed oil
CH₃(CH₂)₁₂COOH	14	myristic acid	58	nutmeg (*Myristica fragrans*)
CH₃(CH₂)₁₄COOH	16	palmitic acid	63	palm oil
CH₃(CH₂)₁₆COOH	18	stearic acid	70	Gk. *stear,* tallow

FOLLOW-UP TO THE CASE IN POINT: Acne and tretinoin

Lana was suffering from the pimples and pustules of common acne, which have long been an embarrassment to millions of adolescents. Many adults are also afflicted by this skin condition, which may affect the shoulders and back as well as the face. Severe acne can be a threat to health, since the lesions can become infected. Even after acne has subsided, it can leave its mark in the form of disfiguring pockmarks. Acne occurs when dead cells and natural oils plug the hair follicles and ducts of oil glands in the skin. The resulting irritation causes pimples that may become infected. The surface of trapped oils may be oxidized by contact with air, forming dark spots known as *blackheads*. People with acne are often advised to keep their skin clean, not to pick at pimples or squeeze blackheads, and to avoid certain foods such as chocolate.

People such as Lana, with severe acne, may require medical intervention. Lana and many other acne sufferers have achieved good results with tretinoin (see figure), a carboxylic acid derived from vitamin A (see A Closer Look: The Chemistry of Vision, on pages 124–125).

Tretinoin
(Retin-A, retinoic acid, vitamin A acid)

Tretinoin is sold in a cream formulation as Retin-A. This formulation is applied topically to the affected area. Two other drugs related to tretinoin, Accutane and Tegison, are administered orally. Tretinoin appears to stimulate the reproduction of the cells of the hair follicles. This activity promotes the extrusion of dead cells and oils that could plug the follicle. The skin of some patients being treated with tretinoin may become very sensitive to sunlight. The use of sunscreens and avoidance of sunlight during treatment usually are recommended. Tretinoin also presents an extremely high risk to a developing fetus. Tretinoin products require a prescription from a physician for these reasons. Since its introduction as an acne medication, tretinoin has been advanced as a kind of miracle drug for the skin. Some studies indicate that tretinoin can reverse sun damage, fade freckles and age spots, reduce the number of fine wrinkles in the skin, and even promote hair growth.

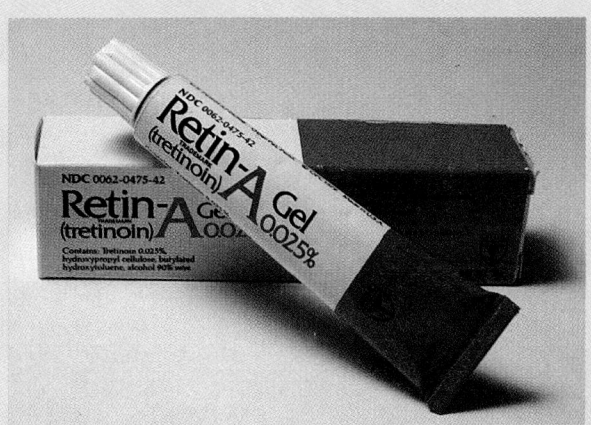

The discomfort and embarrassment of acne often can be relieved with application of tretinoin, a carboxylic acid.

Table 5.2 Unsaturated Aliphatic Carboxylic Acids

Formula	Carbon atoms	Common name	Melting point (°C)
$CH_3(CH_2)_5CH=CH(CH_2)_7COOH$	16	palmitoleic acid	−1
$CH_3(CH_2)_7CH=CH(CH_2)_7COOH$	18	oleic acid	14
$CH_3(CH_2)_4CH=CHCH_2CH=CH(CH_2)_7COOH$	18	linoleic acid	−5
$CH_3CH_2(CH=CHCH_2)_3(CH_2)_6COOH$	18	linolenic acid	−11

The unsaturated dicarboxylic acids maleic acid and fumaric acid are geometric isomers; phthalic acid is an aromatic dicarboxylic acid.

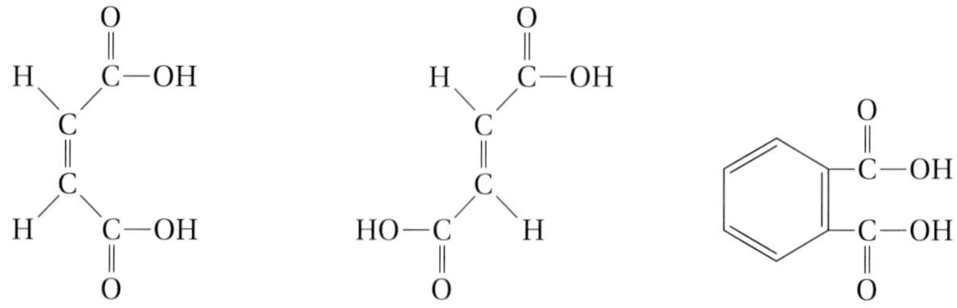

Maleic acid (*cis* isomer; mp 130.5 °C) — Fumaric acid (*trans* isomer; mp 302 °C) — Phthalic acid (mp 210 °C)

Polyfunctional carboxylic acids

Polyfunctional carboxylic acids are carboxylic acids that contain other functional groups as well. Many important acids are polyfunctional. Lactic acid is a **hydroxy acid**—*that is, it contains a hydroxyl and a carboxy group:*

$$CH_3-CH(OH)-COOH$$

Lactic acid

When lactose (milk sugar) is fermented by the bacterium *Lactobacillus,* lactic acid is the only product. It is responsible for the taste of sour milk. Other hydroxy acids are citric acid, tartaric acid, and salicylic acid.

$$HO-C(CH_2-COOH)_2-COOH$$

Citric acid (citrus juice)

$$HO-CH(COOH)-CH(COOH)-OH$$

Tartaric acid (grape juice)

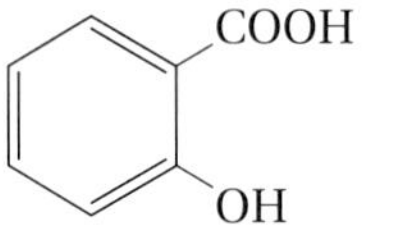

Salicylic acid (willow bark)

Here are some examples of carboxylic acids with functional groups other than hydroxyl:

$$CH_3-C(=O)-COOH$$

Pyruvic acid (a keto acid important in the energy production of living organisms)

$$Cl-CH_2-COOH$$

Chloroacetic acid (a halo acid)

$$NH_2-CH_2-COOH$$

Glycine (an amino acid found in proteins)

EXAMPLE 5.2 **Writing the structure of a carboxylic acid**

Write the structure of 4,4-dibromohexanoic acid.

SOLUTION

Hexanoic acid is the six-carbon straight-chain acid with the carboxyl

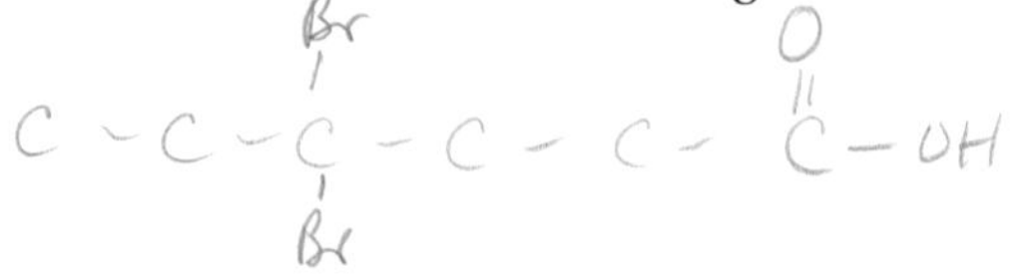

carbon as carbon-1:

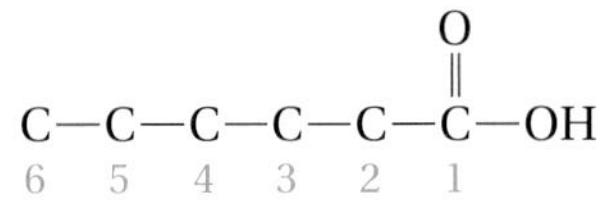

$$\underset{6}{C}-\underset{5}{C}-\underset{4}{C}-\underset{3}{C}-\underset{2}{C}-\overset{\displaystyle \overset{O}{\|}}{\underset{1}{C}}-OH$$

Add two bromine atoms at carbon-4. Then add sufficient hydrogen atoms for each carbon to have four bonds:

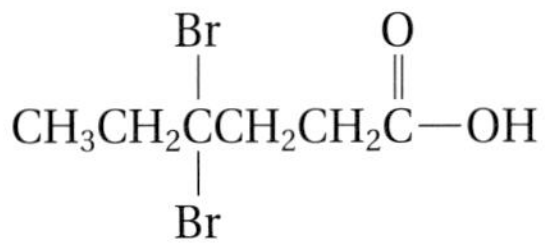

$$CH_3CH_2\overset{\displaystyle Br}{\underset{\displaystyle Br}{C}}CH_2CH_2\overset{\displaystyle \overset{O}{\|}}{C}-OH$$

PRACTICE EXERCISE 5.1

Give the name (IUPAC or common) of the following carboxylic acids.

(a) CH_3CH_2-COOH *propanoic (propionic acid)*

(b) CH_2-COOH
    $\,\,\,\,\,\,|$
    CH_2-COOH *4C → succinic acid*

(c) $\,\,\,\,\,\,\,\,OH$
    $\,\,\,\,\,\,\,\,\,\,|$
    $CH_3CHCH_2CH_2-COOH$ *4-hydroxypentanoic acid*

(d)

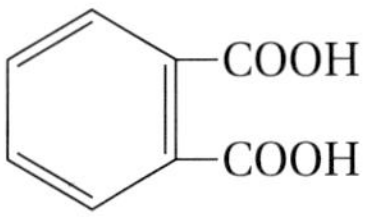

   —COOH
   —COOH *phthalic acid*

b) $CH_2-\overset{\displaystyle \overset{O}{\|}}{C}-OH$
     $\,\,\,\,|$
     $CH_2-\overset{\displaystyle \overset{O}{\|}}{C}-OH$

5.2 The carboxyl group

AIM: To relate the structure of the carboxyl group to the relatively high melting and boiling points, as well as the water solubility, of carboxylic acids.

Like alcohols, carboxylic acids can form intermolecular hydrogen bonds. Dimers—*hydrogen-bonded carboxylic acid pairs*—are found even in the vapor state of the acids with low molar masses.

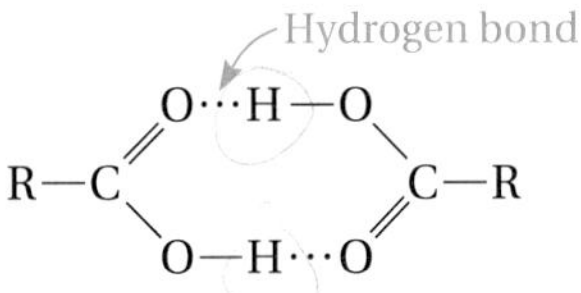

Focus

The carboxyl group avidly forms hydrogen bonds.

Hydrogen bond

$$R-C\overset{\displaystyle O\cdots H-O}{\underset{\displaystyle O-H\cdots O}{}}C-R$$

A carboxylic acid dimer

Because of their hydrogen bonding, carboxylic acids have higher boiling points and higher melting points than other compounds of similar molar

Table 5.3 Boiling Points for Different Compounds with Similar Molar Mass

Class	Compound	Molar mass	Boiling point (°C)
alkane	pentane	74	35
aldehyde	butanal	72	75.7
alcohol	butanol	74	118
acid	propanoic acid	74	141

diethyether 35
butanone 72 80

mass. Table 5.3 illustrates this point. All aromatic carboxylic acids and dicarboxylic acids are crystalline solids at room temperature.

The carboxyl group in carboxylic acids is polar and readily hydrogen bonds to water molecules. Formic, acetic, propionic, and butyric acids are completely miscible with water, but the solubility of carboxylic acids of higher molar mass drops off sharply. Because of their long nonpolar hydrocarbon tails and polar carboxyl group, the fatty acids are interesting. When a small quantity of a fatty acid is dropped onto the surface of water, it spreads out to form a **monomolecular layer**—*a film of the acid one molecule thick.* Figure 5.1 shows how the polar carboxyl groups of acid molecules in a monomolecular layer hydrogen bond to water molecules; the nonpolar tails project out of the water to form their own hydrophobic environment. Most carboxylic acids dissolve in organic solvents such as ethanol, acetone, or ether.

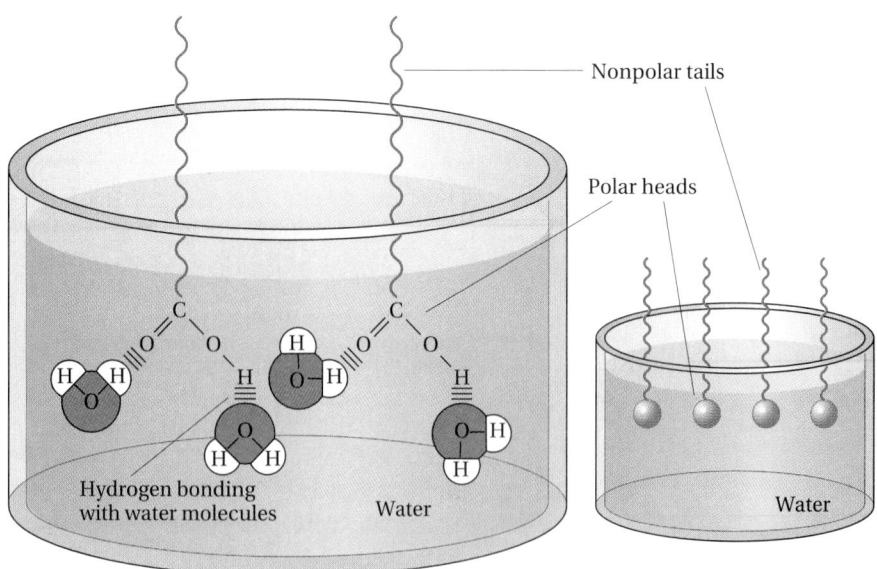

Figure 5.1
The organization of long-chain fatty acids at the surface of water.

5.3 Acidity of carboxylic acids

AIMS: *To explain the acidity of carboxylic acids in terms of K_a values, resonance stabilization, and R groups with different electronegativities. To describe, on a molecular scale, the action of soaps and detergents and explain the effects of hard water on each.*

You may recall from Section 5.1 that carboxylic acids are weak acids because they dissociate to only a slight extent in aqueous solution. The ionization constant K_a is a quantitative measure of the extent to which an acid ionizes. The smaller the K_a value, the weaker is the acid, and the larger the K_a value, the stronger is the acid. Table 5.4 gives the K_a values of some carboxylic acids. As the values show, ionization of carboxylic acids is much greater than that of alcohols and phenols, two classes of compounds that also contain O—H bonds. Moreover, acetic acid ($K_a = 1.8 \times 10^{-5}$) is a much weaker acid than trichloroacetic acid ($K_a = 2.2 \times 10^{-1}$). The acidity of a carboxylic acid heavily depends on its molecular structure. Next, we will investigate the reasons for this dependence.

Acidity and structure of the R group

Why do carboxylic acids lose protons more readily than alcohols? As shown in Figure 5.2, the electronegative carbonyl oxygen pulls electrons toward itself and away from the O—H bonds. This polarizes the O—H bond; the oxygen of the O—H bond draws some of the electrons of the bond to itself

Table 5.4 K_a Values for Some Carboxylic Acids

Name	Structure	K_a(25 °C)
formic acid	HCOOH	1.8×10^{-4}
acetic acid	CH_3COOH	1.8×10^{-5}
chloroacetic acid	$CH_2ClCOOH$	1.5×10^{-3}
dichloroacetic acid	$CHCl_2COOH$	5.0×10^{-2}
trichloroacetic acid	CCl_3COOH	2.2×10^{-1}
benzoic acid	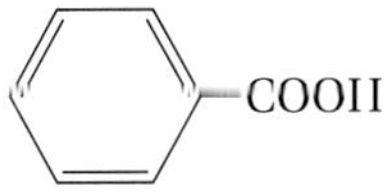 ⬡—COOH	6.0×10^{-5}
phenol*	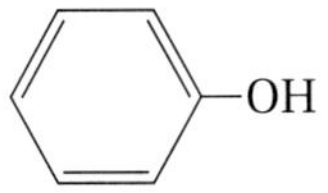 ⬡—OH	1.0×10^{-10}
ethanol*	CH_3CH_2OH	1.0×10^{-16}
water*	HOH	1.0×10^{-14}

*Presented for comparison.

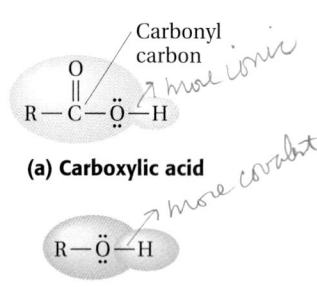

(a) Carboxylic acid

R—Ö—H

(b) Phenol or alcohol

Figure 5.2
The hydroxyl group in a carboxylic acid (a) loses a proton more readily than the hydroxyl group in a phenol or alcohol (b) because the electronegative carbonyl group pulls the bonding electron pair between oxygen and hydrogen away from hydrogen.

and becomes more negative. The hydrogen of the O—H bond loses some of its share of the bonding electrons and becomes more positive. This partial gain of electrons by oxygen and partial loss of electrons by hydrogen makes the O—H bond more ionic and therefore more readily dissociable into a carboxylate ion and a proton. In addition, the carboxylate ion produced by the dissociation of a carboxylic acid is stabilized by resonance—we can draw more than one form of the ion (see Sec. 2.9). Alkoxide ions (R—O⁻) are not stabilized by resonance:

Resonance forms of
carboxylate anion
(two forms can be drawn)

Resonance structures
can be drawn as a
hybrid structure

The R group attached to the carboxyl group may increase or decrease the acidity of a carboxylic acid. Like the carbonyl oxygen of the carboxyl group itself, an R group that is strongly electron attracting polarizes the O—H bond. The proton is released more easily, and the acid is stronger. Chlorine is electronegative. If the hydrogens on the carbon in acetic acid are replaced by one, two, or three chlorines, the resulting acids are successively stronger. Trichloroacetic acid is a much stronger acid than acetic acid.

EXAMPLE 5.3

Identifying the stronger acid

Identify the stronger acid of the following pair, and justify your choice.

$$CCl_3COOH \quad \text{or} \quad CBr_3COOH$$

SOLUTION

The more electronegative the groups attached to carbon-2, the weaker is the O—H bond and the stronger is the acid. Since chlorine is more electronegative than bromine, trichloroacetic acid can more easily lose the carboxyl hydrogen, and it is a stronger acid than tribromoacetic acid.

PRACTICE EXERCISE 5.2

Arrange the following carboxylic acids in order of increasing acidity (you may use Table 5.4):

(a) acetic acid   (b) dichloroacetic acid

(c) benzoic acid   (d) formic acid

Salt formation

Carboxylic acids can be neutralized by bases to produce salts. Organic acid salts are named like inorganic acid salts. First name the cation, then, name the anion by dropping the *-oic acid* or *-ic acid* ending and adding

A number of salts of carboxylic acids are important food additives. Sodium propionate inhibits mold growth in baked goods. Sodium benzoate is used in soft drinks and other acidic foods to inhibit bacterial growth. These additives both preserve and increase the shelf life of many foods.

-ate. For example, the salt of ammonia and formic acid is ammonium formate.

$$H-\overset{\overset{\displaystyle O}{\|}}{C}-OH(aq) + NH_3(aq) \longrightarrow H-\overset{\overset{\displaystyle O}{\|}}{C}-O^-NH_4^+(aq)$$

Formic acid   Ammonia   Ammonium formate

[handwritten: acid + base; cation — anion; salt]

When sodium carbonate or sodium bicarbonate neutralizes carboxylic acids, the solution froths as carbon dioxide is liberated.

$$CH_3-\overset{\overset{\displaystyle O}{\|}}{C}-OH(aq) + NaHCO_3(aq) \longrightarrow CH_3-\overset{\overset{\displaystyle O}{\|}}{C}-O^-Na^+(aq) + H_2O(l) + CO_2(g)$$

Acetic acid   Sodium bicarbonate   Sodium acetate   Water   Carbon dioxide

[handwritten: vinegar; baking soda]

The frothing helps distinguish carboxylic acids from neutral or basic organic compounds.

EXAMPLE 5.4

Neutralization of a carboxylic acid

Write an equation and name the products of the neutralization of benzoic acid with sodium hydroxide.

SOLUTION *[handwritten reaction scheme: benzene ring–C(=O)–OH + NaOH → benzene ring–C(=O)–O⁻Na⁺ + H₂O; labeled "sodium benzoate"]*

The reaction of benzoic acid, the simplest aromatic acid, and sodium hydroxide gives water and a salt. The name of the salt is *sodium* (the cation from the base) *benzoate* (the anion from the acid).

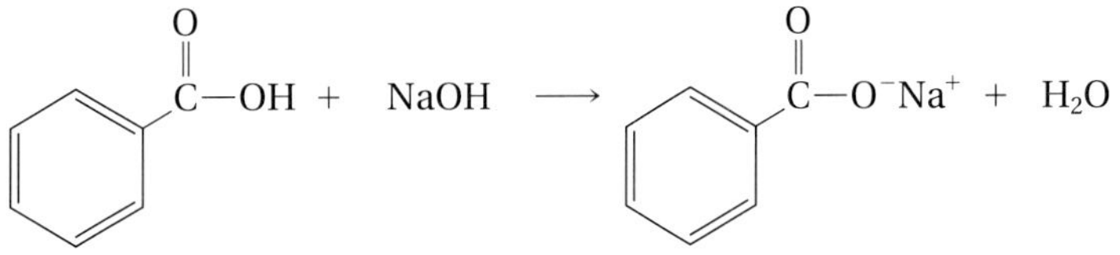

Benzoic acid   Sodium hydroxide   Sodium benzoate (a salt)   Water

PRACTICE EXERCISE 5.3

Write the structure and give the name of the products in each of the following neutralization reactions.

(a) $CH_3CH_2COOH + NaOH \longrightarrow$ *[handwritten: $CH_3CH_2\overset{\overset{\displaystyle O}{\|}}{C}-O^-Na^+ + H_2O$; sodium propanoate]*

(b) $CH_3COOH + Ca(OH)_2 \longrightarrow$ *[handwritten: $(CH_3\overset{\overset{\displaystyle O}{\|}}{C}-O)_2Ca^+ + 2(H_2O)$; calcium acetate]*

Soaps and detergents

Soaps *are the alkali metal (Li, Na, or K) salts of long-chain fatty acids.* Since most dirt is held to surfaces by an oily film, dirt is difficult to remove unless the oily coating is first emulsified with soaps.

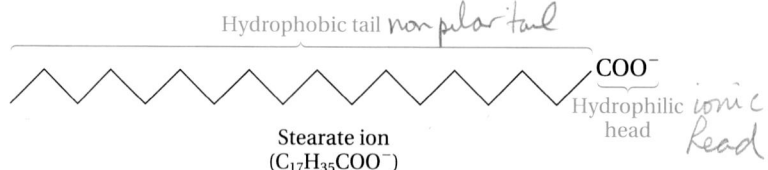

→ neutralized acid = salt

Hydrophobic
hudor (Greek): water
phobos (Greek): fear

Hydrophilic
hudor (Greek): water
philos (Greek): loving

Sodium stearate is a typical soap. Like all soaps, the sodium stearate molecule has two distinct parts—an ionic head (the carboxylate ion) and a long, nonpolar hydrocarbon tail. The charged head is *hydrophilic* ("water-loving"), and the nonpolar tail is *hydrophobic* ("water-hating").

Hydrophobic tail *non polar tail*

COO⁻

Hydrophilic *ionic* head *head*

Stearate ion
($C_{17}H_{35}COO^-$)

Because of their hydrophobic tails, soaps added to water go into colloidal dispersion instead of dissolving. Figure 5.3 shows the dispersed soap molecules arranged in spherical clusters called *micelles*. *In **micelles,** the hydrophobic tails of the soap molecules aggregate to create an oily nonpolar environment protected from water at the center of the cluster.* The charged carboxylate groups form a highly charged, highly solvated shell at the surface of the micelle. The micelles remain dispersed in water because they are like-charged and repel each other. When soapy water contacts an oily surface, the tails of the soap molecules become embedded in the oily film. Micelles then form with small oil droplets at their centers (Fig. 5.4). We might say that the oil dissolves in the soapy water, but it is actually not in true solution. The soap has acted as an emulsifying agent.

Figure 5.3
A micelle. The ionized groups of the soap molecules are in contact with water, and the nonpolar parts of the molecules are protected from water.

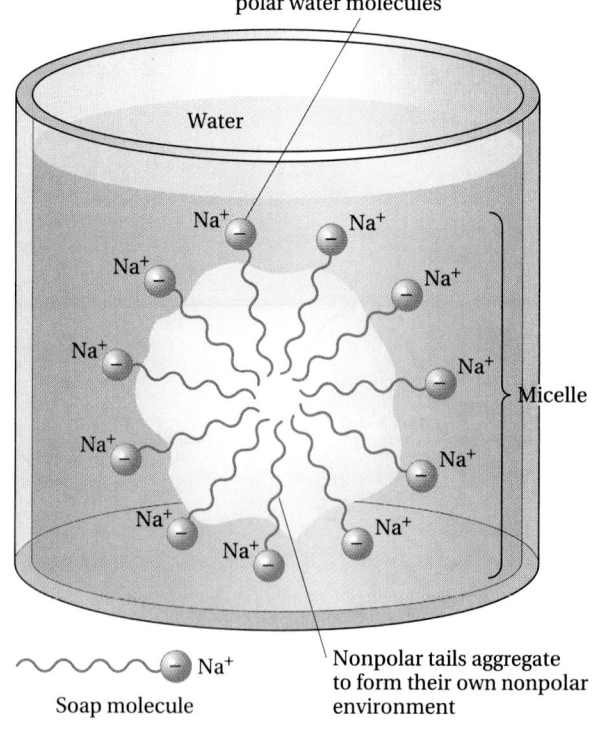

Ionic heads in contact with polar water molecules

Water

Na⁺ Na⁺
Na⁺   Na⁺
Na⁺   Na⁺
                          Micelle
Na⁺   Na⁺
Na⁺   Na⁺
    Na⁺ Na⁺
      Na⁺

⟿ – Na⁺
Soap molecule

Nonpolar tails aggregate to form their own nonpolar environment

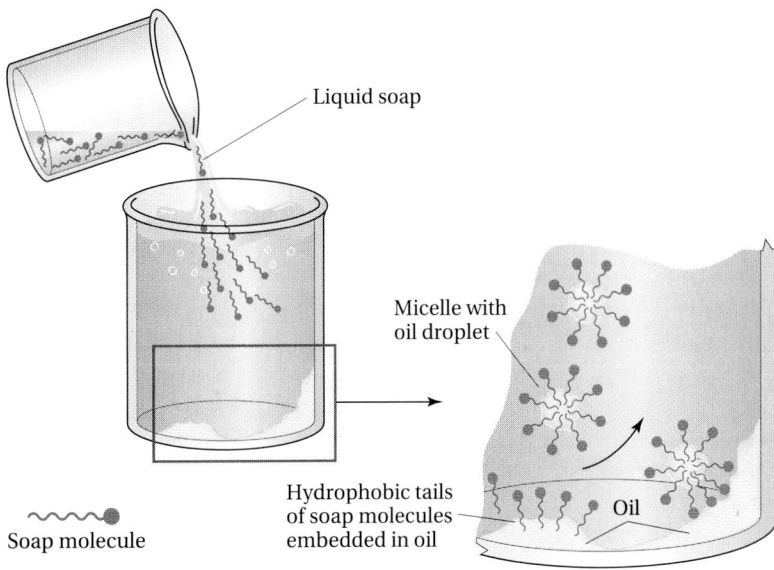

Figure 5.4
Soap "dissolves" oil by packaging it in micelles.

Soaps are essentially useless in acidic water, because in acidic water the carboxylate ions of soap molecules pick up hydrogen ions to form the undissociated fatty acid:

$$RCOO^- \ + \ H^+ \ \rightleftharpoons \ RCOOH$$

| Carboxylate ion | Hydrogen ion | Fatty acid (undissociated) |

The fatty acid precipitates as a scum because it cannot form micelles. Soaps are also essentially useless in **hard water**—*water containing calcium, magnesium, or iron ions.* Carboxylate ions form insoluble salts with these metal ions in the water. With a stearate soap, a typical reaction is

$$2CH_3(CH_2)_{16}COO^- \ + \ Ca^{2+} \ \longrightarrow \ [CH_3(CH_2)_{16}COO^-]_2Ca^{2+}$$

| Stearate ion | Calcium ion | Calcium stearate (precipitate) |

Such precipitates form unsightly scums that stick to clothes and give whites a gray look. A Closer Look: Hard Water and Water Softening describes the treatment of hard water in more detail.

Detergents *are cleaning agents with the desirable properties of soap without soap's scum-forming property.* Detergent molecules are like soap molecules because they have an ionic or polar head and a nonpolar tail. They also act as emulsifying agents for oils and greases, just as soap molecules do. However, they do not form precipitates with the ions present in hard water, so clothes being washed do not get coated with scum. One of

A Closer Look

Hard Water and Water Softening

Hard water can be problematic. It forms precipitates in boilers and hot water systems. When combined with soap, it forms an insoluble scum that coats skin and clings to the bathtub. It also can have a disagreeable taste.

Because natural waters often contain high levels of metal ions, they deliver supplies of hard water to our water systems. Ions are formed when rainwater on its way to these reservoirs leaches calcium from limestone and other metal ions such as magnesium and iron from various mineral rocks.

Hard water exists in two types. *Temporary hardness,* an indication of the presence of hydrogen carbonates (bicarbonates) of calcium and/or magnesium, is produced when rainwater tinged with carbonic acid filters through limestone or chalk rocks. *Permanent hardness* is due to dissolved sulfates and chlorides of calcium and magnesium, whose source is also mineral rock.

Hard water can be softened—the offending ions removed—by several procedures. Temporary hardness can be removed by boiling the water. However, this process produces a precipitate of calcium carbonate that sticks as hard "fur," or scale, on kettles, boilers, and the inside of hot water pipes (see figure). Permanent hardness cannot be removed by boiling, but it can be removed by adding slaked lime, $Ca(OH)_2$, or washing soda (sodium carbonate, Na_2CO_3) to the water. In this process, calcium and magnesium ions, precipitated as insoluble $CaCO_3$ and $Mg(OH)_2$, are replaced in solution by the sodium ions. Another common method of water softening is the addition of sodium triphosphate (Cal-

gon), which combines with the offending ions to form soluble complexes.

Large quantities of hard water are often softened in commercial water softeners by ion exchange. In this process, an undesirable ion in solution displaces an ion, usually sodium, from a solid, insoluble material called a *matrix*. The matrix used in ion-exchange water softeners is usually a natural mineral such as permutite or a synthetic resin. Most of these minerals and resins can be regenerated by pouring strong sodium chloride solution through them to displace the calcium and magnesium ions and replace them with a new stock of sodium ions.

Softening water is not a method of purification, since the ions causing water hardness are made only more soluble or are replaced by other ions. The high sodium ion content in softened water is of concern, especially to people on restricted sodium diets. A solution to the problem is to use soft water for washing and naturally hard water for drinking.

Pipe scale (right) forms when calcium carbonate sticks to the inside of hot water pipes.

the first detergents was the sodium salt of lauryl hydrogen sulfate:

$$CH_3(CH_2)_{10}CH_2-O-\overset{\overset{\displaystyle O}{\|}}{\underset{\underset{\displaystyle O}{\|}}{S}}-O^- Na^+$$

Sodium lauryl sulfate

ester of sulfuric acid
doesn't precipitate in hard water
but sewage treatment plant
cannot deal with it
~ bio degradable

Today's detergents are mostly salts of sulfonic acids:

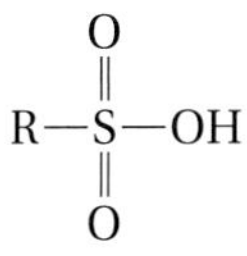

$$R-\overset{\displaystyle O}{\underset{\displaystyle O}{\overset{\|}{\underset{\|}{S}}}}-OH \quad \text{or} \quad RSO_3H$$

Sulfonic acid

Probably the most common detergent is sodium dodecylbenzene sulfonate:

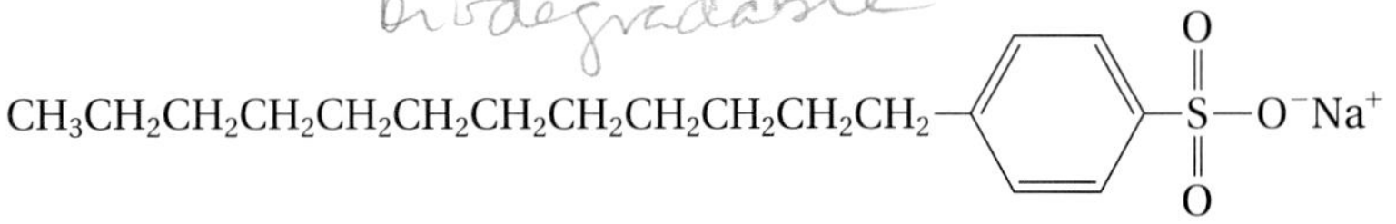

biodegradable

$$CH_3CH_2CH_2CH_2CH_2CH_2CH_2CH_2CH_2CH_2CH_2CH_2-\text{[benzene ring]}-\overset{\displaystyle O}{\underset{\displaystyle O}{\overset{\|}{\underset{\|}{S}}}}-O^-Na^+$$

Sodium dodecylbenzene sulfonate

This molecule has an obvious physical resemblance to a soap molecule—an ionic head and a nonpolar tail. And like a soap, it is **biodegradable**—*it can be broken down by bacteria in the environment into harmless products and does not become a pollutant.* When sudsing must be kept to a minimum, as in automatic washers for dishes and clothing, nonionic detergents are useful. *A* **nonionic detergent** *molecule has a nonpolar hydrocarbon tail and a polar but uncharged head.* One such compound is an ester of palmitic acid, pentaerythrityl palmitate:

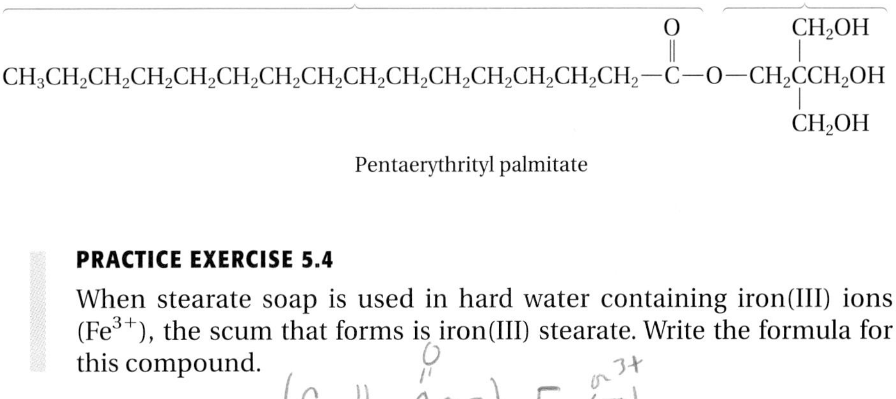

From palmitic acid	From pentaerythritol

$$CH_3CH_2CH_2CH_2CH_2CH_2CH_2CH_2CH_2CH_2CH_2CH_2CH_2CH_2CH_2-\overset{\displaystyle O}{\overset{\|}{C}}-O-CH_2\overset{\displaystyle CH_2OH}{\underset{\displaystyle CH_2OH}{C}}CH_2OH$$

Pentaerythrityl palmitate

> **PRACTICE EXERCISE 5.4**
>
> When stearate soap is used in hard water containing iron(III) ions (Fe^{3+}), the scum that forms is iron(III) stearate. Write the formula for this compound.
>
> $$\left(C_{17}H_{35}\overset{\displaystyle O}{\overset{\|}{C}}O^-\right)_3 Fe(\text{III})$$

5.4 Synthesis of carboxylic acids

AIM: To write equations for the preparation of carboxylic acids from alcohols and aldehydes.

Sometimes a chemist needs to make a carboxylic acid with a particular structure, say, in the synthesis of a new drug. Two common methods for the preparation of carboxylic acids involve the oxidation of primary alcohols or aldehydes (see Sec. 4.4). The starting alcohol or aldehyde must have the same carbon skeleton as the desired acid.

Oxidation of alcohols

The general reaction for oxidation of primary alcohols to carboxylic acids is

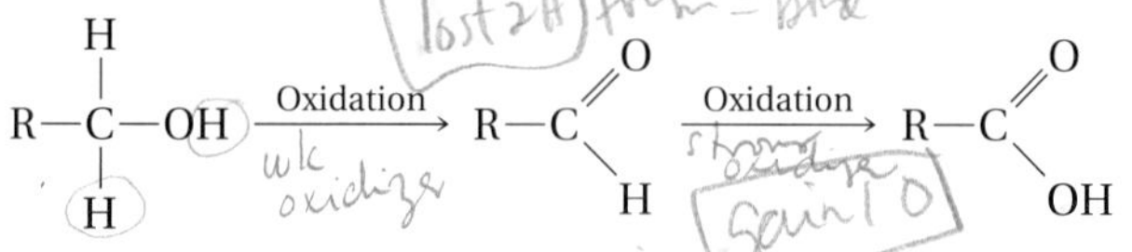

Primary alcohol   Aldehyde   Carboxylic acid

Potassium dichromate ($K_2Cr_2O_7$) and potassium permanganate ($KMnO_4$) are commonly used as oxidizing agents:

$$CH_3CH_2-CH_2OH \xrightarrow{K_2Cr_2O_7} CH_3CH_2-COOH$$

1-Propanol   Propionic acid

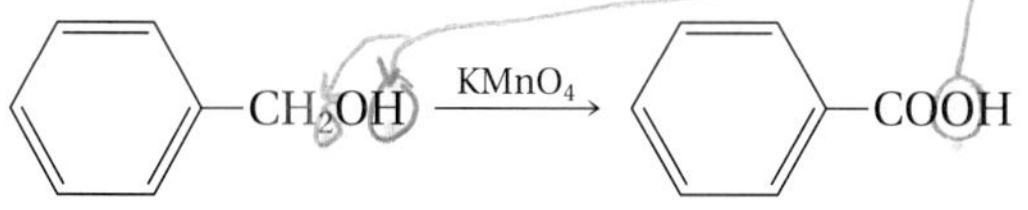

Benzyl alcohol   Benzoic acid

Breaking a C—C bond by oxidation is much more difficult than breaking a C—H bond. The usual oxidizing agents oxidize secondary alcohols to ketones. Tertiary alcohols are not oxidized.

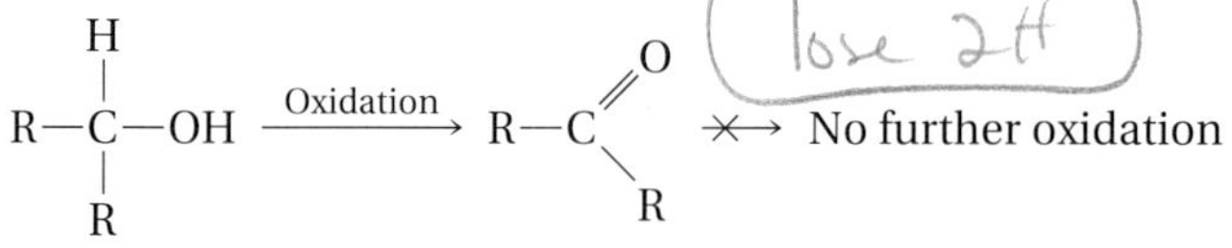

Secondary alcohol   Ketone

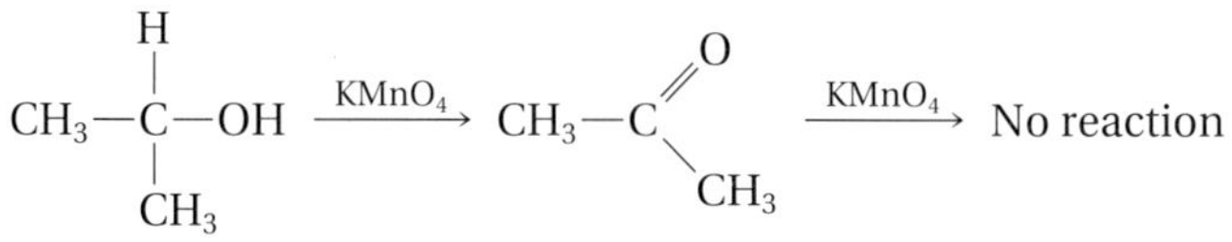

2-Propanol   Acetone

Oxidation of aldehydes

Since an aldehyde is an intermediate in the oxidation of a primary alcohol to a carboxylic acid, it follows that aldehydes can be oxidized to carboxylic acids.

$$CH_3-CHO \xrightarrow{K_2Cr_2O_7} CH_3COOH$$

Acetaldehyde   Acetic acid

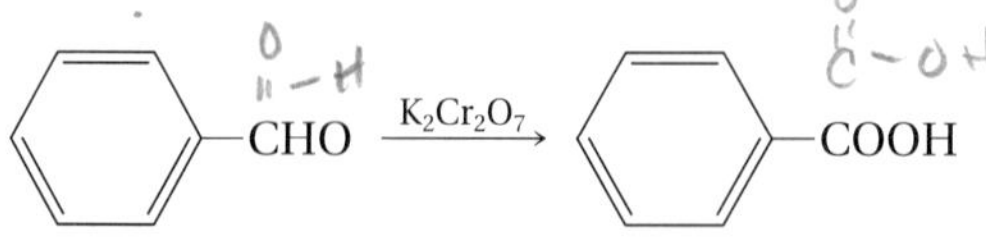

Benzaldehyde   Benzoic acid

PRACTICE EXERCISE 5.5

The following compounds are oxidized with acidified potassium dichromate. Write the structure for the expected product. If no product is expected, write "no reaction."

(a) CH_3CH_2OH (b) CH_3CHOH *lox 2H* $CH_3C\overset{\overset{\displaystyle O}{\|}}{}-CH_3$

primary alcohol + strong oxi $\rightarrow CH_3C$-OH *acid*

        CH_3

secondary alcohol + strong oxidizer → ketone

(c) CH_3CH_2CHO (d) $CH_3\overset{\overset{\displaystyle O}{\|}}{C}CH_3$

aldehyde + strong ox → acid

$CH_3CH_2\overset{\overset{\displaystyle O}{\|}}{C}$-OH *similar*

ketone no rxn   NR

5.5 Carboxylic acid anhydrides

AIMS: To name and write structures of common acid anhydrides. To write equations for the formation of inter- and intramolecular acid anhydrides. To distinguish between an acyl and an acetyl group and write equations for typical acylation reactions.

Carboxylic acid anhydrides *come from two carboxylic acid molecules, or a single dicarboxylic acid molecule, through the loss of a molecule of water.* The name *anhydride* means "without water."

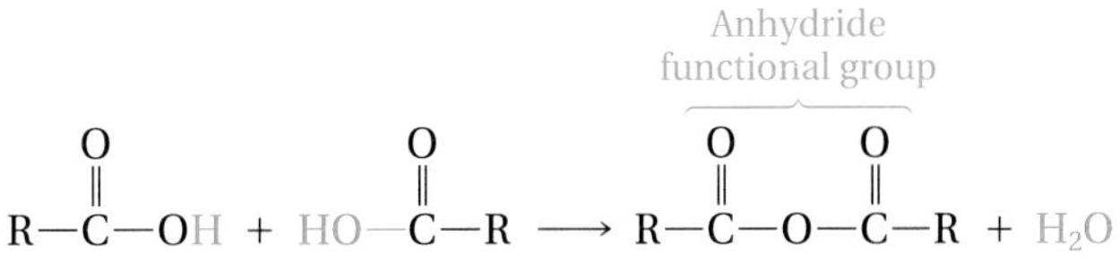

$$R-\overset{\overset{\displaystyle O}{\|}}{C}-OH + HO-\overset{\overset{\displaystyle O}{\|}}{C}-R \longrightarrow R-\overset{\overset{\displaystyle O}{\|}}{C}-O-\overset{\overset{\displaystyle O}{\|}}{C}-R + H_2O$$

Anhydride functional group

Two carboxylic acid molecules   Acid anhydride

loss of water molecule

Names of anhydrides

Anhydrides are named by adding the word *anhydride* after the parent name of the acid. One common acid anhydride is acetic anhydride:

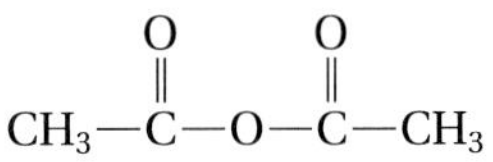

$$CH_3-\overset{\overset{\displaystyle O}{\|}}{C}-O-\overset{\overset{\displaystyle O}{\|}}{C}-CH_3$$

Acetic anhydride
(bp 139.5 °C)

Acetic anhydride, a corrosive liquid, is immensely important as both an industrial chemical and a laboratory reagent.

Certain dicarboxylic acids readily produce anhydrides upon heating if five- or six-membered rings are formed. (Recall that five- and six-membered rings are especially stable.) For example:

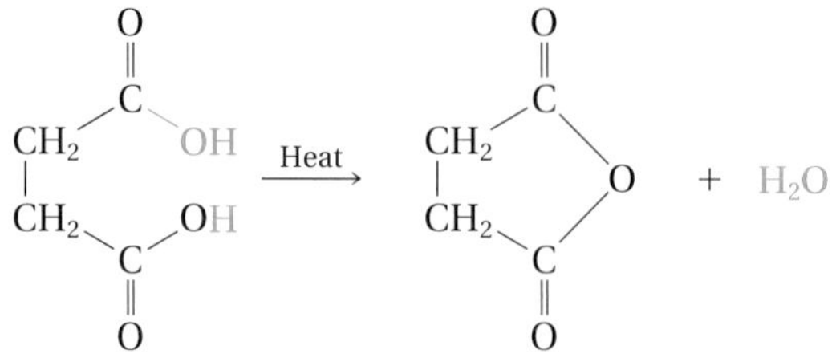

| Succinic acid | Succinic anhydride | Water |

| Phthalic acid | Phthalic anhydride | Water |

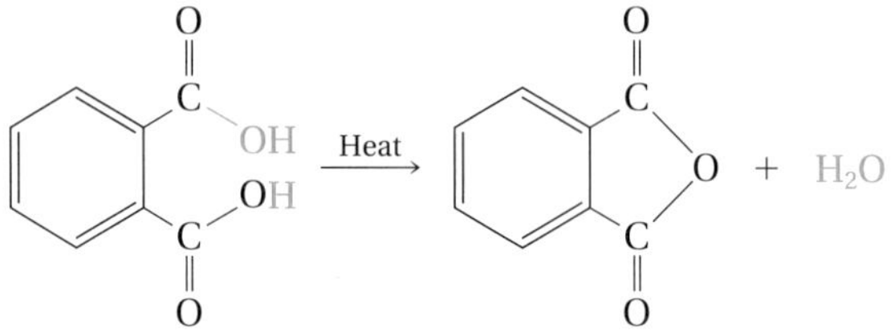

The anhydrides of monocarboxylic acids are liquids. Those of dicarboxylic acids and aromatic carboxylic acids are solids.

Hydrolysis of anhydrides

Processes in which **acyl groups**

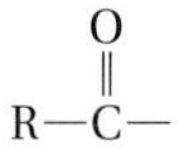

$$\underset{\text{R}-\overset{\displaystyle\text{O}}{\overset{\|}{\text{C}}}-}{}$$

are transferred to another molecule are called **acylation reactions.** Acid anhydrides are very reactive acylating agents. When the acyl group is transferred to water, the reaction is a *hydrolysis.* In organic chemistry, **hydrolysis** *is the splitting of a molecule by water.* When the molecule is split, a hydrogen from water ends up in one product, and the —OH ends up in the other. When acetic anhydride is hydrolyzed, the acetyl group

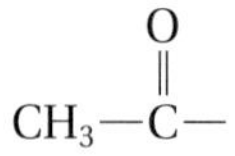

$$\text{CH}_3-\overset{\displaystyle\text{O}}{\overset{\|}{\text{C}}}-$$

is transferred to water. The products of hydrolysis of an anhydride are two molecules of the acid.

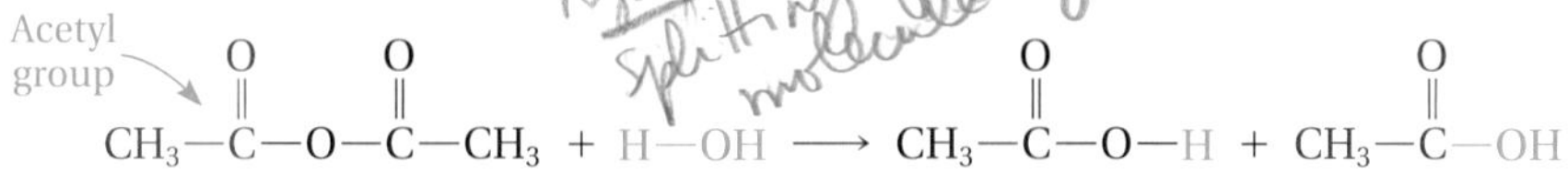

Acetyl group → $\text{CH}_3-\overset{\text{O}}{\overset{\|}{\text{C}}}-\text{O}-\overset{\text{O}}{\overset{\|}{\text{C}}}-\text{CH}_3 + \text{H}-\text{OH} \longrightarrow \text{CH}_3-\overset{\text{O}}{\overset{\|}{\text{C}}}-\text{O}-\text{H} + \text{CH}_3-\overset{\text{O}}{\overset{\|}{\text{C}}}-\text{OH}$

| Acetic anhydride | Water | Two molecules of acetic acid |

PRACTICE EXERCISE 5.6

Name the following compound, and give the product of its hydrolysis.

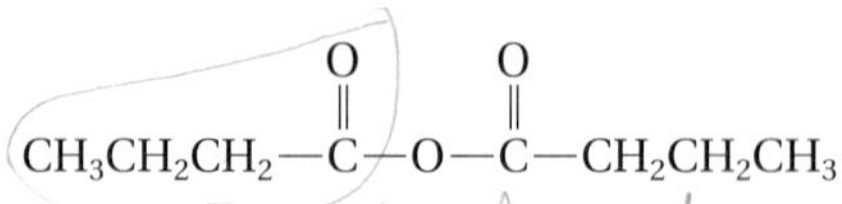

$$\text{CH}_3\text{CH}_2\text{CH}_2-\overset{\text{O}}{\overset{\|}{\text{C}}}-\text{O}-\overset{\text{O}}{\overset{\|}{\text{C}}}-\text{CH}_2\text{CH}_2\text{CH}_3$$

[handwritten: from acid, carbonyl group, ether link to C carbonyl C from alcohol]

5.6 Carboxylic esters

AIMS: To name and write structures of common esters. To write equations for the formation of an ester from an alcohol and a carboxylic acid or an acid anhydride.

Focus

A carboxylic ester is produced when a carboxylic acid or anhydride reacts with an alcohol.

Carboxylic esters *are derivatives of carboxylic acids in which the hydrogen of the* —OH *of the carboxyl group has been replaced by* —R *of an alkyl or aryl group.* They contain a carbonyl group and an ether link to the carbonyl carbon. The general formula is

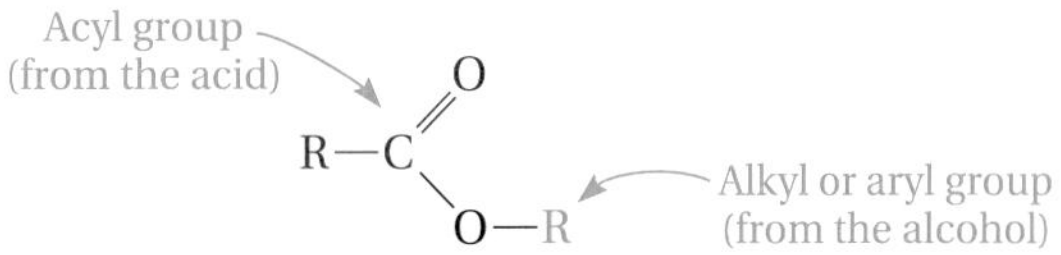

Acyl group (from the acid) → R—C(=O)—O—R ← Alkyl or aryl group (from the alcohol)

The abbreviated formula for a carboxylic ester is RCOOR. The R groups can be short chains or long chains, aliphatic (alkyl) or aromatic (aryl), saturated or unsaturated.

Naming carboxylic esters

Carboxylic esters are named as derivatives of carboxylic acids. The names contain two words, the name of the acid and the name of the —R group of the ether linkage. The *-ic* ending of the acid name is changed to *-ate* and is preceded by the name of the alkyl or aryl group of the ether linkage. This method is illustrated with some derivatives of acetic acid.

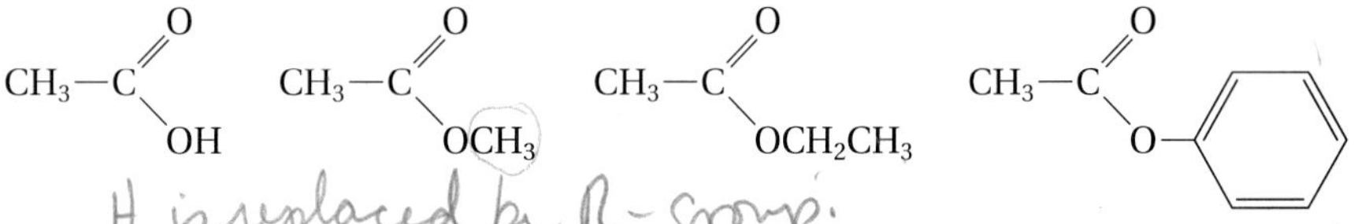

$CH_3—C(=O)OH$ — Ethanoic acid (acetic acid)

$CH_3—C(=O)OCH_3$ — Methyl ethanoate (methyl acetate)

$CH_3—C(=O)OCH_2CH_3$ — Ethyl ethanoate (ethyl acetate)

$CH_3—C(=O)O—$(phenyl) — Phenyl ethanoate (phenyl acetate)

[handwritten: H is replaced by R-group.]

EXAMPLE 5.5

Naming an ester

Give the IUPAC and common names of the following ester.

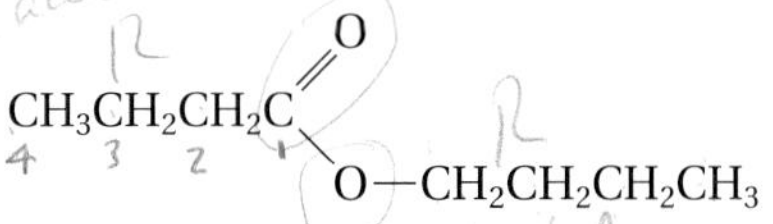

$CH_3CH_2CH_2C(=O)O—CH_2CH_2CH_2CH_3$

[handwritten: butanoic acid, 4 3 2, butyl, butyl butanoate 2 words.]

SOLUTION

In order to find the IUPAC name of the ester, we first inspect the portion of the ester that comes from the carboxylic acid.

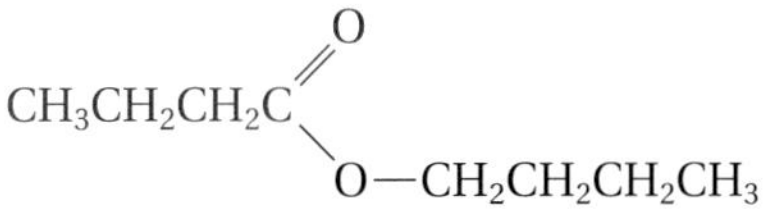

$CH_3CH_2CH_2C(=O)O—CH_2CH_2CH_2CH_3$

The acid portion is four carbons long and is derived from butanoic acid. By removing the *-ic acid* ending and adding an *-ate* ending, we see that the ester is a *butanoate*. We next name the alkyl group bonded to the ether oxygen of the ester.

$$CH_3CH_2CH_2C\overset{O}{\underset{O-CH_2CH_2CH_2CH_3}{\diagup}}$$

This alkyl group, consisting of four carbons in a straight chain, is called *butyl*. We combine both parts of the name in two words to obtain the complete IUPAC name of the ester: *butyl butanoate*.

The common name is derived similarly, except the common name for butanoic acid is *butyric acid*. The common name of the ester is *butyl butyrate*.

PRACTICE EXERCISE 5.7

Name these esters:

(a)

$$CH_3CH_2COO-$$

(b)

$$-COOCH_2CH_3$$

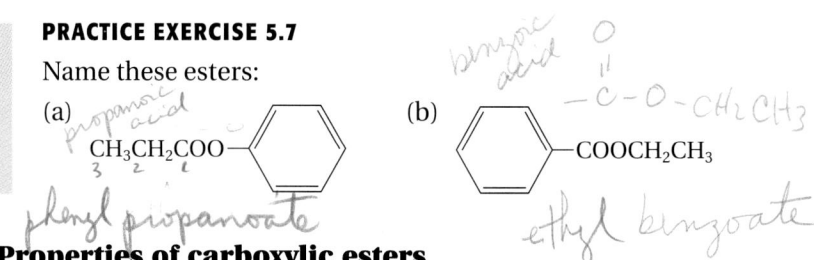

Properties of carboxylic esters

The triesters formed by a molecule of glycerol and three molecules of fatty acids are called *triglycerides*. As we will see in Chapter 16, triglycerides are important energy storage molecules.

Simple carboxylic esters are neutral substances. The molecules are polar but cannot form intermolecular hydrogen bonds. They are much less soluble in water and have much lower boiling points than the carboxylic acids from which they are derived. Esters can hydrogen bond with water. Esters of low molar mass are somewhat soluble, but esters containing more than four or five carbons have very limited solubility in water.

Esters derived from acids and alcohols with low molar masses have delightful odors. They are volatile compounds that occur naturally in fruits and flowers, to which they impart a characteristic taste or fragrance. Table 5.5 lists some of the familiar esters and their characteristic odors. Many of

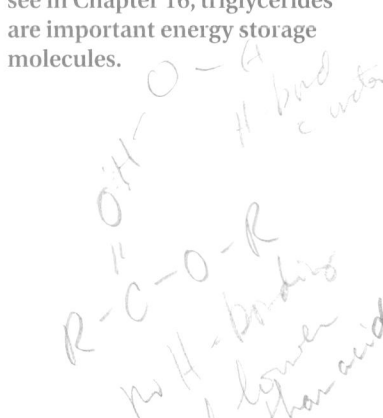

Table 5.5 Flavor Esters

Name	Formula	Odor/flavor
ethyl formate	$HCOOCH_2CH_3$	rum
octyl acetate	$CH_3COO(CH_2)_7CH_3$	oranges
pentyl acetate	$CH_3COO(CH_2)_4CH_3$	bananas
ethyl butyrate	$CH_3(CH_2)_2COOCH_2CH_3$	pineapples
pentyl butyrate	$CH_3(CH_2)_2COO(CH_2)_4CH_3$	apricots
methyl butyrate	$CH_3(CH_2)_2COOCH_3$	apples
methyl salicylate	(see structure)	oil of wintergreen

these esters can be synthesized in the laboratory and are used to flavor foods and drinks artificially. Natural flavors usually consist of more complex mixtures of esters.

Preparation of esters

Esters can be prepared from an acid and an alcohol or from an acid anhydride and an alcohol. **Esterification**—*the formation of esters*—takes place when carboxylic acids are heated with primary or secondary alcohols in the presence of a trace of mineral acid as a catalyst. The reaction is reversible.

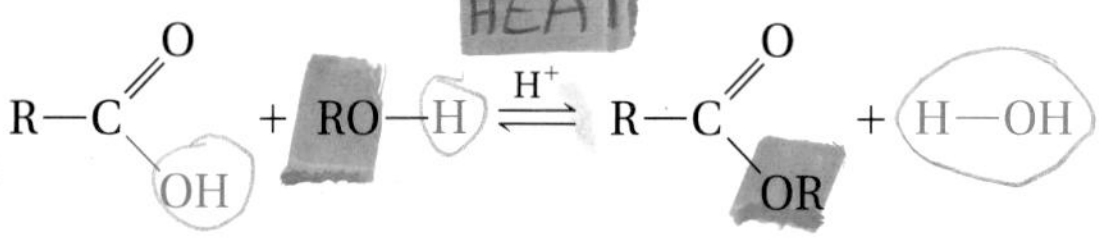

| Carboxylic acid | Alcohol | Carboxylate ester | Water |

Consider this specific example. When salicylic acid and methanol react, the product is methyl salicylate.

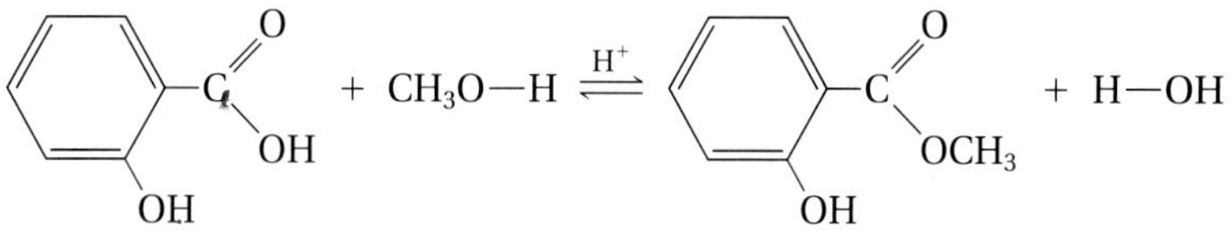

| Salicylic acid | Methanol | Methyl salicylate (oil of wintergreen) | Water |

Methyl salicylate, also called *oil of wintergreen,* is used as a flavoring and in liniments for soothing sore muscles. Many local anesthetics also contain the ester function. See A Closer Look: Ester Local Anesthetics.

There are ways in which the equilibrium in a reversible reaction can be disturbed to improve the product yield. Esterification reactions can be pushed to completion if we use an excess of one of the reactants (acid or alcohol) or if we remove the water from the reaction mixture as it is produced.

The industrial production of esters often involves reacting an acid anhydride with an alcohol. The ester of greatest commercial importance prepared in this way is acetylsalicylic acid, commonly called *aspirin.* Acetylsalicylic acid is obtained from acetic anhydride and salicylic acid:

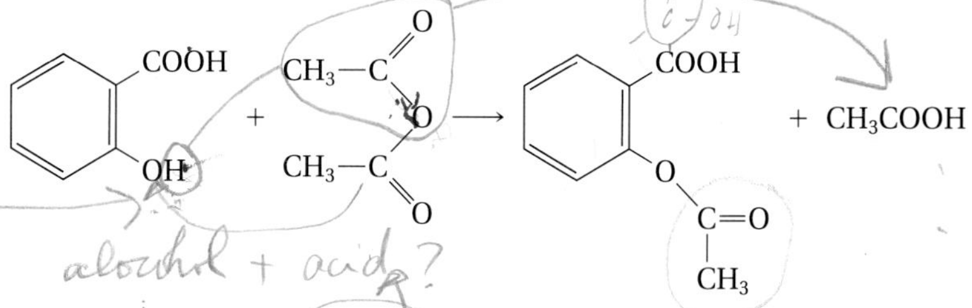

| Salicylic acid | Acetic anhydride | Acetylsalicylic acid (aspirin) | Acetic acid |

A Closer Look

Ester Local Anesthetics

Local anesthetics applied to body surfaces (topical application) or injected around nerves are used primarily to prevent pain during surgical procedures in which the patient stays conscious. Several reliable local anesthetics are available. Some of the most common contain the ester functional group. For example, ethyl-*p*-aminobenzoate (Benzocaine) is a common ingredient in lotions and creams meant to relieve the pain of sunburn.

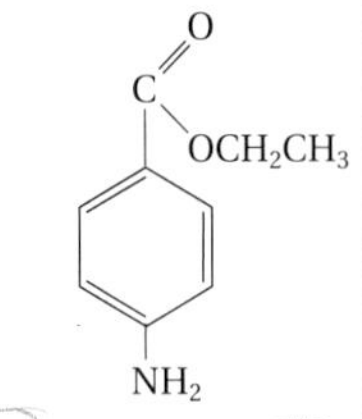

Cocaine, an ester obtained from the coca plant (not to be confused with the cocoa plant), is used in the form of its hydrochloride salt for local anesthesia of the nose, pharynx, and upper respiratory passages. The ability of cocaine to cause constriction of blood vessels is unequaled by other local anesthetics. Cocaine therefore is useful in decreasing

bleeding and shrinking mucous membranes. Cocaine was once used as a local anesthetic in dentistry. Because cocaine is toxic and habit-forming, its use has diminished as other anesthetics have become available.

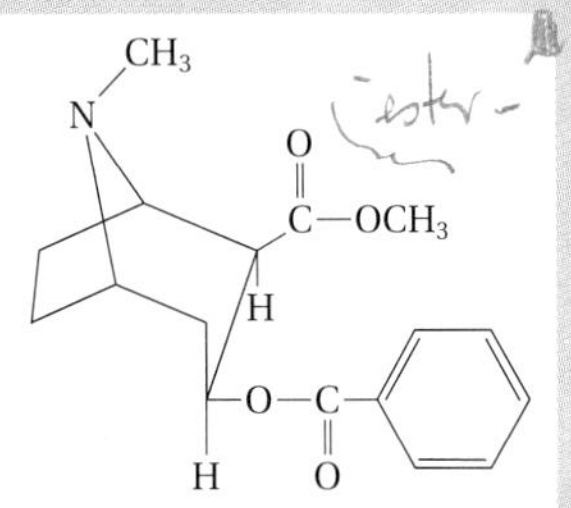

Procaine (Novocain) is an ester of *N,N*-diethylaminoethanol and ethyl-*p*-aminobenzoic acid. Procaine must be injected to produce local anesthesia. It has minimal systemic toxicity and produces no local irritation, and its effects rapidly disappear. These features make it popular for injection into the spinal column to produce a spinal block (see figure). An ester similar to procaine, chloroprocaine (or Nesacaine), also must be injected but is more potent and wears off faster than procaine. Chloroprocaine is rapidly hydrolyzed in plasma and is much less toxic than procaine. It is probably the safest local anesthetic.

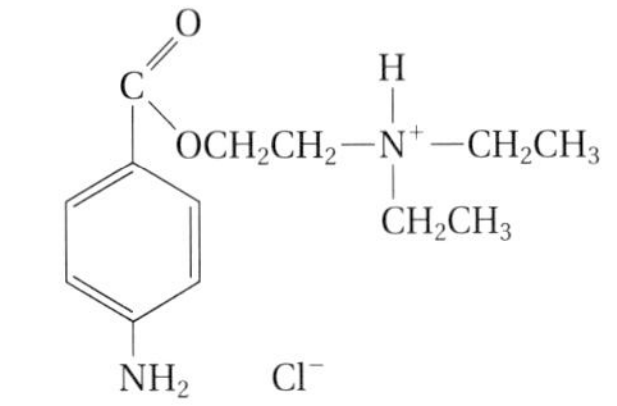

Ethyl-*p*-aminobenzoate (Benzocaine)

Cocaine

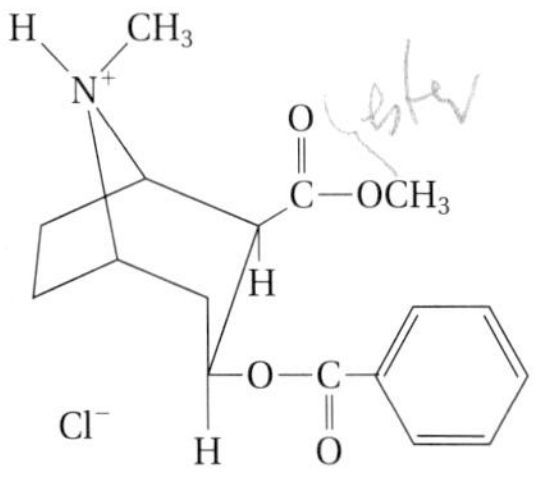

Cocaine (hydrochloride salt)

Procaine (hydrochloride salt) (Novocaine)

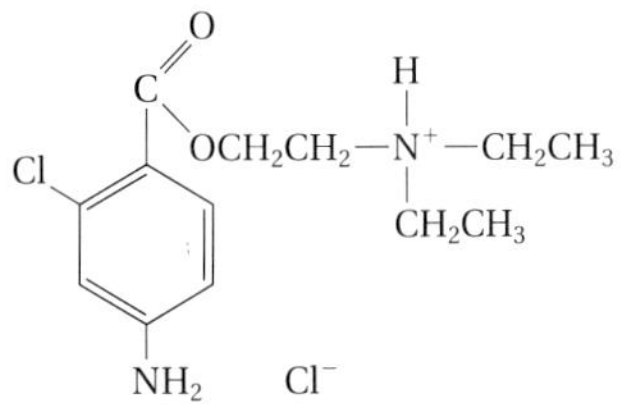

Chloroprocaine (hydrochloride salt) (Nesacaine)

Like the hydrolysis of acetic anhydride, this is an acylation reaction—but in this case the acceptor of the acetyl group is an alcohol instead of water. Salicylic acid is a polyfunctional carboxylic acid because it contains both a carboxyl group and a hydroxyl group. This means that it can react as an acid or as an alcohol depending on the other reactants. In the formation of methyl salicylate, salicylic acid behaves as an acid, and reaction takes place at the carboxyl group. In the production of aspirin, however, salicylic acid acts as an alcohol, and reaction with acetic anhydride occurs at the hydroxyl group. Aspirin is an **antipyretic** (*fever killer*) and an **analgesic** (*pain killer*). It is usually dispensed as its sodium salt, sodium acetyl salicylate.

Another drug in this ester series is tetracaine (Pantocaine, Pontocaine). The potency and duration of action of this compound are higher than those of the other esters. The resulting toxicity, however, is also greater. Tetracaine is the most commonly used anesthetic for spinal anesthesia; it is combined with an equal volume of 10% glucose to increase the specific gravity and thus control the spread of the solution. In 1% to 2% concentrations it is used as a topical anesthetic in the pharynx and the upper respiratory passages.

Tetracaine
(hydrochloride salt)
(Pantocaine, Pontocaine)

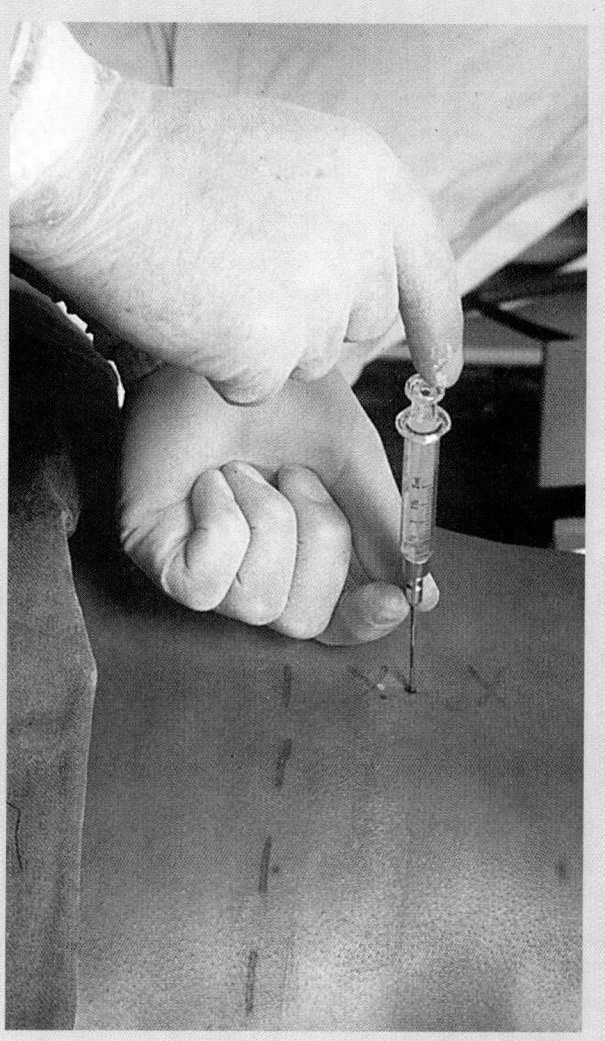

Procaine is injected to produce the local anesthesia of a spinal block.

Sodium acetyl salicylate

See A Closer Look: Aspirin, for more details about this medication.

A Closer Look

Aspirin

How would you like to discover a drug that could relieve pain from headaches and arthritis, prevent heart attacks, and reduce fever? Moreover, your drug would be cheap and available without a prescription. Such a miracle drug already exists. It is called *aspirin.*

In 1763, Edward Stone, an English clergyman, reported that the extract of willow bark would reduce fever. The active ingredient, salicylic acid, was isolated from willow bark in 1860. Because salicylic acid is sour and irritating when taken orally, chemists have modified the structure to remove the undesirable properties while retaining or enhancing its desirable properties. A number of derivatives of the acid have been made and used with success as analgesics and antipyretics. The most common derivative is aspirin (acetylsalicylic acid). A typical aspirin tablet contains about 325 mg acetylsalicylic acid held together with starch, an inert binder. Over 50 billion aspirin tablets are produced annually in the United States. Aspirin, which was introduced in 1899, is still the largest-selling and most widely used medical drug in the world (see figure).

Aspirin seems to work as a pain killer by blocking the synthesis of certain substances called *prostaglandins* (see Sec. 8.11). Some prostaglandins are involved in the inflammation of tissues. At elevated levels, prostaglandins activate pain receptors in tissues, making the tissues more sensitive to pain. This is why aspirin is frequently the drug of choice for the treatment of simple headaches and for arthritis, a condition in which the joints and connective tissues become inflamed.

Fever reduction by aspirin is not completely understood but appears to be due to an action on the source of the body's temperature control, the hypothalamus gland in the brain. In addition, blood vessels expand. The resulting increase in the circulation of the blood helps to dissipate the heat of the fever. Blockage of prostaglandin synthesis also may play a role in the fever-reducing effects of aspirin.

Several recent studies have shown that small daily amounts of aspirin decrease the incidence of heart attack in middle-aged or older individuals. Many times heart attacks result from blood clots that block the coronary arteries. The aggregation of blood cells called *platelets* (see Sec. 13.2) is an important component in the formation of these blood clots. Aspirin helps to prevent these clots by reducing the "stickiness" of blood platelet cells. The clots do not form, thereby reducing the likelihood of a heart attack. The use of aspirin for its "blood thinning" effects is often recommended for patients who have undergone heart bypass surgery.

Like any medication, aspirin use can be abused. Taking an aspirin usually causes a small amount of internal bleeding. This is not a big problem for most people, but those who have ulcers or other bleeding problems are advised to avoid aspirin. Aspirin and salicylates should not be given to children to treat the symptoms of flu or chicken pox. There is a strong correlation between aspirin use and the onset of Reye's syndrome, a disease characterized by weakness and paralysis. Reye's syndrome can be fatal unless treatment is begun immediately.

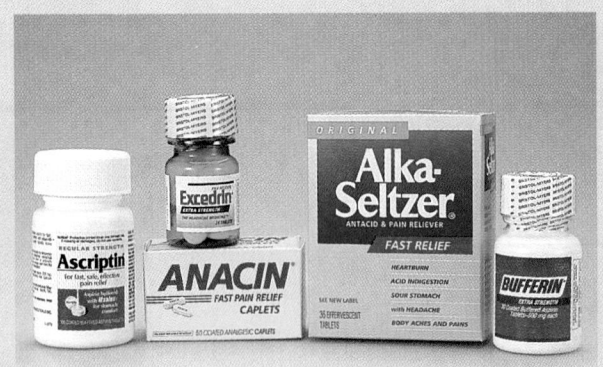

Many medicines contain aspirin or other salicylates.

PRACTICE EXERCISE 5.8

Complete the following reactions by giving the structure and name of the product.

(a) CH_3CH_2COOH + CH_3OH $\xrightarrow[H^+]{H^+}$ *[handwritten:]* $\xrightarrow{\Delta}$ $CH_3CH_2-\overset{O}{\overset{\|}{C}}-O-CH_3$ $+ H_2O$
[handwritten: acid + alcohl $\xrightarrow{H^+}$ ester methyl propanoate]

(b)

⬡—COOH + $CH_3CH_2CH_2OH$ $\xrightarrow[\Delta]{H^+}$ *[handwritten:]* ⬡—$\overset{O}{\overset{\|}{C}}$—O... propylbenzoate $+ H_2O$ ... $CH_2CH_2CH_3$
[handwritten: benzoic acid, propanol]

(c)

$CH_3\overset{O}{\overset{\|}{C}}$—O—$\overset{O}{\overset{\|}{C}}CH_3$ + CH_3CH_2OH ⟶ *[handwritten:]* $CH_3\overset{O}{\overset{\|}{C}}$—O... $+ HO-\overset{O}{\overset{\|}{C}}-CH_3$ acetic acid
[handwritten: acid anhydride acetic + alcohol ethanol → ethyl ethanoate (acetate) CH_2CH_3]

Polyesters

What material can you wear, use to play music, and ride to classes on? The answer is *polyesters*. Chemically speaking, **polyesters** *are polymers consisting of many repeating units of dicarboxylic acids and dihydroxy alcohols joined by ester bonds.* The formation of a polyester can be represented by a block diagram:

$$xHO-\overset{O}{\overset{\|}{C}}-\boxed{}-\overset{O}{\overset{\|}{C}}-OH + xHO-\bigcirc-OH \longrightarrow \left(\overset{O}{\overset{\|}{C}}-\boxed{}-\overset{O}{\overset{\|}{C}}-O-\bigcirc-O\right)_x + 2xH_2O$$

Dicarboxylic acid   Dihydroxy alcohol   Representative polymer unit

Several polyesters are commercially successful. Polyethylene terephthalate (PET) is one of the most important. This polyester is formed from terephthalic acid and ethylene glycol:

$$xHO-\overset{O}{\overset{\|}{C}}-\bigcirc-\overset{O}{\overset{\|}{C}}-OH + xHO-CH_2CH_2-OH \longrightarrow \left(\overset{O}{\overset{\|}{C}}-\bigcirc-\overset{O}{\overset{\|}{C}}-O-CH_2CH_2-O\right)_x + 2xH_2O$$

Terephthalic acid   Ethylene glycol   Representative polymer unit of polyethylene terephthalate (PET)

Figure 5.5
Arterial grafts, fabricated from woven Dacron fibers, are used to repair or replace blood vessels. ©1995 Meadox Medicals, Inc.

The melting point of the polymer is about 270 °C. PET fibers are formed when it is melted and forced through tiny holes in devices called *spinnerettes.* The fibers, marketed as Dacron, Fortrel, or Terylene, depending on the manufacturer, are used for tire cord and permanent-press clothing. Woven Dacron tubing has been used to replace major blood vessels (Fig. 5.5). PET fibers are often the polyesters that are blended with cotton to make permanent-press clothing. Garments made from these blends are more comfortable on hot, humid days than those containing 100% polyester, but they retain the latter's wrinkle resistance. Instead of being forced through spinnerettes, PET melts may be forced through a narrow slit to produce sheets of Mylar, which is used extensively as magnetic tapes for tape recorders and computers.

5.7 Thioesters

AIM: To write an equation for the formation of a thioester.

The carboxylic esters we have seen so far are *oxyesters* because their carboxyl groups contain two oxygen atoms. *In* **thioesters,** *the oxygen in the C—O—C link of an oxyester is replaced by sulfur:*

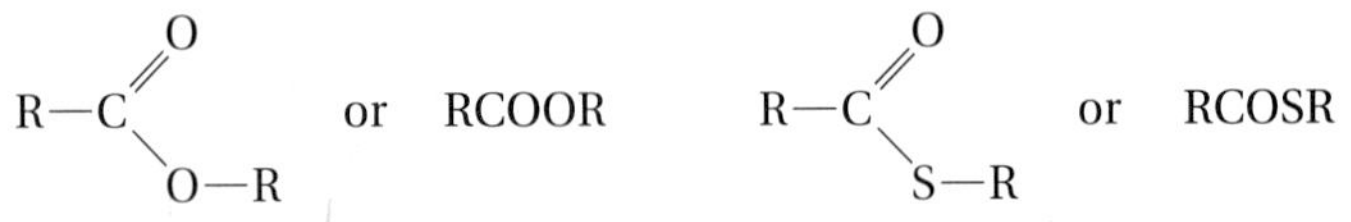

$$
\begin{array}{ccc}
& O \\
& \parallel \\
R-C & & \text{or} \quad RCOOR \\
& \backslash \\
& O-R
\end{array}
\qquad
\begin{array}{ccc}
& O \\
& \parallel \\
R-C & & \text{or} \quad RCOSR \\
& \backslash \\
& S-R
\end{array}
$$

Oxyester   Thioester

Thioesters can be prepared in the same way as oxyesters. A thiol is used instead of an alcohol:

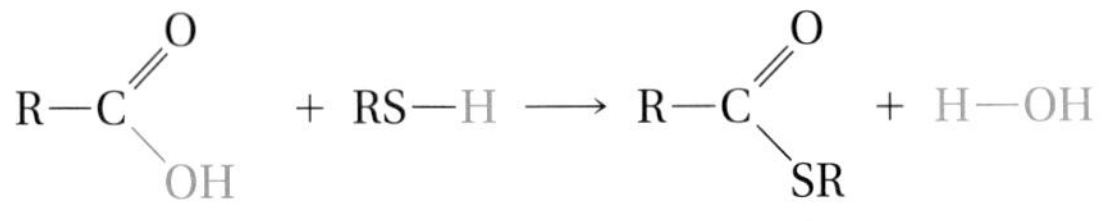

$$
\begin{array}{ccccccc}
& O & & & & O \\
& \parallel & & & & \parallel \\
R-C & & + \; RS-H & \longrightarrow & R-C & & + \; H-OH \\
& \backslash & & & & \backslash \\
& OH & & & & SR
\end{array}
$$

Carboxylic acid   Thiol   Thioester   Water

If an acid anhydride is used in the esterification of a thiol, the process is an acylation:

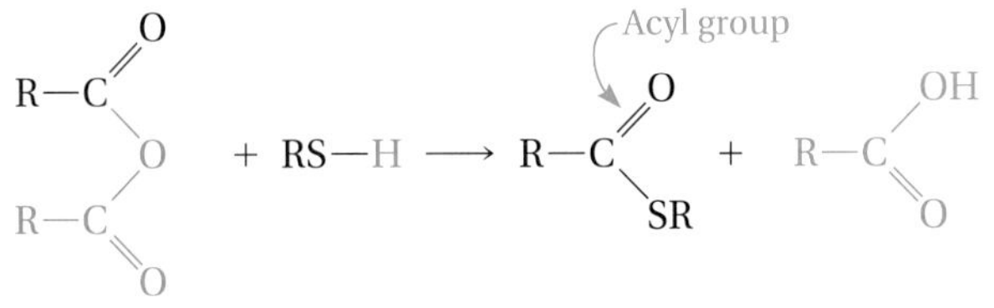

$$
\begin{array}{c}
O \\
\parallel \\
R-C \\
\qquad \backslash \\
\qquad O \qquad + \; RS-H \; \longrightarrow \\
R-C \\
\parallel \\
O
\end{array}
\qquad
\begin{array}{c}
\overset{\text{Acyl group}}{O} \\
\parallel \\
R-C \qquad + \\
\backslash \\
SR
\end{array}
\qquad
\begin{array}{c}
OH \\
\mid \\
R-C \\
\parallel \\
O
\end{array}
$$

Acid anhydride   Thiol   Thioester   Carboxylic acid

Certain thioesters have biological significance. They transfer acyl groups during the synthesis and degradation of such substances as carbohydrates, amino acids, and fatty acids. Acetyl coenzyme A is the most important acyl transfer agent in living organisms. This compound is the thioester of acetic acid and a thiol, coenzyme A:

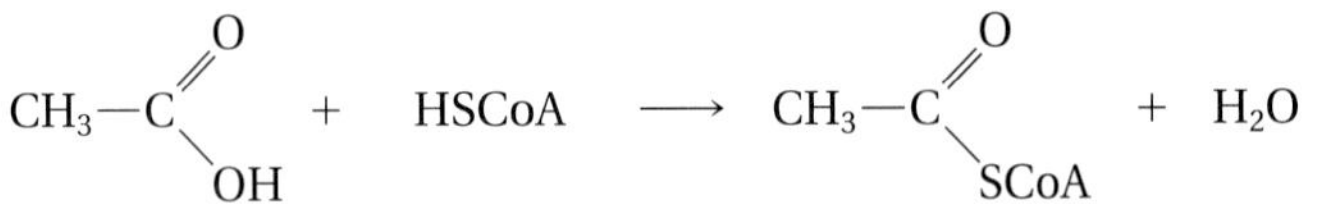

$$
\begin{array}{ccccccc}
& O & & & & O \\
& \parallel & & & & \parallel \\
CH_3-C & & + \;\; HSCoA & \longrightarrow & CH_3-C & & + \; H_2O \\
& \backslash & & & & \backslash \\
& OH & & & & SCoA
\end{array}
$$

Acetic acid   Coenzyme A   Acetyl coenzyme A   Water

5.8 Ester hydrolysis

AIM: To write an equation for the acid- or base-catalyzed hydrolysis of an ester.

If an ester is heated with water for several hours, usually very little happens. In strong acid solutions, however, ester hydrolysis is rapid because it is catalyzed by hydrogen ions: *acid catalyzed hydrolysis of an ester:*

acetic acid *ethyl*

$$CH_3-\overset{\overset{\textstyle O}{\|}}{C}-OCH_2CH_3 + H-OH \overset{H^+}{\underset{\triangle}{\rightleftharpoons}} CH_3-\overset{\overset{\textstyle O}{\|}}{C}-OH + HOCH_2CH_3$$

Ethyl acetate   Acetic acid   Ethanol

Since this hydrolysis process is reversible, a large excess of water pushes the reaction to completion.

An aqueous solution of sodium hydroxide or potassium hydroxide also can be used to hydrolyze esters. Since many esters do not dissolve in water, a solvent such as ethanol is added to make the solution homogeneous. The reaction mixture is usually heated. All the ester is converted to products. At the end of the reaction, the carboxylic acid is in solution as its sodium or potassium salt: *base promoted hydrolysis of an ester: or saponification*

$$CH_3-\overset{\overset{\textstyle O}{\|}}{C}-OCH_2CH_3 + NaOH \overset{NaOH}{\underset{\triangle}{\longrightarrow}} CH_3-\overset{\overset{\textstyle O}{\|}}{C}-O^-Na^+ + HOCH_2CH_3$$

Ethyl acetate   Sodium hydroxide   Sodium acetate   Ethanol

If the reaction mixture is acidified, the carboxylic acid is formed:

$$CH_3-\overset{\overset{\textstyle O}{\|}}{C}-O^-Na^+ + HCl \longrightarrow CH_3-\overset{\overset{\textstyle O}{\|}}{C}-OH + NaCl$$

Sodium acetate   Acetic acid

> Soap is made by the alkaline hydrolysis of naturally occurring triglycerides (triesters of glycerol and fatty acids). The process of making soap is called *saponification* after *sapon* (Latin): soap. Once hydrolyzed, the fatty acid salts are purified, dried, and shaped into bars. Additional ingredients such as dyes for color, perfume for odor, and antiseptics may be added to the soap.

EXAMPLE 5.6

Hydrolysis of an ester

Write the structure and name the products of the acid-catalyzed hydrolysis of isobutyl propanoate.

SOLUTION

$$C-C-\overset{\overset{\textstyle O}{\|}}{C}-O-C-\overset{\overset{\textstyle C}{|}}{C}-C + H_2O \overset{HCl}{\longrightarrow}$$

The ester is hydrolyzed to an alcohol and an acid:

$$CH_3CH_2COOCH_2\underset{\underset{\textstyle CH_3}{|}}{C}HCH_3 + H_2O \overset{HCl}{\longrightarrow} CH_3CH_2COOH + HOCH_2\underset{\underset{\textstyle CH_3}{|}}{C}HCH_3$$

The alkyl part of the name, *isobutyl,* comes from the alcohol, isobutyl alcohol. The acid is *propanoic acid.*

PRACTICE EXERCISE 5.9

Complete the following reactions by writing the names and structures of the expected products.

(a) $CH_3CH_2COOCH_2CH_3 + NaOH$ *(ester + base)*
$$CH_3CH_2\overset{\overset{\textstyle O}{\|}}{C}-O^-Na^+ + HOCH_2CH_3$$
→ salt + alcohol
sodium propionate + ethanol
propanoate

(b)
$$CH_3COO-+ KOH$$
$$CH_3\overset{\overset{\textstyle O}{\|}}{C}-O^-K^+ + HO-\bigcirc$$
ester + base → salt + alcohol
potassium ethanoate + phenol
acetate

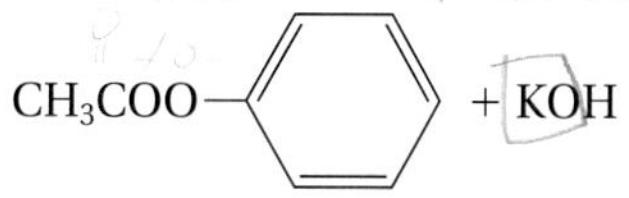

5.9 Phosphoric acids, anhydrides, and esters

AIMS: *To name and draw structures of the various forms of phosphoric acid and phosphate esters. To write equations for the hydrolysis of phosphoric acid anhydrides. To describe a phosphoryl group and write an equation showing a phosphorylation reaction.*

Focus

The chemistry of carboxylic and phosphoric acids is similar.

Recall that phosphorus is one of the important elements found in the human body. This phosphorus is present as phosphate ions, as phosphoric acid anhydrides, and as phosphate esters.

Phosphoric acid is a moderately strong acid. It differs from carboxylic acids because it contains three ionizable protons; a carboxyl function contains only one. The various degrees of ionization are

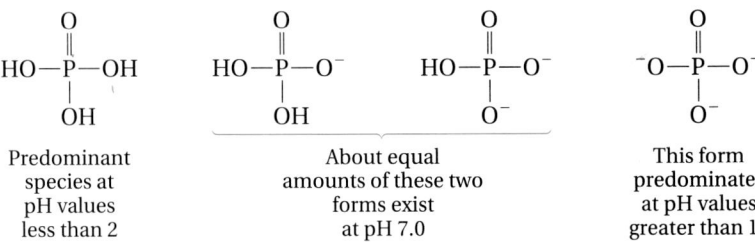

| Predominant species at pH values less than 2 | About equal amounts of these two forms exist at pH 7.0 | This form predominates at pH values greater than 11 |

As shown, the ionization state varies with the pH of the solution. At pH 7.0, the singly and doubly charged phosphate ions are present in about equal amounts. In biochemistry, the symbol P_i (inorganic phosphate) is often used to represent all possible ionization states of phosphoric acid in solution at pH 7.0.

Anhydrides of phosphoric acid

Like carboxylic acids, phosphoric acid can be dehydrated to form the anhydride. Unlike carboxylic acids, however, phosphoric acid can form more than one anhydride bond due to its multiple hydroxyl groups. The simplest anhydride of phosphoric acid is pyrophosphoric acid. The various ionized forms of pyrophosphoric acid (diphosphate ions) at pH 7.0 are often abbreviated PP_i (inorganic pyrophosphate):

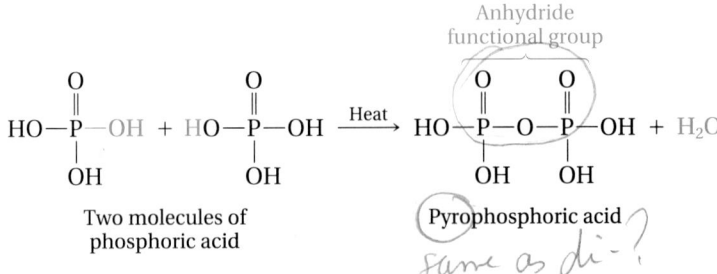

Two molecules of phosphoric acid

Pyrophosphoric acid

Pyrophosphoric acid can react with yet another molecule of phosphoric acid by elimination of water to form triphosphoric acid. Triphosphoric acid is sometimes represented in its various ionized forms (triphosphate ions) at pH 7.0 as PPP$_i$ (inorganic triphosphate):

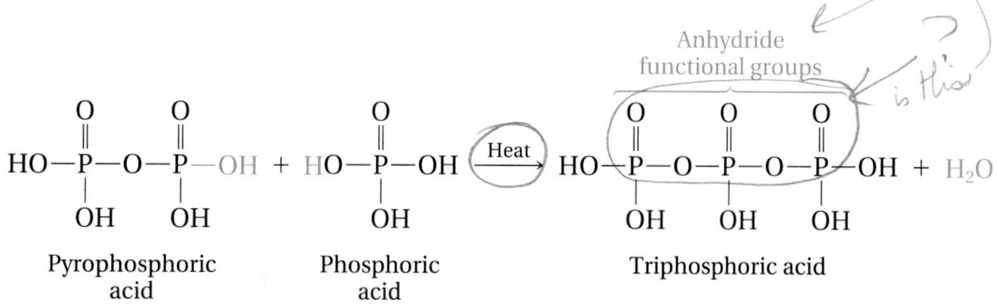

Anhydride
functional groups

is this ←

$$HO-\underset{\underset{OH}{|}}{\overset{\overset{O}{\|}}{P}}-O-\underset{\underset{OH}{|}}{\overset{\overset{O}{\|}}{P}}-OH \;+\; HO-\underset{\underset{OH}{|}}{\overset{\overset{O}{\|}}{P}}-OH \;\xrightarrow{\text{Heat}}\; HO-\underset{\underset{OH}{|}}{\overset{\overset{O}{\|}}{P}}-O-\underset{\underset{OH}{|}}{\overset{\overset{O}{\|}}{P}}-O-\underset{\underset{OH}{|}}{\overset{\overset{O}{\|}}{P}}-OH \;+\; H_2O$$

Pyrophosphoric acid   Phosphoric acid   Triphosphoric acid

Pyrophosphoric acid and triphosphoric acid are the major anhydrides of phosphoric acid. Remember that phosphoric acid anhydrides contain two functional groups: the anhydride and hydroxyl groups. Both these functional groups are important in the reactions we will be discussing.

Esters of phosphoric acid

Phosphate esters are important intermediates in many metabolic pathways. In Chapter 15, for example, we will see that the ester glyceraldehyde-3-phosphate plays an important role in the energy-producing breakdown of glucose.

Just as carboxylic acids and alcohols react to form carboxylic acid esters, phosphoric acid can react with alcohols to form phosphate esters. Because phosphoric acid has three hydroxyl groups, one, two, or all three of these groups can be esterified to form monoesters, diesters, or triesters. This is shown for the reaction of phosphoric acid with methanol.

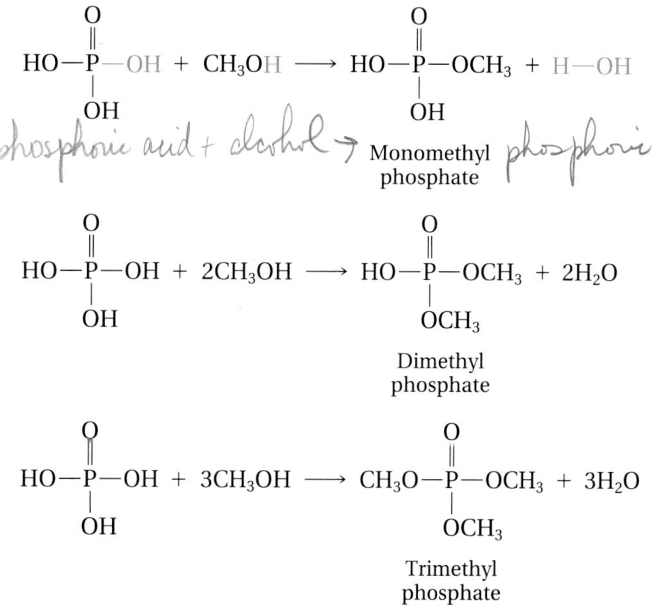

$$HO-\underset{\underset{OH}{|}}{\overset{\overset{O}{\|}}{P}}-OH \;+\; CH_3OH \;\longrightarrow\; HO-\underset{\underset{OH}{|}}{\overset{\overset{O}{\|}}{P}}-OCH_3 \;+\; H-OH$$

phosphoric acid + alcohol → Monomethyl *phosphoric ester + water*
phosphate

$$HO-\underset{\underset{OH}{|}}{\overset{\overset{O}{\|}}{P}}-OH \;+\; 2CH_3OH \;\longrightarrow\; HO-\underset{\underset{OCH_3}{|}}{\overset{\overset{O}{\|}}{P}}-OCH_3 \;+\; 2H_2O$$

Dimethyl
phosphate

$$HO-\underset{\underset{OH}{|}}{\overset{\overset{O}{\|}}{P}}-OH \;+\; 3CH_3OH \;\longrightarrow\; CH_3O-\underset{\underset{OCH_3}{|}}{\overset{\overset{O}{\|}}{P}}-OCH_3 \;+\; 3H_2O$$

Trimethyl
phosphate

The hydroxyl groups of phosphoric acid anhydrides may form esters with alcohols without breaking the anhydride bond. The following example shows a monoester, although more than one hydroxyl group could be

#37
breaks
anhydride
bond

esterified. Again, the methyl esters are used as typical examples:

$$HO-\overset{\overset{\displaystyle O}{\|}}{\underset{\underset{\displaystyle OH}{|}}{P}}-O-\overset{\overset{\displaystyle O}{\|}}{\underset{\underset{\displaystyle OH}{|}}{P}}-OCH_3 \qquad HO-\overset{\overset{\displaystyle O}{\|}}{\underset{\underset{\displaystyle OH}{|}}{P}}-O-\overset{\overset{\displaystyle O}{\|}}{\underset{\underset{\displaystyle OH}{|}}{P}}-O-\overset{\overset{\displaystyle O}{\|}}{\underset{\underset{\displaystyle OH}{|}}{P}}-OCH_3$$

Monomethyl ester of   Monomethyl ester of
pyrophosphoric acid   triphosphoric acid

Hydrolysis and phosphorylation

We have seen that acetic anhydride can undergo cleavage by water in hydrolysis to form acetic acid (see Sec. 5.5) or react with alcohols to form esters in acylation (see Sec. 5.6). Phosphoric acid anhydrides can do the same. Upon hydrolysis, phosphoric acid anhydrides are converted back to phosphoric acid:

$$HO-\overset{\overset{\displaystyle O}{\|}}{\underset{\underset{\displaystyle OH}{|}}{P}}-O-\overset{\overset{\displaystyle O}{\|}}{\underset{\underset{\displaystyle OH}{|}}{P}}-OH + H-OH \longrightarrow HO-\overset{\overset{\displaystyle O}{\|}}{\underset{\underset{\displaystyle OH}{|}}{P}}-OH + HO-\overset{\overset{\displaystyle O}{\|}}{\underset{\underset{\displaystyle OH}{|}}{P}}-OH$$

Phosphoryl group

Pyrophosphoric acid   Two molecules of phosphoric acid

$$HO-\overset{\overset{\displaystyle O}{\|}}{\underset{\underset{\displaystyle OH}{|}}{P}}-O-\overset{\overset{\displaystyle O}{\|}}{\underset{\underset{\displaystyle OH}{|}}{P}}-O-\overset{\overset{\displaystyle O}{\|}}{\underset{\underset{\displaystyle OH}{|}}{P}}-OH + H_2O \longrightarrow 3HO-\overset{\overset{\displaystyle O}{\|}}{\underset{\underset{\displaystyle OH}{|}}{P}}-OH$$

Triphosphoric acid   Three molecules of phosphoric acid

We can consider the hydrolysis of phosphoric acid anhydrides as the transfer of a phosphoryl group ($-PO_3H_2$) to water, just as the hydrolysis of acetic anhydride is a transfer of an acetyl group to water. And just as the transfer of an acetyl group to an alcohol or other functional group is called *acetylation, transfer of a* **phosphoryl group** *to an alcohol or other functional group is called* **phosphorylation.** Here are two typical phosphorylation reactions:

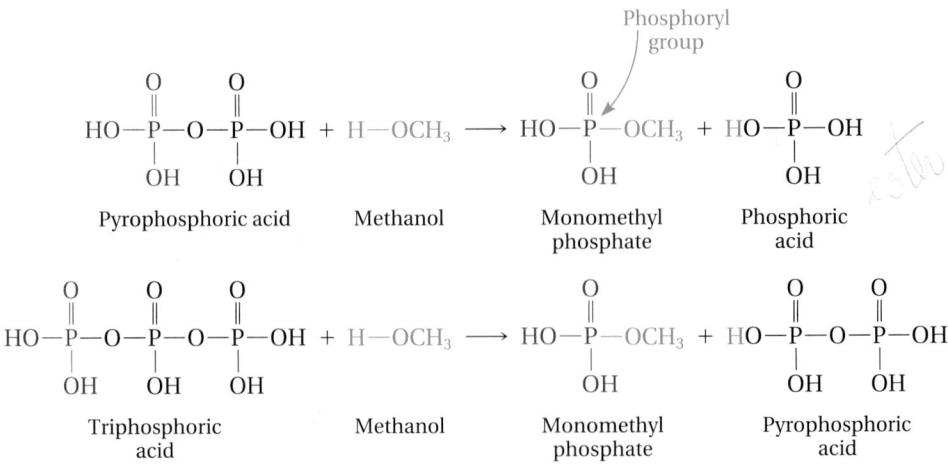

Phosphorylation reactions of this type are enormously important in biochemistry. Adenosine triphosphate (ATP), a monoester of triphosphoric acid, is nature's universal phosphorylating agent. The ATP in living systems is used to phosphorylate water, sugars, proteins, and nucleic acids. Transfers of energy in living organisms come from these phosphorylation reactions.

PRACTICE EXERCISE 5.10

Explain what is meant by the term *phosphorylation*.

5.10 Esters of nitric and nitrous acids

AIM: To state the names and uses of important esters of nitrous and nitric acids.

Focus

Nitrous and nitric acids form important esters.

Alcohols react with nitric acid (HNO_3) and nitrous acid (HNO_2) to produce the alkyl nitrates and alkyl nitrites, respectively. Glycerol and nitric acid produce the ester glyceryl trinitrate, more commonly called *nitroglycerin*.

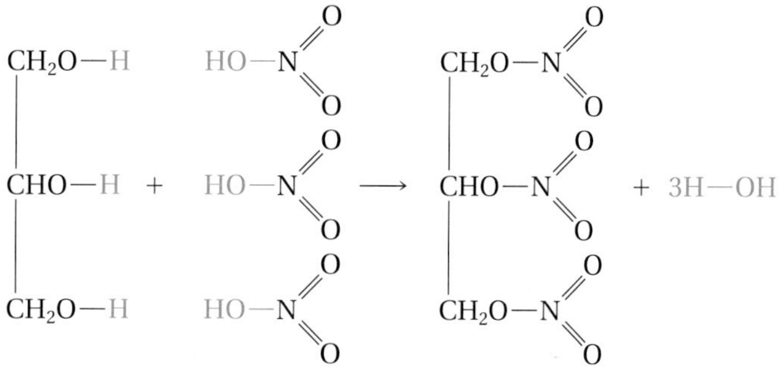

| Glycerol | Three molecules of nitric acid | Glyceryl trinitrate (nitroglycerin) | Water |

Alfred Nobel (1833–1896) conducted experiments to stabilize nitroglycerin on a lake barge. This came after the factory in which he manufactured nitroglycerin was destroyed in an explosion and he was forbidden by the Swedish government to rebuild. Upon his death, he left the bulk of his fortune to establish the Nobel prizes.

Nitroglycerin is an unstable, shock-sensitive, explosive, pale yellow liquid. It was first made in 1846 by the Italian chemist Ascanio Sobrero (1812–1888). The Swedish chemist Alfred Nobel (1833–1896) perfected its synthesis in the mid-19th century and devised a safe method for handling it. Nobel mixed nitroglycerin with *kieselghur*, a claylike absorbent material. The result was dynamite, an explosive that is not sensitive to shock. Since its development, dynamite has been important in the construction of canals, dams, and roads. It also has played a major role in warfare.

Nitroglycerin and isoamyl nitrite, an ester of nitrous acid, have been used as drugs for over a hundred years.

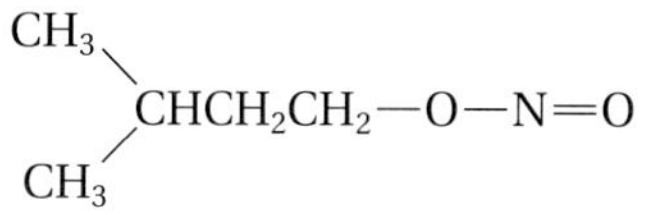

Isoamyl nitrite

Both these esters, when inhaled or taken as tablets, produce immediate relaxation of the smooth muscles of the body and expansion of the blood vessels. They are used to treat people with *angina pectoris*—chest pains caused by an insufficient supply of oxygen to the heart muscle. Isoamyl nitrite is also the active ingredient in the widely abused drug called "poppers."

SUMMARY

Carboxylic acids (RCO_2H) contain the carboxyl functional group ($-CO_2H$). Fatty acids (straight-chain saturated and unsaturated aliphatic carboxylic acids) are widely distributed in nature, as are many other carboxylic acids. Benzoic acid is the parent aromatic carboxylic acid. Fatty acids containing up to four carbons are completely soluble in water. Carboxylic acids of higher molar mass are less soluble, but they tend to form monomolecular layers at water surfaces and micelles within water. Carboxylic acids are weak acids and ionize only to a slight extent in water. The acidity increases if the acid molecules contain electron-withdrawing substituents. Like other acids, carboxylic acids can be neutralized by bases to give salts. Soaps are the alkali metal salts of long-chain carboxylic acids. Detergents are often alkali metal salts of sulfonic acids (RSO_3H). Carboxylic acids may be prepared by either the oxidation of primary alcohols or the oxidation of aldehydes.

Anhydrides (RCO_2COR) and esters (RCO_2R) are important derivatives of carboxylic acids. Anhy-

drides, formed by the dehydration of carboxylic acids, form esters when they react with an alcohol or a phenol, as in the preparation of aspirin. Esters also can be prepared by the reaction of a carboxylic acid and an alcohol. Many naturally occurring esters impart pleasant odors and flavors to fruits. Polyesters (polymeric esters) can be made into synthetic fibers. Esters can be converted to the component carboxylic acid and alcohol by hydrolysis in acidic or basic solution. Certain thioesters, especially esters of the thiol coenzyme A, are biologically important. Esters and anhydrides of phosphoric acid are also biologically important as agents that can transfer phosphoryl groups ($-PO_3H_2$). ATP, an ester of the phosphoric anhydride triphosphoric acid, is the major phosphorylating agent of living cells. Esters of inorganic acids such as nitric and nitrous acids have such varied applications as explosives (nitroglycerin) and medications for the relief of angina pectoris (isoamyl nitrite).

SUMMARY OF REACTIONS

Here are the reactions of carboxylic acids and inorganic acids and their derivatives described in this chapter.

Carboxylic Acids

1. Preparation of carboxylic acids:

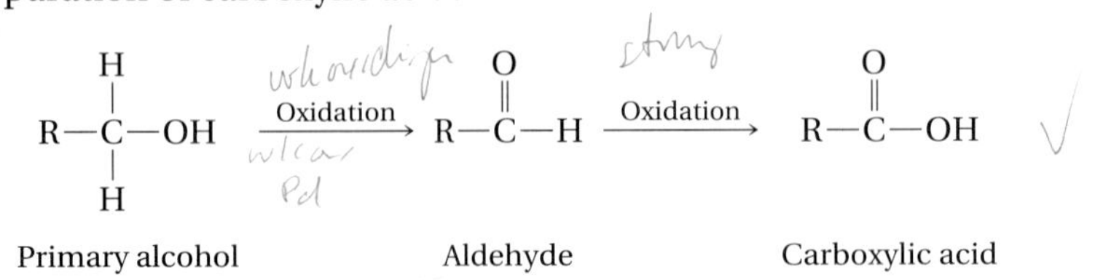

2. Neutralization of carboxylic acids:

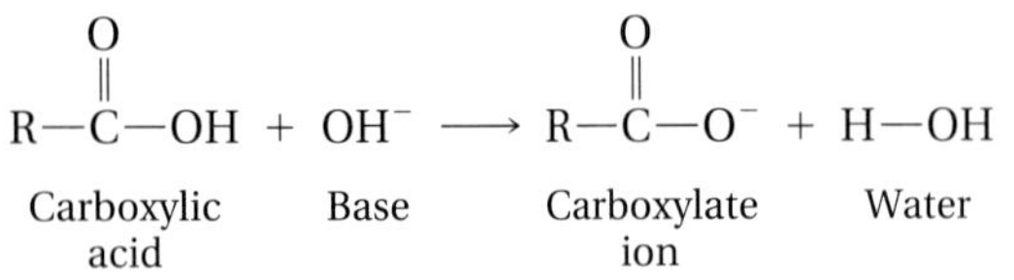

3. Preparation of acid anhydrides:

$$
\underset{\text{Two molecules of acid}}{\overset{\displaystyle O}{R-\overset{\|}{C}-OH}\ +\ \overset{\displaystyle O}{HO-\overset{\|}{C}-R}}\ \longrightarrow\ \underset{\text{Acid anhydride}}{\overset{\displaystyle O\qquad O}{R-\overset{\|}{C}-O-\overset{\|}{C}-R}}\ +\ \underset{\text{Water}}{H-OH}
$$

4. Preparation of esters
   (a) Oxyesters:

$$
\underset{\text{Acid}}{\overset{\displaystyle O}{R-\overset{\|}{C}-OH}}\ +\ \underset{\text{Alcohol}}{RO-H}\ \overset{H^+}{\rightleftharpoons}\ \underset{\text{Ester}}{\overset{\displaystyle O}{R-\overset{\|}{C}-OR}}\ +\ \underset{\text{Water}}{H-OH}
$$

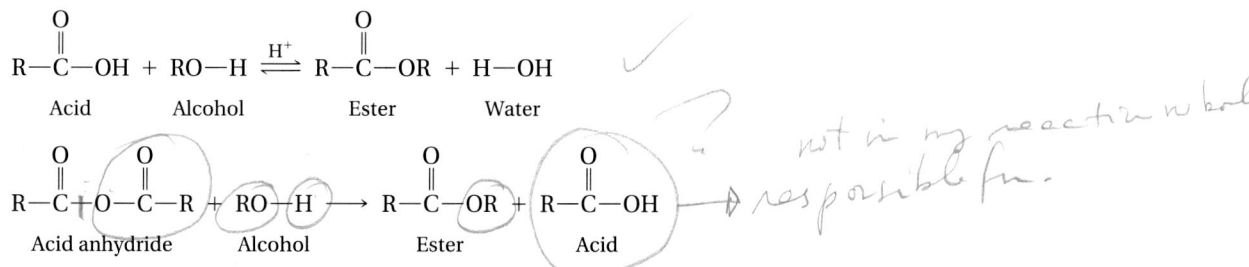

$$
\underset{\text{Acid anhydride}}{\overset{\displaystyle O\qquad O}{R-\overset{\|}{C}-O-\overset{\|}{C}-R}}\ +\ \underset{\text{Alcohol}}{RO-H}\ \longrightarrow\ \underset{\text{Ester}}{\overset{\displaystyle O}{R-\overset{\|}{C}-OR}}\ +\ \underset{\text{Acid}}{\overset{\displaystyle O}{R-\overset{\|}{C}-OH}}
$$

   (b) Thioesters:

$$
\underset{\text{Acid}}{\overset{\displaystyle O}{R-\overset{\|}{C}-OH}}\ +\ \underset{\text{Thiol}}{RS-H}\ \longrightarrow\ \underset{\text{Thioester}}{\overset{\displaystyle O}{R-\overset{\|}{C}-SR}}\ +\ \underset{\text{Water}}{H-OH}
$$

Inorganic Acids

1. Preparation of phosphoric acid anhydrides:

$$
\underset{\substack{\text{Phosphoric}\\\text{acid}}}{\overset{\displaystyle O}{\underset{\displaystyle OH}{HO-\overset{\|}{\underset{|}{P}}-OH}}}\ +\ \underset{\substack{\text{Phosphoric}\\\text{acid}}}{\overset{\displaystyle O}{\underset{\displaystyle OH}{HO-\overset{\|}{\underset{|}{P}}-OH}}}\ \overset{\text{Heat}}{\longrightarrow}
$$

$$
\overset{\displaystyle O\qquad\ O}{\underset{\displaystyle OH\quad OH}{HO-\overset{\|}{\underset{|}{P}}-O-\overset{\|}{\underset{|}{P}}-OH}}\ +\ H_2O
$$

$$
\underset{\substack{\text{Pyrophosphoric}\\\text{acid}}}{\overset{\displaystyle O\qquad\ O}{\underset{\displaystyle OH\quad OH}{HO-\overset{\|}{\underset{|}{P}}-O-\overset{\|}{\underset{|}{P}}-OH}}}\ +\ \underset{\substack{\text{Phosphoric}\\\text{acid}}}{\overset{\displaystyle O}{\underset{\displaystyle OH}{HO-\overset{\|}{\underset{|}{P}}-OH}}}\ \overset{\text{Heat}}{\longrightarrow}
$$

$$
\underset{\substack{\text{Triphosphoric}\\\text{acid}}}{\overset{\displaystyle O\qquad\ O\qquad\ O}{\underset{\displaystyle OH\quad OH\quad OH}{HO-\overset{\|}{\underset{|}{P}}-O-\overset{\|}{\underset{|}{P}}-O-\overset{\|}{\underset{|}{P}}-OH}}}\ +\ \underset{\text{Water}}{H_2O}
$$

2. Preparation of phosphate esters:

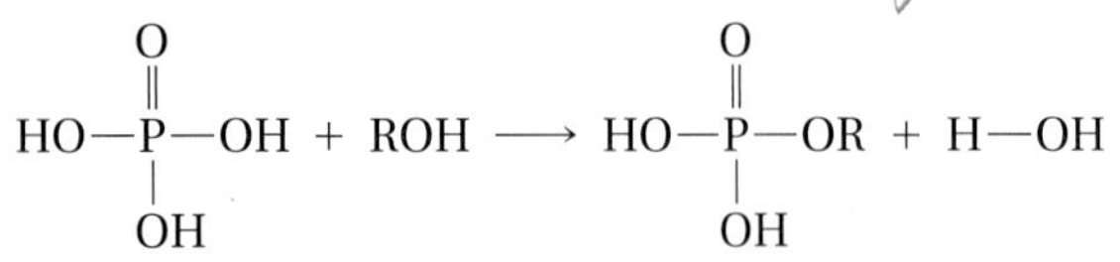

$$\underset{\overset{\displaystyle |}{OH}}{HO-\overset{\overset{\displaystyle O}{\|}}{P}-OH} + ROH \longrightarrow \underset{\overset{\displaystyle |}{OH}}{HO-\overset{\overset{\displaystyle O}{\|}}{P}-OR} + H-OH$$

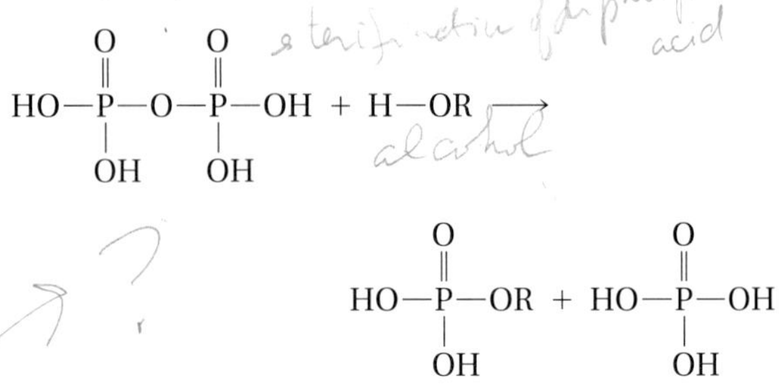

Anhydrides

1. Hydrolysis of acid anhydrides
   (a) Carboxylic acid anhydrides:

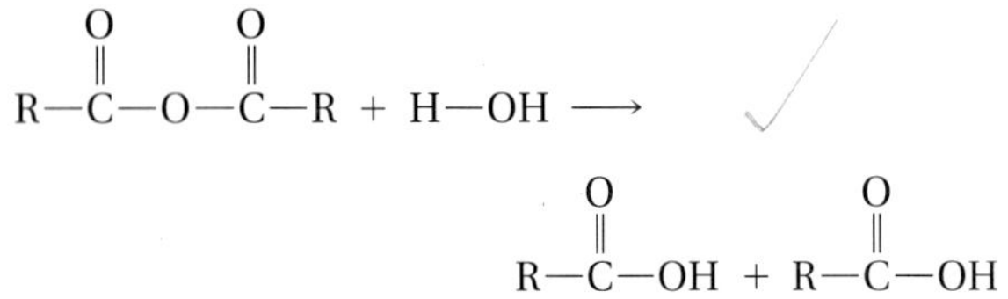

$$\underset{}{R-\overset{\overset{\displaystyle O}{\|}}{C}-O-\overset{\overset{\displaystyle O}{\|}}{C}-R} + H-OH \longrightarrow$$

$$R-\overset{\overset{\displaystyle O}{\|}}{C}-OH + R-\overset{\overset{\displaystyle O}{\|}}{C}-OH$$

   (b) Phosphoric acid anhydrides:

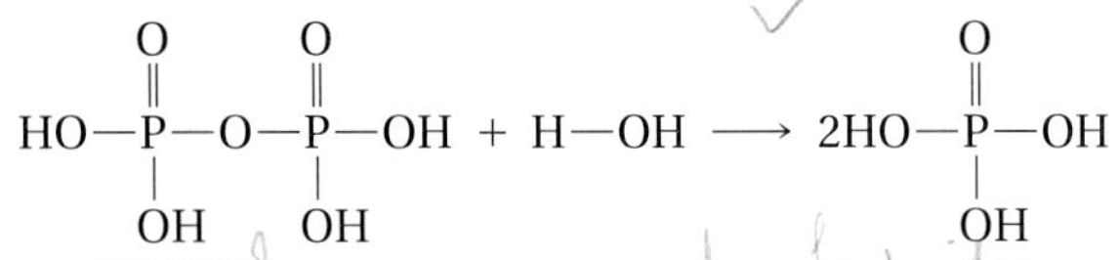

$$\underset{\overset{\displaystyle |}{OH}}{HO-\overset{\overset{\displaystyle O}{\|}}{P}-O}\underset{\overset{\displaystyle |}{OH}}{-\overset{\overset{\displaystyle O}{\|}}{P}-OH} + H-OH \longrightarrow 2\underset{\overset{\displaystyle |}{OH}}{HO-\overset{\overset{\displaystyle O}{\|}}{P}-OH}$$

2. Acylation of alcohols:

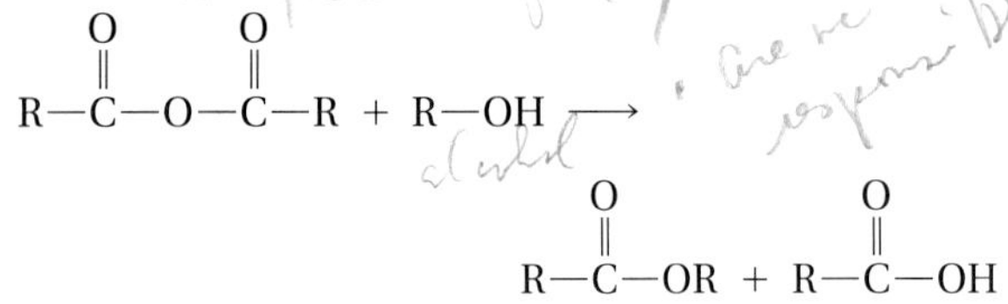

$$\underset{}{R-\overset{\overset{\displaystyle O}{\|}}{C}-O-\overset{\overset{\displaystyle O}{\|}}{C}-R} + R-OH \longrightarrow$$

$$R-\overset{\overset{\displaystyle O}{\|}}{C}-OR + R-\overset{\overset{\displaystyle O}{\|}}{C}-OH$$

3. Phosphorylation of alcohols:

$$\underset{\overset{\displaystyle |}{OH}}{HO-\overset{\overset{\displaystyle O}{\|}}{P}-O}\underset{\overset{\displaystyle |}{OH}}{-\overset{\overset{\displaystyle O}{\|}}{P}-OH} + H-OR \longrightarrow$$

$$\underset{\overset{\displaystyle |}{OH}}{HO-\overset{\overset{\displaystyle O}{\|}}{P}-OR} + \underset{\overset{\displaystyle |}{OH}}{HO-\overset{\overset{\displaystyle O}{\|}}{P}-OH}$$

Esters

1. Hydrolysis of carboxylate esters
   (a) In acid:

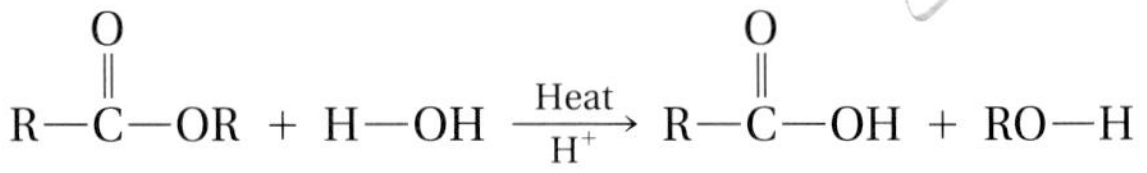

$$R-\overset{\overset{\displaystyle O}{\|}}{C}-OR + H-OH \xrightarrow[H^+]{Heat} R-\overset{\overset{\displaystyle O}{\|}}{C}-OH + RO-H$$

   (b) In base:

$$R-\overset{\overset{\displaystyle O}{\|}}{C}-OR + OH^- \xrightarrow{Heat} R-\overset{\overset{\displaystyle O}{\|}}{C}-O^- + RO-H$$

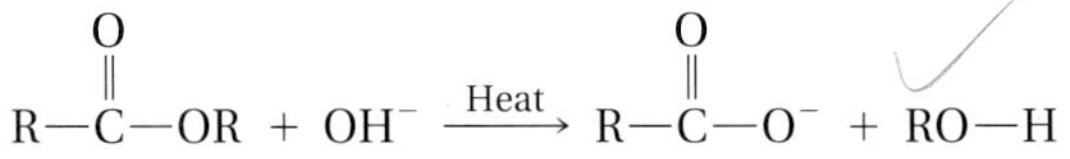

KEY TERMS

Acid anhydride (5.5)	Carboxylic acid (5.1)	Hard water (5.3)	Phosphoryl group (5.9)
Acylation reaction (5.5)	Detergent (5.3)	Hydrolysis (5.5)	Polyester (5.6)
Acyl group (5.5)	Dicarboxylic acid (5.1)	Hydroxy acid (5.1)	Soap (5.3)
Analgesic (5.6)	Dimer (5.2)	Micelle (5.3)	Thioester (5.7)
Antipyretic (5.6)	Ester (5.6)	Monomolecular layer (5.2)	
Biodegradable (5.3)	Esterification (5.6)	Nonionic detergent (5.3)	
Carboxyl group (5.1)	Fatty acid (5.1)	Phosphorylation (5.9)	

EXERCISES

Carboxylic Acids and Their Salts
(Sections 5.1, 5.2, 5.3)

5.11 Give the name (IUPAC or common) or write the structure of each of the following carboxylic acids.
(a) H—COOH

(b)
$$CH_3-\overset{\overset{\displaystyle OH}{|}}{CH}-COOH$$ *lactic acid*

(c) 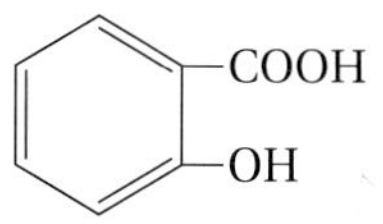
benzene ring with —COOH and —OH substituents

(d)
$$CH_3\overset{\overset{\displaystyle O}{\|}}{C}-COOH$$

(e) stearic acid   (f) lactic acid
(g) *p*-chlorobenzoic acid   (h) malonic acid

5.12 Write the structure or give the name of each of the following compounds.
(a) acetic acid
(b) 2-hydroxypropanoic acid
(c) citric acid
(d) benzoic acid
(e)
$$\overset{\displaystyle COOH}{\underset{\displaystyle COOH}{|}}$$
(f)
$$CH_3CH_2\overset{\overset{\displaystyle Br}{|}}{CH}CH_2CH_2COOH$$
(g) $CH_3(CH_2)_5COOH$
(h)
$$HO-\overset{}{CH}-COOH$$
$$HO-CH-COOH$$

5.13 Which of the acids in Exercise 5.11, if any, are dicarboxylic, polyfunctional, or fatty acids?

5.14 Designate each acid in Exercise 5.12 as a dicarboxylic, polyfunctional, or fatty acid.

5.15 Write the structural formulas for the compounds in Table 5.3, and explain why propanoic acid has the highest boiling point.

5.16 Predict which of the following compounds has the highest boiling point. Molar masses are given in parentheses.
(a) CH_3CHO (44) *aldehyde*   (b) $HCOOH$ (46) *acid O*
(c) $CH_3CH_2CH_3$ (44)   (d) CH_3CH_2OH (46)

5.17 Why are carboxylic acids weak acids?

5.18 Why is trichloroacetic acid a much stronger acid than acetic acid?

5.19 Draw the structure and name the products for each of the following reactions.
(a) $HCOOH + KOH \longrightarrow$
(b)
$$\overset{\displaystyle COOH}{\underset{\displaystyle COOH}{|}} + Ca(OH)_2 \longrightarrow$$
(c) Myristic acid + sodium hydroxide $\longrightarrow$

5.20 Write the structure and name the salt formed in each of the following reactions.
(a) Propanoic acid + sodium bicarbonate $\longrightarrow$
(b) 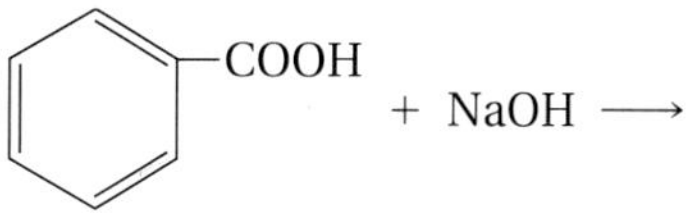
benzene ring with —COOH substituent $+ NaOH \longrightarrow$
(c) $CH_3CH_2CH_2COOH + NH_3 \longrightarrow$

5.21 What is a soap? Describe the two distinct parts of a soap molecule, and explain how soap acts as an emulsifying agent.

5.22 Distinguish between a detergent molecule and a soap molecule, and explain why detergents are better than soaps for washing clothes in hard water.

Preparation of Carboxylic Acids (Section 5.4)

5.23 Complete the following reactions by writing the structure of the expected products.
(a) $CH_3CH_2OH \xrightarrow{K_2Cr_2O_7}$
(b) $CH_3CH_2CHO \xrightarrow{K_2Cr_2O_7}$
(c) $(CH_3)_3COH \xrightarrow{KMnO_4}$

5.24 Write the structure of the acid formed (if any) by the oxidation of the following compounds.
(a) 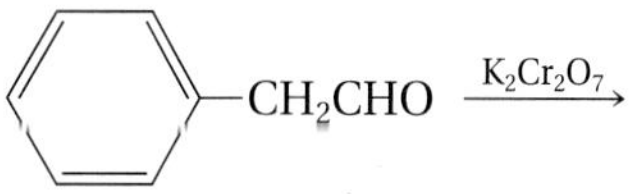
benzene ring with —CH_2CHO substituent $\xrightarrow{K_2Cr_2O_7}$
(b)
$$CH_3CH_2\overset{\overset{\displaystyle O}{\|}}{C}CH_3 \xrightarrow{K_2Cr_2O_7}$$
(c)
$$CH_3CH_2\overset{\overset{\displaystyle}{\underset{\underset{\displaystyle CH_3}{|}}{}}}{CH}OH \xrightarrow{K_2Cr_2O_7}$$

5.25 Name the aldehyde you would oxidize, with acidified $K_2Cr_2O_7$, to produce (a) butanoic acid and (b) formic acid.

5.26 What compound could be oxidized to produce (a) benzoic acid and (b) 3-methylpentanoic acid?

Anhydrides and Esters (Sections 5.5–5.8)

5.27 Draw the general formula for (a) an acid anhydride and (b) an ester.

5.28 What are the general products of (a) hydrolysis of an acid anhydride and (b) hydrolysis of an ester?

5.29 Name the given ester or write the structure and name of the ester that could be produced from the given reactants.
(a) CH_3COOCH_3 (b) $CH_3COOCH_2CH_3$
(c)

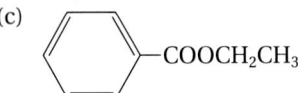$-COOCH_2CH_3$

(d) formic acid + methanol $\longrightarrow$
(e) propionic acid + propanol $\longrightarrow$

5.30 Write the structure and name of the ester that could be produced from the given reactants or name the given ester.
(a) butyric acid + ethanol $\longrightarrow$
(b) acetic acid + butanol $\longrightarrow$
(c) $HCOOCH_3$
(d)

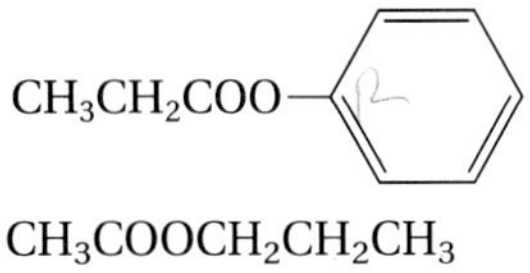

CH_3CH_2COO-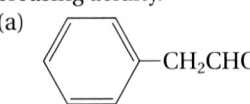

(e) $CH_3COOCH_2CH_2CH_3$

5.31 Show the difference in structure between an oxyester and a thioester.

5.32 Write the structures of the hydrolysis products of the following thioester:

$$\overset{\overset{\displaystyle O}{\|}}{CH_3CH_2C}-S-CH_3$$

5.33 Complete the following reactions by writing the structures of the expected products and by naming each of the reactants and products.
(a) $CH_3COOCH_3 + H_2O \xrightarrow{HCl}$
(b)

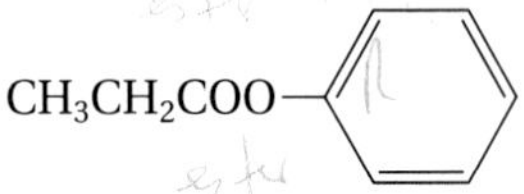

CH_3CH_2COO- ⬡ $+ NaOH \longrightarrow$

(c)

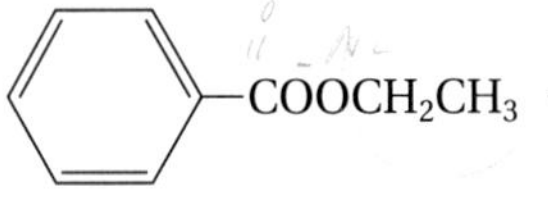

⬡$-COOCH_2CH_3 + NaOH \longrightarrow$

(d) $CH_3CH_2CH_2COOCH_2CH_2CH_3 + NaOH \longrightarrow$
(e) $HCOOCH_2CH_3 + KOH \longrightarrow$
(f)

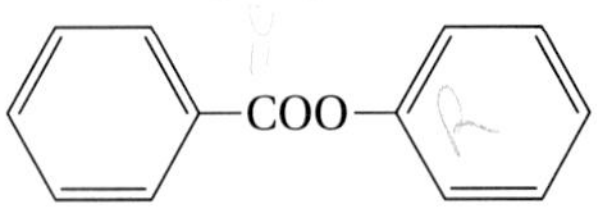

⬡$-COO-$⬡ $+ H_2O \xrightarrow{HCl}$

5.34 Write the structure and name of the expected products for each of the following reactions.
(a) $CH_3COOH + CH_3OH \xrightarrow{H^+}$
(b) $CH_3CH_2CH_2COOCH_2CH_3 + NaOH \longrightarrow$
(c) $CH_3CH_2OH \xrightarrow{K_2Cr_2O_7}$
(d) $CH_3CH_2CH_2COOH + NaOH \longrightarrow$
(e)

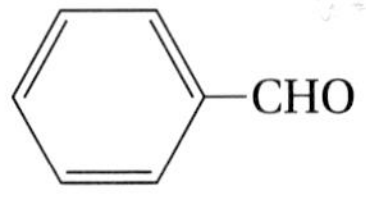

⬡$-CHO \xrightarrow{K_2Cr_2O_7}$

(f)

$$\overset{\overset{\displaystyle O}{\|}}{CH_3C}-O-\overset{\overset{\displaystyle O}{\|}}{CCH_3} + CH_3OH \longrightarrow$$

Acids of Phosphorus and Nitrogen (Sections 5.9, 5.10)

5.35 Write the structures of phosphoric acid to show the various degrees of ionization. Which forms predominate at pH 7?

5.36 Write the structures of (a) pyrophosphoric acid (PP_i) and (b) triphosphoric acid (PPP_i). Why are these compounds called *anhydrides*?

5.37 One molecule of ethanol reacts with one molecule of pyrophosphoric acid.
(a) Give the name and structure of the organic product.
(b) What is the process called?

5.38 Isoamyl nitrite is an inorganic ester whose formula is

$$\overset{\displaystyle CH_3}{\underset{\displaystyle CH_3}{\diagdown}}CHCH_2CH_2-O-N{=}O$$

What will the products of hydrolysis be?

Additional Exercises

5.39 Arrange the following substances in order of decreasing acidity.
(a)

⬡$-CH_2CHO$

(b) CH_3CH_2OH
(c) CH_3COOH (d) H_3O^+

5.40 Write the equilibrium expression for the following reaction:

$$CH_3COOH + H_2O \rightleftharpoons CH_3COO^- + H_3O^+$$

5.41 In the following reaction, which is the stronger base, water or acetate ion? (The K_a of acetic acid is 1.8×10^{-5}.)

$$CH_3COOH + H_2O \rightleftharpoons CH_3COO^- + H_3O^+$$

5.42 List the following compounds in order of increasing acid strength.

(a) CH₃CHCO₂H ③
    |
    Cl

(b) ClCH₂CH₂CO₂H ②

(c) CH₃CH₂CO₂H ①

(d) CH₃CHCO₂H ④
    |
    F

5.43 Write the structural formula of each of the following compounds:

(a) decane
(b) hexanoic acid
(c) dimethyl ether
(d) octanal
(e) ethyl butyrate
(f) 1-pentanol
(g) 2-chlorobutanoic acid
(h) 3-bromo-4-methyl-5-phenyl-1-pentanol

5.44 Write equations for each of the following reactions, naming the organic products formed.

(a) the neutralization of butyric acid
(b) the acylation of methanol with acetic anhydride
(c) the oxidation of 3,3-dimethylbutanal
(d) the hydrolysis of acetic anhydride
(e) the complete ionization of oxalic acid
(f) the esterification of butyric acid with isobutyl alcohol
(g) the hydrolysis of triphosphoric acid
(h) the hydrolysis of isopropyl propionate with base

SELF-TEST (REVIEW)

True/False

1. A secondary alcohol can be oxidized to form a carboxylic acid. *False   get ketone*

2. A carboxyl group contains both a hydroxyl group and a carbonyl group. *True*

3. Carboxylic acids of low molar mass are soluble in water. *true*

4. The first word in the name of an ester comes from the alcohol part of the molecule. *true*

5. Pyrophosphoric acid is an anhydride of phosphoric acid. *true*

6. Ester hydrolysis is catalyzed by both acids and bases. *true*

7. The ester formed between formic acid and butyl alcohol is called *butyl methanoate*. *true*

8. The functional group of a thioester is

    O
    ‖
   —C—S—   *true*

9. Both acylation and phosphorylation yield an ester as a product. *true*

10. Carboxylic acids are able to form intermolecular hydrogen bonds. *true*

Multiple Choice

11. The best name for the compound

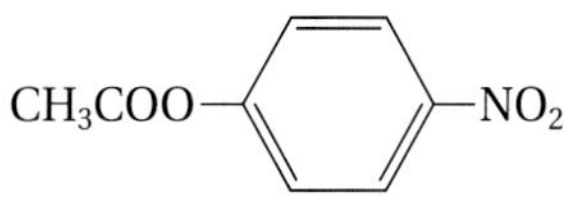

    CH₃COO—⟨benzene ring⟩—NO₂

is
(a) *p*-nitroacetylbenzene.
(b) acetylbenzene nitrite.
(c) *p*-nitrophenyl acetate.
(d) nitrobenzene acetate. *acid + alcohol → ester*

12. The reaction of a carboxylic acid and a thiol results in the formation of a
(a) thioether.   (b) thioester.
(c) thioacetal.   (d) thioanhydride.

13. Which of the following statements about soap molecules is *false?*
(a) The hydrophobic ends of the molecules tend to interact favorably with water. *False*
(b) They can be precipitated by high acid concentration and by calcium ions. *true*
(c) They have both a hydrophilic head and a hydrophobic tail. *true*
(d) They are salts of fatty acids. *true*

14. Carboxylic acids have relatively high melting and boiling points because
    (a) they are generally long-chain compounds.
    (b) they have a high density.
    (c) they have a nonpolar end.
    (d) they form intermolecular hydrogen bonds.

15. Which of the following compounds could be oxidized by potassium permanganate to form propionic acid?
    (a) isopropyl alcohol   (b) propanone
    (c) propane   (d) 1-propanol

16. In acetylsalicylic acid, the acetyl group is
    (a) CH_3-C-O-.   (b) $-COOH$.
    (c) $-COO^-$.   (d) CH_3-CO-.

17. All the following are dicarboxylic acids except
    (a) lactic acid.   (b) succinic acid.
    (c) oxalic acid.   (d) phthalic acid.
    (e) glutaric acid.

18. In the monomethyl ester of pyrophosphoric acid,

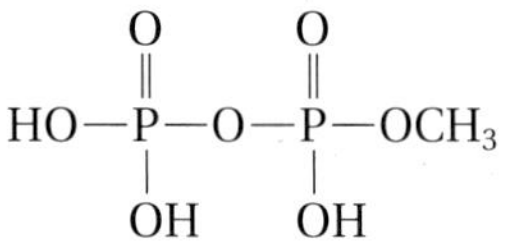

$$HO-\underset{\underset{OH}{|}}{\overset{\overset{O}{\|}}{P}}-O-\underset{\underset{OH}{|}}{\overset{\overset{O}{\|}}{P}}-OCH_3$$

    which of the following functional groups is *not* present?
    (a) hydroxy   (b) anhydride   (c) acid
    (d) ester   (e) all are present

19. Carboxylic acids are weak acids because
    (a) they are always found in low concentration in nature.
    (b) they have large K_a values.
    (c) they do not completely ionize in water.
    (d) they are found in food.

20. Which of the following acids would be least likely to lose a proton?
    (a) acetic acid   (b) chloroacetic acid
    (c) dichloroacetic acid   (d) trichloroacetic acid

21. A carboxylate ester is formed by reacting an alcohol with
    (a) an ether.   (b) a carboxylic acid.
    (c) an anhydride.   (d) another alcohol.
    (e) More than one are correct.

22. A carboxylic acid with six carbons in a straight chain would be named
    (a) succinic acid.   (b) hexanalic acid.
    (c) dimethylbutanoic acid.   (d) hexanoic acid.

23. Which of the following compounds is a product of the hydrolysis of ethyl propionate catalyzed by sodium hydroxide?
    (a) sodium acetate   (b) propanol
    (c) acetic acid   (d) sodium propionate

Amines and Amides

6

Organic Nitrogen Compounds

Many natural and synthetic nitrogen-containing compounds are used as medicinal drugs.

This chapter is about two different classes of organic compounds containing nitrogen: amines and amides. The amines and amides are very significant for the health sciences because they include among their members many important molecules in our bodies. Moreover, many natural and synthetic drugs are organic nitrogen compounds. We will first examine the structures, names, and chemistry of the amines and amides. Then we will learn about some of the major compounds that belong to these categories. Among these compounds are some important drugs that help combat depression, a condition that affected a woman named Maude. Maude's condition and its resolution are described in the following Case in Point.

CASE IN POINT: Antidepressants

Maude always considered herself a person with a sunny disposition. However, over a period of several weeks, and for no discernible reason, Maude was agitated and had difficulty sleeping. After an examination by Maude's physician revealed no physical cause for her symptoms, the doctor suspected that Maude might be suffering from depression and referred her to a psychiatrist. After probing Maude's mood, the psychiatrist suggested that Maude might benefit from an antidepressant drug in addition to taking part in counseling sessions with a group of people similarly affected (see figure). What are the chemical aspects of Maude's depression? We will learn about the use of drugs in the treatment of depression in Section 6.7.

Group sessions along with drug administration are often used in the treatment of psychiatric disorders.

6.1 Amines

AIMS: *To name and write structures of simple aliphatic and aromatic amines. To classify an amine as primary, secondary, or tertiary. To name and write structures of common aliphatic and aromatic heterocyclic amines.*

Focus

Amines may be primary, secondary, or tertiary.

Amines *are derivatives of ammonia (NH_3) in which alkyl or aryl groups (R—) replace hydrogen.* Amines are called *primary, secondary,* or *tertiary* depending on the number of R groups attached to the nitrogen.

$$H-\underset{\underset{\cdot\cdot}{|}}{\overset{\overset{H}{|}}{N}}-H \qquad R-\underset{\underset{\cdot\cdot}{|}}{\overset{\overset{H}{|}}{N}}-H \qquad R-\underset{\underset{\cdot\cdot}{|}}{\overset{\overset{R}{|}}{N}}-H \qquad R-\underset{\underset{\cdot\cdot}{|}}{\overset{\overset{R}{|}}{N}}-R$$

Ammonia   Primary   Secondary   Tertiary
             (1°) amine   (2°) amine   (3°) amine

The terms *primary, secondary,* and *tertiary* are abbreviated 1°, 2°, and 3°, respectively. Here primary, secondary, and tertiary mean something quite different from when they are applied to alcohols. Recall that in alcohols these terms refer to the number of carbon groups attached to the *carbon* that bears the hydroxyl function; in amines they refer to the number of carbon groups attached to the amine *nitrogen.* The carbon groups attached to the nitrogen may be aliphatic, aromatic, or both.

Names of aliphatic amines

Aliphatic amines are named by first listing the alkyl groups attached to the amine nitrogen in alphabetical order. The names of the alkyl groups are followed by the word -*amine;* the complete name is written as one word. We use the prefixes *di-* and *tri-* to indicate more than one alkyl group of the same kind. Here are some examples of the structures and names of aliphatic amines:

$$CH_3-\underset{\underset{\cdot\cdot}{|}}{\overset{\overset{H}{|}}{N}}-H \qquad\qquad CH_3-\underset{\underset{\cdot\cdot}{|}}{\overset{\overset{H}{|}}{N}}-CH_3$$

Methylamine   Dimethylamine
(1°)   (2°)

$$CH_3-\underset{}{\overset{\overset{CH_3}{|}}{CH}}-\underset{}{\overset{\overset{H}{|}}{N}}-CH_2CH_3 \qquad CH_3CH_2-\underset{\underset{\cdot\cdot}{|}}{\overset{\overset{CH_3}{|}}{N}}-CH_3$$

Ethylisopropylamine   Ethyldimethylamine
(2°)   (3°)

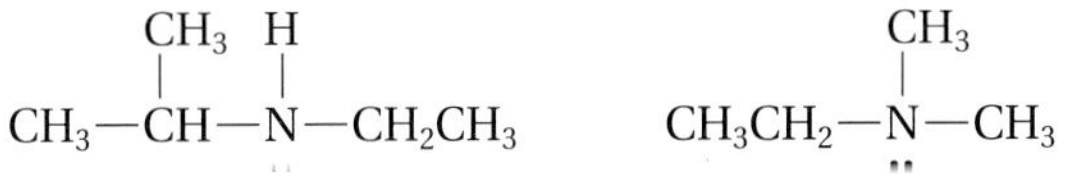

EXAMPLE 6.1

Naming an aliphatic amine

Name the following aliphatic amine, and classify it as primary, secondary, or tertiary.

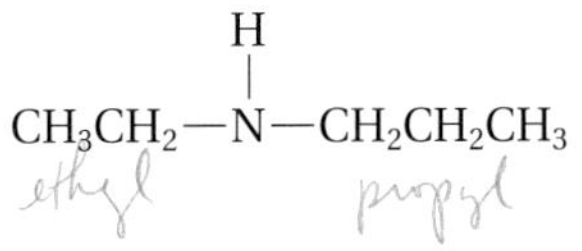

$$CH_3CH_2-\underset{}{\overset{\overset{H}{|}}{N}}-CH_2CH_2CH_3$$

ethyl   *propyl*   *ethylpropylamine (2° amine)*

SOLUTION

There are two alkyl groups attached to the amine nitrogen, so the structure is a secondary amine. In alphabetical order, the alkyl groups are ethyl (CH_3CH_2—) and propyl ($CH_3CH_2CH_2$—). The name of the amine is *ethylpropylamine*.

PRACTICE EXERCISE 6.1

Name the following amines, and classify them as primary, secondary, or tertiary.

(a) $CH_3\overset{\overset{\displaystyle H}{|}}{N}CH_2CH_3$ *ethylmethyl amine 2°*

(b) $CH_3CH_2CH_2NH_2$ *propylamine 1°*

(c) $(CH_3)_2NCH_2CH_3$ *ethyldimethylamine 3°*

(d) $(CH_3CH_2)_3N$ *triethylamine 3°*

Names of aromatic amines

The parent aromatic amine is called *aniline*. Anilines substituted with alkyl groups on the amine nitrogen are named as derivatives of aniline. The prefix *N*- indicates that the substituent is on the amine nitrogen, not on the aromatic ring. For example:

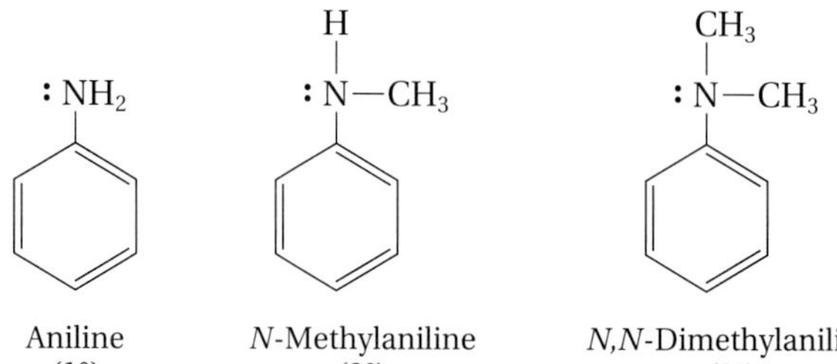

Aniline
(1°)

N-Methylaniline
(2°)

N,N-Dimethylaniline
(3°)

There are exceptions to the aniline naming system. Sometimes, for example, the benzene rings are named as phenyl substituents attached to the amine nitrogen. Two examples are diphenylamine and triphenylamine:

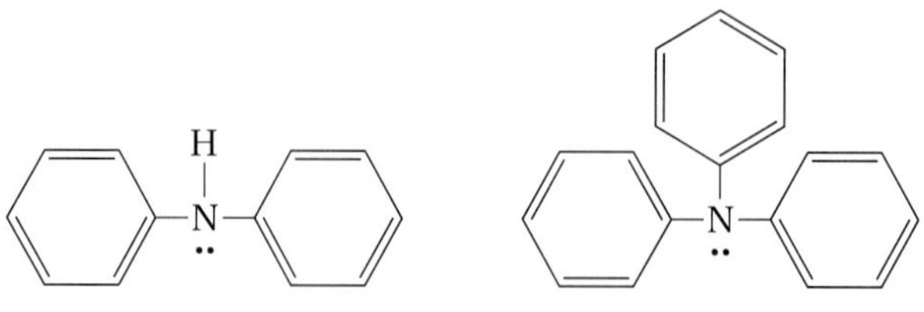

Diphenylamine
(2°)

Triphenylamine
(3°)

Anilines with a single methyl group on the benzene ring are sometimes called *toluidines*. The relative positions of the amino and methyl groups on the aniline rings of the toluidines are indicated by the prefixes *ortho-* (*o-*), *meta-* (*m-*), and *para-* (*p-*). For example:

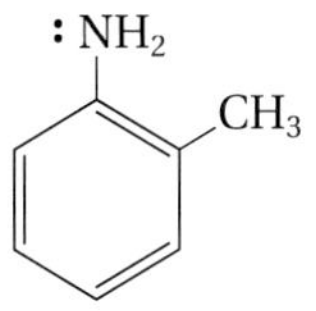

o-Toluidine 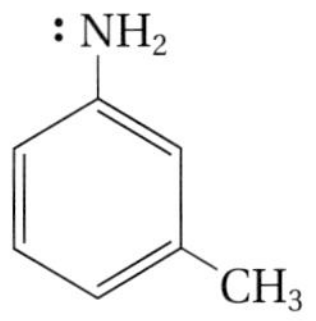 *m*-Toluidine 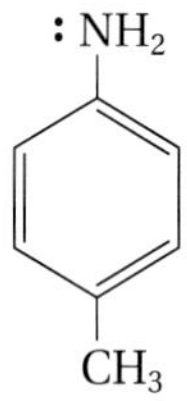 *p*-Toluidine

---

EXAMPLE 6.2 **Naming an aromatic amine**

Name and classify this aromatic amine:

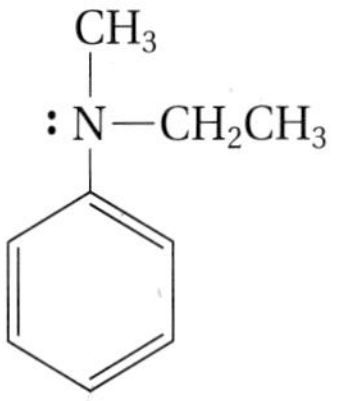

$$CH_3$$
$$: N-CH_2CH_3$$

N, ethyl - N- methylaniline

SOLUTION

This compound has an ethyl group and a methyl group on the amine nitrogen of aniline. It is a tertiary amine named N-*ethyl*-N-*methylaniline*.

PRACTICE EXERCISE 6.2

Name and classify these aromatic amines:

(a) $NHCH_2CH_3$ (b) $N(CH_2CH_3)_2$

N,N-diethylaniline

3°

N-ethylaniline

2°

EXAMPLE 6.3 **Drawing an amine structure when the name is given**

Write the structure for each of the following:
(a) triethylamine (b) *N*-ethylaniline (c) *p*-chloroaniline

SOLUTION

(a) Substitute an ethyl group for each of the three ammonia hydrogens:

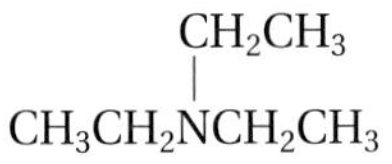

$$CH_2CH_3$$
$$CH_3CH_2NCH_2CH_3$$

a) CH_2CH_3
$CH_3CH_2 - N - CH_2CH_3$

b) $CH_3CH_2 - \ddot{N} - H$

c) $H - \ddot{N} - H$

Cl

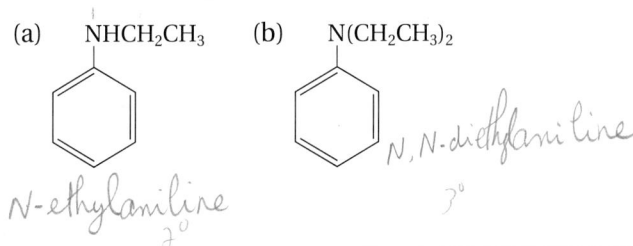

(b) Substitute an ethyl group for one of the hydrogens of the amino group of aniline:

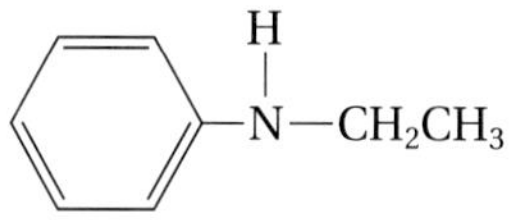

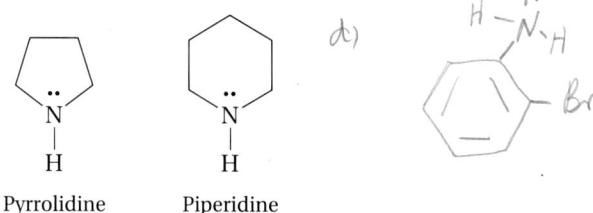

(c) Substitute a chlorine atom for a hydrogen on the phenyl ring of aniline in the *para-* position:

(handwritten annotations in left margin:)

ethylamine

a)

$CH_3CH_2-\overset{..}{N}-H$
|
H

b) N,N-diethylaniline

CH₂CH₃
|
〈O〉-N:
|
CH₂CH₃

c) tripropylamine ← not possible for this

$CH_3CH_2CH_2-\overset{..}{N}-CH_2CH_2CH_3$
|
CH₂
|
CH₂
CH₃   (CH₃CH₂CH₂)₃N

A number of the water-soluble B-complex vitamins contain heterocyclic amine rings. Deficiencies in either thiamine (vitamin B_1), niacin, or pyridoxine (vitamin B_6) can lead to nervous system damage.

PRACTICE EXERCISE 6.3

Write the structure for (a) ethylamine, (b) *N,N*-diethylaniline, (c) tripropylamine, and (d) *o*-bromoaniline.

(handwritten: different ring)

Heterocyclic amines *(handwritten: one or more N in rings)*

In a *heterocyclic amine,* the amine nitrogen is incorporated into a ring. Many aliphatic heterocyclic amines exist. Among the most commonly encountered are pyrrolidine and piperidine.

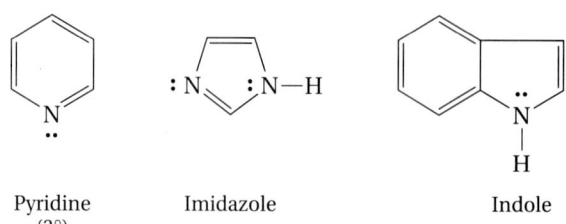

Pyrrolidine (2°)   Piperidine (2°)

(handwritten d): H-N-H ring with Br)

Pyridine, a liquid amine with a nauseating odor, is an example of an aromatic heterocyclic amine; it is a tertiary amine. Several heterocyclic aromatic amines are present in natural products. The ring system of imidazole occurs in the amino acid histidine. The indole ring system is found in the amino acid tryptophan and in many alkaloids, a group of physiologically active amines obtained from plants.

Pyridine (3°)   Imidazole   Indole

Other heterocyclic aromatic ring systems in nature are those of purine, pyrimidine, and pyrrole. Derivatives of purine and pyrimidine are impor-

Chlorophyll is the green pigment found in the chloroplast of plants. The structure of the pyrrole ring system in chlorophyll is very similar to that found in hemoglobin. In photosynthesis, chlorophyll absorbs light energy, which is eventually stored as chemical potential energy in carbohydrate and other nutrient molecules.

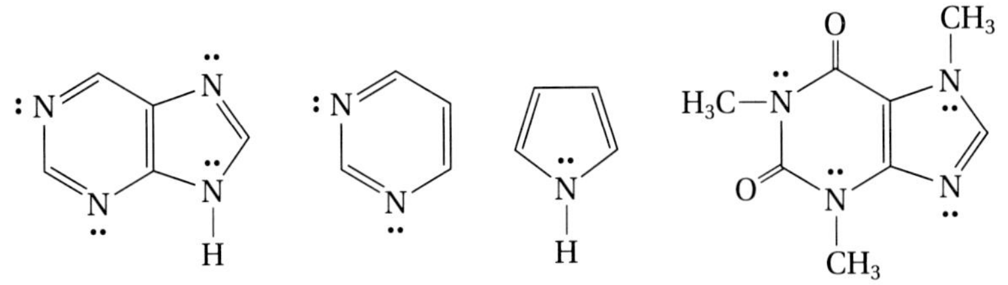

tant components of the nucleic acids, the molecules of heredity. Caffeine, the stimulant in tea, coffee, cocoa, and many soft drinks, is a purine derivative. The hemoglobin of blood is an example of a naturally occurring molecule that contains the pyrrole ring system.

Purine   Pyrimidine   Pyrrole   Caffeine

PRACTICE EXERCISE 6.4

Which structure represents an aromatic heterocyclic amine?

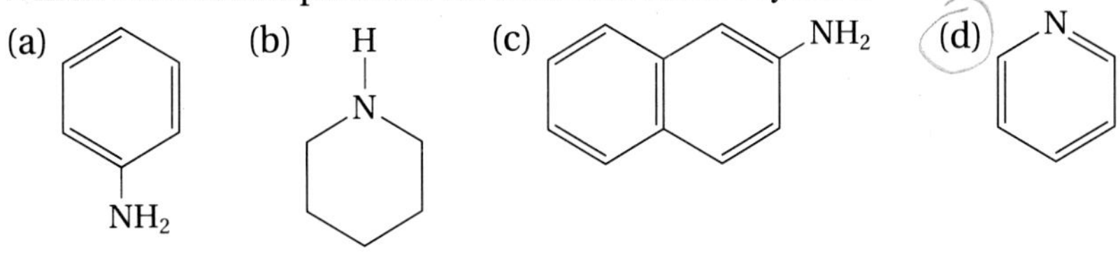

(a)   (b)   (c)   (d)

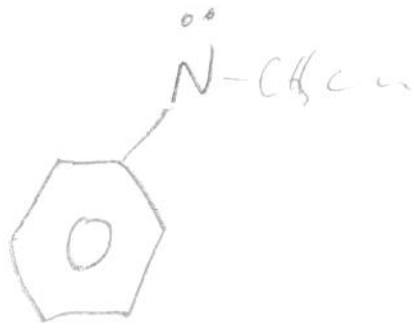

Aliphatic diamines

Aliphatic amines that contain two amino groups in the same molecule are called *aliphatic diamines*. Two aliphatic diamines produced by decaying flesh are putrescine and cadaverine. These amines have characteristic odors. You may judge from their names what the odors are like.

$$H_2N-CH_2CH_2CH_2CH_2-NH_2 \qquad H_2N-CH_2CH_2CH_2CH_2CH_2-NH_2$$

Putrescine   Cadaverine

6.2 Amine basicity

AIM: To show, with equations, how amines act as weak bases.

Focus

Amines are weak bases.

For the weak base ammonia, the unshared electron pair on the nitrogen of primary, secondary, or tertiary amines can accept a hydrogen ion, or proton, from acids. Amines are weak bases of base strength similar to that of ammonia. The general reaction for the protonation of an amine is

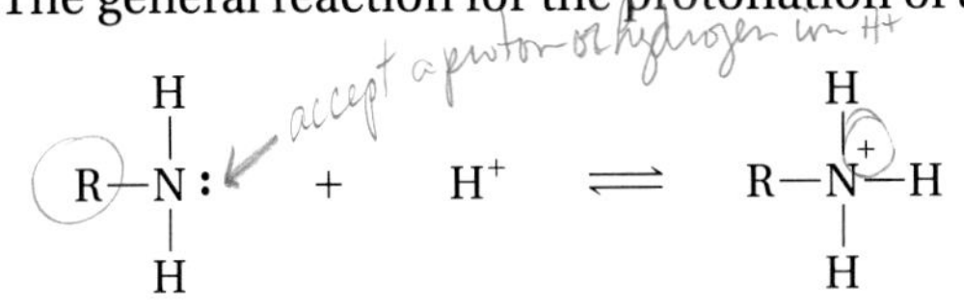

Free amine   Proton   Ammonium ion
(unprotonated amine)   (protonated amine)

An amine that has accepted a hydrogen ion and acquired a positive charge to

become a cation is called a **protonated amine.** *The cations produced by pro-tonation of aliphatic and aromatic amines are called* **alkylammonium ions** *and* **arylammonium ions,** *respectively. An electrically neutral, unproto-nated amine is called a* **free amine.**

Amines are bases, so they can react with acids to form salts. Since amines are derivatives of ammonia, *the salts formed from the reactions of amines and acids are called* **ammonium salts.** Here are three specific examples of the for-mation of ammonium salts from amines and hydrochloric acid:

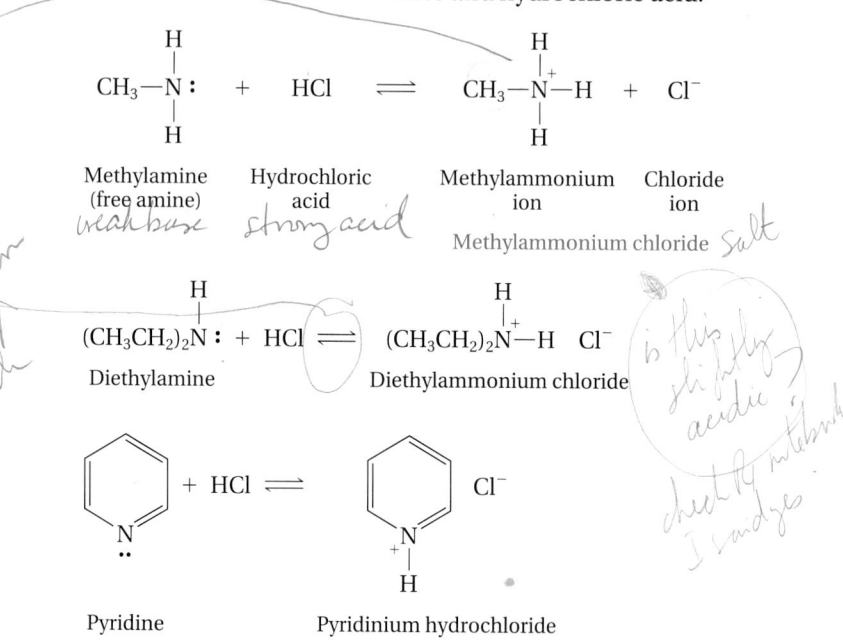

The methylammonium chloride formed in the reaction of methylamine and hydrochloric acid is the salt of a weak base and a strong acid just as ammonium chloride is. The methylammonium ion undergoes slight disso-ciation in water in the same way as for ammonium chloride:

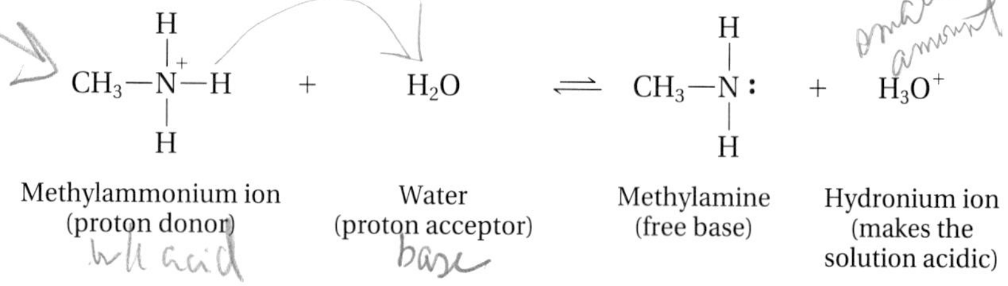

The ammonium salts formed between amines and strong acids make slightly acidic aqueous solutions. Because of their ability to accept a proton and retain it most of the time, amines can act as buffers, or proton sponges, to pro-tect against drastic changes in pH. Many naturally occurring amines serve as pH buffers in the fluids of living organisms, such as blood.

Like alcohols, the solubilities of free amines in water depend on their molecular structure. Amines with large hydrocarbon groups on the amine nitrogen tend to be insoluble. Those with small hydrocarbon substituents, such as methylamine, are very soluble in water. Protonated amines are almost always soluble in water because they are ionic. Many medicinal

[handwritten annotations at top]

a) $(CH_3)_2 \overset{..}{N}H + HCl \rightleftharpoons (CH_3)_2 \overset{\oplus}{N} - H \ Cl^{\ominus}$
weak base + strong acid $\rightleftharpoons$ ammonium salts + cation
free amine dimethylammonium chloride
unprotonated

b/ $CH_3 NH_2 + HNO_3 \rightleftharpoons CH_3 N\overset{\oplus}{H_3} \ NO_3^{\ominus}$
weak base + strong acid ammonium salt + cation
free amine methylammonium nitride

drugs that contain the amino group are supplied as the hydrochloride salts. Because of its greater solubility in water, a hydrochloride salt is usually easier to administer and more readily absorbed into the bloodstream than the free amine.

PRACTICE EXERCISE 6.5

Complete the following reactions, and name the products.

(a) $(CH_3)_2NH + HCl$ (b) $CH_3NH_2 + HNO_3$

(c)

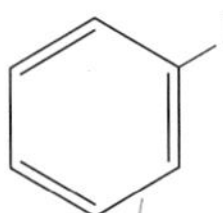

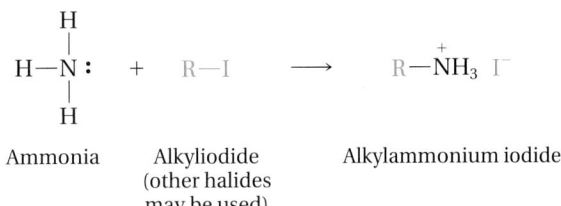

$+ HCl \rightleftharpoons$ *[handwritten:]* $\overset{}{\bigcirc} N\overset{\oplus}{H_3} \ Cl^{\ominus}$

[handwritten:] weak base + strong / aniline acid / anyl ammonium + cation salt / anilinium chloride

(d) $CH_3CH_2NH_2 + H_3PO_4 \rightleftharpoons$ *[handwritten:]* $CH_3CH_2N\overset{\oplus}{H_3} \ H_2PO_4^{\ominus}$
[handwritten:] ethylammonium dihydrogen phosphate

6.3 Preparation of amines

AIMS: To write an equation for the preparation of an amine from a halocarbon and ammonia or an amine nitrogen. To name and write the structure for a quaternary ammonium salt.

Focus

Simple aliphatic amines can be prepared by displacement reactions.

[handwritten: omit]

There are several ways to prepare amines. One common way to prepare aliphatic amines is to displace the halogen in a halocarbon with ammonia or an amine nitrogen. This reaction produces an ammonium salt.

$$H - \underset{\underset{H}{|}}{\overset{\overset{H}{|}}{N}} : \quad + \quad R-I \quad \longrightarrow \quad R-\overset{+}{N}H_3 \ I^-$$

Ammonia   Alkyliodide   Alkylammonium iodide
(other halides
may be used)

The solution containing the ammonium salt can be treated with a base such as sodium hydroxide to produce the free amine.

$$R-\overset{+}{N}H_3 \ I^- \ + \ NaOH \ \longrightarrow \ R-NH_2 \ + \ NaI \ + \ H_2O$$

Ammonium   Sodium   Free   Sodium   Water
salt   hydroxide   amine   iodide

It may be difficult to control the number of alkyl groups attaching to the amine nitrogen. Treatment of ammonia with an alkyl halide usually gives a mixture of primary, secondary, and tertiary amines.

$$NH_3 + R-X \longrightarrow R-NH_2 \quad \text{Primary (1°) amine}$$

$$R-NH_2 + R-X \longrightarrow R-\underset{\underset{}{|}}{\overset{\overset{R}{|}}{N}}H \quad \text{Secondary (2°) amine}$$

$$R-\underset{}{\overset{\overset{R}{|}}{N}}-H + R-X \longrightarrow R-\underset{}{\overset{\overset{R}{|}}{N}}-R \quad \text{Tertiary (3°) amine}$$

Because of its basicity, the nitrogen of a tertiary amine can displace a halogen from a fourth alkyl halide molecule. A nitrogen with *four* R groups attached has a positive charge and is called a *quaternary ammonium ion.* Choline, an important component of some constituents of cell membranes (see Sec. 8.3), is a quaternary ammonium ion.

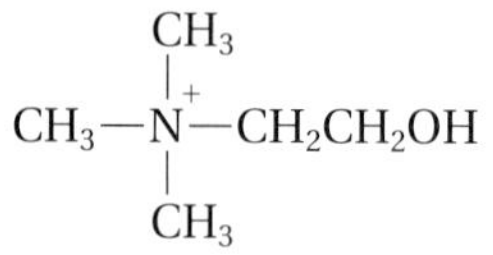

$$CH_3-\overset{\overset{\displaystyle CH_3}{|}}{\underset{\underset{\displaystyle CH_3}{|}}{N^+}}-CH_2CH_2OH$$

Choline

Quaternary ammonium salts having one long alkyl group function as soaps. One quaternary ammonium salt, cetyltrimethylammonium chloride, is found in mouthwashes and is used as a germicide for sterilizing medical instruments.

Salts containing a quaternary ammonium ion are called **quaternary ammonium salts.** The formation of a quaternary ammonium salt can be illustrated by the reaction of trimethylamine with bromomethane.

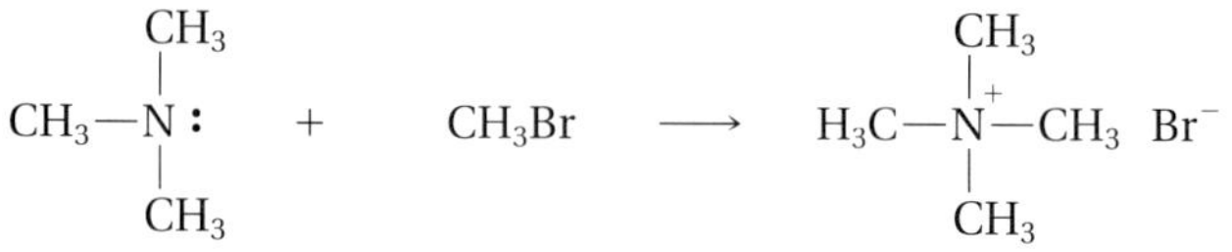

$$CH_3-\overset{\overset{\displaystyle CH_3}{|}}{\underset{\underset{\displaystyle CH_3}{|}}{N}}: \quad + \quad CH_3Br \quad \longrightarrow \quad H_3C-\overset{\overset{\displaystyle CH_3}{|}}{\underset{\underset{\displaystyle CH_3}{|}}{N^+}}-CH_3 \;\; Br^-$$

| Trimethylamine | Bromomethane (methyl bromide) | Tetramethyl ammonium bromide (a quaternary ammonium salt) |

PRACTICE EXERCISE 6.6

An aqueous solution of ammonia is heated with 1-chlorobutane. (a) What is the organic product? (b) Give the structure and name of the free amine that is liberated when the product in part (a) is treated with sodium hydroxide.

6.4 Amides

AIM: To name and write structures of simple amides.

Focus

Amides are ammonia and amine derivatives of carboxylic acids.

Amides *are ammonia or amine derivatives of organic acids.* Amides may be simple, monosubstituted, or disubstituted. For example:

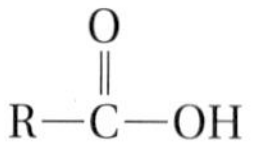

$$R-\overset{\overset{\displaystyle O}{\|}}{C}-OH \qquad \text{Carboxylic acid}$$

$$R-\overset{\overset{\displaystyle O}{\|}}{C}-\underset{\cdot\cdot}{NH_2} \qquad \text{Simple amide}$$

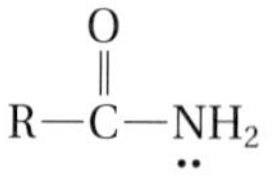

$$R-\overset{\overset{\displaystyle O}{\|}}{C}-\underset{\cdot\cdot}{\overset{\overset{\displaystyle H}{|}}{N}}-R \qquad \text{Monosubstituted amide}$$

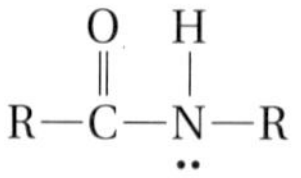

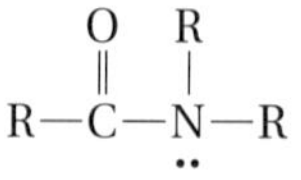

$$R-\overset{\overset{\displaystyle O}{\|}}{C}-\underset{\cdot\cdot}{\overset{\overset{\displaystyle R}{|}}{N}}-R \qquad \text{Disubstituted amide}$$

Although the amides of carboxylic acids are the focus of this chapter, the acid part of the amide need not be a carboxylic acid. The amides of

A Closer Look

The Sulfonamide Antibiotics

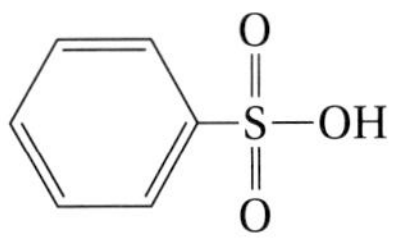

Benzenesulfonic acid

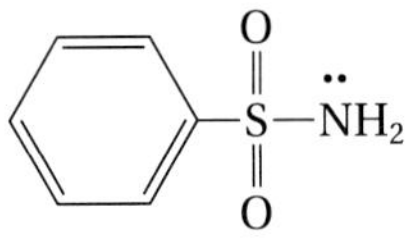

Benzenesulfonamide

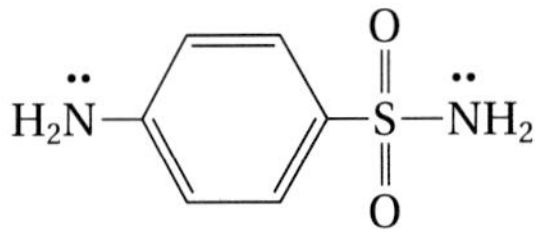

Sulfanilamide

The structures of some sulfa antibiotics.

Certain bacterial infections, especially bladder infections, are often treated with a class of medications called *sulfa drugs*. All of the sulfa drugs are amides of sulfonic acids (RSO_3H). Benzenesulfonamide is an example of an amide of a sulfonic acid; it is the amide of benzenesulfonic acid. The figure above shows the structures of other sulfa drugs used as antibiotics.

Sulfa drugs are effective antibiotics against bacteria that need folic acid to sustain life. These bacteria make folic acid by assembling it from molecular components that include *p*-aminobenzoic acid (see figure below, right).

Sulfanilamide, a compound similar in structure to *p*-aminobenzoic acid, sticks to the bacterial enzyme at a place on the molecule usually reserved for *p*-aminobenzoic acid. The bacterial enzyme cannot use the sulfanilamide to make folic acid, so the bacteria die from folic acid deficiency. Unlike the bacteria, our body cells are relatively unaffected by sulfa drugs. We need folic acid (a B-complex vitamin) in order to stay alive and well, but our cells lack the enzyme needed to make the compound. We obtain our folic acid from the food we eat and from beneficial bacteria that live in our intestines.

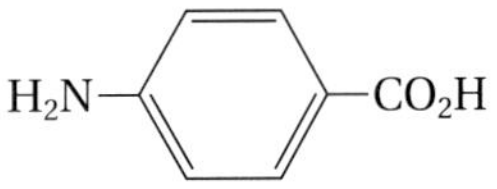

p-Aminobenzoic acid

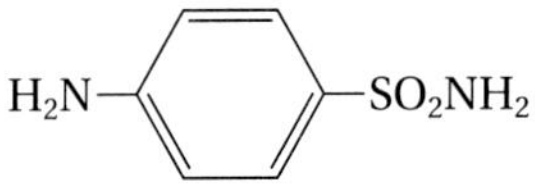

Sulfanilamide

some sulfonic acids are important antibiotics, as discussed above in A Closer Look: The Sulfonamide Antibiotics.

Names of amides

Amides are named by dropping the *-ic* or *-oic* ending from the name of the parent acid and adding the ending *-amide*. Any substituents on the amine nitrogen are named as prefixes preceded by *N*- or *N,N*-. Here are some examples:

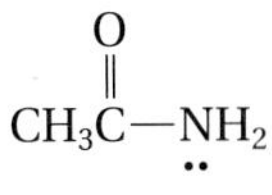

Acetamide

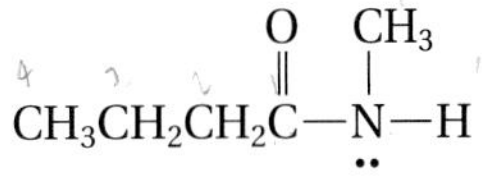

N-Methylbutyramide

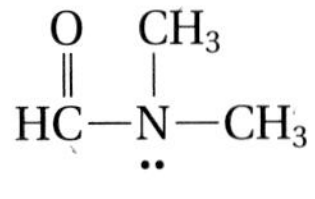

N,N-Dimethylformamide

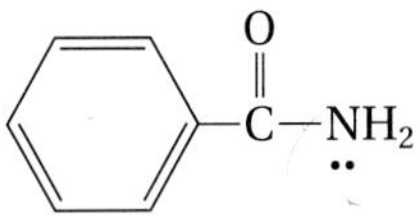

Benzamide

Amides formed from carboxylic acids and aniline are called **anilides.** Acetanilide, formed from acetic acid and aniline, is a simple example. Phenacetin, an analgesic and antipyretic, is also an anilide. APC tablets contain a mixture of aspirin, phenacetin, and caffeine.

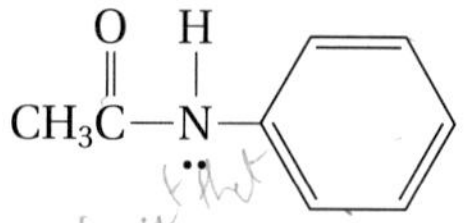

Acetanilide *(handwritten: N-phenylacetamide)*   Phenacetin

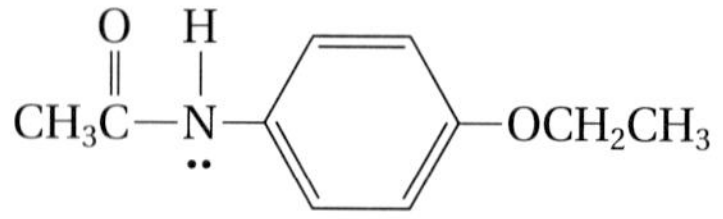

Some amides are used as local anesthetics. See A Closer Look: Amide Local Anesthetics.

EXAMPLE 6.4   Naming amides

Name these amides: (a) $CH_3CH_2CH_2CNH_2$ *(handwritten: butyramide)*   (b) $HCNHCH_3$ *(handwritten: N-methylformamide / methanamide)*

SOLUTION

(a) This amide is derived from butyric acid:

$$CH_3CH_2CH_2COH$$

It is butyramide.

(b) This amide is derived from formic acid: *(handwritten: methanoic)*

$$HCOH$$

It is a formamide. There is also a methyl group attached to the amide N. The name is N-*methylformamide.*

(handwritten: ? If there was a substituent on N, name which subs. first? on N or on C. ↓ first)

PRACTICE EXERCISE 6.7

Name the following amides.

(a) $CH_3CH_2CNH_2$ (b) $CH_3CHCH_2CNH_2$ with CH_3 (c) $CH_3CH_2CNHCH_3$

(handwritten: propanamide   3-methylbutyramide   N-methylpropanamide)

(handwritten left margin: $CH_3-CH_2-C-N\overset{H}{\underset{H}{<}}$ with O above propanamide   amide)

Chemistry of the amide functional group

The electronegativity of oxygen in the amide bond pulls the unshared pair of electrons in amide nitrogen toward the oxygen. Since these electrons are unavailable to accept a proton, *an amide nitrogen is much less basic*

(handwritten: amide functional group   R–C–N–R with O and R)

A Closer Look

Amide Local Anesthetics

Several important local anesthetics contain amide functional groups. Their structures are essentially the same as those of the ester local anesthetics (A Closer Look: Ester Local Anesthetics, page 154) except that nitrogen replaces the ester oxygen. These compounds are less readily metabolized than the esters and tend to accumulate in the plasma.

Lidocaine hydrochloride (Xylocaine) (see figure) has received widespread acceptance since its introduction more than 35 years ago. The major advantages are that it produces rapid anesthesia and is nonirritating and hypoallergenic. Topically it is not as effective as cocaine (A Closer Look: Ester Local Anesthetics, pages 154–155) but is better than procaine. Mepivacaine hydrochloride (Carbocaine) is also an amide. It takes effect faster than lidocaine, and the duration of the anesthesia is somewhat longer. Adverse reactions are few, and tissue irritation is minimal. Bupivacaine hydrochloride (Marcaine) has the same structure as mepivacaine except that a methyl group is replaced by a butyl group. It is more potent and has a longer-lasting action than lidocaine or mepivacaine. This drug has been used occasionally to produce spinal anesthesia.

The most recently introduced local anesthetic with an amide structure is etidocaine (Duranest). It is structurally similar to lidocaine but has greater potency and is longer lasting. Its general toxicologic and pharmacologic actions differ little from the other anesthetics in the group.

Lidocaine (hydrochloride salt)
(Xylocaine)

Bupivacaine (hydrochloride salt)
(Marcaine)

Mepivacaine (hydrochloride salt)
(Carbocaine)

Etidocaine (hydrochloride salt)
(Duranest)

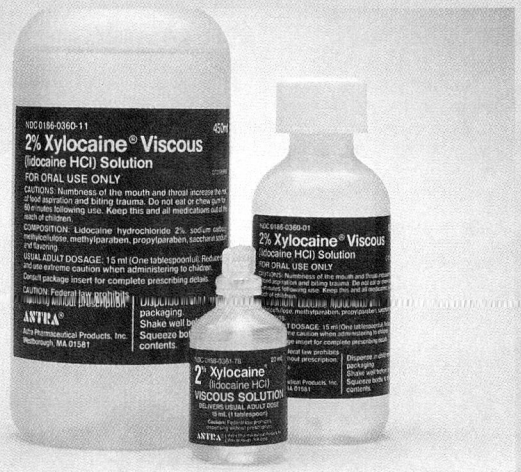

Xylocaine Viscous, a fast-acting, local anesthetic, is sometimes prescribed as a gargle for a severe sore throat.

than an amine nitrogen. It does not usually accept a proton in an acidic solution. However, amides form hydrogen bonds with other amides and with water.

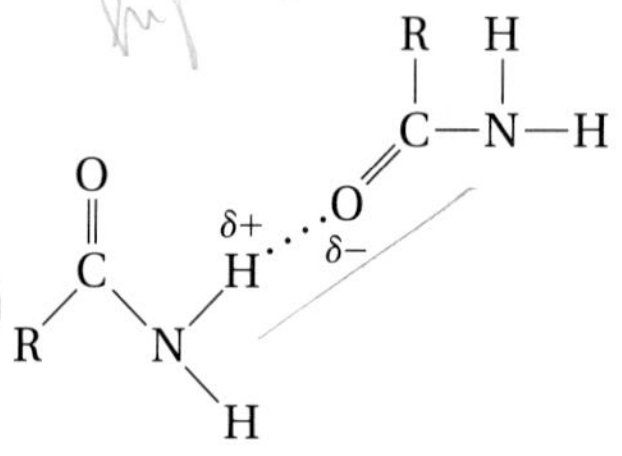

Hydrogen bonding
in amides

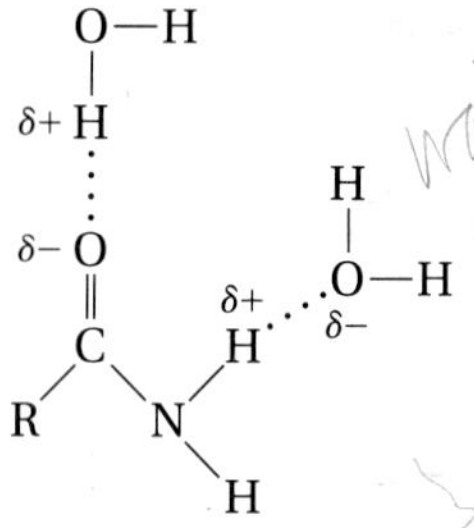

Hydrogen bonding
with water

Such hydrogen bonding is very important in holding protein structures in unique shapes (see Chap. 9).

6.5 Preparation of amides

AIMS: To write equations for the preparation of amides from ammonium salts and carboxylic acid derivatives. To describe the preparation and use of some common polyamides.

Focus

Amides can be prepared from amines and carboxylic acids or their derivatives.

There are many ways to prepare amides. One method is the dehydration of ammonium salts of carboxylic acids; another is the reaction of ammonia or an amine with either an ester or a carboxylic acid anhydride.

Dehydration of ammonium salts

If we mix a carboxylic acid with an amine, we will get an ammonium salt, since the acid is a proton donor and the amine is a proton acceptor.

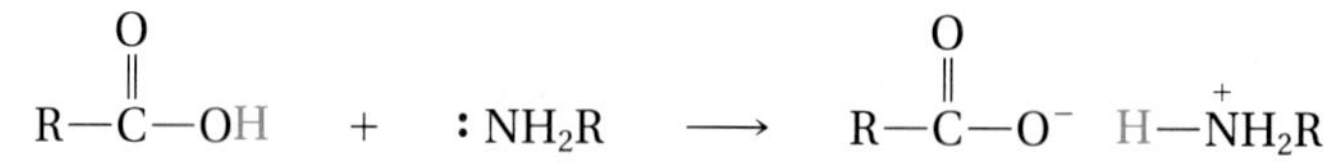

$$R-\overset{O}{\overset{\|}{C}}-OH \quad + \quad :NH_2R \quad \longrightarrow \quad R-\overset{O}{\overset{\|}{C}}-O^- \quad H-\overset{+}{N}H_2R$$

Carboxylic acid
(proton donor)

Amine
(proton acceptor)

Ammonium salt

We can remove a molecule of water from the dry ammonium salt formed from an amine and a carboxylic acid by heating it. This is a dehydration reaction, and the organic product is an amide.

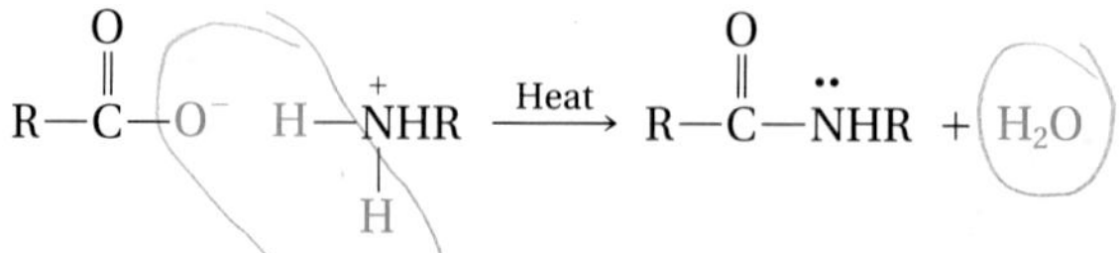

$$R-\overset{O}{\overset{\|}{C}}-O^- \quad H-\overset{+}{N}HR \xrightarrow{\text{Heat}} R-\overset{O}{\overset{\|}{C}}-\overset{..}{N}HR \; + \; H_2O$$

Ammonium salt

Amide

Here are two examples of the preparation of amides by dehydration of ammonium salts of carboxylic acids:

$$CH_3\overset{\displaystyle O}{\overset{\|}{C}}-O^-\quad H-\overset{\displaystyle H}{\underset{\displaystyle H}{\overset{|}{N^+}}}-H\xrightarrow{\text{Heat}} CH_3\overset{\displaystyle O}{\overset{\|}{C}}-\ddot{N}H_2 + H_2O$$

Ammonium acetate   Acetamide

$$CH_3\overset{\displaystyle O}{\overset{\|}{C}}-O^-\quad H-\overset{\displaystyle H}{\underset{\displaystyle H}{\overset{|}{N^+}}}-CH_3\xrightarrow{\text{Heat}} CH_3\overset{\displaystyle O}{\overset{\|}{C}}-\ddot{N}HCH_3 + H_2O$$

N-Methylammonium acetate   *N*-Methylacetamide

PRACTICE EXERCISE 6.8

The salt of propanoic acid with ethylamine is heated. Give the structure and name of the compound that is formed.

Reaction with carboxylic acid derivatives

Amides also can be prepared by the reaction of ammonia or amines with derivatives of carboxylic acids. Esters, especially methyl esters, and acid anhydrides are often the carboxylic acid derivatives.

$$\underset{\text{Methyl benzoate}}{C_6H_5-\overset{\displaystyle O}{\overset{\|}{C}}-OCH_3} + \underset{\text{Ammonia}}{H-\ddot{N}H_2} \longrightarrow \underset{\text{Benzamide}}{C_6H_5-\overset{\displaystyle O}{\overset{\|}{C}}-\ddot{N}H_2} + \underset{\text{Methanol}}{CH_3OH}$$

With an ester as the starting material, an alcohol is formed as a by-product of the reaction. With an anhydride, a carboxylic acid is formed as a by-product.

EXAMPLE 6.5   **Preparing an amide**

Write an equation for the preparation of acetamide from ammonia and an acid anhydride.

SOLUTION

The product, acetamide, has two carbons. We let acetic anhydride react with ammonia. The other product is acetic acid.

$$\underset{\text{Acetic anhydride}}{CH_3\overset{\displaystyle O}{\overset{\|}{C}}-O-\overset{\displaystyle O}{\overset{\|}{C}}CH_3} + \underset{\text{Ammonia}}{H-\ddot{N}H_2} \longrightarrow \underset{\text{Acetamide}}{CH_3\overset{\displaystyle O}{\overset{\|}{C}}-\ddot{N}H_2} + \underset{\text{Acetic acid}}{CH_3\overset{\displaystyle O}{\overset{\|}{C}}-OH}$$

PRACTICE EXERCISE 6.9

Predict the products of the reaction of ethyl acetate with methylamine.

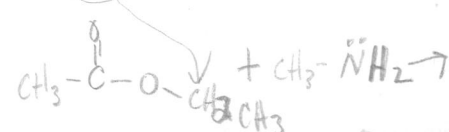

Polyamides → *nylon*

Polyamides *are polymers in which the monomer units are linked by amide bonds.* Various types of nylon, familiar materials to nearly everyone, are polyamides. Of these, nylon 66 was the first polyamide to find wide commercial application. Nylon 66 gets its name from its repeating units of adipic acid, a six-carbon dicarboxylic acid, and 1,6-diaminohexane, a six-carbon diamine. The polymer is prepared by heating the dicarboxylic acid and the diamine; amide bonds are formed as water is removed.

$$x\text{HO}-\overset{\overset{\displaystyle O}{\|}}{C}\,(CH_2)_4\,\overset{\overset{\displaystyle O}{\|}}{C}-OH \;+\; x\text{H}_2\text{N}-CH_2\,(CH_2)_4\,CH_2-NH_2 \longrightarrow$$

Adipic acid   1,6-Diaminohexane

$$\left[\overset{\overset{\displaystyle O}{\|}}{C}\,(CH_2)_4\,\overset{\overset{\displaystyle O}{\|}}{C}-\overset{\overset{\displaystyle H}{|}}{N}-CH_2\,(CH_2)_4\,CH_2-\overset{\overset{\displaystyle H}{|}}{N}\right]_x \;+\; x\text{H}_2\text{O}$$

Representative unit of nylon 66

The finished polymer has a formula weight of about 10,000 and a melting point of 250 °C. The melted polymer can be spun into very fine, strong fibers. Because of this property, nylon 66 was rapidly substituted for the scarcer and more expensive natural silk in women's sheer hosiery.

Today, nylon 66 has been largely replaced by nylon 6, which has similar properties but is even cheaper to produce. The representative polymer unit of nylon 6 is derived from 6-aminohexanoic acid; the long polymer chain is formed by the successive attachment of the carboxyl group of one molecule of the acid to the amino group of the next by formation of an amide bond.

$$x\text{H}_2\text{N}-CH_2\,(CH_2)_4\,\overset{\overset{\displaystyle O}{\|}}{C}-OH \xrightarrow{\text{Heat}} \left[CH_2(CH_2)_4\,\overset{\overset{\displaystyle O}{\|}}{C}-\overset{\overset{\displaystyle H}{|}}{N}\right]_x \;+\; x\text{H}_2\text{O}$$

6-Aminohexanoic acid   Representative polymer
                                      unit of nylon 6

In medicine, nylon was used to make the first synthetic sutures, and it is still used for sutures today. It is also used to make specialized tubing for medical applications (Fig. 6.1). Nylon is spun into fibers for making carpets, tire cord, and textiles and molded into solid objects such as gears, bearings, and zippers.

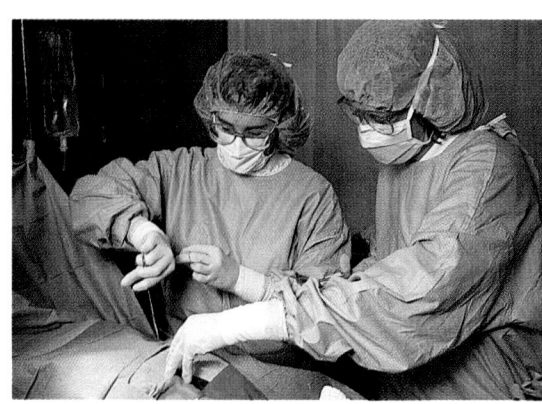

Figure 6.1
Nylon's many uses include surgical applications as well as many other items used by medical personnel.

Aramides, polyamides that contain aromatic rings, have been found to have special properties—bullet resistance and flame resistance. The aromatic rings make the resulting fiber stiffer and tougher. Nomex is a polyamide whose carbon skeleton consists of aromatic rings derived from isophthalic acid and *m*-phenylenediamine.

Isophthalic acid   *m*-Phenylenediamine

Representative unit of Nomex

Figure 6.2
Firefighters and racing car drivers wear fire-resistant clothing made of Nomex

Like nylon, Nomex is a poor conductor of electricity, and since it is more rigid than nylon, it is used to make parts for electrical fixtures. Nomex is also used in the manufacture of flame-resistant clothing for racing car drivers and firefighters (Fig. 6.2). It is also used in the fabrication of flame-resistant building materials. Kevlar has a structure that is similar to that of Nomex. It is a polyamide made from terephthalic acid and *p*-phenylene-diamine.

Terephthalic acid   *p*-Phenylenediamine

Representative unit of Kevlar

Kevlar is used as a replacement for steel in bullet-resistant vests. Since the vests are more flexible and much lighter than their metal counterparts, they can be worn under normal clothing.

Proteins, which are polyamides of naturally occurring amino acids, rank among the most important of all biological molecules. Their structures will be discussed in detail in Chapter 9.

preparation of amines on all notes: not valid reaction. (handwritten)

6.6 Amide hydrolysis

AIM: To predict the products of hydrolysis of an amide.

Compared with esters, amides hydrolyze slowly in water to acid and amine.

$$CH_3\overset{\overset{O}{\|}}{C}-\overset{\overset{H}{|}}{\underset{\cdot\cdot}{N}}-CH_3 + HO-H \xrightarrow[H^+ \text{ or } OH^-]{\text{Heat}} CH_3\overset{\overset{O}{\|}}{C}-OH + H-\overset{\overset{H}{|}}{\underset{\cdot\cdot}{N}}-CH_3$$

N-Methylacetamide   Water   Acetic acid   Methylamine

The hydrolysis can be speeded up by heating the amide in a strongly acidic or basic solution. The hydrolysis of amide bonds is a very important reaction in the digestion of proteins; the hydrolysis is aided by enzyme catalysts.

EXAMPLE 6.6   **Predicting the products of amide hydrolysis**

What are the products when N-methylpropionamide is hydrolyzed?

SOLUTION

The structure of N-methylpropionamide is

$$CH_3CH_2\overset{\overset{O}{\|}}{C}NHCH_3$$

When hydrolyzed with acid, it will give propionic acid and methylamine.

$$CH_3CH_2\overset{\overset{O}{\|}}{C}NHCH_3 + HO-H \xrightarrow[H^+]{\text{Heat}} CH_3CH_2\overset{\overset{O}{\|}}{C}-OH + NH_2CH_3$$

PRACTICE EXERCISE 6.10

What are the products when N-ethylbenzamide is hydrolyzed?

6.7 Important amines

AIM: To name and describe the amines used as neurotransmitters, hallucinogens, decongestants, and antihistamines.

In the central nervous system—the part of the nervous system consisting of the brain, the spinal cord, and the nerves that radiate from the cord—*the most important chemical messengers or* **neurotransmitters** *between nerve cells are three amines: norepinephrine (also called noradrenaline), dopamine, and serotonin.*

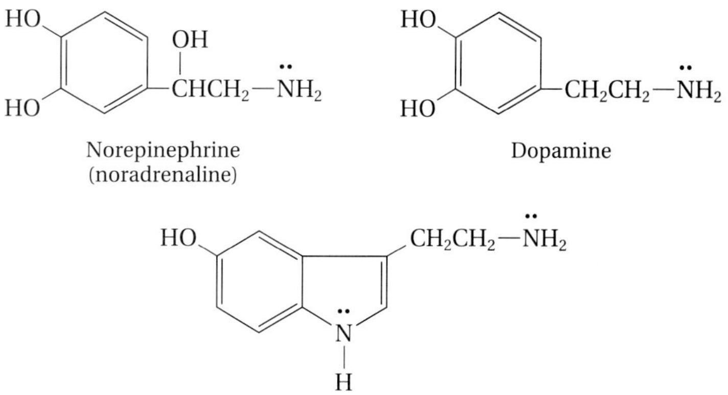

Norepinephrine
(noradrenaline)

Dopamine

Serotonin

Norepinephrine and dopamine are called *catecholamines* because their molecular structures are derived from the dihydroxy phenol known as *catechol*. The catecholamines also have the carbon skeleton 2-phenylethylamine.

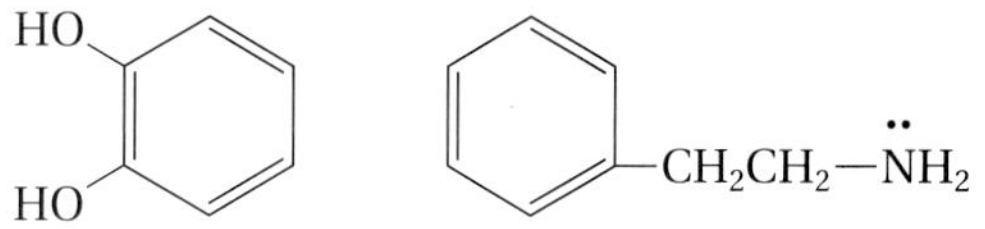

Catechol

2-Phenylethylamine

Dopamine deficiency results in Parkinson's disease, which can cause tremors of the head and extremities, usually in middle-aged and older people. Brain cells of Parkinson's patients have only 5% to 15% of the normal concentration of dopamine. Administration of dopamine does not stem the symptoms of the disease, however, because the amine cannot breach the blood-brain barrier—a natural filter that prevents certain molecules from reaching the brain through the circulation. (How the blood-brain barrier operates is still a puzzle.) Dopa, however, a related carboxylic acid, can pass through the blood-brain barrier. Inside the brain cells, enzymes catalyze the loss of carbon dioxide from the prodrug dopa to produce dopamine.

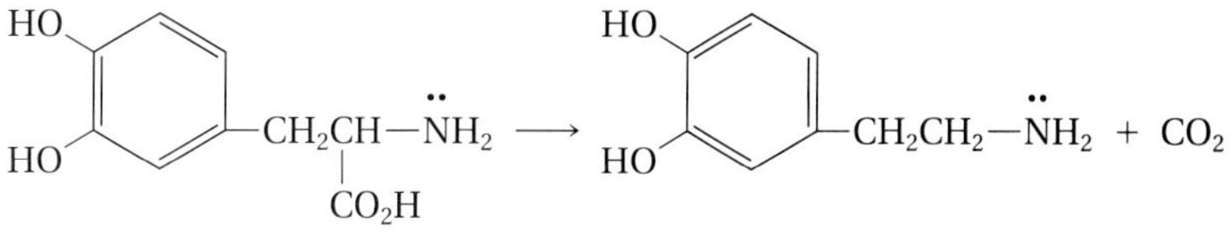

Dopa

Dopamine

The use of dopa to relieve the symptoms of parkinsonism has helped many people resume quite normal, active lives.

Serotonin deficiency has been implicated in mental depression, a disease that strikes millions of people each year. Indeed, depression is the world's most pervasive mental health problem. Our moods appear to rest fundamentally with nerve cells (neurons) in the brain. Signals between neurons are sent by neurotransmitters. Serotonin and norepinephrine are two examples of neurotransmitters that are amines. The neurotransmitters are produced and released by the neurons and then taken up and destroyed

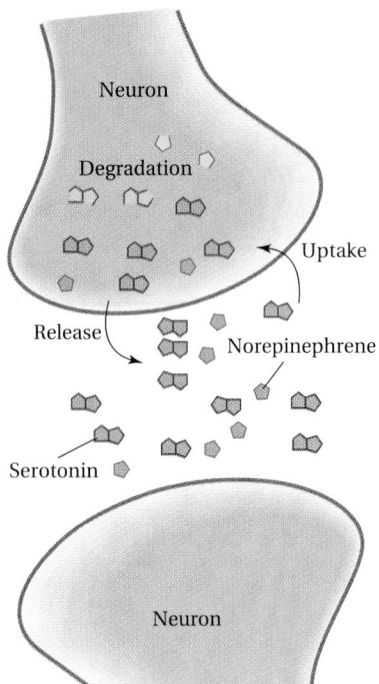

Figure 6.3
Depression can result when the concentration of the neuro-transmitter serotonin between neurons drops to a low level.

by neurons. In a normal individual, the ends of the neurons are bathed by fluid containing the released neurotransmitters (Fig. 6.3). It appears that the concentration of the neurotransmitters between the neurons is important in controlling people's moods. Clinical depression is essentially the result of a chemical imbalance in which certain neurotransmitters drop below normal levels. Over time, three generations of drugs have been developed that help to maintain the levels of neurotransmitters. The earliest class of antidepression drugs to be discovered were the m*onoamine o*x*idase* i*nhibitors* (MAO inhibitors). MAO is an enzyme system responsible for the destruction of neurotransmitters; the MAO inhibitors block the destruction. The mood-elevating effect of iproniazid, the first of the MAO inhibitors, was accidentally discovered while it was being used to treat patients with tuberculosis. Physicians noticed that these patients were exhilarated by the drug. Subsequent studies showed that iproniazid keeps norepinephrine and serotonin at normal levels by inhibiting, or blocking, the degradation of these neurotransmitters by MAO. More than a dozen MAO inhibitors are now available.

Iproniazid
(Euphozid, Marsilid)

A second generation of antidepressants, the *tricyclics,* soon followed the MAO inhibitors. The tricyclics are so called because their chemical structures contain three carbon rings. Amitriptyline is one of several possible examples of tricyclic antidepressive drugs.

Amitriptyline (hydrochloride salt)
(Domenical, Elavil, Larozyl, Saroten)

The tricyclics' antidepressant action is not completely understood, but they are thought to block the uptake of norepinephrine and serotonin by neurons. Because studies indicated that low serotonin is more important than low norepinephrine in depression, scientists began a search for a third generation of drugs that would prevent the uptake of only serotonin. This search for such medications, called *zero-serotonin-uptake drugs,* has been quite successful. You may recall from the Case in Point to this chapter that a

Phencyclidine (PCP) is also known by its street name, "angel dust." PCP is a hallucinogenic drug that contains a piperdine ring. It is a drug of abuse that causes confusion, paranoia, and delirium.

Phencyclidine
(PCP)

FOLLOW-UP TO THE CASE IN POINT: Antidepressants

Maude is one of the more than 12 million people in the United States afflicted by depression. Children and adults alike are affected by this malady, which may include weight loss, insomnia, and general feelings of hopelessness or worthlessness. After a careful analysis of symptoms, Maude's psychiatrist thought that she might be suffering from a chemical imbalance in which the concentration of the neurotransmitter serotonin is depleted. For this reason, the psychiatrist suggested that Maude might benefit from a zero-serotonin-uptake drug.

Maude's psychiatrist prescribed the drug fluoxetine (Prozac), but other zero-serotonin-uptake drugs such as paroxetine (Paxil) and sertraline (Zoloft) are also available. These drugs are the most widely used antidepressant drugs on the market today.

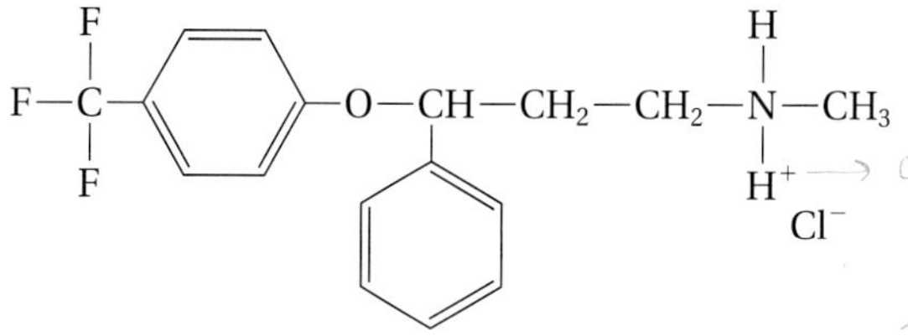

Within a few weeks, Maude began to feel more like her old self. After a few months, her depression had disappeared. She is still attending group counseling sessions but is trying life without her antidepressant medication.

The study of depression is far from over. With the recent explosion of knowledge about the complex chemistry of the brain, the next few years hold the promise of new understanding and treatments for this widespread illness.

Fluoxetine (hydrochloride salt)
(Prozac)

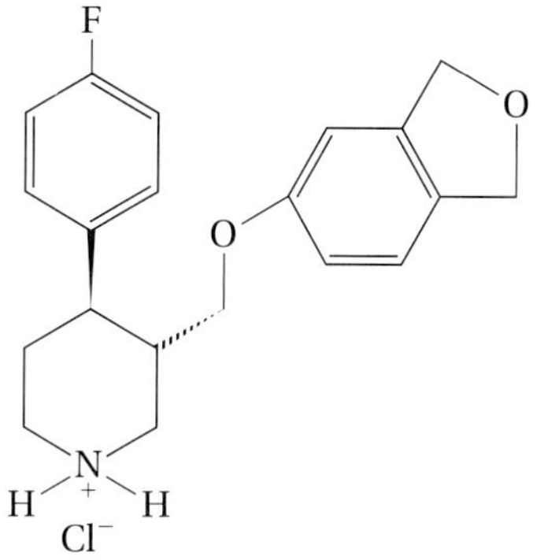

Paroxetine (hydrochloride salt)
(Paxil)

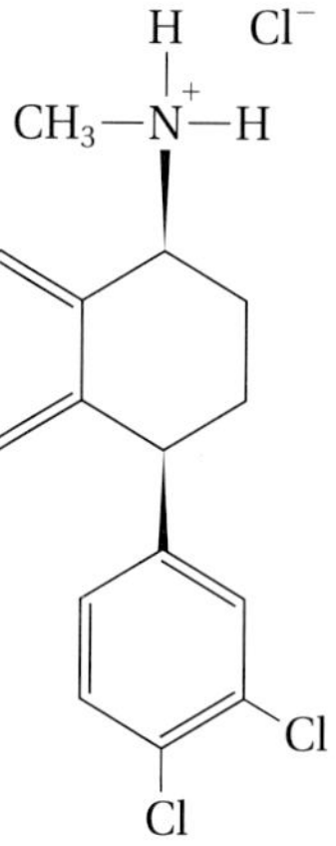

Sertraline (hydrochloride salt)
(Zoloft)

woman named Maude suffered from depression. Above, we learn how she was helped by the third generation of antidepressants.

Epinephrine, commonly called *adrenaline,* is another catecholamine. Its structure is the same as norepinephrine except that epinephrine has an *N*-methyl group.

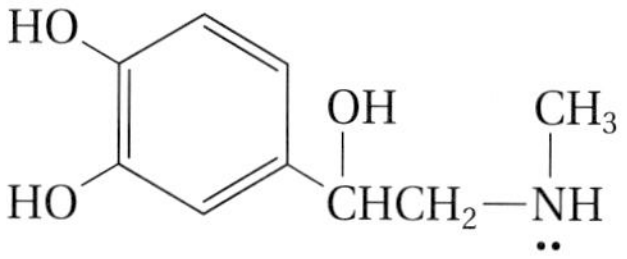

Epinephrine
(adrenaline)

Epinephrine appears to be more important as a hormone than as a neurotransmitter; its effects will be considered in more detail in Chapter 15. Since the catecholamines help determine what stimuli your brain receives, you might very well imagine that blocking these stimuli could approximate the effects of the hallucinogenic drugs and certain narcotics. **Hallucinogens** *are mind-altering drugs that produce visions and distorted views of reality.* Mescaline and DOM (also called STP) are two hallucinogenic drugs whose structures resemble those of the catecholamines.

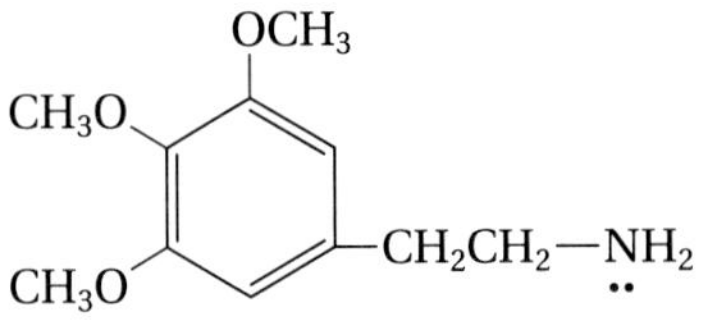

Mescaline
(hallucinogen)

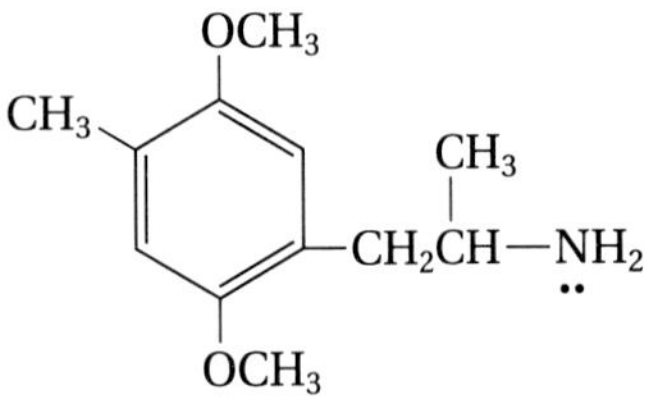

DOM, STP
(hallucinogen)

Mescaline, extracted from the resin pods of the peyote cactus, is used by members of the Native American Church of the Southwest in its rituals.

Scientists have developed several medications using compounds with structures similar to those of the catecholamines. Ephedrine is widely used in cough syrups, while neosynephrine is often an ingredient in nose drops. Both these compounds are **decongestants,** *compounds that cause shrinkage of the membranes lining the nasal passages.*

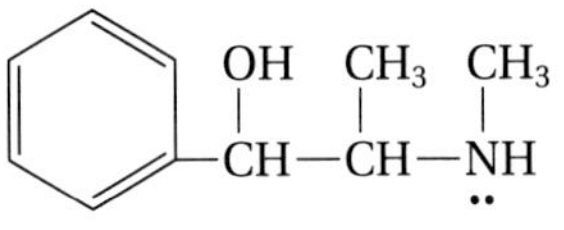

Ephedrine

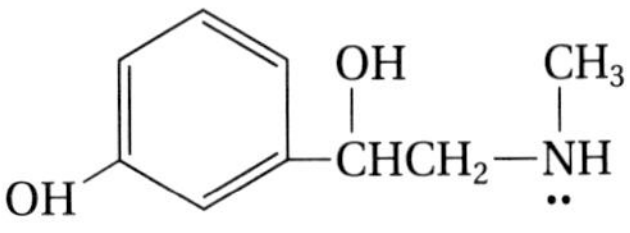

Neosynephrine

Amphetamine is an appetite depressant and nervous system stimulant. Often prescribed for dieters, amphetamines can be abused when the drug is habitually taken as an "upper" for its mood-elevating effects—lessened fatigue, increased confidence, alertness, and increased desire to work.

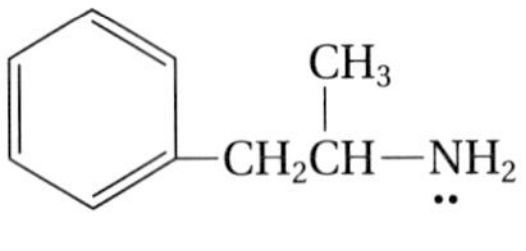

Amphetamine

Do you get hay fever on days when the pollen count is high? If you do, you are allergic to pollen. Your body cells are responding to the invasion of these foreign substances by producing histamine. Histamine is responsible for watery eyes, sniffles, and other hay fever miseries. *The effects of histamine can be relieved to some extent by a class of drugs called* **antihistamines;** two of the most common are pyribenzamine and diphenylhydramine, also called *benadryl.*

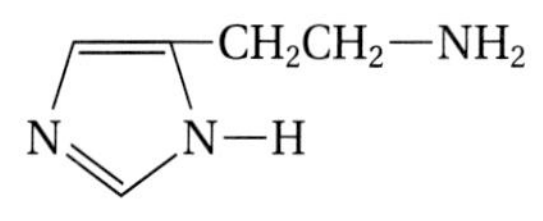

Histamine

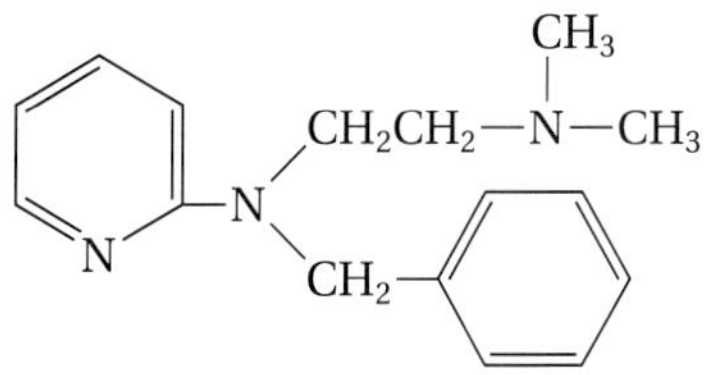

Pyribenzamine

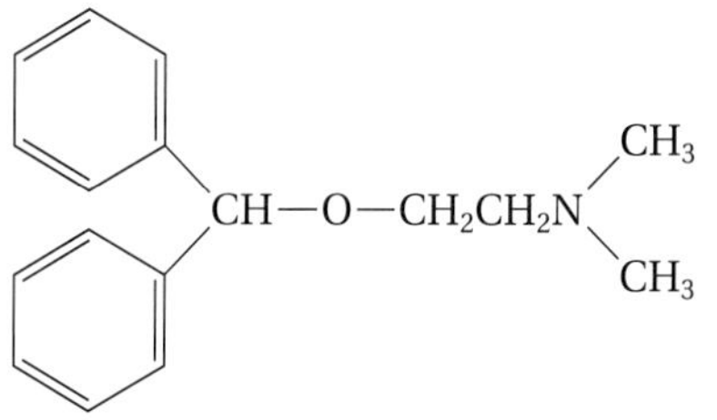

Diphenylhydramine
(benadryl)

Benadryl is also a component of Dramamine, a drug commonly used to prevent motion sickness or nausea.

PRACTICE EXERCISE 6.11

Identify the functional groups in the neosynephrine molecule. What is the physiologic action of this drug?

6.8 Alkaloids

AIM: To name and give examples of uses of medicinal alkaloids.

Focus

Alkaloids are physiologically active amines produced by plants.

Some of the most powerful drugs known are derived from plants. Some have been used for thousands of years. Among the oldest of these drugs are the **alkaloids**—*a group of over 2500 amines obtained from plants.* The name *alkaloid,* meaning "alkali-like," bears testimony to the weakly basic properties of these compounds.

The molecular structures of the alkaloids vary from simple to complex. Nicotine is one of the simplest.

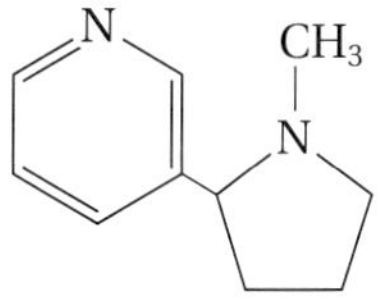

Nicotine

Nicotine is one of several alkaloids present in tobacco. Small doses, such as those obtained by smokers, stimulate the involuntary nervous system. Large doses are toxic and result in nicotine poisoning. Nicotine is a *habituating drug*—one on which people acquire a dependence. An *addicting drug* is one that causes physiologic changes when it is not used for a time.

Several important alkaloids contain the indole ring system. Lysergic acid and reserpine are indole alkaloids. Lysergic acid is produced by ergot, a fungus that grows as brown bodies on infected rye. The diethylamide of lysergic acid is called LSD (lysergic acid diethylamide), a powerful and widely abused hallucinogenic drug. As little as 1 μg of LSD, scarcely enough to see, is sufficient to cause hallucinations. The danger of overdose is high.

Reserpine, an indole alkaloid of the Indian snakeroot (Fig. 6.4a), reduces hypertension (high blood pressure), which, if left unchecked,

Figure 6.4
(a) Indian snakeroot is the source of reserpine. (b) The Cinchona tree yields quinine.

(a)   **(b)**

could lead to a stroke or cardiac arrest. Strychnine, a bitter-tasting compound from the plant *Strychnos nux vomica*, has a long and sometimes sinister history as a poison.

Lysergic acid diethylamide (LSD)

Strychnine

Reserpine

The antimalarial drug quinine is an alkaloid that occurs in the Cinchona (pronounced "sinkona") tree (Fig. 6.4b) of the Andes Mountains of South America. Quinine has the aromatic quinoline ring system as part of its molecular structure. Other drugs that act against malaria have been synthesized, but quinine is still used.

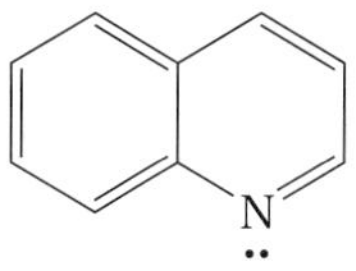

Quinoline 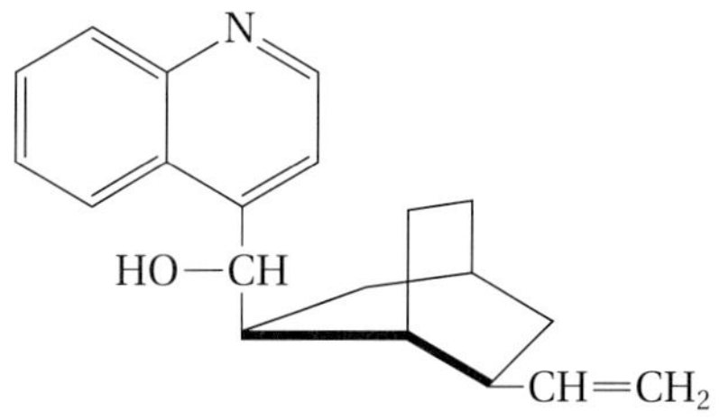   Quinine

The tropane ring system is found in cocaine and atropine. Cocaine stimulates the central nervous system. It is now a widely abused habituating drug. Atropine is an alkaloid obtained from belladonna, hemlane, and deadly nightshade. It has several medicinal purposes, including treatment of certain eye conditions. A tincture of atropine causes dilation of the pupils. Roman women used belladonna to achieve this dilating effect because they thought it attractive. Indeed, belladonna means "beautiful lady."

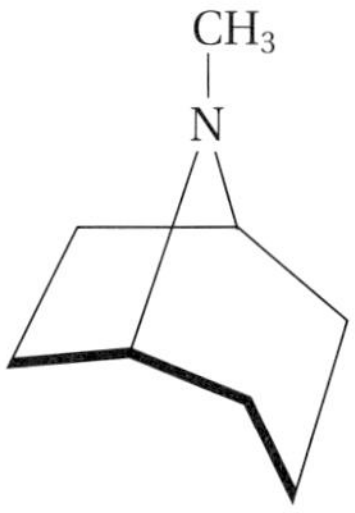

Tropane 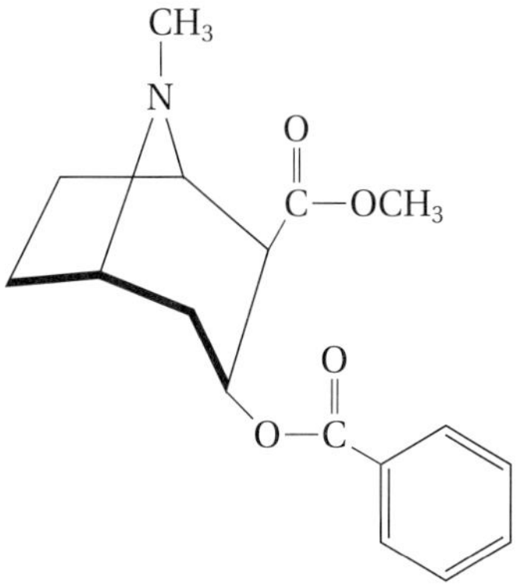   Cocaine 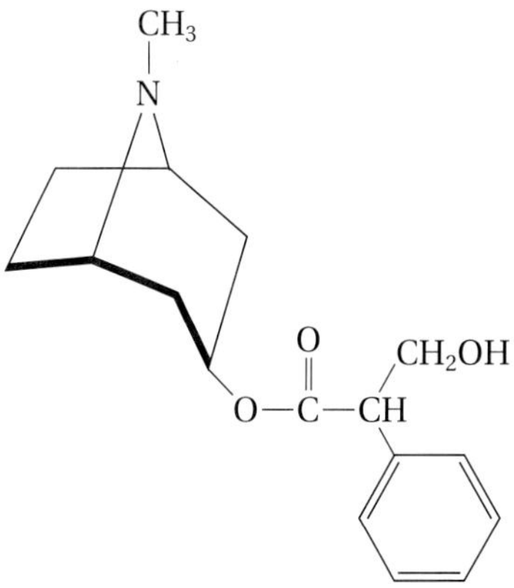   Atropine

Illegally produced "designer" drugs chemically mimic the effects of specific drugs of abuse. Some derivatives of fentanyl are over 1000 times more powerful than morphine. Designer drugs have a high risk of addiction and can cause brain damage and even death.

Opium is the raw resin extracted from the seedpods of the opium poppy (*Papaver somniferum*), not the poppy grown in gardens. When the resin is refined, two important alkaloids, morphine and codeine, can be isolated in pure form. Morphine is the most effective pain-killing drug known. Codeine is a powerful analgesic and cough suppressant. It was an ingredient in cough medicines for many years, but it has been replaced by dextromethorphan, a synthetic alkaloid that is equally effective. Morphine and codeine are addictive drugs.

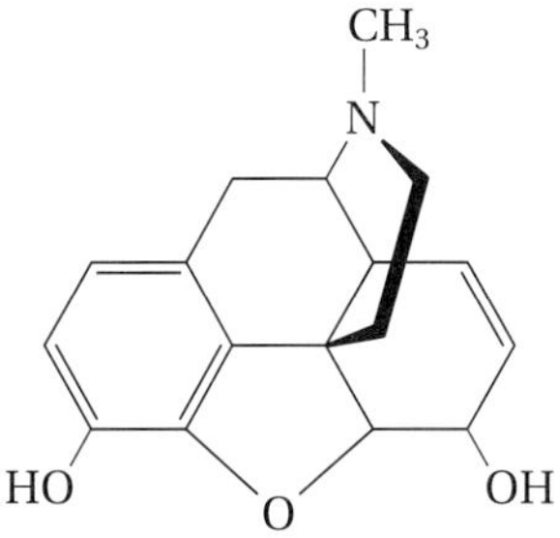

Morphine 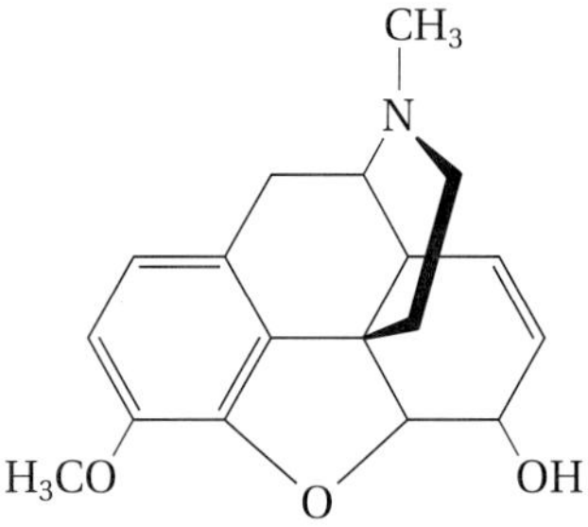   Codeine 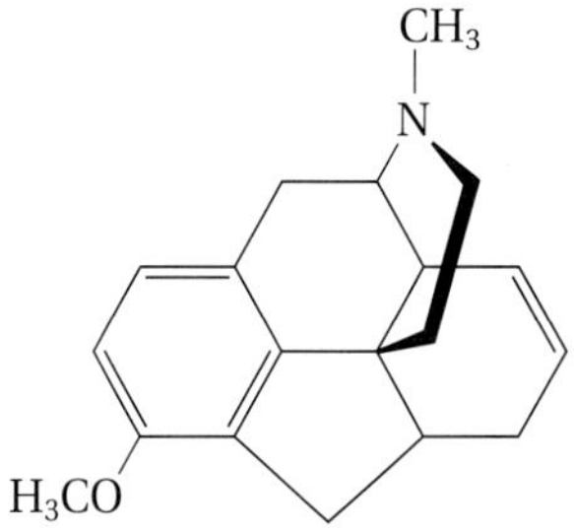   Dextromethorphan

Ironically, heroin, an acetic acid ester derivative of morphine, was introduced because it was thought to be a better analgesic and to lack morphine's addictive properties. It was soon found, however, that heroin is no better than morphine as an analgesic and is even more addictive than other **opiates**—*drugs that produce the psychological and physiologic effects of opium.* Today, heroin is a destructive hard drug. Methadone, a drug that blocks heroin action, has been introduced to wean heroin addicts from their habit. Methadone does not produce the euphoria associated with heroin use. Its use is controversial, since it too is an addictive drug.

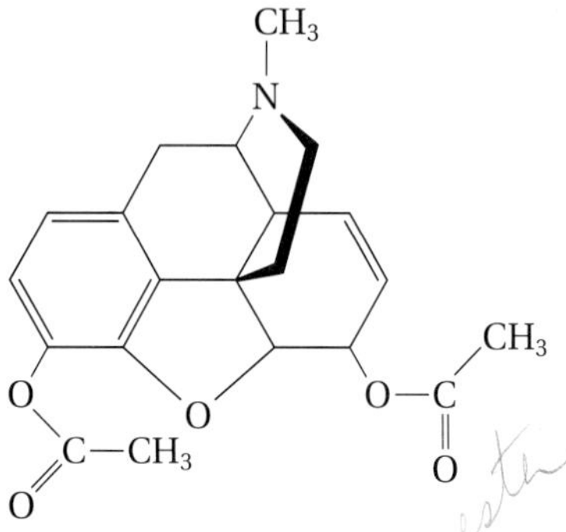

Heroin

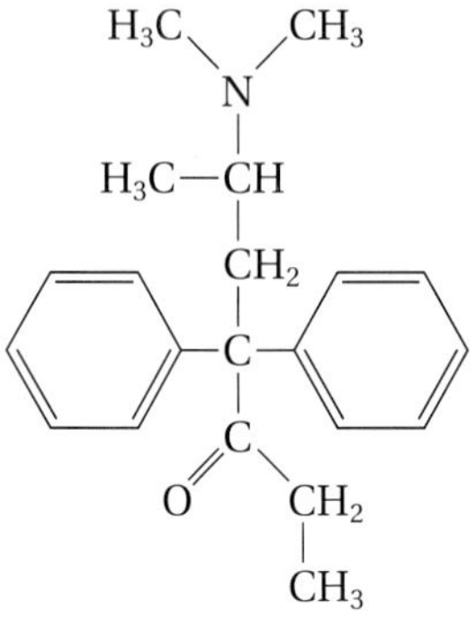

Methadone

PRACTICE EXERCISE 6.12

List some of the physiologic effects and uses of (a) morphine, (b) nicotine, (c) atropine, and (d) quinine.

6.9 Barbiturates *not tested*

AIM: To name and give examples of uses of barbiturate drugs.

Focus

Barbiturate drugs are derivatives of urea.

Urea has a simple molecular structure. In structural terms, we can consider it the diamide of carbonic acid (H_2CO_3):

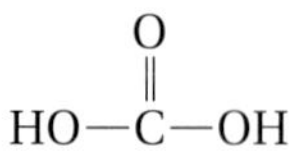

$$HO-\overset{\displaystyle O}{\overset{\|}{C}}-OH \qquad H_2N-\overset{\displaystyle O}{\overset{\|}{C}}-NH_2$$

Carbonic   Urea
acid

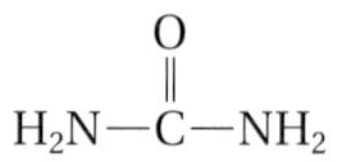

Urea is important in its own right, since it is the form in which our bodies dispose of excess nitrogen in the urine. Its high water solubility (125 g/100 mL of water) and low toxicity make it ideal for this purpose.

Condensation of malonic acid with urea produces barbituric acid, the parent compound of a number of drugs.

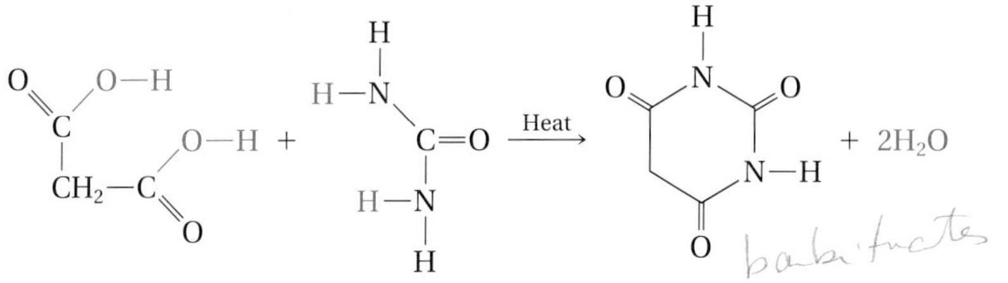

Malonic acid   Urea   Barbituric acid

Note that barbituric acid contains the pyrimidine ring system. It is an acid because the hydrogens on the ring nitrogens readily dissociate in basic solution to form barbiturate salts.

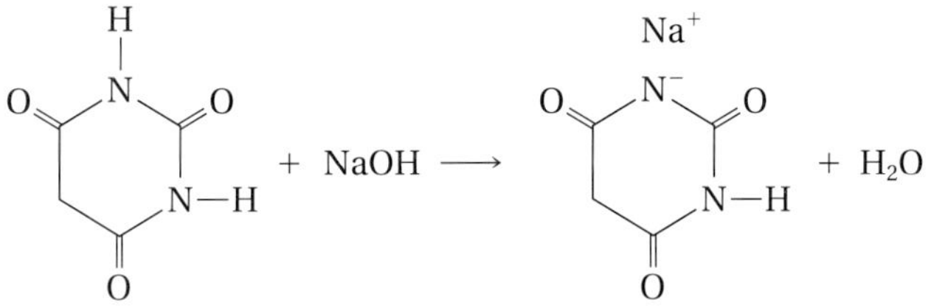

Barbituric acid   Sodium barbiturate

Hypnotic
hupnos (Greek):   sleep

Sedative
sedare (Latin):   to calm

Salts of barbituric acid and its derivatives are called **barbiturates.** The sodium salts are often administered because they ionize in solution and are more water soluble than the acid forms. Barbituric acid is not physiologically active, but some of its substituted derivatives are among the most potent **hypnotics** (*sleep-inducers*) and **sedatives** (*tranquilizers*) known. Barbital was introduced in Germany in 1903 under the name Veronal, and it is still used as a hypnotic. Since the introduction of barbital, medicinal chemists have synthesized thousands of variations on the barbituric acid structure, but only about a dozen are clinically useful.

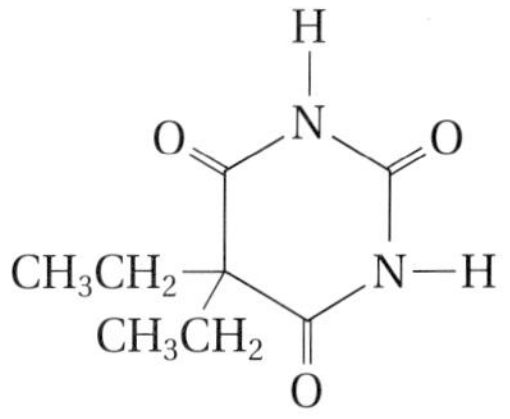

Barbital
(Veronal)

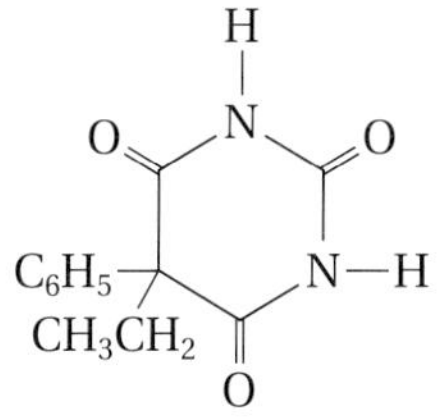

Phenobarbital
(Luminal)

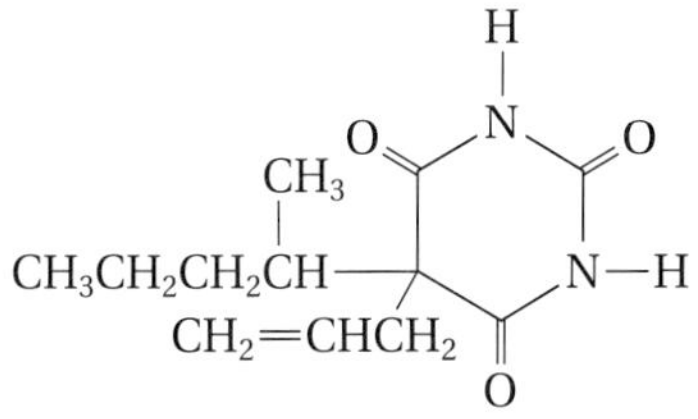

Secobarbital
(Seconal)

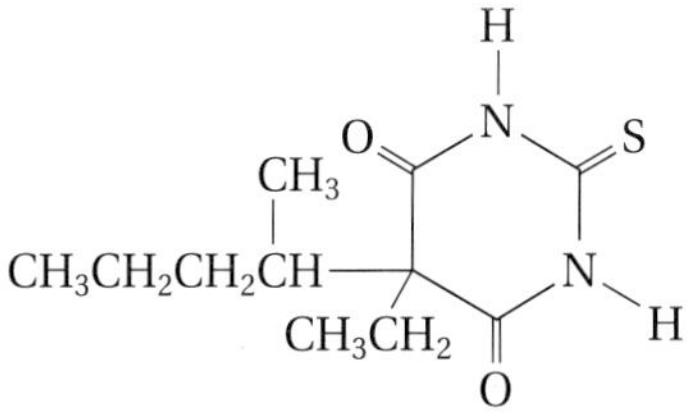

Thiopental

As hypnotics, barbiturates are often classified as long-acting or short-acting. Barbital is long-acting. A 0.3-g dose produces six or more hours of sleep in an adult. Thiopental sodium salt (sodium pentothal), on the other hand, acts only briefly; it is used in surgery to put patients to sleep before a general anesthetic is administered.

PRACTICE EXERCISE 6.13

Write the structure of sodium pentothal.

PRACTICE EXERCISE 6.14

Explain why the hydrogens on the ring nitrogens of barbituric acid and its derivatives are readily lost as protons.

SUMMARY

Amines and amides are major nitrogen-containing classes of organic compounds. Amines are organic derivatives of ammonia. They are classified as primary ($1°$; RNH_2), secondary ($2°$; $RNHR$), and tertiary ($3°$; $RNRR$). A quaternary ammonium salt has four carbon groups attached to the amine nitrogen; the nitrogen is positively charged. Carbon rings that contain amine nitrogen are called heterocyclic amines. The acid-base properties of amines are similar to those of ammonia. That is, the ability of the unshared electron pair of amine nitrogen to accept a proton in acidic solutions makes amines weak bases. Because of their weak basicity, certain amines act as pH buffers in biological fluids.

Amides are derivatives of acids and amines. When formed from carboxylic acids, they have the general formulas $RCONH_2$ (simple amides),

$RCONHR$ (monosubstituted amides), and $RCONRR$ (disubstituted amides). Amides are neutral compounds; the unshared electron pair of the amide nitrogen is pulled toward carbonyl oxygen, rendering it less available for protonation than it is in amines.

Many amines are physiologically important. The catecholamines are neurotransmitters of the central nervous system. Amphetamine and several hallucinogenic drugs are similar in structure to catecholamines. Alkaloids are a class of over 2500 amines isolated from plants. Many have useful medicinal properties. Lysergic acid diethylamide (LSD), reserpine, quinine, atropine, and cocaine are a few alkaloids. Morphine, another alkaloid, is the most effective pain killer known. Barbiturates are derivatives of urea (NH_2CONH_2). A dozen or so barbiturates are used in medicine, mainly as hypnotics and sedatives.

SUMMARY OF REACTIONS

Here are examples of the reactions covered in this chapter.

1. Preparation of amines. (The amine is $1°$ or $2°$.)

Treatment with base liberates the free amine.

2. Preparation of quaternary ammonium salts. (The amine is $3°$.)

3. Preparation of amides.

(a) By dehydration of ammonium salts of acids. (The nitrogen compound is ammonia or a $1°$ or $2°$ amine.)

(b) By reactions of ammonia or amines with esters or anhydrides. (The amine is $1°$ or $2°$.)

4. Hydrolysis of amides. (The amide may be simple, monosubstituted, or disubstituted.)

$$R-\overset{\overset{\displaystyle O}{\|}}{C}-\overset{..}{N}H_2 + HO-H \xrightarrow[\text{H}^+ \text{ or OH}^-]{\text{Heat}} R-\overset{\overset{\displaystyle O}{\|}}{C}-OH + H-\overset{..}{N}H_2$$

KEY TERMS

Alkaloid (6.8)
Amide (6.4)
Amine (6.1)
Alkylammonium ion (6.2)
Ammonium salt (6.2)

Anilide (6.4)
Antihistamine (6.7)
Arylammonium ion (6.2)
Barbiturate (6.9)
Decongestant (6.7)

Free amine (6.2)
Hallucinogen (6.7)
Hypnotic (6.9)
Neurotransmitter (6.7)
Opiate (6.8)

Polyamide (6.5)
Protonated amine (6.2)
Quaternary ammonium
    salt (6.3)
Sedative (6.9)

EXERCISES

Amines (Sections 6.1, 6.2, 6.3)

6.15 Name or write a structural formula for the following amines. Classify them as primary, secondary, or tertiary amines.
  (a) $(CH_3)_2NH$ (b) *p*-chloro-*N*-methylaniline
  (c)   (d) diethylmethylamine

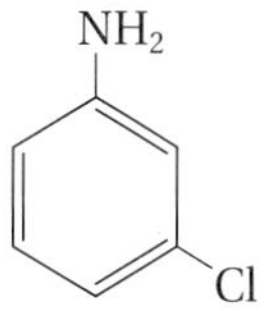

6.16 Write structural formulas and name the following amines. Classify each amine as primary, secondary, or tertiary.
  (a) diethylamine   (b) $(CH_3CH_2)_2NCH_3$
  (c) butylamine   (d)

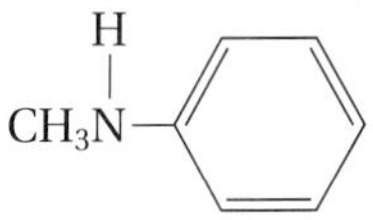

6.17 What is the meaning of the term *heterocyclic amine*?

6.18 Draw the structure and give the name of (a) an aromatic heterocyclic amine and (b) an aliphatic heterocyclic amine.

6.19 Draw structural formulas for (a) pyrimidine and (b) purine. Derivatives of these two compounds are found in what biologically important molecules?

6.20 Draw the structure of pyrrole. List some of the naturally occurring molecules that contain the pyrrole ring system.

6.21 Why are amines weak bases?

6.22 Draw the general formulas for (a) an unprotonated (free) amine and (b) a protonated amine.

6.23 Write the structure and name the expected products of each of the following reactions.
  (a) $CH_3CH_2NH_2 + HCl$ (b) $(CH_3)_2NH + HNO_3$
  (c) $CH_3NH_2 + H_2SO_4$ (d) $(CH_3)_3CNH_2 + HCl$

6.24 Draw the structure and name the organic product for each of the following reactions.
  (a) $CH_3NH_3{}^+I^- + NaOH$
  (b) $(CH_3)_2NH + CH_3Cl$
  (c) $CH_3I + NH_3$
  (d)

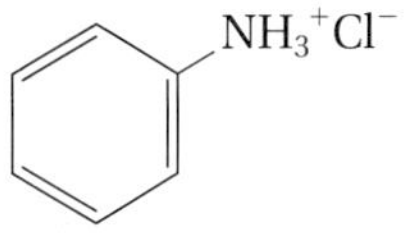

  (e) $(CH_3CH_2)_3N + CH_3CH_2Cl$

6.25 Write an equation for the dissociation of the dimethylammonium ion. Why does an aqueous solution of dimethylammonium chloride test acidic?

6.26 Draw the structure of tetramethylammonium iodide. What happens if this compound is treated with sodium hydroxide?

Amides (Sections 6.4, 6.5, 6.6)

6.27 Name or write structural formulas for the following amides.
  (a) $CH_3\overset{\overset{\displaystyle O}{\|}}{C}NH_2$ (b) $CH_3CH_2\overset{\overset{\displaystyle O}{\|}}{C}NHCH_3$
  (c)

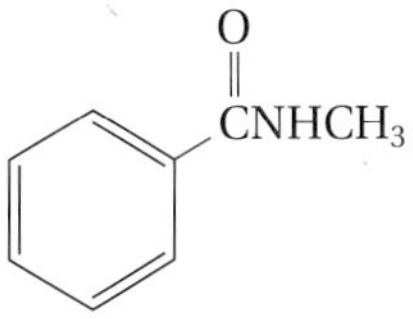

  (d) *N*-ethyl-*N*-methylpropanamide
  (e) acetanilide

6.28 Write structural formulas or name the following.
(a) acetamide   (b) benzamide
(c) *N*-ethylbenzamide

(d) $CH_3CH_2\overset{\overset{\displaystyle O}{\|}}{C}N\underset{\underset{\displaystyle CH_3}{|}}{}CH_3$ (e) $H\overset{\overset{\displaystyle O}{\|}}{C}N\underset{\underset{\displaystyle CH_3}{|}}{}CH_3$

6.29 Write the structure for the product of each of the following reactions.

(a) $CH_3CH_2\overset{\overset{\displaystyle O}{\|}}{C}O^-\ NH_4^+\ \xrightarrow{\text{Heat}}$

(b) $CH_3\overset{\overset{\displaystyle O}{\|}}{C}OCH_3 + NH_3 \longrightarrow$

(c) $H\overset{\overset{\displaystyle O}{\|}}{C}OCH_3 + NH_2CH_2CH_3 \longrightarrow$

6.30 The following amides are hydrolyzed at the conditions shown in the equations. What are the products of each reaction?

(a) $CH_3CH_2\overset{\overset{\displaystyle O}{\|}}{C}-NHCH_3 + H_2O \xrightarrow[H^+]{\text{Heat}}$

(b) 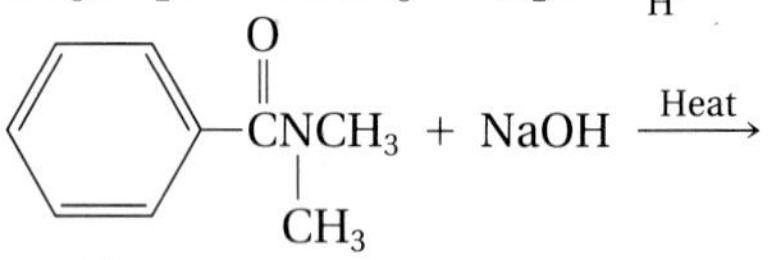 $\langle\!\!\!\bigcirc\!\!\!\rangle -\overset{\overset{\displaystyle O}{\|}}{C}N\underset{\underset{\displaystyle CH_3}{|}}{}CH_3 + NaOH \xrightarrow{\text{Heat}}$

(c) $CH_3\overset{\overset{\displaystyle O}{\|}}{C}-NHCH_2CH_3 + NaOH \xrightarrow{\text{Heat}}$

Important amines (Sections 6.7, 6.8, 6.9)

6.31 Parkinson's disease results when the concentration of dopamine in the brain cells is very low. Explain why dopa is an effective drug for parkinsonism and why dopamine is ineffective.

6.32 What effect does amphetamine have on the body?

6.33 What are the alkaloids? How did these compounds get their name?

6.34 Name the nitrogen-containing functional groups in the LSD molecule.

6.35 Name two opiates that occur naturally.

6.36 What is the structural difference between morphine and codeine? What is codeine used for?

6.37 Is barbituric acid physiologically active? What are some of the physiologic effects of barbiturates?

Additional Exercises

6.38 Give the products of the following reactions. Name the products and reactants whenever possible. If there is no reaction, write "no reaction."

(a) $CH_3CH_2-\overset{\overset{\displaystyle H}{|}}{\underset{\underset{\displaystyle H}{|}}{N}}\!:\ +\ HCl \longrightarrow$

(b) $CH_3-\overset{\overset{\displaystyle H}{|}}{\underset{\underset{\displaystyle H}{|}}{\overset{+}{N}}}-H\ Cl^-\ +\ NaOH \longrightarrow$

(c) $H-\overset{\overset{\displaystyle H}{|}}{\underset{\underset{\displaystyle H}{|}}{N}}\!:\ +\ CH_3I \longrightarrow$

(d) $CH_3-\overset{\overset{\displaystyle O}{\|}}{C}-N\overset{\displaystyle H}{\underset{\displaystyle H}{<}}\ +\ H_2O \xrightarrow{\text{HCl}}$

(e) 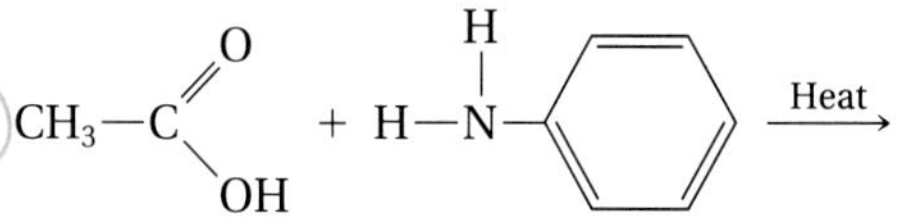 $CH_3-C\overset{\displaystyle O}{\underset{\displaystyle OH}{<}}\ +\ H-\overset{\overset{\displaystyle H}{|}}{N}-\langle\!\!\!\bigcirc\!\!\!\rangle \xrightarrow{\text{Heat}}$

(f) $CH_3-\overset{\overset{\displaystyle H}{|}}{\underset{\underset{\displaystyle CH_3}{|}}{C}}-\overset{\overset{\displaystyle O}{\|}}{C}-\underset{\underset{\displaystyle H}{|}}{N}-CH_2CH_3 \xrightarrow[\text{Heat}]{\text{Acid}}$

(g) $CH_3CH_2C\overset{\displaystyle O}{\underset{\displaystyle NH_2}{<}}\ +\ P_2O_5 \xrightarrow{\text{Heat}}$

6.39 Identify each of these heterocyclic compounds.

(a)

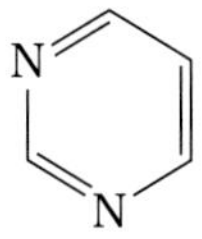

(b)

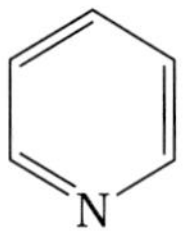

(c)

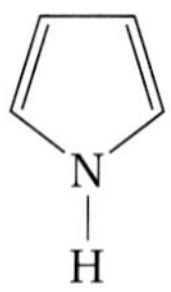

(d)

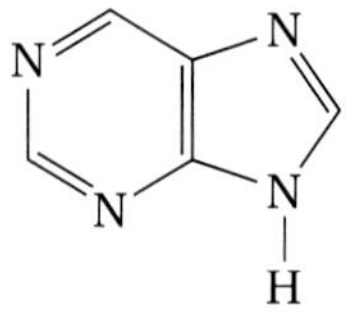

(e)

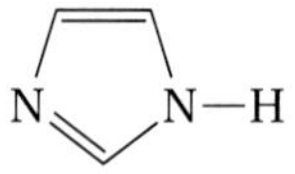

(f)

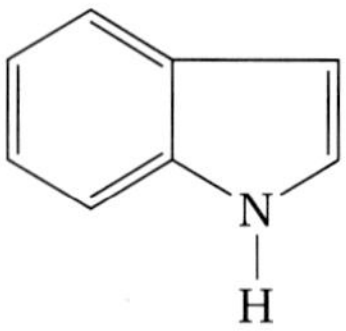

6.40 Many nitrogen-containing compounds of biochemical significance have been mentioned in this chapter. Match the following compounds with the expression that best fits.

(1) Dopa
(2) Serotonin
(3) Mescaline
(4) Amphetamine
(5) Histamine
(6) Benadryl
(7) Nicotine
(8) Quinine
(9) Atropine
(10) Codeine
(11) Morphine
(12) Barbital
(13) Sodium pentothal
(14) Caffeine

(a) responsible for the symptoms of hay fever
(b) a long-acting hypnotic
(c) used as an antimalarial drug
(d) low levels implicated in mental depression
(e) antihistamine drug
(f) drug administered for Parkinson's disease
(g) hallucinogen of similar structure to a neurotransmitter
(h) a short-acting barbiturate
(i) causes dilation of pupils; used to treat certain eye conditions
(j) a common drug of abuse that stimulates the nervous system and depresses the appetite
(k) most effective pain killer known
(l) an alkaloid present in tobacco
(m) a stimulant in tea, coffee, and cocoa
(n) an alkaloid derived from poppies; used as a cough suppressant

6.41 Write structural formulas for the following compounds: (a) triethylmethylammonium bromide, (b) N-propylaniline, (c) N-methylacetamide, and (d) N,N-dimethylbenzamide.

6.42 Name these compounds and classify each as a primary, secondary, or tertiary amine.

(a) [benzene ring]—$N(CH_3)_2$ (b) $CH_3CH_2NH_2$

(c) CH_3—CH—$NHCH_3$ with CH_3 below CH   (d) [two benzene rings connected to N with H above]

(e) $(CH_3)_3C$—NH_2 (f) $CH_3CH_2NHCH_2CH_3$

6.43 Draw structural formulas for the following compounds.
(a) triethylamine   (b) propionamide
(c) N-ethylaniline   (d) N-phenylbenzamide
(e) 3-ethylaniline   (f) N,N-diethylacetamide

6.44 Name the following amides.

(a) CH_3C—$N(CH_3)_2$ with O below C   (b) CH_3CH_2C—NH_2 with O below C

(c) [benzene ring]—C—$NHCH_2CH_3$ with O above C

(d) [benzene ring]—NH—C—CH_2CH_3 with O above C

6.45 Write general reactions to show the formation of a simple, monosubstituted, and disubstituted amide.

6.46 Name the following compounds. Predict their solubility in water. If soluble, will the solution be acidic, basic, or neutral? Explain.

(a) $CH_3CH_2NH_2$ (b) CH_3CH_2C—OH with O below C

(c) CH_3CH_2C—H with O below C   (d) CH_3—C—CH_3 with O below C

(e) CH_3CH_2C—NH_2 with O below C   (f) CH_3—C—OCH_3 with O below C

(g) [benzene ring]—OH (h) [benzene ring]—NH_2

6.47 Using ethene and propanol as your only organic starting material, show how you might prepare each of the following compounds. You may use any inorganic reagents that you need. Name the intermediate products of each reaction.
(a) propylamine   (b) N-ethylpropanamide

6.48 Draw the structure and name the compound described by each of the following statements.
(a) a six-membered aliphatic heterocyclic amine
(b) the simplest aromatic amine
(c) the amine formed by the reaction of dimethylamine with ethyl chloride
(d) an aromatic amine with an ethyl group that is not on the benzene ring

SELF-TEST (REVIEW)

True/False

1. Pyridine is a six-membered aromatic heterocyclic amine. *didn't cover*

2. Amines are relatively weak bases. *true*

3. An amide contains a carbon-nitrogen double bond. *false*

4. A free amine has an unshared electron pair on the amine nitrogen. *true*

5. A tertiary amine is the product of a reaction between a secondary amine and a halocarbon. *didn't cover*

6. The reaction between a carboxylic acid and an amine gives an amide. *true*

7. A secondary amine that contains two carbon atoms is dimethylamine. CH_3-$\overset{H}{N}$-CH_3 *true*

8. In a quaternary ammonium salt the nitrogen atom is bonded to four carbon atoms.

9. Another name for *o*-toluidine is 2-methylaniline. *true*

10. Trimethylamine is a tertiary amine. *true*

Multiple Choice

11. What is the correct name for the following compound?

$$CH_3CH_2\overset{\overset{O}{\|}}{C}-NHCH_3$$

(a) *N*-methylpropanamide
(b) ethylmethylamide   (c) propylmethylamide
(d) methylaminoproprionic acid

12. Which of these is an antimalarial drug?
(a) dopa   (b) quinine   (c) benadryl
(d) atropine

13. The reaction of two moles of ethyl chloride with one mole of ammonia would produce
(a) ethylamine.   (b) diethylamine.
(c) triethylamine.   (d) all of the above.

14. Which of the following is an antihistamine?
(a) mescaline   (b) codeine   (c) caffeine
(d) benadryl

15. Amines are
(a) weak bases.   (b) weak acids.
(c) all physiologically important.
(d) insoluble in water.

16. The acid hydrolysis of *N*-ethylacetamide gives
(a) ethanol and ethylamine.
(b) ethanol and acetic acid.
(c) acetic acid and ethylamine.
(d) ethylamine and acetaldehyde.

17. Which of these is not an aromatic amine?
(a) aniline   (b) diphenylamine
(c) piperidine   (d) *o*-toluidine

18. An amide can be prepared by
(a) heating an ammonium salt.
(b) reacting ammonia with an acid anhydride.
(c) reacting an ester with an amine.
(d) all of the above.

19. Which of these is a tertiary amine?
(a) aniline   (b) pyrrolidine
(c) triphenylamine   (d) *N*-methyldiethylamide

20. An aqueous solution of ethyl ammonium chloride would be
(a) strongly acidic.   (b) weakly acidic.
(c) strongly basic.   (d) neutral.

21. Which of the following terms would *not* describe codeine?
(a) cough suppressant   (b) barbiturate
(c) addictive   (d) analgesic

22. The name of the following compound is

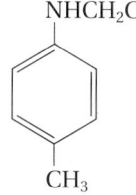

(a) *N*-ethyl-*p*-toluidine.   (b) ethylaminotoluene.
(c) ethylbenzylamide.   (d) ethylphenylaniline.

23. Which of the following terms does not describe the structure of purine?

(a) fused ring   (b) heterocyclic
(c) quaternary   (d) aromatic

24. An anilide would generally have all the following except
(a) a carboxylic acid group.   (b) an amide group.
(c) an aromatic ring.   (d) a carbonyl group.

Carbohydrates

The Structure and Chemistry of Sugars

Honey contains fructose,
an important carbohydrate.

CHAPTER OUTLINE

The class of compounds called *carbohydrates,* or *sugars,* is widespread throughout nature. These versatile molecules provide food for all living creatures. They are also very important structural components of the cell walls of plants and the shells of crustaceans such as crabs and shrimp. In this chapter we will learn what carbohydrates are and then learn to recognize differences in the molecular structures of simple sugars. We also will learn how simple sugars are linked together to form some important polymer molecules. And finally, we will learn a few tests for the identification of sugars.

Despite the importance of sugars as food, not all people can digest a sugar that most people find edible, as we will learn in the following Case in Point.

CASE IN POINT: Lactose intolerance

 Alberto is healthy, but he has always been rather thin. In his sophomore year in college, he tried to gain some weight by lifting weights and increasing his caloric intake. After his first weightlifting session, Alberto and his friends stopped at the campus cafeteria, where Alberto drank two large milkshakes. He awoke in the middle of the night with stomach cramps, gas, and severe diarrhea. Fearing that he had food poisoning, he immediately reported to the college infirmary. After the physician examined Alberto and heard about the suspicious milkshakes, she dismissed the possibility of food poisoning and suggested that Alberto avoid all dairy products until he felt well again. What is the link between ingesting dairy products and the symptoms that Alberto experienced? We will learn more about Alberto's illness and how it is related to the digestion of sugars in Section 7.8.

Lactose, or milk sugar, is present in dairy products.

7.1 Carbohydrate structure and stereochemistry

AIMS: To name and classify a carbohydrate as a monosaccharide, disaccharide, or polysaccharide; as a triose, tetrose, pentose, or hexose; as an aldose or a ketose. To explain what is meant by the handedness of a molecule using the terms asymmetric carbon and stereoisomer. To interpret two-dimensional Fischer projection formulas of sugars as three-dimensional structures.

Carbohydrates abound in nature; among the many forms are starch, cotton, table sugar, and wood. **Carbohydrates** *are polyhydroxy aldehydes or ketones.* The name *carbohydrate* means "hydrate of carbon." The name was coined in the early days of carbohydrate chemistry because many of these compounds have molecular formulas that are multiples of CH_2O, such as $C_6H_{12}O_6$ and $C_5H_{10}O_5$. *Carbohydrates are also called* **sugars.**

Glyceraldehyde is the simplest sugar found in nature. If you have a good understanding of the structure of glyceraldehyde, you will be able to understand more complicated sugars. Here is the structure of glyceraldehyde:

$$\overset{1}{C}HO$$
$$H-\overset{2}{C}-OH$$
$$\overset{3}{C}H_2-OH$$

Glyceraldehyde

The ending *-ose* is a characteristic of the names of many sugars and sugar derivatives. *Sugars that contain an aldehyde functional group (—CHO) are called* **aldoses.** The number of carbons in a sugar gives a more detailed description of the sugar. For example, a sugar containing three carbons and an aldehyde functional group is an *aldotriose.* Glyceraldehyde is an aldotriose. The carbon chains of aldoses are numbered starting with the aldehyde group as carbon 1.

Stereoisomerism of glyceraldehyde

Have you ever noticed that placing an object in front of a mirror can give two different results? If the object is symmetrical—a sphere, say—then its mirror image is indistinguishable from the object. That is, the appearance of the sphere and its reflection are superimposable mirror images—the images can be placed on top of each other to obtain a match. However, if you look at your hands in a mirror, your right hand reflects as a left hand and your left hand reflects as a right hand (Fig. 7.1). Your hands are examples of **nonsuperimposable mirror images**—*mirror images that cannot be placed on top of each other to obtain a match.* Many pairs of familiar objects, such as ears, feet, shoes, and bird wings, are related similarly.

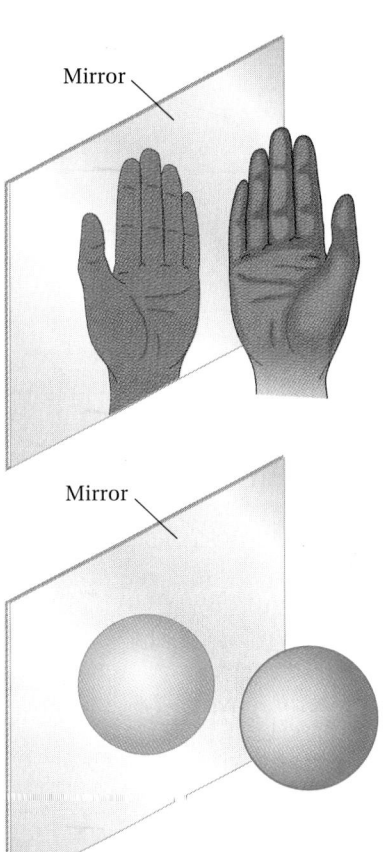

Mirror

Mirror

Figure 7.1
The reflected image of a right hand in a mirror appears as a left hand. Note the relative positions of the thumbs. A sphere and its reflected image are identical.

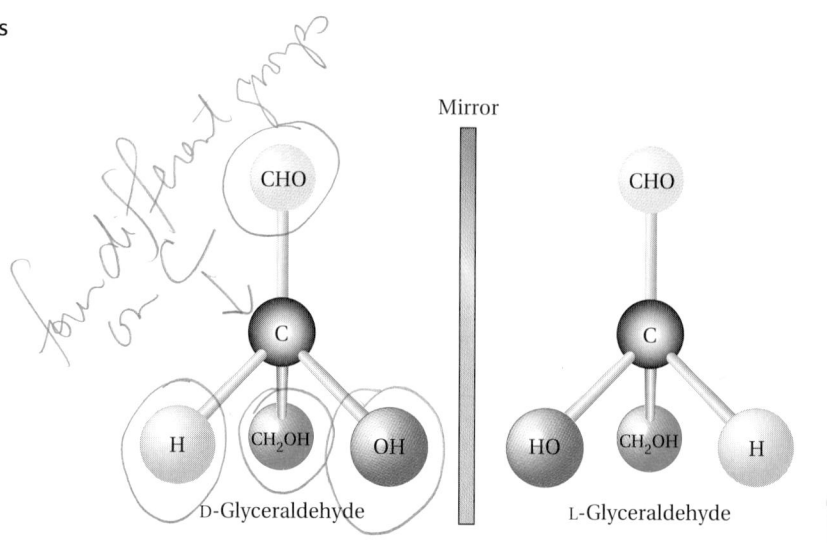

Figure 7.2
Ball-and-stick models of D- and L-glyceraldehyde. These two structures are nonsuperimposable mirror images of one another.

PRACTICE EXERCISE 7.1

Which of the following objects would have a nonsuperimposable mirror image? (Ignore designs or other markings.)

(a) a clam shell   (b) a cup   (c) a ball   (d) a car
(e) a wood screw   (f) a fingerprint   (g) a baseball bat

Looking closely at the glyceraldehyde structure, we see that carbon 2 has four different groups attached: —CHO, —H, —OH, and —CH$_2$OH. *A carbon with four different groups attached is called an* **asymmetric carbon.** Compounds whose molecules contain asymmetric carbons have handedness—that is, there are two kinds of molecules with the same structural formula that are related to one another in much the same way as your hands are related to one another. Glyceraldehyde, with its four different groups attached to the asymmetric carbon in one possible arrangement, is called D-*glyceraldehyde.* Its mirror image is L-glyceraldehyde. Figure 7.2 shows D- and L-glyceraldehyde. Like your hands, D- and L-glyceraldehyde are nonsuperimposable mirror images. No matter how hard you try, there is no way short of breaking bonds to superimpose D- and L-glyceraldehyde so that the four groups attached to the asymmetric carbon of D-glyceraldehyde coincide with the same four groups of L-glyceraldehyde. The D- and L-glyceraldehydes are examples of **stereoisomers**—*molecules whose molecular structure differs only in the arrangement of groups in space.*

Stereochemistry is concerned with the spatial arrangement of the atoms in a molecule. The three-dimensional shapes of molecules are very important in biological systems. In Chapter 10 we will see that the activity of many biological catalysts called *enzymes* are stereospecific—they will work only on a particular molecule that has a particular three-dimensional shape.

EXAMPLE 7.1   **Identifying asymmetric carbons**

Which of the following compounds have an asymmetric carbon?

(a) CH$_3$CHCH$_3$ (b) CH$_3$CHCH$_2$CH$_3$ (c) CH$_3$CHCHO
   | | |
   OH   OH   OH

SOLUTION

An asymmetric carbon has four different groups attached. It may help to draw structures in a more complete form.

(a) The central carbon has one —H, one —OH, but two —CH$_3$ groups attached. It is not asymmetric.

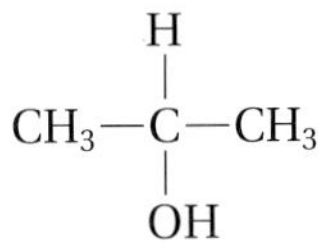

$$CH_3-\underset{\underset{\displaystyle OH}{|}}{\overset{\overset{\displaystyle H}{|}}{C}}-CH_3$$

(b) The central carbon has one —H, one —OH, one —CH$_3$, and one —CH$_2$CH$_3$ group attached. These four groups are different, so the central carbon is asymmetric. It is marked with an asterisk. None of the other carbons in this molecule is asymmetric because no other carbon is attached to four different groups (check to be sure).

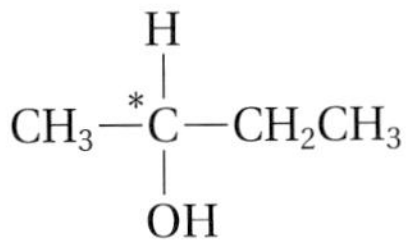

$$CH_3-\underset{\underset{\displaystyle OH}{|}}{\overset{\overset{\displaystyle H}{|}}{\overset{*}{C}}}-CH_2CH_3$$

(c) This molecule also has an asymmetric carbon. It is marked with an asterisk.

$$CH_3-\underset{\underset{\displaystyle OH}{|}}{\overset{\overset{\displaystyle H}{|}}{\overset{*}{C}}}-CHO$$

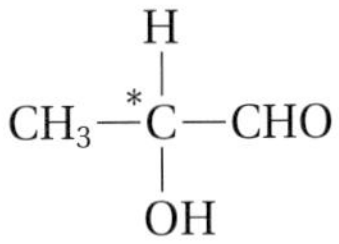

PRACTICE EXERCISE 7.2

Identify the asymmetric carbon, if there is one, in each of the following structures.

(a) CH$_3$CH$_2$CHO   (b) 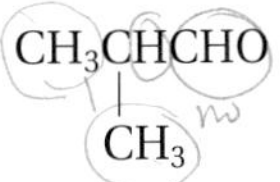 CH$_3$CHCHO   (c) 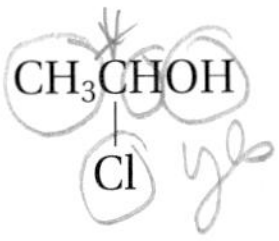 CH$_3$CHOH
                                                | |
                                              CH$_3$ Cl

(d) 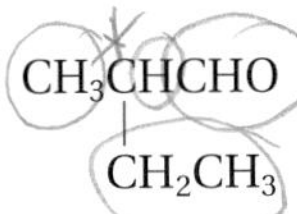 CH$_3$CHCHO
                |
           CH$_2$CH$_3$

EXAMPLE 7.2

Finding nonsuperimposable mirror images

Decide which of the following has a nonsuperimposable mirror image, and draw the stereoisomers.

(a) 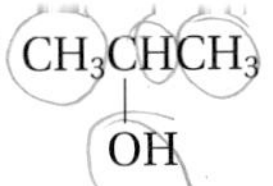 CH$_3$CHCH$_3$ (b) 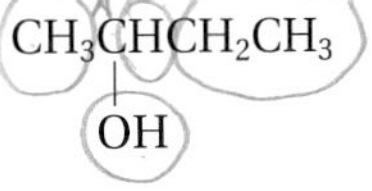 CH$_3$CHCH$_2$CH$_3$
             | |
            OH   OH

SOLUTION

(a) The structure has no asymmetric carbon and exists as only one form— its mirror image is superimposable.

(b) The structure has an asymmetric carbon—its mirror image is non-superimposable. It will exist as a pair of stereoisomers.

One way of writing the structural formulas of stereoisomers is to consider the central carbon—usually the asymmetric carbon if the molecule

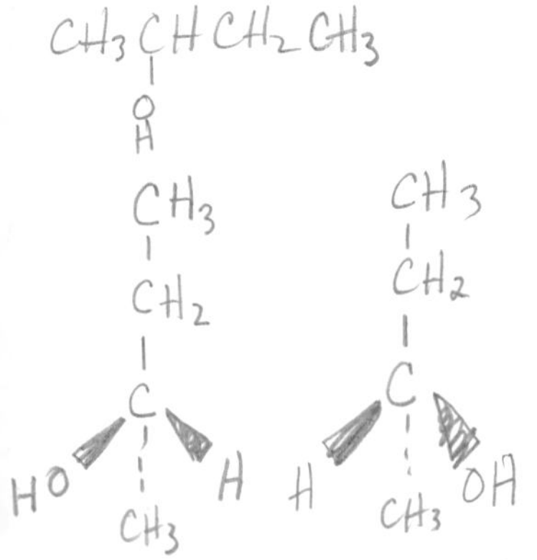

has one—to be in the plane of the paper. The two groups attached to the central atom by *tapered bonds* are understood to be above the plane of the paper; the group attached to the central carbon by a dashed bond is understood to be below the plane. The remaining group is in the plane of the paper. The stereorepresentations of (a) and (b) are as follows:

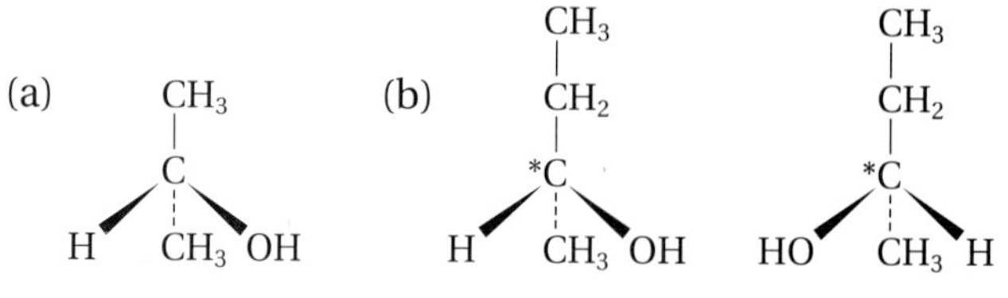

Stereoisomers

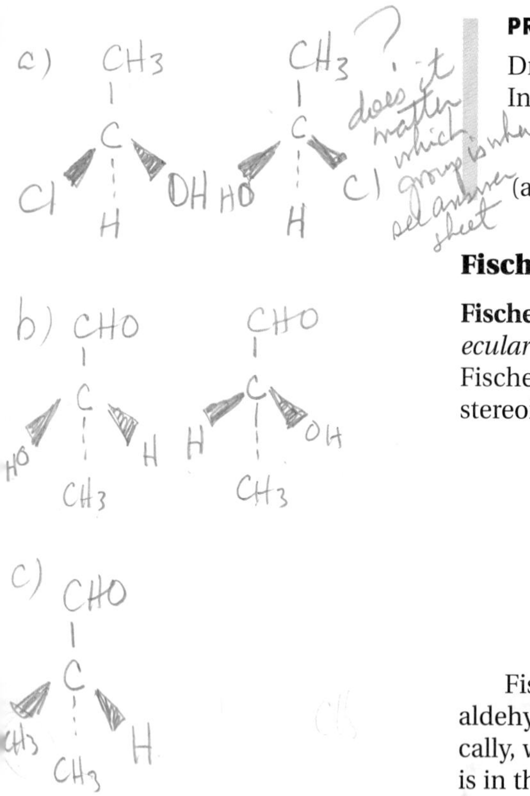

PRACTICE EXERCISE 7.3

Draw stereorepresentations for each of the following compounds. Include stereoisomers if they exist.

(a) $CH_3\overset{\overset{Cl}{|}}{C}HOH$ (b) $CH_3\overset{\overset{OH}{|}}{C}HCHO$ (c) $CH_3\overset{\overset{CH_3}{|}}{C}HCHO$

Fischer projections

Fischer projections *are a way of representing three-dimensional organic molecular structures in two dimensions.* Devised by the German chemist Emil Fischer (1852–1915), they are a convenient means of differentiating stereoisomers. Here are the Fischer projections for D- and L-glyceraldehyde:

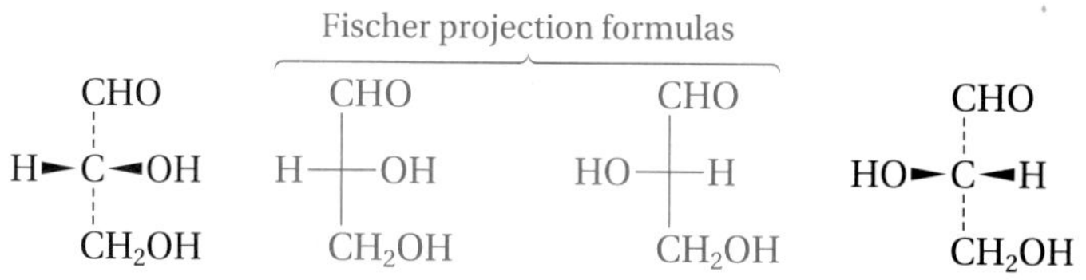

D-Glyceraldehyde L-Glyceraldehyde

Fischer projections place the principal functional group (in this case aldehyde) at the top of the carbon chain. The carbon chain is written vertically, with substituent groups to the left and right. The asymmetric carbon is in the plane of the paper; the chain carbons attached to the asymmetric carbon are below the plane. The groups to the right and left of the asymmetric carbon are above the plane. The Fischer projection formula of glyceraldehyde that shows the hydroxyl group to the right of the carbon chain is D-glyceraldehyde; the Fischer projection formula that shows the hydroxyl group to the left of the carbon chain is L-glyceraldehyde. The naturally occurring form of glyceraldehyde is D-glyceraldehyde. Virtually all sugars found in nature are of the D family.

Optical activity

The D and L forms of stereoisomers are identical in their physical and chemical properties except that they rotate plane-polarized light in opposite

Vibrational pattern of a single light wave in a beam of ordinary light viewed from the side

plain polarized light

When you look at the vibration head on, it looks like this.

Vibrations in a beam of ordinary light are in all directions.

A polarizing filter lets through light in which vibrations are in only one direction.

Polarizing filter   Polarizing filter

Sample tube

D-(+) glyceraldehyde

Rotation to right

L-(−) glyceraldehyde

Rotation to left

Figure 7.3
One of a pair of optical isomers rotates plane-polarized light to the right. Its mirror image rotates the light to the left. The instrument used to measure optical rotation is called a *polarimeter.*

directions. **Plane-polarized light** *is light in which vibrations are in only one direction.* Stereoisomers are often called *optical isomers.* **Optical isomers** *are stereoisomers that rotate plane-polarized light in opposite directions.* For example, D-glyceraldehyde rotates plane-polarized light to the right, and L-glyceraldehyde rotates plane-polarized light to the left. Sometimes you may see the name D-(+)-glyceraldehyde. The D refers to the handedness of the particular glyceraldehyde; the plus sign refers to the direction of rotation of plane-polarized light (Fig. 7.3). There is no connection between the D and the (+); other D sugars may rotate plane-polarized light to the left. Then the sugar would be designated D-(−).

7.2 Monosaccharides

AIMS: **To draw open-chain Fischer projections for the common simple sugars. To identify a sugar as D or L by looking at its Fischer projection formula.**

Focus

Most monosaccharides contain four, five, or six carbons.

The simplest carbohydrate molecules, not bonded to any other carbohydrate, are called **simple sugars** *or* **monosaccharides.** Many of the monosaccharides in nature contain four, five, or six carbons. Sugars containing an aldehyde functional group and consisting of four, five, and six carbons are *aldotetroses, aldopentoses,* and *aldohexoses,* respectively. In this section we will examine some of them.

The four-carbon sugars

D-Threose, with a chain of four carbons, is a naturally occurring tetrose; the aldehyde functional group makes threose an aldotetrose. Two of the carbons of threose, carbon 2 and carbon 3, have four different groups attached, so threose has two asymmetric carbons:

[handwritten annotations: "4C sugar aldotetrose"]

(a)

$$
\begin{array}{c}
\overset{1}{C}HO \\
HO-\overset{2}{C}-H \\
H-\overset{3}{C}-OH \\
\overset{4}{C}H_2OH
\end{array}
$$

Fischer projection formulas

$$
\begin{array}{c}
CHO \\
HO-\!\!\!|-H \\
H-\!\!\!|-OH \\
CH_2OH
\end{array}
\qquad
\begin{array}{c}
CHO \\
H-\!\!\!|-OH \\
HO-\!\!\!|-H \\
CH_2OH
\end{array}
$$

(b)

$$
\begin{array}{c}
\overset{1}{C}HO \\
H-\overset{2}{C}-OH \\
HO-\overset{3}{C}-H \\
\overset{4}{C}H_2OH
\end{array}
$$

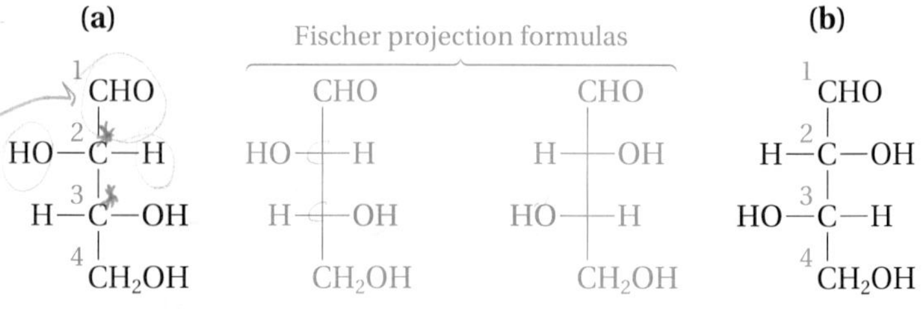

D-Threose   L-Threose

The Fischer projections of the stereoisomer have an —OH group on each of their two asymmetric centers, one pointing to the left and one to the right. If they were not labeled for handedness, how could you tell which is the D and which is the L isomer? First, draw a Fischer projection of the sugar being considered. Then look at the hydroxyl group attached to the last asymmetric carbon in the chain. If the hydroxyl group points to the right, the sugar belongs to the D family; if it points to the left, the sugar belongs to the L family. In the case of threose, the last asymmetric carbon in the chain is at carbon 3, so structure (a) belongs to the D family of sugars. Structure (b) belongs to the L family.

The structures of D- and L-threose are mirror images. However, since there are two asymmetric carbons in a four-carbon sugar molecule, another pair of stereoisomers can exist. These stereoisomers are D- and L-erythrose:

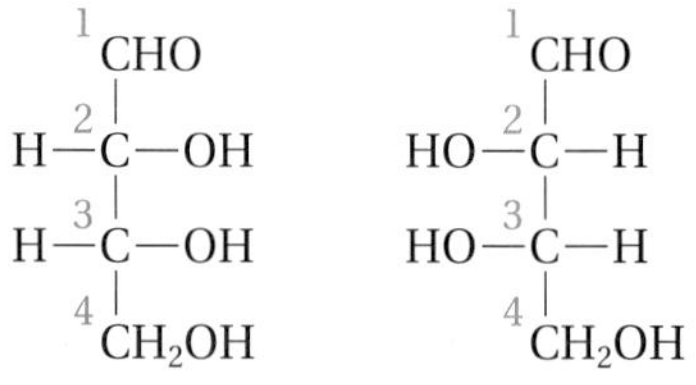

$$
\begin{array}{c}
\overset{1}{C}HO \\
H-\overset{2}{C}-OH \\
H-\overset{3}{C}-OH \\
\overset{4}{C}H_2OH
\end{array}
\qquad
\begin{array}{c}
\overset{1}{C}HO \\
HO-\overset{2}{C}-H \\
HO-\overset{3}{C}-H \\
\overset{4}{C}H_2OH
\end{array}
$$

D-Erythrose   L-Erythrose

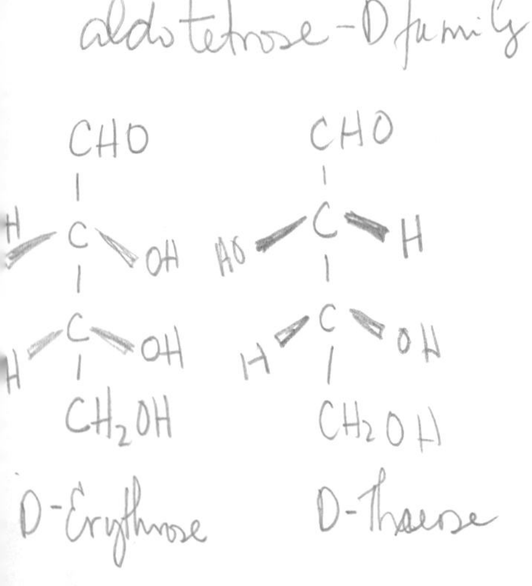

It is easy to calculate the number of possible stereoisomers of a sugar that contains multiple asymmetric carbons. This number is 2^n, where n is the number of asymmetric carbons. Thus glyceraldehyde, with one asymmetric carbon, has 2^1, or 2, stereoisomers; these are the D and L isomers. Four-carbon sugars with two asymmetric carbons can exist as 2^2, or 4, stereoisomers. We can calculate the number of mirror-image pairs by dividing the number of stereoisomers by 2. The four aldotetroses form two pairs of mirror images.

PRACTICE EXERCISE 7.4

Draw Fischer projections for the two aldotetroses that belong to the D family.

[handwritten notes in left margin: "aldotetrose–D family"; two Fischer projections labeled "D-Erythrose" and "D-Threose"]

The five-carbon sugars

Several aldopentoses are found in nature. D-Arabinose and D-xylose are five-carbon sugars produced by plants. D-Arabinose is sometimes called *pectin sugar*. Pectin, the polysaccharide from which it is obtained, forms gels that are useful in making jelly. Because it is isolated from wood, D-xylose is sometimes called *wood sugar*.

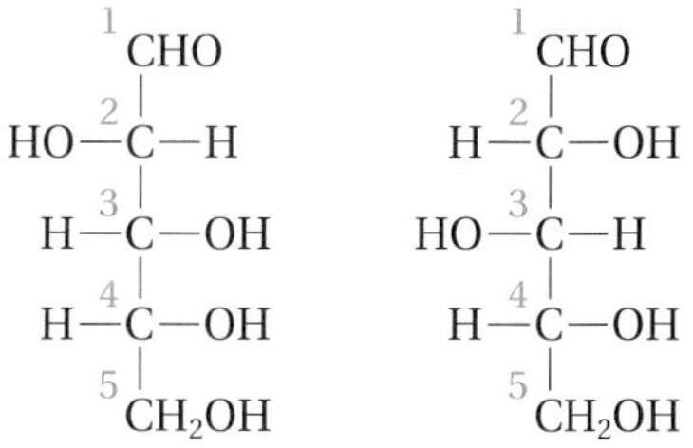

D-Arabinose — *pectin sugar*   D-Xylose — *wood sugar*

Other important aldopentoses are D-ribose and D-2-deoxyribose, a related compound that lacks an —OH group at carbon 2:

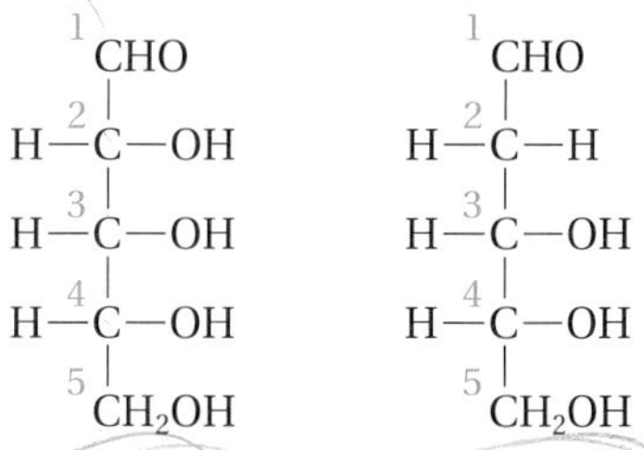

D-Ribose — RNA   D-2-Deoxyribose — DNA → know names

These two sugars are an integral part of the hereditary materials ribonucleic acid (RNA) and deoxyribonucleic acid (DNA). The structures of RNA and DNA are discussed in Chapter 11.

The six-carbon sugars

Only three aldohexoses appear in nature: D-glucose, D-galactose, and D-mannose:

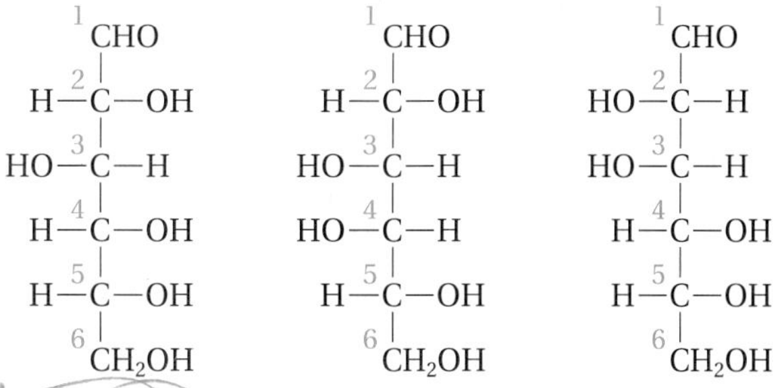

D-Glucose   D-Galactose   D-Mannose

The relative sweetness of sugars and sugar substitutes varies over a wide range. Lactose is about one-sixth as sweet and glucose about three-quarters as sweet as sucrose (table sugar). Fructose is not quite twice as sweet as sucrose. The artificial sweeteners aspartame (NutraSweet) and saccharin are about 150 and 500 times as sweet as sucrose.

The D form of glucose has a central role in the nutrition of virtually all species, including plants and humans. The biochemistry of glucose is so important that Chapter 15 is devoted to it. D-Glucose is abundant in all life forms. Depending on the source, it has been called *grape sugar, corn sugar,* and *blood sugar.* Urine usually contains a trace of D-glucose, but the concentration is greatly increased in the urine of patients with untreated diabetes mellitus. D-Galactose is a constituent of lactose, also called *milk sugar* (see Sec. 7.7). D-Mannose is a major constituent of polymeric molecules called *mannans,* which are found in several plants.

Monosaccharides that contain a ketone functional group $-\overset{\overset{\displaystyle O}{\|}}{C}-$ *are called* **ketoses.** Ketoses containing three, four, five, and six carbons are *ketotrioses, ketotetroses, ketopentoses,* and *ketohexoses,* respectively. No discussion of hexoses would be complete without including D-fructose, a ketohexose because of the presence of a ketone carbonyl group in the molecule at carbon 2. D-Fructose and D-glucose differ in structure only at carbons 1 and 2. The identical stereochemistry at carbons 3, 4, and 5 exists because the breakdown of D-glucose in living systems involves conversion of D-glucose to D-fructose.

$$
\begin{array}{c}
\overset{1}{C}H_2OH \\
| \\
\overset{2}{C}=O \\
| \\
HO-\overset{3}{C}-H \\
| \\
H-\overset{4}{C}-OH \\
| \\
H-\overset{5}{C}-OH \\
| \\
\overset{6}{C}H_2OH
\end{array}
\qquad \textit{ketohexose}
$$

D-Fructose

D-Fructose occurs in a large number of fruits and in honey. It is also the only sugar found in human semen. D-Fructose is one of those sugars which belongs to the D family but rotates plane-polarized light in a left-handed direction.

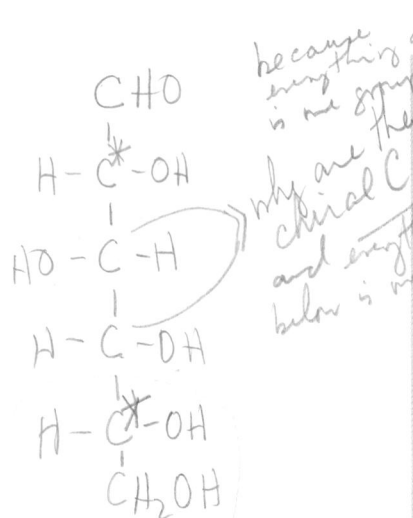

PRACTICE EXERCISE 7.5

Identify each structure as D or L.

(a)
$$
\begin{array}{c}
CHO \\
| \\
H-\!\!\!-OH \\
| \\
HO-\!\!\!-H \\
| \\
CH_2OH
\end{array}
$$
L

(b)
$$
\begin{array}{c}
CHO \\
| \\
H-\!\!\!-OH \\
| \\
CH_2OH
\end{array}
$$
D

(c)
$$
\begin{array}{c}
CH_2OH \\
| \\
=\!\!\!O \\
| \\
HO-\!\!\!-H \\
| \\
H-\!\!\!-OH \\
| \\
H-\!\!\!-OH \\
| \\
CH_2OH
\end{array}
$$
D

(d)
$$
\begin{array}{c}
CHO \\
| \\
HO-\!\!\!-H \\
| \\
HO-\!\!\!-H \\
| \\
HO-\!\!\!-H \\
| \\
CH_2OH
\end{array}
$$
L

PRACTICE EXERCISE 7.6

Draw the Fischer projection formula for D-glucose. Number the carbons, and identify each asymmetric carbon with an asterisk.

7.3 Cyclic structures

AIMS: To describe the bonding that results in cyclic forms of sugars. To classify simple sugars as a pyranose or a furanose and as a hemiacetal or a hemiketal.

Focus

Pentoses and hexoses exist mainly as ring structures.

Until now, saccharides have been depicted as straight-chain compounds. The reality is somewhat different, however, for pentoses and hexoses exist primarily in five- and six-membered rings or cyclic forms. Examples of two such cyclic forms, those of D-glucose and D-fructose, are shown in Figure 7.4.

The five-membered sugar ring system is given the general name **fura-nose** *after the parent cyclic ether furan; the six-membered sugar ring system is considered a derivative of pyran and is called a* **pyranose.**

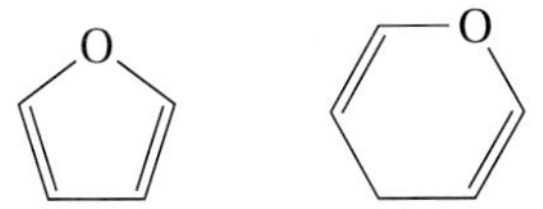

Furan   Pyran

To name a sugar in its cyclic form precisely, remove the *-se* from the sugar name and add *furanose* or *pyranose* according to whether the cyclic form is a five-membered or a six-membered ring. Thus the cyclic form of D-fructose is properly called D-*fructofuranose* and that of D-glucose is called D-*glucopyranose.*

To understand how these cyclic structures are formed from the straight-chain sugars, recall that alcohols can add to carbonyl groups of aldehydes or ketones to form hemiacetals or hemiketals, and five- and six-membered rings are more stable than smaller rings. Hemiacetal formation, discussed previously (Sec. 7.6), is the addition of an alcohol to a carbonyl group of an aldehyde. Hemiketals are formed in a similar way from ketones and alcohols.

Sugars such as D-glucose and D-fructose contain a carbonyl group and several hydroxyl groups in the same molecule. Figure 7.5 illustrates the pos-

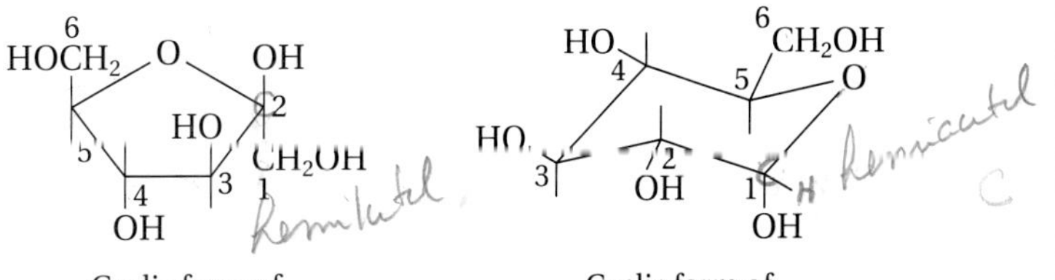

Cyclic form of
D-fructose
(D-fructofuranose)

Cyclic form of
D-glucose
(D-glucopyranose)

Figure 7.4
The cyclic forms of two sugars drawn as in their true shapes. The five- and six-membered rings resemble those of cyclopentane and cyclohexane, respectively.

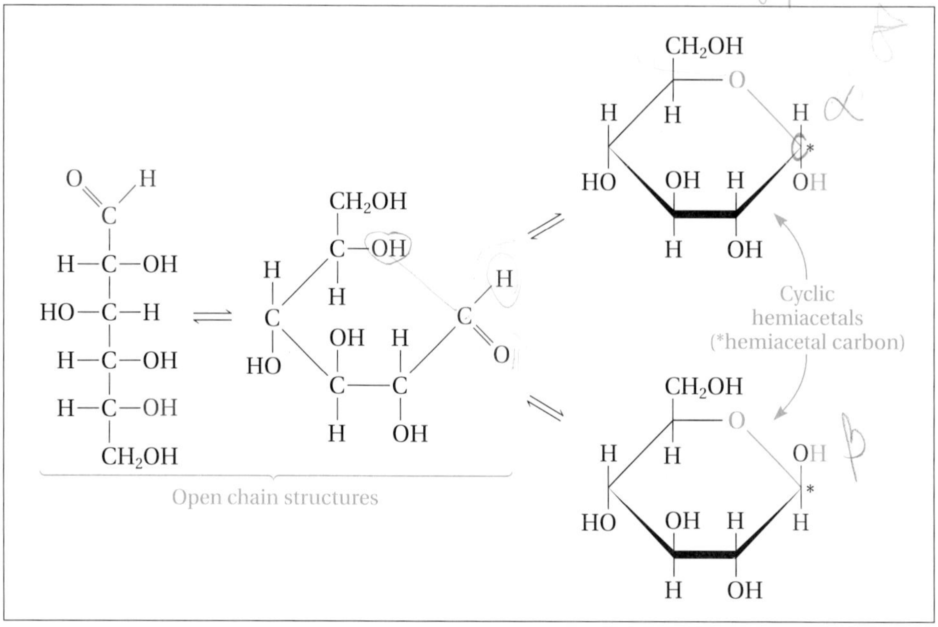

Figure 7.5
An intramolecular reaction between the aldehyde and hydroxyl groups of a sugar, in this case D-glucose, results in formation of a cyclic hemiacetal.

sible internal hemiacetal or hemiketal formation. The internal hydroxyl group that is selected for reaction is one that will give a five- or six-membered ring depending on the saccharide. Smaller rings would be unstable because of ring strain; larger rings are not formed because the more distant ends of the molecule do not often collide in solution.

7.4 Haworth projections

AIMS: *To draw Haworth projections for the common simple sugars. To identify a simple sugar as an alpha or beta anomer. To explain the interconversion of closed-chain forms of sugars.*

Focus

The stereochemistry of the cyclic forms of sugars is depicted by Haworth projections.

The stereochemistry of the cyclic forms of sugars is often represented by their **Haworth projections**—*standardized ways of depicting the positions of hydroxyl groups in space*. In viewing Haworth projections, envision the plane of the ring as tilted perpendicular to the plane of the paper. The attached groups are above and below the plane of the ring. Figure 7.6 shows Haworth projections for D-glucopyranose and D-fructofuranose. Two forms of each of these sugars are possible; these are designated *alpha* and *beta*. Two cyclic forms are possible because, in going from a straight chain to a ring, a new asymmetric carbon is introduced at carbon 1 (the hemiacetal carbon) of aldoses and carbon 2 (the hemiketal carbon) of ketoses. In the

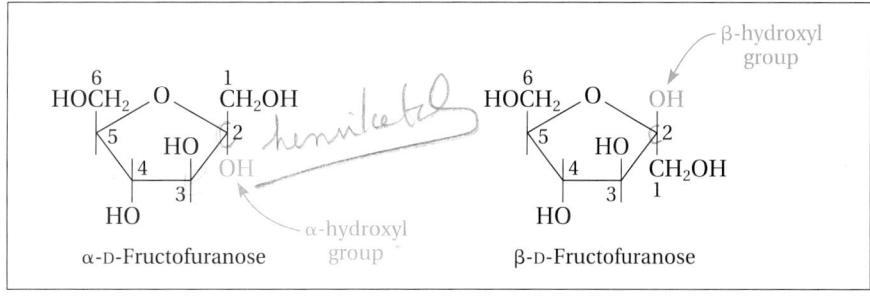

Figure 7.6
Haworth projections of D-glucopyranose and D-fructofuranose. Two different stereoisomers, labeled α and β, may be formed on hemiacetal formation.

Haworth projections of D-glucopyranose, for example, the hydroxyl group at carbon 1, which is formed from the aldehyde functional group of the straight-chain sugar, may end up below (alpha or Greek α) or above (beta or Greek β) the plane of the pyranose ring.

Anomers *are sugars that differ in stereochemistry only at the hemiacetal or hemiketal carbon.* The alpha and beta anomers of the cyclic forms of sugars have different melting points and different abilities to rotate plane-polarized light. Alpha-D-glucose melts at 146 °C, for example, but beta-D-glucose melts at 150 °C. These differences help to demonstrate again how small changes in molecular shape or structure dramatically affect the physical properties of molecules.

How to draw Haworth projections

Conversion of a straight-chain Fischer projection to a ring in a Haworth projection is easily accomplished for D sugars. Hydroxyl groups that point to the left in a Fischer projection of a D sugar point *up* in the Haworth projection. Hydroxyl groups that point to the right in a Fischer projection point *down* in the Haworth projection.

EXAMPLE 7.3   **Drawing a Haworth projection**

Draw the Haworth projection of α-D-glucopyranose.

SOLUTION

The α-D-glucopyranose molecule is the cyclic form of D-glucose. Draw a

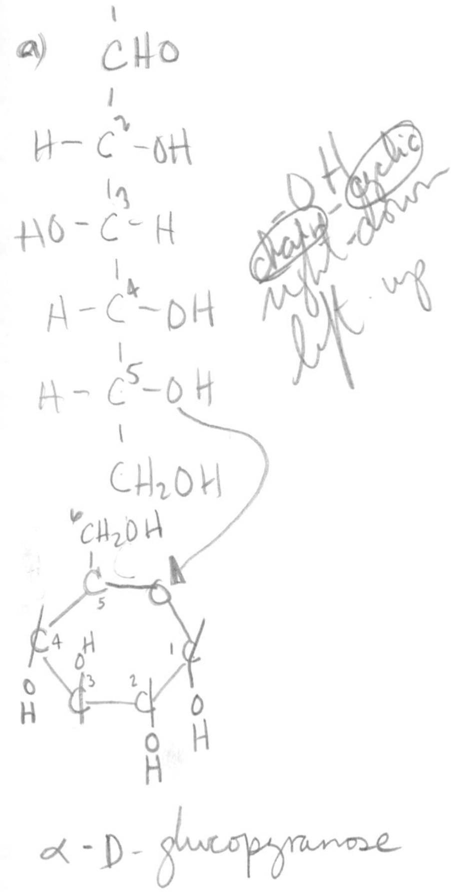

a)

$$CHO$$
$$H-\overset{2}{C}-OH$$
$$HO-\overset{3}{C}-H$$
$$H-\overset{4}{C}-OH$$
$$H-\overset{5}{C}-OH$$
$$CH_2OH$$

(handwritten notes: D-Cyclic, right-down, left-up)

$$CH_2OH$$

(handwritten ring structure)

α-D-glucopyranose

projection of D-glucose in the straight-chain form:

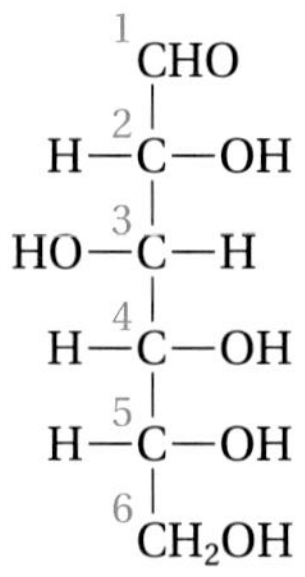

$$\overset{1}{C}HO$$
$$H-\overset{2}{C}-OH$$
$$HO-\overset{3}{C}-H$$
$$H-\overset{4}{C}-OH$$
$$H-\overset{5}{C}-OH$$
$$\overset{6}{C}H_2OH$$

D-Glucose

Draw a six-membered pyranose ring in its abbreviated form as shown:

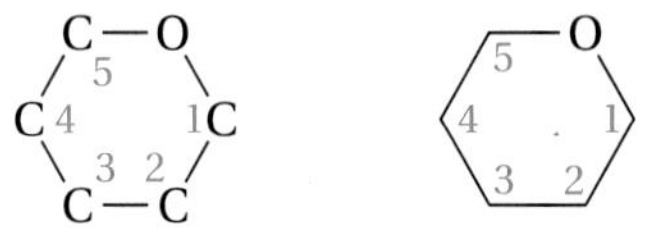

Pyranose ring   Abbreviated pyranose ring

Put in the —CH$_2$OH group of carbon 6 of the hexose. In a D sugar, the sixth carbon is always above the plane of the ring as shown:

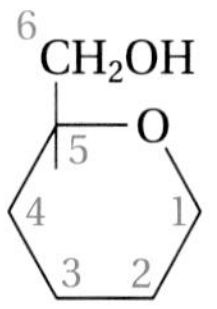

Fill in the —OH groups on carbons 2, 3, and 4. Notice that the oxygen on carbon 5 of the chain form is now in the ring and need not concern us. Hydroxyl groups to the right in the Fischer projection go below the plane of the ring. Those to the left are above the plane. Ring hydrogens are usually omitted for clarity.

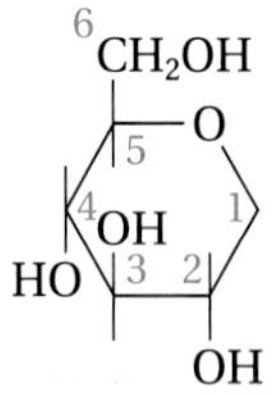

Finally, write the anomeric —OH group at carbon 1—in this case alpha (below the plane of the ring).

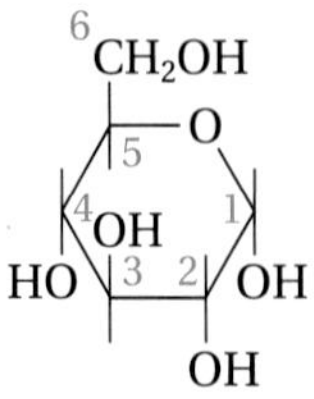

α-D-Glucopyranose

PRACTICE EXERCISE 7.7

Draw the Haworth projection of β-D-glucopyranose. Identify the hemiacetal carbon.

PRACTICE EXERCISE 7.8

Draw the Haworth projection of α-D-ribofuranose. Identify the hemiacetal carbon.

PRACTICE EXERCISE 7.9

Draw the Haworth projection of α-D-fructofuranose. Identify the hemiketal carbon.

Interconversion of straight-chain and ring forms of sugars

The straight-chain sugar forms are in equilibrium with the ring forms. The ring forms are usually quite predominant. For example, if stereochemically pure α-D-glucopyranose is dissolved in an acidic solution, the ring will open and close repeatedly. In reclosing, some β-D-glucopyranose is formed. The final equilibrium mixture consists of about 63% β-D-glucopyranose, about 37% α-D-glucopyranose, and only a tiny amount of the straight-chain aldehyde. From the percentages of products formed, we can say that β-D-glucopyranose is only slightly more preferred than α-D-glucopyranose and that both D-glucopyranose anomers are much more preferred than the straight-chain aldehyde form of D-glucose.

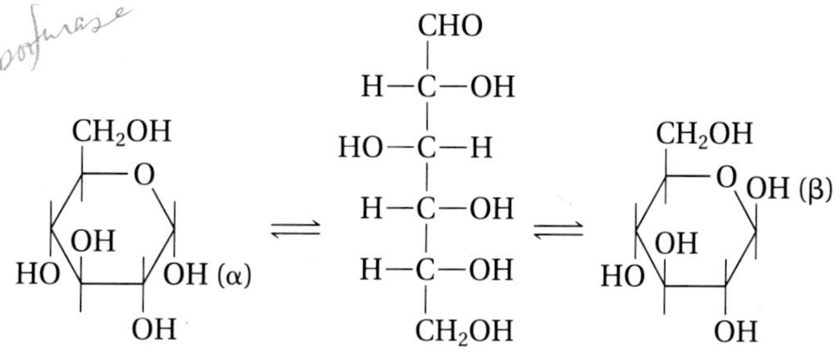

CH₂OH ... (α-D-Glucopyranose, D-Glucose, β-D-Glucopyranose structures)

α-D-Glucopyranose
(about 37%)

D-Glucose
(less than 1%)

β-D-Glucopyranose
(about 63%)

7.5 Glycosides

AIM: To describe the formation of glycosidic bonds and the products of their hydrolysis.

The closed-chain hemiacetal or hemiketal forms of sugars may react with alcohols to form acetals or ketals (see Sec. 4.6). *The acetals or ketals of sugars are called* **glycosides.**

Glycosidic bonds

The covalent ether link between the sugar hydroxyl and the alcohol is a **glycosidic bond.** A simple alcohol such as methanol and a sugar such as α-D-

Focus

Alcohols react with closed-chain forms of sugars to form glycosidic bonds.

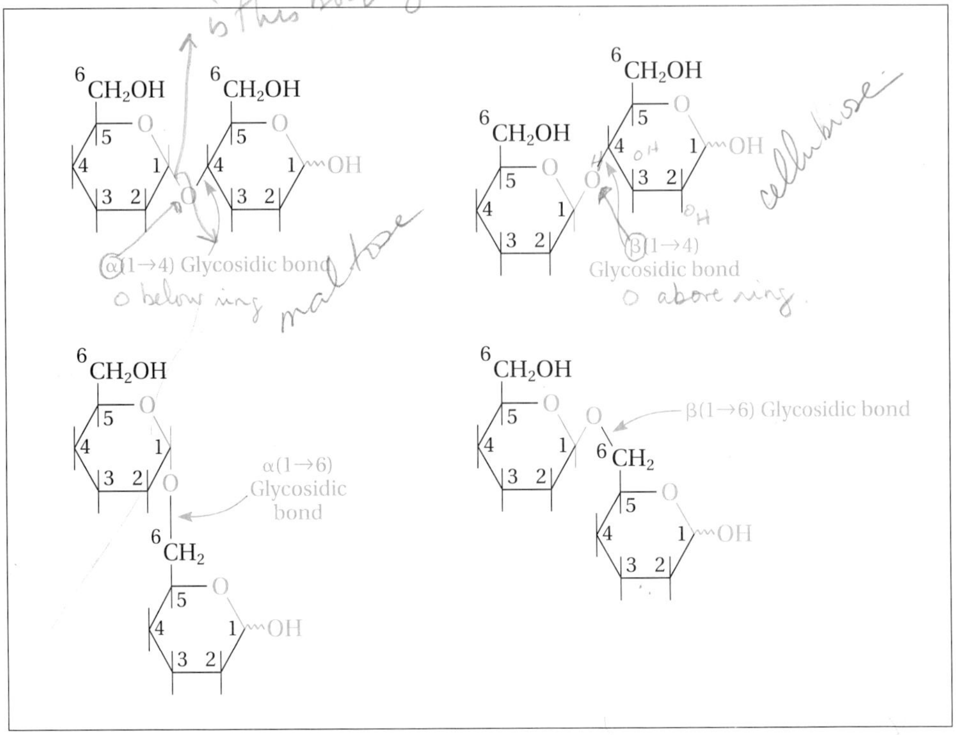

Figure 7.7
Some of the common glycosidic bonds found in polysaccharides. The acetal and hemiacetal portions of the molecules are shown in color. The wavy line connecting the hydroxyl group to carbon 1 indicates that the carbon–oxygen linkage may be either alpha or beta.

glucopyranose produce methyl α-D-glucopyranoside:

α-D-Glucopyranose   Methanol   Methyl α-D-glucopyranoside

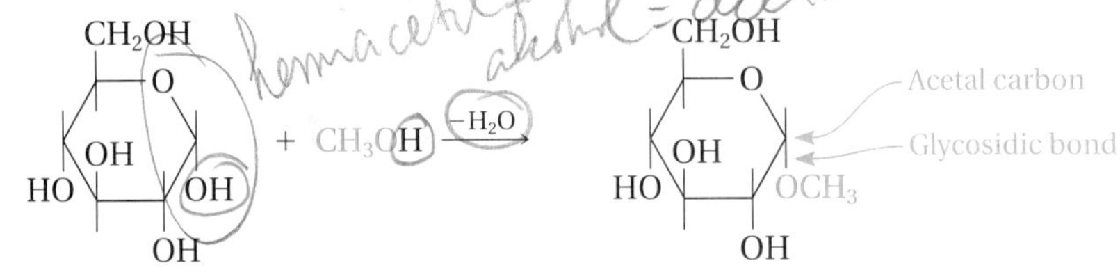

The alcohol used to make a glycosidic bond is often more complex than methanol—in fact, sugars themselves are alcohols. As shown in Figure 7.7, the individual saccharide units are attached through glycosidic bonds. Glycosidic bonds between sugars are designated according to the position numbers of the carbons of the sugars that are linked and also according to the stereochemistry of the linkage. For example, suppose the beta hydroxyl group at carbon 1 in a hexose is linked by a glycosidic bond to carbon 4 of another hexose. This linkage is called a $\beta(1\rightarrow4)$ glycosidic bond. Other common linkages are $\alpha(1\rightarrow4)$, $\alpha(1\rightarrow6)$, and $\beta(1\rightarrow6)$. Once the anomeric —OH group of a sugar is tied up as an acetal, it is no longer free to go from the ring form to the straight-chain form.

PRACTICE EXERCISE 7.10

The hydroxyl group of carbon 1 in α-D-glucopyranose is linked by a glycosidic bond to carbon 4 of another D-glucopyranose molecule. Draw the structure of the glycoside that is formed. Identify the acetal carbon.

Hydrolysis of glycosidic bonds

Glycosidic bonds may be cleaved by hydrolysis reactions. We can take the hydrolysis of an $\alpha(1\rightarrow4)$ glycosidic bond between two hexoses as an example. For simplicity, only the carbon skeleton and the glycosidic bond are shown:

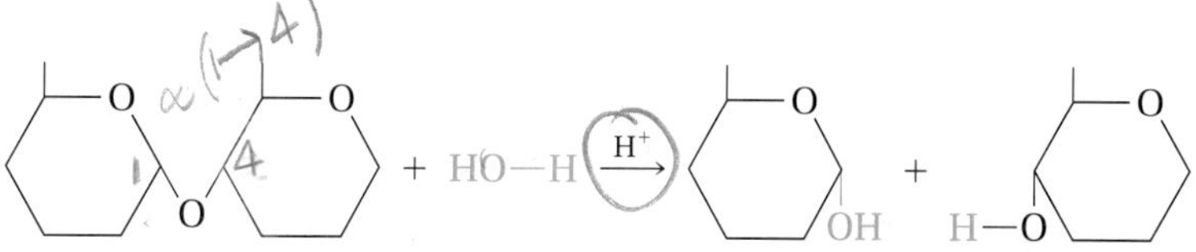

The chemical hydrolysis of most complex sugars can be done by heating an aqueous solution of the carbohydrate. A trace of acid is added as a catalyst. Enzymes act as the catalyst in biological systems. Hydrolysis reactions will be important as we proceed into biochemistry, since they are the means by which sugars, fats, and proteins are broken down to simple materials by digestion.

[handwritten: Hydrolysis of a glucosidic bonds gives the simple sugars monosaccharides]

PRACTICE EXERCISE 7.11

The glycosidic bond in the following compound is hydrolyzed. What are the structures of the products?

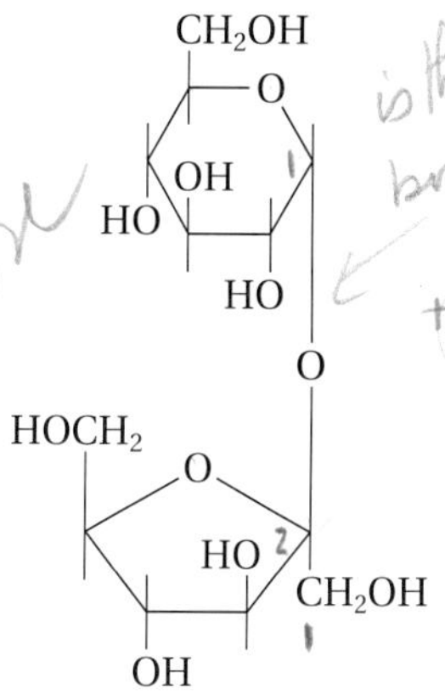

[handwritten annotations: "is the glucosidic bond 1→2? yes", "+ HO - H H⁺", "α D-glucopyranose", "β D-fructofuranose", "glucose"]

7.6 Polysaccharides

AIM: To list the structures, sources, and uses of the following polysaccharides: starch, amylose, amylopectin, glycogen, and cellulose.

Focus

Polysaccharides are composed of many monosaccharide units connected by glycosidic linkages.

Individual sugar units may be connected to one another to form linear, branched, or circular polymers, as shown in Figure 7.8. **Polysaccharides** *have many monosaccharides bonded together to form a long polymer chain.* The bonds connecting the sugar units are glycosidic. The $1\rightarrow4$ and $1\rightarrow6$ linkages are the ones most commonly found in natural polysaccharides consisting of hexoses.

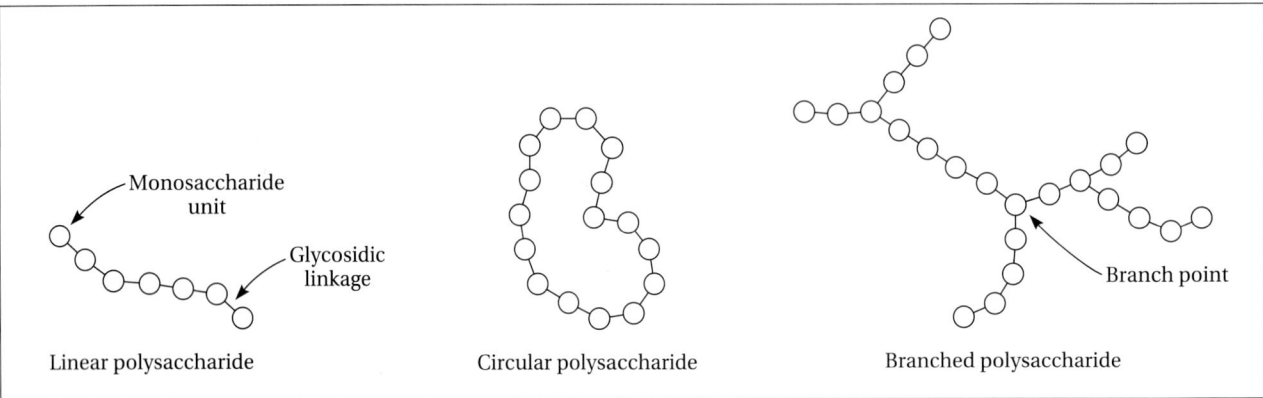

Figure 7.8
Sugar units in a polysaccharide can form linear, circular, or branched structures.

There are four human blood types, A, B, AB, and O. These blood types occur because of differences in composition of relatively small polysaccharide chains on the surface of red blood cells.

Starches *are polymers consisting entirely of* D-*glucose units.* They are the major storage form of D-glucose in plants. Two starches, amylose and amylopectin, are the most common. Amylose is composed of a linear chain of D-glucopyranose molecules linked $\alpha(1\rightarrow4)$ as shown in Figure 7.9. Complete amylose molecules may contain anywhere from a few to about 3000 D-glucopyranose units. Amylopectin consists of chains of D-glucopyranose molecules held together by $\alpha(1\rightarrow4)$ glycosidic linkages, but as shown in Figure 7.10, amylopectin also contains $\alpha(1\rightarrow6)$ cross-linking. These cross-links give amylopectin the appearance of a branched, bushy molecule. There are usually 24 to 30 D-glucopyranose units between the branch points of amylopectin.

Glycogen—*the main storage form of* D-*glucose in animal cells*—has a

Figure 7.9
Partial structure of the starch molecule amylose.

Figure 7.10
Partial structure of an amylopectin molecule.

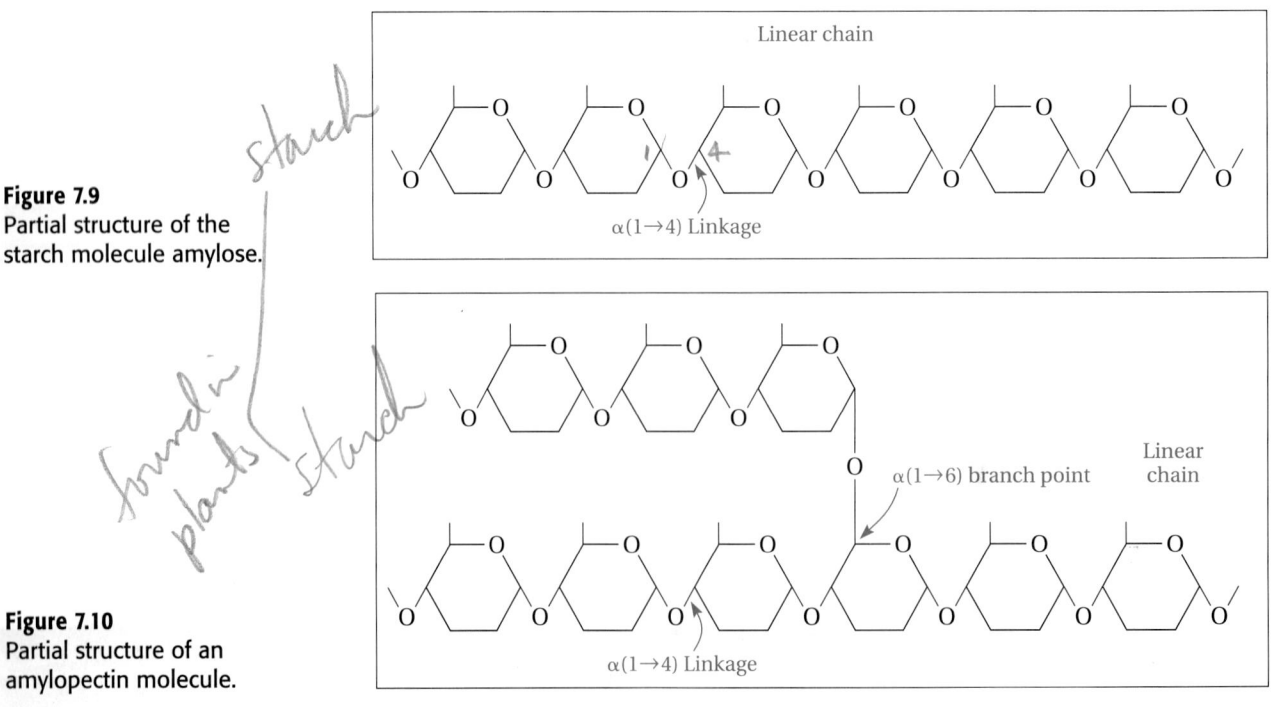

structure similar to amylopectin. The main difference between amylopectin and glycogen is that glycogen is even more highly branched—there are only 8 to 12 D-glucopyranose units between branch points. Glycogen is especially abundant in the liver, where it may amount to as much as 8% of the dry mass of the organ.

Figure 7.11 shows **cellulose,** *the most important structural polysaccharide.* Cellulose is the material of plant cell walls, and it is the major component in wood. Cotton is about 80% pure cellulose. Like amylose, cellulose is a linear polymer consisting of D-glucose units. These D-glucose units, usually from 300 to 15,000 of them, are connected by 1→4 linkages that are beta rather than alpha. The shapes of amylose and cellulose molecules are quite different because of it—$\alpha(1{\to}4)$-linked amylose tends to form loose spiral structures (Fig. 7.12) and $\beta(1{\to}4)$-linked cellulose tends to form straight chains.

The difference in the shapes of amylose and cellulose has a tremendous

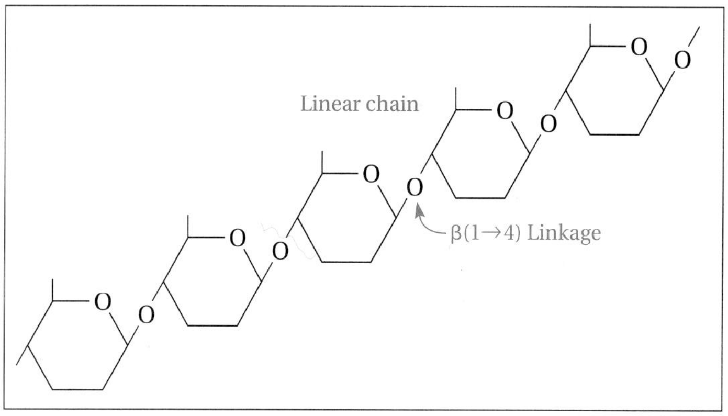

Cellulose is a very important raw material. The wood used in construction, the cotton used in textile production, and the manufacture of paper and paper products all have a large economic impact.

Linear chain

β(1→4) Linkage

Figure 7.11
Partial structure of a cellulose molecule.

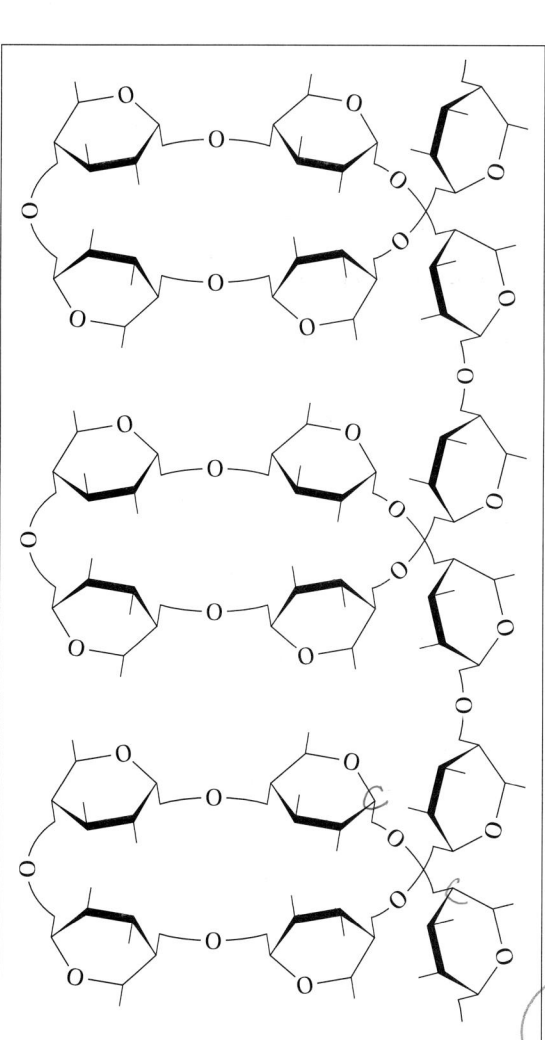

Figure 7.12
Amylose forms loose spiral structures.

[handwritten margin note: because of linear H-bonding between molecules of cellulose gives the wood strength & cotton fibers.]

biological effect. The linear chains of cellulose present a uniform surface of hydroxyl groups. These hydroxyl groups are involved in hydrogen bonding between adjacent cellulose molecules; the large number of these weak interactions give wood and cotton fibers their strength. A second effect of the two different shapes is that enzymes capable of catalyzing the hydrolysis of starch are not capable of hydrolyzing cellulose. Human beings can convert starch to its fuel form, D-glucose, but people lack enzymes to catalyze the hydrolysis of cellulose to glucose. Sawdust has little potential as snack food for people. On the other hand, undigestible dietary carbohydrates do seem to be important for human health as discussed in A Closer Look: Dietary Fiber. The digestive systems of the cud-chewing animals (ruminants) such as cows, sheep, and goats, as well as those of termites, contain microorganisms whose enzymes catalyze the production of glucose from cellulose. These animals can therefore use cellulose as a nutritional source.

PRACTICE EXERCISE 7.12

[handwritten: starch] *[handwritten: starch]*

How does the structure of amylose differ from the structure of amylopectin?

[handwritten: α(1→4)]

[handwritten margin note: starches — amylose: long polymer chains of D-glucose linked α(1→4) pyranose; amylopectin: polymeric chains of D-glucopyranose linked α(1→4) and α(1→6) cross-link]

PRACTICE EXERCISE 7.13

What is the product of the complete hydrolysis of each of the following polysaccharides?

(a) amylose   (b) amylopectin   (c) glycogen   (d) cellulose

[handwritten: since they are all composed of D-glucopyranose monomers, each polymer will yield D-glucopyranose upon hydrolysis.]

A Closer Look
Dietary Fiber

Have you had your fiber today? Many experts recommend that people eat high-fiber diets in order to prevent colon cancer. The nondigestible carbohydrates that we eat constitute dietary fibers. Dietary fibers consist of polysaccharides that cannot be hydrolyzed to monosaccharides and therefore cannot be absorbed into the bloodstream. Cellulose is insoluble fiber. Cellulose from vegetable leaves and stalks provides you with dietary bulk and helps to prevent constipation. There are also soluble fibers. The soluble dietary fibers are noncellulosic. Pectins from fruits, which are used to thicken jams and jellies, are examples of soluble fibers. The pectins are polyhydroxy compounds that have a carboxylic acid group at one end of the molecule and an aldehyde group at the other. Vegetable "gums" are also soluble fibers.

Fruits and vegetables are high in dietary fiber.

Which foods can we eat to ensure that we get enough dietary fiber? Wheat, brown rice, and bran cereals are high in insoluble fiber. Oats, barley, carrots, and fruits are high in soluble fiber. Peas and beans are a good source of both soluble and insoluble fiber.

7.7 Disaccharides

AIM: To write the structures and list the sources and uses of the disaccharides maltose and cellobiose.

Focus

Maltose and cellobiose are derived from the hydrolysis of starch and cellulose.

Disaccharides *are compounds in which two monosaccharides are bonded together.* Disaccharides are glycosides in which the alcohol is a second monosaccharide molecule. Maltose, or malt sugar, is a disaccharide derived from the partial hydrolysis of starch. Diastase, an enzyme found in germinating barley, catalyzes the hydrolysis of starch to maltose. Since starch is composed of many D-glucopyranose units connected by $\alpha(1\rightarrow4)$ glycosidic bonds and maltose is a disaccharide produced from starch, it is not surprising that maltose is composed of two D-glucopyranose molecules connected by an $\alpha(1\rightarrow4)$ glycosidic linkage. Maltose is hydrolyzed to D-glucose by the enzyme maltase.

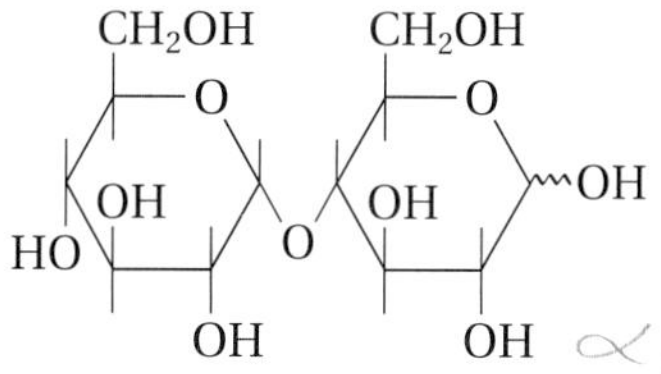

Maltose

Likewise, cellobiose is a disaccharide produced from the partial hydrolysis of cellulose by dilute acid. This disaccharide consists of two D-glucose molecules like maltose, but the $1\rightarrow4$ glycosidic bond that links the two monosaccharides is beta. Cellobiose is not hydrolyzed by the enzyme maltase.

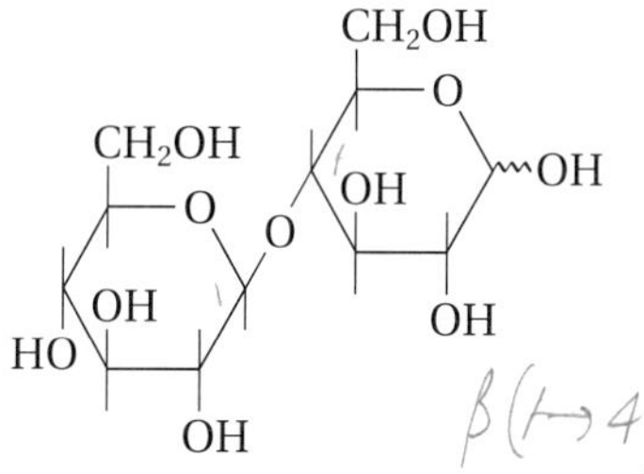

Cellobiose

PRACTICE EXERCISE 7.14

What is the relationship of maltose to cellobiose?

7.8 Sucrose and lactose

AIM: To write the structures and list the sources and uses of the disaccharides sucrose and lactose.

Focus

Sucrose and lactose are the sugars of sap and milk.

The most familiar sugar is sucrose—ordinary table sugar. Sucrose is isolated from the juice or sap of several plants, including sugarcane, sugar beets, and maple trees. The world's production of sucrose from these sources exceeds 7 billion kilograms per year. The sucrose molecule is a *sucrose molecule*: disaccharide composed of α-D-glucose and β-D-fructose:

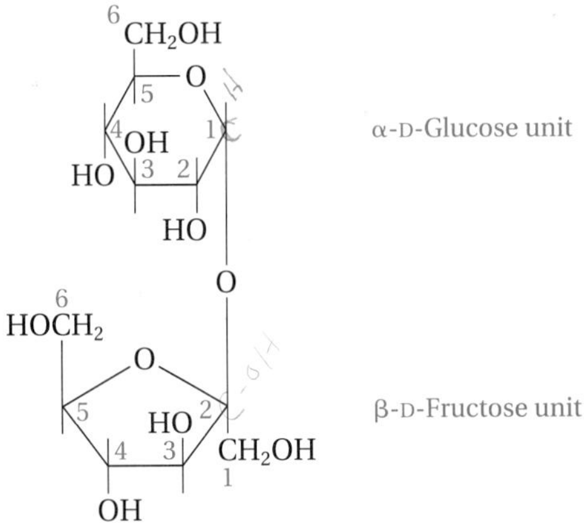

Sucrose

Structurally, the linkage between glucose and fructose in sucrose is unusual among sugars, since both sugars are linked together at the anomeric carbons (carbon 1 of each sugar). Hydrolysis of sucrose by acid or by enzymes gives **invert sugar**—*a mixture of equal molar quantities of glucose and fructose.*

Honey is a rich natural source of invert sugar. Invert sugar is also produced commercially and used when a noncrystalline sweetener is desired. The gooey syrup that traditionally bathes chocolate-covered cherries is one use of invert sugar.

As a food, sucrose has a high caloric value. Many people rely on artificial sweeteners such as saccharin, which is much sweeter than sugar. At one time calcium cyclamate, a chemical 30 times sweeter than sucrose, competed with saccharin as an artificial sweetener; then it was removed from the market as a possible cancer-causing agent. The use of saccharin has been reduced for the same reason.

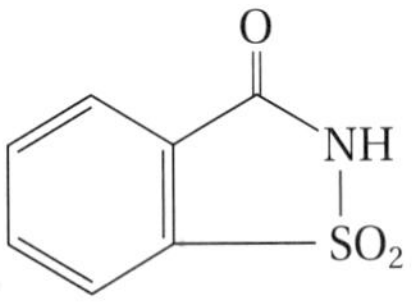

Saccharin 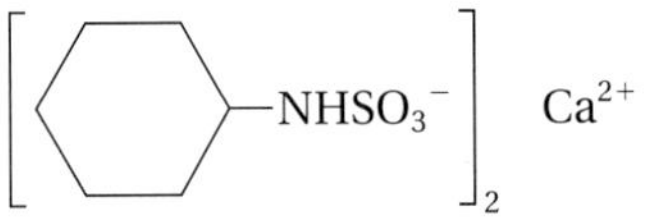   Calcium cyclamate

Lactose, or milk sugar, constitutes 5% of cow's milk and 7% of human milk. Pure lactose is obtained from whey, the watery by-product of cheese production. Lactose is composed of one molecule of D-galactose and one of D-glucose. The linkage between the two sugar units is $\beta(1\rightarrow4)$.

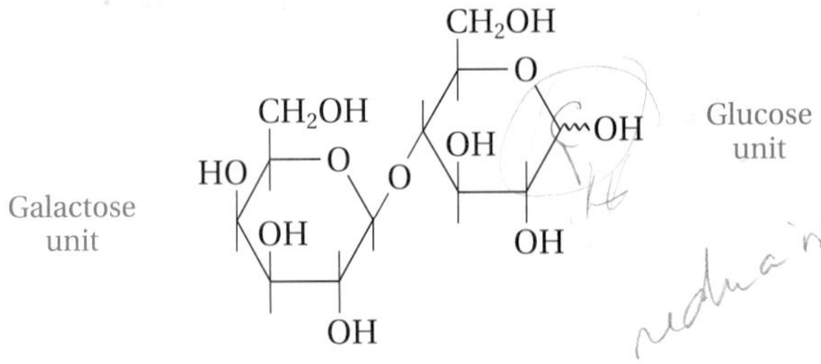

Lactose

Galactose metabolism is impaired in the genetic disease galactosemia. In this disease, galactose builds up to toxic levels, resulting in irreversible nerve damage and early cataract formation. Symptoms do not develop if the intake of galactose and lactose is restricted.

The glycosidic bond between the D-galactose and D-glucose portions of lactose may be cleaved by the enzyme *lactase*. Lactase deficiency is fairly common in people. Lactase deficiency is the reason for the discomfort experienced by Alberto, the student described in the Case in Point earlier in this chapter. In the follow-up, below, we learn more of this deficiency and how it can be treated.

PRACTICE EXERCISE 7.15

Give the names and structures of the compounds produced by the hydrolysis of the glycosidic bond in lactose.

FOLLOW-UP TO THE CASE IN POINT: Lactose intolerance

The physician correctly diagnosed Alberto's stomach cramps, gas, and diarrhea to be the result of lactose intolerance. Like Alberto, many adults and some children lack the enzyme *lactase*. As a result, these people are unable to hydrolyze lactose to its simpler sugar components, D-galactose and D-glucose. The symptoms of lactose intolerance are caused by undigested lactose that sits in the intestinal tract, where it causes cramps and diarrhea. The diarrhea is caused by an imbalance in the osmotic pressure on the walls of the intestine. Lactose draws water through the intestinal wall into the intestines, where the excessive water causes diarrhea, which can lead to dehydration. Some of the lactose can be oxidized by intestinal bacteria, releasing CO_2 into the intestines. The bubbles of gas can cause great discomfort. The effects of lactose intolerance can be avoided by excluding milk and milk products from the diet. Some people with lactose intolerance can eat yogurt, which is a fermented milk product. The fermentation of yogurt by bacteria breaks down the lactose present. In many instances, lactase production diminishes with age. A reduced level of lactase, combined with his consumption of a large amount of milk at one sitting, may have triggered Alberto's illness. Alberto later found that he was able to consume small amounts of dairy products without a repeat of the symptoms of lactose intolerance. Some other sufferers of lactose intolerance—estimated to be as many as one-third of all adult Americans—can avoid the unpleasant

symptoms by adding readily available lactase tablets to their milk (see figure). Lactose-free milk is also available at some groceries.

Lactose intolerance is fairly common in infants, especially among babies of Middle Eastern, Asian, and African descent. Europeans are statistically least susceptible. Dehydration from diarrhea from lactose intolerance or any other cause is always a hazard for babies because, compared with adults, their bodies contain a relatively small mass of water. Switching a lactase-deficient baby from natural milk sources to a commercial milk substitute alleviates the diarrhea and other symptoms.

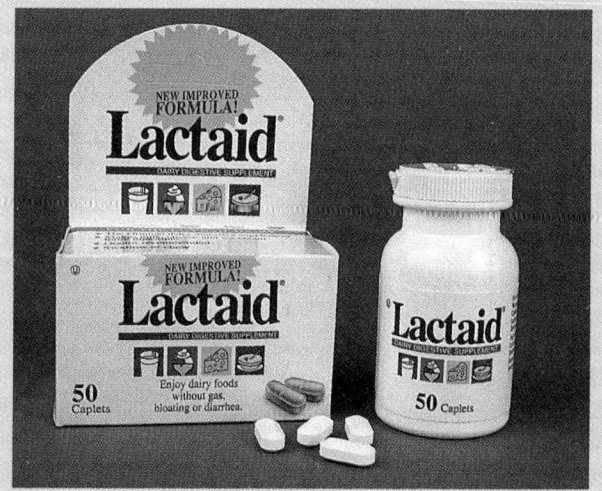

Lactase tablets.

7.9 Reducing and nonreducing sugars

AIM: To predict, on the basis of molecular structure, whether a carbohydrate is reducing or nonreducing.

You may recall from Section 4.5 that Benedict's and Tollens' reagents oxidize aldehydes and alpha-hydroxy ketones. When Benedict's reagent, an alkaline solution of copper(II) sulfate, is the oxidizing agent, the copper(II) ions are reduced to copper(I), and insoluble red copper(I) oxide (Cu_2O) is formed. With Tollens' reagent, silver ions (Ag^+) are the oxidizing agent, and these ions are reduced to silver metal (Ag). The silver deposits as a mirror on the sides of the reaction vessel.

Many sugars behave the same way toward Benedict's and Tollens' reagents as simple aldehydes and alpha-hydroxy ketones. Sugars such as D-glucose and D-fructose give a brick-red color with Benedict's reagent; with Tollens' reagent the silver mirror test is a positive indicator. **Reducing sugars** *give positive tests with Benedict's and Tollens' reagents.* **Nonreducing sugars** *do not give positive tests.* Positive tests are obtained with sugars in which the cyclic hemiacetal or hemiketal forms are in equilibrium with their straight-chain form. For example, even though the equilibrium may greatly favor the hemiacetal, oxidation of all the sugar occurs. As soon as the aldehyde is oxidized by the test reagent, more aldehyde is produced as the equilibrium between hemiacetal and aldehyde is reestablished.

Sugars do not give positive tests with Benedict's and Tollens' reagents if the cyclic and aldehyde forms do not exist in equilibrium with the aldehyde form. Any sugar that is an acetal or a ketal—either because of ether formation of the hemiacetal hydroxyl group with a simple alcohol or by formation of a glycosidic linkage—is nonreducing. Of the simple sugars you have seen in this chapter, only the methyl glycosides and sucrose are nonreducing.

The Benedict's and Tollens' tests for carbohydrates work best for simple sugars. Polysaccharides such as amylose should be reducing sugars, since the hemiacetal form of the terminal sugar unit is in equilibrium with the aldehyde form. If the polysaccharide chain is a long one, however, the number of end groups in a fairly large sample of the material may be so small that a positive Benedict's or Tollens' test is not observed. Thus large polysaccharides such as starch and cellulose are generally not reducing sugars. Chemical and biochemical tests for reducing sugars are very important to patients being treated for diabetes, as discussed in A Closer Look: Tests for Blood Glucose in Diabetes.

EXAMPLE 7.4

Analyzing why a sugar is a reducing sugar

The disaccharide maltose is a reducing sugar. Why?

SOLUTION

Maltose has the structure

The right-hand ring is a cyclic hemiacetal. (The hemiacetal carbon is marked with an asterisk.) This ring can open and close, as you saw earlier for D-glucose (Fig. 7.5), to give an equilibrium between hemiacetal and aldehyde. The aldehyde group can react with Benedict's reagent, reducing copper(II) to copper(I). Therefore, maltose is a reducing sugar.

EXAMPLE 7.5

Deducing why a sugar is nonreducing

Why is sucrose a nonreducing sugar?

SOLUTION

Sucrose (see the structure on page 224) is a disaccharide of α-D-glucose and β-D-fructose. The glucose unit is a cyclic acetal, and the fructose unit is a cyclic ketal. Neither of these two units can ring open, and neither exists in equilibrium with the aldehyde form. Sucrose cannot react with Benedict's reagent and is therefore a nonreducing sugar.

A Closer Look

Tests for Blood Glucose in Diabetes

Patients with diabetes must take steps to keep levels of glucose in the blood as close as possible to the middle of the normal range—4.5 to 8.0 mmol/L. Recent research shows that when patients with diabetes measure their levels of blood sugar at least four times a day and fine-tune their medication, they can avoid the disease's worst complications: kidney disease, blindness, and poor blood circulation that can require limb amputations.

Portable electronic devices called *glucose meters* are available for home testing of blood glucose levels (see figure). The patient must prick his or her finger to get blood for a reading, and the blood glucose level is obtained in less than a minute. These instruments are expensive, however, and some patients may use less expensive methods to check their urine for the presence of glucose. Glucose appears in the urine when the blood glucose concentration rises to 11.0 mmol/L. One kind of tablet called Clinitest uses a reducing sugar test to give the glucose level in the urine. When dissolved in urine, Clinitest tablets produce Benedict's reagent: they contain copper(II) sulfate ($CuSO_4$), sodium carbonate (Na_2CO_3), sodium hydroxide (NaOH), and citric acid ($C_6H_8O_7$). The quantity of glucose in the urine is determined by comparing the color of the test solution against test blocks supplied with the tablets. The Clinitest method is not specific for glucose, and the presence of interfering substances must be considered when interpreting the results. Interferences can be caused by drugs such as aspirin, lactose in nursing mothers, fructose after excessive fruit juice intake, and certain inherited errors of sugar biochemistry. Reagent strips such as Clinistix, BMstix, and Diastix use a color reaction involving enzymes to detect glucose. These products are very specific for glucose in the urine even in the presence of other reducing substances.

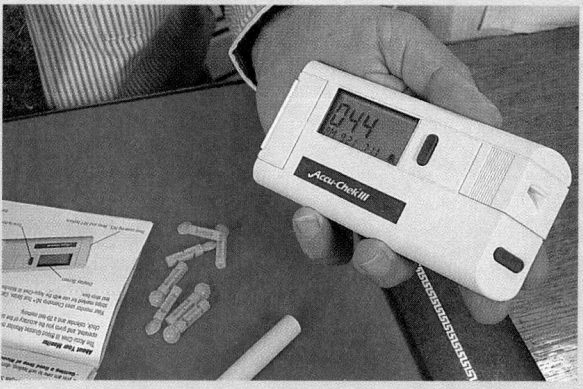

A blood glucose meter can give a result in 20 seconds.

7.10 Chitin, heparin, and acid mucopolysaccharides

AIM: To name and characterize the source of saccharides containing functional groups other than hydroxyl and carbonyl.

Focus

Saccharides may contain functional groups other than carbonyl and hydroxyl.

Two monosaccharides that contain functional groups other than carbonyl and hydroxyl are glucuronic acid and glucosamine. They are related structurally and stereochemically to D-glucose. *N*-Acetylglucosamine is also found in many polysaccharides and is formed from glucosamine.

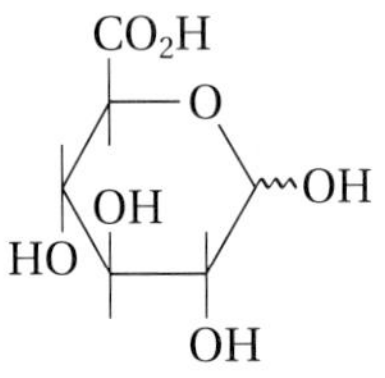

Glucuronic acid

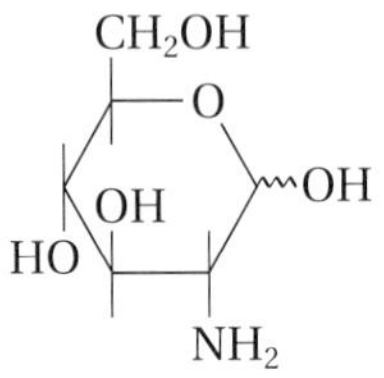

Glucosamine

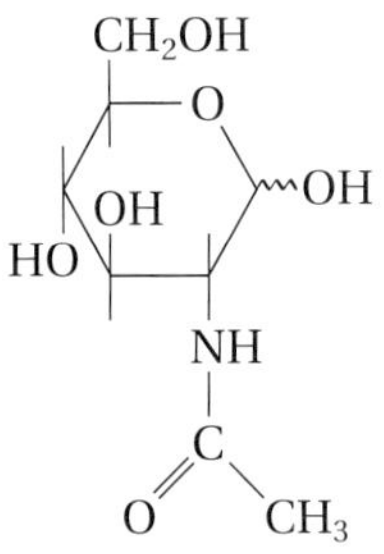

N-Acetylglucosamine

Chitin is an example of a polysaccharide that contains *N*-acetylglucosamine. Chitin forms the shells of crustaceans: crabs, lobsters, and shrimp. It is also a major constituent of the hard skeletons of insects. Chitin is composed of repeating units of *N*-acetylglucosamine:

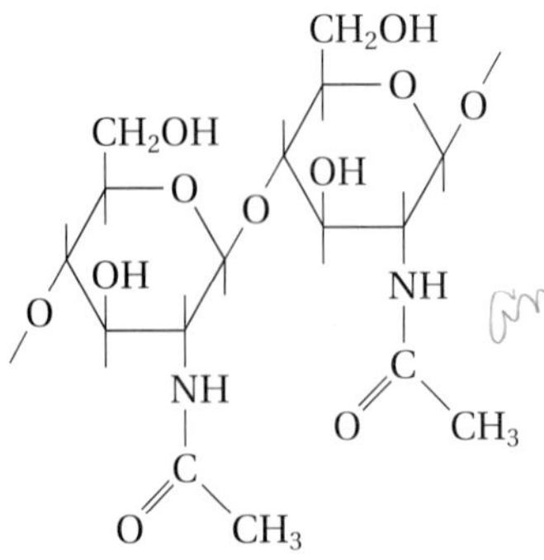

Chitin

Chitin is insoluble in water and is very resistant to hydrolysis of its component saccharides.

Acid mucopolysaccharides *are viscous polysaccharides that contain* N-*acetylglucosamine and glucuronic acid.* Hyaluronic acid is an acid mucopolysaccharide. It consists of repeating units of *N*-acetylglucosamine linked to glucuronic acid. The bonding pattern is

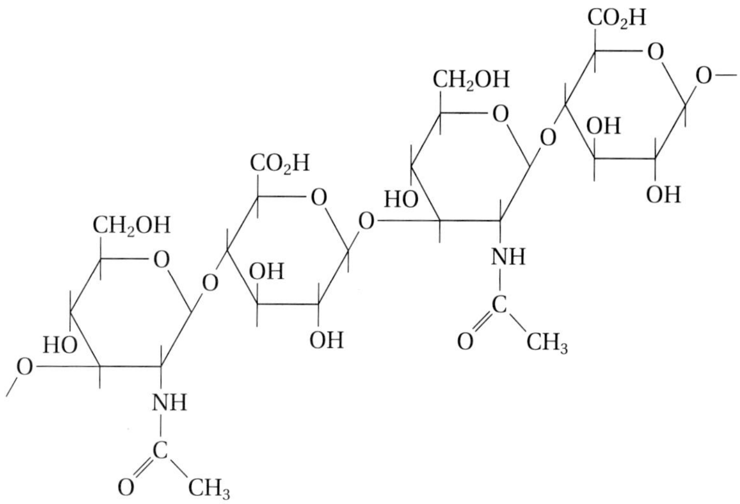

Hyaluronic acid

Hyaluronic acid is found in the connective tissue of animals, where it acts as a glue that helps to hold cells together. Another acid mucopolysaccharide is heparin, an important natural blood anticoagulant:

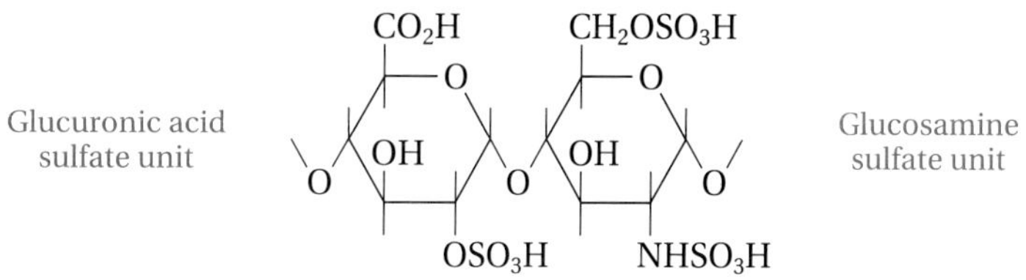

Heparin

Heparin is sometimes used to reduce the possibility of the formation of blood clots in the arteries of people who are at risk of heart attack or who have undergone heart bypass surgery.

The use of anticoagulant drugs such as heparin may cause abnormal bleeding in different parts of the body. Treatment is monitored through the use of periodic blood clotting tests. Using aspirin or alcohol when taking anticoagulant drugs can increase the chance of abnormal bleeding.

PRACTICE EXERCISE 7.16

How many functional groups can you identify in the following structure?

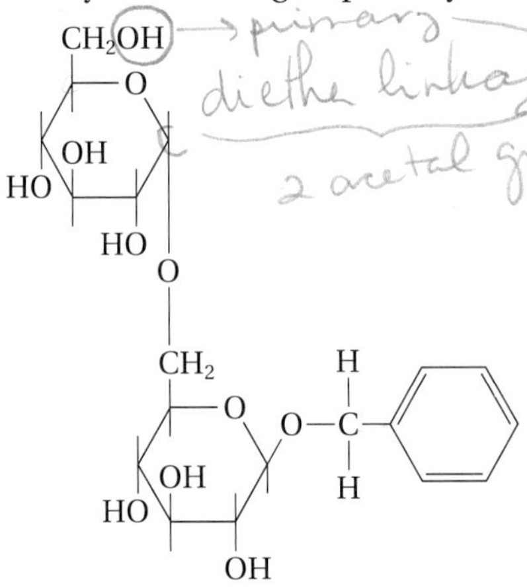

[Handwritten annotations: "→ primary" / "diether linkage → hydroxyl, phenyl" / "2 acetal groups + 6 secondary hydroxyl"]

SUMMARY

Simple sugars, called monosaccharides, are polyhydroxy aldehydes (aldoses) or ketones (ketoses). All sugars contain at least one asymmetric carbon. An important aspect of sugar chemistry is stereochemistry—relating the positions of groups of a compound in space. There is one pair of stereoisomers for compounds containing one asymmetric carbon. Stereoisomers have the same molecular formula, but they are nonsuperimposable mirror images. Stereoisomers may be called optical isomers because they rotate plane-polarized light.

Straight-chain aldoses and ketoses form hemiacetal and hemiketal rings, respectively. The heterocyclic ether rings are five-membered (furanoses) and six-membered (pyranoses). Fischer projections represent straight-chain sugars; Haworth projections represent sugars in ring forms. Cyclization of a straight-chain sugar to a pyranose or furanose ring creates a new asymmetric carbon in the molecule. Molecules whose stereochemistry differs only at the newly created asymmetric carbon are called anomers.

Simple sugars are often linked to other sugars through glycosidic bonds to create disaccharides. Polysaccharides are sugars containing many monosaccharide units.

Starches are polymers of glucose from plants. One form, which is a linear polymer, is amylose. Another, which is a branched polymer, is amylopectin. Glycogen is the animal form of starch; it is more highly branched than amylopectin. Cellulose is another linear polymer of glucose. Cellulose is more resistant to hydrolysis than starch. Many other polymers of monosaccharides are found in nature, where they have many diverse functions—from forming the hard shells of crabs, like chitin, to acting as a blood anticoagulant, like heparin.

Reducing sugars give positive results in the form of a red precipitate with Benedict's reagent or a silver mirror with Tollens' reagent. Nonreducing sugars do not. Most monosaccharides and disaccharides except sucrose are reducing sugars. Most polysaccharides are nonreducing sugars.

KEY TERMS

Acid mucopolysaccharide (7.10)
Aldose (7.1)
Anomer (7.4)
Asymmetric carbon (7.1)
Carbohydrate (7.1)
Cellulose (7.6)
Disaccharide (7.7)

Fischer projection (7.1)
Furanose ring system (7.3)
Glycogen (7.6)
Glycoside (7.5)
Glycosidic bond (7.5)
Haworth projection (7.4)
Invert sugar (7.8)
Ketose (7.2)

Monosaccharide (7.2)
Nonreducing sugar (7.9)
Nonsuperimposable mirror image (7.1)
Optical isomer (7.1)
Plane-polarized light (7.1)
Polysaccharide (7.6)
Pyranose ring system (7.3)

Reducing sugar (7.9)
Simple sugar (7.2)
Starch (7.6)
Stereoisomer (7.1)
Sugar (7.1)

EXERCISES

Carbohydrate Structure and Stereochemistry (Section 7.1)

7.17 Identify the asymmetric carbon (if there is one) in the following compounds.

(a) CH₃CHOH
     |
     CH₂CH₃

(b) CH₃CH₂CHOH
         |
         CH₃

(c) CH₃CHCH₂OH
     |
     OH

(d) CH₃CHCH₂OH
       |
       Cl

7.18 Draw stereorepresentations, and stereoisomers if they exist, for each of the following.

     OH
     |
(a) CH₃CHCHO   (b) CH₃CH₂CHO

7.19 Draw stereorepresentations and Fischer projection formulas of D-glyceraldehyde and L-glyceraldehyde. Explain the meaning of D and L.

7.20 Draw the structural formula for 3-hydroxybutanal, and identify the asymmetrical carbon.

Monosaccharides (Section 7.2)

7.21 Name each monosaccharide, and characterize it as to chain length and functional group.

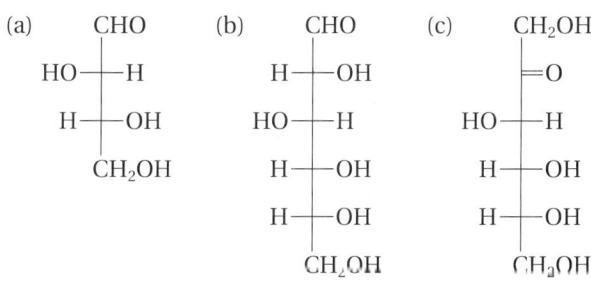

7.22 Draw Fischer projection formulas for D-ribose and D-2-deoxyribose. Number the carbons, and asterisk those which are asymmetrical.

7.23 Identify each of the following monosaccharides as D or L.

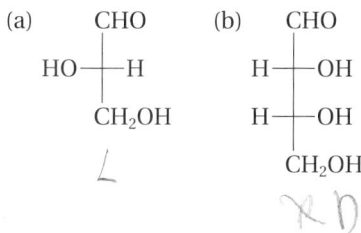

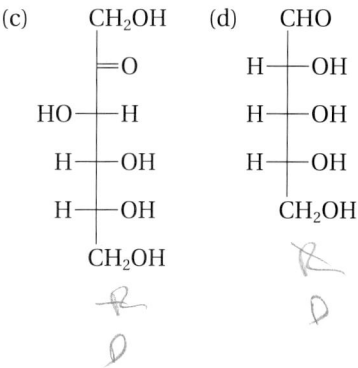

Closed-Chain Structures and Haworth Projections (Sections 7.3, 7.4)

7.24 Draw (a) a structure of the cyclic ether furan and (b) the general structure of a hemiketal.

7.25 Draw (a) the general structure of a hemiacetal and (b) a structure of the cyclic ether pyran.

7.26 Draw the Haworth projections of α-D-glucopyranose and β-D-glucopyranose, the cyclic forms of D-glucose.

7.27 Draw the Haworth projections of α-D-fructofuranose and β-D-fructofuranose, the cyclic forms of D-fructose.

7.28 Draw structures to show the relationship that exists between the two cyclic hemiacetal forms of D-glucose and the open-chain aldehyde form.

Polysaccharides (Sections 7.5, 7.6)

7.29 Write the general structure of (a) an acetal and (b) a ketal.

7.30 Explain the difference between a β(1→4) and an α(1→6) glycosidic bond.

7.31 Give one use for (a) starch, (b) cellulose, and (c) glycogen.

7.32 Name a source of (a) cellulose, (b) glycogen, and (c) starch.

7.33 Describe the structural and stereochemical differences between starch (amylose) and cellulose. What is the biological significance of these differences?

7.34 What is glycogen? How does it differ from amylopectin?

Disaccharides (Sections 7.7, 7.8)

7.35 Give sources, uses, and the products of hydrolysis of (a) sucrose and (b) maltose.

7.36 Give sources, uses, and the products of hydrolysis of (a) lactose and (b) cellobiose.

7.37 Draw a Haworth projection formula of the disaccharide maltose. Label the hemiketal, hemiacetal, ketal, or acetal carbons.

7.38 Draw a Haworth projection formula of the disaccharide sucrose. Label the hemiketal, hemiacetal, ketal, or acetal carbons.

Reducing and Nonreducing Sugars (Section 7.9)

7.39 Explain why lactose is a reducing sugar but sucrose is a nonreducing sugar.

7.40 Why are polysaccharides generally not regarded as reducing sugars?

7.41 Would Benedict's reagent be useful for distinguishing between
(a) glucose and starch
(b) fructose and sucrose
(c) fructose and glucose?

7.42 A solution of invert sugar is treated with (a) Tollens' reagent and (b) Benedict's reagent. What would you expect to happen?

Additional Exercises

7.43 Which of the following compounds exist as optical isomers?

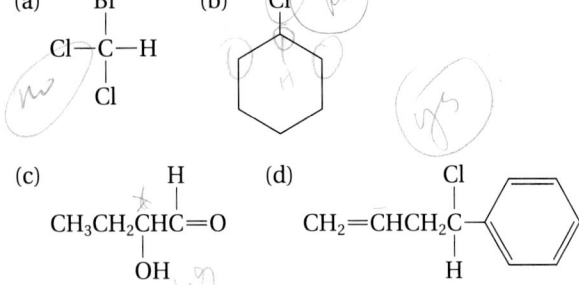

7.44 Draw the Fischer projection formula for D-galactose.

7.45 Draw the Haworth structures for α-D-galactose and β-D-galactose. Identify the anomeric carbon.

7.46 Draw the structure for the disaccharide formed when two D-galactose molecules are joined β(1→4). Would you expect this to be a reducing sugar? Explain.

7.47 Tell the number of asymmetrical carbons in the open-chain structure of (a) L-ribose, (b) D-2-deoxyribose, (c) D-glucose, (d) D-fructose, (e) L-galactose, and (f) D-mannose.

7.48 How would you explain the water solubility of monosaccharides?

7.49 Identify these monosaccharides as D or L.

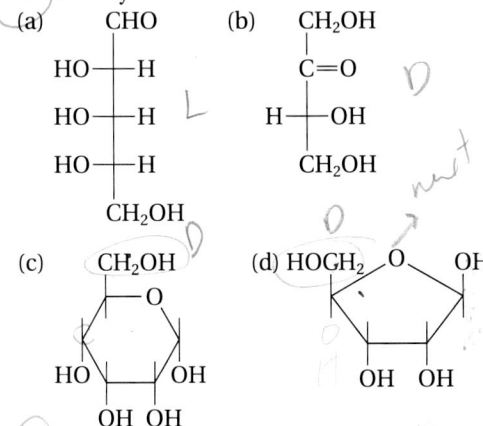

7.50 Match the description on the right with the appropriate polysaccharides. A number may be used more than once.
(a) starch (amylose)
(b) starch (amylopectin)
(c) glycogen
(d) cellulose

(1) found in plants
(2) α(1→4) linkages
(3) β(1→6) linkages
(4) α(1→6) linkages
(5) found in animals
(6) β(1→4) linkages

SELF-TEST (REVIEW)

True/False

1. A compound with optical isomers must have at least one asymmetrical carbon atom.

2. Anomers are stereoisomers that result from ring closure.

3. The cyclic form of an aldose is an intramolecular hemiacetal.

4. In polysaccharides, glycosidic bonds are formed between the alcohol of one sugar unit and the carbonyl group of a different sugar unit.

5. The disaccharide lactose is composed of the monosaccharides D-glucose and D-galactose.

6. A carbohydrate with a free aldehyde group will give a positive Tollens' test and be classified as a reducing sugar.

7. Molecules of D-threose and L-threose are mirror images.

8. A pyranose is a six-membered sugar ring system; a five-membered sugar ring system is a furanose.

9. A molecule of L(−)-glyceraldehyde is identical to a molecule of D(+)-glyceraldehyde.

10. Starch, cellulose, and glycogen are composed of the monosaccharide glucose.

Multiple Choice

11. Which of the following statements is *false* for β-D-fructofuranose?
  (a) It is a reducing sugar. (b) It is a hemiacetal.
  (c) It has a five-membered ring.
  (d) It is a structural isomer of D-glucose.

12. When pure β-D-galactopyranose is dissolved in an acidic solution
  (a) some of it is changed into β-L-galactopyranose.
  (b) all of it changes to the open-chain form.
  (c) some α-D-galactopyranose is formed.
  (d) no change takes place.

13. All the following are aldohexoses except
  (a) galactose.   (b) fructose.   (c) mannose.
  (d) glucose.

14. α-D-Glucopyranose
  (a) has the —OH group on carbon 1 pointing up.
  (b) has a five-membered ring.
  (c) is an intramolecular hemiacetal.
  (d) is an anomer of α-L-glucopyranose.

15. The disaccharide lactose
  (a) is found in milk.
  (b) is nonreducing.
  (c) is made up of the simple sugars D-galactose and D-fructose.
  (d) all of the above are true.

16. All the following are nonreducing sugars except
  (a) starch.   (b) sucrose.   (c) methyl glucoside.
  (d) none of the above.

17. The reaction of ethanol with α-D-ribofuranose would give
  (a) a glycoside.   (b) ethyl α-D-ribofuranoside.
  (c) an acetal.   (d) all of the above.

18. Complete hydrolysis of starch would give
  (a) D-glucose.   (b) amylopectin.   (c) maltose.
  (d) D-fructose.

19. The Fischer projection formula for a stereoisomer of xylose is

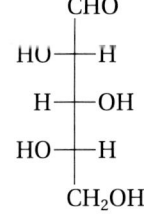

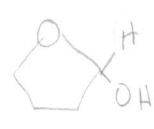

From this formula we know that this stereoisomer is
  (a) D-xylose.   (b) (+)-xylose.   (c) (−)-xylose.
  (d) L-xylose.

20. Which of the following statements is *not* true about an aldotetrose?
  (a) It has four asymmetrical carbons.
  (b) It gives a positive Benedict's test.
  (c) It has an aldehyde group.
  (d) It has three hydroxyl groups.

21. When starch is partially hydrolyzed, the disaccharide obtained is
  (a) maltose.   (b) amylose.   (c) lactose.
  (d) xylose.

22. For a carbohydrate molecule to have a mirror image, it must
  (a) be very complex.
  (b) be an organic compound.
  (c) not have a stereoisomer.
  (d) have an asymmetrical carbon.

23. One thing we can say for certain about the pentose L-ribose is that
  (a) it rotates plane-polarized light to the left.
  (b) it is a mirror image of D-ribose.
  (c) it rotates plane-polarized light to the right.
  (d) it has no asymmetrical carbons.

24. The most branched natural polysaccharide is
  (a) cellulose.   (b) amylose.   (c) glycogen.
  (d) amylopectin.

25. Human beings cannot use cellulose as a nutritional source because
  (a) it is found only in inedible materials.
  (b) it contains β(1→4) glycosidic linkages.
  (c) its structure is too highly branched.
  (d) all of the above are true.

26. Sucrose is unique among the common disaccharides because
  (a) it is not a reducing sugar.
  (b) it has a 1→2 linkage.
  (c) it has no hemiacetal group.
  (d) all of the above are true.

8

Lipids

A Potpourri of Fatty Molecules

Safflower is an Old World herb related to daisies. Its seeds are rich in an edible oil and its large orange-red flowers yield a dye.

L̲ipids are a class of naturally occurring compounds that dissolve in organic solvents such as ether and chloroform. This property sets them apart from carbohydrates, proteins, and nucleic acids, the other great classes of biological molecules. Lipids encompass such a wide variety of molecular structures that their study will allow us to review many of the functional groups we have seen before while we learn about their biological function.

Obesity, or being overweight, occurs when the body stores excessive amounts of one kind of lipid that we discuss in this chapter. Because many people are concerned about controlling their weight, we explore the measurement of obesity and its treatment in this chapter's Case in Point.

CASE IN POINT: Obesity

Fat, fat, fat! Our culture is obsessed with fat. Almost every day we can expect to hear somebody say, "I'm too fat!" While people often want to reduce their weight in order to look better, obesity is a health problem as well as a cosmetic problem. Medical results of being overweight include increases in blood pressure, gallstones, and diabetes and an increased risk of certain cancers. Obese people are also increasing their risk of having a heart attack. How is obesity measured? What are the possible treatments for obesity? We will learn the answers to these questions in Section 8.2.

Besides a reduced-calorie diet, regular aerobic exercise assists in weight loss by increasing the metabolic rate and by burning calories from excess stored fat.

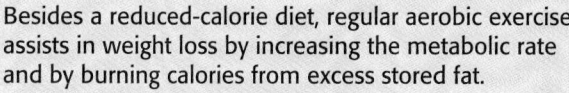

8.1 Waxes

AIM: To describe the general structure, source, and use of waxes.

Lipids *are a large class of relatively water-insoluble compounds found in nature.* Waxes are among the simplest members of the lipid family. **Waxes are esters of long-chain fatty acids and long-chain alcohols.**

$$\boxed{\text{Fatty acid}} \overset{O}{\underset{\|}{-C}}-OH + HO-\boxed{\text{Alcohol}} \xrightarrow{-H_2O} \boxed{\text{Fatty acid}} \overset{O}{\underset{\|}{-C}}-O-\boxed{\text{Alcohol}}$$

You may recall from Section 5.1 that the fatty acids are straight-chain aliphatic carboxylic acids (RCO_2H). (Throughout the first three sections of this chapter you may wish to refer to Tables 5.1 and 5.2 for fatty acid structures.) In waxes, the hydrocarbon chains for both the acid and the alcohol (R—OH) usually contain from 10 to 30 carbons. Waxes are low-melting, stable solids that occur in nature in both plants and animals. A wax coat protects surfaces of many plant leaves from water loss and attack by microorganisms. Carnauba wax, a major ingredient in car wax and floor polish, comes from the leaves of a South American palm tree. Beeswax is largely myricyl palmitate, the ester of myricyl alcohol and palmitic acid.

$$CH_3(CH_2)_{14}-\overset{O}{\underset{\|}{C}}-O-(CH_2)_{29}CH_3$$

Myricyl palmitate

Waxes also coat skin, hair, and feathers and help keep them pliable and waterproof.

8.2 Triglycerides

AIMS: To characterize triglycerides, fats, and oils by source, structure, and use. To distinguish between hydrolytic rancidity and oxidative rancidity and explain how each can be prevented. To describe the production of soap by saponification.

Natural **triglycerides**—*triesters of long-chain fatty acids (C_{12} through C_{24}) and glycerol*—are the major components of animal lipids. The following equation shows the general reaction for the formation of triglycerides:

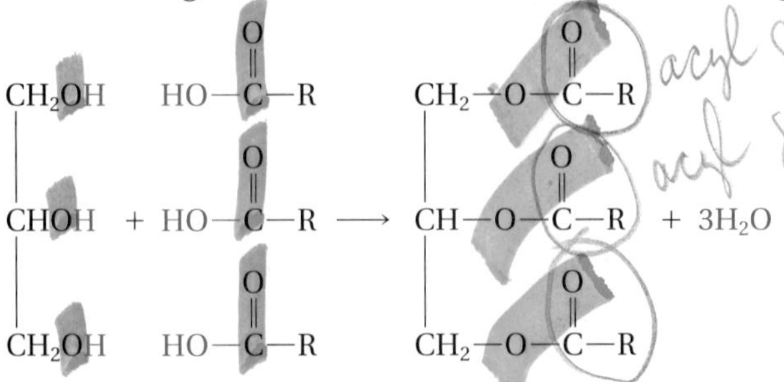

$$
\begin{array}{l}
CH_2OH \\
| \\
CHOH \\
| \\
CH_2OH
\end{array}
+
\begin{array}{l}
HO-\overset{O}{\underset{\|}{C}}-R \\
HO-\overset{O}{\underset{\|}{C}}-R \\
HO-\overset{O}{\underset{\|}{C}}-R
\end{array}
\longrightarrow
\begin{array}{l}
CH_2-O-\overset{O}{\underset{\|}{C}}-R \\
CH-O-\overset{O}{\underset{\|}{C}}-R \\
CH_2-O-\overset{O}{\underset{\|}{C}}-R
\end{array}
+ 3H_2O
$$

Glycerol   3 Fatty acid molecules   Triglyceride (triester of glycerol)   Water

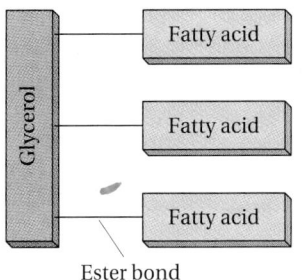

Figure 8.1
Schematic diagram of a triglyceride.

The bonding of the triglyceride building blocks may be represented by a diagram, as shown in Figure 8.1.

The many different types of triglycerides vary with the positions and identities of the fatty acids. **Simple triglycerides** *are triesters made from glycerol and three molecules of one kind of fatty acid.* For example, glycerol and three molecules of stearic acid give a simple triglyceride called *glyceryl tristearate,* or *tristearin:*

$$CH_2-O-\overset{\overset{\textstyle O}{\|}}{C}-(CH_2)_{16}CH_3$$

$$CH-O-\overset{\overset{\textstyle O}{\|}}{C}-(CH_2)_{16}CH_3$$

$$CH_2-O-\overset{\overset{\textstyle O}{\|}}{C}-(CH_2)_{16}CH_3$$

Glyceryl tristearate
(tristearin)

Simple triglycerides are rare. More often, natural triglycerides are **mixed triglycerides**—*triesters of glycerol with different fatty acid components.* The mixed triglycerides in butterfat, for example, contain at least 14 different fatty acids. Triglycerides are the form in which fats are stored in the human body. In the Case in Point earlier in this chapter, we questioned how body fat is measured and how obesity is treated. Answers to these questions can be found in the Follow-up to the Case in Point, below.

Sometimes reactions of glycerol with fatty acids may produce mono- or diesters. For example, see glycerol monostearate, a commonly used emulsifier, in A Closer Look: Cosmetic Creams.

FOLLOW-UP TO THE CASE IN POINT: Obesity

The degree of obesity is measured by the excess of body weight relative to height. Because everyone is different, an *ideal body weight (IBW)* for different heights has been established statistically. The IBW is the weight that gives the lowest incidence of sickness and death in the statistical sample of people of a certain height. The terms *overweight* and *obesity* are based on the IBW. People with a relative weight up to 20% above the IBW are defined as being overweight. Those with a relative weight greater than 20% above the IBW are defined as being obese. Another measure of overweight and obesity is the *body mass index (BMI).* This measure is defined as body mass in kilograms divided by height squared (height is in meters). Overweight is defined as a BMI of 25 to 30 kg/m^2. Obesity is a BMI of >30 kg/m^2.

The causes of most obesity are not very well understood. A few diseases, such as hypothyroidism, tumors of the hypothalamus, and Cushing's disease, can cause obesity, but these diseases are very rare. Inherited and environmental factors may be involved, since about 80% of the offspring of obese parents will be obese. Only about 15% of the offspring of normal-weight parents are obese.

The treatment of obesity usually involves restricting dietary caloric intake and increasing exercise. However, many people go on "crash" diets and lose weight only to gain it back. Permanent reductions in weight often require that, besides diet and exercise, eating habits be changed. No medications are currently available that are effective in the long term. Surgery in which the size of the stomach is restricted can be part of a reducing program for patients who weigh double their IBW.

A Closer Look

Cosmetic Creams

Skin care products such as cold creams, moisturizers, lotions, sunscreens, and cleansing creams are basically emulsions of oil and water (see figure). They are used to protect the skin from the drying effects of detergents and exposure to wind and sun. Emulsifiers are used to stabilize oil-water mixtures. One commonly used emulsifier is the oil glycerol monostearate.

$$CH_2OH$$
$$CHOH$$
$$CH_2O-\underset{\underset{O}{\|}}{C}-(CH_2)_{16}CH_3$$

Glycerol monostearate

Cold creams are oil-in-water emulsions. That is, there is more water than oil in the emulsion. As the water evaporates, it produces a cooling effect on the skin. The thin film of oil that remains is quickly absorbed by the outer skin layers. Water-in-oil emulsions, which contain more oil than water, are also used in cosmetics. Water-in-oil emulsions are sometimes called *warm emulsions* because there is little cooling effect due to water

Glycerol monostearate is an emulsifier commonly used in cosmetic products.

evaporation. These emulsions permit the oil to contact the skin almost immediately. Cleansing creams, which dissolve oily materials on the surface of the skin, and moisturizing lotions, which soften the skin and help replace lost water, are water-in-oil emulsions.

PRACTICE EXERCISE 8.1

Write one structure for the mixed triglyceride that is made from glycerol, stearic acid, palmitic acid, and oleic acid.

PRACTICE EXERCISE 8.2

Name the following triglyceride:

$$CH_2-O-\underset{\underset{O}{\|}}{C}-(CH_2)_{14}CH_3$$
$$CH-O-\underset{\underset{O}{\|}}{C}-(CH_2)_{14}CH_3$$
$$CH_2-O-\underset{\underset{O}{\|}}{C}-(CH_2)_{14}CH_3$$

Hydrogenation of fats and oils

Animal fats tend to be rich in saturated fatty acids; vegetable oils are rich in unsaturated fatty acids!

Two polyunsaturated fatty acids with 18 carbon atoms, linolenic and linoleic acid, are essential fatty acids that cannot be synthesized by the body. These fatty acids are found in fish oils and in many plants. Some studies suggest that the consumption of linolenic acid may reduce the risk of heart disease.

The distinction between fats and oils is usually based on melting point. *At room temperature, **fats** are solid triglycerides and **oils** are liquid triglycerides.* The melting point of a fat or oil depends on its structure, usually increasing with the number of carbons. The number of carbon-carbon double bonds in the fatty acid component also has an effect: Triglycerides rich in unsaturated fatty acids, such as oleic and linoleic, are generally oils; triglycerides rich in saturated fatty acids, such as stearic and palmitic, are generally fats. The triglycerides in liquid olive oil contain mostly unsaturated oleic acid, for example, but those of solid beef fat, or tallow, contain mostly saturated stearic acid. Hydrogenation (see Sec. 3.2) converts vegetable oils into fats. It is commercially important in the manufacture of margarine and lard substitutes. Powdered metallic nickel (which is removed later) is dispersed in the hot oil as a catalyst. Hydrogen adds to some of the double bonds in the unsaturated fatty acid carbon chains, saturating them and thereby converting the oil to a fat. For example, the hydrogenation of triolein (melting point $-17\,°C$) produces tristearin (melting point $55\,°C$):

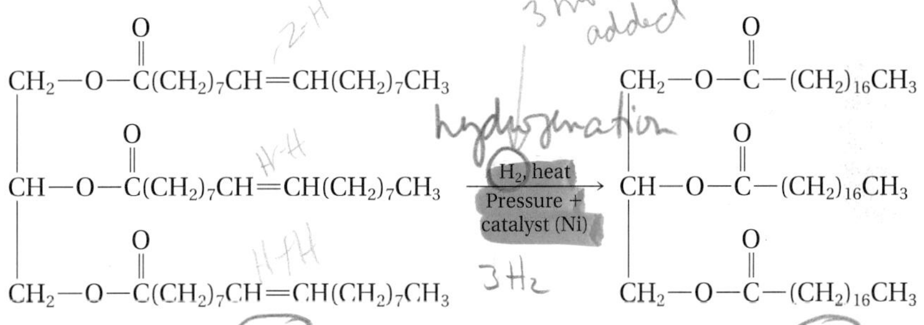

Triolein (an oil)   Tristearin (a fat)

The consistency of the fat is controlled by the degree of hydrogenation; the higher the degree of saturation of the fat by hydrogenation, the harder is the product at room temperature.

PRACTICE EXERCISE 8.3

A mixed triglyceride with the following structure is hydrogenated. What is the hydrogenation product?

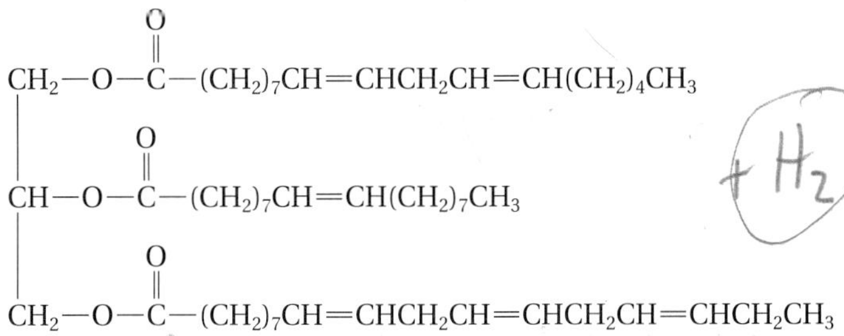

Rancidity

Triglycerides soon become **rancid,** *developing an unpleasant odor and flavor on exposure to moist air at room temperature.* The release of volatile fatty acids (particularly butyric acid) from butterfat causes the disagreeable odor of rancid butter. These acids are formed either by the hydrolysis of ester bonds or by the oxidation of double bonds. The hydrolysis of a fat or oil is often catalyzed by enzymes, called *lipases,* present in airborne bacteria. *Hydrolytic rancidity* is prevented or delayed by storing food in a refrigerator. The unwelcome odor of sweat results when bacterial lipases catalyze the hydrolysis of oils and fats on the skin.

Oxidative processes, however, not hydrolysis, are the major cause of rancidity in food. Warmth and exposure to air induce oxidative rancidity. In *oxidative rancidity,* double bonds in the unsaturated fatty acid components of triglycerides rupture, forming low-formula-weight aldehydes with objectionable odors. The aldehydes then oxidize to the equally offensive low-formula-weight fatty acids. Oxidative rancidity shortens the shelf life of cookies and similar foods. Antioxidants are compounds that delay the onset of oxidative rancidity. Two naturally occurring substances often used as antioxidants are ascorbic acid (vitamin C) and α-tocopherol (vitamin E).

Saponification of fats and oils

Saponification
sapon (Latin):   soap

Fats and oils are simply esters. Like other esters, they are easily hydrolyzed in the presence of acids and bases (Sec. 5.8). *Hydrolysis of oils or fats by boiling them with aqueous sodium hydroxide is called* **saponification.** This process is used to make soap. Soaps are the alkali metal (Na, K, or Li) salts of fatty acids.

Soap is made by heating beef tallow or coconut oil in large kettles with an excess of sodium hydroxide. When sodium chloride is added to the saponified mixture, the sodium salts of the fatty acids separate as a thick curd of crude soap. If the fat is tristearin, the soap is sodium stearate:

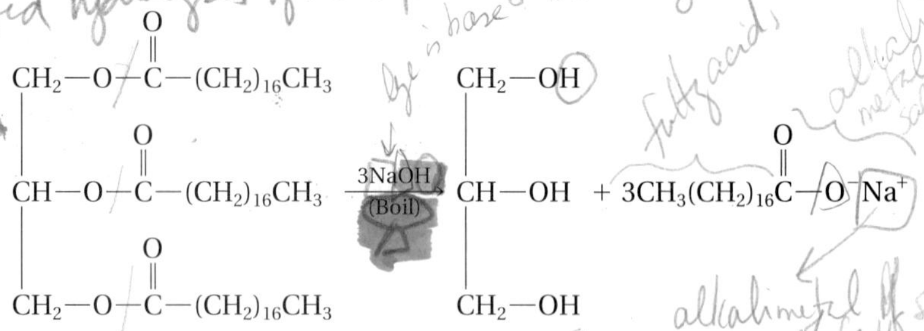

Tristearin
(a triester)   Glycerol   Sodium stearate
(soap)

Glycerol is an important by-product of the reaction. It is recovered by evaporating the water layer. The crude soap is then purified, and coloring agents and perfumes are added according to market demands.

PRACTICE EXERCISE 8.4

What are the products when a mixed triglyceride with the following structure is saponified?

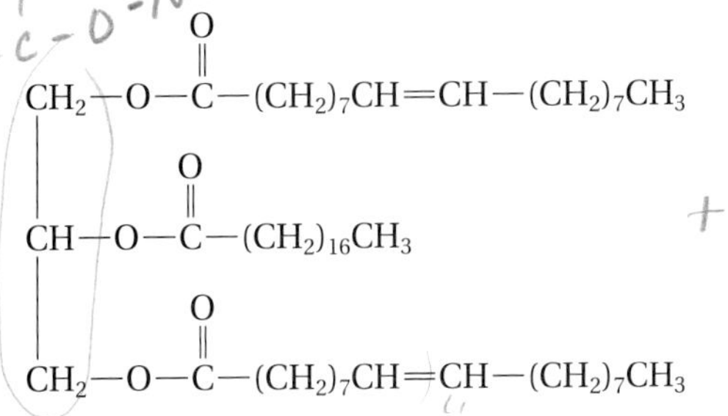

[handwritten notes:]

CH_2-OH
$CH-OH$ + $CH_3(CH_2)_7CH=CH(CH_2)_7$ sodium oleate (2 mols)
CH_2-OH
glycerol (1 mol)
$CH_3(CH_2)_{16}-C-O-Na^+$ sodium stearate (1 mol)

[typed structure with handwritten annotations:]

$CH_2-O-C-(CH_2)_7CH=CH-(CH_2)_7CH_3$
$CH-O-C-(CH_2)_{16}CH_3$
$CH_2-O-C-(CH_2)_7CH=CH-(CH_2)_7CH_3$

$+ 3NaOH \xrightarrow{\Delta \text{ boil}}$

8.3 Lipid composition of cell membranes

AIM: To recognize the general structures of the following three types of lipid molecules: phosphoglycerides, sphingomyelins, and glycolipids.

It is possible to break cells, empty them of their contents, and isolate the *cell membranes. The* **cell membrane** *is the "sack" that holds the contents of cells and acts as a selective barrier for the passage of certain substances in and out of the cell.* The interior of cells also contains membrane structures, as described in A Closer Look: Cells. Chemical analysis of the isolated membranes shows that lipids are the major components. These lipids are not triglycerides, but another group of compounds called *complex lipids.* **Complex lipids** *contain parts made from substances besides fatty acids and glycerol; some contain no glycerol.* The complex lipids fall into two categories: *phospholipids* and *glycolipids.*

Phospholipids *are lipids that are esters of phosphoric acid.* There are two main types of phospholipid molecules in cell membranes: phosphoglycerides and sphingomyelins.

Phosphoglyceride *molecules are built from long-chain fatty acids (14 to 24 carbons), glycerol, and phosphoric acid.* Two fatty acids are covalently bonded to adjacent hydroxyl groups of glycerol by ester linkages. The phosphoric acid is bonded through phosphate ester linkages to the remaining hydroxyl function of glycerol. The resulting molecule is called a *phosphatidic acid,* which is a phosphoglyceride.

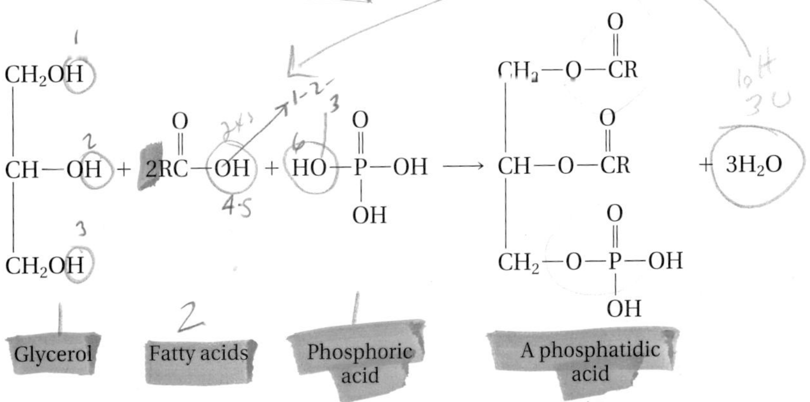

$$CH_2OH + CH-OH + 2RC-OH + HO-P-OH \longrightarrow CH_2-O-CR + 3H_2O$$

Glycerol	Fatty acids	Phosphoric acid	A phosphatidic acid

[handwritten:] ester phosphoglyceride

A Closer Look

Cells

The two major cell designs are *prokaryotic* and *eukaryotic*. The former is the more ancient of the two. Microscopic examination of fossilized remains shows that prokaryotes were present on Earth at least 3 billion years ago, whereas eukaryotes did not appear until 2 billion years later. In the modern world, the prokaryotic cell design is limited to bacteria and blue-green algae. The cells of other cellular organisms, including green plants and people, are eukaryotic.

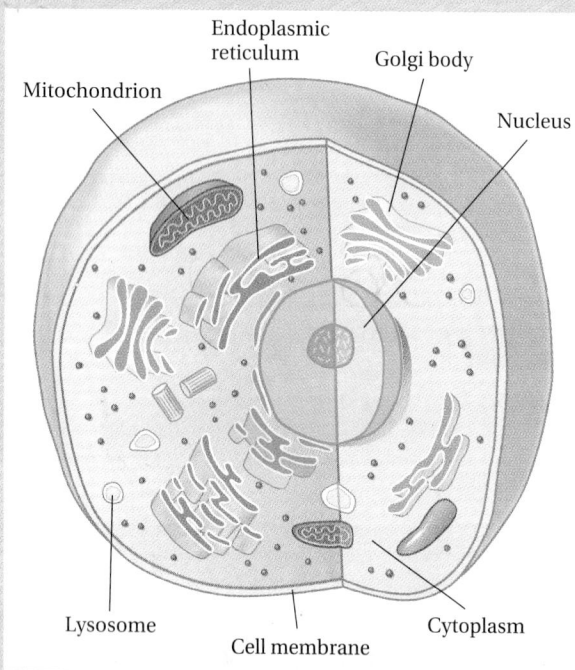

The eukaryotic cell

Both types of cells are essentially packages of chemicals necessary for life encased in a cell membrane. Eukaryotic cells are considerably larger and somewhat more complicated than prokaryotes, but the chemical processes carried out by both types of cells are very similar, and both are exceedingly efficient chemical factories. The major feature that distinguishes prokaryotes from eukaryotes is the latter's *organelles* ("little organs")—small membrane-enclosed bodies suspended in the interior cellular fluid or cytoplasm (see figure). The organelles are the sites of many specialized functions in eukaryotes. The most prominent membrane-encased organelles and their functions are as follows:

Organelle	Function
Nucleus	Cell reproduction
Mitochondrion	Production of most cellular energy in cells using oxygen for respiration
Golgi body	Processing of proteins into glycoproteins
Lysosome	Digestion of cellular wastes and substances taken into cells

Yet another membrane structure in eukaryotes is the highly folded, netlike endoplasmic reticulum (ER). Among its various functions, the ER serves as an attachment site for ribosomes—small organelles that are not membrane-encased but are the sites where proteins are made. Endoplasmic reticula with and without attached ribosomes are called, respectively, *rough ER* and *smooth ER* because of their appearance under a microscope.

Living cells contain little or no free phosphatidic acid. Usually, the phosphorus of the phosphatidic acid is linked to the hydroxyl group of a second alcohol. Choline, a common amino alcohol constituent of cell membrane phospholipids, is an example. In Figure 8.2, the hydroxyl group of choline is attached to the phosphorus of the phosphatidic acid through a phosphate ester bond. This phospholipid is a phosphoglyceride

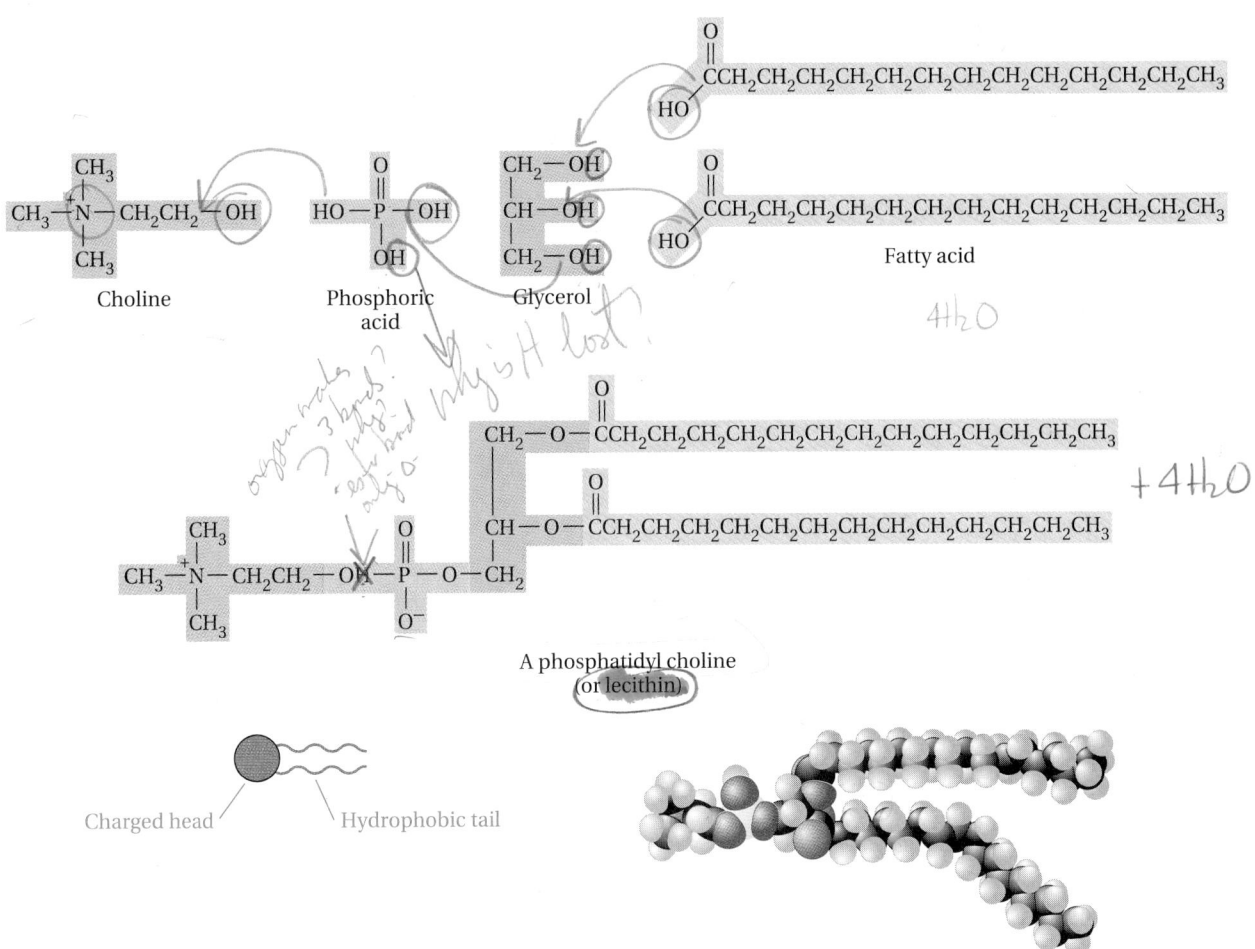

Choline

Phosphoric acid

Glycerol

Fatty acid

$4H_2O$

$+4H_2O$

A phosphatidyl choline (or lecithin)

Charged head

Hydrophobic tail

Figure 8.2
A phosphatidyl choline (phosphoglyceride) or lecithin molecule is constructed from the amino alcohol choline, phosphoric acid, glycerol, and two fatty acid molecules. A simplified representation of the molecule (lower left) has been adopted for all membrane lipids. The hydrophilic (charged) head is shown as a purple sphere, and the hydrophobic tails as red wavy lines. A space-filling model of a phosphatidyl choline molecule is shown at the lower right of the figure.

because it contains phosphorus and has a backbone of glycerol. It also may be called a *phosphatidyl choline* because it is an ester of phosphatidic acid and choline. *An older name for phosphatidyl choline is* **lecithin.** It is important to recognize that the names *phosphoglyceride, phosphatidic acid,* and *phosphatidyl choline* are only general names for classes of compounds. The lengths of the hydrocarbon chains of the fatty acids may vary and these chains may be saturated or contain one or more double bonds.

[handwritten margin notes: "why unprotonated chlorine / protonated ↓", "amino alcohol", "H—N—CH₂CH₂—OH + HO—P—OH + CH₂—OH + R—C—OH →", "AHOH", "phosphoric acid + glycerol", "+N—C—R→long chain", "O=C—O—R→long chain", "CH—O—C—R", "CH—O—C—R", "CH₂—O—C—R", "OH / N—CH₂CH₂—O—P—O—CH₂ / OH", "OH / H—N—CH₂CH₂—O—P—O / OH", "why not O⁻ like / because it in equilibrium and choline too!", "both ways"]

PRACTICE EXERCISE 8.5

When the phosphorus of a phosphatidic acid is linked to the hydroxyl group of the amino alcohol ethanolamine, $NH_2CH_2CH_2OH$, the compound formed is phosphatidyl ethanolamine. Phosphatidyl ethanolamine, also called *cephalin*, is found in brain tissue and is important in blood clotting. Draw the general structure for cephalin.

The second type of phospholipid molecules encountered in cell membranes is **sphingomyelins.** Sphingomyelins do not contain glycerol. Instead, they contain sphingosine, a long-chain unsaturated amino alcohol. Only one fatty acid is attached to sphingosine, as shown in Figure 8.3, through an amide linkage. The structure of the nonpolar end of sphin-

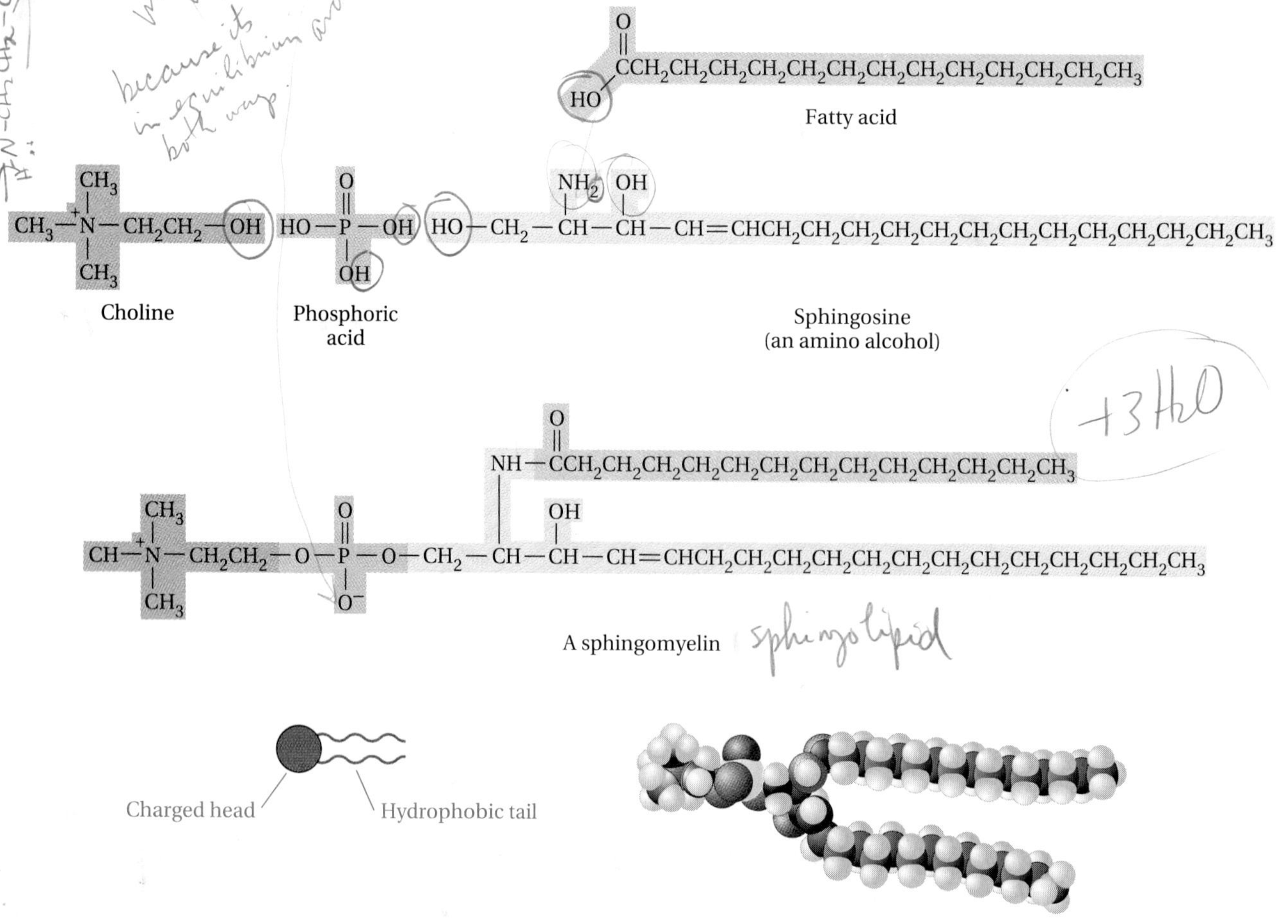

[Figure 8.3 chemical structures]

Fatty acid:
$HO—CCH_2CH_2CH_2CH_2CH_2CH_2CH_2CH_2CH_2CH_2CH_2CH_2CH_3$ (with $=O$)

Choline: $CH_3—N^+(CH_3)—CH_2CH_2—OH$ with CH_3

Phosphoric acid: $HO—P—OH$ with $=O$ and OH

Sphingosine (an amino alcohol): $HO—CH_2—CH—CH—CH=CHCH_2CH_2CH_2CH_2CH_2CH_2CH_2CH_2CH_2CH_2CH_2CH_2CH_3$ with NH_2 and OH

[handwritten: "+3 H₂O"]

A sphingomyelin:
$CH_3—N^+(CH_3)—CH_2CH_2—O—P—O—CH_2—CH—CH—CH=CHCH_2CH_2CH_2CH_2CH_2CH_2CH_2CH_2CH_2CH_2CH_2CH_2CH_3$ with CH_3, O^-, $=O$, and $NH—CCH_2CH_2CH_2CH_2CH_2CH_2CH_2CH_2CH_2CH_2CH_2CH_3$ (with $=O$) and OH

[handwritten: "sphingolipid"]

Charged head — Hydrophobic tail

Figure 8.3
A sphingomyelin molecule consists of one molecule each of choline, phosphoric acid, sphingosine, and a fatty acid. A space-filling model of a sphingomyelin molecule is shown at the bottom of the figure.

The abnormal metabolism and accumulation of certain types of lipid molecules occur in a number of genetic diseases. For example, a glycolipid accumulates and damages the brain in Tay-Sachs disease.

gomyelins may differ somewhat, depending on the length and degree of saturation of the fatty acid attached to the sphingosine amino group. Sphingomyelins are the only phospholipids that are not built on glycerol. Large amounts of sphingomyelins are found in brain and nervous tissue and in the myelin sheath, the protective coat of nerves.

Glycolipids *are lipid molecules that contain carbohydrates, usually simple sugars such as glucose or galactose.* Figure 8.4 shows a glycolipid that consists of sphingosine, a fatty acid, and a sugar. These are called *cerebrosides* because of their abundance in the brain. The cerebrosides are not phospholipids, because they do not contain phosphorus.

The classification of lipids is summarized in Figure 8.5.

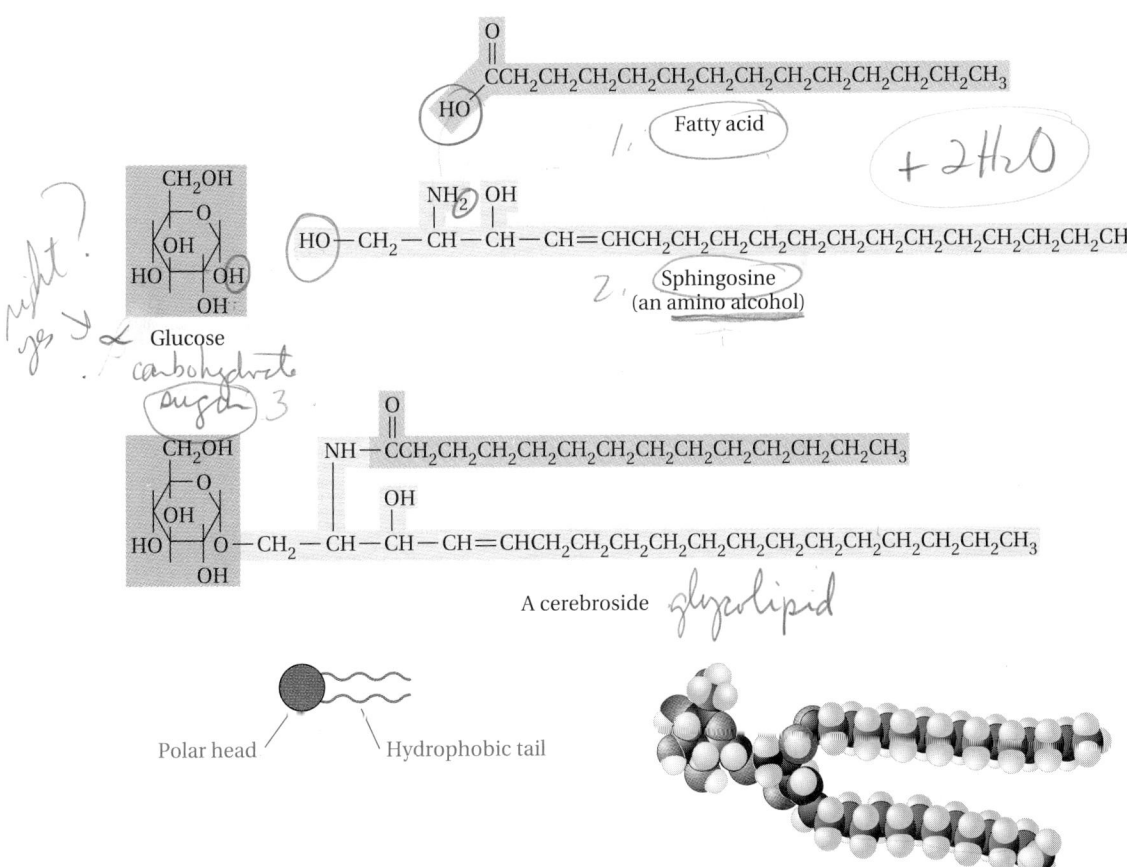

Figure 8.4
A cerebroside (glycolipid) consists of a sugar, a sphingosine, and one fatty acid molecule. The sugar unit shown is glucose, but it may be galactose. A space-filling model of a cerebroside is shown at the bottom of the figure.

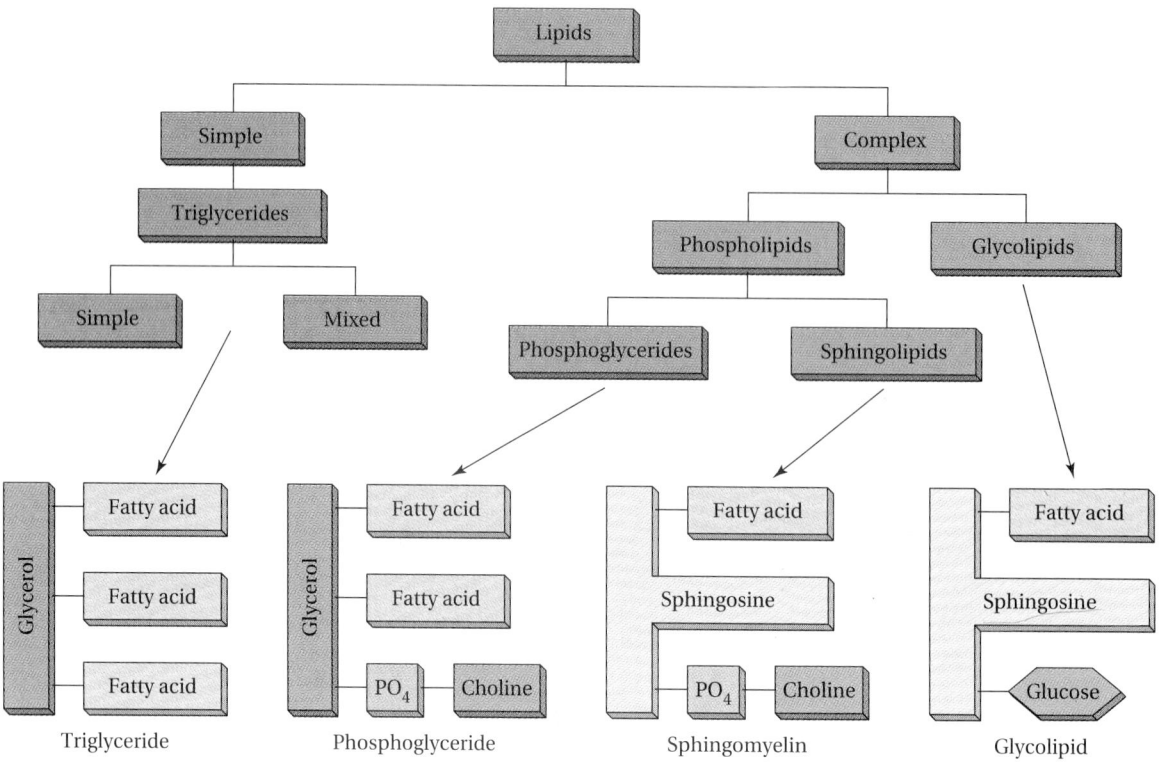

Figure 8.5
Lipid classification diagram.

8.4 Structure of liposomes and cell membranes

AIMS: To sketch sections of the liposomal bilayer in water, labeling the polar end of the lipid molecules. To explain the relationship between the degree of unsaturation in phospholipid molecules and membrane flexibility. To use the fluid mosaic model to describe the movement of lipid molecules in membranes.

lecithin

Phosphatidyl choline is a typical membrane phospholipid. It contains a charged head consisting of negatively charged phosphate and positively charged choline attached through glycerol to two hydrophobic fatty acid tails. If we vigorously shake a mixture of phosphatidyl choline and water, the lipid molecules form microscopic spheres rather than dispersing evenly in water. *These lipid spheres, or* **liposomes,** *are packages of water surrounded by a* **lipid bilayer**—*a two-layer-thick wall of phosphatidyl choline.* Figure 8.6 shows a cross section of a liposome. The lipid molecules of the liposomal bilayer are more ordered than the sulfonic acid molecules in detergent micelles (Sec. 5.3).

All the hydrophobic hydrocarbon tails of the lipids are protected from water, and all the hydrophilic phospholipid heads interact with water. Lipo-

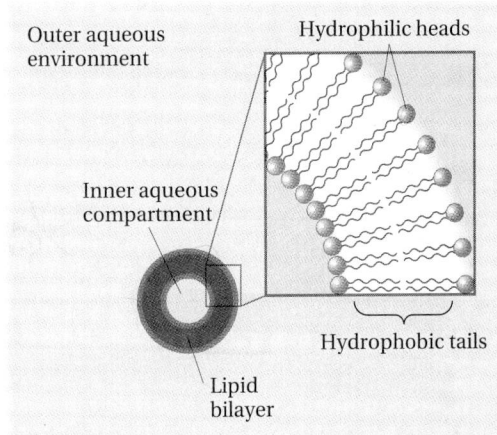

Figure 8.6
Cross-section of a liposome.
The whole liposome is spherical.

somes are stable structures. When broken to expose the lipid hydrocarbons to water, liposomes spontaneously reseal. Lipid bilayers are tight, so any leakage of the contents of a liposome to the outside is quite slow.

PRACTICE EXERCISE 8.6

Predict the result if the liposome experiment (shaking phosphatidyl choline in a solvent) is repeated using a nonpolar solvent (such as carbon tetrachloride) instead of water. *will not form micelles.*

Structure of cell membranes

Cell membranes also consist of lipid bilayers similar to those described for liposomes. Some cell membranes are more flexible than others, due to differing properties of the different fatty acids of the lipids that make the lipid bilayers. Lipid bilayers made of phospholipids having a high percentage of unsaturated fatty acids are more flexible than those having a high percentage of saturated fatty acids. As shown in Figure 8.7, unsaturated fatty acid chains are bent and fit into bilayers more loosely than saturated fatty acid chains. The looser the packing in the bilayer, the more flexible is the membrane. A high percentage of glycolipids also tends to increase membrane flexibility.

Figure 8.7
The space-filling models show that the molecule of the saturated fatty acid palmitic acid (a) is straighter than that of the unsaturated fatty acid oleic acid (b). The highly ordered packing of three molecules of palmitic acid (c) is disrupted by the presence of a molecule of oleic acid between two molecules of palmitic acid (d). For this reason, the phospholipids rich in saturated fatty acids make more tightly packed, less flexible membranes (e) than those rich in unsaturated fatty acids (f).

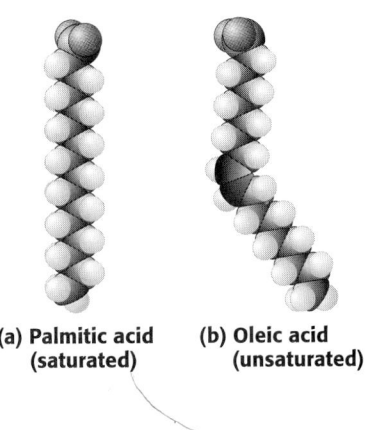

**(a) Palmitic acid
(saturated)**   **(b) Oleic acid
(unsaturated)**

palmitic acid

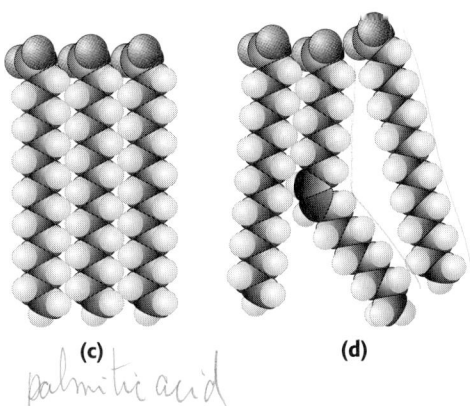

(c)   **(d)**

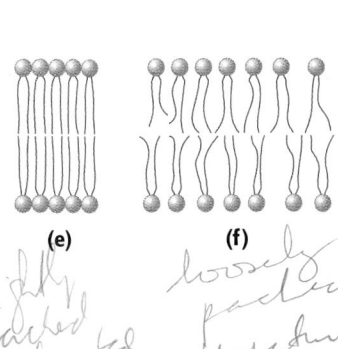

(e)   **(f)**

tightly packed saturated   *loosely packed unsaturated*

The fluid mosaic model of membrane structure

*The **fluid mosaic model** of membrane structure proposes that lipids of the bilayer are in constant motion, gliding from one part of their bilayer to another at high speed.* Although lipids move freely within their own layer of the bilayer, they cannot easily cross or flip-flop to the other lipid layer (see Fig. 8.8).

Stuck in the mass of moving lipids are protein molecules, some moving and some apparently anchored in place. Figure 8.9 shows two kinds of membrane proteins: peripheral and integral. **Peripheral proteins** *perch on either side of the lipid bilayer.* They can usually be removed from the membrane by high salt concentrations. **Integral proteins** *may be partially embedded in one side of the bilayer or jut all the way through.*

Many integral proteins are *glycoproteins*—protein molecules with attached carbohydrates. The carbohydrate portion of the embedded glycoproteins is found at the surface of the bilayer, where its hydroxyl groups can hydrogen bond with water. The protein portion of membrane glycoproteins is usually hydrophobic. This ensures a favorable interaction of the protein with the hydrophobic lipid tails of the bilayer. Many membrane proteins facilitate the transport of ions and molecules into and out of the cell.

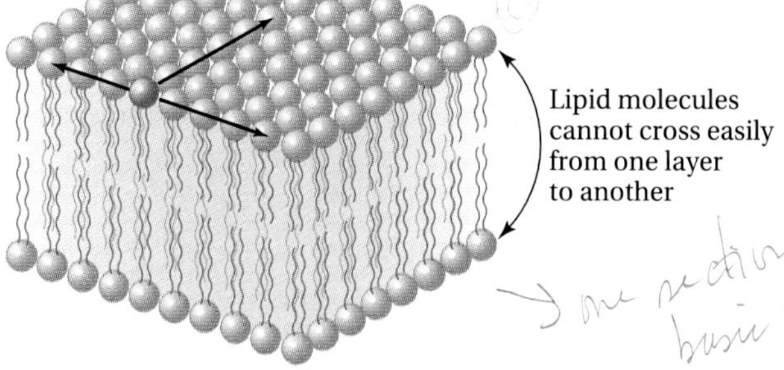

Figure 8.8
Lipid molecules move easily within their own layer of the bilayer but do not readily flip-flop to the other layer.

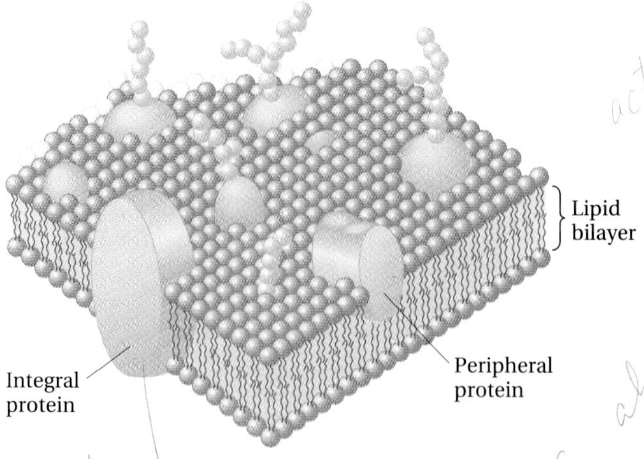

Lipid molecules cannot cross easily from one layer to another

Figure 8.9
Fluid mosaic model. Most membrane proteins are integral; some membrane proteins are peripheral. The beadlike structures attached to the proteins represent the carbohydrate portions of glycoproteins.

Lipid bilayer

Integral protein

Peripheral protein

8.5 Steroids

AIMS:** **To draw the fundamental chemical structure of all steroid molecules. To describe the function of cholesterol.

Steroid
stereos (Greek):   solid

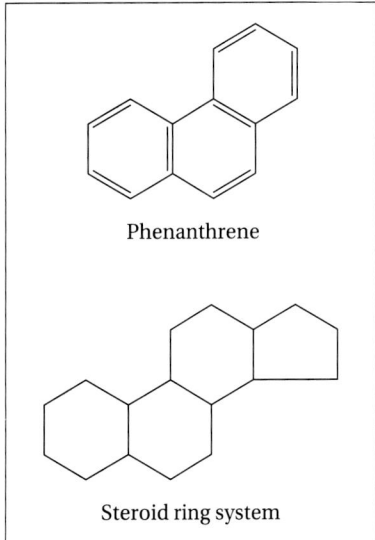

Figure 8.10
The phenanthrene and steroid ring system.

The amount of artery-plugging cholesterol in the blood is closely related to the dietary intake of saturated fats. Cooking oils, which contain less saturated triglyceride, may be a healthier alternative to the hydrogenated products.

The next class of lipids we will examine is **steroids**—*a family of lipids found in plants and animals.* The connection between steroids and other classes of lipids is mainly historical. Scientists first isolated steroids along with fats by extracting plant and animal tissues with chloroform or ether. Like other lipids, steroids are insoluble in water but soluble in organic solvents. As we will soon see, however, the structures of steroids are quite different from those of triglycerides, phospholipids, and glycolipids. Medical research has revealed that certain steroids, the *steroid hormones,* are enormously important to the proper functioning of the human body and, when misused, to the dysfunctioning of the human body. Hence it will be worth our while to examine the chemical structure and biological function of steroids.

All steroid molecules are saturated derivatives of phenanthrene, a tricyclic aromatic hydrocarbon. The fundamental structure of every steroid molecule also contains a fused cyclopentane ring. Figure 8.10 compares the structures of phenanthrene and the steroid ring system.

All steroid molecules are formed from the steroid ring system by replacing ring hydrogens with hydrocarbon chains or functional groups or by introducing carbon-carbon double bonds into one or more of the cyclohexane rings. *Steroid molecules containing a hydroxy function and no carboxyl or aldehyde groups are called* **sterols.** Cholesterol, a waxy solid, is an example of a sterol:

Cholesterol is found only in animal cells; plants do not make it. A typical animal cell membrane contains about 60% phospholipid and 25% cholesterol. Cholesterol appears to impart rigidity to cell membranes. The higher the percentage of cholesterol, the more rigid is the membrane.

PRACTICE EXERCISE 8.7

What structural features of cholesterol permit cholesterol molecules to become part of cell membrane structures?

8.6 Steroid hormones

AIM: To state the source and at least one function of cortisone, prednisone, and aldosterone.

Hormones *are chemical messengers produced by the endocrine glands in the human body.* Figure 8.11 shows the locations and names of the endocrine glands. Hormones produced by these glands are secreted directly into the bloodstream. Some of the most important hormones in the human body are steroids. Cholesterol is not a steroid hormone, but many steroid hormones are probably made from it.

Steroid hormones *are produced and secreted at two major places in the human body: the adrenal glands and the gonads (the testes in males and the ovaries in females).* Adrenal glands are small mounds of tissue at the top of each kidney. An outer layer of the gland, the adrenal cortex, produces a number of potent steroid hormones. *The adrenal cortex steroids are often called adrenal corticoids or simply* **corticoids.** The chemical structures of more than 30 corticoids are known, but only a few seem to function as hormones. Active corticoids fall into three categories according to function: *glucocorticoids, mineralocorticoids,* and *secondary sex hormones.* We will cover the glucocorticoids and mineralocorticoids in this section. Secondary sex hormones will be discussed with other male and female hormones in Sections 8.7 and 8.8, respectively.

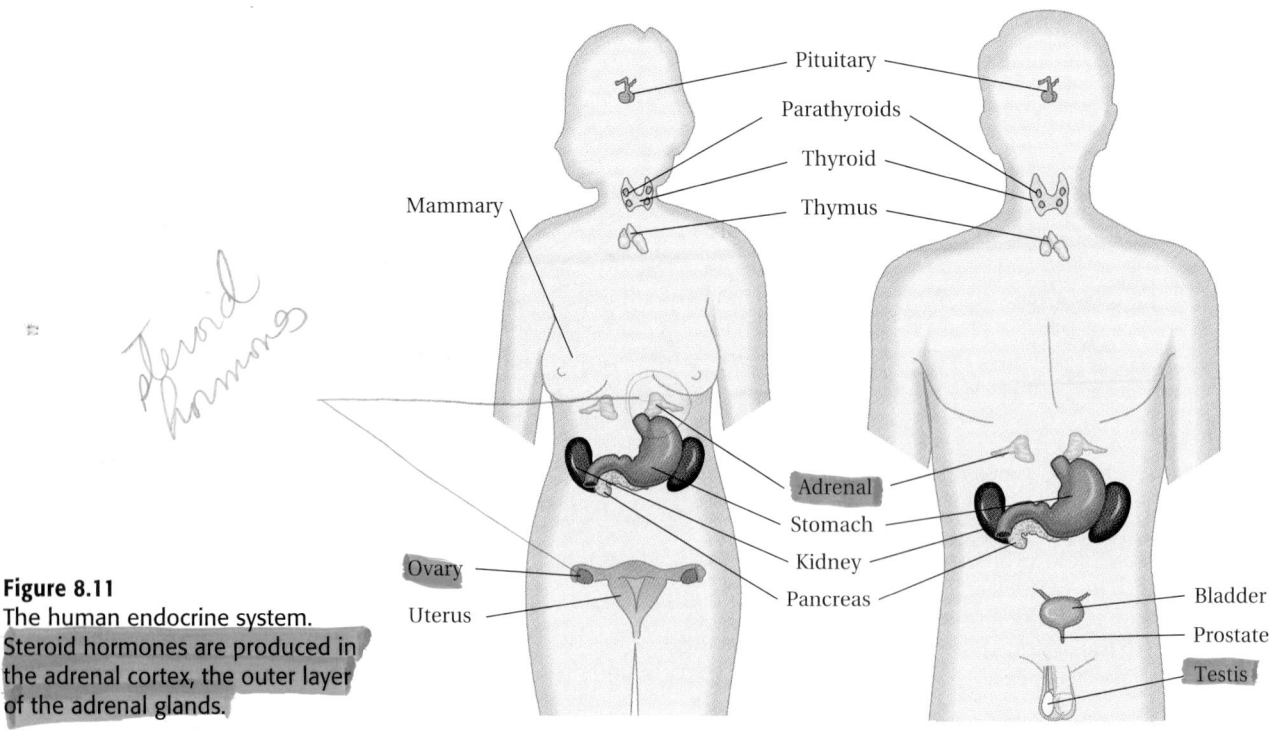

Figure 8.11
The human endocrine system. Steroid hormones are produced in the adrenal cortex, the outer layer of the adrenal glands.

Glucocorticoids

Glucocorticoids *regulate the body's use of glucose—how much is burned as fuel to provide energy and how much is stored as glycogen for future energy needs.* Glucocorticoids tend to prevent the uptake of glucose by tissues and therefore promote its storage as glycogen (see Sec. 7.6). Other nonsteroid hormones, especially insulin, promote the uptake of glucose by tissues. Together, the glucocorticoids and insulin maintain the body's delicate balance between too much and too little glucose. Glucocorticoids also affect body protein and fat utilization in a similar way. Cortisone, cortisol, and corticosterone are the major glucocorticoids:

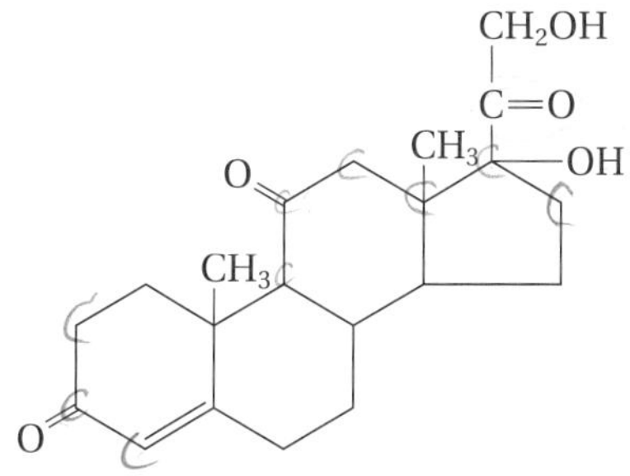

Cortisone

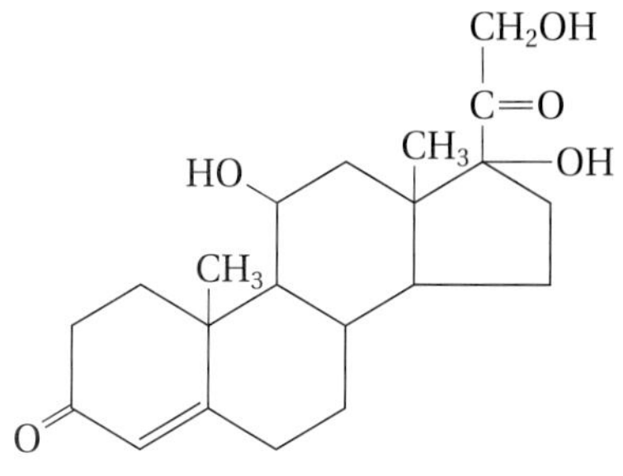

Cortisol

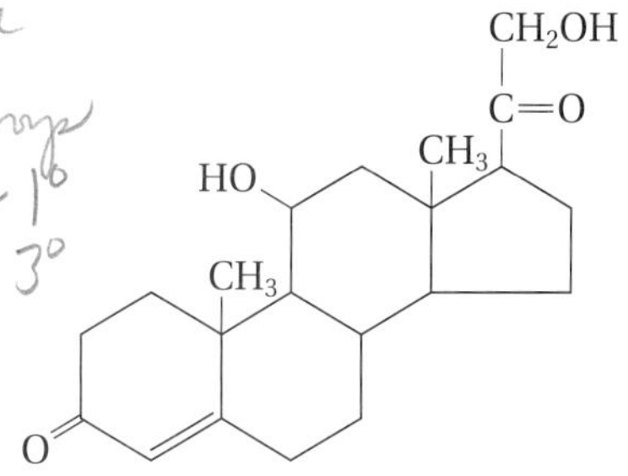

[handwritten annotations: 1 C–C double bond / 3 ketone carbonyl group / 2 hydroxy group (1°, 3°)]

Corticosterone

Quite a few years ago scientists learned that cortisone reduces the symptoms of rheumatoid arthritis, allergic diseases, and inflammation. Many people hoped it would become a new wonder drug. However, since the side effects of cortisone injections are often severe, the drug is rarely used for these purposes. Fortunately, medicinal chemists have been able to synthesize new steroids that give many of the advantages of cortisone therapy with fewer undesirable side effects. Two of the main cortisone substitutes in use today are prednisone and prednisolone:

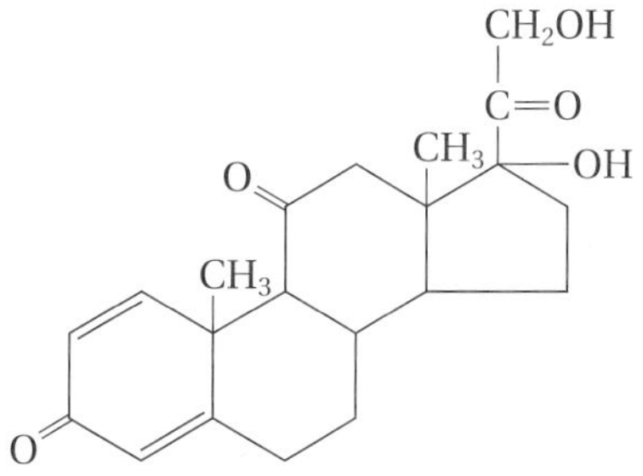

Prednisone

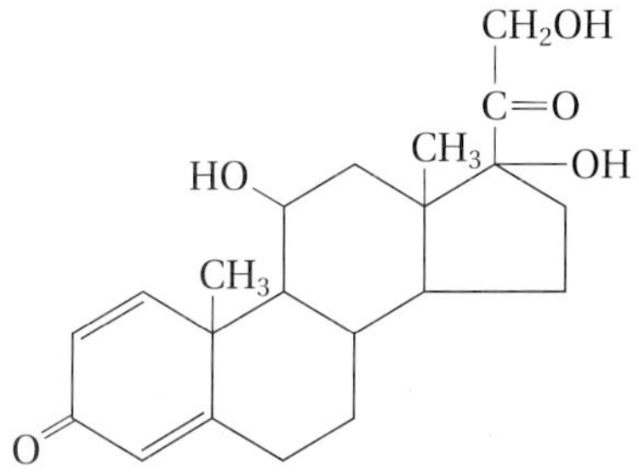

Prednisolone

Glucocorticoids reduce the body's immunity to the introduction of foreign materials. (This may be why these compounds are effective in the treatment of allergic diseases, which are a severe response to foreign materials.) For this reason, physicians sometimes give injections of cortisone or related drugs to organ transplant patients to help suppress their immune system's response to the new tissue or organ, thereby preventing rejection.

Mineralocorticoids

Mineralocorticoids *help retain sodium chloride and maintain fluid balance in the body.* All corticoids affect salt and water balance to some extent, but aldosterone is by far the most important mineralocorticoid. Aldosterone's sodium chloride–retaining activity is about 1000 times the activity of cortisol, for example. Aldosterone is the principal hormone in maintaining life in an animal whose adrenals have been removed.

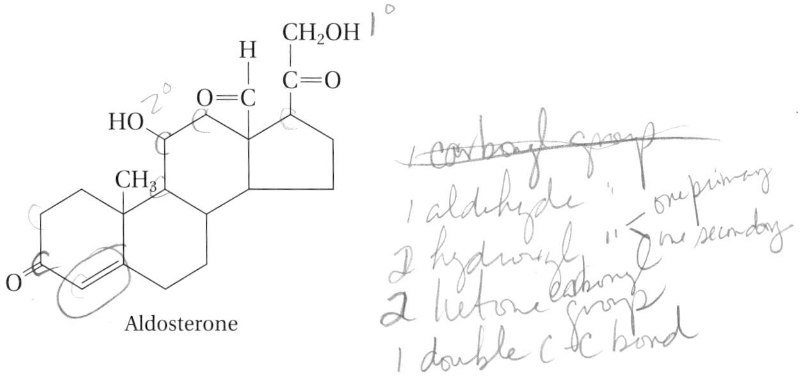

Aldosterone

PRACTICE EXERCISE 8.8

Name the functional groups in (a) aldosterone and (b) cortisone.

8.7 Male sex hormones

AIM: To state the source and at least one function of testosterone and androgens.

Focus

The primary male sex hormone is produced in the testes.

Testosterone *is the principal sex hormone in males and is produced in the testes (testicles).*

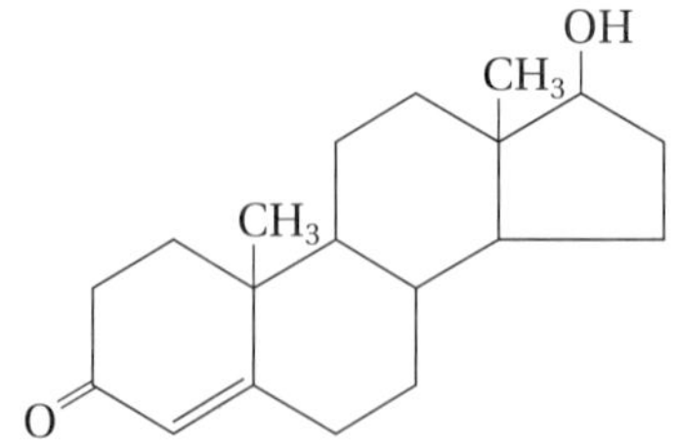

Testosterone

The testes perform two functions: They produce sperm, and they produce testosterone. At puberty, testosterone promotes the maturation and growth of the male sex organs. It also aids in the development of male secondary sex characteristics such as deep voices and beards, and it contributes to the greater muscular development and bone growth of males compared with females.

Women also produce testosterone. Masculinization does not occur in normal women, however, because chemical reactions in a woman's body rapidly convert testosterone into female hormones. The adrenal cortex also produces sex hormones—*androgens* in males and *estrogens* in females. Beginning at puberty, androgens and estrogens produce the secondary sex characteristics considered masculine or feminine: deep voices and beards in males, higher voices, breast development, and lack of facial hair in females. The major androgens are dehydroepiandrosterone and androstenedione.

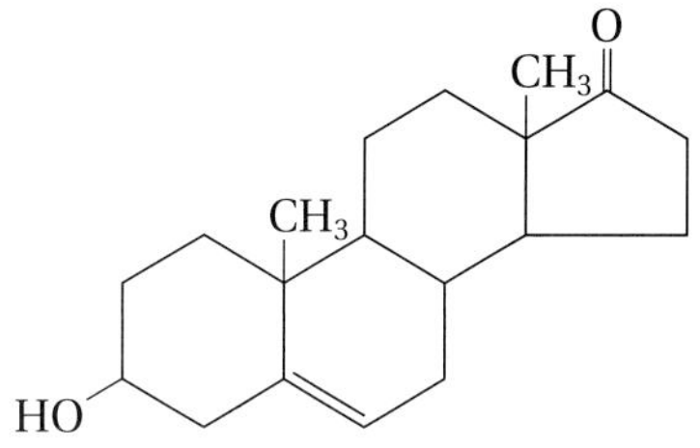

Dihydrotestosterone is a hormone in which the double bond of testosterone has been reduced by enzyme reactions in the body. Dihydrotestosterone appears to be involved in two conditions experienced by men: male-pattern baldness and growth of the prostate, an exclusively male gland that controls the flow of urine from the bladder.

Dehydroepiandrosterone 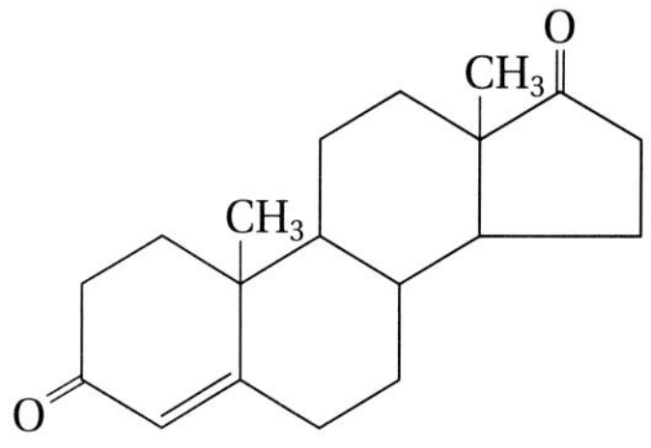   Androstenedione

Overproduction of androgens can cause masculinization, also called *virilization*, in females. The "bearded ladies" sometimes seen in circus sideshows usually suffer from excessive androgen production.

Certain world-class athletes have used synthetic steroids structurally related to testosterone. These drugs, the anabolic steroids, promote muscular development. Norancholane is one of the anabolic steroids.

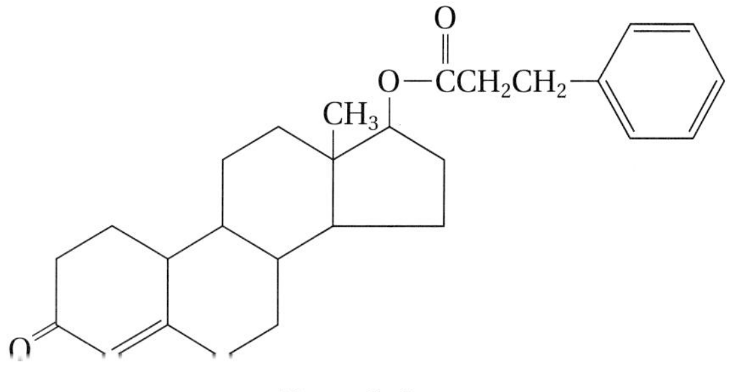

Norancholane

The use of any drug for this purpose is officially banned by the world's amateur athletic unions. Anabolic steroids can be detected in the urine for only a few days after their use is discontinued, however; so this rule is hard to enforce. In the short term, there are reports of testicular atrophy in some users of anabolic steroids. The long-term effects of synthetic bodybuilding steroid drugs are not well understood but may include cancer.

8.8 Female sex hormones

AIM: To state the source and at least one function of estrogens and progesterone.

The primary female sex hormones are produced by the ovaries. *Two estrogens—estrone and estradiol—and progesterone are the principal female sex hormones.* An apparently small change in the molecular structure of a steroid causes a profound change in its effect on the body. Testosterone and progesterone, two steroid hormones that help divide all humanity into male and female, differ by only two atoms of carbon and two of hydrogen.

Estrone   Estradiol   Progesterone

Estrogens are produced in the adrenal cortex and contribute to the secondary sex characteristics of women—breast development, high voices, and lack of facial hair. They are also extremely important as primary sex hormones. Estrogens are important to the development of the egg, the ovum, in the ovaries. Progesterone causes changes in the wall of the uterus that prepare it for pregnancy and that maintain pregnancy.

Figure 8.12 shows how the levels of estrogens and progesterone affect egg and uterine wall development in the human female's ovarian cycle. The cycle usually takes 28 days; the first day of the menstrual period is considered the beginning of the cycle. Near the end of menstruation, development of a new egg begins in the follicle, a small body inside the ovary. Estrogens promote growth of the follicle and also stimulate thickening of the uterine lining. By about the fourteenth day of the cycle, the follicle, now called the *graafian follicle,* has matured. At ovulation, the matured follicle ruptures, releasing the egg, which enters the oviduct (fallopian tube), a channel between the ovary and the uterus.

Once the egg is released, the ruptured follicle, now called the *corpus luteum,* begins to secrete progesterone. The action of progesterone maintains the thickness of the uterine lining and stimulates the lining's final development. If the egg is not fertilized, the corpus luteum begins to shrink and secrete less progesterone. Eventually, the concentration of progesterone is too small to maintain the uterine lining. The lining begins to slough off, and hemorrhagia takes place. This marks the onset of menstruation and a new ovarian cycle. A woman loses an average of 50 to 150 mL of blood during menstruation. If the egg is fertilized, the corpus

Levels of estrogen decline in women who have reached menopause. These women are often given replacement estrogen. This therapy helps to prevent osteoporosis—the loss of calcium from bones.

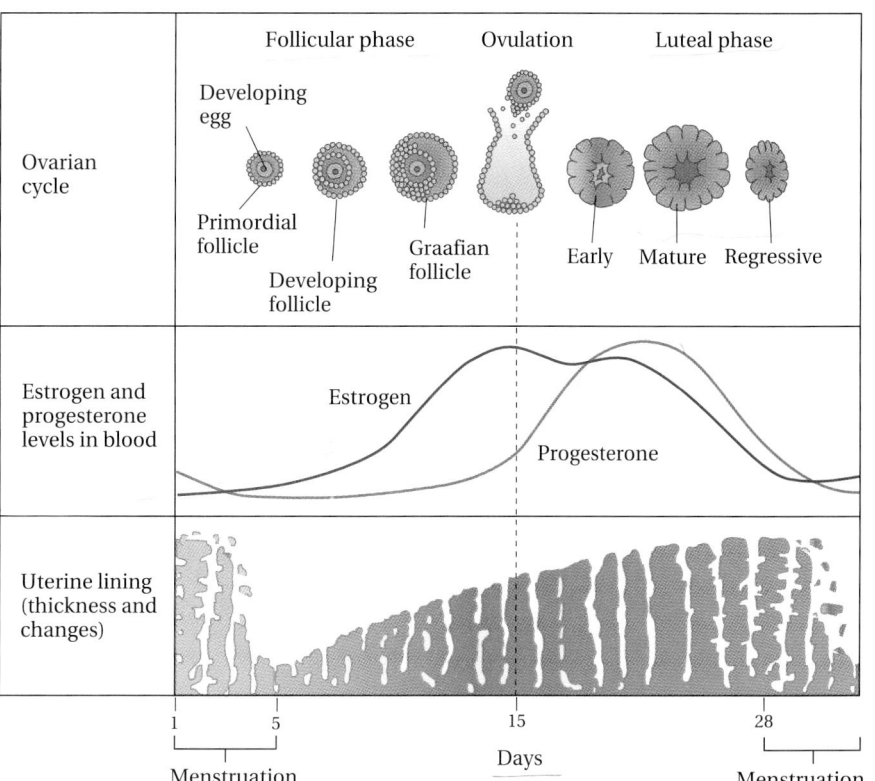

Figure 8.12
Effects of blood levels of estrogens and progesterone on the development of the egg and the uterine lining during the ovarian cycle.

luteum continues to produce progesterone throughout the course of pregnancy. The progesterone makes implantation of the fertilized egg on the uterine wall possible, promotes the development of the mammary glands, and prevents new follicles and eggs from maturing.

8.9 The pill

AIM: To explain how birth control pills prevent conception.

Focus

The original oral contraceptive pill prevented ovulation.

Since progesterone prevents maturation and release of the egg, this hormone would seem to be a good candidate for a birth control drug—a *contraceptive*—that could be taken by mouth. However, progesterone itself is an ineffective oral contraceptive, because a woman's body breaks it down before it reaches the ovaries. In the late 1940s, scientists began a search for a synthetic oral drug that mimicked the hormonal effect of progesterone. Several such drugs are now on the market; they are collectively called "the pill." Many early preparations of the pill contained rather large doses (up to 1.5 mg) of norethindrone or norethynodrel. A synthetic estrogen, mestranol (usually about 0.06 mg), is also included in this class of drugs.

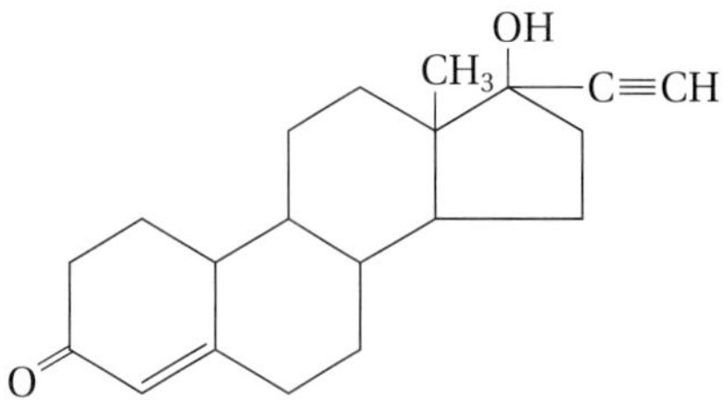

Norethindrone
(Norlutin)

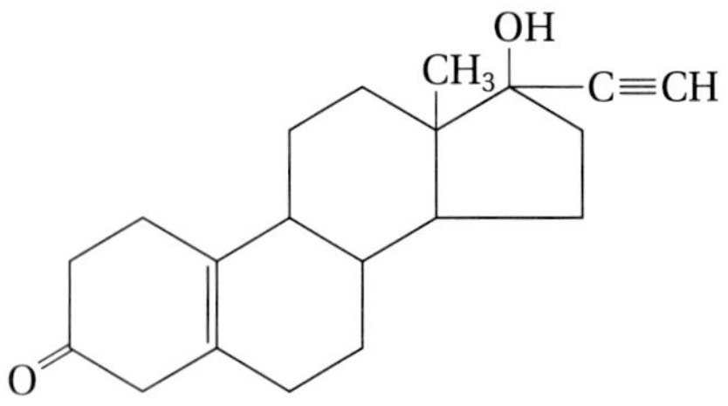

Norethynodrel
(Enovid ketone)

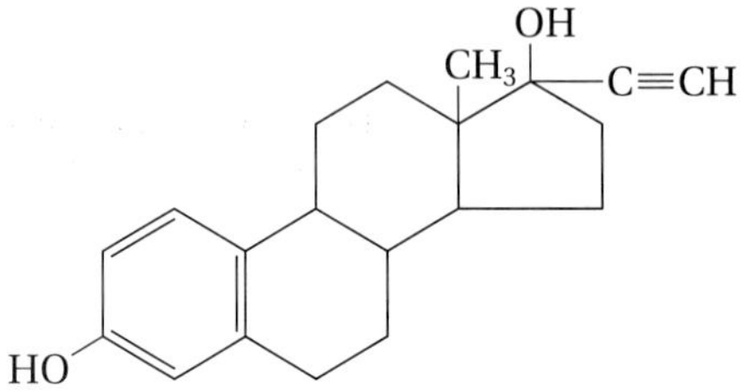

Mestranol
(synthetic estrogen)

Women who take the pill often experience the symptoms of pregnancy—dizziness, headaches, vomiting, and increases in weight and breast sensitivity—because of the high levels of progesterone-like drug in their bodies. Put another way, women's bodies have no way of "knowing" that they are not pregnant.

The "minipill" was introduced around 1971. Minipills contain much smaller amounts (0.1 to 0.2 mg) of the synthetic progesterones and no synthetic estrogen. The minipill was introduced in the hope that its users would experience fewer side effects. The action of the newer generations of the pill appears to be somewhat different from that of the ancestors. Minipills seem to stop conception by preventing sperm from entering the oviduct, by preventing release of the egg, and by making the uterus less receptive to any fertilized egg that reaches it.

Scientists have had somewhat less success in developing a "morning-after pill." Diethylstilbestrol (DES), a nonsteroid compound that mimics the effects of the estrogens, is effective in preventing pregnancy if taken in large doses for 3 days after intercourse; DES prevents implantation of a fertilized egg by causing excessive thickening of the uterine lining.

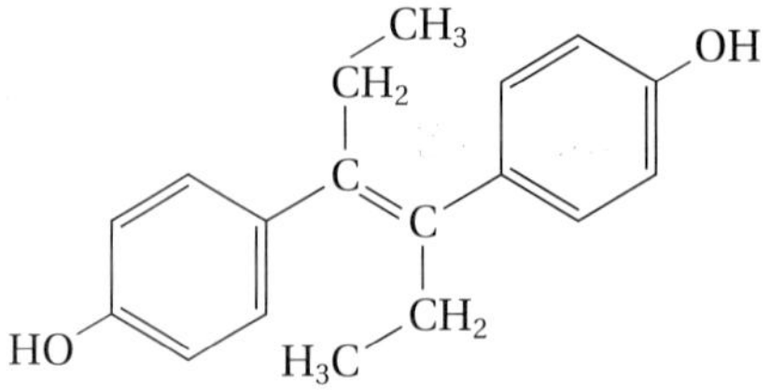

A steroid-like drug called RU-486 is an *abortifacient*, a drug used to terminate pregnancy.

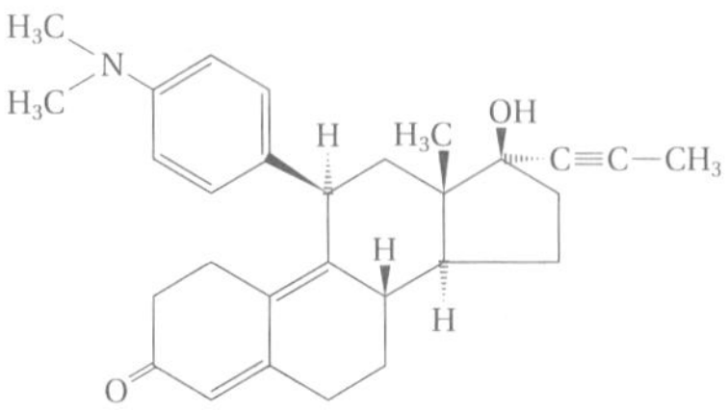

RU-486

Diethylstilbestrol
(DES)

The undesirable side effects of DES include vomiting, nausea, and exces-

"5"

sive vaginal bleeding. In the 1940s and early 1950s, DES was routinely used to prevent spontaneous abortions (miscarriages). This practice was stopped when several studies showed that daughters born to women who took DES have a higher-than-average incidence of vaginal cancer after reaching puberty.

8.10 Plant steroids

AIM: To state the source and at least one function of digitoxin.

Focus

Digitalis is a mixture of plant steroids used to treat heart disease.

Many plants produce steroids also. Nobody knows, though, what function many of these compounds have in plants. Many plant steroids are toxic to humans. Nevertheless, one of the oldest drugs, in use since 1785, is in fact a mixture of toxic steroids. The drug is *digitalis,* obtained from the common foxglove plant. Digitalis is potentially deadly, but in very small doses it improves the tone of heart muscles and is widely used to treat congestive heart failure. Digitoxin is one of the major components of digitalis. Digitoxin has a steroid ring system, but it is unlike animal steroids in that the steroid part of the molecule is attached to a carbohydrate part.

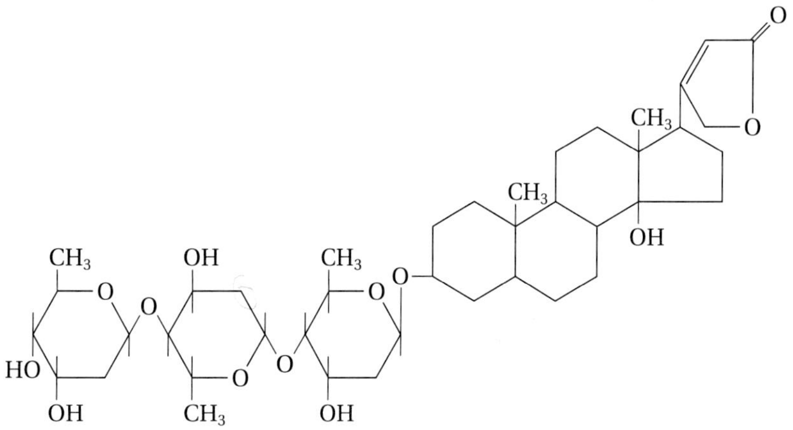

Digitoxin

8.11 Prostaglandins and leukotrienes

AIM: To recognize prostaglandins and leukotrienes and state several of their biological effects.

Focus

Prostaglandins and leukotrienes influence many body functions.

Prostaglandins *are a class of fatty acid derivatives that have a wide range of physiologic activity.* These substances have hormone-like effects. Unlike hormones, however, they are not transported to their site of action in the bloodstream but are synthesized in the same environment in which they act. Prostaglandins are involved in the control of pain and fever, acid secretion into the stomach, relaxation and contraction of smooth muscle, blood pressure, and the wake/sleep cycle. They also stimulate many inflammatory responses, particularly of the joints (rheumatoid arthritis), skin (psoriasis), and eyes. It seems likely that prostaglandins play a role in almost every stage of reproduction.

Since their initial discovery in the 1930s, more than a dozen different prostaglandins have been identified. Slight structural differences account for their distinct biological effects. Prostaglandins are synthesized in the body by oxidation and cyclization of unsaturated continuous-chain fatty acids containing 20 carbons. Arachidonic acid is one such acid.

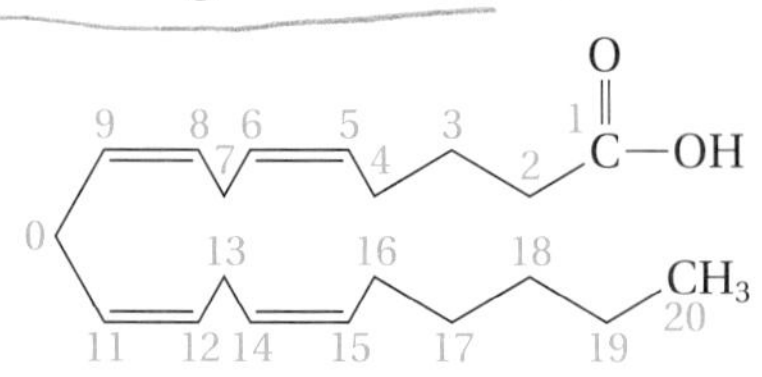

Arachidonic acid

This acid is converted into a prostaglandin structure when the eighth and twelfth carbons fuse to make a cyclopentane ring. Every prostaglandin molecule contains 20 carbons arranged in the basic structure of prostanoic acid.

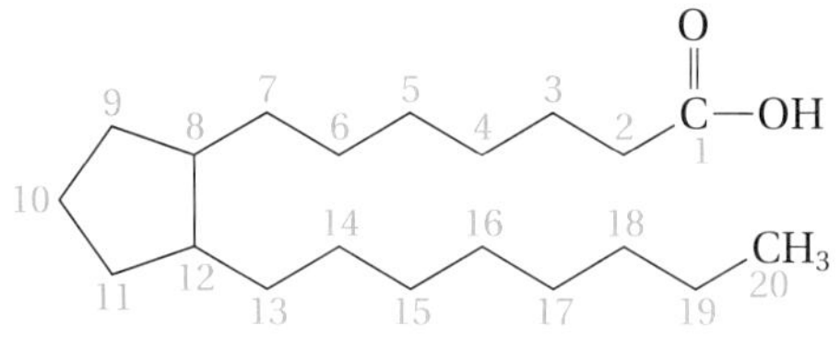

Prostanoic acid

Prostanoic acid is so named because prostaglandins were first isolated from human semen, and scientists thought that they were produced in the prostate gland. It is now known that they are present in minute amounts in nearly all animal tissues and fluids.

Prostaglandin E_2 demonstrates the widespread effects of prostaglandins on a variety of body functions. Prostaglandin E_2 does not itself cause pain, but it enhances the intensity of the pain induced by other pain-producing chemicals in the body. This prostaglandin also induces the signs of inflammation such as swelling, redness, and heat in the inflamed body part. It also lowers arterial blood pressure and promotes blood clotting. Prostaglandin E_2 and prostaglandin $F_{2\alpha}$ have been used extensively as drugs in reproductive medicine, where they can be employed to induce labor and terminate pregnancy.

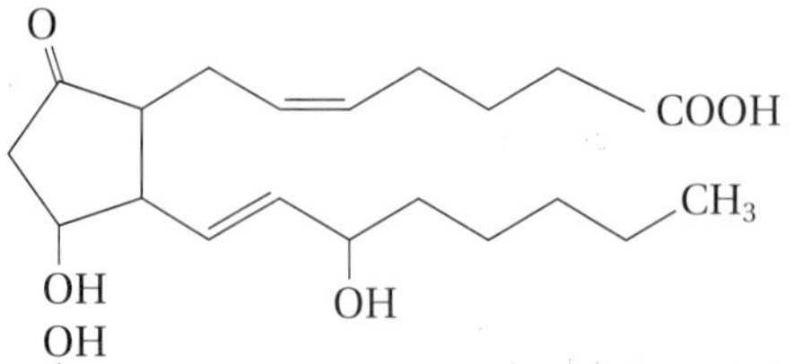

Prostaglandin E_2

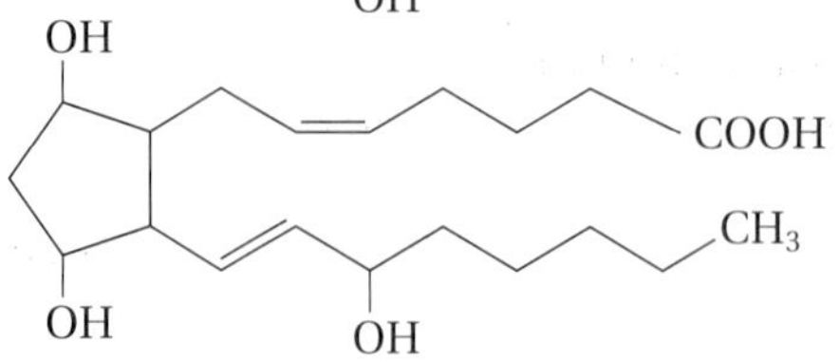

Prostaglandin $F_{2\alpha}$

Aspirin blocks the synthesis of prostaglandins that induce pain and fever. It also has been reported that low doses of aspirin reduce the danger of heart attack and stroke by blocking the synthesis of prostaglandins that promote blood clotting. Aspirin's inhibition of the formation of these prostaglandins accounts for its effectiveness as an analgesic (pain killer), antipyretic (fever killer), and anticlotting agent.

Leukotrienes *are another class of hormone-like substances synthesized from arachidonic acid.* Leukotriene B_4 and leukotriene E_4 are examples of leukotrienes. Unlike the prostaglandins, the leukotrienes contain three carbon-carbon double bonds in a row in their molecular structures.

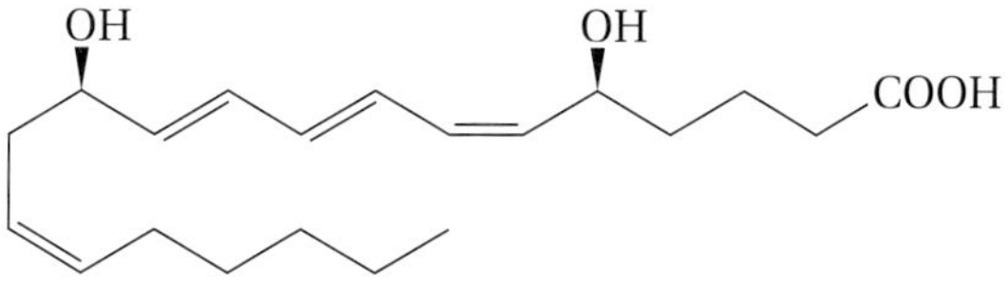

Leukotriene B_4

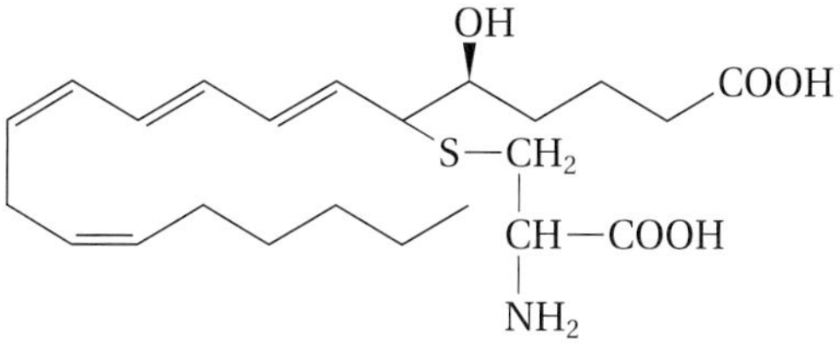

Leukotriene E_4

Certain leukotrienes constitute what is referred to as *the slow-acting substances of anaphylactic shock.* Anaphylactic shock is a drastic allergic response of the body that can be fatal. Severe allergies to foods such as shellfish or peanuts can trigger anaphylactic shock in susceptible people. The symptoms of anaphylactic shock include closing of the throat, blocking the airways. Heart stoppage also can occur, leading to death. Leukotrienes are also implicated in less severe allergic reactions, asthma, inflammations, and heart attacks. Recent studies suggest that dietary fish oils can reduce levels of leukotrienes and reduce the risk of heart attack.

SUMMARY

Lipids are a broad class of naturally occurring, relatively water-insoluble molecules. Triglycerides—triesters of glycerol and fatty acids—are the most abundant lipids in animal tissue.

The membranes of all cells are composed of lipids. The major lipids of cell membranes are the phospholipids, the glycolipids, and in animal cells, cholesterol. Phospholipids have a backbone of glycerol (phosphoglycerides) or of sphingosine (sphingomyelins). Cerebrosides, containing glucose or galactose, with a backbone of sphingosine are common glycolipids. Phospholipid and glycolipid molecules have polar heads and hydrophobic tails. The lipid bilayer is the fundamental structure of liposomes and the membranes of cells and organelles. Its behavior is best described by a fluid mosaic model. Proteins, which may be peripheral or integral, are associated with cell and organelle membranes. Membrane proteins are often glycoproteins.

Steroid molecules have no structural features in common with other lipids. They are often classified as lipids, however, because, like fatty acid esters, they are extractable from plant and animal tissues with organic solvents. Aldehyde and ketone functions as well as hydroxyl groups appear frequently in steroid molecules, which sometimes contain a double bond.

Cholesterol is the most abundant steroid in animals. Many steroids present in lesser amounts in the animal body act as hormones. The adrenal cortex, the outer layer of the adrenal gland, produces three groups of steroid hormones: the glucocorticoids, the mineralocorticoids, and the secondary sex hormones. Glucocorticoids (cortisone, cortisol, corticosterone) regulate the body's use of glucose and affect salt and water balance. Mineralocorticoids also affect the body's salt and water balance; aldosterone is the most important mineralocorticoid. Secondary sex hormones produce secondary sex characteristics of the male and female. They are the androgens in the male and the estrogens in the female. Primary sex hormones produced by the gonads are responsible for the maturation of the reproductive system.

Prostaglandins and leukotrienes, two groups of fatty acid–related molecules, exert many hormone-like effects on the body. Prostaglandins are used for inducing labor and terminating pregnancy and also have been used for controlling blood pressure and reducing inflammation.

KEY TERMS

Cell membrane (8.3)
Complex lipid (8.3)
Corticoid (8.6)
Estrogen (8.8)
Fat (8.2)
Fluid mosaic model (8.4)
Glucocorticoid (8.6)
Glycolipid (8.3)
Hormone (8.6)

Integral protein (8.4)
Lecithin (8.3)
Leukotriene (8.11)
Lipid (8.1)
Lipid bilayer (8.4)
Liposome (8.4)
Mineralocorticoid (8.6)
Mixed triglyceride (8.2)
Oil (8.2)

Peripheral protein (8.4)
Phosphoglyceride (8.3)
Phospholipid (8.3)
Progesterone (8.8)
Prostaglandin (8.11)
Rancid (8.2)
Saponification (8.2)
Sex hormone (8.5)
Simple triglyceride (8.2)

Sphingomyelin (8.3)
Steroid (8.5)
Steroid hormone (8.6)
Sterol (8.5)
Testosterone (8.7)
Triglyceride (8.2)
Wax (8.1)

EXERCISES

Waxes and Simple Lipids (Sections 8.1, 8.2)

8.9 Write the general formula for an ester.

8.10 Name the two classes of organic compounds that are produced when waxes are hydrolyzed.

8.11 (a) Name the following fatty acids, and state whether they are saturated or unsaturated. (b) Which of them are solids and which are liquids at room temperature?

(a) $CH_3(CH_2)_{10}COOH$
(b) $CH_3(CH_2)_4CH{=}CHCH_2CH{=}(CH_2)_7COOH$
(c) $CH_3(CH_2)_{14}COOH$
(d) $CH_3(CH_2)_7CH{=}CH(CH_2)_7COOH$

8.12 Decide whether the following fatty acids are most abundant in animal fat or vegetable oil: (a) stearic acid, (b) oleic acid, and (c) linoleic acid.

8.13 What is a triglyceride?

8.14 Write the structure for glyceryl trioleate, a simple triglyceride. Would you expect this compound to be an oil or a fat? Why?

8.15 (a) What happens, chemically, when an oil is hydrogenated? (b) Why is this process of great commercial importance?

8.16 Why do animal fats and vegetable oils become rancid when exposed to moist warm air?

8.17 Palmolive soap is mostly sodium palmitate. Write the structure of this compound.

8.18 The following compound is hydrolyzed by boiling with sodium hydroxide. What are the saponification products?

$$
\begin{array}{l}
CH_2-O-\overset{\displaystyle O}{\overset{\|}{C}}-(CH_2)_{14}CH_3 \\[1mm]
CH-O-\overset{\displaystyle O}{\overset{\|}{C}}-(CH_2)_{10}CH_3 \\[1mm]
CH_2-O-\overset{\displaystyle O}{\overset{\|}{C}}-(CH_2)_7CH{=}CH(CH_2)_7CH_3
\end{array}
$$

Complex Lipids and Cell Membranes (Sections 8.3, 8.4)

8.19 Name the two categories of complex lipids.

8.20 Write a general structure for the following complex lipids: (a) a phosphoglyceride and (b) a sphingomyelin.

8.21 What is the difference, structurally, between sphingomyelins and phosphoglycerides?

8.22 Draw structural formulas for the products of complete hydrolysis of phosphatidyl choline.

8.23 Use diagrams to show the difference between a liposome and a micelle.

8.24 What is the role of phospholipids in cell structure?

8.25 Explain how the degree of unsaturation in membrane lipids affects the flexibility of cell membranes.

8.26 Describe the major features of the fluid mosaic model of membrane structure.

8.27 What are the functions of membrane proteins?

8.28 Describe the two types of membrane proteins.

Steroids (Sections 8.5–8.10) *may questions.*

8.29 Draw the fundamental chemical structure that applies to all steroid molecules.

8.30 What steroid forms a large part of animal cell membranes but is not found in plants?

8.31 What is a hormone, and where are steroid hormones produced?

8.32 Cortisone is a hormone. Where is it produced in the body, and what is its biological function?

8.33 Explain why a physician might prescribe cortisone shots before a patient has a kidney transplant operation.

8.34 What is (a) an androgen and (b) an estrogen?

8.35 Name (a) the primary male sex hormone and (b) the three primary female sex hormones.

8.36 Compare the structural formulas of testosterone and progesterone. How are these two molecules different?

8.37 Comment on the use and action of anabolic steroids.

8.38 What are the functions of the estrogens and progesterone in the female ovarian cycle?

8.39 Why is progesterone ineffective as an oral contraceptive?

8.40 What is the action of each of the following as an oral contraceptive?
(a) the pill   (b) the minipill   (c) DES

8.41 How does the structure of digitoxin (the major component of the drug digitalis) differ from that of all the other steroids discussed?

8.42 What is the medicinal use of digitalis?

Prostaglandins and Leukotrienes (Sections 8.10, 8.11)

8.43 Draw the basic structure for a prostaglandin molecule.

8.44 How are prostaglandins synthesized in the body?

8.45 Prostaglandins are among the most potent biological compounds known. What are some of their physiologic effects?

Additional Exercises

8.46 Match the following.
(a) phosphoglyceride   (1) long-chain esters
(b) lipid bilayer 3   (2) absent in plants
(c) testosterone   (3) fluid mosaic model
(d) lecithin   (4) body's salt balance
(e) saponification   (5) congestive heart failure
(f) fats   (6) phosphate ester
(g) wax   (7) fat hydrolysis
(h) digitalis 5   (8) highly saturated
(i) cholesterol 2   (9) male sex hormone
(j) aldosterone   (10) phosphatidyl choline

8.47 Which type of lipids are found in cell membranes?

8.48 Explain the relationship between the nature of the fatty acids of complex lipids (that is, whether they are saturated or unsaturated) and the fluidity of cell membranes.

8.49 Name all the functional groups in the following compounds:
(a) cortisone
(b) phosphatidyl choline
(c) testosterone
(d) diethylstilbestrol
(e) prostaglandin E_2
(f) sphingosine
(g) digitoxin
(h) choline
(i) triolein.

8.50 Distinguish between a fat and an oil.

8.51 Explain why saturated fatty acids have higher melting points than unsaturated fatty acids.

SELF-TEST (REVIEW)

True/False

1. A triglyceride contains three ester bonds. T
2. Integral proteins of membranes help move small molecules through membranes. T
3. The major component of most membranes is carbohydrate. F
4. Cholesterol is a sterol found in virtually all plants and animals. F
5. Increasing the percentage of unsaturated fatty acids in a membrane increases its flexibility. T
6. Because the lipid molecules in one layer of a membrane can move about freely, membranes are very permeable. F
7. The sex hormone testosterone is produced only by males. F
8. Triglycerides containing only saturated fatty acids are often spoiled by oxidative rancidity. F
9. In a liposome, the hydrophobic hydrocarbon tails form the juncture of the lipid bilayer. T
10. Aldosterone is a mineralocorticoid hormone that helps maintain the body's salt and water balance. T

Multiple Choice

11. The polar end of a lipid molecule
    (a) is characterized by long hydrocarbon chains.
    (b) is hydrophobic.
    (c) interacts favorably with water.
    (d) Both (a) and (b) are correct.

12. The chemical structure common to all steroid molecules is

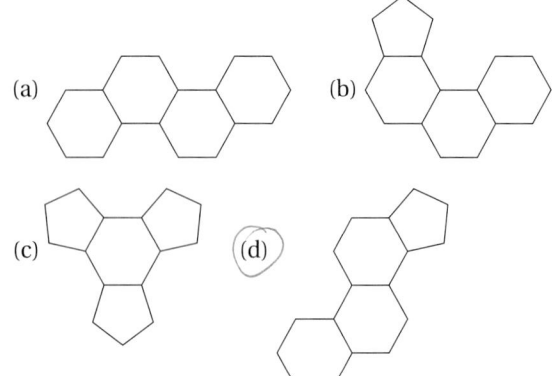

13. Hydrolysis of a mixed triglyceride may
    (a) give three different alcohols.
    (b) give ethylene glycol.
    (c) give three different fatty acids.
    (d) give phosphoric acid.

14. A lipid molecule composed of glycerol, phosphoric acid, and two fatty acids is called a
    (a) glycolipid.   (b) phosphoglyceride.
    (c) cholesterol.   (d) sphingomyelin.

15. Prostaglandins are involved in all the following regulatory functions *except*
    (a) timing of menstrual cycle.
    (b) body temperature regulation.
    (c) muscle contraction.
    (d) stomach acid secretion.

16. Which of the following statements is *not* true about a sphingomyelin?
    (a) It has an amide bond.
    (b) It has a phosphate ester bond.
    (c) It has a glycerol backbone.
    (d) Choline is found at the polar head of the molecule.

17. Sources of hormones in the body include
    (a) the adrenal glands.   (b) the ovaries.
    (c) the testes.   (d) all of the above.

18. The adrenal corticoids called *glucocorticoids*
    (a) produce the body's sex characteristics.
    (b) are used to reduce the risks of organ transplant operations.
    (c) include the primary sex hormones.
    (d) all contain at least one carbohydrate molecule.

19. Liquid oils are changed into solid fats by
    (a) hydrogenation.   (b) solidification.
    (c) hydration.   (d) rancidization.

20. Waxes are formed from long-chain
    (a) acids and amines.   (b) amines and alcohols.
    (c) alcohols and aldehydes.   (d) acids and alcohols.

21. Which of the following is *not* primarily a female sex hormone?
    (a) progesterone   (b) estradiol   (c) mestranol
    (d) estrone

22. In hydrolytic rancidity of triglycerides, which part of the molecule is attacked?
    (a) double bond   (b) alcohol group
    (c) acid group   (d) ester group

23. Which of the following drugs is an important steroid produced by a plant?
    (a) androgen   (b) digitalis   (c) progestin
    (d) estrogen

24. Alkaline hydrolysis of a fat to make soap is called
    (a) glycolysis.   (b) saponification.
    (c) hydration.   (d) anabolication.

25. The "power stations" of a eukaryotic cell are the
    (a) nucleolus.   (b) Golgi bodies.
    (c) mitochondria.   (d) peroxisomes.

26. Lysosomes function in
    (a) breaking down complex molecules.
    (b) cellular reproduction.
    (c) transporting proteins across cell membranes.
    (d) protein synthesis.

Amino Acids, Peptides, and Proteins

9

Molecular Structures and Biological Roles

Proteins are the source of amino acids. Egg whites are rich in protein.

This chapter begins with a discussion of the structures and chemistry of the most common naturally occurring amino acids. Then we will see how combinations of the natural amino acids form large molecules called *peptides* and *proteins* (see chapter opener figure). Along the way we will encounter a few important peptides and proteins that highlight the relationship between molecular structure and biological function. The Case in Point for this chapter involves a class of molecules called the *enkephalins,* which promote feelings of comfort and sometimes exhilaration.

CASE IN POINT: Runner's high

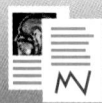 Karla, a college student, does not consider herself to be athletic. A few months ago, however, she decided to try running for exercise. Her first few times out she was able to run for only a few minutes. Gradually, she began to cover more and more distance, until she could run about 5 miles in 35 minutes, three times a week. In addition to her increased endurance, Karla has lost a few pounds and feels mentally more alert. To her surprise, Karla has become such a dedicated runner that she becomes edgy if her schedule forces her to miss a running session. Part of Karla's attraction to running is that during her runs she sometimes experiences a feeling that can only be described as exhilaration. Karla's exhilaration is "runner's high." What causes Karla's runner's high? We will find out in the discussion of certain peptide hormones in Section 9.4.

Physical exercise can be exhilarating.

9.1 The amino acids *are L* [handwritten]

AIM: To classify the 20 common amino acids according to their side-chain structures.

Focus

There are 20 common natural amino acids.

An **amino acid** *is any carboxylic acid (RCO_2H) that contains an amino group ($—NH_2$) in the same molecule.* To most chemists and biochemists, however, the term is usually reserved for the 20 amino acids that are found and used in living organisms. All 20 amino acids common in nature have a skeleton consisting of a carboxylic acid group and an amino group covalently bonded to a central atom. *The central atom of amino acids is called the* **alpha carbon.** The remaining two groups on the alpha carbon are hydrogen and an *R group—the* **amino acid side chain.**

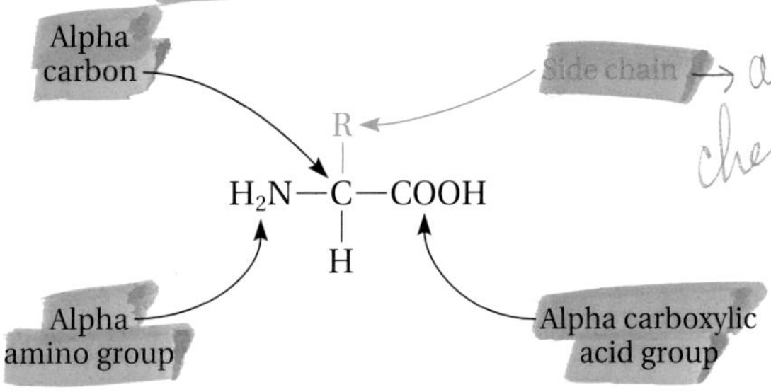

Alpha carbon

side chain → *accounts for dif chemical nature properties* [handwritten]

$$H_2N—\underset{\underset{H}{|}}{\overset{\overset{R}{|}}{C}}—COOH$$

Alpha amino group

Alpha carboxylic acid group

The chemical nature of the side chains accounts for differences in amino acid properties. The 20 common naturally occurring alpha amino acids are divided into seven categories according to their side-chain structures.

Aliphatic side chains   This group consists of amino acids having hydrocarbon (aliphatic) side chains. Glycine, the simplest amino acid, has a hydrogen rather than an aliphatic side chain, but it is still placed in this category. Alanine, with a methyl side chain, is the smallest true member of this group. Other aliphatic R groups are those of the amino acids valine (isopropyl), leucine (isobutyl), and isoleucine (*sec*-butyl). Leucine and isoleucine have the same molecular formulas and are structural isomers. Proline is the only amino acid with its alpha amino group incorporated into a ring. The amino group of proline is a secondary amine.

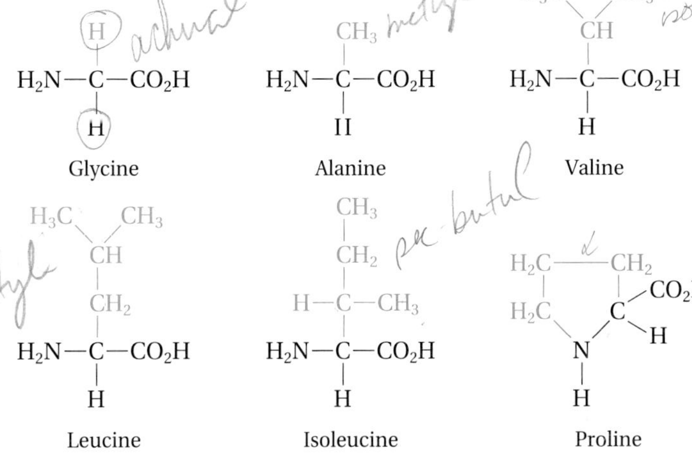

Glycine
glycos (Greek): sweet
(The amino acid glycine has a sweet taste.)

Glycine

Alanine

Valine

Leucine

Isoleucine

Proline

Hydroxylic side chains The amino acids in this group are serine and threonine. Both have aliphatic side chains containing hydroxy functional groups.

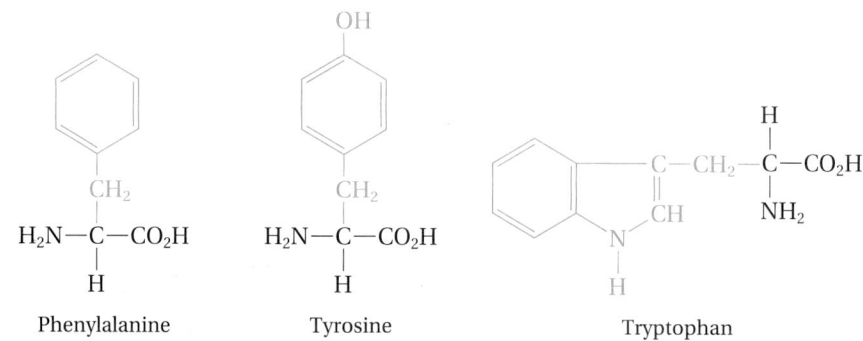

Serine   Threonine

Eight of the amino acids, isoleucine, leucine, lysine, methionine, phenylalanine, threonine, tryptophan, and valine, are essential amino acids. These amino acids cannot be synthesized by the body but are essential to its proper functioning. They must be supplied in our diet.

Aromatic side chains There are three amino acids that have aromatic rings in their side chains. Phenylalanine is an amino acid in which one of the methyl hydrogens of alanine is replaced by a phenyl group. In tyrosine, the phenyl ring of phenylalanine is substituted in the *para* position with a phenolic hydroxyl group. The aromatic ring system of tryptophan is derived from indole (see Sec. 6.1).

Phenylalanine   Tyrosine   Tryptophan

Acidic side chains Aspartic and glutamic acids have side chains that are terminated by carboxylic acid groups. At the usual biological pH, slightly above pH 7, these carboxylic acid groups are ionized. For this reason, aspartic acid and glutamic acid are often referred to as their carboxylate ions, aspartate and glutamate.

Aspartic acid   Glutamic acid

Amide side chains Asparagine and glutamine are the amides of aspartic and glutamic acids, respectively. Their side chains are electrically neutral at pH 7.0.

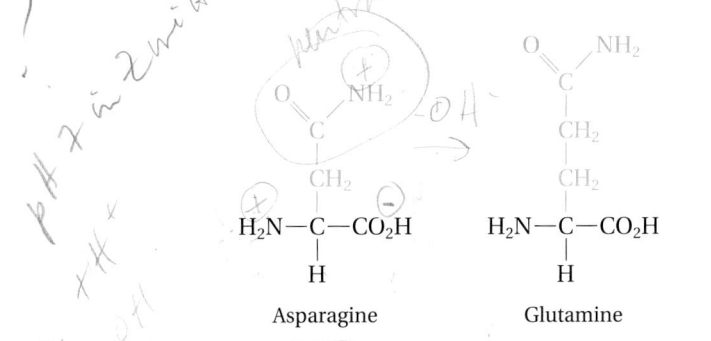

Asparagine   Glutamine

Basic side chains   This group consists of three amino acids containing weakly basic nitrogens. The nitrogens on the side chains of lysine and arginine are such sufficiently strong bases that they remove a proton from water at neutral pH. At pH 7.0, therefore, the side chains of arginine and lysine are positively charged. The nitrogens of the side chains of histidine are more weakly basic than those of lysine and arginine. In solution at pH 7.0, the side chains of only about half of the histidine molecules are positively charged at any moment.

Lysine   Arginine   Histidine

Sulfur-containing side chains   Methionine and cysteine are two common sulfur-containing amino acids. Cysteine is often found linked to another cysteine through formation of a disulfide bond (—S—S—) to form the amino acid *cystine*. Cystine is not considered one of the 20 common amino acids, since it is formed from two molecules of cysteine.

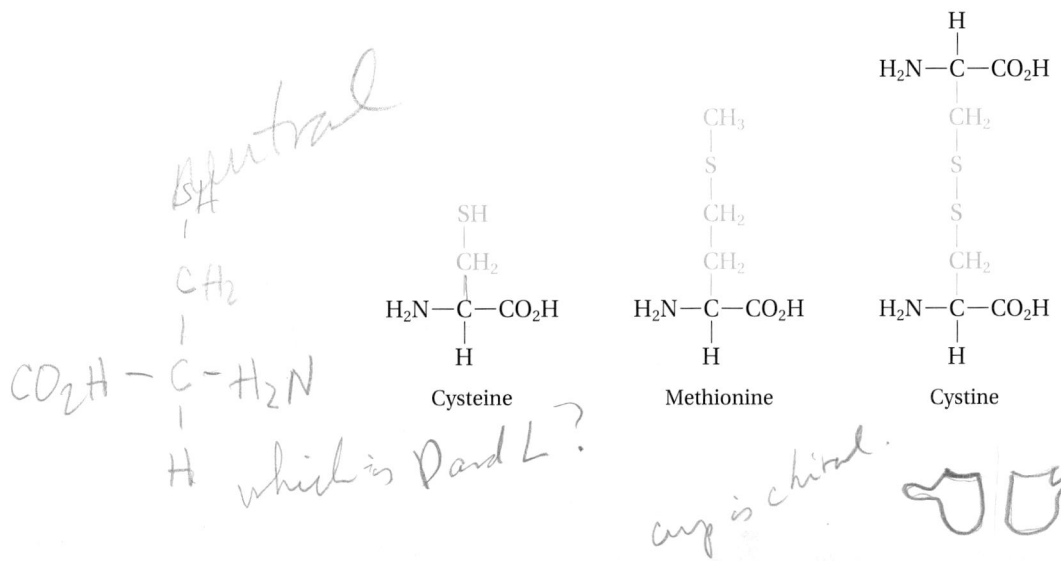

Cysteine   Methionine   Cystine

PRACTICE EXERCISE 9.1

Categorize the following amino acids according to their side-chain groups:

(a) leucine   (b) lysine   (c) serine   (d) tyrosine

9.2 Stereoisomers of amino acids

AIMS: To state the handedness of amino acids found in nature. To draw Fischer projection formulas for alpha amino acids.

Focus

All the amino acids but glycine have handedness.

A glance at the structures of common amino acids will show that, except for glycine, there are four different groups attached to the alpha carbons. This is a criterion for handedness in organic molecules (see Sec. 7.1). Amino acids fit this rule and may exist as D or L forms that are nonsuperimposable mirror images of each other. The handedness of amino acids is related to that of D- and L-glyceraldehyde as follows:

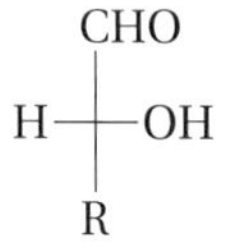

CHO
H——OH
R
D-Glyceraldehyde

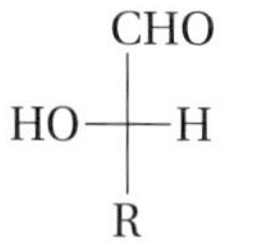

CHO
HO——H
R
L-Glyceraldehyde

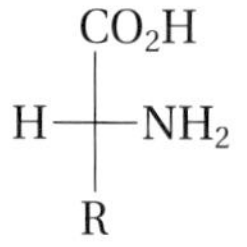

CO_2H
H——NH_2
R
D-Amino acid

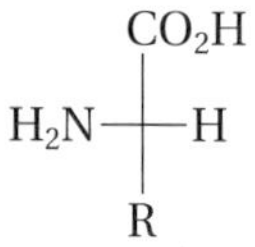

CO_2H
H_2N——H
R
L-Amino acid

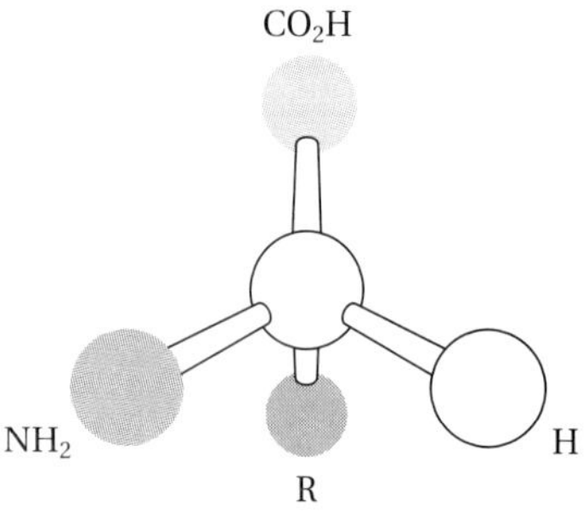

CO_2H

NH_2 R   H

Configuration of L-amino acids

Although the amino acids found in nature are mostly the L forms, the D forms of a few amino acids are found in the cell walls of certain bacteria.

Nature shows a marked preference for the L forms of amino acids. With few exceptions, the amino acids found in living systems belong to the L family.

EXAMPLE 9.1   **Drawing Fischer projection formulas**

Draw the Fischer projection formula of L-alanine.

SOLUTION

Place the carboxylic acid group at the top and the methyl group (the R group) at the bottom. For the L isomer, the amino group is on the left.

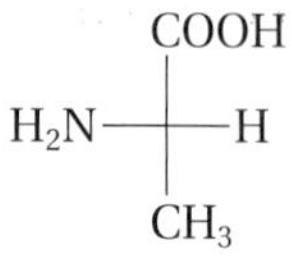

COOH
H_2N——H
CH_3

[Handwritten margin notes, top left:]

Fischer projections drawn by hand showing L-cysteine and D-cysteine structures with NH_2-C-H / $H-C\ NH_2$, CH_2, SH groups, labeled (L-cysteine) and D-cysteine.

[Handwritten top right:] SH / CH_2 / $NH_2-C-C-OH$ ...

PRACTICE EXERCISE 9.2

Give (a) the Fischer projection formula for L-cysteine and (b) the stereorepresentation of D-cysteine. (c) Which isomer predominates in nature?

9.3 Zwitterions

AIM: To describe the formation of zwitterions and their effect on the properties of amino acids.

Focus

Amino acids form internal salts called *zwitterions*.

Consider the proximity of the carboxylic acid and amino groups in the glycine molecule. The weakly acidic proton of the carboxylic acid group easily transfers to the weakly basic amino group to form an internal salt.

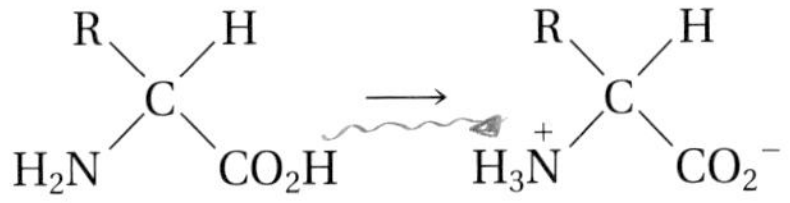

Amino acid → Internal salt (zwitterion)

[Handwritten:] carboxylic acid is a weak acid; amino group is a weak base

Amino acids are really ionic compounds. And as you may recall, ionic compounds generally have much higher melting points than molecular compounds. Glycine is a solid with a melting point of 233 °C—much greater than the temperature required to boil water. To see how the internal salt formation affects such properties as melting point, we can compare the melting point of glycine with those of structurally similar amines and carboxylic acids. Glycine is a two-carbon compound with amino and carboxylic acid groups, so we might expect glycine to be a liquid with a melting point between those of ethylamine ($CH_3CH_2NH_2$), −80.6 °C, and acetic acid (CH_3CO_2H), 16.6 °C. The much higher melting point of glycine is the result of internal salt formation.

$H_2NCH_2CH_3$ CH_3CO_2H $H_2NCH_2CO_2H$

Ethylamine   Acetic acid   Glycine
(mp −80.6 °C)   (mp 16.6 °C)   (mp 233 °C)

[Handwritten:] because ionic compound

Zwitterion
zwitter (German):   hybrid

The internal salts of such compounds as amino acids are called **zwitterions.** In the pure solid state and in aqueous solution near neutral pH, the amino acids exist almost completely as zwitterions.

[Handwritten:] at pH 6 basic R units as cation; at pH 6-7 acidic R units anion

Isoelectric points

The isoelectric point of the milk protein casein is 4.6. Casein is least soluble and will precipitate at this acidic pH. This acidic condition is met in the cheese-making process by the action of certain bacteria in producing lactic acid.

In zwitterions of amino acids with uncharged side chains, the positive and negative charges cancel one another. There is a net zero charge on the molecule. *Any amino acid in which the positive and negative charges are balanced is at its* **isoelectric point.** *For amino acids in aqueous solution, the pH at which this balancing of charges occurs is the* **isoelectric pH.** The isoelectric point of amino acids with uncharged side chains occurs near pH 7 in aqueous solution. An amino acid tends to be least soluble at its isoelectric pH because of the net zero charge. At lower pH values (where more protons

[Handwritten:] would be: sulfur containing.

are available), the solubility of the amino acid increases because the carboxylate ion of the zwitterion picks up a proton from the solution; the amino acid acquires a net positive charge.

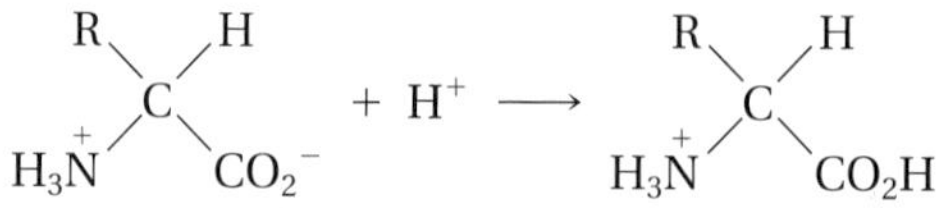

The formation of zwitterions is not limited to amino acids. For example, *p*-aminosulfonic acid, the parent acid of the antibiotic sulfanilamide, exists mainly in the zwitterion form.

$$\underset{\text{Zero net charge}}{\underset{H_3\overset{+}{N}}{\overset{R}{\diagdown}}\underset{CO_2^-}{\overset{H}{\diagup}}C} + H^+ \longrightarrow \underset{\text{Positive charge}}{\underset{H_3\overset{+}{N}}{\overset{R}{\diagdown}}\underset{CO_2H}{\overset{H}{\diagup}}C}$$

(handwritten: pH 2-3 cation)

At higher pH values, solubility of the amino acid also increases because a proton is removed from the positively charged ammonium ion of the zwitterion and the net charge of the amino acid is negative.

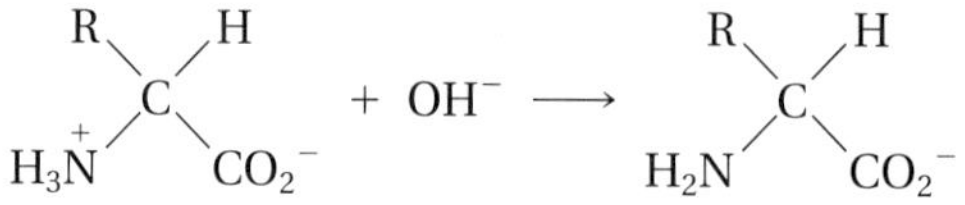

$$\underset{\text{Zero net charge}}{\underset{H_3\overset{+}{N}}{\overset{R}{\diagdown}}\underset{CO_2^-}{\overset{H}{\diagup}}C} + OH^- \longrightarrow \underset{\text{Negative charge}}{\underset{H_2N}{\overset{R}{\diagdown}}\underset{CO_2^-}{\overset{H}{\diagup}}C}$$

(handwritten: pH 10)

EXAMPLE 9.2   Drawing formulas of zwitterions

Draw the structural formula for the amino acid alanine at (a) pH 2 and (b) pH 11.

SOLUTION

(a) At pH 2 the amino group and the carboxylate ion will be protonated.

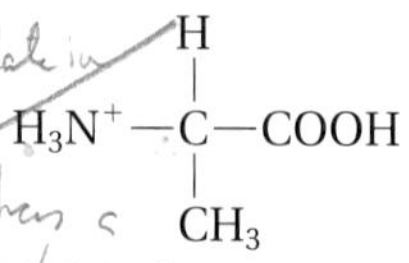

$$H_3N^+-\overset{\overset{H}{|}}{\underset{\underset{CH_3}{|}}{C}}-COOH$$

At this pH the amino acid has a net positive charge.

(b) At pH 11 the functional groups are the unprotonated amino group and the carboxylate ion.

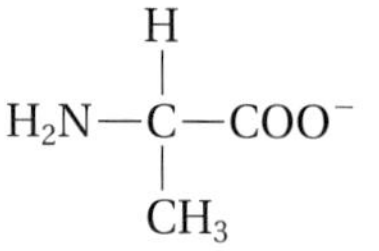

$$H_2N-\overset{\overset{H}{|}}{\underset{\underset{CH_3}{|}}{C}}-COO^-$$

At this pH the amino acid has a net negative charge.

PRACTICE EXERCISE 9.3

Consider the amino acids glycine and aspartic acid. Draw structural formulas for the species that predominate in aqueous solution at

(a) pH 2   (b) pH 10

9.4 Peptides

AIMS: To name and describe the bond that links amino acids together. To draw complete structural formulas for simple peptides. To contrast the biological functions of some peptide hormones.

Focus

Amino acids are linked to form peptides.

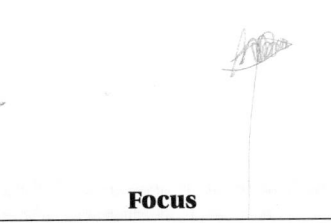

$NH_2-C-C-OH$
H
amino acid.

A **peptide** *is any combination of amino acids in which the alpha amino group (—NH$_2$) of one acid is united with the alpha carboxylic group (—CO$_2$H) of another through an amide bond.*

$$H_2N-C-C-OH + H-N-C-C-OH \rightarrow$$

Amino acid   Amino acid

peptide bond.

$$H_2N-C-C-N-C-C-OH + H_2O$$

Peptide

The amide bonds formed in peptides always involve the alpha amino and alpha carboxylic acid groups and never those of side chains. More amino acids may be added in the same fashion to form chains such as those in Figure 9.1. *The amide bond between the carbonyl group of one amino acid and the nitrogen of the next amino acid in the peptide chain is called a* **peptide bond** *or peptide link. Amino acids that have been incorporated into peptides are called* **amino acid residues.** As more amino acid residues are added, a backbone common to all peptide molecules is formed. The amino acid residue with a free amino group at one end of the chain is the *N-terminal residue;* the residue with a free carboxylic acid at the other end of the chain is the *C-terminal residue.* The number of amino acid residues in a peptide is often indicated by a set of prefixes for peptides of up to 10

Figure 9.1
Parts of a peptide—in this case, a tetrapeptide. The peptide bonds of the zigzag backbone are shown in color. Note that the C-terminal is at the right and the N-terminal is at the left.

Amino acid residues

1   2   3   4

N-terminal residue   C-terminal residue

H_2N ... COOH

Peptide bonds

alanyl-valyl-alanyl-glycine
Ala-Val-Ala-Gly

Table 9.1 Prefixes for Short Peptides

Residues	Prefix
2	*di-*
3	*tri-*
4	*tetra-*
5	*penta-*
6	*hexa-*
7	*hepta-*
8	*octa-*
9	*nona-*
10	*deca-*

Table 9.2 Three-Letter Abbreviations for Amino Acids

Amino acid	Abbreviation
alanine	Ala
arginine	Arg
asparagine	Asn
aspartic acid	Asp
cysteine	Cys
glutamine	Gln
glutamic acid	Glu
glycine	Gly
histidine	His
isoleucine	Ile
leucine	Leu
lysine	Lys
methionine	Met
phenylalanine	Phe
proline	Pro
serine	Ser
threonine	Thr
tryptophan	Trp
tyrosine	Tyr
valine	Val

residues, as shown in Table 9.1. *We call any peptide with more than 20 amino acid residues a* **polypeptide.** In theory, the process of adding amino acids to a peptide chain may be continued indefinitely. Names of peptides are derived from names of amino acid residues. By convention, names of peptides are always written from left to right starting with the N-terminal end; a peptide that contains N-terminal glycine, followed by a histidine, followed by C-terminal phenylalanine is named *glycyl-histidyl-phenylalanine*. The sequence is extremely important; glycyl-histidyl-phenylalanine is a different molecule from phenylalanyl-histidyl-glycine. The methyl ester of the dipeptide aspartyl phenylalanine is an artificial sweetener (see A Closer Look: Aspartame).

Structural formulas or even full word names for large peptides become very unwieldy and time-consuming to write. To simplify matters, chemists write peptide and protein structures by using three-letter abbreviations for the amino acid residues with dashes to show peptide bonds. Table 9.2 lists the abbreviations. Peptide structures written with these abbreviations always start with the N-terminal group to the left and end with the C-terminal group to the right. For example, we can write glycyl-histidyl-phenylalanine as Gly-His-Phe and phenylalanyl-histidyl-glycine as Phe-His-Gly.

EXAMPLE 9.3   Naming a peptide

Write the (a) the three-letter amino acid name and (b) the name using full abbreviations for the peptide in Figure 9.1 assuming that R_1 is methyl, R_2 is isopropyl, R_3 is methyl, and R_4 is hydrogen.

SOLUTION

(a) The full name, starting at the N-terminal end, is alanyl-valyl-alanyl-glycine.
(b) Using abbreviations: Ala-Val-Ala-Gly.

PRACTICE EXERCISE 9.4

Glutathione, a tripeptide that is widely distributed in all living tissues, is named glutamyl-cysteinyl-glycine. (a) Draw the complete structural formula for this peptide. (b) Write the amino acid sequence of the peptide using the three-letter abbreviations.

PRACTICE EXERCISE 9.5

Write three-letter amino acid sequences for all possible tripeptide structures that contain one residue each of glutamic acid, cysteine, and glycine.

Chemists have isolated over 200 peptides important to the smooth functioning of the human body. The peptide hormones oxytocin and vasopressin are two examples showing that apparently minor differences in the order of amino acid residues can result in profoundly different biological actions.

A Closer Look

Aspartame

Aspartame, an artificial sweetener with the brand name NutraSweet®, is the methyl ester of the dipeptide aspartylphenylalanine.

phenylalanine O *aspartic acid residue*

$$\text{[benzene ring]}-CH_2-CH-NH-\overset{\displaystyle O}{\overset{\|}{C}}-CH-NH_2$$
$$\qquad\qquad\quad | \qquad\qquad\qquad |$$
$$\qquad\qquad CO_2CH_3 \qquad\qquad CH_2CO_2H$$

Aspartylphenylalanine methyl ester (aspartame)

name seems to be backwards

Aspartame is a substitute for sucrose and saccharin.

Aspartame is more than 50 times sweeter than sucrose. Now approved for use in more than 30 countries, this product of amino acid chemistry has found wide acceptance in the food industry as a substitute for both sucrose and saccharin (see figure). The main advantage of aspartame over saccharin is its taste, which is very similar to that of cane sugar. Its chief disadvantage is its instability. Aspartame is not recommended for the preparation of foods in which cooking temperatures exceed 150 °C. High temperatures and extremes of pH can cause aspartame in solution to hydrolyze to the unesterified dipeptide, aspartylphenylalanine and methanol, with a simultaneous loss of sweetness. Methanol is a toxic alcohol, and there is disagreement among scientists whether the breakdown of the small amounts of aspartame used for sweetening produces sufficient amounts of methanol to cause harm. Food scientists have found that aspartame's stability is improved by using it in combination with saccharin. This combination is now used in a number of soft drinks. Products that contain aspartame, however, must carry a warning label for people who suffer from the hereditary disease phenylketonuria. People who have phenylketonuria are unable to break down phenylalanine and must therefore limit its intake, as we will see in Section 17.8.

Oxytocin and vasopressin are formed in the hypothalamus (pituitary gland) and enter the bloodstream. Each hormone is a nonapeptide (contains nine amino acid residues) with six of the amino acid residues drawn into a loop by a disulfide bond. The disulfide bond is formed by the coupling of cysteine residues in the first and sixth positions of their peptide chains, as shown in Figure 9.2. *In peptides and proteins, disulfide bonds formed between two cysteine —SH groups that draw a single peptide chain into a loop or hold two peptide chains together are called* **disulfide bridges.**

Figure 9.2
Oxytocin and vasopressin. These peptide hormones differ by two amino acid residues (shown in color). The C-terminal residues have amide functional groups rather than carboxylic acid groups.

```
      2      1                          2      1
    Tyr — Cys                         Tyr — Cys
     |      |                          |      |
   3 Ile    S                        3 Phe    S
     |      |                          |      |
   4 Gln    S                        4 Gln    S
     |      |                          |      |
    Asn — Cys — Pro — Leu — GlyNH2    Asn — Cys — Pro — Arg — GlyNH2
     5     6     7     8      9        5     6     7     8      9
          Oxytocin                          Vasopressin
```

peptide hormones

Although the amino acid composition differs at only the third and eighth positions of their peptide chains (counting from the N-terminal end), the biological roles of these two peptides are different. Oxytocin stimulates milk ejection in females and contraction of the smooth muscle of the uterus in labor. Oxytocin has been called the "cuddle drug," because in females it stimulates sensations during lovemaking and produces feelings of relaxed satisfaction and attachment. Vasopressin is an **antidiuretic**—*it helps to maintain a proper water balance in both sexes by helping to retain water.* Defective production of vasopressin results in diabetes insipidus, characterized by the production of massive volumes of urine. Injections of the hormone control the volume of urine produced.

Another example of how the sequence of amino acid residues affects biological function can be seen when we compare the blood pressure–controlling activities of two peptides, bradykinin and boguskinin. Bradykinin is a nonapeptide formed directly in the bloodstream when a fragment is chopped from a large protein, α-2-globulin. Boguskinin is a synthetic octapeptide that lacks only the proline residue at position 7 of bradykinin.

$$\underset{1}{\text{Arg}}-\underset{2}{\text{Pro}}-\underset{3}{\text{Pro}}-\underset{4}{\text{Gly}}-\underset{5}{\text{Phe}}-\underset{6}{\text{Ser}}-\underset{7}{\text{Pro}}-\underset{8}{\text{Phe}}-\underset{9}{\text{Arg}}$$

Bradykinin

$$\underset{1}{\text{Arg}}-\underset{2}{\text{Pro}}-\underset{3}{\text{Pro}}-\underset{4}{\text{Gly}}-\underset{5}{\text{Phe}}-\underset{6}{\text{Ser}}-\underset{7}{\text{Phe}}-\underset{8}{\text{Arg}}$$

Boguskinin

Bradykinin is partially responsible for triggering pain, welt formation (as in scratches), movement of smooth muscle, and lowering of blood pressure. Blood pressure is lowered when, in response to a signal, bradykinin and related peptides relax muscles of blood vessel walls. Blood vessels dilate, or expand, and blood flows into the expanded volume, lowering blood pressure. Less than 1 μg of bradykinin lowers blood pressure in an average-sized adult. Boguskinin, on the other hand, is completely inactive—hence the name *bogus,* meaning "false."

Parts of the brain contain **enkephalins**—*peptides involved with feelings of emotion and sensation of pain.* Two major enkephalins are methionine enkephalin and leucine enkephalin, which differ in structure by only one amino acid residue:

Tyr—Gly—Gly—Phe—Met   Tyr—Gly—Gly—Phe—Leu

Methionine enkephalin   Leucine enkephalin

These two pentapeptides are messengers in brain processes associated with emotional euphoria and relief of pain—the same processes affected by morphine, heroin, and other opiate drugs. Researchers hope that administration of these peptides, or similar synthetics, will bring relief to people with chronic pain without the danger of addiction. Karla, the jogger in the Case in Point earlier in this chapter, experiences the effects of these peptide hormones during her workouts.

Karla sometimes experiences runner's high during her runs. What is causing this effect? Many sports physiologists believe that runner's high comes from increases in the levels of enkephalins in the brain. These increases, which can cause remarkable feelings of well-being, appear to be stimulated by endurance exercises such as distance running. The effects of enkephalins appear to last beyond the period of exercise to increase all-around mental alertness and a sense of comfort. Exercise is a good thing, but like most good things, it can be overdone, since the body may suffer from exhaustion and other breakdowns when overtaxed without sufficient time to recover. Enkephalins may be responsible in part for the potentially harmful addiction of some people to excessive exercise.

9.5 Primary structure of proteins

AIM: To characterize the primary structure of a protein.

Focus

The primary structure is the first level of organization of a protein.

Protein
proteios (Greek):   of first
importance

When the number of amino acid residues becomes greater than about 40, a naturally occurring peptide is called a **protein.** On average, a peptide molecule containing 100 amino acid residues has a molar mass of about 10,000 g. Proteins are so vital to living organisms that many of the remaining topics in this book concern them. The structures of proteins are usually studied at four levels of organization: *primary structure, secondary structure, tertiary structure,* and *quaternary structure. The* **primary structure** *of a peptide or protein is the order in which the amino acid residues of a peptide or protein molecule are linked by peptide bonds.* The primary structure of a protein molecule also includes any disulfide bridges that the molecule contains. Just as we saw for peptides in the preceding section, differences in the chemical and biological properties of proteins result from differences in the order of amino acids in the polypeptide chain (primary structure). The secondary, tertiary, and quaternary structures of proteins are discussed in the following three sections.

9.6 Secondary structure of proteins

AIMS: To describe the bonding and structure in the following secondary structures of proteins: alpha helix, beta-pleated sheet, and collagen helix. To distinguish among the following fibrous proteins: alpha keratin, beta keratin, and collagen.

Focus

The alpha helix, beta-pleated sheet, and collagen helix are three types of secondary structures found in proteins.

The polypeptide chains of many proteins contain specific, repeating patterns of folding of the peptide backbone. *These specific, repeating patterns of folding of the peptide backbone constitute the protein's* **secondary structure.** Three commonly found patterns are the *alpha helix,* the *beta-pleated sheet,* and the *collagen helix.*

Figure 9.3
A corkscrew must be turned in a right-handed, or clockwise, direction to penetrate a cork.

Alpha helix

In some proteins, regions of the backbone of the peptide chain are coiled into a spiral shape called an **alpha helix,** *similar to a corkscrew.* As Figure 9.3 shows, a corkscrew must be turned in a right-handed, or clockwise, direction to penetrate a cork. The alpha helixes of proteins are always right-handed. The helixes are held together by hydrogen bonds, shown in Figure 9.4, formed between the hydrogen of an N—H of a peptide bond and the carbonyl oxygen of another peptide bond group four residues away in the same peptide chain.

The tightness of coiling is such that 3.6 amino acid residues of the peptide backbone make each full turn of the alpha helix. There are no amino acid residue side chains inside the alpha helix; they are located on the outside. The cyclic amino acid proline does not fit well into the peptide backbone of alpha helixes. Alpha helixes in long protein chains often end at a place where proline residues occur in the primary structure. Proline is

Figure 9.4
A peptide chain twisted into a right-handed alpha helix constitutes a protein's secondary structure. The N-terminal to C-terminal direction is from top to bottom. The dotted lines show the hydrogen bonds between the carbonyl oxygen of one amino acid residue and the N—H hydrogen of another, four amino acid residues further down the chain.

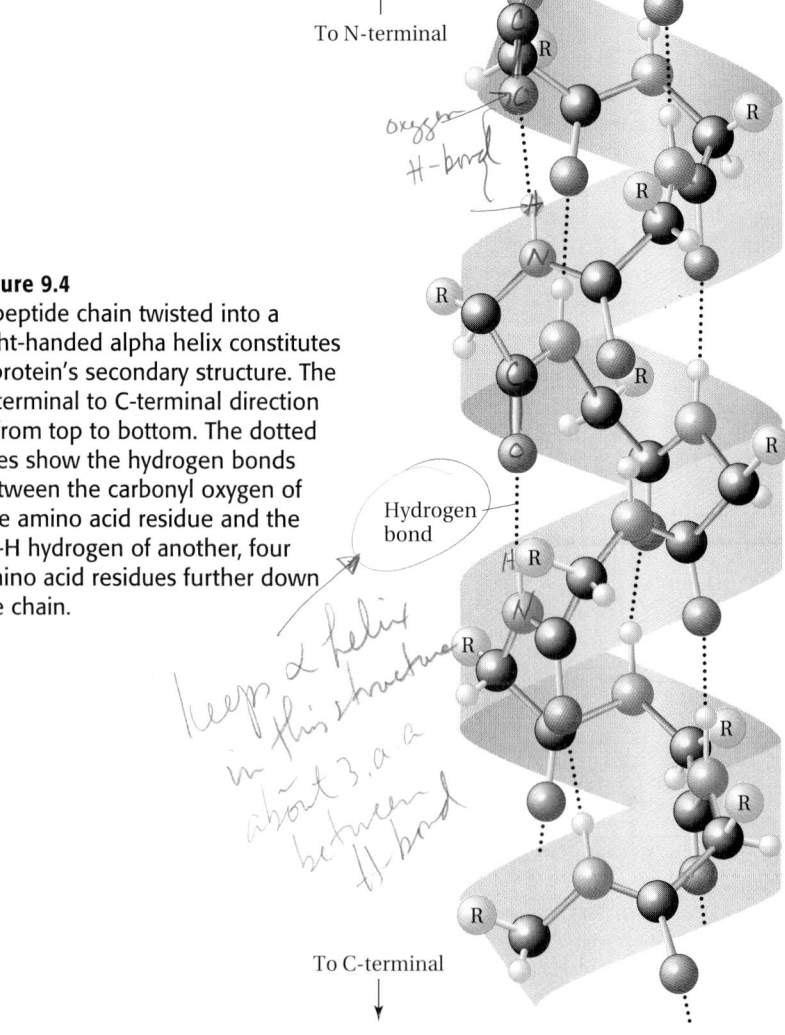

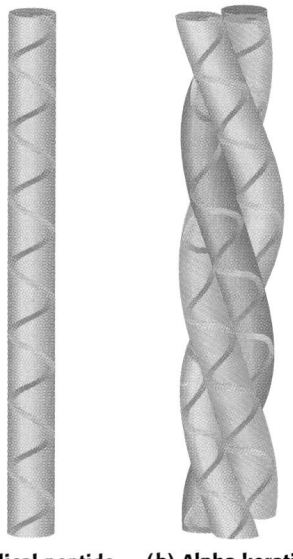

(a) Helical peptide chain **(b) Alpha keratin**

tightly wiled

Figure 9.5
Three helical peptide chains, like the one shown in (a), are twisted or supercoiled to form a rope in alpha keratin (b).

sometimes called an *alpha helix disrupter* for this reason. *The inability of proline to fit into the alpha helix is one of many pieces of evidence that the secondary structure of a protein is determined by its primary structure.*

Much of what is known about protein alpha helixes comes from studies of fibrous proteins. **Fibrous proteins** *tend to be long, rod-shaped molecules with great mechanical strength.* Such proteins are usually insoluble in water, dilute salt solutions, and other solvents. *The polypeptide chains of a class of fibrous proteins called* **alpha keratins** *consist mainly of alpha helixes.* Alpha keratins form the hard tissue of hooves, horns, outer skin layer (epidermis), hair, wool, and nails of mammals. We see in Figure 9.5 complex protein molecules consisting of long alpha helixes that are twisted together—*supercoiled*—in ropes of three or seven strands. It takes many supercoiled ropes to make a strong but elastic wool fiber. If you have ever washed a wool sweater, you know that warm, wet wool fibers can be stretched, but they eventually return to their original length. This is because the alpha helixes of the damp fibers are easily pulled into an extended form. The extended form is less stable than the alpha helix. It will, in time, return to the original alpha helix. Disulfide bridges (see Sec. 9.1) between alpha helixes help to make alpha keratins rigid; the alpha keratins of a hard hoof have more disulfide linkages than relatively elastic epidermis.

Beta-pleated sheet

Figure 9.6 shows the beta-pleated sheet, another kind of secondary structure commonly found in proteins. *The* **beta-pleated sheet** *consists of peptide chains arranged side by side to form a structure that resembles a piece of paper folded into many pleats.* The carbonyl groups of one zigzag peptide backbone are hydrogen bonded to peptide N—H hydrogens of adjacent peptide chains that run in opposite directions in the N-terminal to C-ter-

Figure 9.6
A beta-pleated sheet is another kind of secondary structure. Hydrogen bonds (shown as dotted lines) hold adjacent strands of the sheet together.

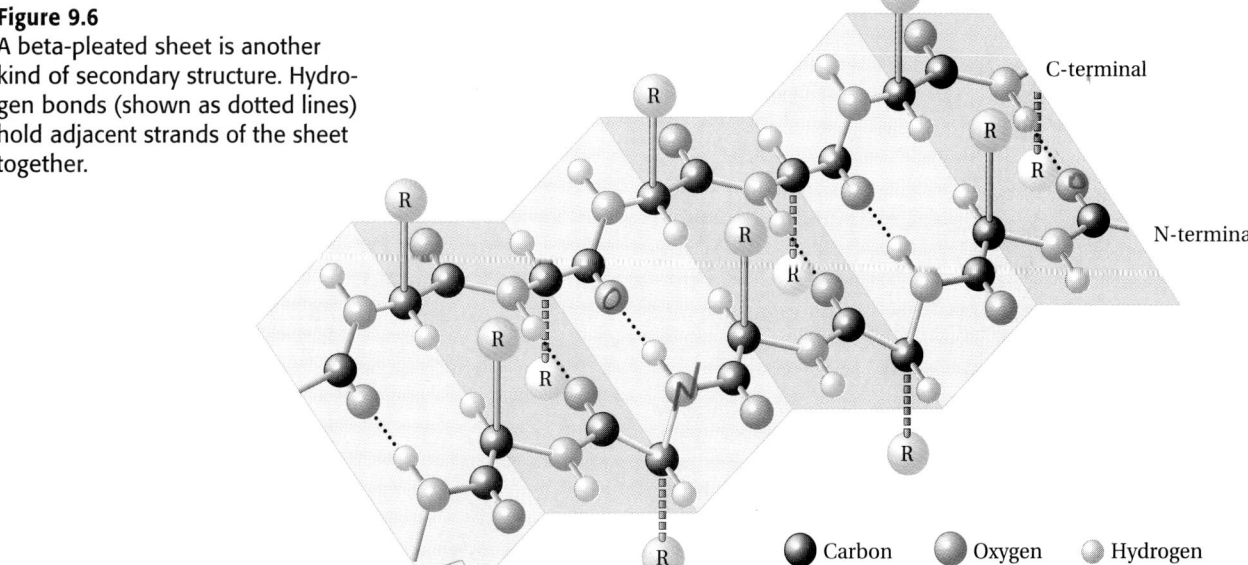

Carbon Oxygen Hydrogen
Nitrogen R Side chain Hydrogen bond

Figure 9.7
Spider webs are made of fibroin, a protein that exists mainly as beta-pleated sheets.

Figure 9.8
Silk fibroin consists of stacked beta-pleated sheets. The small R groups of glycyl and alanyl residues permit the stacking to occur.

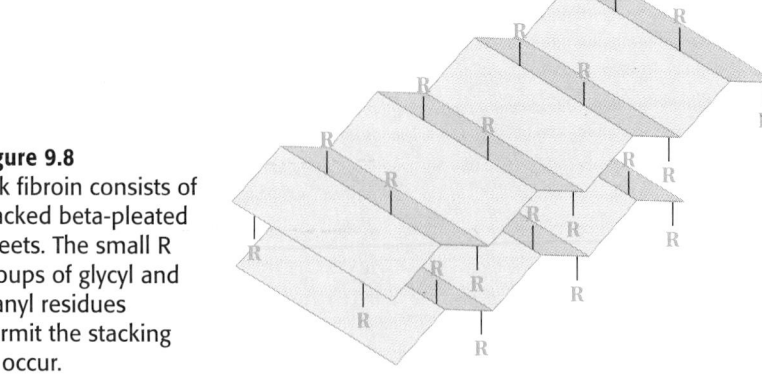

minal sense. Beta-pleated sheets are formed by separate strands of protein or by a single chain looping back on itself.

Beta keratins *are a class of fibrous proteins that consist mainly of beta-pleated sheets.* The long, thin fibers secreted by silkworms and spiders (Fig. 9.7) are composed of fibroin, a well-studied beta keratin. Fibroin consists mainly of layers of beta-pleated sheets, as shown in Figure 9.8. Again, the primary structure of a protein determines its secondary structure. Fibroin peptide chains are particularly rich in glycyl and alanyl residues. The small side chains of these residues are important to the organization of fibroin, since larger side chains would interfere with the packing of the sheets in layers.

Collagen helix

Collagen is yet another kind of fibrous protein. **Collagen** *is the most abundant protein in human beings and many other animals with spinal columns (vertebrates).* About one-third of the protein in the human body is present as the collagen of bones, teeth, inner skin layer (dermis), tendons, and cartilage. The inner material of the eye lens is almost pure collagen. Collagen occurs in all organs, where it imparts strength and stiffness.

Collagen is formed from three peptide chains, each a helix, wound into a rope. There are important differences between alpha helixes and collagen helixes, however. Collagen is rich in proline (40%), which does not fit into regular alpha helixes, and collagen helixes are not as tightly coiled as alpha helixes. Unlike alpha helixes, which are right-handed, collagen helixes are left-handed. Collagen contains large quantities of glycine (35%) and the unusual amino acids 4-hydroxyproline (5%) and 5-hydroxylysine (1%).

4-Hydroxyproline

5-Hydroxylysine

In addition, collagen is a glycoprotein (see Sec. 9.11). Sugar units (usually disaccharides of glucose and galactose) are covalently attached to the peptide chains.

9.7 Tertiary structure of proteins

AIMS: To describe the forces that help determine the tertiary structure of proteins. To define a conjugated protein. To describe the tertiary structure of myoglobin.

Focus

Folding of polypeptide chains produces tertiary structure.

The polypeptide chains of proteins fold to form three-dimensional structures. *The folding of a polypeptide chain of a protein molecule into a relatively stable three-dimensional shape constitutes the protein's* **tertiary structure.** Like the formation of a secondary structure, the formation of the tertiary structure of a protein is determined by its primary structure. The tertiary structure may contain within it regions of secondary structure, such as alpha helix and beta-pleated sheet. The overall folding of a polypeptide into its tertiary structure results from interactions between the side chains of the amino acid residues within the primary structure of the protein. For example, disulfide bridges can hold two regions of the same polypeptide chain together (Fig. 9.9). Weak attractions between amino acid side chains are also important. A polypeptide chain folds in the way that maximizes energetically favorable hydrogen bonds. *Ionic bonds, also called* **salt bridges,** *are usually formed between the negatively charged carboxylate ion side-chain groups of aspartate or glutamate and the positively charged amino side-chain groups of lysine or arginine.* One major factor in how a protein folds is the presence of aliphatic or aromatic amino acid residues in the peptide chain. The hydrophobic side chains of these residues tend to face the interior of protein molecules, much as the hydrophobic tails of soap molecules exclude water by forming micelles. Protein molecules that have different primary structures experience different side-chain interactions and therefore fold into different tertiary structures. Conversely, protein molecules that have the same primary structure experience the same side-chain interactions and therefore fold into the same tertiary structure.

Forces that stabilize III structure.

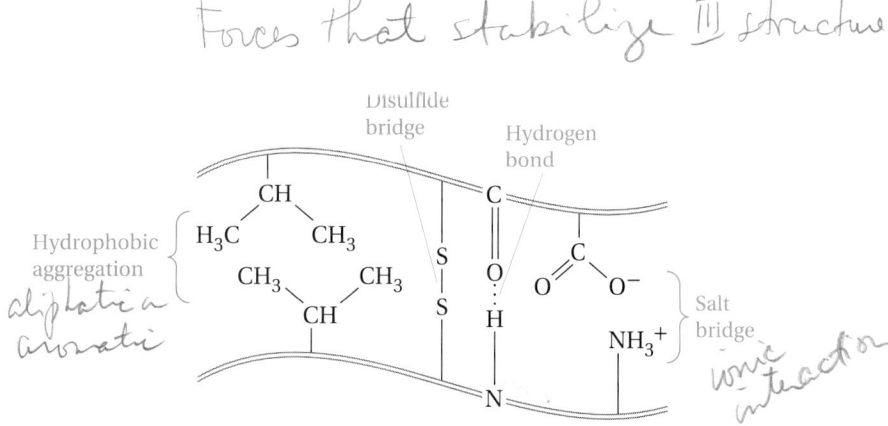

Figure 9.9
Many forces stabilize the tertiary structure of proteins.

The unusually large amount of myoglobin in whale muscle enables whales to remain submerged in water for long periods of time without rising to the surface for oxygen.

Myoglobin—*the oxygen-storage protein of mammalian muscle*—was the first protein to have its tertiary structure determined. Myoglobin is a *globular protein*. In contrast to rod-shaped fibrous proteins, **globular proteins** *have more or less spherical shapes*. Like many other globular proteins, myoglobin is soluble in water and dilute salt solutions. Myoglobin consists of a single peptide chain of 153 amino acids. This relatively small protein contains 1260 atoms, not counting hydrogens. Myoglobin is an example of a *conjugated protein*. **Conjugated proteins** *have structures that incorporate nonprotein portions called* **prosthetic groups.** Prosthetic groups are necessary for the protein to perform its biological function. They are usually organic molecules that are permanently attached to the protein by covalent bonds. The myoglobin molecule contains a prosthetic group called a *heme group*. The heme group contains an iron atom in the iron(II) or ferrous (Fe^{2+}) state (Fig. 9.10). Oxygen is bound to the heme iron to form oxymyoglobin.

In 1957, John C. Kendrew of the Medical Research Council laboratories in Cambridge, England, and his colleagues determined the three-dimensional structure of myoglobin by a method called *X-ray crystallography.* (See A Closer Look: X-ray Crystallography, for more about the X-ray method for determining protein structures.) The X-ray crystallographic picture of myoglobin in Figure 9.11 shows a globular structure in which almost three-fourths of the peptide chain is alpha helix. The tertiary structure of myoglobin's single chain consists of segments of alpha helix between turns of the polypeptide chain. Because of the turns, the molecule takes on a globular

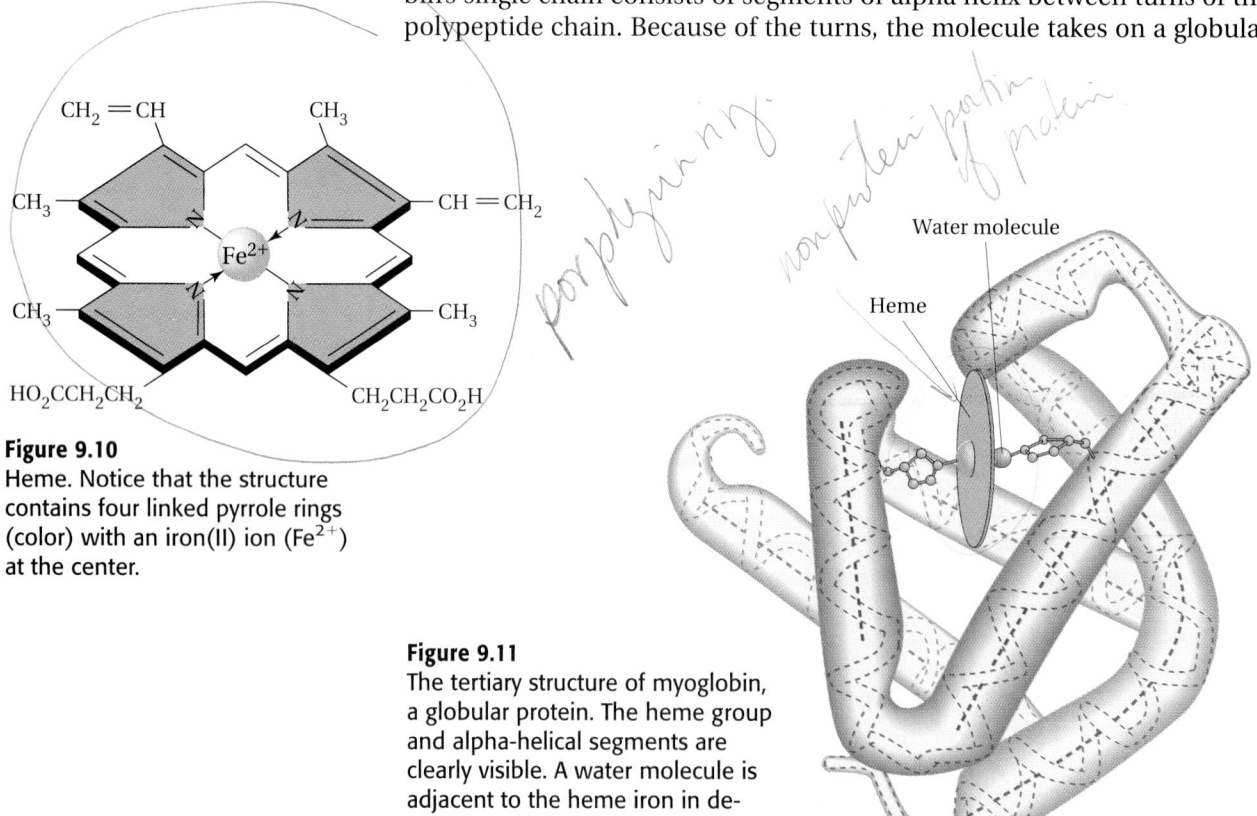

Figure 9.10
Heme. Notice that the structure contains four linked pyrrole rings (color) with an iron(II) ion (Fe^{2+}) at the center.

Figure 9.11
The tertiary structure of myoglobin, a globular protein. The heme group and alpha-helical segments are clearly visible. A water molecule is adjacent to the heme iron in de-oxymyoglobin, but it is replaced by oxygen in oxymyoglobin.

A Closer Look

X-ray Crystallography

The determination of the three-dimensional structure of a biologically interesting molecule is fascinating and practical. The fascination comes from learning how the atoms in the molecule are arranged in space and trying to figure out how this arrangement enables the molecule to perform its biological function. The practical aspect is that this knowledge, once obtained, can sometimes be used to design compounds that will either mimic or block the action of the biological molecule.

X-ray crystallography is a very powerful tool for determination of the three-dimensional structures of molecules, including large molecules such as those of proteins and deoxyribonucleic acids (DNA). Exposing a crystal of a substance to a narrow beam of X rays causes the X rays to be scattered, or diffracted, by their interactions with the electrons of the atoms in the crystal. Hundreds or thousands of diffraction patterns are recorded from different angles all around the crystal.

The spacing and intensity of the spots on diffraction patterns contain information about the electron densities in molecules. Computers are used to translate the diffraction information into three-dimensional maps of the electron densities all over the molecule (see figure). The electron-density maps give very accurate, complete molecular structures that pinpoint the positions of the atoms in space. For example, a region in the molecule that has a high electron density must be a

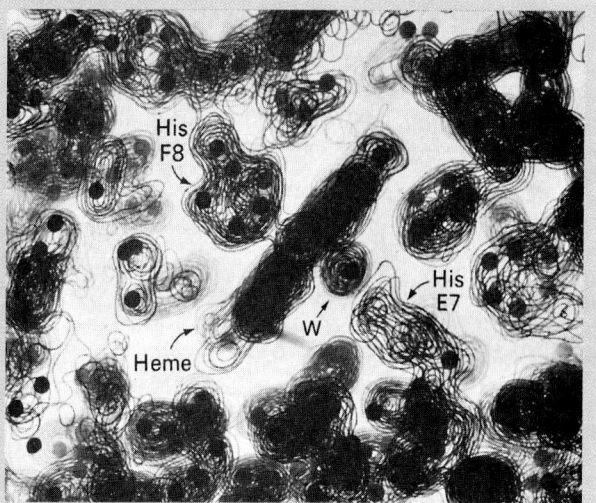

Electron-density map of myoglobin. The heme group is seen edge-on with its two associated histidine (His) residues and a water molecule (W).

covalent bond, because electrons are concentrated in covalent bonds. Conversely, a region that contains essentially no electron density must be a nucleus of an atom, because electrons are not found in the nucleus. The identity of a chemical bond, say, a carbon-carbon bond, is confirmed if the distance between two nuclei connected by a region of high electron density is 0.15 nm, the length of an ordinary C—C bond. Hydrogen atoms are too small to be directly detected by X-ray crystallography. However, the positions of hydrogens can be inferred once the positions of the heavier atoms in a molecule are known.

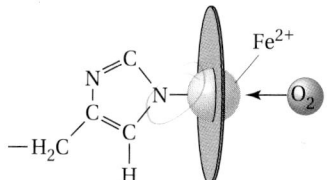

Figure 9.12
Heme is attached to myoglobin through a bond from the iron of the heme group to the nitrogen of a histidyl residue.

shape with the heme group resting in a "basket" formed by segments of alpha helix. There are other interesting features. The entire molecule is very compact—no more than two water molecules will fit on the inside. All the hydrophobic amino acid residues, those with aliphatic and aromatic side chains, are turned toward the interior of the molecule so that they are not exposed to water. All the charged side chains are exposed to water on the exterior of the molecule. Finally, as Figure 9.12 shows, the heme group is attached to the protein by a bond between the iron(II) ion of the heme group and a nitrogen of a histidyl residue of the polypeptide chain.

9.8 Quaternary structure of proteins

AIMS: To define the terms **subunit** *and* **quaternary** structure. *To describe the quaternary structure of hemoglobin. To distinguish among oxyhemoglobin, deoxyhemoglobin, and methemoglobin.*

Focus

In some proteins, polypeptide chains aggregate to form quaternary structures.

Some proteins consist of more than one polypeptide chain. *These individual chains are called* **subunits** *of the protein. Proteins composed of subunits are said to have* **quaternary structure.** Many proteins have structures that contain subunits. Proteins consisting of dimers (two subunits), tetramers (four subunits), and hexamers (six subunits) are fairly common. The proteins that comprise the individual subunits may be identical, or they may be different. Like the secondary and tertiary structures, the quaternary structure of a protein is determined by its primary structure. The polypeptide chains of subunits are held in place by the same forces that determine tertiary structure—hydrogen bonds, salt bridges, and sometimes disulfide bridges—except the forces are *between* the polypeptide chains of the subunits instead of *within* them. Hydrophobic aliphatic and aromatic side chains of subunits can aggregate to exclude water.

Hemoglobin—*the globular oxygen-transport protein of blood*—is an example of a protein that has a quaternary structure. Max Perutz, also of the Medical Research Council laboratories, determined the structure of horse blood hemoglobin in 1959. Hemoglobin is a larger molecule than myoglobin. The hemoglobin molecule has a molar mass of 64,500. It contains about 5000 individual atoms, excluding hydrogens, in 574 amino acid residues.

The quaternary structure of hemoglobin consists of four peptide subunits. Two of the subunits are identical and are called the *alpha subunits.* The remaining two subunits, called the *beta subunits,* are identical to each other but different from the alpha subunits. Figure 9.13 shows the four subunits of hemoglobin interlocked in a compact globular structure held together by ionic and hydrogen bonds between the amino acid side chains of the polypeptide subunits.

A chromoprotein has a color because of a colored prosthetic group. Hemoglobin is a chromoprotein that gets its red color from the heme group. Plants have their green color because of chlorophyll, a porphyrin ring with a structure similar to the heme group.

A heme prosthetic group containing an iron(II) ion capable of carrying one oxygen molecule is associated with each of the four protein subunits of hemoglobin. **Globin** *is the protein from which the heme groups have been removed. The hemoglobin complex with oxygen is* **oxyhemoglobin;** *that without oxygen is* **deoxyhemoglobin.** Oxyhemoglobin, the major hemoglobin of arterial blood, is bright red; deoxyhemoglobin of venous blood is purplish. *Long exposure to oxygen will convert the heme iron of hemoglobin to the iron(III), or ferric state (Fe^{3+}), to give* **methemoglobin.** The brown color of dried blood results from the conversion of hemoglobin to methemoglobin. Oxygen does not form complexes with methemoglobin. *In* **methemoglobinemia,** *a hereditary blood disease, the heme iron of hemoglobin, in either the alpha or the beta chains, exists in the iron(III) state.* People with iron(III) ions in both the alpha- and the beta-chain heme groups would be unable to live because their blood could not transport oxygen.

The folding of the polypeptide chains of the alpha and beta chains of the hemoglobin subunits is very similar. Moreover, the folding of

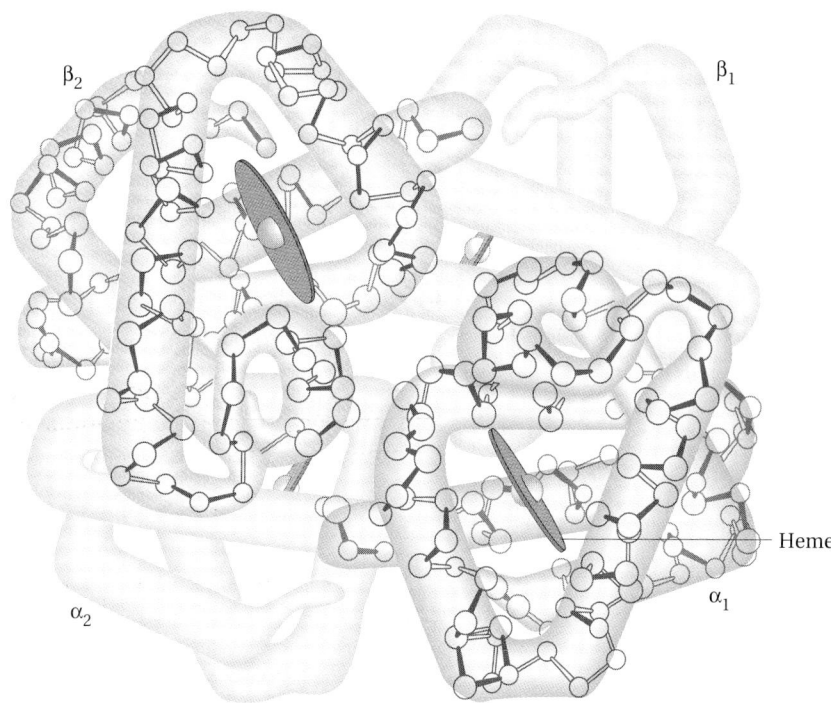

β₂

β₁

Heme

α₂

α₁

Figure 9.13
The quaternary structure of hemoglobin, showing the arrangement of alpha and beta chains.

myoglobin is very similar to the folding of the subunits of hemoglobin. This surprised biochemists, since the primary structures of sperm whale myoglobin and horse hemoglobin are very dissimilar. It is now clear that the tertiary structures devised by nature for myoglobin and the hemoglobin subunits are crucial to oxygen transport by these proteins.

9.9 Hemoglobin function

AIM: To describe the mechanism of oxygen transport by hemoglobin.

Focus

The environment of the heme groups gives myoglobin and hemoglobin the ability to carry oxygen.

Nature has gone to considerable trouble to construct complicated proteins such as myoglobin and hemoglobin for the apparently simple task of binding oxygen. In oxygen transport, however, as in many biological processes, a delicate balance must be maintained. The binding between iron(II) ions and oxygen must be strong enough that the oxygen can be stored or transported yet weak enough to provide a means for releasing the oxygen where it is needed.

Free iron(II) ions in water form very unstable complexes with molecular oxygen. Complexes of oxygen with iron(II) ions contained in heme groups are only slightly more stable. In myoglobin and hemoglobin, however, the heme groups are tucked into hydrophobic folds in the globin peptide chains. This hydrophobic environment fosters formation of relatively stable complexes between molecular oxygen and heme iron. It is the hydrophobic environment of the heme group that makes myoglobin and

A hereditary disease giving symptoms similar to sickle cell anemia is thalassemia, which is often found in people whose family origins are from Mediterranean countries. In thalassemia, however, either the alpha or beta chains of hemoglobin are not made (which is fatal) or are made in reduced amounts (thalassemia trait). Like sickle cell trait, thalassemia trait confers resistance to malaria.

Why is sickle cell anemia found almost exclusively in people of African descent? Although patients with sickle cell anemia often die very young, persons with sickle cell trait have a high resistance to malaria, a disease prevalent in certain parts of Africa. Apparently, ancestral Africans with sickle cell trait had a distinct survival advantage over those who lacked it. Sickle cell trait and sickle cell anemia are passed from generation to generation, accounting for the high incidence of the two conditions today.

The electrophoresis of hemoglobin obtained from red blood cells is a simple and effective technique for screening individuals for sickle cell trait or sickle cell anemia. **Electrophoresis** *is a method for the separation of ions according to their charge,* as described in A Closer Look: Electrophoresis. At pH 8.4, HbA migrates more rapidly than HbS. Since both HbA and HbS are red, they are readily visible without staining.

A Closer Look

Electrophoresis

Since they are charged, ions in solution move in electrical fields. Anions migrate to the anode (the positively charged electrode), and cations migrate to the cathode (the negatively charged electrode). The more highly charged an ion, the more rapidly it migrates. The difference in migratory rates of differently charged ions is the basis for electrophoresis, a powerful tool for the separation of mixtures of proteins.

Electrophoresis succeeds as a protein-separation method because different kinds of protein molecules behave as complex ions with slightly different charges. The size of the charge on a given protein molecule depends not only on the number and kind of acidic and basic side chains on the amino acid residues of the protein but also on the pH of the solution containing the protein. In acidic solutions, most proteins are positively charged, because the amino groups are present as positively charged ions; carboxylic acid groups are also protonated and are therefore electrically neutral. In basic solutions, most proteins are negatively charged, because the amino groups are unprotonated and are electrically neutral; carboxylic acid groups are unprotonated and negatively charged. At the isoelectric pH of a protein, the positive and negative charges balance, and the protein behaves as if it had no charge—it will not migrate in an electrical field.

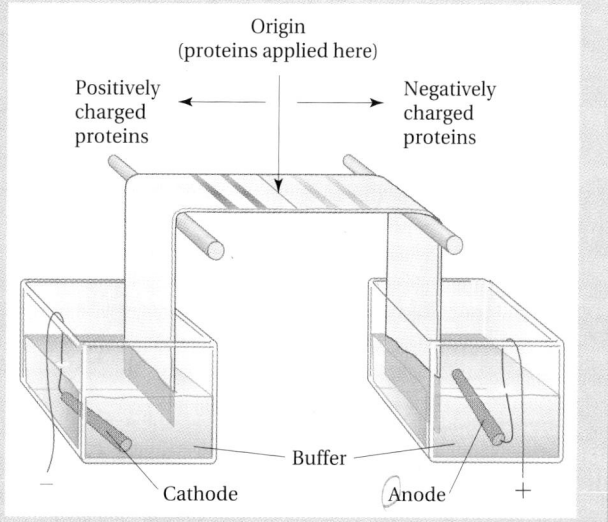

A procedure called *electrophoresis* is used to separate a mixture of proteins.

Biochemists perform electrophoresis experiments by applying a sample of solution containing a mixture of proteins to the center of a strip of porous material such as thick filter paper. Both ends of the paper are dipped like wicks into troughs containing a buffer solution of appropriate pH, and an electric current is passed through the system (see figure). Charged proteins separate by migrating at different rates toward the electrode of opposite charge. The separated proteins can be seen directly if they are colored, but often it is necessary to make them visible by staining with a suitable dye.

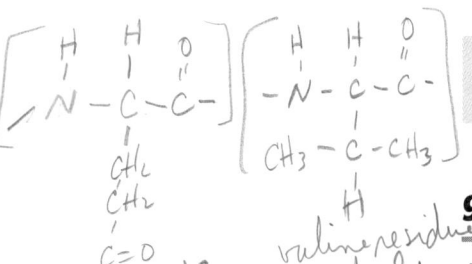

glutamic acid residue, pH 7, (C=O, OO), CH₂, CH₂

valine residue, neutral at all values, hydrophobic, CH₃—C—CH₃, H

PRACTICE EXERCISE 9.6

Draw the structure of a glutamic acid residue, and compare it with a valine residue. How do their side chains differ?

9.11 Glycoproteins

AIM: To characterize glycoproteins.

Focus

Carbohydrates are sometimes found attached to proteins.

Glycoproteins *are protein molecules that have sugar molecules (glycosides) covalently bonded to them.* Mucous secretions consist in part of mucoproteins, a special class of glycoproteins. Some of the glycoproteins we mention are many cell membrane proteins (see Chap. 8), collagen (see Sec. 9.9), fibrinogen of blood clotting (see Chap. 10), and immunoglobulin G (see Chap. 13).

The measurement of certain glycoproteins in the blood of patients with diabetes can be an important indicator of how well the level of blood sugar is being controlled over a period of time, as described in A Closer Look: Glycoproteins: Control of Blood Glucose in Diabetes.

9.12 Denaturation

AIM: To state three ways to denature proteins.

Focus

Proteins are easily denatured.

A protein that is folded into its normal, biologically active structure is in its **native state.** *Denaturation* *occurs when a native protein unfolds owing to the disruption of weak attractive forces or cleavage of disulfide bridges,* as shown in Figure 9.14. Denaturation can disrupt the secondary, tertiary, and

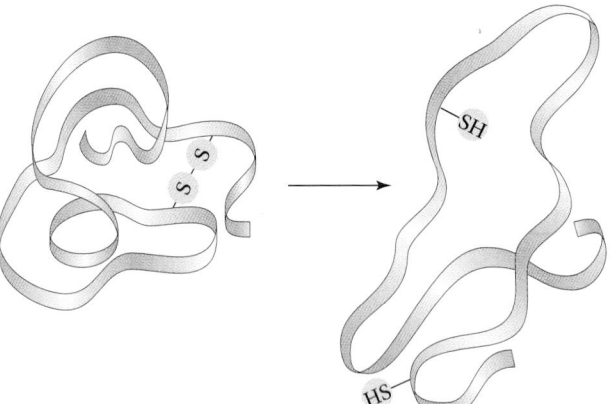

Figure 9.14
Protein denaturation results from disruption of weak forces and, in some instances, cleavage of disulfide bridges.

Glycoproteins: Control of Blood Glucose in Diabetes

Glycoproteins play an important role in control of the level of blood glucose in patients with diabetes. The goal of those patients is usually to maintain blood glucose as near as possible to the midpoint of the nondiabetic range, about 6 to 7 mmol of glucose per liter of blood. The absence of glucose in urine tests (A Closer Look: Test for Glucose in Diabetes) or a normal amount of blood glucose from blood tests done four times a day does not mean that the level of blood glucose is constant at all times. The most accurate way to find out how the blood glucose concentration changes over time would be to monitor it constantly. This is usually impractical, but a test for a variant of normal adult hemoglobin (HbA) can indicate the degree of control of blood glucose levels. Normally, between 5% and 8% of HbA has a glucose molecule attached to it. This glycoprotein is called *hemoglobin A₁ (HbA₁)*. People with untreated diabetes have much higher levels of HbA₁ than healthy people. The blood levels of HbA₁ reflect the average blood glucose level during the previous 6 to 8 weeks. Therefore, the percentage of HbA₁ in the blood can be used to estimate the degree of control of blood sugar during this period. Blood levels of HbA₁ below 7% repre-

sent good control, levels of 7% to 11% represent moderate control, and levels greater than 11% represent poor control. This information is useful for adjusting the diet or the amount and frequency of insulin injections.

Evidence suggests that glycoproteins are responsible for circulation problems sometimes encountered in diabetes. A test used to monitor the amount of glycoproteins and glucose control involves albumin, the major protein of the fluid part of the blood. Similar to hemoglobin's reaction with glucose, albumin and other proteins react with a sugar in the blood to make glycoproteins. The sugar is fructosamine, an amine derivative of fructose. Fructosamine reacts with albumin and other proteins in direct proportion to its concentration in the blood. The amount of fructosamine in the blood therefore provides a measure of the proportion of glycoproteins being made in the body. Since the body makes fructosamine from glucose, the amount of circulating fructosamine also reflects the amount of glucose in the blood over a period of time. Fructosamine lasts a shorter time in the body than hemoglobin, so the level of the sugar reflects the control of blood glucose for the previous period of only 3 to 6 weeks instead of the 6 to 8 weeks for HbA₁. The normal fructosamine level is less than 2.8 mmol/L. Higher levels in a patient with diabetes indicate poor control of blood glucose.

quaternary structures of a protein, but it does not break the peptide chains, or in any other way alter the primary structure.

Protein denaturation may or may not be reversible. The proteins of egg white unfold and congeal into a rubbery mass when we boil an egg. This denaturation is irreversible because the protein can never return to its original state. Not all proteins are so heat-sensitive. Thermolysin, a protein-cutting enzyme of *Bacillus thermoproteolyticus,* a microorganism that lives in hot springs, resists unfolding even in boiling water. Heat, extremes of pH, and many chemicals, especially organic solvents, cause the irreversible

Heat sterilization of medical instruments kills bacteria because the heat irreversibly denatures their proteins.

denaturation of proteins by disrupting weak attractive forces. Figure 9.15 shows disulfide bonds cleaved in the presence of oxidants and reductants.

In reversible denaturation, the protein unfolds in the presence of a denaturing agent, such as a concentrated urea solution, but is restored to its native state on removal of the agent. Whether a protein undergoes reversible or irreversible denaturation varies with the kind of protein reacting and the conditions of reaction.

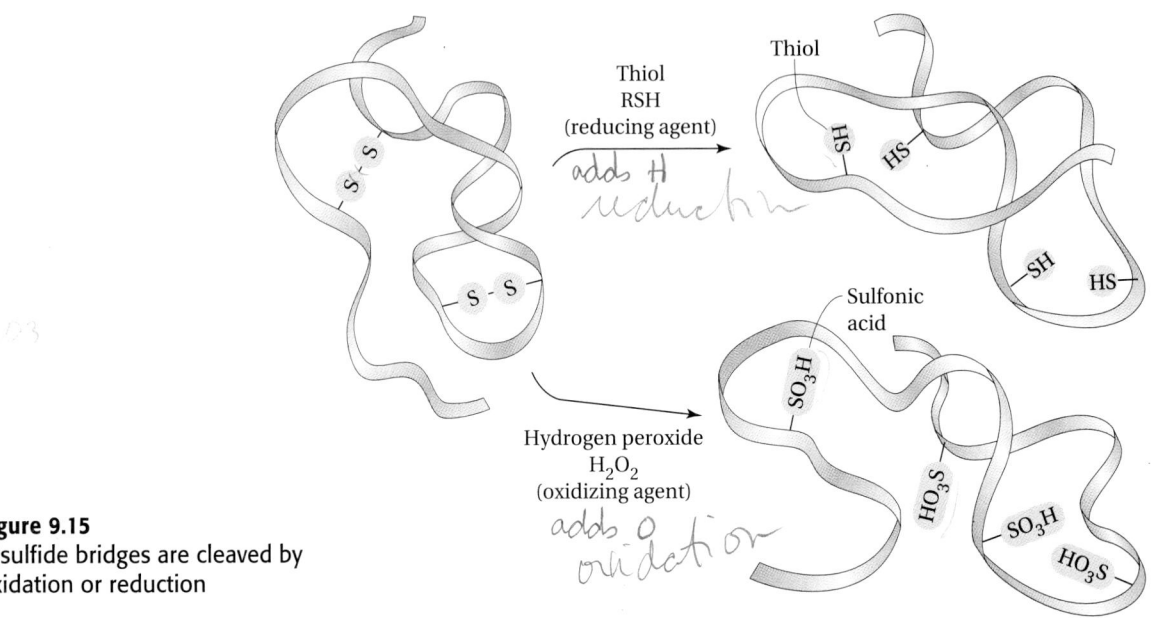

Figure 9.15
Disulfide bridges are cleaved by oxidation or reduction

SUMMARY

Twenty alpha amino acids commonly occur in nature. These compounds are grouped according to the chemical properties of their side chains. All the amino acids but glycine belong to the L family of stereoisomers. At biological pH, most amino acids exist as electrically neutral zwitterions, in which the positive charge of the protonated alpha amino group is compensated by the negatively charged alpha carboxylate ion.

The formation of an amide bond (peptide bond) between the alpha amino group of one amino acid and the alpha carboxylic acid group of another produces a peptide. Peptides have a common backbone and N-terminal and C-terminal ends. The order in which amino acids are linked in a peptide is the peptide's amino acid sequence. The peptide's amino acid sequence determines its biological role.

Proteins are peptides of greater than about 100 amino acid residues. Structural proteins are often fibrous, water-insoluble molecules; other proteins are globular and water-soluble. Conjugated proteins have nonprotein portions called prosthetic groups, which are often relatively small organic molecules.

Protein molecules fold into unique shapes. The chemical nature of the amino acid side chains and the order of their occurrence within the peptide chain (primary structure) govern the type of folding. Two patterns of chain folding (secondary structure) are found in many proteins; these are the alpha helix and the beta-pleated sheet. Collagen forms a secondary structure of collagen helixes. Each molecule of a particular kind of protein has the same overall three-dimensional folding (tertiary structure). Forces that hold proteins in their respective foldings include hydrophobic aggregation, hydrogen bonds, ionic bonds, and disulfide bridges.

Myoglobin is a conjugated globular protein that stores oxygen in muscle. The folding of its single-peptide chain is mostly alpha helix. The heme prosthetic group is tucked into folds in the chain. Hemoglobin, the oxygen carrier of blood, consists of four single-chain subunits comprising its quaternary structure—two alpha chains and two beta chains. Sickle cell anemia is an example of how a minor modification of the primary structure of a protein affects the protein's biological function.

Peptide chains of proteins are unfolded (denatured) by chemical agents, heat, or extremes of pH. Denaturation may be reversible or irreversible.

KEY TERMS

Alpha carbon (9.1)
Alpha helix (9.6)
Alpha keratin (9.6)
Amino acid (9.1)
Amino acid residue (9.4)
Amino acid side chain (9.1)
Antidiuretic (9.4)
Beta keratin (9.6)
Beta-pleated sheet (9.6)
Carboxyhemoglobin (9.9)
Collagen (9.6)

Conjugated protein (9.7)
Cyanohemoglobin (9.9)
Denaturation (9.12)
Deoxyhemoglobin (9.8)
Disulfide bridge (9.4)
Electrophoresis (9.10)
Enkephalin (9.4)
Fibrous protein (9.6)
Globin (9.8)
Globular protein (9.7)
Glycoprotein (9.11)

Hemoglobin (9.8)
Isoelectric pH (9.3)
Isoelectric point (9.3)
Methemoglobin (9.8)
Methemoglobinemia (9.8)
Myoglobin (9.7)
Native state (9.12)
Oxyhemoglobin (9.8)
Peptide (9.4)
Peptide bond (9.4)
Polypeptide (9.4)

Primary structure (9.5)
Prosthetic group (9.7)
Protein (9.5)
Quaternary structure (9.8)
Salt bridge (9.7)
Secondary structure (9.6)
Subunit (9.8)
Tertiary structure (9.7)
Zwitterion (9.3)

EXERCISES

Amino Acids (Sections 9.1, 9.2, 9.3)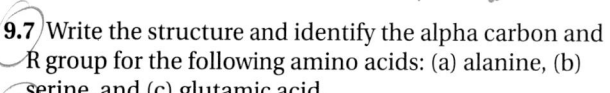

9.7 Write the structure and identify the alpha carbon and R group for the following amino acids: (a) alanine, (b) serine, and (c) glutamic acid.

9.8 Categorize the following amino acids according to their side-chain groups.
(a) valine   (b) glutamine   (c) cysteine
(d) phenylalanine

9.9 Draw a stereorepresentation of (a) D-serine and (b) L-alanine.

9.10 Identify the following compounds represented by their Fischer projection formulas.

(a) CO_2H (b) CO_2H (c) CO_2H

H—NH$_2$ H—NH$_2$ H$_2$N—H

    CH$_2$OH   H   CH$_2$SH

9.11 Define the term *zwitterion*. Draw the amino acid leucine as a zwitterion.

9.12 Consider the amino acids valine, glutamic acid, and lysine. Draw structural formulas for the species that predominate in solution at (a) pH 2 and (b) pH 10.

9.13 Define *isoelectric pH*.

9.14 At which pH will alanine be least soluble, (a) pH 2, (b) pH 7, or (c) pH 10? Why?

Peptides (Section 9.4)

9.15 In peptide chemistry, what is the meaning of the word *residue?*

9.16 What is the name given to the amide bond in a peptide chain?

9.17 Consider the tripeptide seryl glycyl phenylalanine. (a) Draw the complete structural formula, and (b) write the three-letter abbreviation. (c) How many peptide bonds does this molecule have?

9.18 Translate the following three-letter abbreviations: (a) Ala-Ser-Gly and (b) Gly-Ser-Ala. Are the structures of these tripeptides the same? Explain your answer.

9.19 Explain the biological functions of the peptide hormones (a) oxytocin and (b) vasopressin.

9.20 Name two pentapeptides that act as the body's own opiates.

Primary and Secondary Structures of Proteins (Sections 9.5, 9.6)

9.21 What is meant by the *primary structure* of a protein?

9.22 Define the *secondary structure* of a protein.

9.23 Describe three common repeating patterns that are found in the secondary structure of proteins.

9.24 Consider the structure of the alpha helix. Are the amino acid residue side chains all inside the helix, part of the helix, or all outside the helix?

9.25 What is the function of collagen in the body?

9.26 Compare the molecular structures of collagen and alpha keratin with regard to secondary structure and amino acid composition.

Tertiary and Quaternary Structures of Proteins (Sections 9.7, 9.8)

9.27 Define what is meant by a *conjugated protein.*

9.28 (a) In what ways are the molecular structures of myoglobin and hemoglobin similar? (b) How are they different?

9.29 With the aid of diagrams, describe the factors that contribute to the folding of peptide chains in the tertiary structures of proteins.

9.30 Describe the heme group and explain its function.

9.31 What is the oxidation state of iron in (a) oxyhemoglobin and (b) methemoglobin?

Hemoglobin Function (Section 9.9)

9.32 Discuss the role of hemoglobin in the transport of oxygen in the body.

9.33 Why are carbon monoxide and cyanide ions poisons?

Sickle Cell Anemia (Section 9.10)

9.34 What is the basic structural difference between normal hemoglobin (HbA) and sickle cell hemoglobin (HbS)?

9.35 Why is exposure to low oxygen levels potentially dangerous for a person who has sickle cell trait?

9.36 What technique is used for screening individuals for sickle cell trait or sickle cell anemia?

9.37 The symptoms of sickle cell anemia can be treated by hyperbaric oxygenation. Why does this procedure offer relief for the patient?

Glycoproteins and Denaturation (Sections 9.11, 9.12)

9.38 Name and describe the biological functions of at least two glycoproteins.

9.39 What happens when a protein is denatured?

9.40 Describe one way a protein can be denatured irreversibly and one way a protein can be denatured reversibly.

Additional Exercises

9.41 Match the following.

(a) lysine	(1) fibrous protein
(b) vasopressin	(2) antidiuretic hormone
(c) denaturation	(3) basic amino acid
(d) peptide link	(4) protein of tendon
(e) globular protein	(5) unfolding
(f) beta keratin	(6) net charge equals zero
(g) zwitterion	(7) spherical shape
(h) collagen	(8) amide bond

9.42 (a) Write the three-letter abbreviation and draw the structural formula of the tripeptide glutamyl-cysteinyl-glycine. (b) Draw the different ionic species of this tripeptide at low (pH = 2), neutral (pH = 7), and high pH (pH = 11). (c) A disulfide bridge is formed when this tripeptide is treated with a mild oxidizing agent. Draw the structural formula of the resulting hexapeptide.

9.43 Write the three-letter abbreviations of all the different tripeptides that can be made if the peptides contain one amino acid residue each of histidine, methionine, and glutamic acid.

9.44 For each of the following fibrous proteins, match all the items at the right that apply. Some answers may be used more than once, some not at all.

(a) beta keratin	(1) spider webs
(b) collagen	(2) cartilage
(c) alpha keratin	(3) predominantly alpha helixes
	(4) rich in proteins
	(5) a glycoprotein
	(6) found in hair
	(7) rich in glycine and alanine

9.45 Draw the structural formulas of the expected products of acid hydrolysis of (a) Cys-Gly-Ala, (b) Glu-Leu-Val-Pro, and (c) Ser-His-Phe-Tyr-Trp.

9.46 What unique structural features do each of the following amino acids have that help to distinguish them from all the others?
(a) serine
(b) phenylalanine
(c) glycine
(d) cysteine
(e) tyrosine
(f) proline

9.47 Define (a) salt bridge, (b) prosthetic group, (c) tertiary structure, (d) globular protein, (e) beta-pleated sheet, (f) secondary structure, (g) alpha-helix, and (h) primary structure.

9.48 Distinguish among alpha keratin, beta keratin, and collagen by biological function and molecular structure.

9.49 Match the following. Each answer will be used once only.
(a) fibrous protein *6*
(b) native state *4*
(c) disulfide bridge *7*
(d) peptide bond *9*
(e) globular protein *1*
(f) denaturation *2*
(g) prosthetic group *3*
(h) salt bridge *5*
(i) conjugated protein *8*

(1) water-soluble protein, easily denatured
(2) can be reversible or irreversible
(3) the metal ion of a conjugated protein
(4) a protein in its normal, biologically active form
(5) ionic bond between carboxylate ion and protonated amine group
(6) important structural protein
(7) covalent bond that can hold two peptide chains together
(8) protein that has a prosthetic group
(9) covalent bond that holds two amino acids together

SELF-TEST (REVIEW)

True/False

1. Amino acids found in nature are generally the L forms.
2. At the isoelectric pH of an amino acid, its solubility in water is maximum.
3. A basic amino acid such as asparagine would have a net negative charge at high pH.   *amide side chain - neutral*
4. The amino acid glycine is the C-terminal amino acid of the pentapeptide Gln-Asp-Pro-Val-Gly.
5. In aqueous solutions, the hydrocarbon side chains of a protein tend to point inward, avoiding interaction with the water.
6. Protein denaturation is always irreversible.
7. The amino acid proline disrupts the alpha-helical secondary structure of a protein.
8. Myoglobin is used for oxygen storage in the muscle.
9. Proteins are composed of amino acids.
10. All protein molecules have a quaternary structure.

Multiple Choice

11. All the following amino acids have aliphatic side chains *except*
(a) valine.   (b) alanine.   (c) serine.
(d) leucine.

12. Disulfide bridges can be broken by
(a) dissolving the protein in water.
(b) cooling the protein.
(c) oxidizing agents.
(d) dehydrating agents.

13. Proteins that tend to be water soluble and generally have spherical shapes are
(a) conjugated proteins.   (b) globular proteins.
(c) beta keratins.   (d) fibrous proteins.

14. Which of the following contributes most importantly to the secondary structure of a protein?
(a) hydrogen bonds   (b) disulfide bridges
(c) salt bridges   (d) none of the above

15. A disruption of the secondary and tertiary structures of a protein
    (a) is never reversible.
    (b) is caused only by heat.
    (c) results in a conjugated form.
    (d) is called *denaturation*.

16. The peptide Phe-Glu-Ala-Val
    (a) has valine at the N-terminal end.
    (b) contains four peptide bonds.
    (c) is the same as the peptide Val-Ala-Glu-Phe.
    (d) has four amino acid residues.

17. Two commonly occurring peptide secondary structures in proteins are
    (a) alpha helix and beta-pleated sheet.
    (b) protofibril and globular.
    (c) globular and beta-pleated sheet.
    (d) protofibril and alpha helix.

18. The most abundant protein in higher vertebrates is
    (a) collagen.   (b) oxytocin.
    (c) alpha keratins.   (d) hemoglobin.

19. Amino acids found in biological systems are usually
    (a) in the left-handed, or L, form.
    (b) in the D form.
    (c) found in an uncombined state.
    (d) beta amino acids.

20. The heme group in a hemoglobin molecule
    (a) is absent in sickle cell anemia.
    (b) is called a *prosthetic group*.
    (c) is held jointly by the four peptide subunits.
    (d) can never contain Fe(III) ions.

21. The complex present in carbon monoxide poisoning is
    (a) cyanohemoglobin.   (b) deoxyhemoglobin.
    (c) carboxyhemoglobin.   (d) methemoglobin.

22. Myoglobin, a globular, conjugated protein,
    (a) has most of its aliphatic side chains on the exterior surface of the molecule.
    (b) would be expected to be water soluble.
    (c) carries oxygen from the lungs to the tissues.
    (d) has no prosthetic group.

23. A structural protein found in mammalian hair is
    (a) alpha keratin.   (b) collagen.
    (c) beta keratin.   (d) fibroin.

24. A molecule of the amino acid valine in a strongly acid solution would be a
    (a) zwitterion.
    (b) negatively charged species.
    (c) positively charged species.
    (d) neutral molecule.

25. Which of these is *not* usually disrupted in the denaturation of a protein?
    (a) ionic bonds   (b) hydrogen bonds
    (c) peptide bonds   (d) hydrophobic aggregation

26. The major type of hemoglobin of arterial blood is
    (a) cyanohemoglobin.   (b) oxyhemoglobin.
    (c) methemoglobin.   (d) deoxyhemoglobin.

27. The correct general formula for a dipeptide at neutral pH is

    (a)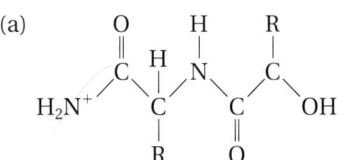

    (b)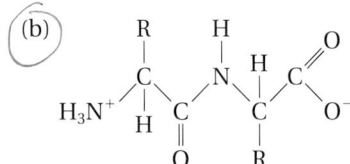

    (c)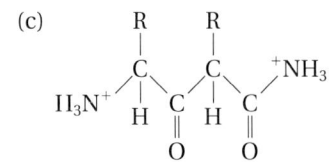

    (d)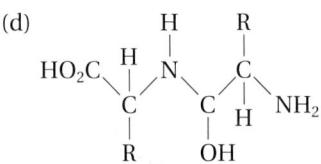

28. Which type of bonding holds proteins in the beta-pleated sheet?
    (a) ionic bonds   (b) hydrogen bonds
    (c) hydrophobic aggregation   (d) covalent bonds

Enzymes

Catalysis of the Reactions of Life

Enzymes of clinical significance are derived from many unusual sources, including the venom of snakes.

The operation of living cells may be compared with the operation of highly efficient chemical factories. *Enzymes,* substances that catalyze biological reactions, are the workers in these cell factories, busily disassembling raw materials (nutrients taken into the cell) and reassembling them into products used to help the cell survive and grow. Depending on the tasks they perform, particular enzymes may be distributed throughout the cell or limited to a specific location.

In this chapter we will learn about the structure of enzymes and see how they function. Enzymes are proving useful as medical drugs, as we will learn in the Case in Point.

CASE IN POINT: Enzyme therapy for heart attacks

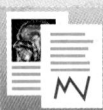

Art, a 45-year-old stockbroker, awoke early Monday morning. The markets were not performing well, and he was feeling a great deal of stress over his clients' and his own financial losses. By 11 o'clock in the morning, the markets had dropped further. Art began to feel a tightness in his chest and pains running down his left arm. Fifteen minutes later a coworker found Art slumped over his desk. After a quick examination, the emergency room physician was certain that Art had suffered a heart attack. Among other steps, the physician ordered an intravenous drip of a solution containing an enzyme. What enzyme did the doctor order? What effect did the enzyme have on Art's recovery? We will learn the answers to these questions in Section 10.12.

Clinical laboratory technicians use automated equipment to measure enzyme concentrations in blood serum. This procedure helps in the diagnosis of medical problems such as heart attacks.

10.1 Enzymes

AIM: To state three properties that show that enzymes are catalysts.

Enzymes *are biological catalysts.* In 1926, American chemist James B. Sumner reported the first purification of an enzyme. The enzyme he isolated was urease, which is able to hydrolyze urea, a constituent of urine, into ammonia and carbon dioxide. The reaction is this:

$$NH_2-\overset{\overset{\displaystyle O}{\|}}{C}-NH_2 \;+\; H_2O \;\xrightarrow{\text{Urease}}\; 2NH_3 \;+\; CO_2$$

Urea   Water   Ammonia   Carbon dioxide

Sumner demonstrated that urease is a protein. Since Sumner's pioneering work, hundreds of proteins that serve as enzymes have been isolated and structurally characterized. Until the 1980s, it was believed all enzymes were proteins. An exciting recent finding is that certain ribonucleic acids (RNAs) are capable of cleaving out sections of other long, polymeric RNA molecules. These catalytic RNA molecules, called *ribozymes*, also can splice the ends of two RNA molecules together. The existence of ribozymes means that some enzymes are not proteins.

Enzymes constitute a substantial portion of the total protein of the cell. A typical cell contains about 3000 different enzymes and many molecules of each kind. Enzymes can speed up chemical reactions, while other proteins cannot. Besides being able to speed up reactions, enzymes have two other properties of true catalysts. First, they are unchanged by the reaction they catalyze. Second, and very important, even though they speed up reactions, enzymes cannot change the normal position of a chemical equilibrium. In other words, an enzyme can help a reaction-product to be formed faster, but the same amount of product is eventually formed whether or not an enzyme is present. (Few reactions in cells ever reach equilibrium because the products are rapidly converted to another substance in a further enzyme-catalyzed reaction. The removal of a reaction product pulls the reaction toward completion.)

People have used the catalytic power of enzymes to suit their desires since prehistoric times. The fermentation of fruit sugars to alcohol by yeast enzymes was a very early discovery. Herdsmen who made canteens from the stomachs of goats and sheep found that when they filled them with milk instead of water the milk soon clumped into cheese. The agent for this transformation is rennin, an enzyme produced in the stomachs of cud-chewing animals (ruminants). Yogurt, an ancient food with modern popularity, is prepared by the action of enzymes produced by several bacteria. Brewing beer from grain, leavening bread with yeast, and fermenting apple cider to vinegar are other practical applications of the catalytic power of enzymes.

10.2 Names of enzymes

AIM: To identify the function of an enzyme from its name.

Focus

Many enzymes are named for the reactions they catalyze.

A class of enzymes responsible for energy-producing reactions consists of the oxidoreductases. These enzymes catalyze oxidation-reduction reactions—reactions that involve the transfer of electrons. Many of these enzymes are found in the mitochondria of a cell.

Proteases are used as meat tenderizers, in laundry detergents as stain removers, in beer production to remove cloudiness, and in leather tanning.

Biochemists often name enzymes by taking the name of the compound undergoing change and adding the ending -*ase*. The ending -*ase* serves as a signpost that says that the substance in question is an enzyme. In the hydrolysis of urea, the substance undergoing change is urea, and the name given to the enzyme that catalyzes this change is *urease*. Sometimes the enzyme name also reflects the kind of chemical transformation that occurs. For example, an enzyme catalyzes the removal of hydrogen from ethanol (CH_3CH_2OH) to give acetaldehyde (CH_3CHO). The enzyme that catalyzes this dehydrogenation reaction is called *alcohol dehydrogenase*. In another example, *enzymes that catalyze the hydrolysis of one or more peptide bonds of proteins are given the general name* **protease** *or* **peptidase.** Some proteases are digestive enzymes that have names established long ago: *pepsin, trypsin, chymotrypsin. Thrombin* and *plasmin* are peptidases involved in blood clotting.

EXAMPLE 10.1   **Discerning the function of an enzyme**

What is the role of the enzyme sucrase? *enzyme used in the hydrolysis*
SOLUTION *of a disaccharide into 2 monosaccharides sucrose*

Based on the name of the enzyme, the substance undergoing a change is sucrose. Sucrose is a disaccharide. Most likely this enzyme catalyzes the breakdown (hydrolysis) of the disaccharide into two monosaccharides.

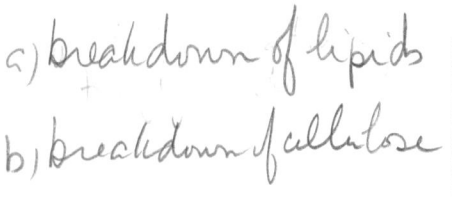

a) breakdown of lipids
b) breakdown of cellulose

PRACTICE EXERCISE 10.1

Predict the function of the following enzymes: (a) lipase and (b) cellulase.

PRACTICE EXERCISE 10.2

Suggest general names for enzymes that catalyze (a) an oxidation reaction involving a gain of oxygen and (b) a hydrolysis reaction.
oxydase   hydrolase

10.3 Enzyme specificity

AIM: To explain what is meant by the specificity of an enzyme.

Focus

Enzymes are specific catalysts.

Most of the chemical changes that occur in the cell are catalyzed by enzymes. **Substrates** *are reactants that are transformed to products by the catalytic action of enzymes.* As in nonenzymatic reactions, substrates are transformed into products by bond-making and bond-breaking processes. In an enzymatic reaction, these processes occur through interactions of the enzyme with its substrate. One property of enzymes is **specificity**—*catalyzing one chemical reaction with only one substrate.*

Enzymes exhibit stereospecificity. They will act on only one stereo-isomer of a compound, for example, the L form of an amino acid but not the D form.

To get some idea of the power and specificity of enzymes, consider carbonic anhydrase. This enzyme catalyzes only one reversible reaction, the breakdown of carbonic acid to water and carbon dioxide.

$$H_2CO_3 \underset{\text{anhydrase}}{\overset{\text{Carbonic}}{\rightleftharpoons}} CO_2 + H_2O$$

At ideal conditions, a single molecule of carbonic anhydrase is capable of catalyzing the breakdown of about 36 million molecules of carbonic acid in 1 minute. The next section examines how enzymes do their work.

10.4 Enzyme-substrate complexes

AIM: To use the lock-and-key model and the induced-fit model to explain binding and specificity in enzyme action.

Focus

Lock-and-key and induced-fit models of binding of substrates to enzymes help explain enzyme specificity.

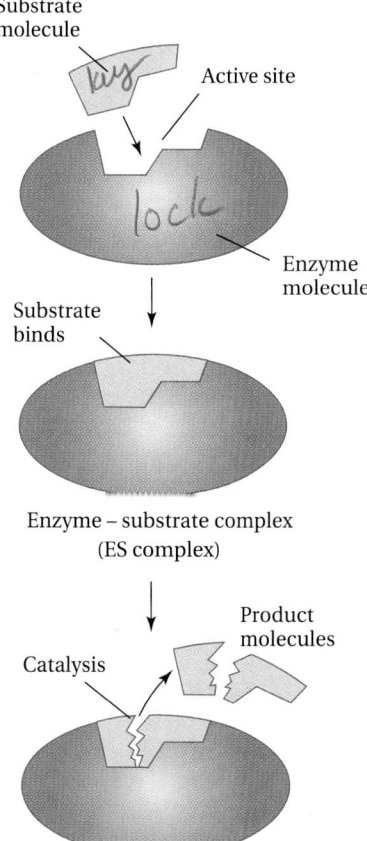

Substrate molecule

Active site

Enzyme molecule

Substrate binds

Enzyme – substrate complex (ES complex)

Catalysis

Product molecules

A substrate molecule must contact an enzyme molecule before it can be transformed into products. Once the substrate has made contact, it must bind to the enzyme at the **active site**—*a region of the enzyme where the processes that convert substrates to products can take place.* The active site is usually a dimple, pocket, or crevice formed by folds in the tertiary structure of the protein. One model that explains the binding of substrates to enzymes is the **lock-and-key model**—*the active site of a given enzyme is shaped so that only the substrate molecule for that enzyme will fit into it, much as only one key will fit into a certain lock.* The interaction of a substrate molecule with an enzyme molecule is illustrated in Figure 10.1. *When the substrate key fits the enzyme lock, an* **enzyme-substrate complex** *is formed.*

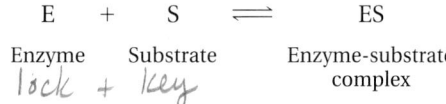

$$E \quad + \quad S \quad \rightleftharpoons \quad ES$$

Enzyme   Substrate   Enzyme-substrate
complex

The formation of an enzyme-substrate complex is an example of *complementarity.* The **complementarity,** *or fit,* between enzyme and substrate in binding helps to produce enzyme specificity. Without complementarity between the enzyme and its substrate, the enzyme would be unable to exert its catalytic power and transform substrate to products. Complementarity between enzyme and substrate is governed by factors such as the overall shapes of the active site and the substrate. Electric charge also can be important. The substrate may be positively charged and the enzyme active site negatively charged, or vice versa. Hydrogen bonds and other weak forces also may hold the enzyme-substrate complex together. Moreover, hydrophobic regions of the substrate associate with similar regions on the

Figure 10.1 (left)
Interaction of a substrate molecule with an enzyme molecule. The substrate fits the active site of the enzyme as a key fits a lock. The result is an enzyme-substrate complex. The bond-breaking and bond-making processes that transform substrates to products occur while the substrate is bound to the active site.

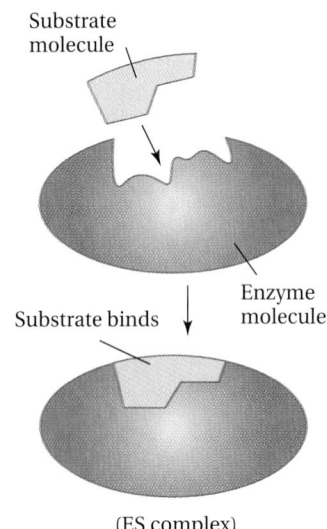

Substrate molecule

Enzyme molecule

Substrate binds

(ES complex)

Figure 10.2
The shape of the enzyme molecule changes to accommodate the substrate in the induced-fit model of enzyme action.

active site of the enzyme. For most enzymes, hydrophobic association and contributions from shape, charge, and a large number of weak forces are responsible for the formation and strength of the enzyme-substrate complex.

The lock-and-key analogy for the formation of enzyme-substrate complexes suffers from an important fact that we have not considered yet: Locks and keys are rigid objects, but substrates and enzymes are flexible. The chains and functional groups of substrate molecules are vibrating and rotating. Enzymes in solution "breathe"—undergo small changes in tertiary structure. Because of this flexibility, the *induced-fit model* is a better model for the formation of enzyme-substrate complexes than the lock-and-key model. *In the* **induced-fit model** *of enzyme-substrate binding, the substrate and enzyme make continuous structural adjustments as the substrate approaches the active site of the enzyme.* The complementarity between enzyme and substrate achieves perfection only at the moment of contact. Figure 10.2 illustrates the induced-fit model of enzyme-substrate complex formation. We should recognize that the lock-and-key and induced-fit models for the formation of enzyme-substrate complexes are just that—models. In reality, the detailed processes of enzyme-substrate complex formation are as numerous as the number of substrates and enzymes themselves.

10.5 Active sites

AIM: To describe what happens at the active site of an enzyme.

We have seen that enzyme binding specificity is analogous to a lock-and-key relationship between enzyme and substrate. But what accounts for the marvelous efficiency of enzymes in transforming substrates to products? The beauty of the precise binding of substrates to enzymes is that it achieves two important ends. First, binding brings specific substrates and enzymes together. This increases the effective concentration of the substrate at the enzyme's active site. The speed of chemical reactions depends on the concentrations of the reactants. Even in dilute solution, the substrate and enzyme are in a one-to-one correspondence at the active site. Second, when the substrate is bound to the active site, it is in precisely the right position for the substrate to undergo the bond making and bond breaking catalyzed by the enzyme. Both these features contribute to the speed of enzyme-catalyzed reactions, which may be millions of times faster than the same reaction in the absence of the enzyme.

Active sites are usually only a small portion of the enzyme protein. Most proteins have molar masses of at least 25,000 g/mol, but only a small part, the active site, is involved in catalysis. Perhaps proteins are large because a firm foundation is needed for the active site to ensure precise alignment of substrate and enzyme. Many mechanical devices are based on this principle. A precision lathe, for example, may be used to turn out small metal spindles. If close tolerances are required, the framework of the lathe is apt to be large and made of cast iron to prevent errors due to vibration.

when substrate is bound to active site of an enzyme.

PRACTICE EXERCISE 10.3

Describe two features that make enzyme-catalyzed reactions go very fast. *① high concentration of substrate and enzyme ② substrate is "locked" into ideal position for bond-making & bond-breaking to occur.*

10.6 Cofactors

AIM: To describe the functions of cofactors.

Focus

Cofactors are essential to the catalytic activity of some enzymes.

Some enzymes consisting only of polypeptide chains can catalyze transformations of substrates, but others need nonprotein *cofactors,* also called *coenzymes,* to assist the transformation. *A* **cofactor** *(coenzyme) is a nonprotein portion of an enzyme necessary for the enzyme's function.* Cofactors may be metal ions of elements such as magnesium, potassium, iron, or zinc, but they also may be small organic molecules. Usually, a cofactor is easily removed from an enzyme molecule, whereas a prosthetic group (see Sec. 9.7) is covalently attached. *An enzyme that contains its bound cofactor is called a* **holoenzyme** *("whole enzyme"); if the cofactor is missing, the enzyme is called an* **apoenzyme.** As Figure 10.3 shows, the job of the enzyme superstructure is to bring the substrate into proximity with the cofactor. The cofactor participates in the chemical transformation catalyzed by the enzyme. Many cofactors are members, or are synthesized from members, of the vitamin B complex group (see Chap. 12).

10.7 Enzyme assay

AIM: To discuss the importance of assaying enzyme activity in the diagnosis of disease.

Focus

Assays of enzyme activity are useful in the diagnosis of disease.

Enzyme activity *is expressed as the rate at which the enzyme catalyzes the conversion of a substrate to products. Experiments in which enzyme activity is measured are called* **enzyme assays.** Many clinical testing procedures consist of assays for enzymes in blood serum—the fluid that remains when blood has clotted. The concept that makes these testing procedures useful is simple: Slight leakage of enzymes from tissues and organs to the circulatory system is normal; additional leakage is abnormal and indicates damage or disease. For instance, damage to the heart muscle in heart attacks

Figure 10.3
Many enzymes contain binding sites for cofactor and substrate molecules. The closeness of these sites in the protein superstructure of the enzyme brings the cofactor and substrate together for reaction.

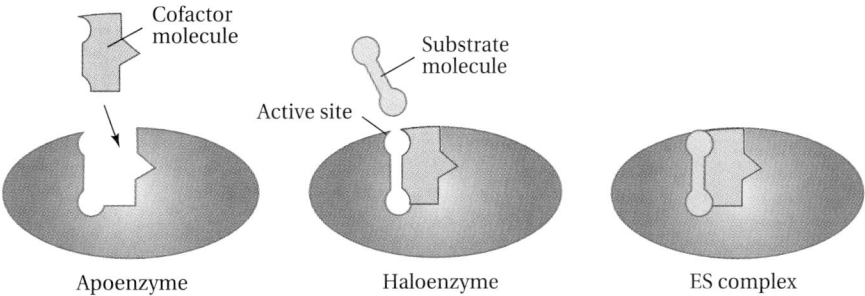

Apoenzyme   Haloenzyme   ES complex

Some enzymes, such as lactate dehydrogenase (LDH), have a number of different forms that differ slightly structurally. The relative amounts of these different forms in the blood have diagnostic value. A large elevation of LDH_5 is linked to acute hepatitis, while a large elevation of LDH_1 typically occurs after a heart attack.

results in the release of the enzyme *creatine phosphokinase* (CPK) into the bloodstream. Comparison of the range of CPK levels found in normal patients with that of the victim of a possible heart attack can indicate whether an attack has occurred and, if it has, the extent of the damage.

Over 40 other blood serum enzymes can be subjected to clinical testing. Information about a few selected groups of serum enzymes is given in Table 10.1. Enzyme activities are often measured by following the appearance of colored products or the disappearance of colored substrates with time. Alternatively, reactions that involve the uptake or release of protons can be followed by measuring changes in the pH of the assay solution with time.

Because of an increased demand for clinical tests, the assay procedures for many serum enzymes have been automated and computerized. The enzyme activities are recorded on a form that also shows ranges for normal patients.

The activity levels of enzymes are almost always reported in *international units*. The international unit defines a standard level of enzyme activity that will convert a specified amount of substrate to product within a certain time. *The internationally agreed on value of* **1 international unit** *(1 IU) is that quantity of enzyme that catalyzes the conversion of 1 micromole (1μmol) of substrate per minute at a specified set of reaction conditions.* For example, an enzyme preparation with a value of 40 IU presumably contains an amount of the enzyme 40 times greater than the standard. It is the level of enzyme activity with respect to normal activity that is significant in the diagnosis of disease.

Table 10.1 Some Serum Enzymes Used to Diagnose Disease

Enzyme	Site	Physiologic function	Elevated serum levels in
amylases	salivary glands, pancreas	starch digestion	mumps, pancreatic obstruction or inflammation
peptidases	digestive tract, tissue cells	protein digestion	tissue injury, shock, fever, anemia
lipases	pancreas	fat digestion	pancreatic disorders
alkaline phosphatases	bone marrow, liver	cleavage of phosphate ester bonds at alkaline pH	bone inflammation (Paget's disease), bone softening (osteomalacia), hepatitis, obstructive jaundice
acid phosphatases	prostate	cleavage of phosphate ester bonds at acidic pH	prostate cancer
transaminases	heart, liver	control of nitrogen balance	hepatitis, myocardial infarction
dehydrogenases	heart muscle, skeletal muscle	oxidation-reduction reactions	myocardial infarction, hepatitis, acute and chronic leukemia
creatine phosphokinase (CPK)	heart muscle, skeletal muscle	formation of creatine phosphate in muscle	myocardial infarction

EXAMPLE 10.2

Interpreting enzyme assays

What disease is indicated in a patient with elevated blood serum levels of both transaminase and alkaline phosphatase enzymes? *hepatitis, liver disease.*

SOLUTION

Although other factors would enter into a diagnosis, elevation of these two enzymes occurs in the disease hepatitis.

PRACTICE EXERCISE 10.4

A patient was vomiting, complained of stomach pains, and had a fever. An enzyme assay of the patient's blood revealed abnormally high levels of serum amylase and serum lipase. Suggest a possible diagnosis of the patient's illness. *pancreatic disorder*

10.8 Effects of pH, temperature, and heavy metals

AIM: To interpret, on a molecular scale, changes in enzyme activity that occur as a result of changing pH and changing temperature.

Focus

Enzyme activity is sensitive to environmental conditions.

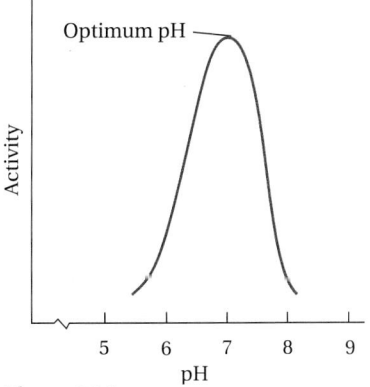

Figure 10.4
Typical pH-activity profile for an enzyme. The pH value at which the maximum activity is observed is called the enzyme's pH optimum. Here, the enzyme has its optimum pH near 7.

Enzyme assays must be carried out at identical conditions for comparisons of activity to be valid. Particular attention must be paid to pH and temperature, since small variations in these two conditions often cause large changes in the activity values obtained. Suppose, for example, that the anionic form of a single carboxylic acid side chain of aspartic acid is involved in binding or catalysis by a certain enzyme. At pH values near or above neutrality, this group is completely dissociated into a carboxylate anion and a proton, and the enzyme has the greatest possible activity. If the pH of the solution is gradually lowered, the fraction of carboxylate anion decreases, and the fraction of undissociated carboxylic acid increases. At low pH, where all the carboxylic acid groups are in the undissociated form, the enzyme would become completely inactive.

The variation of an enzyme's activity with changes in pH is shown in a **pH-activity profile.** Figure 10.4 gives an example. Most enzymes exhibit maximum activity at a pH near 7, the pH of most body fluids. *The pH where maximum activity occurs is called the* **pH optimum** *of the enzyme.* The pH-activity profiles of different enzymes may be very complex if ionization of more than one group on the enzyme is important to binding or catalysis or if substrate ionizations are also important.

The pH-activity profiles of enzymes generally reflect reversible changes in activity of enzymes. That is, a decline in the activity of an enzyme, say, with a decrease in pH, is reversed if the pH is increased to its original value. Like other proteins, however, most enzymes are stable only within a narrow range of pH. Extremes of pH will irreversibly denature most enzymes, and a

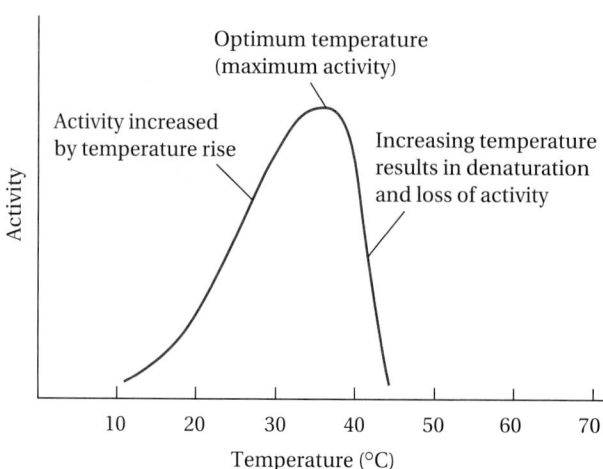

Figure 10.5
The activity of an enzyme depends on the temperature. If the temperature is raised past a certain point, heat denaturation causes a sharp decrease in enzyme activity.

Denaturation leaves peptide bonds intact but disrupts the secondary, tertiary, and quaternary structures of a protein molecule. If the protein is an enzyme, denaturation will damage the active site and destroy the activity of the enzyme.

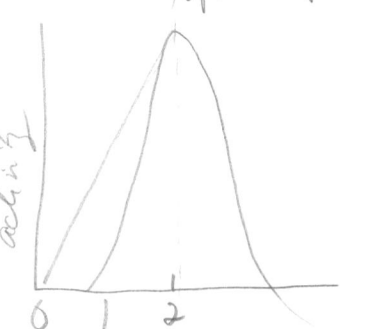

return to the pH optimum will not restore the activity. What constitutes an extreme of pH depends on the particular enzyme being studied.

Enzyme activity usually varies with temperature. At temperatures near 20 °C, most enzymes are active and relatively stable. As with nonbiological catalysts, an increase in temperature increases the reaction rate. Figure 10.5 illustrates that activity can rapidly decrease beyond the optimal temperature. The decrease in activity is usually due to heat denaturation of the enzyme.

Although some enzymes contain metal ion cofactors, ions of heavy metal ions such as those of lead and cadmium are often damaging to enzymes, as discussed in A Closer Look: Lead Poisoning.

PRACTICE EXERCISE 10.5
Enzyme assays must be done at a specified pH and specified temperature. Why?

PRACTICE EXERCISE 10.6
Pepsin, an enzyme found in gastric juice, has an optimal pH of about 2. Draw its pH-activity profile.

10.9 Induction and degradation of enzymes

AIM: To explain how induction and degradation can control enzyme concentrations in a cell.

Focus

Induction and degradation help control enzyme concentrations in cells.

The concentrations of various body chemicals must be strictly controlled. To assert this control, cells need to slow down or speed up the production of cellular products. Cells use several means to control levels of cellular products. Two of these means involve enzyme **induction**—*the synthesis of enzymes on demand*—and **degradation**—*the hydrolysis of enzymes to their constituent amino acids.*

Each enzyme molecule, given a sufficient supply of raw materials, will catalyze the formation of products at a rapid, steady pace. One way to

A Closer Look

Lead Poisoning

Lead is a heavy metal poison that affects the functioning of the brain, blood, liver, and kidneys. Lead poisoning often leads to mental retardation and neurologic disorders through damage to the brain and central nervous system. Some of the harmful effects of lead poisoning result from the effects of lead(II) ions (Pb^{2+}) on the action of enzymes. Many enzymes contain thiol groups (—SH groups) from the amino acid cysteine. Lead ions, as well as other heavy metal ions, bind tightly to —SH groups and denature the enzymes. This renders the enzymes inactive. Lead ions also form strong bonds with the side-chain carboxylate ions of acidic amino acid residues with similar results.

Compounds of lead are widespread in the environment. Leaded gasoline was once a major contributor to environmental lead. This problem source has been eliminated by the introduction of lead-free gasoline. However, another more insidious source of environmental lead pollution—lead paint—still exists.

Lead poisoning is a current major problem with children. The interior woodwork of most homes built before 1950 was coated with paint that contained up to 50% lead. Lead paint poses little hazard unless it flakes and powders. Children may eat the sweet-tasting chips of lead-based paint. These same children also may breathe lead-containing paint dust, which further increases their body burden of lead.

It is estimated that more than 3 million children in the United States have blood levels of 10 μg of lead or more per deciliter. Such blood levels are considered a cause for concern by the U.S. Centers for Disease Control and Prevention. Children with these blood levels of lead are at risk of

The presence of lead paint in many older buildings makes its abatement essential in preventing lead poisoning, especially in children.

underdeveloped nervous systems and lowered intelligence.

A child with lead poisoning is often constipated and vomits. A radiograph of the child's abdomen will reveal the presence of radiopaque ingested paint fragments. A blood sample can be taken and measured for lead content. Blood tests also reveal a low red blood cell count and low hemoglobin levels.

Chronic lead poisoning can be treated by giving the patient calcium EDTA (ethylenediaminetetraacetic acid) intravenously. The EDTA complexes (binds tightly) with the lead ions in the body and allows them to be excreted as the Pb-EDTA complex in the urine. During treatment, the lead content of the urine is monitored. A rise in the lead content of the patient's urine indicates that lead is being eliminated from the body.

increase production is for the cell to increase the number of enzyme molecules. *Enzymes that are synthesized in response to a temporary need of the cell are called* **inducible enzymes.** Cells grown in a medium that is deficient in a certain nutrient may not produce the enzymes required to transform that nutrient to useful products. If the cells are transferred to a medium that contains the nutrient, they begin to produce the necessary enzymes. When the cells are returned to the original medium, the induced enzymes soon disappear, degraded by the digestive machinery of the organism.

10.10 Control of enzyme activity

***AIMS:** To compare the control mechanisms of competitive inhibition and enzyme modulation. To define an allosteric enzyme and a pacemaker enzyme.*

Induction and degradation of enzymes are important ways for the cell to economize on the variety of enzymes it produces. Used exclusively, however, they would be a rather coarse way of controlling production of substances in the cell. A finer control for the slowdown of cellular processes is provided by *inhibitors.*

Competitive inhibitors *are molecules that are similar in shape or charge to substrate molecules and capable of binding to enzyme active sites.* However, competitive inhibitors are not transformed into products. If the inhibitor in Figure 10.6 binds at the active site, for example, it will block further use of the enzyme by substrates for as long as it remains bound. Certain enzymes normally present in the body are complexed with natural competitive inhibitors. When a need for such an enzyme arises, events occur that cause the complex to dissociate. The enzyme is then free to go about its appointed task. A Closer Look: HIV Protease and Its Inhibition, describes one way that scientists are trying to enlist enzyme inhibitors in the fight against HIV.

Cells have other ways of using competitive inhibitors to control the activity of enzymes. A fine slowdown control of enzyme activity built into some enzymes relies on **feedback inhibition**—*the concentration of a product at the end of a series of steps builds up and then "feeds back" to inhibit the enzyme in a preceding step.* Assume that substrate *A* is converted by a team of three enzymes, E_1, E_2, and E_3, to a product *D* as shown in this equation:

$$A \xrightarrow{E_1} B \xrightarrow{E_2} C \xrightarrow{E_3} D$$

Feedback

Each of the enzymes works at transforming the product made by the previous enzyme into a new product, finally turning out the end product *D*. Now suppose that enough *D* has been produced to satisfy the needs of the cell. A slowdown of the production line will occur if the product *D*, now present in a relatively high concentration, is a competitive inhibitor of the enzyme E_1.

More often, an enzyme such as E_1 is *not* competitively inhibited by a product such as *D*. Instead, enzymes that catalyze the early stages of a long

Figure 10.6
Competitive inhibitors block the active site of an enzyme so that the substrate molecule cannot enter. Naturally, a substrate molecule that cannot enter the active site cannot be transformed to products.

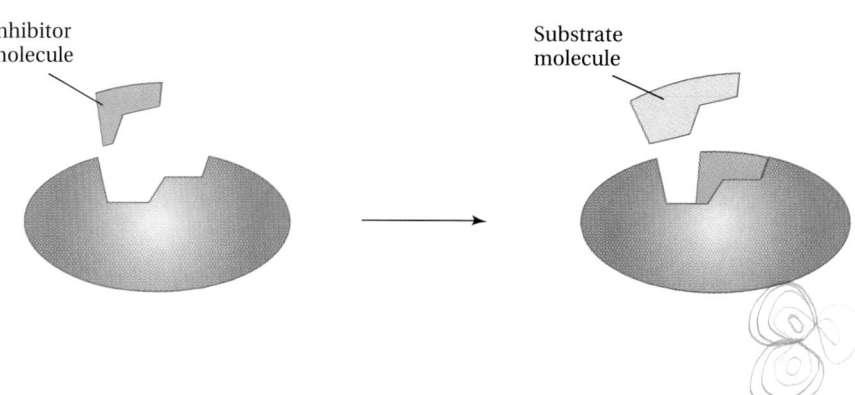

HIV Protease and Its Inhibition

The human immunodeficiency virus (HIV) is the causative agent in acquired immunodeficiency syndrome (AIDS). The destruction of cells of the immune system by HIV heightens the susceptibility of patients with AIDS to infections and tumors that inevitably lead to death. Responsible for one of only four enzyme activities produced by the virus, *HIV protease* is a target for the development of therapeutic drugs for AIDS. We can describe the enzyme and the path that research on the HIV protease is taking, but it is too early to say whether the work will lead to effective ways of slowing or stopping the progress of AIDS.

The three-dimensional structure of HIV protease has been determined by X-ray crystallography (see figure). The enzyme is a protein dimer consisting of two identical subunits. Each subunit contains only 99 amino acid residues. HIV protease binds to large proteins produced during reproduction of the virus and splits them into smaller polypeptides. These smaller polypeptides are needed for the assembly of new, infectious viruses. The biological function of HIV protease provides an approach to the design of medicines to treat AIDS. If HIV protease could be inhibited in an HIV-infected person, only incomplete, noninfectious viruses would be formed. Very good inhibitors of HIV protease have been discovered. Some of the best can be considered "decoy" molecules. These decoy molecules consist of peptides similar to the peptide substrates cleaved by HIV protease, except the peptide bond that would ordinarily be cleaved by the protease is replaced by a carbon-carbon or other hydrolysis-resistant bond.

Like ducks drawn to decoys, the protease is attracted to the peptide parts of the inhibitor molecules and binds them tightly to its active site. However, the enzyme cannot cleave the hydrolysis-resistant bond of the decoy molecule. The protease is rendered unable to process viral proteins as long as the decoy inhibitor remains bound to the active site. Other compounds are also being tried in order to block the action of HIV protease. Many of these molecules have no obvious structural resemblance to the natural protein substrate of HIV protease. For example, a chemically modified version of buckyball has been found to block the active site of the enzyme.

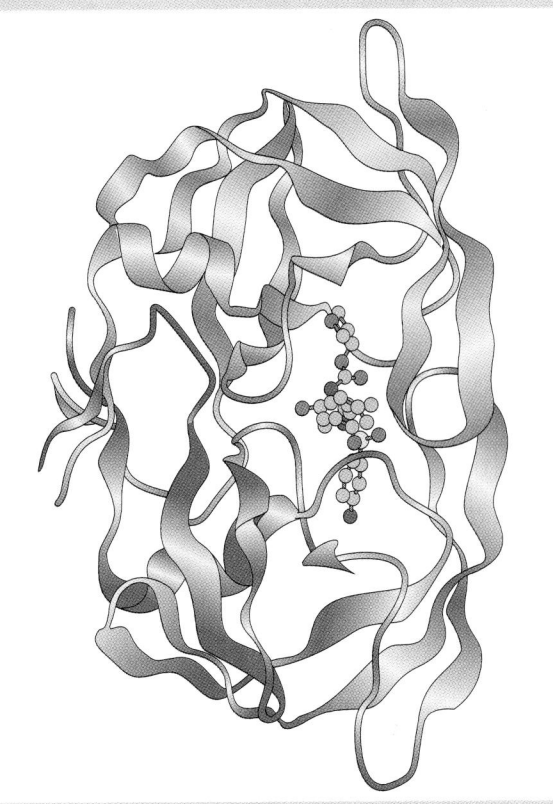

The dimer of HIV protease with bound inhibitor. The identical subunits of the dimer are colored in blue and red. Each subunit contains a short section of alpha helix (shown as a curly ribbon) and several regions of beta-pleated sheet structure (shown as sections of parallel ribbons).

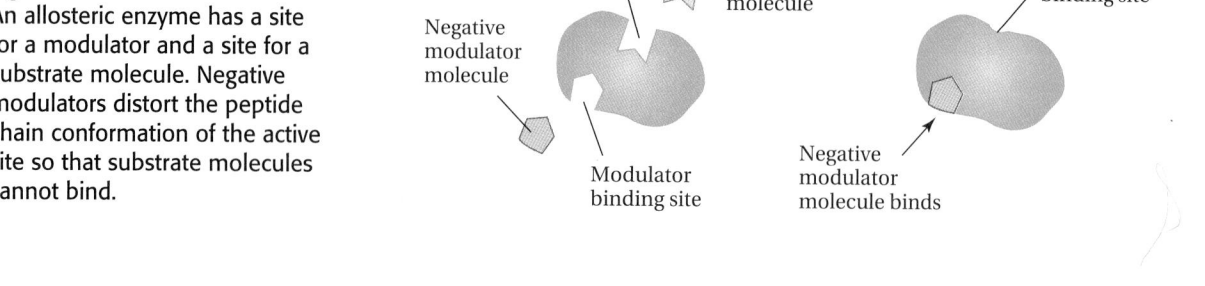

Figure 10.7
An allosteric enzyme has a site for a modulator and a site for a substrate molecule. Negative modulators distort the peptide chain conformation of the active site so that substrate molecules cannot bind.

Allosteric
allo (Greek): other
stereos (Greek): site or space

Allosteric enzymes are used as control points in many metabolic pathways. For example, a high concentration of a metabolic intermediate can serve as a negative modulator of an enzyme that catalyzes a reaction producing more of the intermediate.

sequence of reactions often contain another binding site in addition to the active site. *This second site is called the* **modulator binding site.** *Enzymes containing both an active site and a modulator binding site are called* **allosteric enzymes.** *A molecule that binds to the modulator binding site and slows down a reaction is a* **negative modulator.** In our example, if E_1 is an allosteric enzyme, product D will often serve as a negative modulator.

Binding of a negative modulator to the modulator binding site of an allosteric enzyme usually distorts the enzyme's active site (Fig. 10.7). This distortion hampers the formation of the enzyme-substrate complex. The enzyme's activity is decreased when the negative modulator is bound to the modulator binding site. *Some allosteric enzymes have* **positive modulators**—*modulators that increase the enzyme activity.* In these enzymes, the active site may be distorted in the *absence* of the modulator. Figure 10.8 shows a positive modulator binding to the modulator site and inducing refolding of an enzyme to the active form.

Pacemaker enzymes *control the rates of cellular processes.* Often allosteric enzymes, pacemakers are usually found near the start of a sequence of enzyme-catalyzed reactions. Negative modulation of allosteric pacemaker enzymes by end products of the sequence is important to the economy of the cell. After all, there is not much sense in expending the material and energy resources of the cell by going through a series of complex reactions to make a final product that already is present in sufficient quantities to meet cellular needs. On the other hand, positive modulation provides a means to speed up a process in a way that is more rapid, more

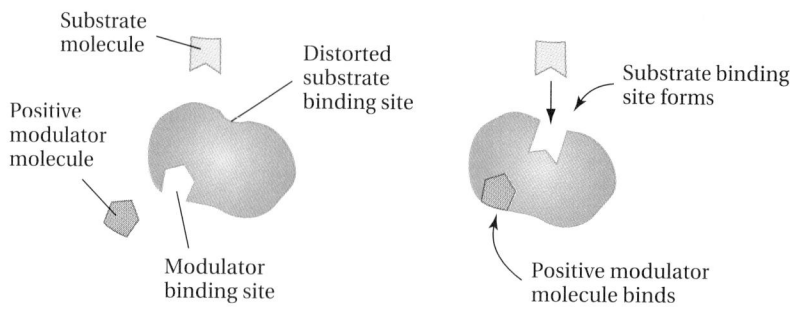

Figure 10.8
Positive modulators change the peptide chain conformation of the active site so that substrate molecules can bind.

[handwritten marginal notes at top:]
positive modulator: a molecule that binds to the modulator binding site of allosteric enzyme and increases enzyme activity
negative modulator: a molecule that binds to the modulator binding site of allosteric enzyme and slows down enzyme activity

[left margin handwritten:]
they affect the activity of an enzyme by changing the conformation of its peptide chain

direct, and more economical than the synthesis of whole new enzyme molecules.

PRACTICE EXERCISE 10.7

Explain the difference between a positive modulator and a negative modulator.

PRACTICE EXERCISE 10.8

Distinguish between a cofactor and a positive modulator.

[handwritten:]
- an enzyme cofactor takes part in the chemical reaction that the enzyme catalyzes
- pos. modulator is essential for activating the enz. but it does not take part in the actual chemical Rx

10.11 Zymogens

AIMS: To describe the action of a peptidase on a zymogen to form an active enzyme. To describe at least two physiologic processes that rely on activation of zymogens.

Focus

Active enzymes are sometimes produced from zymogens.

Some enzymes are synthesized in biologically inactive forms that are transformed into active enzymes as the organism needs them. *These inactive precursors are called* **zymogens** *or* **proenzymes.** Along with induction and positive modulation, the activation of zymogens represents a way to control enzyme activity in an organism. A peptidase is usually required for the conversion of a zymogen to an active enzyme. The peptidase catalyzes the hydrolysis of one or more specific peptide bonds in the zymogen. Once these peptide bonds are broken, the remaining protein chains are free to fold into a catalytically active conformation.

Table 10.2 lists many peptidases that are formed from zymogen precursors. *Pepsin,* for example, a protein-digesting enzyme of the stomach, is produced by the action of stomach acid on the zymogen *pepsinogen.* Pepsin itself is also capable of converting pepsinogen into the active enzyme. *The process by which an enzyme molecule catalyzes the activation of its own zymogen is called* **autoactivation.**

Zymogens of some other peptidases are formed in the pancreas; these include *trypsinogen, chymotrypsinogen,* and *procarboxypeptidase.* As the zymogens are formed, they are packaged as **zymogen granules**—*zymogens in coats of lipid and protein.* The zymogen granules are then stored in the

Table 10.2 Peptidases Formed from Zymogen Precursors

Zymogen ⟶ active enzyme	Site of zymogen formation	Site of zymogen activation	Activating agent
pepsinogen ⟶ pepsin	stomach cell walls	stomach	stomach acid or autoactivation
trypsinogen ⟶ trypsin	pancreatic cells	small intestine	enterokinase or autoactivation
chymotrypsinogen ⟶ chymotrypsin	pancreatic cells	small intestine	trypsin
procarboxypeptidase ⟶ carboxypeptidase	pancreatic cells	small intestine	trypsin

pancreatic cells and, when required, are released into the small intestine. Trypsinogen is converted to trypsin by chopping small peptides from the zymogen. (See A Closer Look: Trypsin: Anatomy of an Enzyme.) These peptide cleavages are catalyzed by *enteropeptidase*. Autoactivation of trypsinogen also may occur. Trypsin then activates chymotrypsinogen and procarboxypeptidase to their active enzymes by peptide bond cleavages.

Once activated, pepsin, trypsin, and chymotrypsin help to digest food proteins. All three enzymes are **endopeptidases**—*peptidases that catalyze the hydrolysis of peptide bonds in the interior of peptide chains but not those at the C-terminal or N-terminal end. Carboxypeptidase is also a digestive enzyme, but it is an* **exopeptidase**—*a peptidase that catalyzes the removal of amino acid residues, one at a time, from the C-terminal or N-terminal end of peptide chains.* Carboxypeptidase starts from the C-terminal end. Working in concert, the four digestive peptidases are capable of completely degrading a protein molecule to its constituent amino acids.

EXAMPLE 10.3   **Determining the C-terminal group of a peptide**

You may recall from Chapter 9 that methionine enkephalin, Tyr-Gly-Gly-Phe-Met, is one of nature's pain killers. What products will be produced if a solution of methionine enkephalin is treated for a short time with carboxypeptidase? → *exopeptidase*

SOLUTION

Carboxypeptidase is an exopeptidase that catalyzes the hydrolysis of peptide bonds, one at a time, starting at the C-terminal end of the peptide. After a short exposure of the methionine enkephalin to carboxypeptidase, the products would be the C-terminal amino acid methionine (Met) and the tetrapeptide Tyr-Gly-Gly-Phe.

EXAMPLE 10.4   **Determining the N-terminal group of a peptide**

Aminopeptidase is an exopeptidase that catalyzes the hydrolysis of peptide bonds, one at a time, starting at the N-terminal end of the peptide. When we treat a tripeptide with aminopeptidase and isolate the products, we obtain the amino acid leucine (Leu) and the dipeptide Asp-Gly. What is the sequence of amino acids in the original tripeptide?

SOLUTION

Aminopeptidase cleaves peptide bonds from the N-terminal end of peptides, so the leucine obtained from the hydrolysis must have been the N-terminal residue of the tripeptide. The original tripeptide must be Leu-Asp-Gly.

EXAMPLE 10.5   **Enzyme hydrolysis of peptides**

Chymotrypsin preferentially catalyzes the hydrolysis of peptide bonds in which the amino acid residue on the C-terminal side of the cleaved peptide bond has an aromatic side chain. What are the products if a solution of the following hexapeptide is treated with chymotrypsin?

Arg-Gly-Phe-Gly-Gly-Phe

SOLUTION

The only hydrophobic side chains in the hexapeptide are two phenylalanine (Phe) residues: at the third position from the N-terminal end and at the C-terminal end. Since chymotrypsin is an endopeptidase, it will not remove the C-terminal Phe residue. It will, however, cleave the Phe-Gly peptide bond. The products will be two tripeptides: Arg-Gly-Phe and Gly-Gly-Phe.

PRACTICE EXERCISE 10.9

Consider the general structure for a hexapeptide with its five peptide bonds:

$$NH_2-\overset{\displaystyle H}{\underset{\displaystyle R}{C}}-\overset{\displaystyle O}{\underset{}{C}}\underset{1}{-}N-\overset{\displaystyle H}{\underset{\displaystyle R}{C}}-\overset{\displaystyle O}{\underset{}{C}}\underset{2}{-}N-\overset{\displaystyle H}{\underset{\displaystyle R}{C}}-\overset{\displaystyle O}{\underset{}{C}}\underset{3}{-}N-\overset{\displaystyle H}{\underset{\displaystyle R}{C}}-\overset{\displaystyle O}{\underset{}{C}}\underset{4}{-}N-\overset{\displaystyle H}{\underset{\displaystyle R}{C}}-\overset{\displaystyle O}{\underset{}{C}}\underset{5}{-}N-\overset{\displaystyle H}{\underset{\displaystyle R}{C}}-CO_2H$$

Which peptide bonds could be hydrolyzed by (a) endopeptidases and (b) exopeptidases?

1 & 2 (at the ends)   *2-3-4 (not at the ends)*

Activation of digestive peptidases from their proenzymes is a way to control levels of peptidase activity in digestion, but it is also a necessary protective device for the pancreas. Premature activation of trypsinogen and chymotrypsinogen occurs in one form of acute pancreatitis. In this disease, the patient's pancreas is destroyed by its own protein-digesting enzymes.

10.12 Blood clotting

AIM: To describe the function of zymogens and enzymes in the blood-clotting and clot-dissolution mechanisms.

Focus

Blood clot formation and dissolution are controlled by activation of zymogens.

Along with protein digestion, blood clotting is another important process that involves the activation of zymogens. Blood clotting must be precisely controlled, because blood must clot when it is shed, but it must not clot in the blood vessels. The early steps of the process leading to blood clotting are extremely complex. At the end of the process, the cleavage of peptides from the zymogen *prothrombin* to give *thrombin* sets the stage for actual clot formation.

Prothrombin ⟶ Thrombin

(Zymogen)   (Active enzyme)

Thrombin is a peptidase that accelerates clot formation by catalyzing the

Trypsin: Anatomy of an Enzyme

Most of our knowledge about how enzymes work comes from detailed studies of a relatively few enzymes, one of which is trypsin. Over the years, biochemists have probed nearly every facet of trypsin's action and structure, ranging from its pH dependence to its tertiary structure (see figure, part a). Trypsin is a globular protein containing 245 amino acid residues with five disulfide bridges. Trypsin aids digestion by catalyzing the

(a)

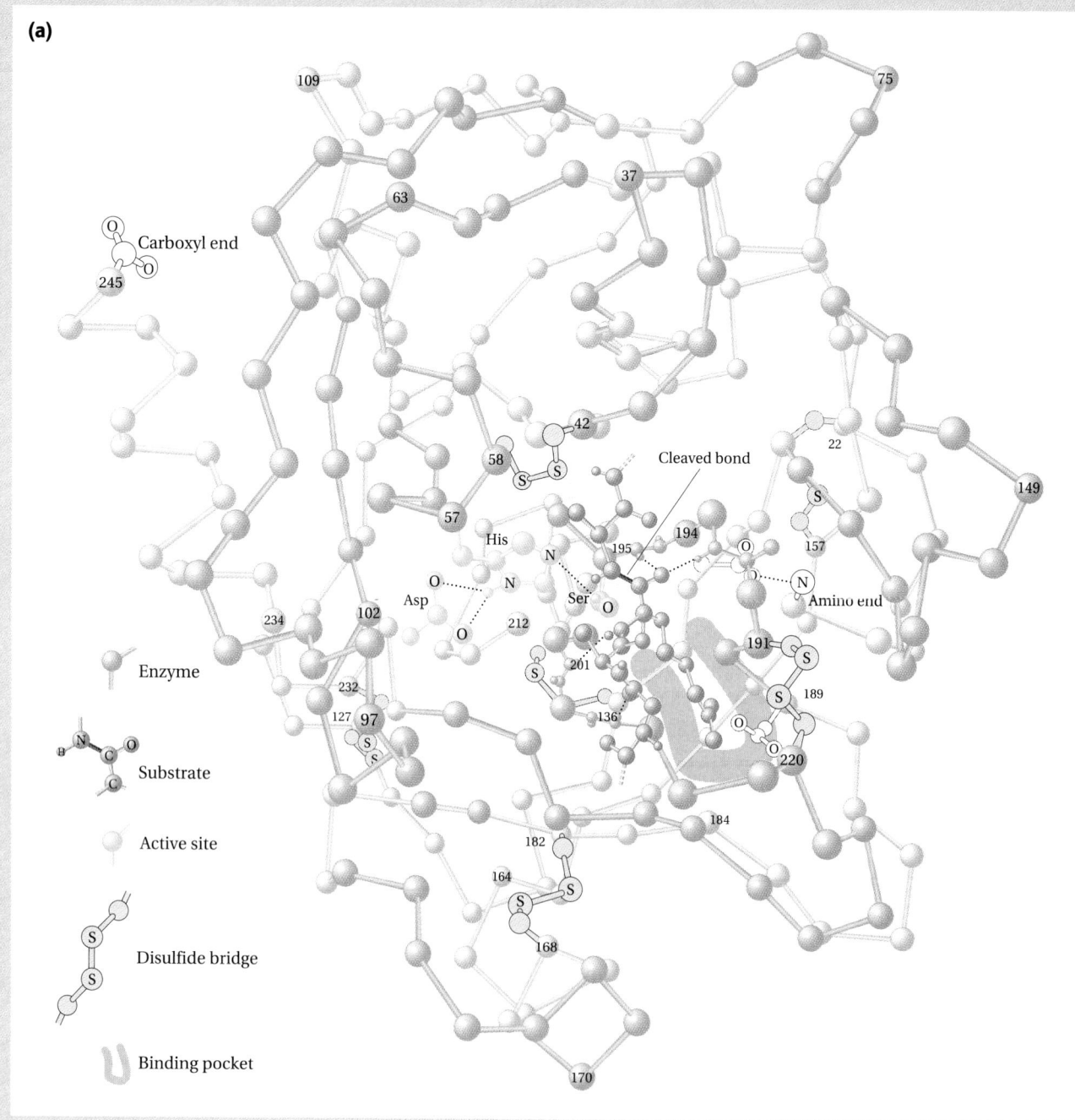

cleavage of peptide bonds of dietary proteins and peptides. However, the specificity of trypsin is such that it cleaves only peptide bonds in which the carboxyl group of the peptide bond to be cleaved comes from lysine and arginine, two residues whose side chains are positively charged at the enzyme's optimum pH.

As shown in part (b) of the figure, a lysine side chain of the polypeptide binds selectively to the active site pocket. It is drawn there by the negatively charged carboxylate ion of an aspartate residue at position 189 of the amino acid sequence of the enzyme's peptide chain. Except for lysine and arginine, the side chains of other amino acid residues do not bind because of a lack of complementarity of size or charge.

Once the substrate is bound in the proper orientation, the catalytic groups of the enzyme go to work. Three side-chain groups on the enzyme seem to be necessary, but the one involved most directly is the hydroxyl group of serine 195. The cleavage of the substrate's peptide bond occurs in two stages.

First, the hydroxyl group of serine 195 displaces the amide nitrogen of the peptide bond of the substrate. As a result of the displacement, the substrate's peptide bond is broken, and an ester bond forms between enzyme and substrate. This intermediate is called an *acyl enzyme*. The piece of substrate protein just past the bond cleavage diffuses away.

Second, a water molecule at the enzyme active site hydrolyzes the ester bond of the acyl enzyme. This reaction is also catalyzed by the enzyme. The released peptide fragment floats away, and the free enzyme is able to do its catalytic work on yet another peptide bond.

Although the process is complex, one trypsin molecule can cleave more than 100 peptide bonds per second under ideal conditions. Several other enzymes that catalyze hydrolysis processes also employ a crucial serine hydroxyl group at the active site, although the binding sites of these enzymes are quite different from trypsin's. Included in this family of trypsin-like enzymes are chymotrypsin, thrombin, plasmin, and acetylcholinesterase.

(a) (left) The chain conformation of trypsin as deduced by X-ray crystallography. Only the positions of the alpha carbons of the amino acid residues of the enzyme's peptide chain are shown. The numbers refer to the positions of some of these residues in the enzyme's peptide chain. The shaded area at the lower right of the molecule is the active site pocket; part of a bound peptide substrate is shown in dark blue. The arrow marks the position of the catalytically important serine 195. Residues of histidine 57 and aspartate 102 also contribute to the speed of trypsin's catalysis of certain peptide bonds.

(b) (below) The mechanism of cleavage of a peptide bond at the active site of trypsin.

(b)

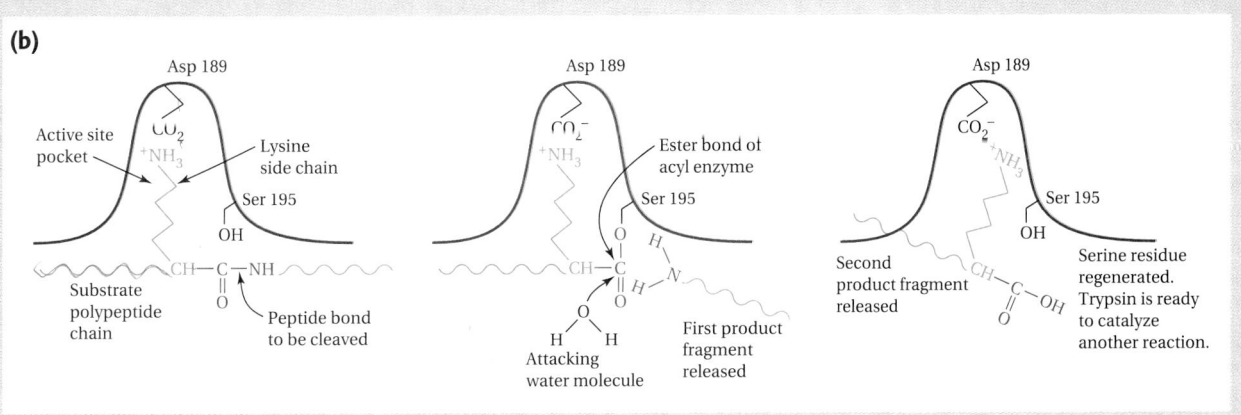

311

Fibrinogen is not a zymogen. Here the *-ogen* ending simply indicates that fibrinogen is the precursor molecule to fibrin, which is a fibrous protein and *not* an enzyme.

conversion of *fibrinogen* to *fibrin*. Clot formation begins when thrombin cleaves the peptide bonds of four Arg-Gly pairs in fibrinogen, a large (molar mass 340,000 g/mol), soluble blood protein. Cleavage of these Arg-Gly bonds releases four highly charged peptides that are responsible for maintaining the solubility of fibrinogen. Following the release of these peptides, the resultant fibrin molecules form long, insoluble fibers that precipitate as a mass held together by hydrogen bonds. The eventual formation of covalent bonds between fibrin molecules makes a stable clot.

After blood clots serve their function, they are dissolved and reabsorbed into the bloodstream. *Plasmin*, a peptidase activated from its zymogen *plasminogen*, is responsible for clot breakdown. A clot that prematurely breaks loose from the point of formation and enters a blood vessel may cause an *embolism*, a blockage of the flow of blood to tissues and organs. Embolisms also may be caused by air bubbles, clumps of bacteria, or other foreign matter, including bullets.

Especially severe damage or death results when the blockage causes a decreased oxygen supply to the brain (stroke), lung (pulmonary embolism), or heart (myocardial infarction). Potentially damaging blood clots are prevented from forming in the bloodstream of a normal individual by the presence of inhibitors that prevent completion of the clotting activation sequence. Among these inhibitors is heparin, a complex polysaccharide concentrated in the lungs and arterial cell walls (see Sec. 7.10). You may recall from the Case in Point earlier in this chapter that a stockbroker named Art suffered a heart attack. The administration of a solution containing an enzyme was among his treatments, which are discussed below in the Follow-up to the Case in Point.

PRACTICE EXERCISE 10.10

How is blood clot formation controlled by zymogen activation?

FOLLOW-UP TO THE CASE IN POINT: Enzyme therapy for heart attacks

Many heart attacks involve the formation of blood clots that clog (infarct) the arteries of the heart. These clots block the flow of blood into the heart, starving the heart tissues for oxygen. Failure to rapidly restore the flow of blood to the heart muscle results in permanent damage to the heart or even death. What can be done to dissolve the clots? In Art's case, the doctor ordered intravenous administration of a solution containing 100 mg of an enzyme called *tissue plasminogen activator* (TPA). The TPA solution was administered over a 2-hour period. TPA is a peptidase that activates plasminogen, the zymogen of plasmin. The plasmin then goes about its natural work of dissolving the fibrin in blood clots that may be blocking the coronary arteries. Studies show that the survival of victims of coronary infarctions is significantly increased if steps are taken to dissolve the clots within about 2.5 hours of the onset of the heart attack. Thanks to the early administration of TPA and other excellent medical care, Art is now recuperating from his heart attack; his prospects for a full recovery are good. TPA is not the only enzyme used for the dissolution of clots in heart attacks. *Streptokinase*, a peptidase obtained from certain bacteria, also can be used with good results.

10.13 Antibiotics and other therapeutic drugs

AIM: To discuss the action of antibiotics and other therapeutic drugs as enzyme inhibitors.

Knowledge of how enzymes and inhibitors work has been of enormous benefit to medical scientists in their search for ways to cure or control diseases. **Antibiotics**—*drugs used to control harmful bacteria*—show how enzyme inhibitors can be put to therapeutic use. You may recall that sulfanilamide and other sulfa drugs work as antibiotics because they inhibit the synthesis of folic acid from *p*-aminobenzoic acid (A Closer Look: The Sulfonamide Antibiotics). Bacteria require folic acid to survive but cannot use the inhibitor to make it, so the microorganism dies from folic acid deficiency.

In good antibiotic design, the drug must interfere with the metabolism of the infecting organism but not with that of the infected individual. In 1928, Alexander Fleming, working at St. Mary's Hospital in London, was searching for a universal antibiotic. He discovered that a growth culture of a pus-producing bacterium disappeared in an area of the culture where green mold was growing. Figure 10.9 shows the mold component responsible—*penicillin.* In 1941, the drug was first applied to severe bacterial infections in humans. The results were dramatic. The patients who received penicillin made rapid, complete recoveries. It is now believed that one action of peni-

Figure 10.9
Below, penicillin mold and, right, the structure of penicillin and some of its commonly used derivatives.

cillin is to inhibit an enzyme that catalyzes the synthesis of cell walls of reproducing bacteria. The new cells have defective cell walls, the cell contents leak out, and the cell dies. Since the cell wall material is not present in humans, penicillin can be used to treat bacterial infections in people.

As we will learn in A Closer Look: The Toxicity of Pesticides, some enzyme inhibitors are very effective poisons.

Since the discoveries of sulfanilamide and penicillin, scores of new antibiotics have been developed. Indeed, the search for new antibiotics has stepped up in recent years because of the evolution of the so-called superbugs. Because of the widespread use of antibiotics, these highly antibiotic-resistant bacteria have evolved. Whatever antibiotics are used now and in the future, probably no antibiotic will ever be capable of winning the war against infecting bacteria. Instead, antibiotics fight a holding action that gives the infected organism's natural defenses time to mobilize against the invader.

Many therapeutic drugs besides antibiotics are also enzyme inhibitors. Several drugs introduced to lower blood pressure, including captopril and linisopril, are inhibitors of a peptidase, the *angiotensin-converting enzyme* (ACE).

Captopril (Capoten)

Linisopril (Prinivil)

ACE catalyzes the release of the octapeptide angiotensin II from the decapeptide angiotensin I.

$$\text{Asp}-\text{Arg}-\text{Val}-\text{Tyr}-\text{Ile}-\text{His}-\text{Pro}-\text{Phe}-\text{His}-\text{Leu} \xrightarrow{\text{ACE}}$$

Angiotensin I

$$\text{Asp}-\text{Arg}-\text{Val}-\text{Tyr}-\text{Ile}-\text{His}-\text{Pro}-\text{Phe} + \text{His}-\text{Leu}$$

Angiotensin II

An assay for elevated levels of angiotensin-converting enzyme in the blood can be used to detect *sarcoidosis*—an accumulation of certain tissue called *granuloma*. Sarcoidosis usually affects the lungs, where it causes coughing and shortness of breath, but it may appear in other parts of the body such as the joints.

Angiotensin II causes the constriction of arteries. Since the heart still tries to pump the same amount of blood through the narrowed arteries, blood pressure increases. The ACE inhibitors help lower blood pressure by preventing ACE from converting angiotensin I to angiotensin II. You may recall from A Closer Look: HIV Protease and Its Inhibition that scientists are also working to find therapeutically useful inhibitors of enzymes produced by the human immunodeficiency virus (HIV), the causative agent for the acquired immunodeficiency syndrome (AIDS). Whether the quest will eventually lead to new drugs that delay the onset of AIDS or control the disease is still open to question.

The Toxicity of Pesticides

Substances that inhibit bacterial enzymes while leaving human enzymes unaffected have become important antibiotics, such as sulfa drugs. However, several powerful insecticides, including parathion and malathion, are inhibitors of human *and* insect enzymes.

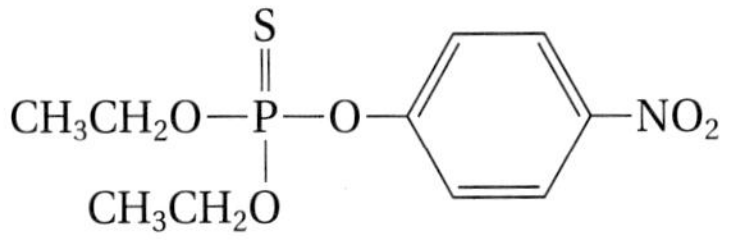

Parathion

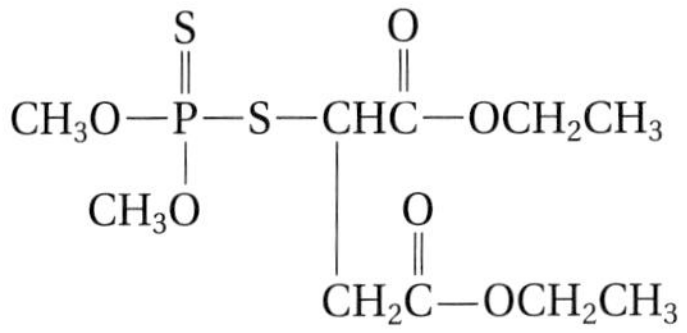

Malathion

Poisons, such as these insecticides, inhibit the enzyme acetylcholinesterase, which is responsible for a crucial event in the process of muscle contraction. Its inhibition results in paralysis of the body's muscles and eventually death. Inhibitors of this type were originally called "nerve gases," although nowadays many liquids and solids that have a similar effect are known. Muscle contraction begins with signals from the nervous system that are communicated to muscles by acetylcholine (the ester of acetic acid and the nitrogen-containing alcohol choline).

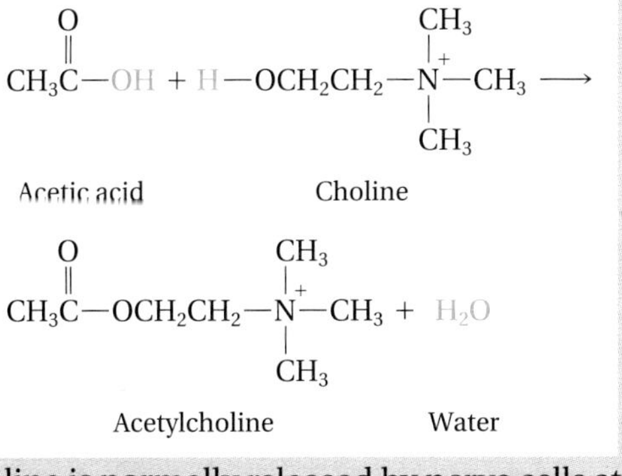

Acetic acid   Choline

Acetylcholine   Water

Acetylcholine is normally released by nerve cells at their junction with muscle. It flows into the muscle cells and triggers a chain of events leading to muscle contraction. In order for muscles to alternately contract and relax, such as in the functioning of the

Crops are often sprayed with insecticides.

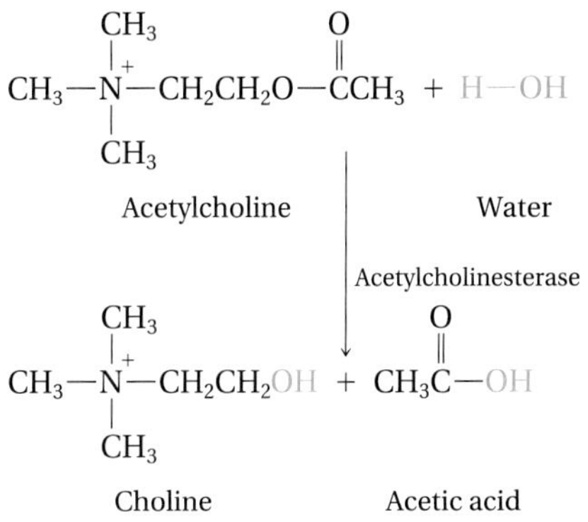

cardiorespiratory system, acetylcholine must be released and destroyed alternately. Acetylcholinesterase is the destroyer, promoting the hydrolysis of acetylcholine back to acetic acid and choline.

Acetylcholine   Water

Acetylcholinesterase

Choline   Acetic acid

Without acetylcholinesterase, muscles cannot relax, and paralysis sets in. The action of parathion is similar to that of many pesticides. Parathion blocks the action of acetylcholinesterase by forming a stable phosphate ester with the hydroxyl group of a serine residue of the enzyme. This hydroxyl group is crucial for the enzyme to do its catalytic work, just as it is in trypsin.

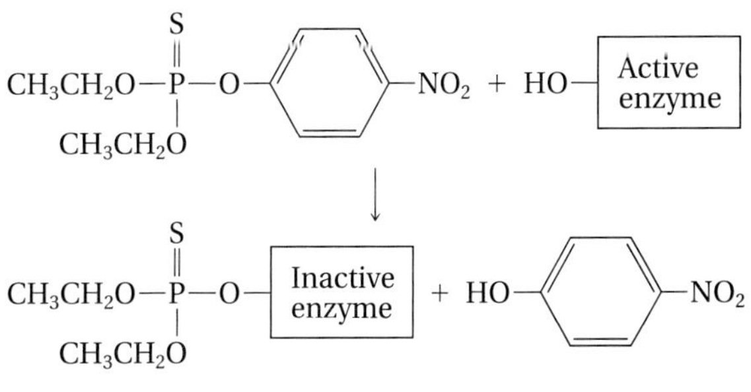

A dose of parathion sufficient to paralyze the muscles of the respiratory system would be lethal.

SUMMARY

RNA

Enzymes, which except for ribozymes are proteins, serve as catalysts of chemical reactions that take place in biological systems. Enzyme-catalyzed reactions begin when a substrate molecule binds to an enzyme. In binding, the substrate fits a region of the folded peptide chains of the enzyme called the active site. The fit of substrate to enzyme occurs through complementarity of shape, charge, or weaker forces.

Conversion of bound substrates to products often results from chemical reactions involving the substrate and amino acid side chains of the enzyme. In many instances, enzymatic reactions involve a cofactor, a small organic molecule, or a metal ion contained within the enzyme active site.

Compared with their nonenzymatic counterparts, enzyme-catalyzed reactions are more rapid and occur at milder conditions of temperature and pH. They are also very specific; only one kind of product is formed by interaction of one kind of substrate with a particular kind of enzyme. Enzyme activity levels in body fluids may be used to diagnose diseases. In certain illnesses, the patient's levels of blood serum enzymes are higher than those of a normal individual.

Enzymatic reactions in living systems may be subject to sensitive control. Such reactions may be speeded up by induction, activation of zymogens or proenzymes, and positive modulation. Enzymatic reactions may be slowed down by enzyme degradation, inhibition by a substrate mimic, or negative modulation. Enzymes whose activities are controlled by positive or negative modulation, or both, are usually allosteric, or second-site, enzymes. Allosteric enzymes are often pacemakers, appearing at the beginning of a sequence of several reactions and controlling the manufacture of products for the entire sequence. Protein digestion and blood clotting are examples of biological events in which zymogens must be converted to active enzymes.

Antibiotics permit people to fight off bacterial infections until their natural defenses can take effect. Many antibiotics are designed to inhibit the normal enzymatic processes of bacteria.

KEY TERMS

Active site (10.4)
Allosteric enzyme (10.10)
Antibiotics (10.13)
Apoenzyme (10.6)
Autoactivation (10.11)
Cofactor (10.6)
Competitive inhibitor (10.10)
Complementarity (10.4)
Degradation (10.9)

Endopeptidase (10.11)
Enzyme (10.1)
Enzyme activity (10.7)
Enzyme assay (10.7)
Enzyme-substrate complex (10.4)
Exopeptidase (10.11)
Feedback inhibition (10.10)
Holoenzyme (10.6)
Induced-fit model (10.4)

Inducible enzyme (10.9)
Induction (10.9)
International unit (10.7)
Lock-and-key model (10.4)
Modulator binding site (10.10)
Negative modulator (10.10)
Pacemaker enzyme (10.10)
Peptidase (10.2)
pH-activity profile (10.8)

pH optimum (10.8)
Positive modulator (10.10)
Proenzyme (10.11)
Protease (10.2)
Specificity (10.3)
Substrate (10.3)
Zymogen (10.11)
Zymogen granule (10.11)

EXERCISES

Enzymes (Sections 10.1–10.6)

10.11 Describe what an enzyme is and what it does.

10.12 Define the terms (a) *substrate* and (b) *specificity*.

10.13 What type of reaction is catalyzed by each of the following enzymes? (a) esterase (b) lactase (c) dehydrogenase (d) amylase

10.14 Suggest a name for the enzyme that catalyzes (a) the hydrolysis of maltose and (b) the hydrolysis of a lipid.

10.15 Explain how the lock-and-key model describes the action of an enzyme.

10.16 What is an *enzyme-substrate complex?*

10.17 What is the *active site* of an enzyme?

10.18 Explain why enzyme molecules are very large structures but their active sites are small.

10.19 Why do some enzymes need cofactors (coenzymes)?

10.20 List some typical cofactors.

Enzyme Assay (Section 10.7)

10.21 Define enzyme activity.

10.22 Why is an enzyme assay a valuable tool in clinical testing?

Effects of pH, Temperature, and Heavy Metals (Section 10.8)

10.23 What effect does an enzyme have on the energy of activation of a reaction?

10.24 Explain the term *optimal pH* of an enzyme.

10.25 The optimal temperature for a given enzyme is 37 °C. What will probably happen to the enzyme and its activity when (a) the temperature is lowered to 0 °C and (b) the temperature is raised to 100 °C?

10.26 Trypsin, an enzyme present in pancreatic juice, has an optimal pH of 8.2. Draw a pH-activity profile for trypsin.

Control of Enzyme Activity (Sections 10.9–10.13)

10.27 What is an *inducible* enzyme?

10.28 An enzyme that has its maximum activity at 37 °C is denatured by raising the temperature to 100 °C. The temperature is then lowered to 37 °C. Comment on the activity of the enzyme.

10.29 What is a competitive inhibitor, and how does it work?

10.30 How does feedback inhibition control enzyme activity?

10.31 What type of enzyme contains both a substrate binding site and a modulator binding site?

10.32 Distinguish between a competitive inhibitor and a negative modulator.

10.33 What is a *zymogen?*

10.34 Why is a peptidase usually required to convert a zymogen into an active enzyme?

10.35 Explain how soluble fibrinogen is converted to insoluble fibrin—the blood clot.

10.36 Describe how blood clots are dissolved.

10.37 How does penicillin kill bacteria?

10.38 Inhibitors of the enzyme acetylcholinesterase are extremely effective poisons. Briefly describe how they work.

Additional Exercises

10.39 Describe one way in which poisons work.

10.40 List at least three ways in which the activity of an enzyme is regulated.

10.41 Describe what is meant by the *specificity* of an enzyme.

10.42 What is the difference between (a) urea and urease, (b) lactose and lactase, and (c) pepsin and pepsinogen?

10.43 Explain the relationship of a zymogen to the corresponding enzyme.

10.44 Discuss how the concentration of enzymes in blood serum can give useful information about specific disease conditions.

10.45 Define (a) competitive inhibitor, (b) positive modulator, (c) allosteric enzyme, and (d) feedback inhibition.

SELF-TEST (REVIEW)

True/False

1. The strength of the enzyme-substrate complex is determined entirely by the fit of the substrate into the active site.

2. The shape of a substrate molecule does not affect the rate of an enzyme-catalyzed reaction.

3. In the clinical assay for blood serum enzymes, disease or tissue injury is indicated by increased enzyme activity as compared with normal.

4. An increase in the temperature of an enzyme always increases its activity.

5. Both competitive inhibitors and negative modulators affect the active sites of the enzymes they regulate.

6. Trypsinogen is the zymogen of the endopeptidase trypsin

7. An exopeptidase hydrolyzes the terminal peptide bonds of a peptide chain.

8. Antibiotics are produced by the body to combat bacterial infections.

9. The product of a series of consecutive enzyme-catalyzed reactions may function in an allosteric manner in feedback inhibition.

10. Plasmin is responsible for the formation of blood clots.

Multiple Choice

11. A sample of blood serum to be assayed for enzyme *Z* is mixed with a known amount of substrate for this enzyme. Which of the following conditions must be controlled?
    (a) pH   (b) amount of blood serum
    (c) temperature   (d) all of these

12. The demand for a particular substance in the body varies widely with time. The production of this substance is most likely controlled by a(n)
    (a) changeable enzyme.   (b) allosteric enzyme.
    (c) proenzyme.   (d) accelerin enzyme.

13. A patient's lactate dehydrogenase (LDH) and glutamic-oxaloacetic transaminase (SGOT) levels are both elevated. The patient's creatine phosphokinase (CPK) level is normal. Which of the following is a probable diagnosis? (Use Table 10.1.)
    (a) myocardial infarction   (b) hepatitis
    (c) pancreatic disorder   (d) leukemia

14. Which of the following substances is most directly involved in the formation of a blood clot?
    (a) fibrinogen   (b) prothrombin
    (c) heparin   (d) thrombin

15. In an enzyme-catalyzed reaction, a substrate is changed to products
    (a) at a number of different locations on the enzyme.
    (b) at the active site of the enzyme.
    (c) after the enzyme-substrate complex is formed.
    (d) More than one are correct.

16. The enzyme maltase would show specificity for
    (a) all carbohydrates.
    (b) all disaccharides.
    (c) maltose.
    (d) more than one are correct.

17. Which of the following statements is *not* generally true of enzymes?
    (a) They are not changed by the reaction they catalyze.
    (b) They are proteins.
    (c) They shift the equilibrium toward the product side of the equation.
    (d) They speed up chemical reactions.

18. The activity of some enzymes is dependent on the availability of necessary
    (a) cofactors.   (b) inhibitors.
    (c) antigens.   (d) inductors.

19. Peptidase proenzyme can be activated by the action of
    (a) cofactors.   (b) a peptidase.
    (c) an inductor.   (d) an acinar cell.

20. In the lock-and-key model of enzyme action, a competitive inhibitor
    (a) is like a key that will not fit the lock.
    (b) is like the wrong lock for the available key.
    (c) is like a key that will not work but still fits the lock.
    (d) more than one are correct.

21. The function of an antibiotic such as penicillin is best described as
    (a) phagocytosis.   (b) positive modulation.
    (c) enzyme inhibition.   (d) immune response.

22. Enzyme-catalyzed reactions in the cell can be slowed down by
    (a) negative modulation.   (b) enzyme degradation.
    (c) inhibition.   (d) all of these.

Nucleic Acids

The Molecular Basis of Heredity

These twins are
genetically identical.

Mice beget mice, cats beget cats, and people beget people. The observation that organisms reproduce their own species is widely known and self-evident. Yet the mystery of how this occurs is one of the toughest cases that scientific sleuths have had to crack. The study of the chemical composition and biological functions of the molecules of heredity—*nucleic acids*—has already led to enormous advances in our understanding of human heredity and the origins and treatment of inherited diseases. Our Case in Point, a true story, illustrates one of these advances.

CASE IN POINT: Treatment for an inherited disease

In September 1990, a 4-year-old girl named Ashanthi was treated by a revolutionary new method for the cure of inherited diseases. When she was a baby, Ashanthi's parents and physician were concerned about her repeated infections. Eventually, they learned that the infections resulted from a lack of the enzyme adenosine deaminase (ADA). ADA-deficient patients are subject to all kinds of infections. Because even a simple infection can be fatal, ADA-deficient patients traditionally have been kept in isolation. What revolutionary treatment was Ashanthi given, and what was its outcome? We will find out in Section 11.11.

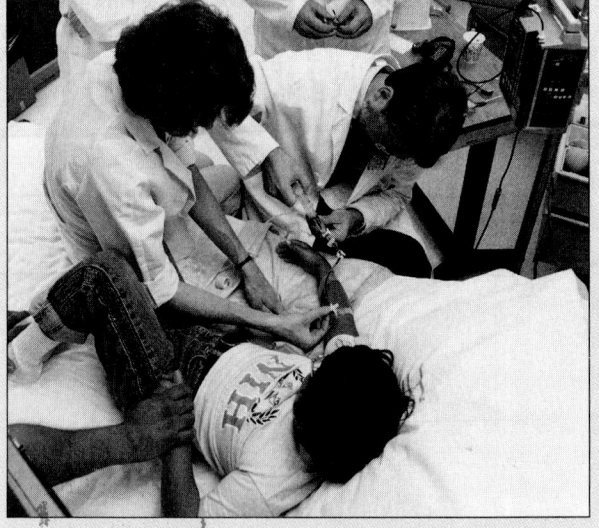

Ashanthi received a newly developed treatment for ADA deficiency when she was very young.

11.1 Nucleic acids, nucleotides, and nucleosides

AIMS: To give the names and one-letter symbols for the five major nitrogen bases found in nucleic acids. To state two differences between the molecular composition of DNA and RNA. To name and draw structures of nucleosides and nucleotides. To describe the bonding that joins nucleotides together in nucleic acids.

Focus

Nucleotides are the fundamental building blocks of nucleic acids.

Cells contain two types of **nucleic acids: deoxyribonucleic acid (DNA)** *and* **ribonucleic acid (RNA).** *DNA and RNA molecules are* **polynucleotides—** *polymers composed of many repeating units of nucleotides. Each* **nucleotide** *consists of a nitrogen base, a sugar unit, and a phosphate group attached to the sugar unit. The combination of a base and a sugar unit makes a* **nucleoside.** Adding a phosphate group to the sugar unit of a nucleoside makes a nucleotide.

$$\text{Base} \ + \ \text{Sugar} \ \longrightarrow \ \underbrace{\text{Base} \ - \ \text{Sugar}}_{\text{A nucleoside}}$$

$$\underbrace{\text{Base} \ - \ \text{Sugar}}_{\text{A nucleoside}} \ + \ \text{Phosphate} \ \longrightarrow \ \underbrace{\text{Base} \ - \ \text{Sugar} \ - \ \text{Phosphate}}_{\text{A nucleotide}}$$

Structure of the sugar units

The sugar unit of the nucleotides strung together to make RNA is β-D-ribose—hence the name *ribonucleic acid.* The sugar unit of DNA is β-D-2-deoxyribose—hence the name *deoxyribonucleic acid.*

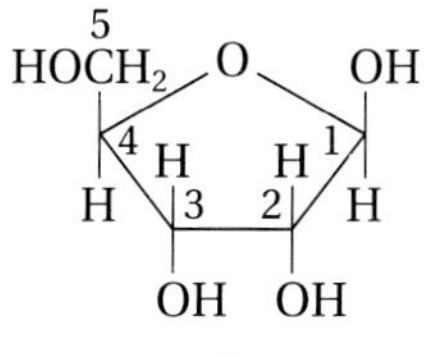

β-D-Ribose 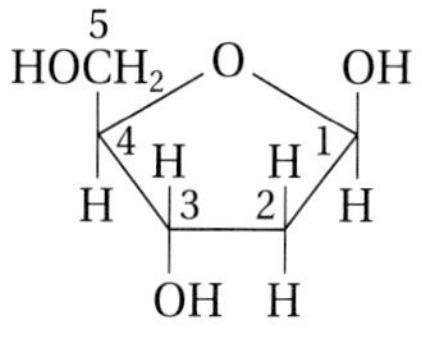   β-D-2-Deoxyribose

Structure of the base units

The molecular mass of a DNA molecule can be as high as several billion. RNA molecules are much smaller, often falling in the range of 20,000 to 40,000.

Four different nitrogen bases (heterocyclic amines) are found in DNA. (Recall from Section 6.2 that amines are weak bases.) Two of these bases, adenine (A) and guanine (G), are derivatives of purine; the other two, thymine (T) and cytosine (C), are derivatives of pyrimidine (Fig. 11.1). *Adenine and guanine are the* **purine bases** *of DNA; thymine and cytosine are the* **pyrimidine bases** *of DNA.* Except for thymine, these same bases are found in RNA. The pyrimidine base uracil (U) is found instead of thymine in RNA.

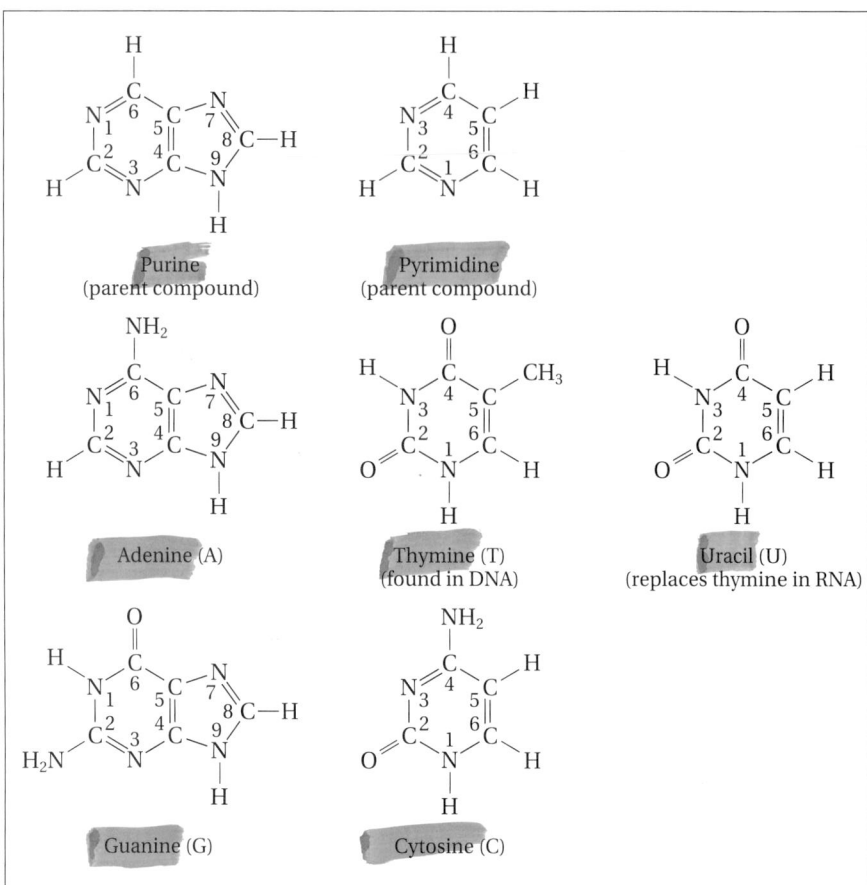

Figure 11.1
Structures of the nitrogen bases found in DNA and RNA.

Structure of nucleosides

The base and sugar units of nucleosides are held together by a covalent bond between a nitrogen of the purine or pyrimidine base and a ring carbon of the sugar unit, as shown for the deoxyribonucleosides of DNA in Figure 11.2. Ribonucleosides (the nucleosides found in RNA) are similar in structure to the deoxyribonucleosides, except that ribose rather than deoxyribose is the sugar, and uracil replaces thymine. The name for these compounds—*nucleosides*—reflects their role as constituents of nucleic acids (*nucleo-*) and their carbohydrate character (*-oside*, as in *glycoside*).

Since both the ring atoms of the sugar unit and the nitrogen bases of nucleosides are numbered, the conventional way to name nucleosides is to number the atoms of the base unit as 1, 2, 3, 4, and so forth. Carbons of the sugar ring are numbered 1′, 2′, 3′, 4′, and 5′ (read as "one prime," "two prime," and so forth). In Figure 11.2 we see that purine nucleosides are formed by a covalent bond between nitrogen 9 of the base and carbon 1′ of the sugar unit. The base and sugar of pyrimidine nucleosides are joined together by a covalent bond between nitrogen 1 of the base and carbon 1′ of the sugar.

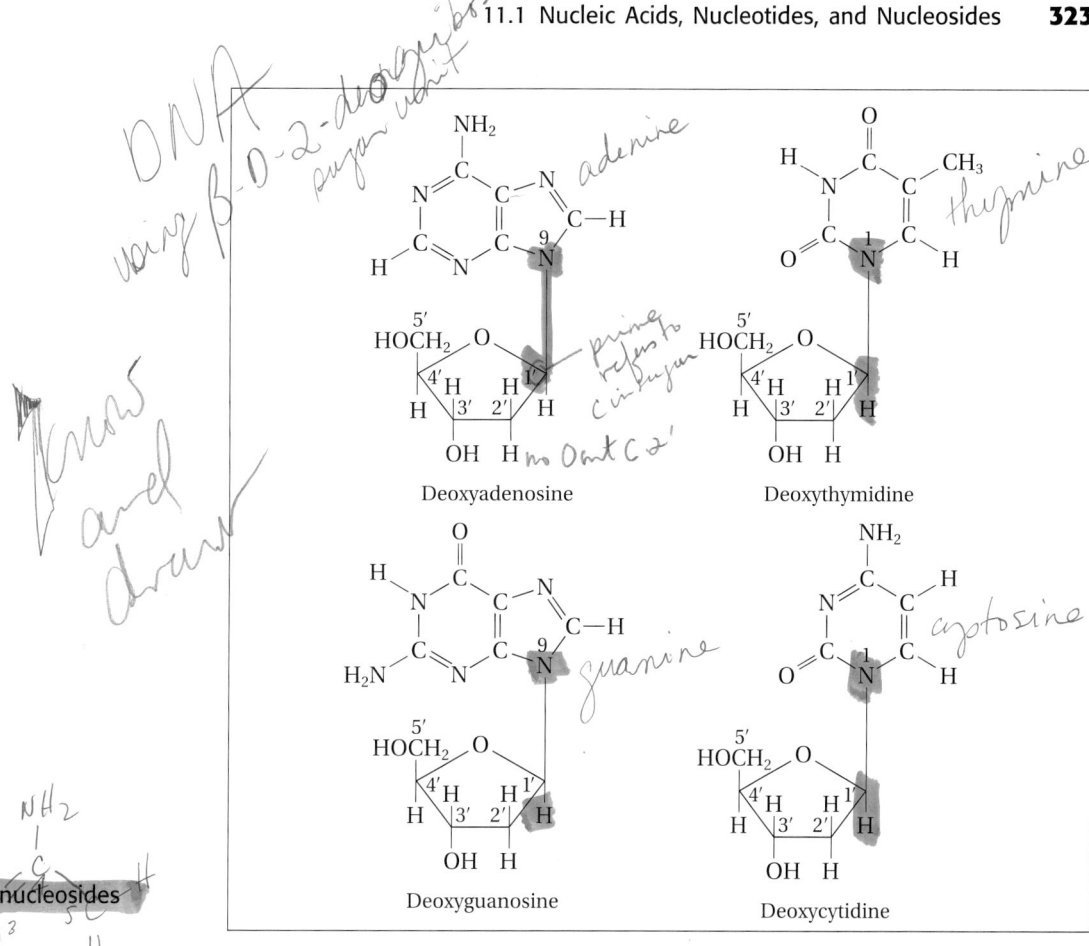

Deoxyadenosine

Deoxythymidine

Deoxyguanosine

Deoxycytidine

Figure 11.2
Structures of the nucleosides found in DNA.

PRACTICE EXERCISE 11.1

Draw the structure of the nucleoside formed by the combination of the nitrogen base cytosine and the sugar β-D-ribose. Name this nucleoside.

Structure of nucleotides

Addition of a phosphate group to a sugar hydroxyl group of a nucleoside forms a *nucleotide*. In other words, nucleotides are phosphoric acid esters of nucleosides. (See Section 5.9 to review phosphate esters.) In nucleotides of DNA and RNA, the phosphate ester is formed at the 5′-hydroxyl group of the nucleoside. Table 11.1 summarizes the names of the nucleosides and nucleotides of DNA and RNA. In the following structure of the nucleotide, the base is A, T, G, or C in DNA and A, U, G, or C in RNA.

Phosphate ester   (or OH in RNA)

A nucleotide

Table 11.1 Names of Bases, Nucleosides, and Nucleotides Found in DNA and RNA

Base	Nucleoside	Nucleotide
DNA		
adenine (A)	deoxyadenosine	deoxyadenosine 5′-monophosphate (dAMP)
guanine (G)	deoxyguanosine	deoxyguanosine 5′-monophosphate (dGMP)
thymine (T)	deoxythymidine	deoxythymidine 5′-monophosphate (dTMP)
cytosine (C)	deoxycytidine	deoxycytidine 5′-monophosphate (dCMP)
RNA		
adenine (A)	adenosine	adenosine 5′-monophosphate (AMP)
guanine (G)	guanosine	guanosine 5′-monophosphate (GMP)
uracil (U)	uridine	uridine 5′-monophosphate (UMP)
cytosine (C)	cytidine	cytidine 5′-monophosphate (CMP)

Figure 11.3 shows how DNA and RNA molecules consist of nucleotides linked together through phosphate ester bridges between the 3′-hydroxyl group of one nucleotide and the 5′-hydroxyl group of the next nucleotide in the chain. *Since the phosphate ester bridges holding the nucleotides together each contain two phosphate ester linkages, these bridges are called* **phosphodiesters.** Notice that even though two of the four oxygens attached to the phosphorus of each bridge are tied up as phosphate esters and one is present in a phosphorus-oxygen double bond, one oxygen is free to lose a proton as in the ionization of other phosphoric acid esters (see Sec. 5.9).

$$
\begin{array}{ccc}
& O & & O \\
& \parallel & & \parallel \\
RO-P-OR & \rightleftharpoons & RO-P-OR + H^+ \\
& | & & | \\
& OH & & O^-
\end{array}
$$

It is the presence of many such dissociating groups that gives DNA and RNA their highly acidic character.

PRACTICE EXERCISE 11.2

A phosphate group is added to the 5′-hydroxyl group of the nucleoside in Practice Exercise 11.1 to form a nucleotide. What is the structure and name of this nucleotide?

Shorthand structures for RNA and DNA nucleotide sequences

Most DNA and RNA molecules are too long to write out in full, so biochemists often use a shorthand form. The shorthand shows the sequence of their bases. The sugar units and phosphate groups are identical in all the nucleotides. Thus we can describe the structure of an RNA molecule by ignoring the sugars and phosphodiester bridges and writing only the sequence of the nitrogen bases. By convention, we start at the left with the end of the molecule that has a free 5′-hydroxyl group (not attached to another nucleotide) and work toward the end that has the free 3′-hydroxyl

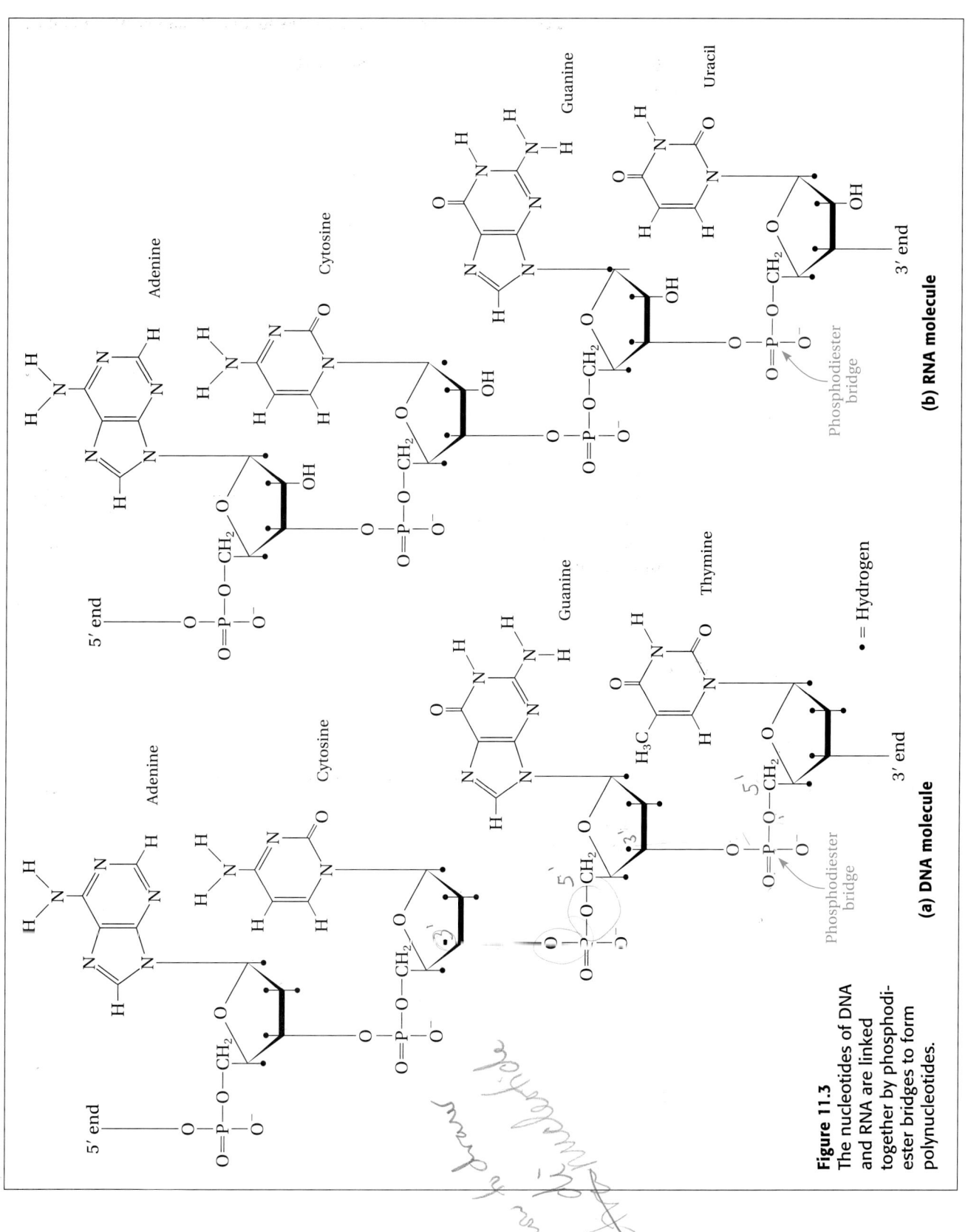

Figure 11.3
The nucleotides of DNA and RNA are linked together by phosphodiester bridges to form polynucleotides.

group (not attached to another nucleotide). Another way of saying this is that we work in the 5′ to 3′ direction.

$$U—A—G—C—U—G—C—C$$
$$5'———————————→3'$$

We can write abbreviations for the base sequence in DNA by putting a lowercase *d* in front of the base sequence. This indicates that all the sugar units of the sugar-phosphodiester backbone of the molecule are deoxyribose in DNA.

$$dA—A—T—G—T—C—A—C$$
$$5'———————————→3'$$

EXAMPLE 11.1 | **Writing base sequence abbreviations**

[handwritten: dA -T-G-G-C; 5'————→3']

Write the base sequence abbreviation of a strand of DNA composed of the bases adenine, thymine, guanine, guanine, and cytosine. Assume adenine is not bonded at the 5′ position.

SOLUTION

Use the one-letter abbreviation of each base starting with adenine that has the free 5′ end. Since this is DNA, begin the sequence with a lowercase *d*.

$$dA—T—G—G—C$$

[handwritten: a) adenine, cytosine, guanine, uracil (RNA)]
[handwritten: b) thymine, adenine, cytosine, guanine (DNA)]

PRACTICE EXERCISE 11.3

Translate the following base sequence abbreviations:

(a) A—C—G—U   (b) dT—A—C—G

PRACTICE EXERCISE 11.4

What is a phosphodiester bridge, and where is it found?

[handwritten: phosphodiester bridge links 3' hydroxyl group of one nucleotide to the 5' hydroxyl group of another nucleotide]

11.2 The DNA double helix

AIM: To discuss the significance of A = T and G = C as it relates to the formation of the double-helical structure of DNA.

Focus

Strands of DNA form a double helix.

The three-dimensional structure of DNA was unknown until 1953, when James D. Watson and Francis Crick proposed that these polymeric molecules form a **DNA double helix**—*a double-stranded DNA molecule in which the deoxyribose-phosphate backbones of the DNA strands are wound in a helix.* Figure 11.4 depicts the structure of the now-famous DNA double helix. In the double helix, adenine can form two hydrogen bonds to a

Figure 11.4 (below)
The DNA double helix.

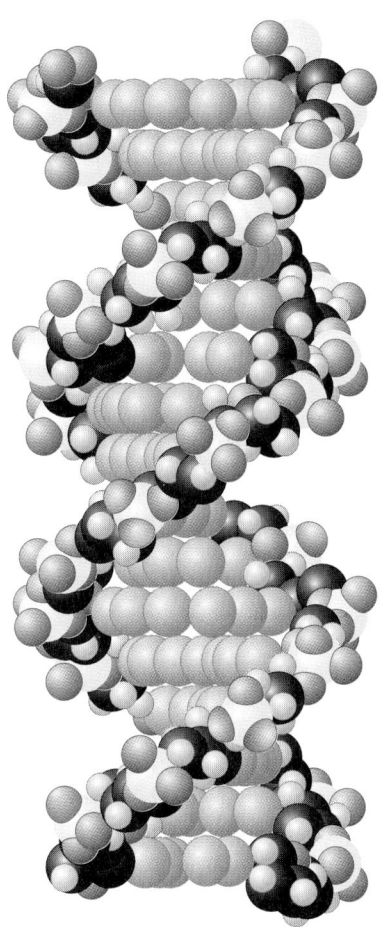

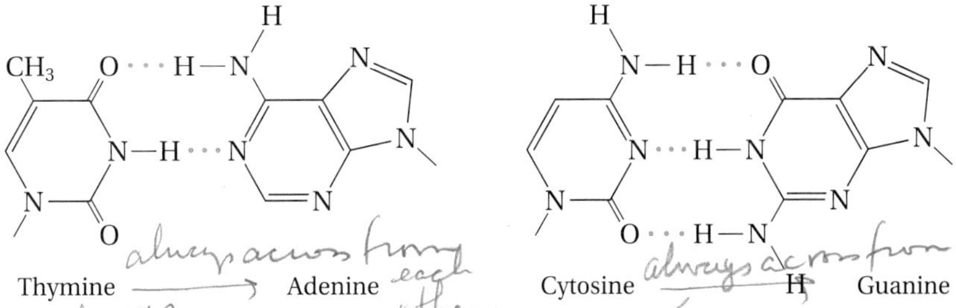

Thymine ──────→ Adenine
always acros from each other

always bonded together
(a) AT pair

Cytosine *always across from* Guanine

always bonded together because lowest energy configuration.
(b) GC pair

Figure 11.5
Complementary base pairing:
(a) adenine (A) forms two hydrogen bonds with thymine (T);
(b) guanine (G) forms three hydrogen bonds with cytosine (C).

In 1962, James D. Watson and Francis Crick shared the Nobel prize in medicine and physiology with Maurice Wilkins for their discovery of the structure of DNA.

thymine on the opposing DNA chain (Fig. 11.5). Guanine can form three hydrogen bonds to cytosine on an opposing chain. *This specific matching of opposing base pairs by hydrogen bonding is termed* **complementary base pairing.** In any DNA molecule, A = T and G = C.

The strands of DNA constituting the DNA double helix run in opposite directions, much as traffic on a two-way street moves in opposite directions. In other words, *the two DNA strands of the double helix are* **antiparallel.** The sugar phosphate units of the nucleotide are on the outside of the DNA molecule, with the complementary base pairs stacked neatly inside the molecule. The hydrogen bonds between the base pairs hold the double helix together. A and T each constitute about 30% of the bases of human DNA. The bases C and G each constitute about 20%. (Since there are only four different bases in DNA, these numbers always add up to 100%.)

PRACTICE EXERCISE 11.5

Describe the DNA double helix in your own words.

the two strands of the double helix run in opposite directions.

a double-stranded DNA molecule in which the deoxyribose-phosphate backbones of the DNA strands are wound in a helix; H-bonding between the opposing N base pairs holds the molecule together; and there is complementary base pairing.

11.3 The central dogma

AIMS: To name the three processes that constitute the central dogma of molecular biology. To describe a unique characteristic of all retroviruses.

DNA is the bearer of genetic information in all living organisms. The entire basis of information storage and transmission in cells is embodied in three steps:

1. *DNA is replicated.* **Replication** *means copying;* it is the way DNA is supplied to new cells formed by cell division and to a new organism formed by reproduction.

2. *The information contained in DNA is passed to a form of RNA called* **messenger RNA** *(mRNA) by* **transcription** *("rewriting").*

3. The mRNA directs **translation,** *the synthesis of proteins.* In everyday parlance, the word *translate* means "to change from one language to another." The meaning in molecular biology is quite similar: The "language" of DNA that is transcribed to mRNA is translated to a completely different language—the primary structure of a protein.

Replication, transcription, and translation can be summarized as

$$\text{DNA} \xrightarrow{\text{Replication}} \text{DNA} \xrightarrow{\text{Transcription}} \text{mRNA} \xrightarrow{\text{Translation}} \text{Protein}$$

Replication, transcription, and translation are so important to an understanding of the relationship between heredity and protein synthesis that they have been called the **central dogma** *of molecular biology.* The enormous diversity of living creatures ultimately resides in the central dogma. The information needed to make enzymes and other proteins is stored in DNA. The base composition of the DNA of various species is different, but the base composition of all the DNA from an individual is the same. Differences among species result from differences in the proteins constructed with this information. The cells of trees contain enzymes to make cellulose for cell walls, but humans lack these enzymes. At least some of the proteins in cats—hemoglobin, for example—are not found in bacteria. Moreover, the hemoglobin of cats differs slightly in structure from the hemoglobin of mice and humans. Differences in proteins are also found among individuals of the same species. Your eyes may be brown and your friend's eyes blue because your friend lacks enzymes needed to manufacture the pigment that makes eyes brown. It is the immense number of possible variations in our protein makeup that turns each of us into a unique individual.

The steps of replication, transcription, and translation are followed in every kind of organism except the *retroviruses. In* **retroviruses,** *the genetic material is RNA.* In order for the retroviruses to reproduce, the message contained in the viral RNA must be transcribed—or more accurately, *reverse transcribed*—into DNA.

$$\text{RNA} \xrightarrow{\text{Reverse transcription}} \text{DNA}$$

Once formed, the retroviral DNA is transcribed and translated much as it is for other organisms. The viruses responsible for many human diseases,

Retrovirus
retro (Latin): backward
virus (Latin): poison, slime

such as the common cold, poliomyelitis, measles, influenza, and some tumors, are retroviruses. The human immunodeficiency virus (HIV), which causes AIDS, is also a retrovirus. Because of its centrality in the reproduction of HIV, *reverse transcriptase*, the enzyme that catalyzes the transcription of the virus's RNA into DNA, is the target of researchers seeking treatments for AIDS. This work is described in A Closer Look: HIV Reverse Transcriptase.

11.4 Replication

AIM: To outline the process of replication.

Focus

DNA is replicated on a DNA template.

Watson and Crick also proposed a model for DNA replication. They suggested that each strand of DNA in the double helix acts as a **template**—*a pattern*—for the synthesis of its complement, the other strand. Since DNA is double-stranded, complementary replication would produce two DNA molecules, each a double helix containing a strand of the original DNA and a new strand complementary to it. In other words, suppose the structure of the double helix is

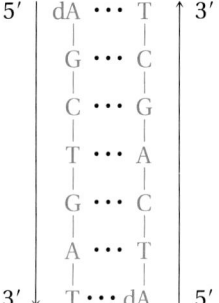

Then complementary replication of each strand produces

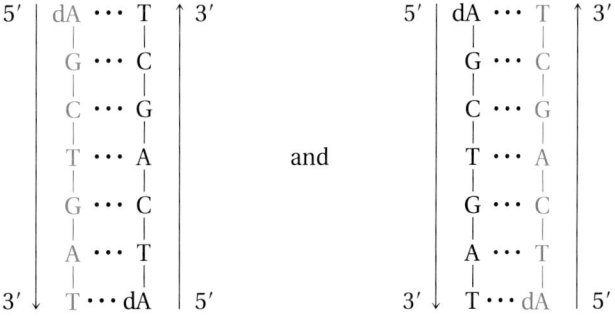

This type of replication is called *semiconservative replication* because the double-stranded DNA in each round of replication contains one strand from the previous round (the conserved strand) and one new strand. Figure 11.6 illustrates semiconservative replication in a slightly different way. The growth of the DNA chain requires **DNA polymerase**—*a large enzyme that catalyzes the formation of DNA*. The synthesis of DNA uses as starting mate-

Parent strands

First
generation

Second
generation

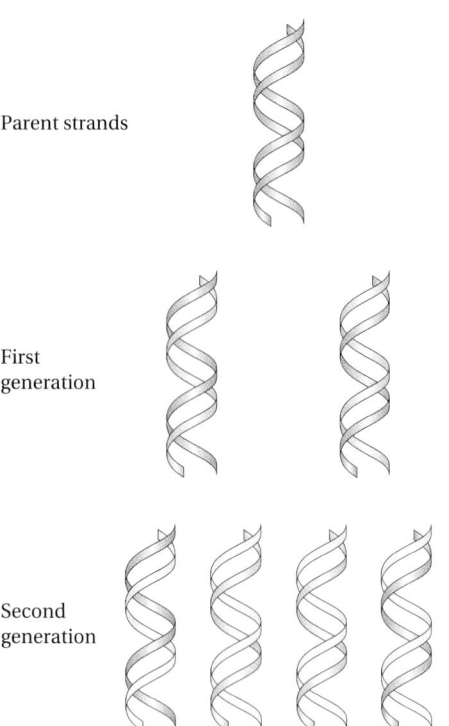

Figure 11.6 (above)
Semiconservative replication.
In semiconservative replication,
the DNA of each succeeding
generation of cells contains one
strand from the generation before
it and one new strand that is
complementary to the old strand.

Figure 11.7 (right)
General structure of
nucleoside 5′-triphosphates.
In the nucleoside triphosphates
required for the synthesis of DNA,
the sugar units are 2′-deoxyribose
and the bases are A, T, G, and C.
The nucleoside 5′-triphosphates are
named after the parent nucleoside
(Table 11.1) by adding the word *5′-
triphosphate*—as in deoxyguanosine
5′-triphosphate (dGTP) or adenosine 5′-triphosphate (ATP).

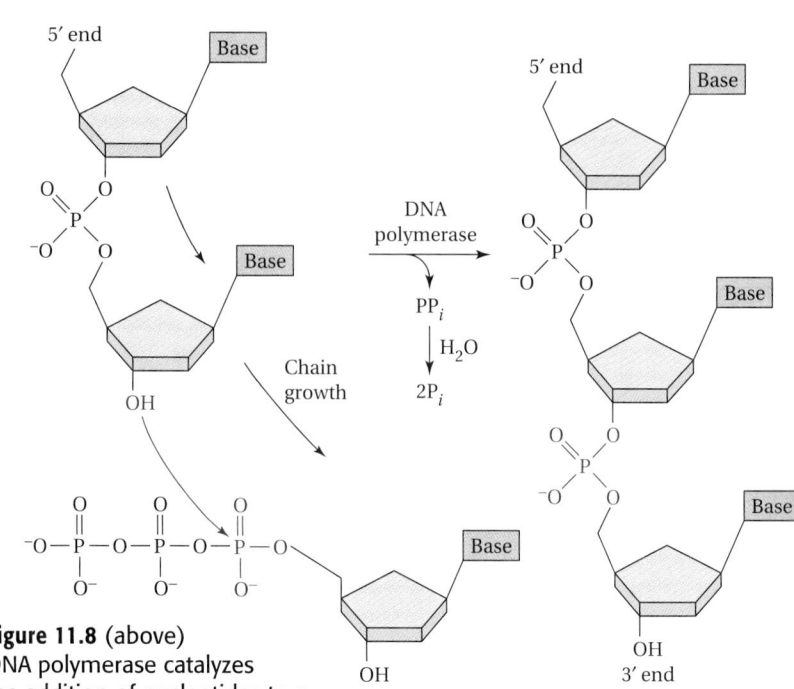

Figure 11.8 (above)
DNA polymerase catalyzes
the addition of nucleotides to a
growing DNA chain. A "high energy"
phosphate bond is broken in the formation of the phosphodiester linkage,
releasing a pyrophosphate ion (PP$_i$). Notice that the direction of chain growth
is in the 5′ to 3′ direction. Triphosphates are anhydrides of phosphoric acid
and therefore phosphorylate alcohols to make phosphate esters. Thus the
esterification reaction is spontaneous and proceeds with a large release of
free energy. The hydrolysis of the PP$_i$ to 2P$_i$ also releases free energy, which
further aids in driving esterification to completion.

rials the nucleoside 5′-triphosphates of the bases found in DNA (Fig. 11.7).
DNA must be present for new DNA to be formed, which agrees with Watson
and Crick's idea that DNA serves as a template for DNA synthesis. DNA
strands elongate when an incoming nucleoside triphosphate phosphory-
lates the 3′-hydroxyl group of ribose at the end of a growing DNA chain (Fig.
11.8). Stated another way, the direction of chain growth is always from the
5′ end to the 3′ end of the elongating DNA molecule.

HIV Reverse Transcriptase

HIV is a retrovirus—it contains RNA as the instructions for its reproduction. In addition to the protease discussed in A Closer Look: HIV Protease and Its Inhibition, on page 305, the HIV virus produces a reverse transcriptase (see figure). HIV reverse transcriptase is needed to make DNA from the RNA. The information in the DNA is then transcribed into mRNA. These instructions are in turn translated into proteins needed to assemble a mature, infectious virus. Like HIV protease, HIV reverse transcriptase is an important target for medical drugs to slow down or prevent the reproduction of HIV. Zidovudine (AZT) and DDI are two drugs used in the treatment of AIDS. These drugs are designed to impede HIV reverse transcriptase from synthesizing DNA from viral RNA. As the chemical structures of these drugs show, they are chemical modifications of the deoxynucleosides normally found in DNA.

Zidovudine and DDI bind to the active site of HIV reverse transcriptase. They are added to a growing DNA copy of the viral RNA. However, they lack a 3′-hydroxyl group. This group is needed to lengthen the DNA chain, so chain growth stops before the DNA has reached its full length. Incomplete, defective proteins are made, and infectious viruses cannot be assembled. The deoxynucleoside analogues are helpful to patients with symptoms of AIDS, but none cures the disease. High toxicity is one reason for the limited effectiveness of these antiviral drugs; toxicity limits the size of the doses that can be administered, so not all viral reproduction can be stopped. More important, rapid changes (mutations) in HIV produce versions of the reverse transcriptase that are insensitive to the drugs.

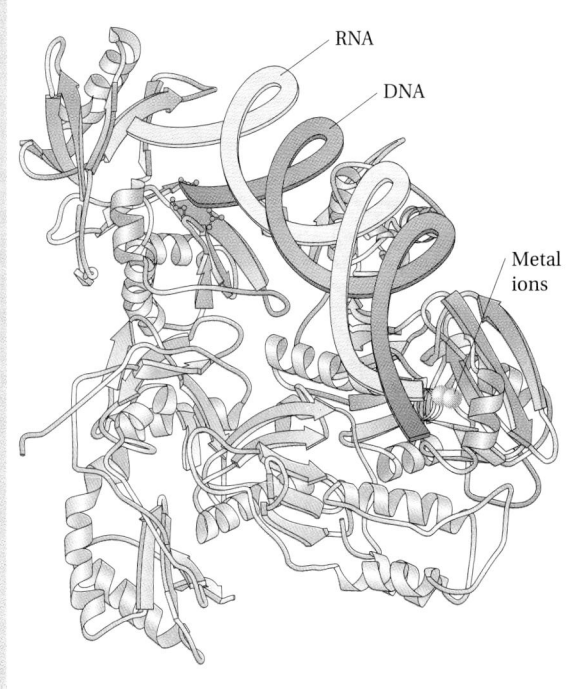

The structure of HIV reverse transcriptase. The structure is a dimer consisting of a subunit of molar mass 66,000 g/mol (blue) and a subunit of molar mass 51,000 g/mol (green). The molecule is shown with viral RNA (gold) and a strand of DNA (brown) bound to the enzyme. Two metal ions (blue spheres) are thought to be important in the reverse transcription of RNA to DNA.

3′-Azido-3′-deoxythymidine
(Zidovudine, AZT)

2′, 3′-Dideoxyinosine (DDI)

CHAPTER 11 Nucleic Acids

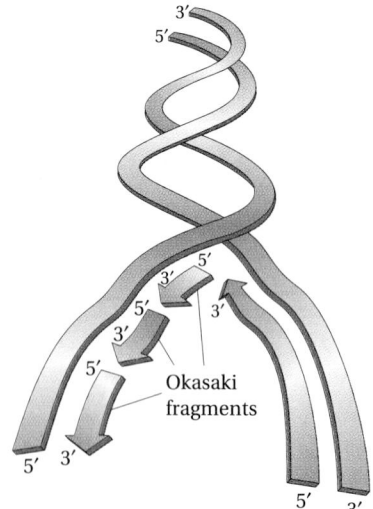

Figure 11.9
DNA is replicated in the 5′ to 3′ direction. Since the strands of DNA run in opposite directions, one complementary strand is synthesized in a continuous chain in the 5′ to 3′ direction. The other is also synthesized in the 5′ to 3′ direction, but in sections of about 1000 nucleotides that are connected by an enzyme. The short sections are called *Okazaki fragments.*

Although replication seems to proceed essentially by the method described by Watson and Crick, there are other considerations. The bases in the double helix are inside the DNA molecule. How are the bases of the strands of DNA exposed so that replication can occur? It appears that the DNA strands unwind as replication proceeds. *Special proteins, termed* **unwinding proteins,** *catalyze the unwinding of double-stranded DNA.*

Another problem concerns the direction of growth of the new DNA chains. As we have seen, the strands of DNA are antiparallel, yet when double-stranded DNA unwinds during replication, only one new strand can grow continuously in the 5′ to 3′ direction. How is the other new strand formed? In 1968, R. Okazaki showed that the other new strand is formed in the 5′ to 3′ direction in pieces that are subsequently connected by the action of an enzyme (Fig. 11.9). *These pieces of DNA, which have only a fleeting existence independent of each other, are called* **Okazaki fragments.**

EXAMPLE 11.2   Writing a complementary DNA strand

Write the abbreviated base sequence (written 5′ to 3′) for the strand of DNA complementary to the following strand:

dA—G—G—A—C—T—C—C

SOLUTION

In DNA replication, A pairs with T, G pairs with C. Using the given strand as a template, and remembering that the complementary strand is antiparallel, gives

3′ dT—C—C—T—G—A—G—G 5′

Written 5′ to 3′, the complementary strand is

5′ dG—G—A—G—T—C—C—T 3′

PRACTICE EXERCISE 11.6

Complete the following segment of a DNA double helix. Write symbols for the missing bases, and draw the hydrogen bonds.

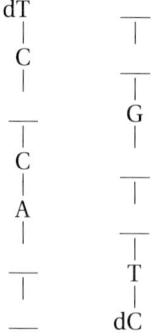

11.5 Genes

AIMS: To characterize a gene. To describe the function of exons and introns.

Focus

A gene is a DNA segment that directs the synthesis of a single kind of protein.

Table 11.2 Number of Nucleotide Base Pairs in the DNA of Various Organisms

Organism	Nucleotide base pairs (millions)
Homo sapiens	5500*
birds	2000
higher plants	2300
sponges	100
fungi	20
bacteria	2†
complex virus	0.17
simple virus	0.05

*1 million genes.
†100 genes.

Early geneticists defined a *gene* as the smallest unit of heredity that would lead to an observable trait. In 1943, George Beadle and Edward Tatum made the connection between heredity and protein synthesis when they showed that hereditary differences are the result of differing abilities to synthesize proteins. They then suggested a refined definition of a gene as the unit of heredity that directs the synthesis of an enzyme. This definition is known as the *one-gene, one-enzyme hypothesis.*

Two later discoveries somewhat changed this definition. First, scientists discovered that the synthesis of all proteins, not just enzymes, is governed by genes. Second, they found that information for protein synthesis resides in the number and sequence of bases in the DNA molecule. *We can regard a* **gene** *as a segment of the DNA molecule that directs the synthesis of one kind of protein molecule.* The proteins expressed (produced) in protein synthesis are enzymes, hemoglobin, collagen, and so forth.

DNA molecules are among the largest molecules known. For example, the DNA found in one human cell contains about 5.5 billion nucleotide base pairs that make up perhaps 1 million genes (Table 11.2). If we could make each nucleotide base pair of human DNA one book page and make the pages into books of 1000 pages each, our library would contain 5.5 million volumes. The number of nucleotide base pairs of a bacterial DNA is much smaller, about 2 million, but they still would occupy 2000 volumes. Despite the enormous number of base pairs, scientists are determined to learn the base sequence of human DNA, as discussed in A Closer Look: The Human Genome Project.

Introns and exons

The genes of eukaryotic cells differ from those of prokaryotic cells in an important respect. Essentially 100% of the bases of DNA in bacteria (prokaryotes) are used as a source of information for making proteins or for regulating the making of proteins. In contrast, only about 5% of the DNA of human cells (eukaryotes) is used in protein synthesis or its regulation. *The base sequences of the genes that contain the information (code) for protein synthesis in eukaryotic cells are called* **exons.** The remaining 95% of DNA of eukaryotic cells interrupts the coding sequences of genes (Fig. 11.10). *The noncoding base sequences that interrupt eukaryotic genes are called intervening sequences or* **introns.** A large human gene, such as the gene for the synthesis of collagen, may have as many as 50 introns interspersed between coding segments of DNA. Introns often contain much more DNA than the genes they interrupt. Scientists do not understand why human genes contain so much "junk" DNA. The base sequences of introns are somewhat similar among families and ethnic

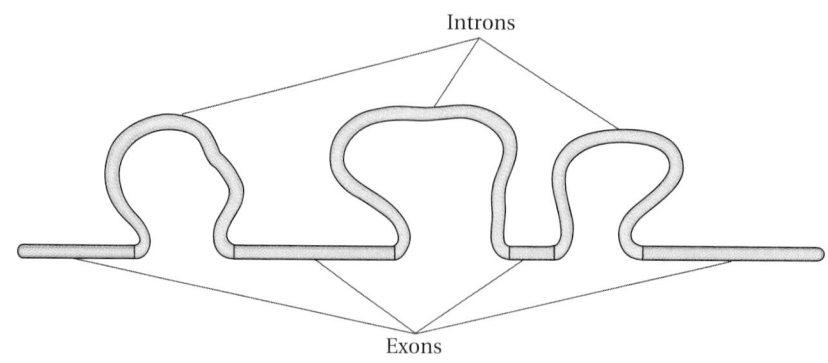

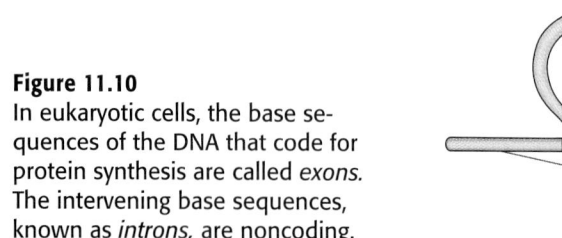

Figure 11.10
In eukaryotic cells, the base sequences of the DNA that code for protein synthesis are called *exons.* The intervening base sequences, known as *introns,* are noncoding.

groups. Since each individual inherits portions of these base sequences from both parents, a person's DNA structure is characteristic of that person. The genetic individuality of people makes it possible to identify a

The Human Genome Project

The long-term goal of scientists involved in a venture called the *Human Genome Project* (HGP) is the determination of the complete base sequence of human DNA (the human genome). Long before this goal is accomplished, these scientists expect to map signposts at close intervals within the genome. They also will have sequenced short regions of interest in the human genome as well as in such organisms as bacteria, yeast, fruit flies, and mice.

Why is it important to know about human DNA, down to the last base pair? Scientific curiosity is one perfectly valid reason. Like a mountain to be climbed, the sequence of human DNA is of interest because it exists. Supporters of the HGP are also quick to point out that much fundamental information about human heredity is still lacking. For example, the number of human genes is not known with certainty. Furthermore, the biological function of many genes is still unknown. Undoubtedly, the HGP will produce evidence of previously unknown genes with previously unknown functions. Supporters of the HGP correctly point out that many diseases are of genetic origin. Knowledge of the human genome will help to

identify genes that are linked to about 4000 hereditary diseases. In time, the knowledge gained from the HGP might be applied to the cure of these diseases.

Despite the many excellent justifications advanced by its supporters, the HGP remains somewhat controversial. When resources are limited, critics say, satisfying scientific curiosity is not a particularly good reason to start a grand project like the HGP. Critics of the project are also concerned that its cost will divert scientists and dollars from other equally worthy research projects. Originally, it was estimated that a DNA base pair could be sequenced at an average cost of about $1. This estimate, however, assumed great improvements in efficiency relative to today's sequencing technology. With today's technology, it costs about $5 to determine the sequence of a base pair. Some critics also say that it is unclear that advancements toward gene therapies could not be made by simpler, less expensive methods for studying genes. An alternative plan might be better. For example, it might be more efficient to first identify a gene of interest before going to the trouble of sequencing the DNA of the gene. This criticism is supported by evidence suggesting that only about 5% of the base pairs in DNA play a functional role in human cells.

person with great certainty by examination of the base sequences of his or her DNA, as described in A Closer Look: DNA Fingerprinting.

DNA Fingerprinting

Violent criminals often leave evidence in the form of blood and hair at the scene of their crimes. Perpetrators of sexual assaults may leave evidence as seminal fluid. A method called *DNA fingerprinting,* or *DNA profiling,* is helping to solve crimes. DNA fingerprinting also can be used to establish family relationships. For example, the paternity of a child may be at issue.

Except for identical twins, the base sequences of introns and, to a lesser extent, exons are slightly different for different individuals. These small variations among individuals form the basis for DNA fingerprinting. To make a DNA fingerprint, enzymes called *restriction enzymes* are used to cleave DNA at specific base sequences (see Sec. 11.11). Because of the differences in the base sequences among individuals, the action of a restriction enzyme may cut a fragment of 1000 bases from the DNA of one person and 3000 bases from the DNA of another person. The DNA fragments produced by the restriction enzymes can be separated by electrophoresis. The pattern of separated DNA fragments is the DNA fingerprint (see figure). The variations in DNA sequences among individuals make it very improbable that two people will have the same DNA fingerprint. In forensic work, DNA fingerprints from evidence taken at the scene of a crime are compared with the "fingerprint" obtained from DNA provided by a suspect. In principle, the chance that a DNA fingerprint is unique to one individual can be as high as 10^{19} to 1. This is a very high degree of certainty, since the world contains fewer than 6×10^9 people. In practice, the chance of identifying an individual from a DNA fingerprint is usually lower. This has caused controversy in the use of DNA finger-

A scientist compares DNA fingerprints.

printing for criminal proceedings. Is a 100 to 1 chance, a 1000 to 1 chance, or a 10,000 to 1 chance of having identified a murderer adequate to convict an accused person? Despite the controversy, more than half the states permit the use of evidence based on DNA fingerprints.

11.6 Classes of RNA

AIM: To describe the function of the three major classes of RNA: tRNA, rRNA, and mRNA.

In addition to DNA, the nucleic acids of cells include three major classes of RNA molecules: *ribosomal RNA, messenger RNA,* and *transfer RNA.*

Ribosomal RNA

Ribosomal RNA *(rRNA) molecules are an integral part of* **ribosomes,** *the organelles where proteins are synthesized* (refer to A Closer Look: Cells, on page 242). Ribosomes are composed of two subunits, one light subunit and one heavy subunit (Fig. 11.11). Each subunit is composed of about 65% rRNA and 35% protein. The rRNA maintains the structure of the ribosome and provides sites for the binding of messenger RNA. Ribosomes are self-assembling units. When the proteins and rRNA needed to make a complete ribosome are mixed together in solution, the ribosome forms spontaneously.

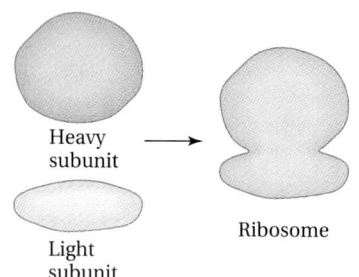

Ribosomal subunits

Figure 11.11
Heavy and light subunits combine to form a complete ribosome.

Messenger RNA

Messenger RNA *(mRNA) molecules carry the information for protein synthesis to ribosomes in prokaryotic and eukaryotic cells.* In prokaryotes, which have no nucleus, mRNA is transcribed from DNA, and proteins are synthesized on ribosomes in the cytoplasm. In eukaryotes, the mRNA is transcribed from DNA in the nucleus. The mRNA then carries the information for protein synthesis to ribosomes in the cytoplasm. In contrast to DNA, which remains intact and unchanged throughout the life of the cell, mRNA has a short lifetime—usually less than an hour. Then it is rapidly degraded back to the constituent nucleotides.

Transfer RNA

Molecules of **transfer RNA** *(tRNA) deliver amino acids to mRNA during translation.* Cells have at least one type of tRNA for each of the 20 common amino acids found in proteins. These tRNA molecules are relatively small; most contain fewer than 100 nucleotides. Moreover, scientists have found some bases other than the usual A, U, G, and C in the nucleotides that make up the tRNA molecules. Unlike DNA molecules, most RNA molecules are single-stranded. However, a single strand of RNA may sometimes loop back on itself to make "hairpin turns" held in place by hydrogen bonds between complementary bases on the same strand. The maximum number of complementary base pairs of this type are found in tRNA molecules. Figure 11.12(a) shows in a schematic way how the alanine-carrying tRNA of yeast is shaped like a cloverleaf. The tRNAs for all amino acids are quite similar in appearance. The cloverleaf shape is present but twisted in the three-dimensional structures (see Fig. 11.12b).

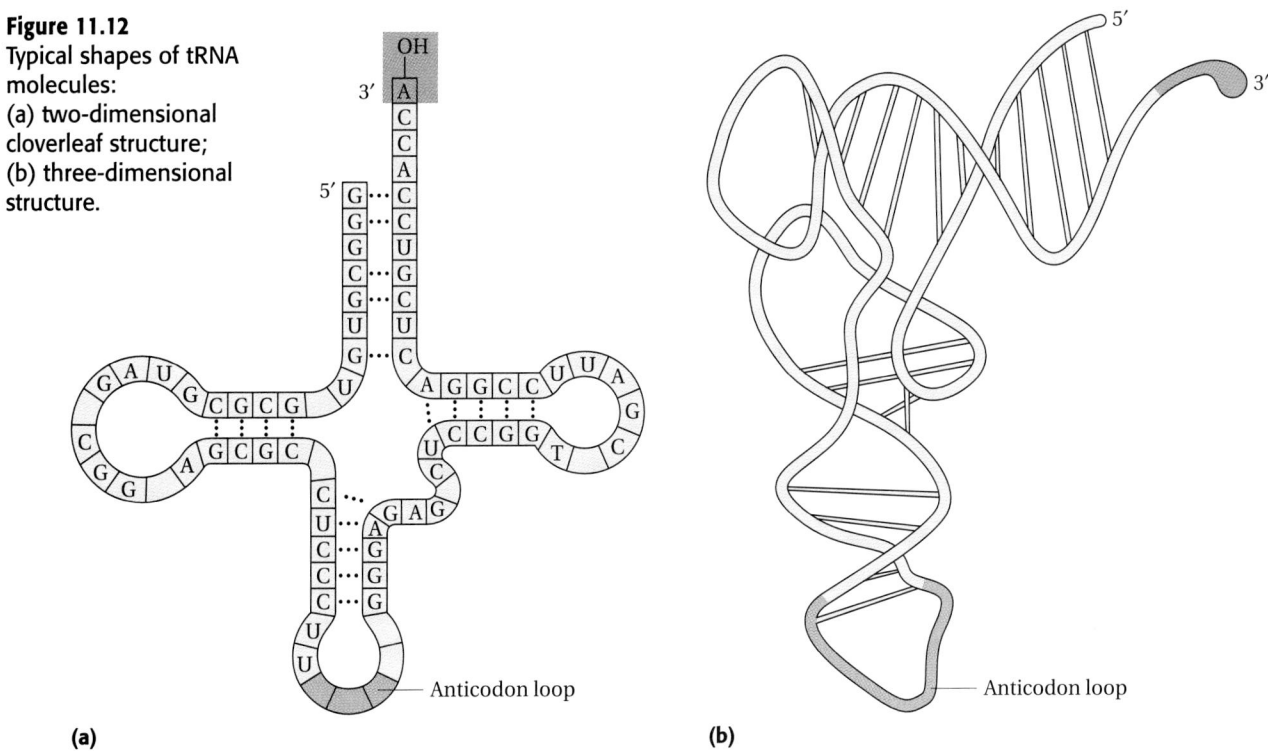

Figure 11.12
Typical shapes of tRNA molecules:
(a) two-dimensional cloverleaf structure;
(b) three-dimensional structure.

(a)

(b)

11.7 Transcription

AIM: To describe the process of transcription.

Focus

RNA is made on a DNA template.

Now we will turn our attention to *transcription*—the means by which information for protein synthesis passes from DNA to mRNA. Molecules of tRNA and rRNA are made by a similar process. The mechanics of transcription are quite similar to the mechanics of replication because DNA is the template upon which RNA is formed. The major differences are that *transcription requires an enzyme called* **RNA polymerase** rather than DNA polymerase, that the sugar units of the nucleoside triphosphates used to make RNA are ribose rather than deoxyribose, and that U substitutes for T in RNA. Suppose, for example, the order of bases in a segment of DNA is this:

5′   dT
      │
      A
      │
      A
      │
      G
      │
      C
      │
3′   T

Then the DNA molecule and its complementary RNA strand formed in transcription are

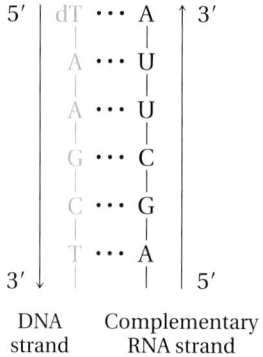

| DNA | Complementary |
| strand | RNA strand |

In prokaryotic cells, each mRNA molecule is transcribed from start to finish in a continuous chain. Because of introns, eukaryotic cells handle the transcription of DNA into mRNA somewhat differently than prokaryotic cells. You may recall that the genes of eukaryotic DNA are interrupted by stretches of introns that do not code for proteins, and only the exons contain the information needed to synthesize proteins. The transcription of DNA in eukaryotic cells is a three-step process (Fig. 11.13). First, the entire

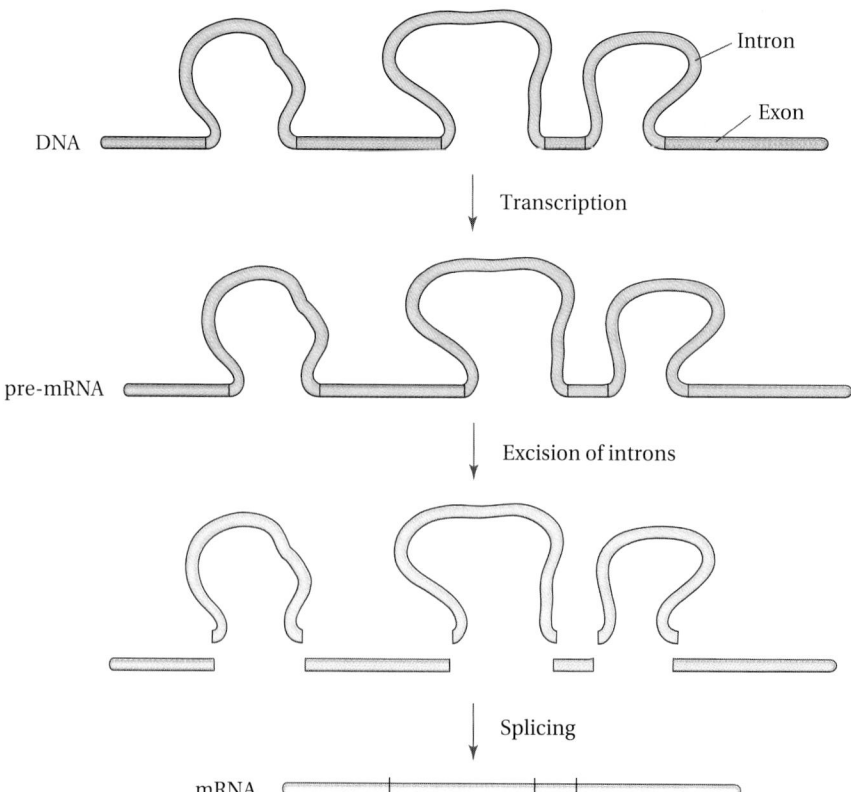

Figure 11.13
Formation of mRNA of eukaryotic cells involves transcription of genes containing exons and introns, excision of the introns of the RNA transcript, and splicing the exons to make the final mRNA.

gene containing both the introns and the exons is transcribed into an RNA molecule called a *pre-mRNA,* much as we just described. Second, the segments of the pre-mRNA molecules that correspond to the introns are cut out by *ribonuclease* (RNase) enzymes. Third, the portions of pre-mRNA that correspond to the exons of the DNA are spliced together by *ligase* enzymes to make a finished mRNA. The finished mRNA contains the information needed for translation into a protein.

| EXAMPLE 11.3 | Writing a complementary RNA strand |

Consider the following segment of DNA:

    5′ dACCGTCCAAGATAAT 3′

Write the sequence of bases in the complementary segment of RNA.

SOLUTION

In forming RNA, G pairs with C, C with G, T with A, and A with U. The lowercase *d* is dropped because this is RNA.

    3′ UGGCAGGUUCUAUUA 5′

PRACTICE EXERCISE 11.7

The sequence on a DNA strand is

5′ dA—T—C—G—A—A 3′

Write the sequence of the complementary mRNA strand.

11.8 Translation

AIMS: **To outline, with the aid of sketches, the steps of protein synthesis, indicating the function of tRNA, mRNA, ribosomes, and enzymes. To write the anticodon for a given codon.**

Focus

Translation involves tRNA, mRNA, ribosomes, and enzymes.

The translation of the information contained in mRNA into protein requires a supply of amino acids, tRNA molecules, mRNA, ribosomes, and a number of enzymes. As we shall see, translation involves four steps: *activation* of tRNA, *initiation, elongation,* and *termination* of a polypeptide chain.

Activation of tRNA

In translation, tRNA molecules carry amino acids to mRNA bound to ribosomes. *A tRNA molecule carrying its specified amino acid is termed an* **activated tRNA.** Activation of tRNAs occurs in two steps. In the first step, the amino acid reacts with adenosine triphosphate (ATP) to form a mixed carboxylic-phosphoric acid anhydride.

Amino acid Adenosine triphosphate (ATP) Acylamino adenosine monophosphate (acylamino AMP)

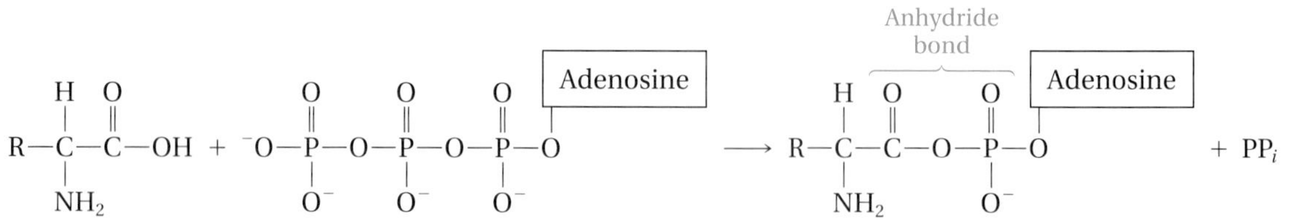

The mixed anhydride bond is extremely energetic and reacts readily with the 3′-hydroxyl group on the ribose ring at the 3′ end of the tRNA to form an ester.

Acylamino adenosine monophosphate (acylamino AMP) tRNA Acylamino tRNA (activated tRNA) Adenosine monophosphate (AMP)

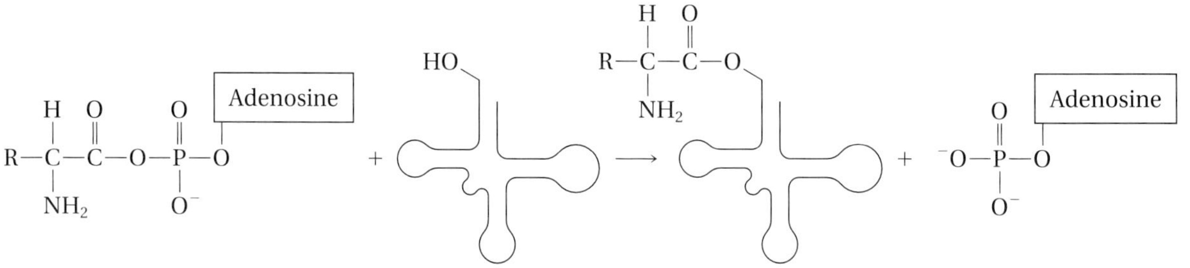

The structures of all tRNA molecules appear quite similar. How then does a given tRNA molecule recognize the amino acid it is to carry to mRNA on the ribosome? Cells contain an entire set of enzymes—one for each amino acid tRNA combination—that match tRNA molecules to their proper amino acids. These enzymes are very specific for both the structure of the amino acid and the tRNA. Almost never does an amino acid attach to the wrong tRNA.

Initiation

The beginning step in the growth of a polypeptide chain of a protein is called **initiation.** For initiation to begin, molecules of mRNA bind at a ribosomal site. A tRNA molecule carrying the N-terminal amino acid (AA 1) of the protein to be synthesized meets the mRNA at the first of two **tRNA binding sites**—*specific locations where tRNA molecules bind to ribosomes* (Fig. 11.14). The first amino acid incorporated into proteins in eukaryotic cells is always methionine. The activated tRNA carrying methionine must land on the correct spot on mRNA or else the protein that is synthesized will have the wrong sequence of amino acids. The mRNA landing point for the activated tRNA consists of three adjacent bases at the 5′ end of the mRNA molecule. *These three bases or* **base triplets** *on mRNA are called a* **codon.** *A loop in each tRNA contains three complementary bases called* **anticodons.** The anticodon of the tRNA molecule forms three complementary base pairs with the first triplet or codon of the mRNA molecule (Fig. 11.15). As in the

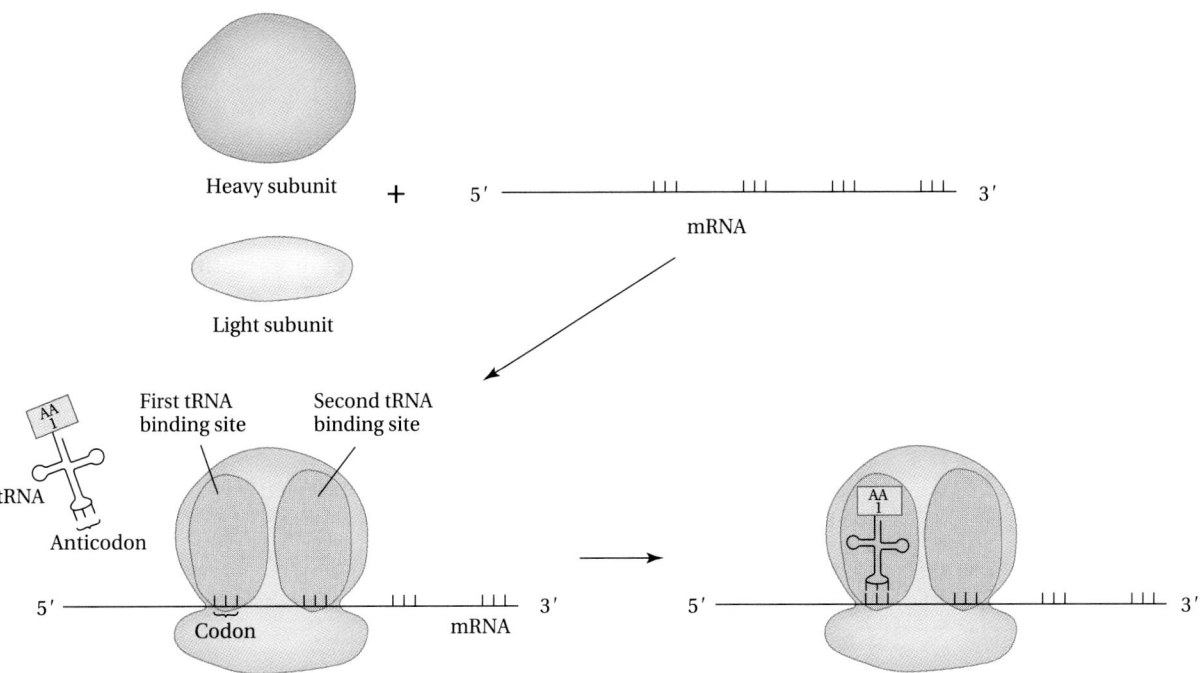

Figure 11.14
Initiation of translation.
Subunits of ribosome combine
with mRNA to form a ribosome-
mRNA complex. Activated tRNA
carrying the first amino acid
(methionine) to be incorporated
into the polypeptide chain enters
the first tRNA binding site. The
anticodon of the tRNA forms
three complementary base
pairs with the codon of mRNA.

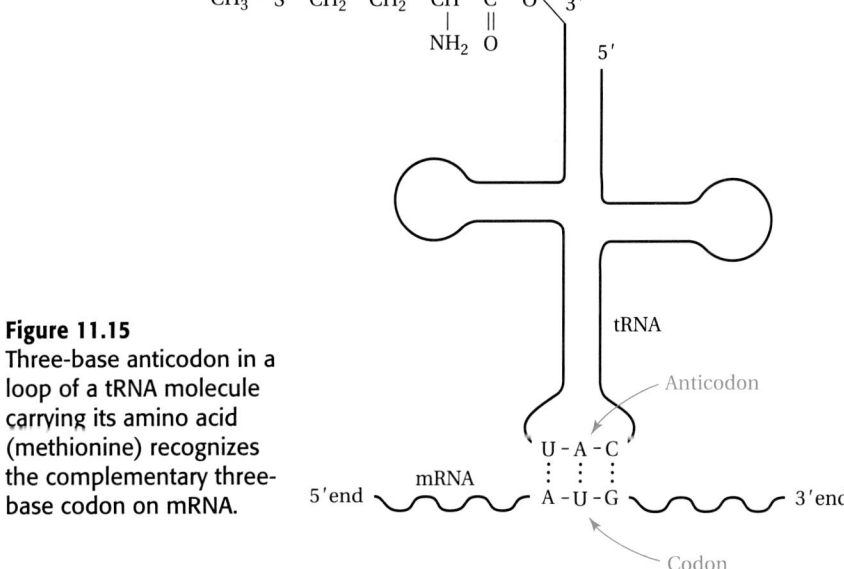

Figure 11.15
Three-base anticodon in a
loop of a tRNA molecule
carrying its amino acid
(methionine) recognizes
the complementary three-
base codon on mRNA.

DNA double helix, a stable complex between tRNA and mRNA forms only
when the bases of the codon and anticodon are complementary. In the initiation of translation, a codon for methionine on mRNA signals the proper
starting place for the initiation of translation. Later in this chapter we will
learn about the codons for each amino acid.

Elongation of the polypeptide chain

A number of antibiotics interfere with protein synthesis in bacteria. The tetracyclines, for example, block binding at the second tRNA binding site. Erythromycin blocks the shift to expose a new base triplet once a peptide bond has been formed.

Once the growth of a polypeptide chain has been initiated, the next step is *elongation* of the peptide chain. *In* **elongation,** *the polypeptide chain grows an amino acid at a time.* The tRNA molecule bearing the second amino acid (AA 2) to be incorporated into the protein enters the second tRNA binding site on the ribosome (Fig. 11.16). The anticodon of this tRNA molecule forms complementary base pairs with the second base triplet of the mRNA molecule. An enzyme promotes the formation of a peptide bond between the first (N-terminal) and the second amino acids. The first tRNA molecule leaves its binding site. The mRNA and the remaining tRNA, which now bears the dipeptide, shift to the first tRNA binding site (to the left in the figure). The shift exposes a new base triplet of mRNA to the second tRNA binding site. A tRNA molecule carrying the third amino acid (AA 3) to be incorporated into the protein being synthesized enters the second tRNA binding site. These steps are repeated with appropriately activated tRNA molecules until a complete protein is synthesized.

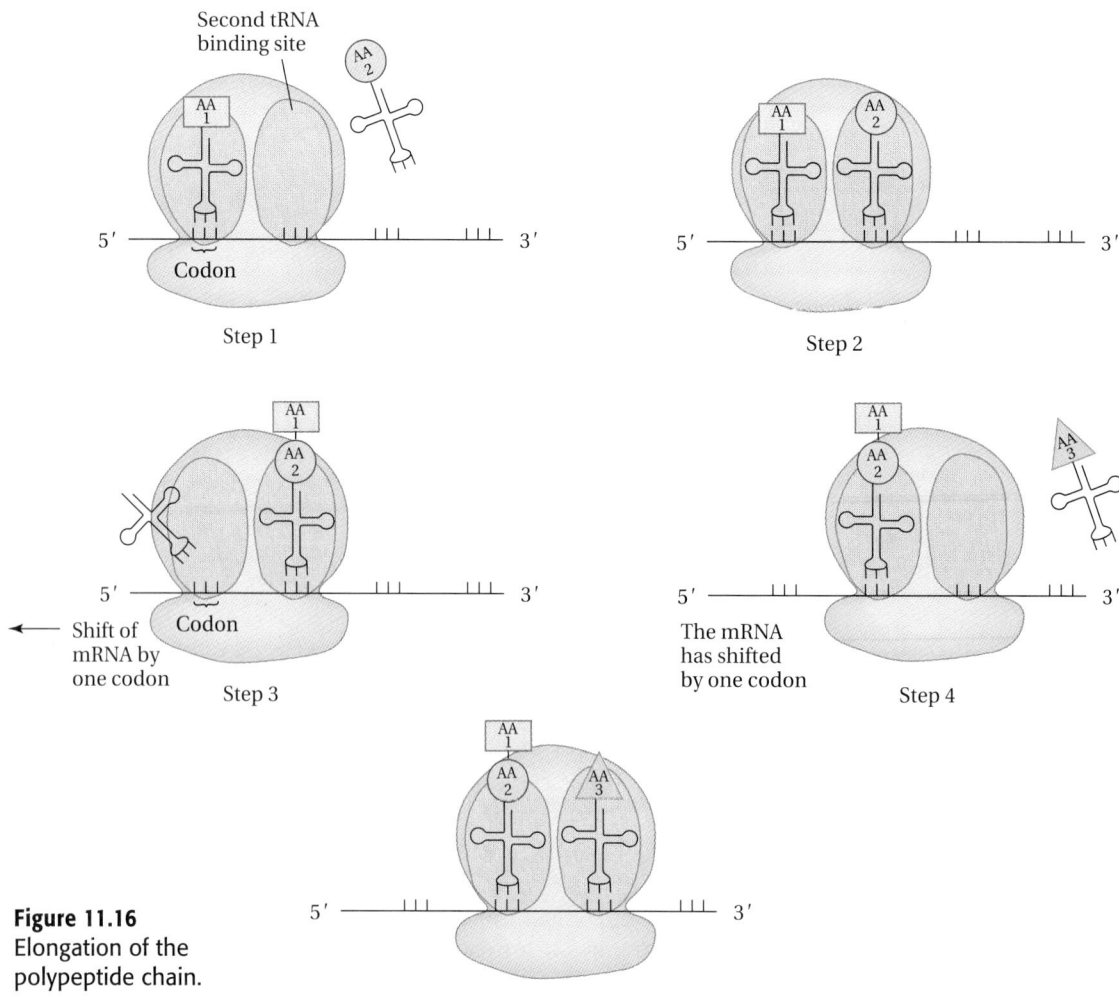

Figure 11.16
Elongation of the polypeptide chain.

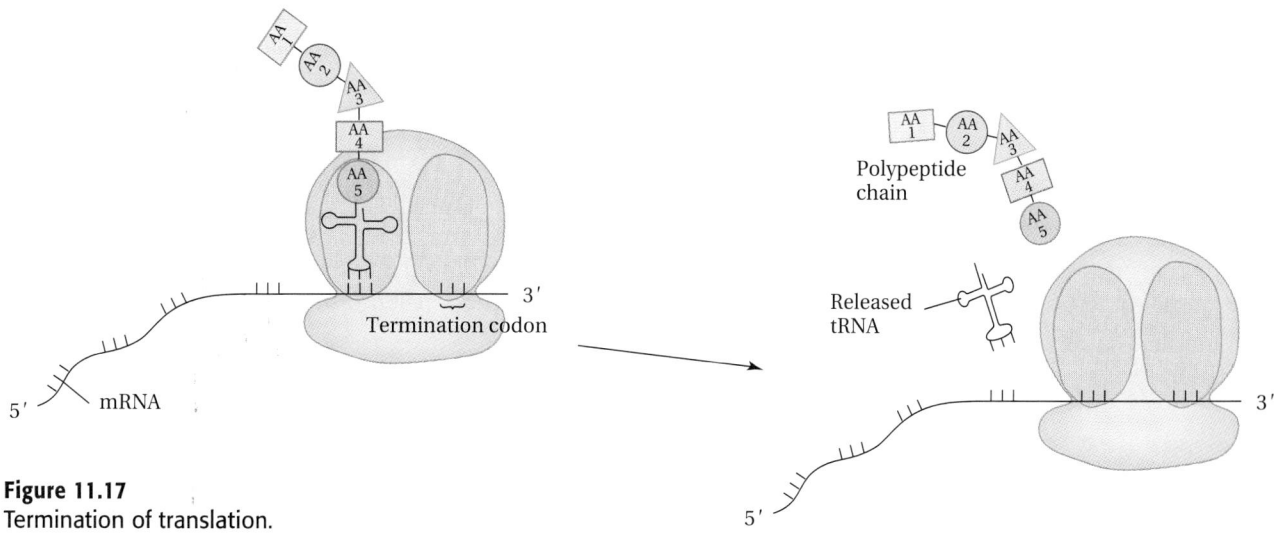

Figure 11.17
Termination of translation.
A growing polypeptide chain
terminates when it encounters
a noncoding base triplet
(termination codon) on mRNA.
The tRNA is cleaved from the
completed polypeptide chain,
which diffuses away from the
ribosome.

Termination

Termination *is the cessation of growth of the polypeptide chain.* The protein
chain undergoes termination when incoming tRNA molecules encounter a
termination codon—*a base triplet of mRNA that does not code for an amino
acid.* On termination of a protein chain, an enzyme cleaves the bond
between the protein and tRNA. The protein diffuses away from the ribosome to complete translation (Fig. 11.17).

Posttranslation processing

Although the translation of proteins in eukaryotic cells is initiated with
methionine as the N-terminal amino acid, few finished proteins have
methionine as the N-terminal amino acid residue. The reason is that most
new proteins undergo **posttranslation processing**—*modification subsequent to translation.* Cleavage of N-terminal methionine from proteins is
part of posttranslation processing. Other modifications may occur as well.
Necessary disulfide (—S—S—) bonds are formed, for example. If the protein is to be a glycoprotein, the carbohydrate molecules are attached. Other
posttranslation processes include methylations, hydroxylations, and
attachment of coenzymes or prosthetic groups. With the completion of
posttranslation processing, the new protein molecule is ready to make its
contribution to the life of the cell.

11.9 The genetic code

*AIMS: To write the amino acid sequence of a peptide given the
DNA or RNA base sequence. To explain why the genetic
code is termed* **degenerate.**

Focus

The genetic code is a dictionary
of three-letter words.

The proper amino acid sequence of a protein molecule depends on the formation of a complex between a base triplet, or codon, on mRNA and a complementary triplet, or anticodon, on tRNA. What are the codons for the 20
common amino acids? We will find out in this section.

It is sometimes helpful to compare mRNA to a dictionary. Actually, DNA is the master dictionary, and mRNA is a copy, but nature protects the DNA of eukaryotic cells by locking it in the nucleus. We, with our copy of the dictionary, or the cell with its mRNA, can work just as well from the copy as from the original. Whereas all the words in an English dictionary are composed from 26 letters, the words in the mRNA dictionary are composed from only 4 letters: the bases A, U, G, and C. The words in the mRNA dictionary have 20 meanings, the 20 common amino acids. *The collection of words or codons in the mRNA dictionary that is translated from one language to another—nucleic acid language to protein language—is called the* **genetic code.**

A sequence of three bases is required to code for an amino acid. Table 11.3 lists the amino acids and their mRNA codons. Three codons do not code for an amino acid. These codons—*UAA, UAG, and UGA*—are the *termination codons.* No anticodon on a tRNA molecule is complementary to them, so they signal "stop" at the end of a newly synthesized chain.

Over the years, scientists have examined the spellings of genetic code words in many different organisms, and so far the code has been the same in almost all of them. The genetic code appears to be nearly *universal.* As the genetic code was being deciphered, it became apparent that the code is

Table 11.3 Three-Letter Code Words for the Amino Acids*

		U		C		A		G	
U		UUU	Phe	UCU	Ser	UAU	Tyr	UGU	Cys
		UUC	Phe	UCC	Ser	UAC	Tyr	UGC	Cys
		UUA	Leu	UCA	Ser	UAA	End	UGA	End
		UUG	Leu	UCG	Ser	UAG	End	UGG	Trp
C		CUU	Leu	CCU	Pro	CAU	His	CGU	Arg
		CUC	Leu	CCC	Pro	CAC	His	CGC	Arg
		CUA	Leu	CCA	Pro	CAA	Gln	CGA	Arg
		CUG	Leu	CCG	Pro	CAG	Gln	CGG	Arg
A		AUU	Ile	ACU	Thr	AAU	Asn	AGU	Ser
		AUC	Ile	ACC	Thr	AAC	Asn	AGC	Ser
		AUA	Ile	ACA	Thr	AAA	Lys	AGA	Arg
		AUG	Met	ACG	Thr	AAG	Lys	AGG	Arg
G		GUU	Val	GCU	Ala	GAU	Asp	GGU	Gly
		GUC	Val	GCC	Ala	GAC	Asp	GGC	Gly
		GUA	Val	GCA	Ala	GAA	Glu	GGA	Gly
		GUG	Val	GCG	Ala	GAG	Glu	GGG	Gly

* Codes are read in the 5′ to 3′ direction. "End" denotes a termination codon.

degenerate—*that is, there is more than one code word for most of the amino acids.* The term *degenerate* does not imply that the code is defective, but only that different code words can have the same meaning. On the other hand, no codon specifies more than one amino acid. Of the 64 possible combinations of triplets formed by 4 bases, 61 code for an amino acid, and 3 are termination codons.

EXAMPLE 11.4	**Writing an amino acid sequence**

Write the amino acid sequence formed by the strand of DNA in Example 11.3.

SOLUTION

The mRNA rewritten in the $5' \rightarrow 3'$ direction from the solution to Example 11.3 is

5′ AUUAUCUUGGACGGU 3′

Divide the mRNA into 3-base codons, and use Table 11.3 to find the corresponding amino acids.

5′ AUU   AUC   UUG   GAC   GGU 3′
Ile——Ile——Leu——Asp——Gly

PRACTICE EXERCISE 11.8

Use Table 11.3 to write a base sequence for mRNA that codes for the tripeptide Ala-Gly-Ser.

PRACTICE EXERCISE 11.9

Determine the amino acid sequence of a tetrapeptide if the corresponding base sequence on DNA is

5′ dA—G—T—G—T—T—T—C—T—C—C—T 3′

11.10 Gene mutations and molecular diseases

AIMS: *To show how the addition, deletion, or substitution of a nucleotide can result in a gene mutation. To explain the relationship between a gene mutation and a molecular disease using sickle cell anemia as an example.*

Focus

Gene mutations are changes in the base sequence of DNA.

Substitutions, additions, or deletions of one or more nucleotides in the DNA molecule are called **gene mutations.** If mutated DNA is faithfully transcribed, the mutation is reflected in a change in the nucleotide sequence of mRNA. As a result, the primary structure of a protein may be changed, or a given protein may not be synthesized at all. The functioning of genes involved in the multiplication and growth of cells must be kept under strict cellular controls. When gene mutations cause the failure of these controls, the result can be cancer, as described in A Closer Look: Oncogenes and Tumor Suppressor Genes.

A Closer Look

Oncogenes and Tumor Suppressor Genes

Researchers have made significant advances in understanding the molecular basis of cancer. Essentially a family of diseases, cancer is characterized by uncontrolled cell division and cell growth. Because DNA is involved in the regulation of cell multiplication, cancer may be thought of as a genetic disease. The formation of tumors is related to mutations in DNA that cause unregulated growth of cells. If the DNA in a cell undergoes mutations that cause inappropriate multiplication, the result may be formation of a tumor. The cells in an entire tumor are often the descendants of a single cell that has started unregulated and unrestrained multiplication. Not every tumor is cancerous. Cells of malignant tumors invade surrounding normal tissue and metastasize, or spread to new sites, to start new tumors. Benign tumors stay where they begin and do not invade surrounding tissues.

How do mutations in DNA lead to cancers? The answer to this simple question is complex and still incomplete. Certainly part of the answer lies in the activities of two types of genes normally found in cells: *oncogenes* and *tumor suppressor genes.*

Oncogenes are genes whose normal activity is necessary for cells to divide and grow. In normal cells of adults, the activity of these genes is strictly controlled, since most cells in organs and other tissues need to divide and grow rather slowly, if at all. If the controls on these genes are turned off, cell division and growth begins and continues unabated. Many of the known oncogenes have been shown to encode the information for a group of polypeptides called *growth factors.* The controls for the production of growth factors are encoded in the same or a nearby gene. Mutation of the DNA in the control portion of the gene inactivates the control, and unrestrained cell growth results.

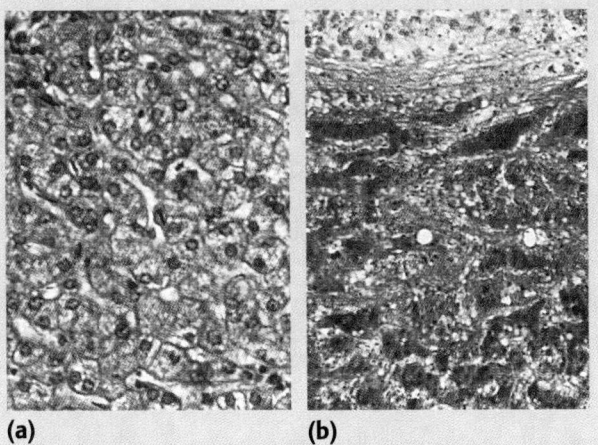

(a)　　　　　　　　　　　　　(b)

Oncogenes appear to trigger the processes that convert normal cells (a), to cancerous cells (b).

Another family of oncogenes consists of those which encode certain *protein kinase enzymes.* The protein kinases stimulate many activities of cells, including growth. One or more protein kinases important for cellular growth are normally active only if they are bound to growth factors. However, mutations in the normal DNA that encodes the kinases can cause the enzymes to be active even in the absence of growth factors. The result is unrestrained cell growth.

The normal job of tumor suppressor genes is to prevent unrestrained cellular growth. In an inherited disease called *retinoblastoma,* children lack both copies of a gene called *RB.* As a result of the missing tumor-suppressing *RB* gene, the children develop tumors in both eyes.

In addition to the effects of both oncogenes and tumor suppressor genes, additional steps are probably required to transform abnormal, rapidly growing cells into cancer cells. However, the nature of the transformation is still murky. Knowledge of the role of DNA mutation and cellular transformation may or may not lead to better treatments for diseases, but improved treatments will probably not be devised without it.

PRACTICE EXERCISE 11.10

Explain how gene mutations occur.

When a human cell divides and replicates its DNA, approximately 4×10^9 bases are copied. It has been estimated that there are perhaps 2000 errors made during each replication.

Albinism is a genetic disease in which the enzyme tyrosinase is not produced. Tyrosinase is necessary for the production of melanin, the pigment of the skin, hair, and eyes. Persons affected by albinism are very sensitive to sun exposure.

Mutations may be harmful or beneficial to the survival of a species. Sickle cell trait and sickle cell anemia illustrate both the harm and the benefit. As mentioned in Section 9.10, the difference between normal adult hemoglobin (HbA) and sickle cell hemoglobin (HbS) is the replacement of a glutamic acid residue by a valine residue in the sixth position of the beta chain of the protein.

HbA: Val—His—Leu—Thr—Pro—Glu—Glu—Lys——————
HbS: Val—His—Leu—Thr—Pro— Val—Glu—Lys——————
        1   2   3   4   5   6   7   8

The code words for glutamic acid and valine are

Glutamic acid	Valine
GAA	GUA
GAG	GUG
	GUU
	GUC

Substitution of the middle base, A, by U changes the meaning of the code words for glutamic acid to two code words for valine. People with sickle cell trait have immunity to malaria, so in this sense the mutation is beneficial. People with sickle cell anemia are very sick and often die young, which certainly is a harmful effect of the mutation. The benefits and the harm of hemoglobin S come from the substitution of a single nucleotide base in the gene that directs the synthesis of the beta chain of hemoglobin.

Among the hardest mutations to recognize are those that cause no detectable change in the activity of enzymes. At the other extreme, a mutation allowing no active protein at all is among the easiest to detect. A cell may be unable to synthesize a certain protein if a mutation changes a codon for an amino acid into one of the termination codons: UAA, UGA, or UAG. For example, one of the codons for tyrosine is UAU. A change of the last base from U to G produces the termination codon UAG. Translation of the message for a polypeptide chain stops when the termination codon is reached. A complete protein chain is usually necessary to produce an active protein. The single base change prevents an active protein from being synthesized.

The idea that certain diseases stem from faulty genes originated long before anyone knew about DNA's role in heredity. About 1900, Archibald Garrod, an English physician, proposed that certain lifelong diseases arise because an enzyme necessary to good health is defective or missing. Garrod's concept has been extended to include other proteins as well. Over 4000 **molecular diseases**—*inborn errors*—are known. Although they have not been labeled previously as such, we have already met two molecular diseases: methemoglobinemia (Sec. 9.9) and sickle cell anemia (Sec. 9.10).

EXAMPLE 11.5

Finding peptide sequences from mRNA base sequences

Consider the following segment of mRNA:

     5′ CGGGGUUGCAAU 3′

(a) What is the amino acid sequence formed by translation? (b) What amino acid sequence would result from the substitution of adenine for the second uracil?

SOLUTION

(a) Divide the mRNA sequence into 3-base codons, using Table 11.3.

5′ CGG   GGU   UGC   AAU 3′

Arg——Gly——Cys——Asn

(b) The substitution gives a different amino acid sequence:

5′ CGG   GGU   AGC   AAU 3′

Arg——Gly——Ser——Asn

PRACTICE EXERCISE 11.11

The following base sequence is a portion of an mRNA:

5′ A—C—G—G—U—C—A—G—C—G—A—G—C—C—C 3′

(a) Translate this genetic message to give the amino acid sequence of a pentapeptide. (b) Uracil is replaced by adenine in a mutation. What is the new amino acid sequence? (c) Suppose uracil is deleted. What is the peptide composition?

11.11 Recombinant DNA and gene therapy

AIM: Describe how recombinant DNA technology can be used to treat a molecular disease.

Focus

Scientists have learned to transplant genes.

Since the late 1970s, scientists have taken giant steps in **recombinant DNA technology**—*transplantation of genes from one organism into another. Recombination* consists of cleaving DNA chains, inserting a new piece of DNA into the gap created by the cleavage, and resealing the chains. When the recombination is successful, the transplanted gene will express (synthesize) its normal protein product. For example, bacteria that receive human genes can be induced to express human proteins of value for research or for the treatment of disease.

Therapeutic drugs from recombinant research

Recombinant DNA research has produced several notable proteins for human therapy. Human insulin produced by bacteria is on the market for use by diabetics. This product is said to be preferable to insulin from hogs, which is very similar in amino acid composition to human insulin but which sometimes causes an adverse allergic response. A protein called *beta-interferon* (see Sec. 13.2), made by incorporation of the human interferon gene into bacteria, has been approved for treatment of multiple sclerosis patients. The protein has been shown to reduce the number and severity of attacks that strike patients. Another human protein produced in large quantities by recombinant techniques and of therapeutic value is tissue plasminogen activator (TPA; see Case in Point in Chapter 10). You may recall that this enzyme converts the zymogen plasminogen to plasmin, the enzyme that dissolves blood clots (see Sec. 10.12), and is useful in the treatment of heart attacks. The number of protein products of recombinant DNA entering testing for therapeutic use seems to increase almost daily, and hundreds more are the subject of intense research. Undoubtedly, the surface of the potential for new pharmaceuticals from recombinant DNA has been barely scratched.

Gene therapy

The explosion of knowledge about the structure, function, and manipulation of the human genome has led scientists to the dawn of a new era—the era of *gene therapy*—in the treatment of disease. **Gene therapy** *is the transfer of new genetic material to the cells of an individual with resulting therapeutic benefit to the individual.* Ashanthi, the 4-year-old girl described in the Case in Point earlier in this chapter, was the first human recipient of gene therapy. In the Follow-up to the Case in Point, below, we will learn the outcome of this therapy.

Procedures for gene therapy are now being worked out for a number of hereditary diseases such as hemophilia, in which clotting factors are defective or missing. In 1994, researchers successfully applied gene therapy to a woman to reduce the symptoms of familial hypercholesteremia,

FOLLOW-UP TO THE CASE IN POINT: Treatment for an inherited disease

In September of 1990, the first gene therapy was done on Ashanthi, a 4-year-old girl who suffered from an inherited disease called *adenosine deaminase (ADA) deficiency*. Adenosine deaminase (ADA) plays an important role in the functioning of a type of cells called *T cells*. One role of T cells is to destroy virus-infected cells, parasites, and cancer cells. Another function is to activate other cells that are responsible for the destruction of infecting bacteria. In the pioneering gene therapy treatment, Ashanthi received some of her own T cells, which had been removed and infected with a virus. The viral gene was cleverly modified; a gene for the production of ADA had been inserted into the viral gene. The virus was of a type that attaches its own genetic material to the DNA of the cells it infects. It was hoped that when the virus infected the isolated T cells, the DNA necessary to make ADA also would be inserted into the T cell gene, along with the usual viral gene.

The gene therapy for ADA deficiency seems to have worked. The T cells Ashanthi received produced ADA and strengthened her ability to ward off

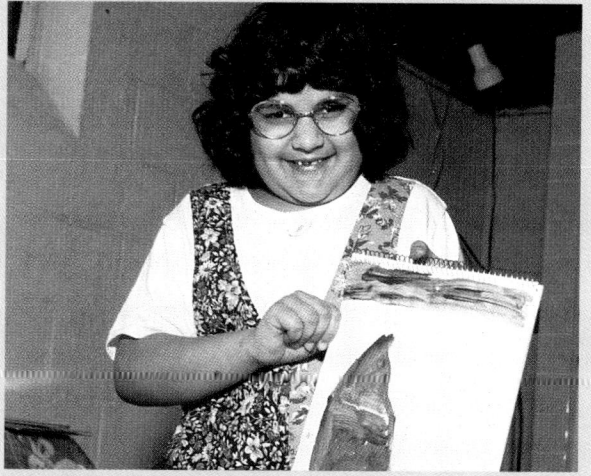

Recovered from her ADA deficiency, Ashanthi now attends public school.

infections. The procedure also was effective for a second girl who started treatment in January 1991. Both girls are enrolled in public schools and have had no more than the average number of infections.

an inherited disease whose symptoms are characterized by high blood cholesterol and clogging of the arteries. The researchers removed some of the woman's liver cells, which contained a flawed gene for a key enzyme that processes cholesterol. They added a correct form of the gene to the DNA in the liver cells and returned these cells to the liver. The woman now has cleaner arteries and a lower blood cholesterol level.

Gene therapy also could prove useful for treating a number of non-hereditary diseases. For example, new research might make it possible to insert into cells certain genes that would produce factors to shrink tumors or stop the progression of AIDS. In the AIDS work, researchers are using an approach that allows them to insert a gene for a ribozyme (see Sec. 10.1) into cells infected with HIV. The scientists hope that this catalytic RNA will disable HIV by cleaving its RNA.

The demonstrated feasibility of gene transplantation in humans has aroused the attention of scientists and other concerned citizens, for it opens the door to profound ethical problems. Human control of the human genome offers unlimited opportunities for moral or immoral choices. Who will benefit and who will not? And for what reasons? Biomedical ethicists generally agree that the cure of diseases of existing human beings is an ethical use of gene therapy. However, the techniques of gene therapy are potentially applicable to the creation of humans to suit various purposes: superintelligent people to rule the world, tall people for basketball, workers who would not be bored by repetitive tasks or harmed by dirty environments. For this reason, the genetic manipulation of sperm or egg (germ line cells) is more ethically problematic than the use of gene therapy to cure diseases.

PRACTICE EXERCISE 11.12

Why do you think recombinant DNA research is controversial? What are your personal thoughts on the matter?

SUMMARY

Ribonucleic and deoxyribonucleic acids (RNA and DNA) consist of giant molecules consisting of nucleotides bound into a polymer through phosphodiester linkages. The sugar units of the nucleotide units in RNA are ribose; in DNA they are deoxyribose. The nucleotide base groups of DNA are two purines, adenine (A) and guanine (G), and two pyrimidines, thymine (T) and cytosine (C). Uracil (U) substitutes for T in RNA.

The complete DNA molecule exists as a double-stranded helix—two single strands of DNA running in opposite directions. The bases of one DNA strand in the helix are paired through hydrogen bonding with complementary bases from the other; A is always paired with T, and G is always paired with C.

DNA acts as an instruction manual for the synthesis of all the proteins made by the cell. The process by which DNA is copied is called replication. The information contained in DNA can be copied (replicated) and passed to a kind of RNA, messenger RNA (mRNA), in transcription. Information contained in mRNA can be used to direct synthesis of specific proteins (translation).

Protein synthesis requires molecules of transfer RNA (tRNA). The tRNA molecules, at least one kind for each amino acid, carry amino acids to the ribosomes for protein synthesis. Molecules of tRNA recognize their appropriate positions on mRNA by an anticodon—a base triplet that is complementary to a triplet on mRNA. The triplet code words for the amino acids have been worked out; the genetic code is degenerate and nearly universal.

The fidelity of replication, transcription, and translation is remarkable, but sometimes mutations can occur. Mutations can result in the synthesis of a fully active protein, a less active protein, or no protein at all. Some mutations are harmful and lead to death or illnesses called hereditary diseases, molecular diseases, or inborn errors. Others help increase the organism's ability to survive. Recombinant techniques in the past few years have increased the scientist's ability to manipulate genes. In gene therapy, a faulty gene responsible for a disease is replaced by a normal gene.

KEY TERMS

Activated tRNA (11.8)
Anticodon (11.8)
Antiparallel strands (11.2)
Base triplet (11.8)
Central dogma (11.3)
Codon (11.8)
Complementary base pair (11.2)
Degenerate code (11.9)
Deoxyribonucleic acid, DNA (11.1)
DNA double helix (11.2)
DNA polymerase (11.4)

Elongation (11.8)
Exon (11.5)
Gene (11.5)
Gene mutation (11.10)
Gene therapy (11.11)
Genetic code (11.9)
Initiation (11.8)
Intron (11.5)
Messenger RNA (11.3)
Molecular disease (11.10)
Nucleic acid (11.1)
Nucleoside (11.1)
Nucleotide (11.1)

Okazaki fragments (11.4)
Phosphodiester (11.1)
Polynucleotide (11.1)
Posttranslation processing (11.8)
Purine bases (11.1)
Pyrimidine bases (11.1)
Recombinant DNA technology (11.11)
Replication (11.3)
Retrovirus (11.3)
Ribonucleic acid, RNA (11.1)

Ribosomal RNA (11.6)
Ribosome (11.6)
RNA polymerase (11.7)
Template (11.4)
Termination (11.8)
Termination codons (11.8)
Transcription (11.3)
Transfer RNA (11.6)
Translation (11.3)
tRNA binding site (11.8)
Unwinding proteins (11.4)

EXERCISES

Nucleic Acids and the Double Helix
(Sections 11.1, 11.2)

11.13 Cells contain two types of nucleic acids. What are they called?

11.14 What is the structural difference between the sugar unit in RNA and the sugar unit in DNA?

11.15 Explain the difference between a nucleoside and a nucleotide.

11.16 Explain the difference between a nucleotide and a nucleic acid.

11.17 (a) Name the nitrogen bases found in DNA. (b) Identify them as derivatives of purine or pyrimidine.

11.18 Are the nitrogen bases in RNA the same as those found in DNA? Explain.

11.19 (a) Draw the structure of the nucleoside formed by the combination of the nitrogen base adenine and the sugar β-D-2-deoxyribose. (b) What is its name?

11.20 How are nucleotides linked together in RNA and DNA molecules?

11.21 What gives DNA and RNA their highly acidic character?

11.22 (a) What does the following abbreviation mean?

U—G—C—A—G

(b) Which end is 3′ and which end is 5′?

11.23 What type of bonding holds a DNA double helix together?

11.24 Which of the following base pairs are found in a DNA molecule?
(a) A—A   (b) A—T
(c) C—G   (d) G—C
(e) G—A   (f) A—U

11.25 The two strands of DNA in the DNA double helix are *antiparallel*. What does this mean?

11.26 Consider the double helix. Are the base pairs
(a) part of the backbone structure
(b) inside the helix
(c) outside the helix?

The Central Dogma (Section 11.3)

11.27 Define the three steps that make up the central dogma of molecular biology.

11.28 What is a common characteristic of all retroviruses?

Replication and Genes (Sections 11.4, 11.5)

11.29 Describe the process of DNA replication.

11.30 A segment of a DNA strand has the following base sequence: dC—G—A—T—C—C—A. Write the base sequence that appears on its complementary strand in the Double Helix. Label the 3′ and 5′ ends of the sequence.

11.31 What is the function of DNA polymerase?

11.32 Why does the replication of DNA require a DNA template?

11.33 Distinguish between a gene and a polypeptide.

11.34 Contrast the function of exons and introns in a gene.

Classes of RNA and Transcription (Sections 11.6, 11.7)

11.35 Describe the function of (a) rRNA and (b) mRNA.

11.36 What is the function of tRNA? Sketch the shape of a typical tRNA molecule.

11.37 What components are needed for the synthesis of RNA?

11.38 Write the base sequence for the complementary RNA strand when the DNA base sequence is as follows:

5′ dA—T—C—G—C—T—A 3′

Label the ends of the RNA strand.

Translation and the Genetic Code (Sections 11.8, 11.9)

11.39 Give the location and describe the general structure of (a) a codon and (b) an anticodon.

11.40 Does every base triplet of mRNA code for an amino acid in protein synthesis? Explain.

11.41 The codon for methionine is AUG. What is the anticodon?

11.42 (a) Name the first amino acid that is always incorporated into proteins in eukaryotic cells. (b) Predict what would happen if this amino acid were missing from the diet.

11.43 The base sequence of a DNA fragment is a gene:

5′ dA—G—C—T—G—G—G—A—C 3′

(a) Write the base sequence of the mRNA strand.
(b) Write the amino acid sequence of the tripeptide. (Remember to start reading codons at the 5′ end of mRNA.)

11.44 Describe the changes that may occur in posttranslation processing.

Gene Mutations and Molecular Diseases (Section 11.10)

11.45 What causes a gene mutation?

11.46 (a) Name some molecular diseases. (b) Can these diseases be cured? Explain.

Recombinant DNA and Gene Therapy (Sections 11.11)

11.47 What are some outcomes of recombinant DNA research?

11.48 Describe a possible use of gene therapy.

Additional Exercises

11.49 Match the following.

(a) anticodon	(1) RNA → protein
(b) A, T, G, C	(2) links nucleotides
(c) tRNA	(3) inborn error
(d) transcription	(4) triplet on mRNA
(e) ribosome	(5) cloverleaf shape
(f) phosphodiester bridge	(6) bases of DNA
(g) codon	(7) DNA → RNA
(h) degenerate	(8) site of protein synthesis
(i) molecular disease	(9) genetic code
(J) translation	(10) triplet on tRNA

11.50 Answer the questions below based on the following strand of DNA:

5′ T T A G A T T C G C A T C C G 3′

(a) What is the structure of the mRNA formed from this DNA template?
(b) What polypeptide would be formed from this strand of mRNA?
(c) What polypeptide would be formed
   (1) if the first base at the 5′ end of mRNA is deleted?
   (2) if the first G from the 5′ end in the mRNA sequence is replaced by an A?
   (3) if the terminal C at the 5′ end of the mRNA is replaced by a G?

11.51 To make the tetrapeptide Phe-Ser-Ala-His, (a) what is the sequence of bases on the mRNA strand? (b) What is the sequence of bases on the complementary segment of the DNA molecule?

11.52 The genetic code is *degenerate*. What does this mean?

11.53 What are the different types of RNA in the cell? Describe the functions of each in protein synthesis.

11.54 Why are nucleic acids *acidic?*

11.55 One of the following triplets cannot be a codon: UCG, ATC, GCU. Explain.

11.56 In which direction is the DNA molecule (a) synthesized continuously? (b) synthesized discontinuously (using Okazaki fragments)?

11.57 What are the possible consequences if a DNA molecule fails to replicate itself *every time* without error?

11.58 Match one of the processes at the right to each of the phrases on the left.
(a) utilizes codon-anticodon interaction
(b) *not* carried out in the nucleus
(c) produces mRNA
(d) uses tRNA and mRNA
(e) produces a DNA copy
(f) results in single-stranded RNA
(g) produces a specific protein
(h) semiconservative in nature

(1) replication
(2) transcription
(3) translation

SELF-TEST (REVIEW)

True/False

1. RNA and DNA are structurally the same except that uracil is found in place of thymine in RNA.

2. The backbone of a DNA molecule is held together by phosphodiester bridges.

3. RNA and DNA are termed *nucleic acids* because their nitrogen bases can donate protons.

4. Gene mutations are always harmful to the organism.

5. In any organism the A + G to T + C ratio in DNA equals 1.

6. In the double helix of DNA, the largest number of stabilizing double bonds is achieved when purines oppose purines and pyrimidines oppose pyrimidines.

7. Protein synthesis takes place in the nucleus, where the genetic information is found.

8. The tRNA molecules are responsible for the sequence of amino acids in a protein.

9. An anticodon on a tRNA molecule must be complementary to a codon on the mRNA for a stable tRNA-mRNA structure to form.

10. The replacement of one letter in the genetic code by another letter always leads to a gene mutation.

Multiple Choice

11. Human beings are different from monkeys because of
(a) differences in types of nucleotides.
(b) a different A + G/T + C ratio.
(c) differences in proteins.
(d) a different genetic code.

12. The *ordered* sequence of steps in the central dogma of genetic information transmission is
(a) translation, replication, transcription.
(b) replication, translation, transcription.
(c) translation, transcription, replication.
(d) replication, transcription, translation.

13. The RNA strand transcribed from the strand of DNA molecule (-G-A-T-T-C-G-) would be
(a) -C-U-A-A-G-C-.   (b) -C-T-U-U-G-C-.
(c) -T-U-G-G-A-T-.   (d) -C-T-A-A-G-C-.

14. The type of RNA molecule that brings amino acids to the site of protein synthesis is
(a) rRNA.   (b) aRNA.   (c) mRNA.
(d) tRNA.

15. A segment of mRNA is composed of the alternating bases adenine and guanine, A-G-A-G-A-G-A-G. The protein formed from this molecule
(a) would probably contain only one amino acid.
(b) would probably contain two different amino acids.
(c) would contain three or more different amino acids.
(d) would be the same as one from one with an mRNA sequence of G-A-G-A-G.

16. Which of the following statements is *incorrect* concerning the structure of DNA?
(a) The structure is a double helix.
(b) The bases are arranged on the outside of the molecule.
(c) The two DNA strands are antiparallel.
(d) The two strands are held together by hydrogen bonds.

17. A nucleic acid is made up of
    (a) a sugar, a nucleoside, and a nitrogen base.
    (b) a nitrogen base, a phosphate group, and a sugar.
    (c) a protein, a sugar, and a phosphate group.
    (d) a nitrogen base, an amino acid, and a sugar.

18. When the DNA molecules of two different species are compared, it is found that
    (a) the percent of guanine in each is always equal.
    (b) the lengths of the molecules are the same.
    (c) the A-to-T ratios are the same.
    (d) the age of the subjects compared must be the same, or results can vary.

19. A nitrogen base bonded to a sugar gives
    (a) a nucleoside.   (b) a nucleotide.
    (c) a nucleic acid.   (d) all of these.

20. The following units are found in RNA:
    (a) β-D-2-deoxyribose, A, T, G, C.
    (b) β-D-2-deoxyribose, C, U, A, G.
    (c) β-D-ribose, A, T, G, C.
    (d) β-D-ribose, A, G, U, C.

21. A segment of a DNA molecule that directs the synthesis of a particular protein molecule is a(n)
    (a) gene.   (b) deoxyribozyme.
    (c) codon.   (d) operon.

Digestion and Nutrition

 12

Materials for Living

Eating a selection from the recommended food groups provides a well-balanced diet.

CHAPTER OUTLINE

In previous chapters we have introduced the four major classes of large biological molecules: carbohydrates, lipids, proteins, and nucleic acids. We have learned that enzymes are vital to our bodies' proper functioning. The final chapters of this book explore the ways in which the human body maintains its internal environment, produces energy, and makes molecular building blocks for growth. Our bodies must be supplied with substances through dietary intake to perform these tasks, and these substances must be processed by digestion so that they can be used by our bodies. In this pivotal chapter we will examine nutrition and digestion.

Everyone knows that the building of strong bones in children is important. However, the maintenance of bone strength is often a problem for older adults, particularly women, as we will learn in the Case in Point.

CASE IN POINT: Osteoporosis

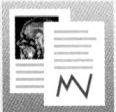

Bessie lived on a farm her entire life where she was admired by her family and neighbors for her capacity for hard work. In her eighty-sixth year she planted, hoed, weeded, and carried water to a large garden beside the farmhouse. One evening as she stood on the porch surveying her day's work, she fell suddenly. Like many elderly women, Bessie had broken a hip. What was the cause of Bessie's fracture, and how was it related to the special nutritional needs of older women? We will learn the answers to these questions in Section 12.9.

Many women suffer from osteoporosis in later life.

12.1 Essential needs

AIM: To list five essential needs of the human body.

Our bodies depend on external sources for five essential requirements:

1. *Oxygen. Cells that require oxygen to live are called* **aerobic cells.** Human cells are aerobic. Oxygen is not a nutrient, but human cells require it for energy production. The length of time human cells can survive without oxygen depends on the type of cell. The cells of heart muscle survive only about 30 seconds without oxygen, and brain cells die within about 5 minutes without it. Skeletal muscle cells live much longer than either heart muscle cells or brain cells in the absence of oxygen.

2. *Water.* Like oxygen, water is not a nutrient. But it is indispensable as the medium in which all body functions take place.

3. *Carbon compounds.* Carbon compounds supply energy and raw materials for growth and maintenance through the diet—mostly in the form of carbohydrates, lipids, proteins, and nucleic acids.

4. *Vitamins.* **Vitamins** *are compounds of carbon needed for growth and health that must be supplied in the diet.* Vitamins are divided into two major categories: fat-soluble and water-soluble (or B-complex) vitamins. Vitamin C is also water-soluble.

5. *Salts or minerals.* The various salts or minerals needed by the body for the maintenance of the acid-base balance, the electrolyte balance, and the growth of teeth and bones must be supplied through the diet.

The focus of this chapter is carbon compounds, vitamins, and salts or minerals. We begin this chapter by discussing the intake and digestion of dietary carbon compounds.

12.2 Digestion

AIM: To name the type of chemical reaction that is common to the digestion of carbohydrates, proteins, lipids, and nucleic acids.

The bulk of the food we eat is composed of complex carbohydrates (such as starch and glycogen), complex lipids, proteins, and nucleic acids, most of which are molecules too large to be absorbed by body tissues without further processing. Organs of the digestive system (Fig. 12.1) are sites of hydrolysis of these substances into simpler molecules that are absorbed into the bloodstream and transported to body cells.

When food is chewed, it mixes thoroughly with **saliva,** *the fluid secreted by the salivary glands.* An average adult secretes almost 1.5 L of saliva per day. Saliva is about 99% water and has a pH slightly above 7. It also contains small amounts of salts and two proteins: *mucin* and the enzyme *amylase* (ptyalin). Mucin is a glycoprotein that gives saliva its viscosity and stringiness. Amylase promotes the hydrolysis of starch and

---

Focus

The human body has five essential needs.

Vitamin
vita (Latin): life

Focus

Digestion begins in the mouth.

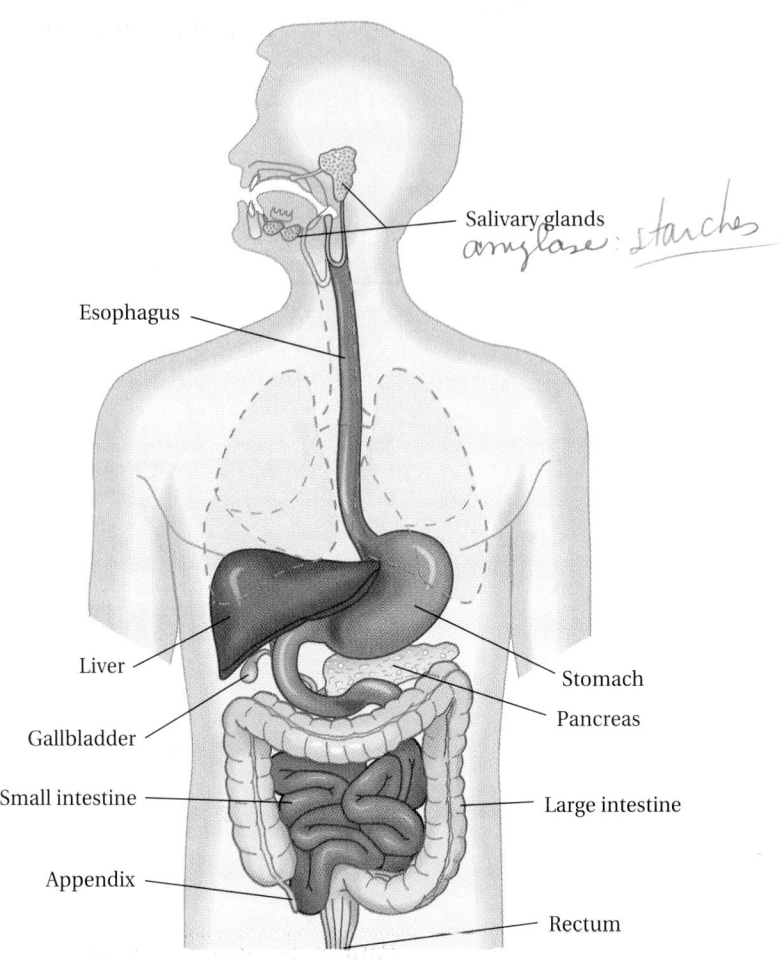

Salivary glands
amylase: starches

Esophagus

Liver

Gallbladder

Small intestine

Appendix

Stomach

Pancreas

Large intestine

Rectum

Figure 12.1
Organs of the human digestive system.

glycogen to the disaccharide maltose. No digestion of fats, proteins, or nucleic acids takes place in the mouth.

12.3 The stomach

AIM: To name the enzyme that hydrolyzes protein molecules in the stomach and characterize the environment in which it works.

Focus

Protein digestion begins in the stomach.

The mass of food mixed with saliva is swallowed and enters the stomach. The environment of the stomach is quite different from that of the mouth. The fluid of the stomach is very acidic owing to the presence of **gastric juice,** *a fluid secreted by cells that line the stomach*. The gastric juice is about 0.1 M hydrochloric acid. (A 0.1 M solution of a strong acid has a pH of 1.)

Most enzymes have a pH optimum in solutions near pH 7. Pepsin, however, the protease of the stomach, is an exception; it is most active at the pH of gastric juice. Pepsin, formed from pepsinogen by autoactivation or by the action of hydrochloric acid, begins to catalyze the hydrolysis of large proteins in the acid environment of the stomach, breaking them down to smaller peptides.

Considering that it contains strong acid and a voracious protease, the stomach usually holds up well. Pepsin and hydrochloric acid do not digest the cells that line the stomach because the outer membranes of these cells are mainly lipids, not proteins. Moreover, cells of the stomach lining rapidly slough off and are completely replaced every 3 days or so. Bacterial infection, nervous tension, and excessive alcohol or aspirin consumption can cause the stomach lining to break down in places. The breakdown may result in peptic ulcers, a form of self-digestion of the stomach.

> **PRACTICE EXERCISE 12.1**
>
> Explain why the frequent ingestion of large doses of an antacid, such as milk of magnesia, can affect protein digestion in the stomach.

12.4 The small intestine

AIM: To outline the digestive process for carbohydrates, proteins, lipids, and nucleic acids, citing locations and enzymes.

Focus

Digestion is completed in the small intestine.

The contents of the stomach, now thick and creamy, enter the upper part of the small intestine. Secretions from the pancreas and gallbladder flow into the upper part of the small intestine. These secretions are alkaline, and they neutralize stomach acid entering the small intestine.

Digestion and absorption of proteins and sugars

Pancreatic juice—*an aqueous fluid with a high bicarbonate ion concentration and a slightly basic pH (about pH 8.0)*—contains several proteases, including chymotrypsin, trypsin, and carboxypeptidase. These proteases, formed from their zymogens, catalyze the hydrolysis of proteins into a mixture of small peptides and individual amino acids. Digestion of proteins is completed within a few hours, but without proteases, the process under the same conditions would take about 7 years.

Digestion of sugars also continues in the small intestine. Maltose is hydrolyzed to glucose by maltase, sucrose is hydrolyzed to fructose and glucose by sucrase, and lactose is hydrolyzed to glucose and galactose by lactase (unless a person has a lactase deficiency). The absorption of monosaccharides, amino acids, and a few small peptides into the bloodstream starts as the products of digestion begin their long journey through the small intestine.

The concentrations of sugars and amino acids in the small intestine are lower than in the cells of the intestinal wall. Since the normal flow of substances through membranes is from higher to lower concentration, these compounds must be transported into intestinal cells by means of energy-requiring cellular pumps. From the cells of the small intestine, the sugars and amino acids make their way into the bloodstream. The liver extracts from the blood the sugars and amino acids it needs for its cells. The rest remain in the bloodstream for delivery to other tissues.

Digestion and absorption of lipids

Hydrolysis of lipids also begins in the small intestine. The lipids we consume as part of our diet are mainly the triglycerides, cholesterol, and the complex phospholipids—the phosphatidyl cholines, sphingomyelins, and cerebrosides of cell membranes. You may recall that cholesterol comes entirely from animal products.

Fats form small, insoluble globules in water rather than dissolving in it. The fats in these globules are relatively inaccessible to enzymes. The globules are broken up by being mixed with bile when they enter the duodenum. **Bile** *is a soaplike fluid produced in the liver and stored in the gallbladder.* The digestion of fats by enzymes requires bile to form water-soluble micelles with dietary fats just as the cleansing action of soaps depends on micelle formation with greasy dirt. Bile consists of cholesterol, *bile salts,* and **bile pigments**—*a group of colored compounds.* **Bile salts** *are derivatives of cholesterol;* Figure 12.2 shows the major ones.

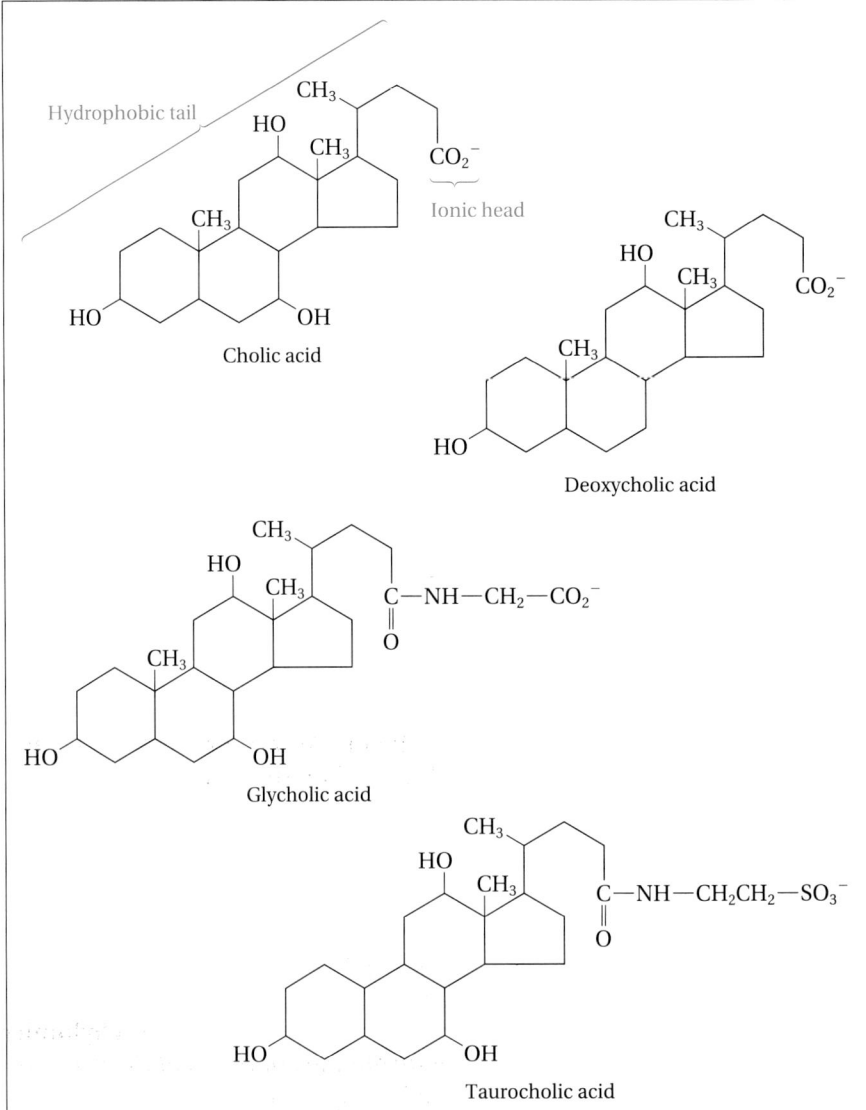

Figure 12.2
The bile salts are soaplike molecules derived from cholesterol that help dissolve dietary lipids. Notice that the bile salts, like soaps, have ionic heads and hydrophobic tails.

Dietary lipids that enter the small intestine are hydrolyzed into their simpler component molecules. *Lipase, a pancreatic enzyme, promotes the hydrolysis of lipids.* Lipase hydrolyzes only lipids that are solubilized by forming micelles with bile salts. The breakdown of tripalmitin, the triester of palmitic acid and glycerol, is a typical example of the hydrolysis of a triglyceride. The hydrolysis products are glycerol and the 16-carbon fatty acid palmitic acid.

$$
\underset{\text{Tripalmitin}}{
\begin{array}{c}
CH_3{+}CH_2{)}_{\overline{14}}\overset{\displaystyle O}{\overset{\|}{C}}{-}O{-}CH_2 \\[2pt]
CH_3{+}CH_2{)}_{\overline{14}}\overset{\displaystyle O}{\overset{\|}{C}}{-}O{-}CH \\[2pt]
CH_3{+}CH_2{)}_{\overline{14}}\overset{\displaystyle O}{\overset{\|}{C}}{-}O{-}CH_2
\end{array}}
\; + \; 3H_2O
\; \xrightarrow[\text{lipase}]{\text{Pancreatic}} \;
\underset{\text{Glycerol}}{
\begin{array}{c}
HO{-}CH_2 \\
HO{-}CH \\
HO{-}CH_2
\end{array}}
\; + \;
\underset{\text{Palmitic acid}}{3CH_3{+}CH_2{)}_{\overline{14}}\overset{\displaystyle O}{\overset{\|}{C}}{-}OH}
$$

Likewise, phospholipids such as the phosphatidyl cholines are hydrolyzed to their component parts.

$$
\underset{\substack{\text{A phosphatidyl choline} \\ \text{(lecithin)}}}{
\begin{array}{c}
R{-}\overset{\displaystyle O}{\overset{\|}{C}}{-}O{-}CH_2 \\[2pt]
R{-}\overset{\displaystyle O}{\overset{\|}{C}}{-}O{-}CH \\[2pt]
\overset{\textstyle CH_3}{\underset{\textstyle CH_3}{CH_2{-}\overset{+}{N}{-}CH_2{-}CH_2{-}O{-}\underset{\textstyle O^-}{\overset{\displaystyle O}{\overset{\|}{P}}}{-}O{-}CH_2}}
\end{array}}
\; + \; 4H_2O
\; \xrightarrow[\text{lipase}]{\text{Pancreatic}}
$$

$$
\underset{\text{Fatty acid}}{2R{-}\overset{\displaystyle O}{\overset{\|}{C}}{-}OH}
\; + \;
\underset{\text{Glycerol}}{
\begin{array}{c}
HO{-}CH_2 \\
HO{-}CH \\
HO{-}CH_2
\end{array}}
\; + \;
\underset{\text{Choline}}{\overset{\textstyle CH_3}{\underset{\textstyle CH_3}{CH_3{-}\overset{+}{N}{-}CH_2{-}CH_2{-}OH}}}
\; + \;
\underset{\substack{\text{Phosphate} \\ \text{ion}}}{P_i}
$$

The breakdown products of dietary lipids are absorbed into intestinal cells as is cholesterol. The intestinal cells contain another lipase—one that catalyzes the synthesis of lipids. Resynthesized lipids form as a result of the action of this lipase (Fig. 12.3). Because of their relative nonpolarity, cholesterol and other lipids are too insoluble in body fluids to be transported by the blood to other body tissues. However, insoluble cholesterol and other lipids are solubilized by interacting with **lipoproteins**—*proteins designed to bind to lipids*—in the intestinal lining. The lipoproteins with their bound lipids form bodies called *chylomicrons*. The structures of chylomicrons are similar to those of micelles. *In* **chylomicrons** *the nonpolar portions of lipids are surrounded by a film of the polar lipoprotein, a combination that is soluble in body fluids.*

Figure 12.3
The path taken by dietary lipids in going from the small intestine to the bloodstream. Triglycerides and complex lipids are hydrolyzed to their simpler components. These components pass into cells of the intestinal mucosa, where they are resynthesized into lipids. The newly synthesized lipids complex with lipoproteins and cholesterol to make soluble bodies called *chylomicrons.* The chylomicrons enter the bloodstream through the lymphatics, a network of vessels that carries substances from body tissues to the blood.

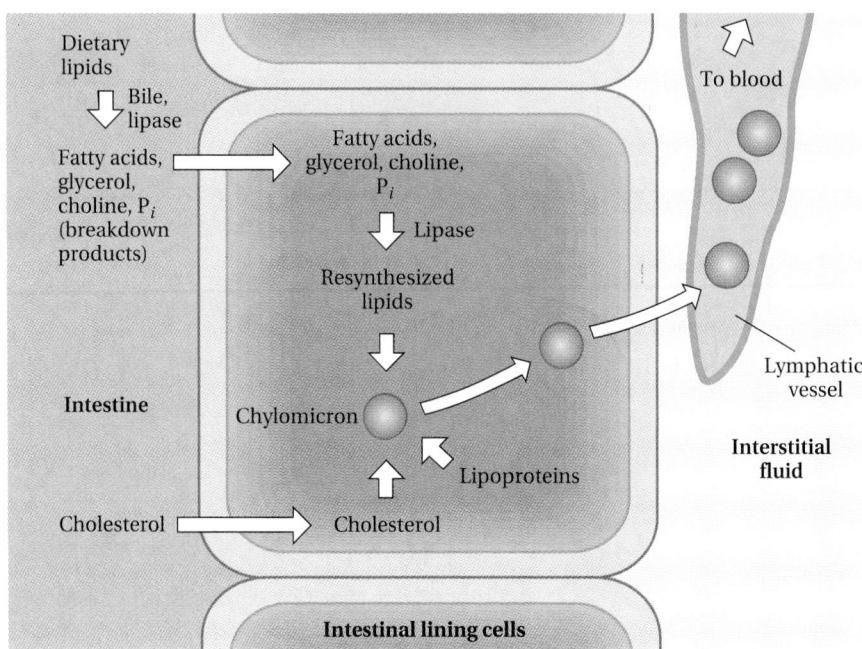

The incorporation of triglycerides into chylomicrons prepares the lipids for transport through the bloodstream, much as a ship transports passengers who cannot swim. Chylomicrons cannot enter the blood through capillaries, probably because they are too large, but they can enter lymphatic vessels. After a meal heavy in fats, the lymph may acquire a milky white appearance because of the high concentration of triglycerides. The fat laden lymph enters the bloodstream through the thoracic duct (see A Closer Look: Blood Lipoproteins and Heart Disease).

Digestion of nucleic acids

A group of enzymes known as **nucleotidases** *catalyze the hydrolysis of nucleic acids to their component nucleotides.* Nucleotidases are produced in the pancreas and transported to the small intestine, where they perform their catalytic work. The nucleotides produced by the action of nucleotidases are then absorbed into intestinal cells. The nucleotidase that catalyzes the hydrolysis of DNA is called *deoxyribonuclease (DNase); ribonuclease (RNase)* catalyzes the hydrolysis of RNA. An interesting therapeutic use of DNase is in cystic fibrosis, as described in A Closer Look: Therapies for Cystic Fibrosis, on page 364.

Food is almost completely absorbed by the time it reaches the end of the small intestine. A watery mixture of undigested materials, the feces, passes into the large intestine (the colon). There, much of the water is reabsorbed into the body. The reabsorption gives the feces their characteristic semisolid texture. The color of feces comes from bile pigments that have escaped the small intestine.

A Closer Look

Blood Lipoproteins and Heart Disease

Chylomicrons are the largest lipid-carrying particles of blood, but they are usually found in the blood only after a fatty meal. Blood also contains smaller particles consisting of lipoproteins and their associated lipids. The role of these particles is to transport lipids between tissues independent of the chylomicrons.

One general method of classifying lipoproteins is by density. Since lipids are less dense than proteins, particles containing lipoproteins associated with large amounts of lipid are less dense than those associated with smaller amounts. The lipoproteins of chylomicrons haul more lipids than any other lipoproteins. The protein content of chylomicrons is often as little as 1% of the total mass of the particle, and chylomicrons have densities of less than 0.96 g/cm^3. On the other hand, the mass of proteins in the so-called high-density lipoproteins (HDLs) may be as much as 60% of the total mass of the particles. The HDL particles have densities from 1.06 to 1.21 g/cm^3. Two other types of lipoproteins, the very-low-density lipoproteins (VLDLs) and low-density lipoproteins (LDLs), have densities between these two extremes.

Biomedical scientists studying the link between blood cholesterol levels and the risk of heart disease have become interested in the cholesterol content of HDLs and LDLs. So far they have learned that the risk of heart disease increases as the amount of LDL cholesterol (see figure) rises in proportion to the amount of HDL cholesterol. As the following table shows, the LDL cholesterol/HDL cholesterol ratio is presently the best method for assessing the risk of heart disease.

Risk	LDL/HDL ratio
Men	
$\frac{1}{2}$ average	1.00
average	3.55
2 times average	6.25
3 times average	7.99
Women	
$\frac{1}{2}$ average	1.47
average	3.22
2 times average	5.03
3 times average	6.14

Several inherited diseases are connected with defects in lipoprotein synthesis or function. Depending on the defect, these diseases may result in either hypolipoproteinemia or hyperlipoproteinemia (low or high concentrations of lipoprotein in the blood, respectively). Many individuals with such diseases as diabetes mellitus and atherosclerosis (hardening of the arteries) also show one or the other condition. The treatment depends on the cause of the disease and nature of the condition, but it often consists of adjusting the amount of fat in the diet.

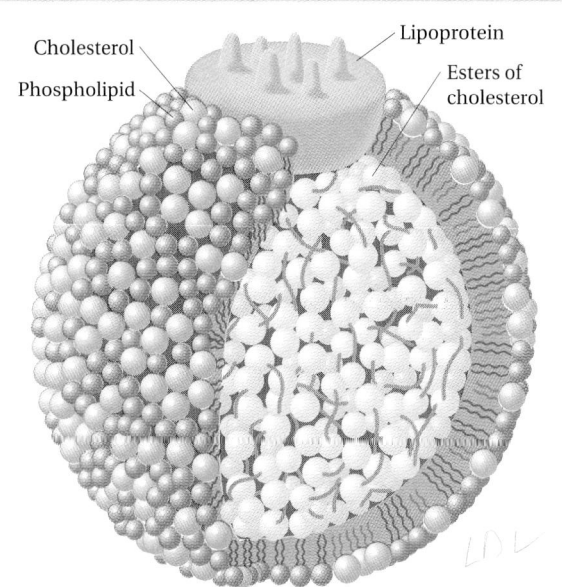

Cholesterol — Lipoprotein — Phospholipid — Esters of cholesterol

Schematic diagram of an LDL particle. As the level of LDL increases in the blood, so does the deposition of atherosclerotic plaque. For this reason, LDL is often called "bad cholesterol."

Therapies for Cystic Fibrosis

A treatment for cystic fibrosis uses deoxyribonuclease (DNase), the enzyme normally used in the body to hydrolyze DNA to its constituent nucleotides. Cystic fibrosis is a genetic disease that affects 20,000 Americans annually. The disease, which historically has killed 50% of its victims by age 20, is characterized by the thick, infected mucous secretions that plug air passages in the lungs. These secretions cause inflammation and damage to lung tissue and ultimately death. Traditional treatment for cystic fibrosis has involved sessions of thumping the patient on the back and chest to clear the mucus from the passages and administering antibiotics to control infections. The high viscosity of the mucus of cystic fibrosis results from large amounts of DNA in the secretions. The DNA comes from the nuclei of white blood cells that have gathered in the lungs to fight infections and have died there. This unwanted DNA is the target of the DNase used in the treatment of cystic fibrosis. A solution of DNase is inhaled into the lungs through an atomizer, which produces a fine spray (see figure). In the lungs, the DNase catalyzes the hydrolysis of the DNA to thin the mucous secretions. The thinned secretions are much more easily expelled from the lungs, improving lung function and reducing infections. DNase treatment is not a cure for cystic fibrosis, but it can improve life for many patients with the disease. The DNase used for cystic fibrosis treatment is a human enzyme obtained by recombinant DNA technology and marketed as Pulmozyme. Since it is identical to the enzyme in our bodies, the recombinant DNase does not cause an

An inhalant spray of deoxyribonuclease is effective in treating the symptoms of cystic fibrosis.

allergic response. Its major side effect appears to be hoarseness in some patients.

In addition to the DNase treatment, there has been progress toward the treatment or cure for cystic fibrosis by gene therapy. The secretions of cystic fibrosis are caused by mutations in the gene for cystic fibrosis transmembrane conductance regulator (CFTR). These mutations lead to a marked decrease in chloride transport across membranes of airway cells. Researchers have used a virus carrier to deliver the normal CFTR gene to the nasal cells of patients with cystic fibrosis. They found that normal chloride transport was achieved in treated cells for a few days, with no adverse effects. Many questions remain to be answered before this promising gene therapy can be used on a large scale.

Undigested materials are subjected to the action of bacteria in the large intestine. One action of these bacteria is to degrade the amino acid tryptophan into skatole and indole.

Skatole

Indole

These two heterocyclic amines are largely responsible for the odor of feces. Feces are stored in the rectum, where more water reabsorption occurs, until they are excreted.

PRACTICE EXERCISE 12.2

Describe the digestion and absorption of carbohydrates in the body.

PRACTICE EXERCISE 12.3

Outline the digestion and absorption of dietary lipids.

12.5 Balanced diet

AIM: To explain why basal metabolism varies among individuals.

Recall that the amount of energy that can be obtained from a molecule is related to the energy contained in the molecule's chemical bonds. When an organic compound oxidizes, the amount of energy released is independent of the method of oxidation. Both combustion and oxidation in aerobic cells ultimately oxidize carbon compounds to carbon dioxide and water. The same amount of energy is released by burning a food substance in air as is released when the food is oxidized in steps by a living organism. It is possible to burn various foods and measure the heat evolved. *The quantity of heat evolved per gram of food burned is that food's* **caloric value.** Each gram of carbohydrate or protein burned produces about 4 kcal of heat energy. Fats produce more energy; just over 9 kcal is produced for each gram burned. Nutritionists usually use the dietary Calorie (note the capital C) in place of the kilocalorie; 1 Calorie equals 1 kcal or 1000 cal.

Focus

A balanced diet provides for basal metabolism, activity, and growth.

Heats of combustion of foods are determined using a bomb calorimeter. A food sample is burned inside a well-insulated container surrounded by a known quantity of water. The energy released by the combustion of the food is calculated from the increase in temperature of the water.

EXAMPLE 12.1

Calculating the caloric value of food

A college student eats a snack that contains 12 g of carbohydrate, 7 g of protein, and 3 g of fat. What is the caloric value of the snack in dietary Calories?

SOLUTION

Each of the 19 total grams of carbohydrates and proteins supplies 4 kcal of energy. Each of the 3 grams of fat supplies 9 kcal. The total caloric value of this snack is:

$$(19 \ g \times \frac{4 \text{ kcal}}{1 \ g}) + (3 \ g \times \frac{9 \text{ kcal}}{1 \ g}) = 103 \text{ kcal}$$

When we are at rest, our body cells are using nutrients to produce energy at a level near the minimum to sustain life. *This minimum level of energy production is called the* **basal metabolism.** (Chapters 14 through 17 present a more detailed picture of metabolism.)

The basal metabolism of healthy people of the same age, size, and physical condition is remarkably constant. It varies, however, according to age, size, and sex: Children have a higher basal metabolism than adults; obese people have a higher basal metabolism than thin people; males have a higher basal metabolism than females. A typical healthy 35-year-old male of average size and weight has a basal metabolism requiring about 1700 Calories per day. Such a man needs a daily intake of food with a caloric value of 1700 Calories even if he never wiggles a finger. The caloric requirements of women are slightly less than those of men of comparable size, age, and weight.

Most of us do eat and wiggle our fingers, or even lift a leg or two, in a typical day. These activities require the expenditure of energy beyond that required for basal metabolism. Adults in sedentary occupations require a food intake equivalent to 2000 to 2500 Calories per day. Very active people may need as many as 4000 to 6000 Calories per day to meet their energy needs. Children need more food than comparably active adults to provide for new tissue growth.

When we eat less than our daily minimum caloric requirement, our bodies call on storage reserves of fat to supply energy, and we lose weight. When we eat too much, energy reserves are stored as fat, and we gain weight. Only when our caloric intake equals our caloric output do we stay the same weight. However, the diet of every human being must be well balanced with regard to intake of carbohydrates, fats, proteins, minerals, and vitamins, since the body needs some of each to efficiently produce substances needed to stay healthy. In other words, getting enough calories is not necessarily good nutrition. Guidelines for a balanced diet have been issued, as described in A Closer Look: The Nutrition Pyramid.

Table 12.1 Essential and Nonessential Amino Acids

Essential	Nonessential
isoleucine	alanine
leucine	arginine*
threonine	asparagine
lysine	aspartic acid
methionine	cysteine
phenylalanine	glutamic acid
tryptophan	histidine*
valine	glutamine
	glycine
	proline
	serine
	tyrosine

*Needed in the diet of growing children.

Focus

Dietary protein must contain all the essential amino acids.

Recombinant DNA research in plants holds promise for producing corn and wheat that contain all essential amino acids. Success would greatly help to solve the world's food problems.

12.6 Proteins and amino acids in the diet

AIMS: *To distinguish between essential and nonessential amino acids. To distinguish between complete and incomplete proteins. To distinguish between nutritional marasmus and kwashiorkor.*

The amino acids synthesized by the body are called the **nonessential amino acids.** Amino acids not synthesized by the body must be obtained in the diet and are called the **essential** *(or indispensable)* **amino acids.** Table 12.1 lists 8 essential and 12 nonessential amino acids for humans. Two of the amino acids listed as nonessential, arginine and histidine, are nonessential for adults but are not synthesized in sufficient amounts to satisfy the needs of growing children.

Not all foods contain all the amino acids. This means that we must pay attention to the sources of protein eaten. We must make certain that the proteins we consume contain enough of the essential amino acids to satisfy our bodies' needs. **Complete proteins** *are those which contain all the essential amino acids.* Eggs, dairy products, kidneys, and liver are sources of complete proteins. Zein, the principal corn protein, is incomplete because it contains no lysine or tryptophan. Wheat protein is lacking in lysine also.

A Closer Look

The Nutrition Pyramid

In order to promote public health, the U.S. Department of Agriculture issues guidelines on the content and amount of food in a well-balanced diet. The latest guidelines, issued in 1992, replace an earlier set nearly 50 years old. The new guidelines are summarized in a pyramid, with the number of daily servings of the various food groups decreasing from the bottom to the top of the pyramid (see figure). At the base of the pyramid are grains—bread, cereal, rice, and pasta. At the top are sweets and fats, which should be eaten only sparingly. Many people are confused by what constitutes a daily serving, since this is not given on the chart. The following list may be helpful.

Breads, cereals, rice, and pasta (6 to 11 servings): A serving could be one slice of bread, half a bun or bagel, 1 ounce of dry cereal, or a half-cup of cooked cereal, rice, or pasta.

Vegetables (3 to 5 servings): A serving is 1 cup of raw, leafy greens or a half-cup of any other vegetable.

Fruit (2 to 4 servings): A serving could be one medium apple, banana, or orange; a half-cup of fresh,

cooked, or canned fruit; or three-quarters of a cup of fruit juice.

Dairy products (2 to 3 servings): A serving could be 1 cup of milk, 8 ounces of yogurt, 1.5 ounces of natural cheese, or 2 ounces of processed cheese.

Meat, poultry, dry beans, eggs, and nuts (2 to 3 servings): A total of 5 to 7 ounces of lean cooked meat, poultry, or fish a day. Count a half-cup of cooked beans, one egg, or 2 tablespoons of peanut butter as 1 ounce of meat.

The food guide pyramid.

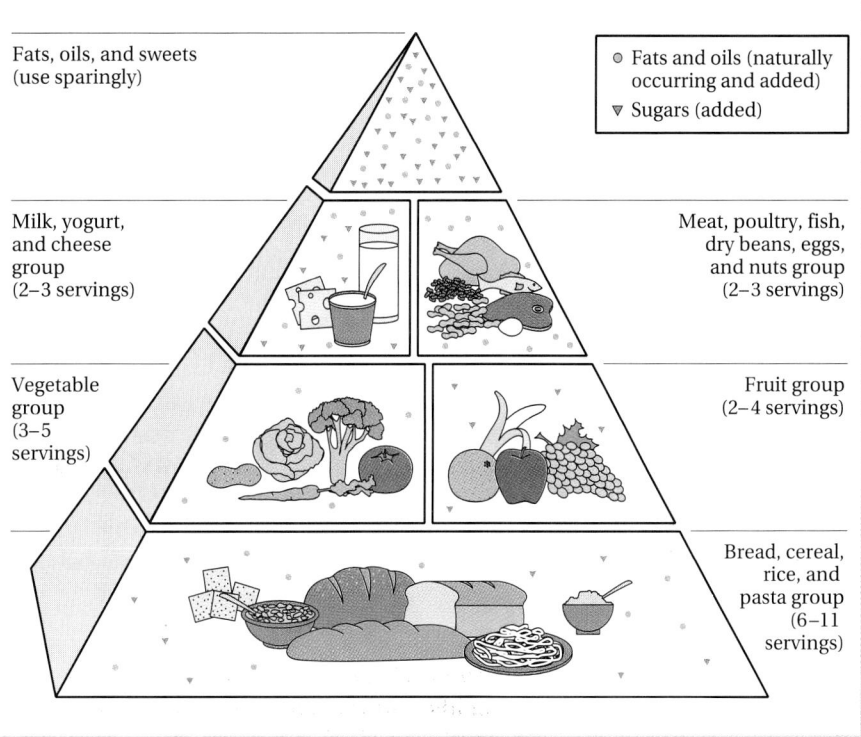

Marasmus
marasmos (Greek): a wasting away.

Gelatin, another incomplete protein, contains no tryptophan. Incomplete corn, wheat, and gelatin protein are easily converted to complete protein by adding the missing amino acid. Soybeans, peanuts, and potatoes are sources of complete protein, as are poultry, fish, and red meats. A combination of several foods containing protein is more likely to contain sufficient amounts of all essential amino acids than a limited selection.

A lack of total dietary calories and a lack of sufficient protein are the two most important nutritional problems that an individual may

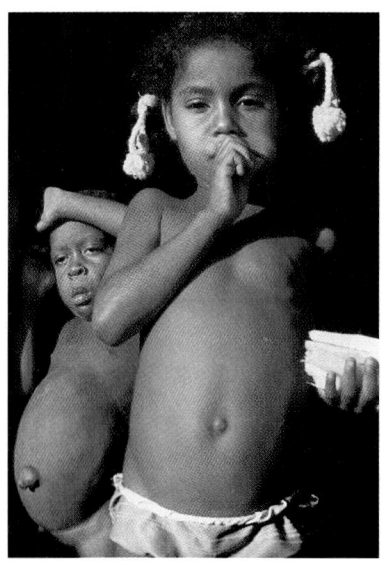

Figure 12.4
The disease kwashiorkor is caused by a protein deficiency. The symptoms include bloating of the belly and stunted growth.

Focus

Vitamins A, D, E, and K are the fat-soluble vitamins.

The term *vitamine* was coined in 1912 by a Polish biochemist, Casimir Funk, who believed that all vitamins were amines. When it was discovered that not all vitamins were amines, the "e" was dropped.

Retinoic acid (13-*cis*-retinoic acid), a derivative of vitamin A, appears promising as a drug for treating severe acne and other skin conditions.

---

encounter. *Nutritional* **marasmus**—*starvation*—*is the name given to malnutrition caused by an insufficient total food intake. The most common protein-deficiency disease is* **kwashiorkor.** The name of this disease is an African word that means "weaned child." Kwashiorkor is most prevalent in certain parts of Africa and Asia. As the name implies, kwashiorkor often occurs in infants who have just left their mother's breast. These children are often fed a diet consisting solely of thin, starchy gruel that contains virtually no protein. Their plump faces often look healthy enough, but their bodies have underdeveloped muscles owing to a lack of protein. Moreover, such children have swollen bellies (Fig. 12.4). The blood protein serum albumin is important in maintaining a balance between blood fluid and tissue fluid volumes. The swelling of kwashiorkor is caused by the escape of water from the blood to the tissues owing to a lack of sufficient serum albumin to maintain the osmotic pressure of the blood.

12.7 Fat-soluble vitamins

AIMS: To name the fat-soluble vitamins, indicating their dietary sources and biologic functions. To identify fat-soluble vitamins responsible for particular vitamin deficiency diseases.

Even an intake of food of adequate caloric value and protein quality does not guarantee good nutrition. We also need vitamins and minerals. The definition of a *vitamin* has two aspects. First, vitamins are organic compounds absolutely necessary, usually in small amounts, for an animal's growth and health. And second, they are substances the animal cannot synthesize and must therefore be supplied in the diet.

Fat-soluble vitamins essential to humans are our chief interest. These are vitamins A, D, E, and K. Their chemical structures have little in common, but they are all insoluble in water, dissolve in fat, and are stored in body fat, especially liver fat. A person who has a problem absorbing dietary fats will probably also have a problem absorbing enough fat-soluble vitamins. On the other hand, taking excessive amounts of fat-soluble vitamins can cause **hypervitaminosis**—*a dangerous overaccumulation of vitamins in the tissues.*

Vitamin A   Vitamin A (retinol) occurs only in animal tissues. It is an unsaturated primary alcohol and was first isolated from fish oils. The double bonds of the unsaturated carbon chain are all of the *trans* configuration.

Vitamin A

Polar bear liver is toxic to humans because of the high concentration of vitamin A it contains.

Cod liver oil is one of the richest sources of vitamin A, but dairy products usually provide enough in a normal diet. Certain vegetables are another excellent source of vitamin A, but only indirectly. Carrots and spinach contain a yellow compound, called β-*carotene*, which the body can transform to vitamin A.

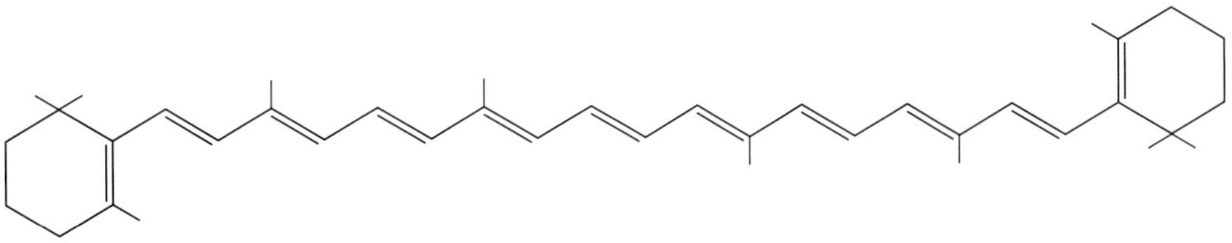

β-Carotene

One of the earliest signs of vitamin A deficiency in humans is a loss of the ability to see in dim light called *night blindness*, or *nyctalopia*. Vitamin A keeps the mucous membranes of the eye and the respiratory, digestive, and urinary tracts in healthy condition. When vitamin A is deficient, *mucous membranes become hard and dry—a process called* **keratinization.** When tear ducts become keratinized, tear secretions stop, and the outer surface of the eye becomes dry and scaly. Serious eye infections are then likely because tears normally remove bacteria from the eye. If left untreated, keratinization can cause blindness.

Vitamin D The substances called *vitamin D* are actually a number of different compounds. Of these compounds, vitamin D_2 is most critical to human health. This vitamin promotes the uptake of calcium and phosphorus, both of which help form strong teeth and bones in growing children. Vitamin D is also involved in the maintenance of bone mass in adults. Ergosterol, a steroid abundant in yeast and many fungi, is the starting material, or precursor, for the formation of vitamin D_2. Irradiation by light transforms ergosterol into the vitamin by cleavage of one carbon-carbon bond in the ergosterol steroid nucleus. You may have noticed that the labels on some cartons of vitamin D–fortified milk say "contains irradiated ergosterol."

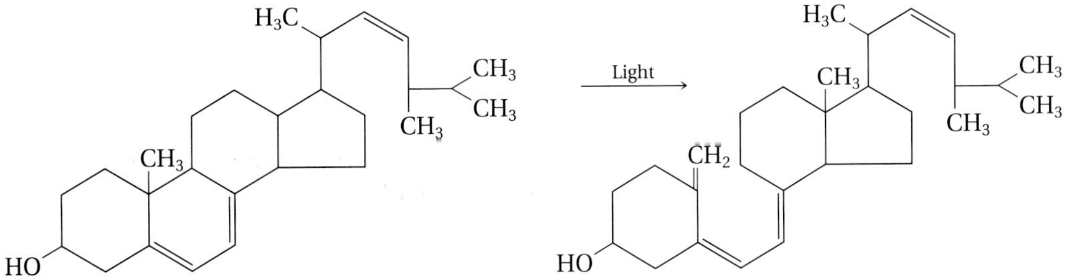

Ergosterol   Vitamin D_2

Steroids in the skin can be converted to vitamin D by sunlight, and children who play in the sun usually get enough vitamin D this way. This is why vitamin D is called the "sunshine vitamin."

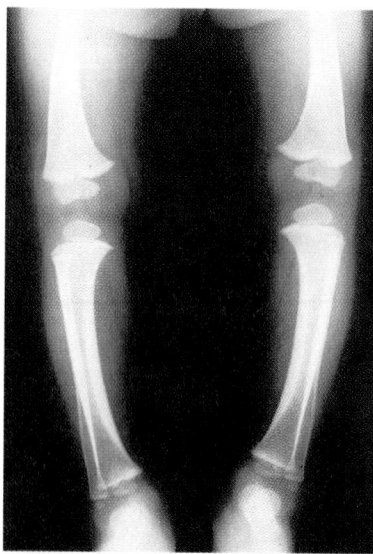

(a)

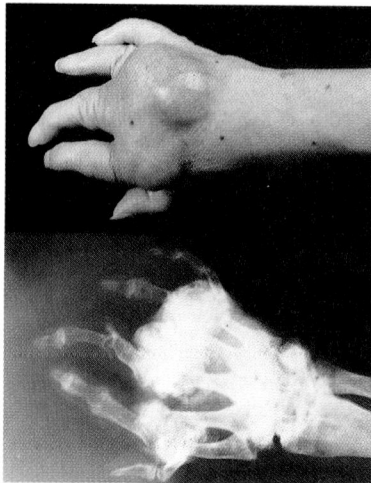

(b)

Figure 12.5
The distortion of the leg bones in rickets (a) is due to a deficiency of vitamin D; the deposition of additional bonelike material in joints (b) is caused by too much vitamin D.

Rickets, *a disease of children, is caused by a vitamin D deficiency.* Some symptoms of severe rickets are bowed legs (Fig. 12.5), an underdeveloped and abnormal formation of the ribs called "pigeon breast," and poor tooth development. These symptoms can be completely overcome by adding vitamin D to the diet. The best natural sources of vitamin D are cod, tuna, and other fish liver oils, as well as sardines and herring.

Vitamin E   Several naturally occurring compounds have vitamin E activity. Of these, α-tocopherol is the most potent. α-Tocopherol is abundant in fish oils and vegetable oils; wheat germ oil is the richest source.

α-Tocopherol

Vitamin E is used as a food additive to prevent oxidation—an antioxidant—and thereby prevent food spoilage. It extends the shelf life of foods containing unsaturated fatty acids that otherwise quickly become rancid. Vitamin E also functions as an antioxidant in the body by inhibiting the oxidation of unsaturated fats in much the same way it prevents food spoilage. An important use for vitamin E in the body appears to be as an antioxidant for vitamin A. In the absence of vitamin E, symptoms of vitamin A deficiency appear. Vitamin E is also necessary for the proper development and operation of the membranes of muscle cells and red blood cells.

Vitamin K   Vitamin K is known as the coagulation or antihemorrhagic vitamin because it is essential for the production of prothrombin, the precursor of the blood-clotting enzyme thrombin.

Vitamin K

Spinach and leafy green vegetables are good sources of vitamin K. Deficiencies are not usually due to a lack of the vitamin in the diet, because bacteria in the large intestine synthesize this vitamin, and people absorb it from the bacteria. Certain drugs such as warfarin and dicoumarin are vitamin K *antagonists*—drugs that interfere with the action of the vitamin. They are used in anticoagulation therapy to prevent potentially harmful blood clots. Both these drugs have structures similar to vitamin K and probably interfere with the synthesis of prothrombin. Warfarin is also used as a rat poison; the poisoned rodents die of severe internal bleeding.

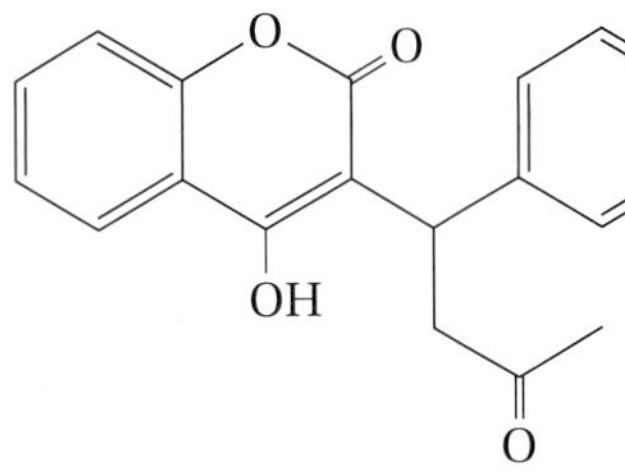

Warfarin 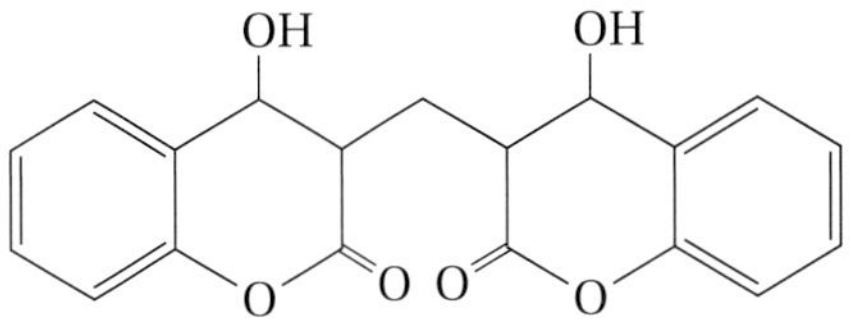   Dicoumarin

12.8 Water-soluble vitamins

AIMS: To name the water-soluble vitamins, indicating their dietary sources and biologic functions. To identify water-soluble vitamins responsible for particular vitamin deficiency diseases.

Focus

The B-complex vitamins are water-soluble.

The B-complex vitamins have only one property in common: They are soluble in water. As we will see, their biological functions are as diverse as their number. From Chapter 10 you may recall that many of the B-complex vitamins are important as enzyme cofactors. Vitamin C is also water-soluble, and we will consider it along with the B vitamins.

Because of their water solubility, the B vitamins and vitamin C are rapidly eliminated by our bodies. Hypervitaminosis is not usually a problem with these substances, as it may be with the fat-soluble vitamins. Because they are not stored in body tissue like the fat-soluble vitamins, however, the intake of water-soluble vitamins in the diet must be more frequent to maintain adequate levels.

Ascorbic acid (vitamin C)   Vitamin C is essential for the formation of the collagen in tendons, ligaments, teeth, bones, and skin.

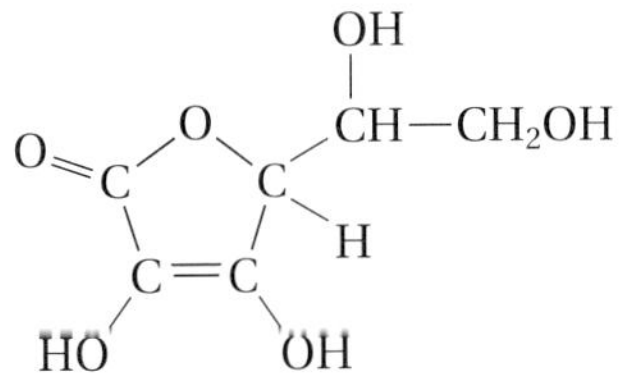

Ascorbic acid
(vitamin C)

Although vitamin C is abundant in citrus fruits and fresh vegetables, it is slowly oxidized by air and destroyed in appreciable amounts during cooking.

A deficiency of vitamin C results in **scurvy**—*symptoms include skin lesions, loose teeth, and rotting gums.* Scurvy is easily prevented by including sources of vitamin C, especially citrus fruits, in the diet. Once the bane of seafarers who had no fresh fruits and vegetables on long voyages, scurvy is still reported in the United States. Today it exists mostly as the "bachelor's scurvy" sometimes found among older widowers and bachelors, a group of individuals whose nutritional needs are often neglected. Vitamin C also promotes the absorption of dietary iron in the body.

In 1970, Linus Pauling claimed that large doses of vitamin C (1 to 1.5 g per day) could prevent the common cold. A fierce controversy ensued that is still going on.

Thiamine (vitamin B$_1$) Meat, yeast, and unpolished grain are the main sources of thiamine in the diet. A modified form of thiamine is a cofactor in the breakdown of fatty acids.

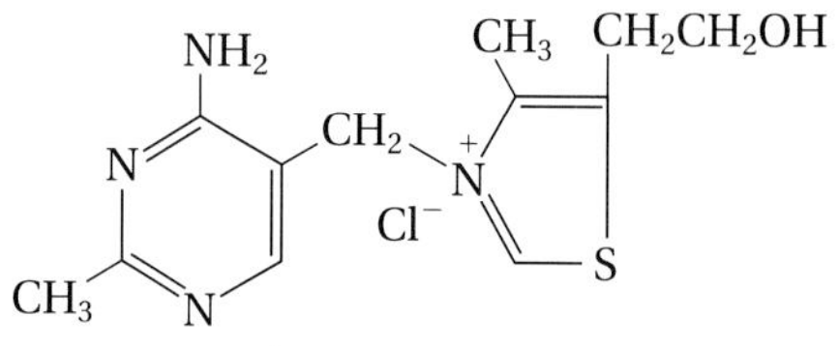

Thiamine chloride
(vitamin B$_1$)

A thiamine deficiency causes **beriberi**—*a disease that affects the nervous system and the heart.* Natives of Java coined the name *beriberi,* which means "sheep" in Javanese, because they thought that its victims walked like sheep. Typical symptoms of beriberi include pain in the arms and legs, weak muscles, and distorted skin sensations. Since most foods are low in thiamine, good nutrition is necessary to prevent beriberi. Even someone who eats sensibly may have a slight thiamine deficiency.

Addition of thiamine to some processed foods, such as white bread, has helped to lower the incidence of beriberi in the general population of the United States. It is sometimes seen in people who abuse alcohol and other malnourished people. Beriberi is still a problem in parts of Asia, where the chief food staple is rice. Rice is low in thiamine, and polished rice has none at all, since all the thiamine is in the outer layer.

Nicotinic acid (niacin) Milk, yeast, and meat are important dietary sources of nicotinic acid. Many enzymes that catalyze oxidation-reduction reactions use a modified form of this vitamin (nicotinamide) as a cofactor.

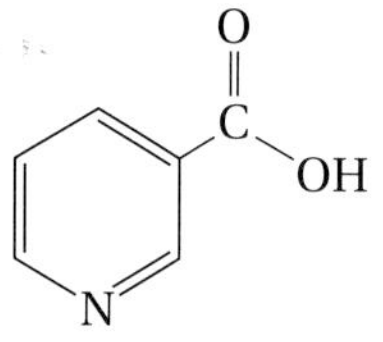

Nicotinic acid
(niacin)

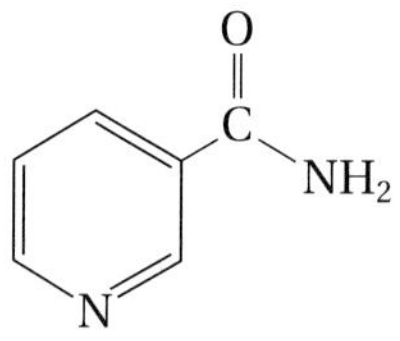

Nicotinamide
(niacinamide)

A niacin deficiency results in **pellagra,** *a disease whose symptoms include damage to the nerves as well as to the linings of the skin and intestines.* The vitamin is sometimes administered as nicotinamide, the amide of nicotinic acid. Despite the similarity in name and structure, nicotine is not converted to nicotinic acid in the body. Indeed, nicotine may interfere with the proper utilization of nicotinic acid.

Riboflavin (vitamin B₂)   Cheese, eggs, liver, and milk are good sources of riboflavin. A modified form of riboflavin is an enzyme cofactor in certain biological oxidation-reduction reactions.

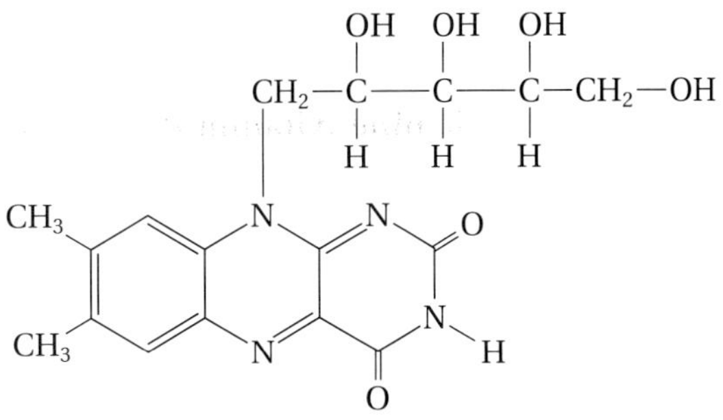

Riboflavin
(vitamin B₂)

The symptoms caused by a deficiency of riboflavin are rather vague, but they include general weakness, eye damage, and a reddening of the tongue.

Pyridoxine (vitamin B₆)   Pyridoxine is the recognized form of vitamin B₆. In our bodies, however, the related aldehyde (pyridoxal) and the related amine (pyridoxamine) are the active forms. Modified forms of pyridoxal and pyridoxamine are cofactors for several enzymes needed in the breakdown of amino acids. Wheat germ, yeast, peanuts, corn, and meat are good sources of this vitamin.

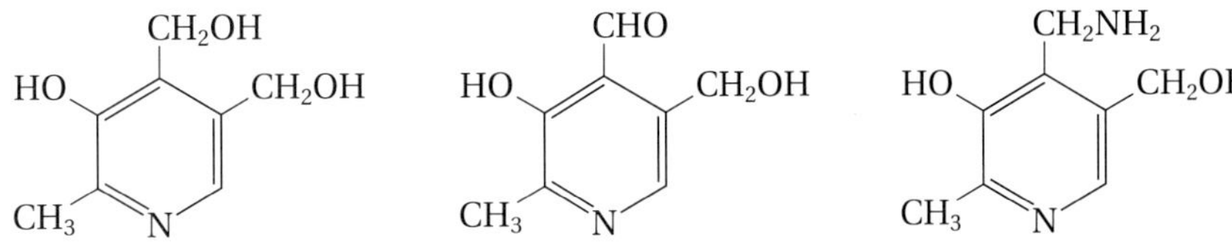

Pyridoxine   Pyridoxal   Pyridoxamine

Deficiencies of pyridoxine result in nervous system damage.

Folic acid   Folic acid is a yellow compound that is only slightly soluble in water. One good dietary source is leafy green vegetables. Bacteria also manufacture the vitamin, and it is possible for humans to absorb it from intestinal bacteria. Folic acid is the cofactor for a group of enzymes that catalyze the transfer of methyl groups (CH_3—) between biological molecules. In this role, it is particularly important for the synthesis of the purine and pyrimidine bases of nucleic acids.

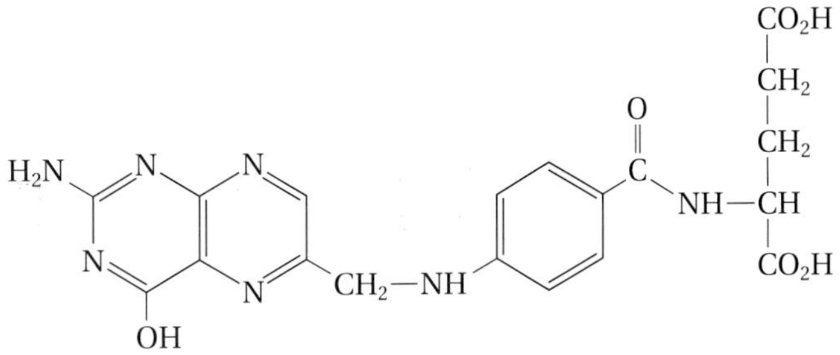

Folic acid

Anemia, a deficiency of red blood cells, results from a severe folic acid deficiency.

Pantothenic acid Pantothenic acid is a component of a cofactor that plays an important role in carbohydrate and lipid biochemistry. Pantothenic acid is abundant in such foods as liver, egg yolk, cabbage, fruits, and sweet potatoes.

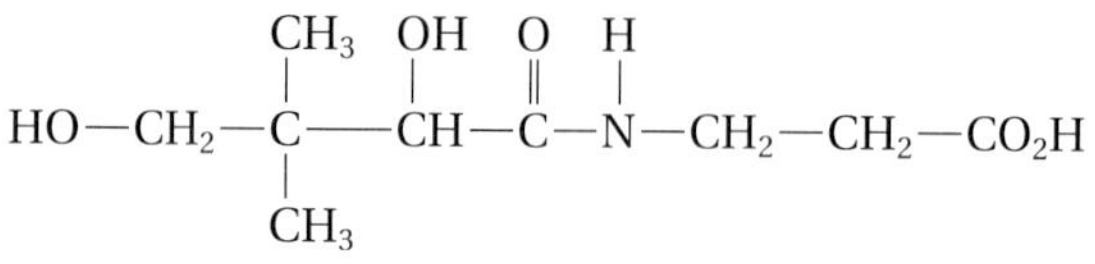

$$HO-CH_2-\underset{\underset{CH_3}{|}}{\overset{\overset{CH_3}{|}}{C}}-\underset{\underset{}{|}}{\overset{\overset{OH}{|}}{CH}}-\overset{\overset{O}{\|}}{C}-\underset{\underset{}{|}}{\overset{\overset{H}{|}}{N}}-CH_2-CH_2-CO_2H$$

Pantothenic acid

A deficiency results in fatigue, muscle spasms, and intestinal disturbances.

Biotin Biotin is a human vitamin, yet deficiencies hardly ever occur—apparently, enough biotin from intestinal bacteria is absorbed through the intestinal walls. Liver, kidney, milk, and molasses are excellent sources. Biotin serves as a cofactor for enzymes that use carbon dioxide to synthesize certain biological molecules.

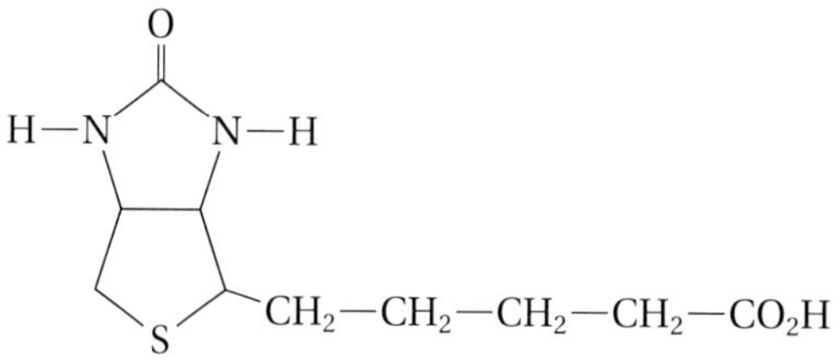

Biotin

Cobalamin (vitamin B_{12}) This vitamin is unusual because it is the only cobalt-containing organic compound found in nature (Fig. 12.6). The cobalt is present as a cobaltous or Co(I) ion bound to a corrin ring—a ring that is similar to the heme ring of hemoglobin.

Vitamin B_{12} is essential for the synthesis of red blood cells. Liver is the chief source of vitamin B_{12}, although the vitamin is also found in eggs, milk, meat, and seafood. We need only about 2×10^{-5} g of vitamin B_{12} daily. Normally, there is no difficulty in obtaining this minute amount in the diet. **Pernicious anemia** *is most often caused by an inability to absorb vitamin B_{12}.* Victims of pernicious anemia suffer from general fatigue and weakness. Before 1926, older people often died from pernicious anemia. Their red blood cells are immature, very large, and low in number. In 1926 it was discovered that people with pernicious anemia could benefit from eating half a pound of lightly cooked liver every day. Some 20 years later it was shown that the active ingredient in the liver is vitamin B_{12}. Daily injections of 1×10^{-6} g (1 μg) of pure vitamin B_{12} control pernicious anemia. Since vitamin B_{12} is formed principally by bacteria, there is little or

Pernicious anemia
pernecare (Latin): to kill
anaimia (Greek): without blood

We now know that pernicious anemia results from immunologic damage to the stomach lining, preventing the secretion of a substance called intrinsic factor, which aids the absorption of vitamin B_{12}.

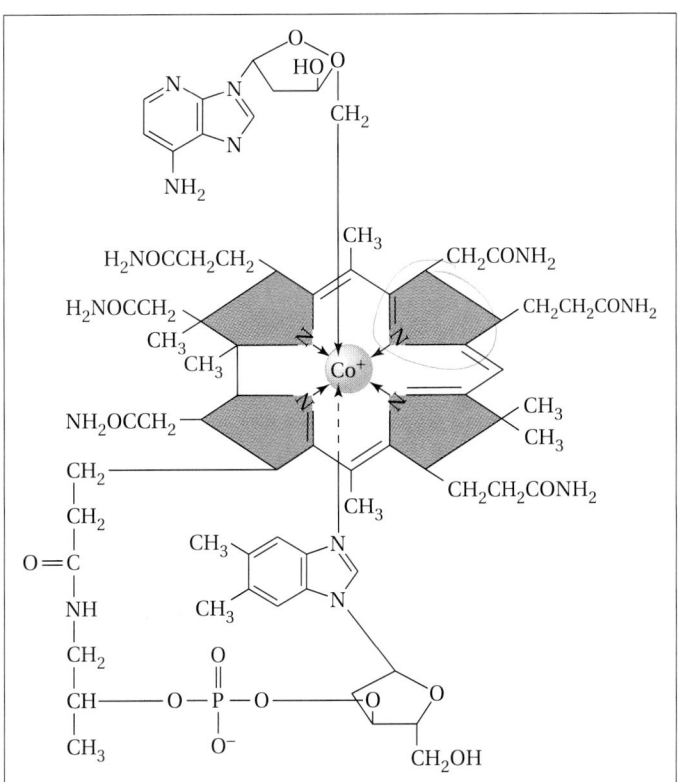

Figure 12.6
The structure of vitamin B_{12}. The corrin ring system (brown) contains a cobalt(I) ion (Co^+) (blue/gray).

none in most plants, and strict vegetarians may develop symptoms of pernicious anemia.

EXAMPLE 12.2

Determining the number of molecules of vitamin B_{12} needed daily

The 2×10^{-5} g of vitamin B_{12} we require daily is a small mass, but how many molecules of the vitamin does this mass represent? (Use a molar mass for vitamin B_{12} of 1400 g/mol.)

SOLUTION

Remember that 1 mole of any substance is 6.02×10^{23} molecules. Use the molar mass to find the number of moles in the given mass. Then calculate the number of molecules.

$$2 \times 10^{-5}\,g \times \frac{1\ mol}{1400\ g} \times \frac{6.02 \times 10^{23}\ \text{molecules}}{1\ mol} = 9 \times 10^{15}\ \text{molecules}$$

PRACTICE EXERCISE 12.4

The recommended daily allowance (RDA) of niacin (molar mass = 122 g/mol) is 18 mg. How many molecules is this?

? molecules : 18 mg . $\dfrac{1 g}{1000 mg}$. $\dfrac{1\,mol}{122 g}$ $\dfrac{6.02 \times 10^{23}\ \text{molecules}}{1\,mol}$ =

0.0008882×10^{23}
8.9×10^{19}

12.9 Minerals

AIM: To list the macronutrients and micronutrients by name and chemical formula, indicating dietary sources, functions in the body, and consequences of deficient or excessive intake.

Hydrogen, oxygen, carbon, and nitrogen are the most abundant elements in living organisms. Together they account for 99.4% of all the atoms in the human body. Most of the hydrogen and oxygen is combined as water; the remainder, together with carbon, nitrogen, sulfur, and phosphorus, makes up the molecular building blocks of life: sugars, fatty acids, amino acids, and nucleotides.

Experiments have demonstrated which **mineral elements**—*elements other than H, O, C, and N*—are essential for life. One procedure involves incineration. Technicians burn samples of plants and dead animals and analyze the elemental composition of the ashes. Scientists then vary the diets fed to test animals, excluding these elements one at a time. A detrimental effect on health, growth rate, or life span establishes the excluded element as essential. The essential elements found in relatively large amounts include metals and nonmetals. *The metals calcium, potassium, sodium, and magnesium and the nonmetals phosphorus, sulfur, and chlorine are called* **macronutrients.** *Trace amounts of many other elements are also found; those essential for life are called* **micronutrients.**

Table 12.2 lists the approximate abundance of the elements in the human body. Although these essential ingredients are described as elements, no free element, either metal or nonmetal, is present in the body. Rather, they exist as simple ions, polyatomic ions, or in covalent molecules. Besides the elements listed in Table 12.2, lead, mercury, silver, cadmium, barium, and antimony are often found in the body in trace amounts. All these metals are highly toxic. None of them seems to have any beneficial role in the human body.

Macronutrients

Calcium and phosphorus   Dairy products such as milk and cheese are good sources of calcium and phosphorus. Nuts, beans, egg yolk, and shellfish also contain calcium, but the calcium in milk is more readily absorbed from the digestive tract than the calcium in vegetables. Meats and wholewheat flour are other sources of phosphorus. Nutrition experts believe that the most desirable ratio of calcium to phosphorus in the diet is 1:1—the ratio present in milk and cheese.

Human infants need extra calcium in the first weeks of life. The demand is met by drinking mother's milk, which has a calcium/phosphorus ratio of 2:1. (A newborn infant fed only cow's milk can develop a calcium deficiency.) The recommended daily intake of calcium for young adults ages 11 to 24 and nursing and pregnant women is 1.2 g every day, the amount obtained by drinking a liter of milk. People under age 11 and over age 25 require about 0.8 g of calcium daily. Some experts suggest that women over 50 should consume 1.2 g of calcium each day.

Table 12.2 The Elemental Composition of the Body

Element	Percentage of total mass of body	Percentage of total number of atoms in body	Kilograms per 70-kg person
most abundant			
oxygen	65.0	25.5	45.500
carbon	18.0	9.5	12.600
hydrogen	10.0	63.0	7.000
nitrogen	3.0	1.4	2.100
macronutrients			
calcium	1.5	0.31	1.050
phosphorus	1.0	0.22	0.700
potassium	0.35	0.06	0.245
sulfur	0.25	0.05	0.175
chlorine	0.15	0.03	0.105
sodium	0.15	0.03	0.105
magnesium	0.05	0.01	0.035
micronutrients			
iron	0.006	0.05	0.004
zinc	0.003	0.01	0.002
copper	0.0001	< 0.01	0.0001
manganese cobalt chromium selenium iodine molybdenum	< 0.0001	< 0.01	< 0.0001
*trace amounts**			
tin vanadium nickel fluorine silicon arsenic	< 0.0001	< 0.01	< 0.0001

*Need has not been established.

About 90% of the calcium and 80% of the phosphorus in the body are present in the bones and teeth. Bone is a combination of inorganic salts and collagen. *A network of collagen fibers forms the basic structure of bone: the* **bone matrix.** A complex salt of calcium phosphate with a composition similar to hydroxyapatite $[Ca_{10}(PO_4)_6(OH)_2]$ deposits as crystals around the collagen matrix. The fibers lend flexibility and toughness; the salt lends hardness and rigidity. The dentine and enamel of teeth are similar to bone. Dentine is moderately hard and has a mineral content of 70%; enamel is harder, with a mineral content of 98%.

A calcium deficiency results in bones and teeth that do not form properly. In children, a calcium deficiency causes rickets. In the elderly, calcium is sometimes released from bone to keep blood calcium levels constant. *As a result of the loss of calcium, bones can become brittle and porous, a condition called* **osteoporosis.** In the Case in Point earlier in this chapter we learned that Bessie, an elderly farm woman, had broken her hip. In the Follow-up to the Case in Point, below, we will learn more about these fractures.

Calcium and phosphorus are found in the body not only in bones and teeth. They are present, for example, in the blood as calcium and phosphate ions. Calcium ions are required for clotting blood, maintaining heartbeat rhythm, and forming milk. Phosphorus is a component of DNA and RNA. Another phosphorus-containing molecule, adenosine triphosphate (ATP), transmits energy for nearly all body functions.

Sodium, potassium, and chlorine Sodium ions (Na^+), potassium ions (K^+), and chloride ions (Cl^-) are three of the principal electrolytes (conduc-

FOLLOW-UP TO THE CASE IN POINT: Osteoporosis

It is common to have a great-aunt or grandmother who, like Bessie, has broken a hip. Most elderly victims of broken hips suffer from osteoporosis, a crippling disease characterized by the loss of calcium from bones. The bones of the spine, hips, and wrists are in general affected mostly in older people, postmenopausal women in particular. Initially, health scientists thought that raising the dietary intake of calcium would cure the disease or at least prevent it from occurring. However, increased calcium intake does not always slow the loss of calcium from bone. Instead, recent research shows that the cause of osteoporosis is closely related to regulation of the release and uptake of calcium in bones. Several hormones are involved in regulation of the calcium levels in the blood and bones. Of these, the principal one is *parathyroid hormone* (PTH). When the concentration of calcium in the blood is low, PTH is released. PTH stimulates the release of calcium from bone into the bloodstream. PTH also stimulates calcium retention in the kidneys and the adsorption of dietary calcium in the intestines. When the blood level of calcium is high, the thyroid gland secretes the hormone calcitonin. The effects of calcitonin are the opposite of the effects of PTH. Consequently, in response to the calcium level in the blood, bones are constantly being broken down and built up. Osteoporosis results if more calcium from bone is lost than is deposited (see figure, part a). Postmenopausal women may lose bone at a rate of more than 1% per year.

Load-bearing exercises such as walking and running help prevent the bone depletion of osteoporosis. Eating foods that are rich in calcium and vitamin D also is helpful, since vitamin D aids in the absorption of dietary calcium. The steroid hormone estrogen, which declines in postmenopausal women, also plays a role in calcium deposition in bone. Many postmenopausal women are aided by estrogen replacement therapy, which helps prevent bone depletion. The current belief is that healthful habits begun in the teens and continued throughout adulthood will develop dense bones. As a result, the chance of developing osteoporosis in later life will be reduced.

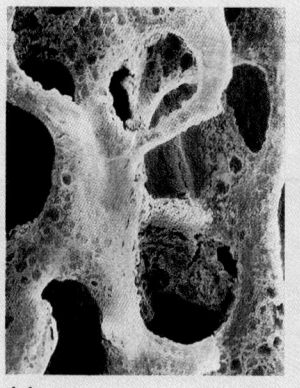 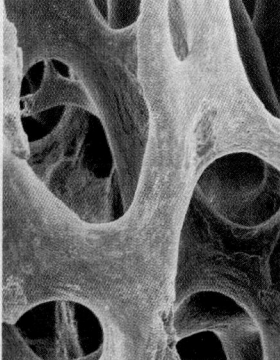

(a) **(b)**

The bone of a person with osteoporosis (a) has a lower density than normal bone (b).

tors of electricity in aqueous solutions) in the body. Electrolytes help keep the volume of body fluids constant. Sharp changes in body fluid volume or electrolyte concentration often indicate illness. Abnormal loss of water (*dehydration*) and abnormal retention of water (*edema*) are examples of the loss of control of body fluid volumes. Although sodium and potassium ions are similar in their physical and chemical properties—both are Group 1 metals—they cannot replace each other in the body. Sodium ions are the principal cations of blood, and potassium ions are the principal cations inside cells. Both sodium and potassium ions help maintain nerve responses at normal levels; potassium ions exert a relaxing effect on heart muscle between heartbeats.

Sodium chloride is the main source of sodium in the diet. Bread, cheese, carrots, celery, eggs, milk, oatmeal, and clams all are high in sodium. The recommended daily intake for people who do not have a history of high blood pressure is about 5 g of sodium chloride per day, which is about one-half the amount many people consume. For people with high blood pressure, a daily intake of less than 1 g of sodium chloride is usually recommended. Most of the sodium entering the body is absorbed in the small intestine; about 95% of the sodium ions excreted are in the urine.

Potassium occurs in all food, so a dietary deficiency is uncommon. The daily requirement for an adult is about 1 g. Foods high in both sodium and potassium are not always suitable, however, because people who need potassium are often on low-sodium diets. Foods high in potassium but low in sodium include beef and beef liver, chicken, pork, bananas, dried apricots, orange juice, broccoli, potatoes, and pineapples.

The main source of chlorine in the diet is table salt (sodium chloride). In both ingestion and excretion, sodium ions and chloride ions are inseparable; a low-salt diet produces decreased excretion of both sodium ions and chloride ions in the urine. Excessive sweating and diarrhea produce a simultaneous loss of both sodium ions and chloride ions from the body. A very important use of chloride ions by the body is in the production of hydrochloric acid in the stomach.

Magnesium   Most of the magnesium (Mg^{2+}) in the body is combined with calcium and phosphorus in the bone. High concentrations are found in muscles and red blood cells, while the remainder is distributed throughout body tissues and fluids. A normal diet provides an adequate supply of magnesium. Foods rich in this element include green vegetables, soybeans, nuts, dried peas and beans, and whole-grain cereals. The recommended dose is 300 mg per day.

Magnesium is essential in nerve impulse transmission and muscle contraction. A magnesium deficiency causes muscle tremors, twitches, and convulsions, sometimes accompanied by behavioral disturbances. Alcohol consumption leads to an increased loss of magnesium from the body. Since magnesium is a depressant, intravenous injection of magnesium salts produces anesthesia and even paralysis, but it also can reduce convulsions. Magnesium ions are not readily absorbed through the intestinal wall, so they pull water into the intestine from adjacent tissues by osmosis. Compounds such as Epsom salts ($MgSO_4 \cdot 7H_2O$) are used as laxatives for this reason.

Sulfur Sulfur is present in every cell of the body and in most food proteins. A normal diet provides an adequate supply. In proteins, sulfur is a component of certain amino acids. The proteins of hair, fingernails, and feathers are rich in sulfur-containing amino acids, which is why these materials smell so offensive when burned. Sulfur is also present as sulfate ions in the blood and other body fluids. In the breakdown of food in the body, sulfur is eventually oxidized to sulfate and excreted in the urine. Low concentrations of sulfuric and phosphoric acids give normal urine a slightly acidic pH.

Micronutrients

Iron The human body contains less than 5 g of iron. The greatest dietary need is during the first 2 years of life and for women during childbearing years. The recommended allowance is 10 mg per day for an adult male and 20 mg per day for an adult female. The best sources of iron are liver, kidney, and green vegetables. Other sources are egg yolk, brewer's yeast, fish, whole wheat, nuts, oatmeal, molasses, and beans. Absorption of iron as iron(II) ions (Fe^{2+}) takes place in the stomach and upper part of the small intestine. The presence of vitamin C, which promotes the reduction of dietary iron(III) ions (Fe^{3+}) to iron(II) ions (Fe^{2+}) assists the absorption. Nearly all the iron in the body is continually reabsorbed and reused.

Iron ions are cofactors of the cytochromes, proteins that transport electrons in cellular respiration.

A healthy adult loses about 1 mg of iron every day in the urine, sweat, and feces. Ordinarily, this loss is easily made up by the diet. However, women who have excessive loss of menstrual blood or who are in the late stages of pregnancy or persons suffering from iron-deficient anemia need to take 100 mg of iron daily, as iron(II) salts, to correct the deficiency. Surgical removal of the stomach or intestines results in *diminished iron absorption leading to* **iron-deficient anemia,** *which is characterized by fatigue.* High concentrations of iron in the body are harmful and may lead to congestive heart failure or cirrhosis of the liver.

Carbonic anhydrase, carboxypeptidase, DNA polymerase, and RNA polymerase are a few of the enzymes that require Zn^{2+} ions for catalytic activity.

Zinc, copper, manganese, molybdenum, and cobalt Zinc is essential for normal growth, life span, and reproduction in plants and animals. About 2 g of zinc is present in the body of an adult. Most of it is concentrated in the skin; bones and teeth contain lesser amounts. Zinc is also present as a cofactor in many enzymes. The recommended daily allowance for healthy adults is 15 mg of zinc per day, which is easily obtained in a normal diet. Liver, eggs, meat, milk, whole-grain products, and shellfish are rich sources of zinc. Zinc deficiency results in an impaired sense of taste and a poor appetite; it also causes slow healing of wounds and in severe cases produces dwarfism.

The human body contains about 100 mg of copper, distributed mainly in the muscles, bone, liver, and blood. The copper concentration in the blood of an adult is normally in the range 100 to 200 μg/100 mL of blood. The amount of copper in the human body, though minute, is an essential component of several vital enzymes. One such enzyme is involved in the formation of melanin, a skin pigment that helps protect us from harmful rays of the sun. The proper formation and maintenance of the protective coat of nerves also depend on copper. In instances of severe copper deficiency, this coat becomes defective, and degeneration of the nervous system occurs.

The formation of hemoglobin in the body depends on traces of both copper and iron. A decrease in the formation of hemoglobin occurs when there is a normal intake of iron but a deficiency of copper. *People with an inherited disorder called* **Wilson's disease** *accumulate copper in the liver and brain.* These people excrete copper in their urine, but copper is often undetectable in the blood. The large deposits of copper in the liver can cause damage, and cirrhosis often develops. A balanced diet supplies the recommended 2 mg of copper per day for an adult. Foods that are a good source of copper include liver, kidney, raisins, nuts, dried peas and beans, and shellfish.

The functioning of the central nervous system and the thyroid gland, as well as the formation of normal bone and cartilage, depends on a small amount of manganese in the body. The total manganese content of the body is only about 15 mg, most of which is stored in the liver and kidney. An average daily intake of about 5 mg is satisfactory.

Molybdenum is probably absorbed into living systems as the molybdate ion (MoO_4^{2-}) from such foods as yeast, liver, kidney, beans, and peas. If the levels of molybdenum or zinc in the body are high, the absorption of copper decreases, and symptoms of copper deficiency appear.

Cobalt, as we have seen, is an essential component of vitamin B_{12}.

Selenium and chromium   Selenium is a nonmetal of the same chemical group as oxygen and sulfur. Although selenium is extremely toxic; in minute amounts it is an essential human nutrient. Chromium is an essential trace element in all plant and animal tissues. The estimated amount of chromium in the body of an adult is only about 6 mg. Chromium deficiency in animals has been shown to impair growth and reduce life span. Some nutrition experts believe that the impaired ability of many middle-aged people to use glucose might be reversed by chromium supplements.

Iodine   Nearly all the iodine in the body is concentrated in the thyroid gland. Iodine is needed only for the synthesis of iodine-containing thyroid hormones. One such hormone, thyroxine, is involved in controlling body growth and regulating basal metabolism.

Thyroxine

A normal diet usually provides the daily requirement for iodine, about 100 μg. The need for iodine is highest during adolescence and in pregnancy. People who have an iodine deficiency may develop **goiter**—*an enlarged thyroid gland.* Simple goiter occurs as the gland grows larger in an attempt to compensate for the low iodine content in the diet. The widespread use of *iodized salt*—table salt to which a small amount of sodium iodide has been added—has virtually eliminated simple goiter.

12.10 Trace elements

AIM: To name the trace elements found in the body, and state some of their possible functions.

Many other elements are present in trace quantities in living systems, but it is hard to establish whether they are essential or merely passing through. Fluorine, nickel, tin, vanadium, silicon, and arsenic are essential trace elements in animals and also may be required by humans, though their exact functions have not been established. Deficiencies of these elements in test animals result in reduced growth rates. Since nickel deficiency also results in damage to liver cells, it has been suggested that nickel might play a role in the structure of liver cell membranes. Tin and vanadium appear to have an effect on fat biochemistry, silicon seems to be involved in the structure of skin and connective tissue, and arsenic appears to influence the ability to reproduce.

Fluorine accumulates in bones and teeth as fluoride ion. The concentration there depends on a person's age and fluoride ion intake. Fluoride ions are rapidly absorbed by the body and are distributed mostly in the fluid surrounding cells, though some are retained in the bones and teeth. The fluoride ions that are not retained are rapidly excreted in the urine.

Years of research have established that fluorine, as the fluoride ion, is effective in protecting teeth against dental caries (cavities). Today, many municipalities add fluoride to drinking water at a level of about 1 ppm. This concentration corresponds to a daily intake of about 1 or 2 mg. The fluoride ion replaces a small fraction of the hydroxide ions (OH^-) in the hydroxyapatite mineral of the bones and teeth; the modified hydroxyapatite has much greater mechanical strength than the non-fluorine-containing mineral.

Although fluoride ion intake of about 1 to 2 mg per day is beneficial for the maintenance of strong, healthy teeth, higher or lower amounts are detrimental. A regular daily intake of about 2 to 4 mg per day of fluoride causes teeth to become discolored or mottled; the teeth are hard but brittle. When the daily intake is about 10 to 20 mg per day, fluorosis occurs. **Fluorosis** *is characterized by severe mottling of the teeth (dental fluorosis) and abnormal changes in bone growth (skeletal fluorosis).* If fluoride ion is absent from drinking water or present below 0.5 ppm, there is a noticeable increase in the incidence of dental caries, but there is no discoloration or mottling.

PRACTICE EXERCISE 12.5

Match the following:

Medical condition

(a) simple goiter 4
(b) pancreatitis 6
(c) pellagra 1
(d) kwashiorkor 8
(e) poor coagulation of blood 2
(f) night blindness and 10
    keratinization of mucous
    membranes
(g) Wilson's disease 3
(h) fluorosis 9
(i) osteoporosis 11
(j) scurvy 5
(k) pernicious anemia 7

Cause or result of condition

(1) niacin deficiency
(2) vitamin K deficiency
(3) copper accumulation in liver
    and brain, but no copper in
    blood
(4) iodine deficiency
(5) vitamin C deficiency
(6) premature activation of
    pancreatic enzymes
(7) vitamin B_{12} deficiency
(8) protein deficiency
(9) excessive fluoride ion in diet
(10) vitamin A deficiency
(11) excessive loss of bone calcium

SUMMARY

Sugars, lipids, proteins, and nucleic acids, the sources of energy and materials for growth in living organisms, are absorbed by the human body in digestion. All digestion processes involve enzyme-catalyzed hydrolysis reactions: hydrolysis of complex carbohydrates to monosaccharides, hydrolysis of lipids to fatty acids and other products, hydrolysis of proteins to amino acids, and hydrolysis of nucleic acids to nucleotides. Bile salts aid the digestion of fats and lipids by emulsifying them so that they are susceptible to attack by hydrolyzing enzymes called lipases.

A balanced diet requires proper amounts of the carbon compounds mentioned above, as well as vitamins and minerals. The protein consumed must contain sufficient amounts of the eight essential amino acids, since these acids are not synthesized by humans. Moreover, human beings do not make vitamins, the organic molecules needed for health and growth. The vitamins are divided into two categories: fat-soluble and water-soluble. The fat-soluble vitamins required by humans are A, D, E, and K. The water-soluble vitamins include the B-complex vitamins and vitamin C. The water-soluble vitamins are precursors of cofactors of enzymes that catalyze many important chemical reactions in the body.

Four elements—hydrogen, oxygen, carbon, and nitrogen—account for more than 99% of all the atoms in the human body. Twenty-two other elements, known as the mineral elements, make up the remainder. Although the mineral elements constitute a very small portion of the total number of atoms in the body, they are nevertheless essential—life cannot exist without them. The mineral elements are classified as either macronutrients or micronutrients. The seven macronutrients are the metals calcium, magnesium, sodium, and potassium and the nonmetals phosphorus, sulfur, and chlorine. Ten elements (iron, copper, zinc, manganese, cobalt, molybdenum, chromium, selenium, iodine, and fluorine) are all essential but are present only in trace amounts in the body—these are the micronutrients. Other trace elements that are present include nickel, tin, vanadium, arsenic, and silicon. These elements are essential in test animals and may be essential to humans.

KEY TERMS

Aerobic cell (12.1)
Basal metabolism (12.5)
Beriberi (12.8)
Bile (12.4)
Bile pigment (12.4)
Bile salt (12.4)
Bone matrix (12.9)
Caloric value (12.5)
Chylomicron (12.4)

Complete protein (12.6)
Essential amino acid (12.6)
Fluorosis (12.10)
Gastric juice (12.3)
Goiter (12.9)
Hypervitaminosis (12.7)
Iron-deficient anemia
  (12.9)
Keratinization (12.7)

Kwashiorkor (12.6)
Lipoprotein (12.4)
Macronutrient (12.9)
Marasmus (12.6)
Micronutrient (12.9)
Mineral element (12.9)
Nonessential amino acid
  (12.6)
Nucleotidase (12.4)

Osteoporosis (12.9)
Pancreatic juice (12.4)
Pellagra (12.8)
Pernicious anemia (12.8)
Rickets (12.7)
Saliva (12.2)
Scurvy (12.8)
Vitamin (12.1)
Wilson's disease (12.9)

EXERCISES

Digestion (Sections 12.1–12.4)

12.6 What is digestion, and where does it begin?

12.7 Name the enzyme contained in saliva, and describe its function. What is mucin?

12.8 What makes gastric juice acidic?

12.9 What is the principal zymogen in gastric juice?

12.10 How is pepsinogen converted to pepsin?

12.11 Why is pepsin called a *peptidase?* At what pH is pepsin most active?

12.12 What classes of compounds are digested in the small intestine?

12.13 Name the enzymes secreted in pancreatic juice.

12.14 Why must fats be emulsified before they can be digested?

12.15 (a) What is bile? (b) Where is it produced? (c) What is its function in lipid digestion?

Dietary Needs (Sections 12.5, 12.6)

12.16 Explain basal metabolism.

12.17 Which nutrients have the highest caloric value?

12.18 What other items do we need in our diet apart from sufficient calories?

12.19 Distinguish between essential and nonessential amino acids.

12.20 What are complete proteins?

12.21 Which foods are a source of (a) complete protein and (b) incomplete protein?

12.22 Describe the difference between nutritional marasmus and kwashiorkor.

Vitamins (Sections 12.7, 12.8)

12.23 List the fat-soluble vitamins, and describe an important function of each.

12.24 Why is it dangerous to take excessive amounts of fat-soluble vitamins?

12.25 Why must we have a frequent intake of water-soluble vitamins?

12.26 The following diseases are all caused by a vitamin deficiency. Name the vitamin that is responsible, and identify it as fat-soluble or water-soluble.
(a) rickets   (b) pellagra   (c) scurvy
(d) beriberi   (e) nyctalopia
(f) pernicious anemia

Macronutrients, Micronutrients, and Trace Elements (Sections 12.9, 12.10)

12.27 List the seven macronutrients, and write the formula for each.

12.28 Identify the macronutrients as metallic elements or nonmetallic elements.

12.29 Name three principal electrolytes in the body.

12.30 Respond to this statement: "A potassium deficiency in the diet can be compensated by an increased sodium intake."

12.31 Explain why the calcium/phosphorus ratio in the diet is an important factor in calcium absorption.

12.32 What is the result of (a) an excessive intake of calcium and (b) an excessive intake of phosphorus?

12.33 In what way is growth affected if there is an inadequate supply of both calcium and phosphorus in the diet?

12.34 Organic compounds, such as oxalic acid, form insoluble substances with calcium. Why should this reaction lead to a calcium deficiency?

12.35 What are the effects of rickets?

12.36 What is the major difference between sodium ions and potassium ions with respect to their occurrence in the body?

12.37 List the nine micronutrients, and write the chemical symbol for each element.

12.38 Identify the micronutrients as metallic elements, nonmetallic elements, or metalloids.

12.39 What foods are rich in iron?

12.40 Iron exists as Fe^{2+} and Fe^{3+}. (a) Which of these forms is most readily absorbed by the body? (b) How does the presence of vitamin C help the absorption?

12.41 Why do we need iron? What happens if our diet is deficient in iron?

12.42 Comment on this statement: "The majority of iron in the diet is absorbed by the body and used in the synthesis of hemoglobin."

12.43 What metallic element, besides iron, is responsible for hemoglobin formation? Give the name of the inherited disease that causes an abnormal distribution of this element in the body.

12.44 What is the function of iodine in the body?

12.45 Explain why a lack of iodine in the diet causes simple goiter.

12.46 What is the major source of fluorine in the diet? Is fluorine an essential element?

12.47 In what respect is fluorine beneficial? What is the cause of mottled teeth?

12.48 Describe one function of each of the following trace elements.
(a) molybdenum   (b) zinc   (c) manganese
(d) cobalt   (e) chromium   (f) selenium
(g) nickel   (h) tin   (i) vanadium
(j) silicon   (k) arsenic

Additional Exercises

12.49 Name at least two sources of folic acid.

12.50 What are the end products of the complete digestion of
(a) proteins?   (b) carbohydrates?
(c) triglycerides?   (d) nucleic acids?

12.51 Name the enzymes or zymogens in (a) saliva, (b) gastric juice, and (c) pancreatic juice.

12.52 Describe, briefly, how proteins and sugars are digested and absorbed in the intestinal tract.

12.53 Discuss the digestion and absorption of lipids.

12.54 Give a definition and example for the following.
(a) macronutrient   (b) micronutrient
(c) mineral element

12.55 Name the water-soluble vitamins.

12.56 Describe the disease condition that results in a deficiency of each of the following:
(a) thiamine   (b) vitamin C   (c) vitamin B_{12}
(d) vitamin D   (e) vitamin A

12.57 What is the major function in the body of each of the following mineral elements?
(a) cobalt   (b) chromium   (c) copper
(d) zinc   (e) manganese   (f) magnesium

SELF-TEST (REVIEW)

True/False

1. Potassium ions can substitute for sodium ions in the body. F
2. Some micronutrients can be toxic to the body. T
3. A common function of vitamins is to act as an enzyme cofactor. T
4. Most grains are sources of high-quality protein. F
5. A person can never take too much of a vitamin. F
6. Marasmus and kwashiorkor describe the same condition. F
7. Essential amino acids cannot be synthesized in the body. T
8. The four most abundant elements in the body are carbon, hydrogen, oxygen, and calcium. F
9. Some trace elements function as enzyme cofactors. T
10. A lipase catalyzes the breakdown of lipid molecules. T

Multiple Choice

11. The enzyme responsible for the hydrolysis of proteins in the stomach is
(a) pepsin.   (b) chymotrypsin.
(c) amylase.   (d) hydrolase.

12. Compounds of the following macronutrient are used as laxatives.
(a) iron   (b) sulfur
(c) magnesium   (d) calcium

13. An iodine deficiency causes
(a) simple goiter.   (b) Wilson's disease.
(c) pellagra.   (d) osteoporosis.

14. An incomplete protein is one that
(a) has a low amino acid concentration.
(b) is a result of a defect in amino acid metabolism.
(c) lacks one or more essential amino acids.
(d) is produced in the urea cycle.

H – C – N – O

15. The minimum energy output for a person to survive
    (a) is the same regardless of the person's age or sex.
    (b) is the basal metabolism.
    (c) is greater the older the person gets.
    (d) more than one are correct.

16. Which of the following is *not* a fat-soluble vitamin?
    (a) vitamin A   (b) vitamin K
    (c) vitamin C   (d) vitamin D

17. Most nutrients are absorbed into the bloodstream in the
    (a) stomach.   (b) large intestine.
    (c) small intestine.   (d) colon.

18. Which of the following is *not* a trace element?
    (a) tin   (b) arsenic
    (c) vanadium   (d) sulfur

19. A patient who has her gallbladder removed should control the dietary intake of
    (a) carbohydrates.   (b) fats.
    (c) proteins.   (d) all of these.

20. Carboxypeptidase is associated with the digestion of
    (a) protein.   (b) fats.
    (c) carbohydrates.   (d) nucleic acids.

21. Lipids are transported to body tissues
    (a) through the lymphatic system only.
    (b) in a combined state with lipoprotein.
    (c) in bodies called *chylomicrons*.
    (d) more than one are correct.

22. The following fat-soluble vitamin is important in the blood-clotting process.
    (a) vitamin K   (b) vitamin B_1
    (c) vitamin D   (d) folic acid

23. Vitamin B_{12}
    (a) contains a cobalt(II) ion.
    (b) aids in the synthesis of red blood cells.
    (c) deficiency causes pernicious anemia.
    (d) all of the above are correct.

Body Fluids

Maintaining the Body's Internal Environment

Replenishing liquids to maintain proper fluid balance is important for everyone.

The maintenance of a consistent internal environment in the human body is an interactive and dynamic process. Organs and organ systems working together continuously adjust imbalances that occur. A breakdown in any part of the collaborating systems of the body can cause serious illness or even death. What role does chemistry play in these systems? Consider Larry's plight in the following Case in Point.

CASE IN POINT: A diabetic imbalance

Larry, a 32-year-old teacher, collapses on the steps of the public library. The paramedics discover that he wears a medical alert tag identifying him as a diabetic. Upon his arrival at the hospital, the attending physician immediately orders several lab tests. One of these tests is measurement of the pH of Larry's blood. What imbalance does Larry's doctor suspect? And what does the chemical concept of pH have to do with Larry's illness? We will find out in Section 13.8.

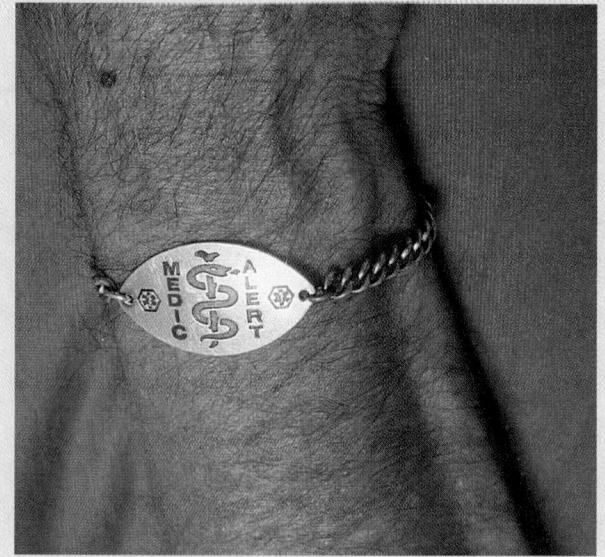

In case of an emergency, many people display their medical condition on a Medic Alert tag.

13.1 Body water

AIMS: To describe the distribution of body water and compare the electrolyte compositions of blood plasma, interstitial fluid, and intracellular fluid. To define homeostasis and cite some examples of homeostatic control in the body.

Focus

Tissue cells are bathed in fluid inside and out.

The average adult body contains 42 L (42 kg) of water. (Remember that the mass of 1 L of water is 1 kg.) This 42 L of water is compartmentalized in three regions of the body. The majority, about 28 L, is located in the **intracellular fluid**—*the fluid inside cells.* About 11 L of water is in the **interstitial fluid**—*fluid that fills the space between tissue cells. Lymph,* a fluid discussed later, may be considered part of the interstitial fluid. Most of the remaining 3 L of water is present in the fluid portion of the blood. Most body fluids consist of water and its dissolved substances. The retention of these fluids is important. See A Closer Look: Artificial Skin.

Figure 13.1 shows the composition of the major body fluids. As the figure shows, the salt composition of *blood plasma*—the fluid portion of the

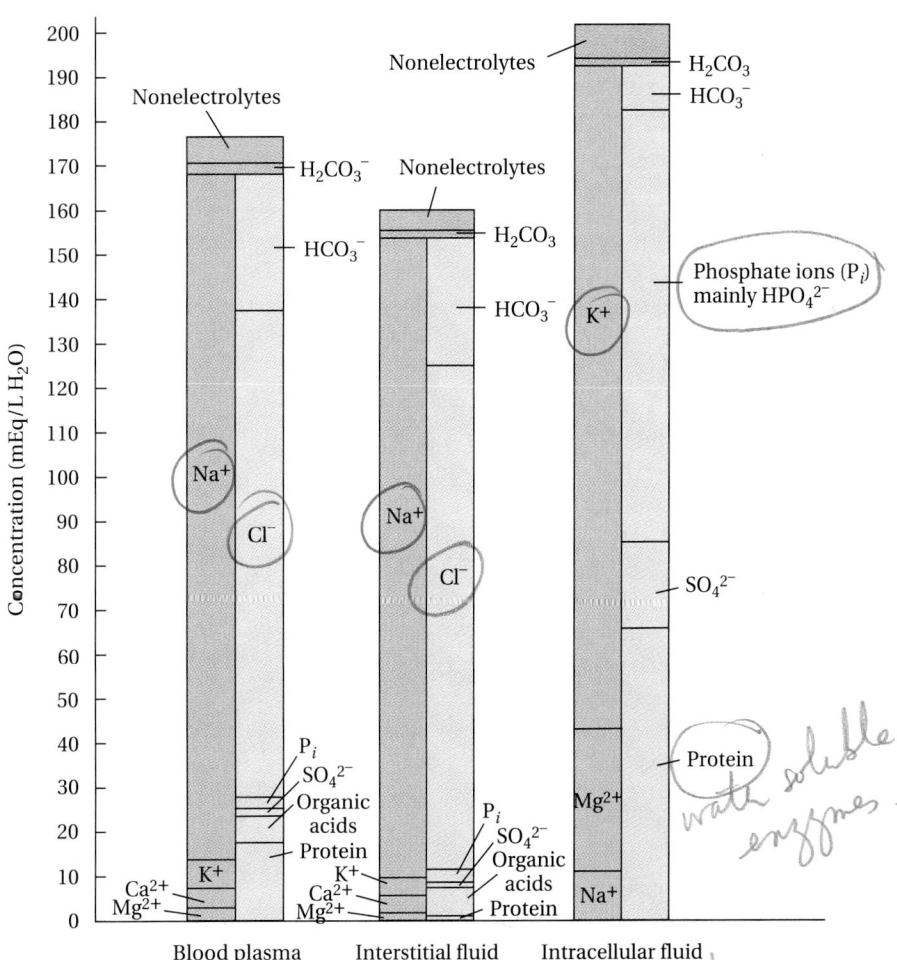

Figure 13.1
Composition of the major body fluids.

Artificial Skin

Many burn victims lose large areas of skin. These people often die because vital fluids ooze out of their bodies, and they are too weakened to fight off bacterial infection. The essential requirement in treating burns is covering the damaged area quickly. When there is too little unburned skin on the patient's body for grafting, surgeons use skin from human cadavers or pigs. These grafts are usually rejected by the body after a few days and must be replaced.

An artificial skin that shows encouraging results was first used in the 1980s. The body does not recognize the artificial skin as foreign, and drugs that prevent rejection are not required. The wounds heal with little scarring.

Like natural skin, artificial skin has two layers (see figure). The bottom layer is a blend of a complex carbohydrate obtained from shark cartilage and collagen (a protein) extracted from cowhide. These components are added to water and acidified to produce short white fibers. The mixture is poured into a shallow pan and freeze-dried to remove the water. The fibers form a thin white sheet of material that is light and highly porous. The sheet is then baked in an oven to preserve its shape. A top layer is added. This layer, made from a thin sheet of a rubber-like plastic, is bonded with an adhesive to the fibrous sheet. The completed sheet of synthetic skin is soft and pliable as natural skin and about as thick as a paper towel. It is freeze-dried again and stored in a sterile container at room temperature until required. Ten square feet of material, enough to cover wounds over 50% of an adult body, can be made in a few days.

Artificial skin reacts with human flesh as does natural skin. The fibrous bottom layer, which is placed next to the burned area, is porous and provides scaffolding into which body cells migrate to make more collagen. Over a period of months, collagen in the artificial skin breaks down in the same way as collagen in healthy skin and sloughs off. Nerve fibers, which are still alive, and blood vessels in the flesh grow up into the new material. The upper plastic layer does not become part of the body but serves only as a protective flexible covering. Within a week or so after grafting, small areas of this layer are removed and replaced with thin patches of the patient's own skin. The need to cover the artificial graft with natural skin, although inconvenient, is not a serious drawback because it can be done later when the patient's condition is improved.

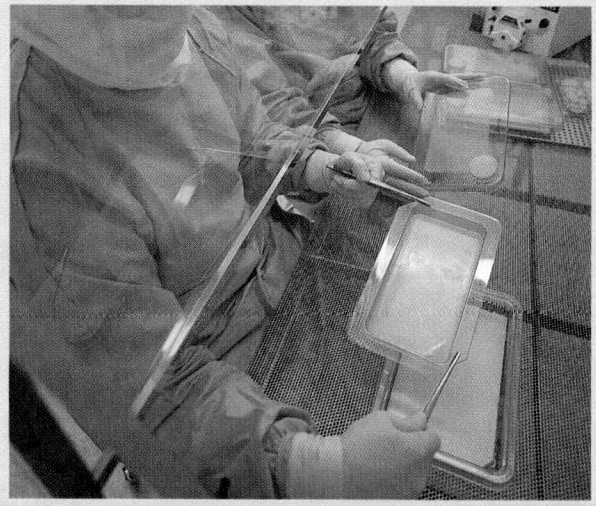

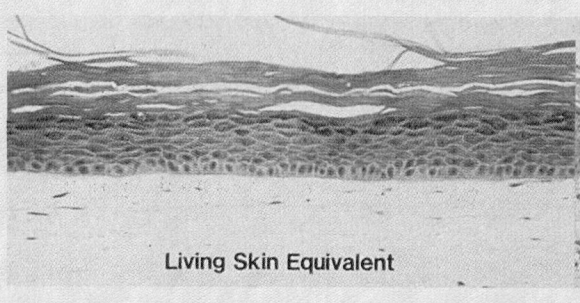

Living Skin Equivalent

A skin graft using artificial skin can be a lifesaving procedure.

blood—and interstitial fluid is nearly identical. Sodium ions and chloride ions are the major electrolytes of these two fluids. In contrast, the electrolyte composition of intracellular fluid is quite different from that of the extracellular fluids. Potassium ions and HPO_4^{2-} ions are the major inorganic electrolytes of the intracellular fluid—a reflection of the tendency of cells to transport potassium ions and HPO_4^{2-} ions from low concentrations in the extracellular fluid to higher concentrations in the intracellular fluid (cytoplasm). Cells also transport sodium ions from a low concentration in the cell to higher concentrations in the extracellular fluid. The higher protein concentration of intracellular fluid compared with the extracellular fluids results from the high concentration of water-soluble enzymes in the cytoplasm of cells.

Temperature, energy production, pH of fluids, and the levels of salts, water, and nutrients all must be carefully controlled to maintain an environment in which body cells can operate at peak efficiency. **Homeostasis** *is the process that tends to keep all these crucial factors in balance.*

The composition of seawater and interstitial fluid are very similar. Does this give credence to the theory that the first living cells originated in the sea?

Homeostasis
homo (Greek): the same
stasis (Greek): condition

13.2 Blood

AIM: To list the functions of blood, the formed elements of blood, and the plasma proteins and distinguish between blood serum and blood plasma.

Focus

Blood has many important functions.

Supplies of food and oxygen are carried to tissues and wastes are carried away from tissues by circulating blood. Fluid received from the blood bathes tissues and maintains their pH at slightly alkaline values. Blood helps to equalize body temperature by evenly distributing heat generated by active cells. Blood carries hormones that help coordinate the activities of the different body organs. Cells and proteins of the blood help defend the body against infecting organisms.

Formed elements of blood

On average, 1 mL of blood from an adult contains about 5×10^6 red cells, 7×10^3 white cells, and 250×10^3 platelets.

Within the blood are red blood cells, white blood cells, and platelets. **Erythrocytes** *(red blood cells) carry oxygen to tissues and remove carbon dioxide.* Hemoglobin, the chief carrier of oxygen and carbon dioxide in the blood, is packed into the erythrocytes. Each erythrocyte contains about 70 million molecules of hemoglobin. Blood also contains *leukocytes* and *thrombocytes.* **Leukocytes** *(white blood cells) perform many functions in cells, including the destruction of foreign organisms.* **Thrombocytes** *(platelets) are cells that play an important role in blood clotting. Erythrocytes, leukocytes, and thrombocytes are called the* **formed elements** *of the blood.*

Soluble elements of blood

Plasma *is unclotted blood from which the formed elements have been removed.* The plasma contains clotting agents and other proteins, as well as various salts and glucose. When whole blood is permitted to stand in a test

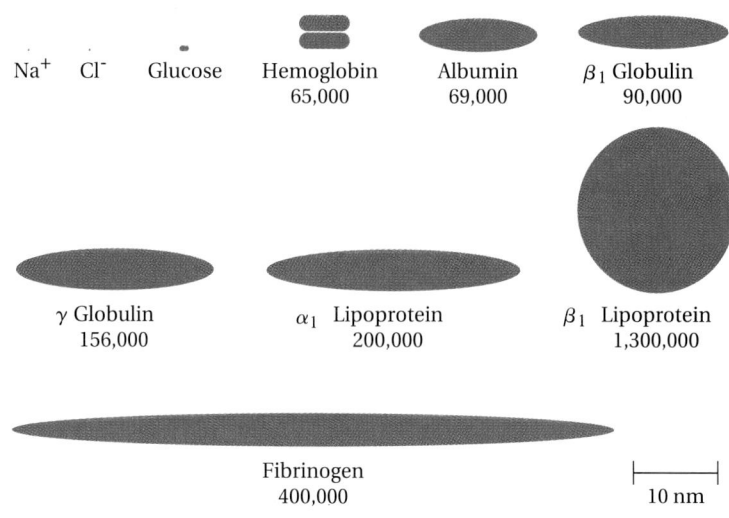

Figure 13.2
Relative shapes and sizes of blood proteins. The molar masses are given below the names of the proteins. Relative sizes of a sodium ion, a chloride ion, and a glucose molecule are shown for comparison.

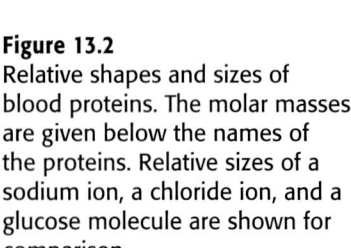

Blood drawn for some medical tests is prevented from clotting by the addition of the natural polysaccharide heparin, citrate ion, or oxalate ion. The anticlotting agent that is added depends on the test to be performed.

tube, it clots. *The straw-colored liquid that separates from the clot is* **serum.** Serum lacks the formed elements and clotting agents.

The major plasma proteins are *fibrinogen,* the *albumins,* the *globulins,* and the *lipoproteins* (Fig. 13.2). Recall that fibrinogen is essential for clot formation (see Sec. 10.12). *Albumins* help to maintain plasma volume by increasing the osmotic pressure of the blood. When the concentration of serum albumin is too low, water flows from the blood to the tissues. Owing to the force of gravity, the excess tissue fluid often collects in the lower extremities, leading to swelling (edema). *Globulins* perform many functions, such as defense against infecting organisms and the transport of metal ions such as iron and copper. *Lipoproteins* carry lipids in the bloodstream, and most plasma lipids are found associated with them (refer to A Closer Look: Blood Lipoproteins and Heart Disease, page 363). Recent concerns about the safety of blood supplies have led to new research to find blood replacements and to prevent the loss of blood during medical procedures: See A Closer Look: Blood: Risks and Replacements.

The circulatory system

The heart, the blood vessels, and the lymphatic network constitute the major organs of the circulatory system (Fig. 13.3). The heart pumps oxygen-carrying blood from the lungs through the large blood vessels, the arteries. Oxygen-laden hemoglobin is red, and arterial blood has a bright scarlet hue. Like the trunk of a tree, arteries separate into thinner vessels, the branchlike arterioles and the twiglike capillaries.

A vast network of capillaries enmeshes every tissue, and no tissue cell is very far from a capillary. Scientists estimate that the length of capillaries in an average person's body is in the neighborhood of 100,000 km (over 60,000 miles). The microscopic capillaries are so thin that erythrocytes must squeeze through single file. Hemoglobin molecules of erythrocytes exchange their loads of oxygen for loads of carbon dioxide in the capillaries.

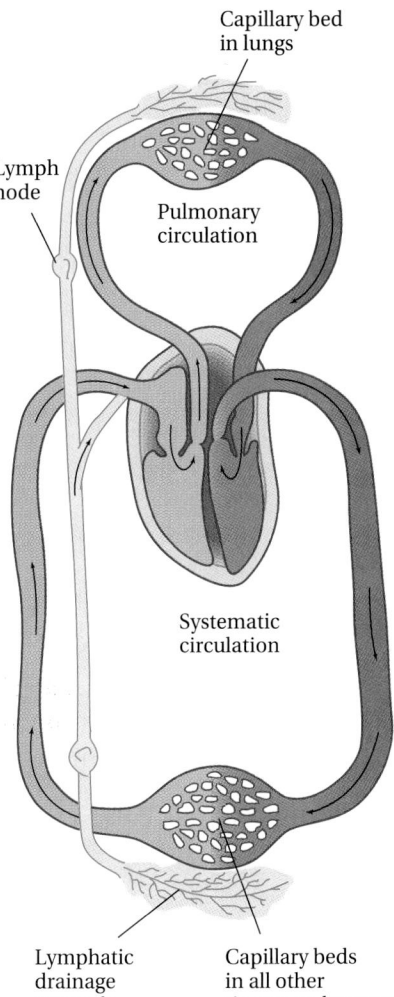

Figure 13.3
The circulatory and lymphatic systems.

Capillary bed in lungs

Lymph node

Pulmonary circulation

Systematic circulation

Lymphatic drainage network

Capillary beds in all other tissues and organs

Blood leaves the capillaries by entering small veins that lead in succession to larger veins. The large veins carry the blood back to the heart. Venous blood—which has a bluish cast—is pumped into the capillaries of the lungs, where carbon dioxide is lost. A new load of oxygen is picked up, and the circulatory cycle begins again.

The lymphatic system consists of the lymphatics, the thymus, the spleen, and the lymph nodes. The lymphatics are tubes through which interstitial fluid drains into the bloodstream. **Lymph**—*the fluid that enters small capillary-sized lymphatics*—flows to progressively larger lymphatics and drains into the bloodstream at the thoracic duct or the right lymphatic duct.

PRACTICE EXERCISE 13.1

Describe the functions of the blood and the lymph, and explain how these two fluids function in the circulatory system.

A Closer Look

Blood: Risks and Replacements

Some loss of blood in surgical procedures must be expected. However, the loss of too much blood can be fatal. Transfusions of blood obtained from blood bank programs have saved many lives, but there are disadvantages to the use of donated blood. The blood type of donors and recipients must be carefully matched to avoid a potentially fatal allergic response caused by transfusing mismatched blood. There is also a possibility of catching diseases from blood transfusions. In the early 1980s, many hemophiliacs who were treated with whole blood or blood products became infected with HIV, the virus that causes AIDS. Blood is now heat-treated to kill HIV. Today, the chances of being infected with HIV by a transfusion of blood pooled from many donors are small, but they are finite—from 1 in 45,000 to 1 in 225,000. The risk of catching hepatitis is about 1 in 3500. One way that patients are avoiding these risks is to bank their own blood before surgery. Hospitals are trying to use less blood and transfuse only as absolutely necessary. Blood lost during surgery is also recovered and returned to the patient.

Because of real risks and people's fears about blood transfusions, scientists are trying to find synthetics to replace natural blood. There will likely never be a substitute to perform all of blood's many functions, so current research is geared to synthetics that can temporarily perform the oxygen-carrying function of red blood cells. One promising line of research employs perfluo-

rocarbons as the oxygen-carrying materials. *Perfluorocarbons* are organic compounds in which all the hydrogens are replaced by fluorines. A unique property of perfluorocarbons is their ability to dissolve oxygen. One perfluorocarbon product, Fluosol DA, is at present the only product to win approval in the United States for clinical use as a temporary blood replacement. Since 1989, Fluosol DA has been approved for the single purpose of oxygenating heart muscles during artery-widening balloon angioplasty. Fluosol DA is a milky aqueous emulsion containing 14% perfluorodecalin with 6% perfluorotripropyl amine as the emulsifier. The product carries oxygen to tissue cells and transports carbon dioxide from tissues cells like blood. It appears to be safe except for flulike symptoms in some patients. It can be transfused without delay for blood typing because it contains no antigens. One disadvantage is that the concentration of perfluorocarbons in the emulsion is low, and thus has a low oxygen-carrying capacity. However, emulsions containing more of the perfluorocarbon tend to separate into aqueous and organic layers and are not stable enough to use. In order to improve the stability of the emulsions, scientists are trying new fluorocarbons and new emulsifiers such as lecithin, egg yolk phospholipids, and triglycerides.

Another line of blood-replacement research involves the use of hemoglobin, the natural oxygen carrier of blood that has been removed from its red blood cells. The hemoglobin research is in relative infancy, and we do not know what its future holds.

13.3 Antibodies and interferons

AIM: To compare and contrast phagocytosis and the immune response.

Focus

Antibodies and interferons form an important natural line of defense against infection.

The body uses blood components in two major defenses against infectious organisms or other foreign matter. The first of these mechanisms, *phagocytosis,* is a process by which certain leukocytes called *macrophages* engulf and subsequently destroy foreign objects. The pus that forms around a cut on an infected finger or at the root of an abscessed tooth is composed of macrophages that have fought and died in the battle with the foreign invader. Phagocytosis forms an immediate first line of defense in infec-

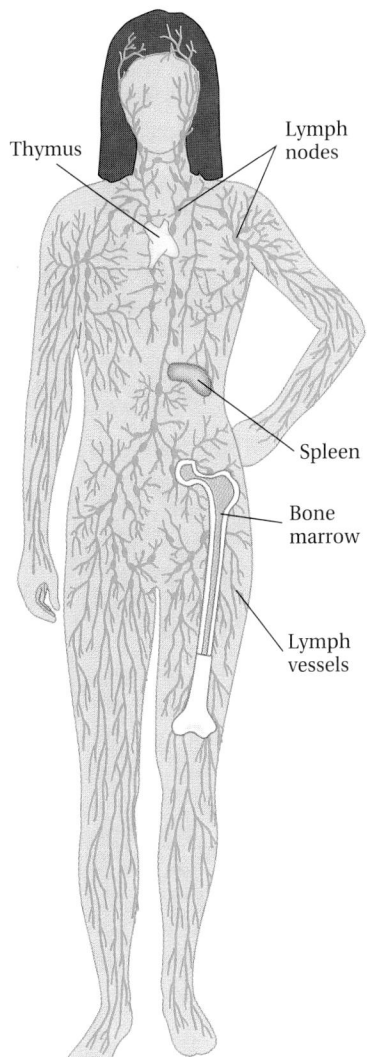

Thymus

Lymph nodes

Spleen

Bone marrow

Lymph vessels

Figure 13.4
Antibody-producing cells, the lymphocytes, are produced in the thymus, the spleen, the lymph nodes, and the bone marrow.

tions. The second line of defense takes about a week to achieve its full power following an infection. This second defense is the *immune response.* In the immune response, special cells of the body are called on to produce proteins called *antibodies.* Another class of proteins, the *interferons,* is produced specifically to defend against infections by viruses.

Antibodies

Antibodies *are large protein molecules whose job is to react with infecting agents.* The antibody-producing cells are *lymphocytes.* Lymphocytes are produced in bone marrow and in the other lymphatic tissues illustrated in Figure 13.4: the thymus, the spleen, and the lymph nodes. The lymph nodes act as filters of lymph, trapping bacteria and other foreign substances. When infecting agents are trapped in a lymph node, lymphocytes begin to produce antibodies against them. The antibodies then enter the bloodstream by means of the lymphatic system. *The infecting agents that are acted on by antibodies are called* **antigens.** In a normal individual, antigens may be bacteria, viruses, proteins, carbohydrates, or just about anything the body recognizes as foreign.

The antibody proteins released into the bloodstream make up a part of blood serum called **immunoglobulins.** *There are five classes of immunoglobulins, of which the most common are immunoglobulins G (IgG or gamma globulins).* The number of antibodies produced by the lymphocytes is enormous. Since you began reading this section, your body has produced over a million billion antibody molecules. Even more astonishing, the different antibodies you have produced probably number in the millions. Despite the great numbers of different antibodies being continually produced, all antibody molecules have certain structural features in common. They all consist of two identical large-protein portions (heavy chains) and two identical smaller-protein portions (light chains). Figure 13.5(a) shows the Y shape of IgG molecules.

The peptide chains of all antibody molecules have very similar primary structures. However, small regions of the light and heavy chains differ in primary structure from antibody to antibody. These small regions are called the *variable regions* of the light and heavy chains. The interaction of the antibody molecule with the antigen occurs in these variable regions, as

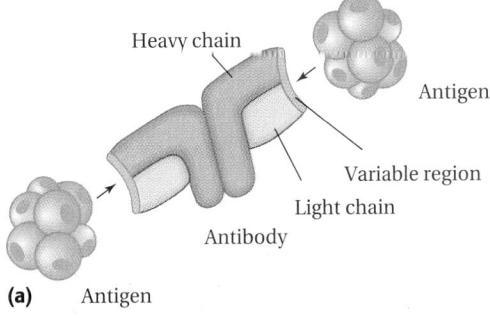

Heavy chain

Antigen

Variable region

Light chain

Antibody

(a) Antigen

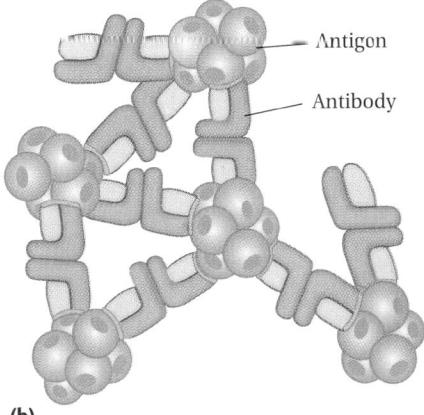

Antigen

Antibody

(b)

Figure 13.5
(a) Representation of an immunoglobulin G antibody molecule (IgG) with antigens. Each light chain of IgG has a molar mass of about 25,000, and each heavy chain has a molar mass of about 50,000. (b) An antibody-antigen complex is formed when antibody molecules interact with antigens.

Multiple sclerosis is an example of an *autoimmune disease*—the body mistakenly recognizes a normal component as foreign and generates an immune response to destroy the component. Other autoimmune diseases include insulin-dependent diabetes mellitus, rheumatoid arthritis, myasthenia gravis, and systemic lupus erythematosis.

illustrated in Figure 13.5(b). This interaction occurs by a lock-and-key or "sticky patch" mechanism similar to that which binds substrates to enzymes. It is likely that only a small part of the total antigen is recognized by the antibody. Once antibody-antigen complexes have been formed, they are destroyed by phagocytosis. A system called the *complement system* also contributes to the destruction of foreign cells and viruses. *The* **complement system** *consists of at least 11 plasma proteins that are responsible for the lysis (breakage) of the cell membrane and the death of foreign cells.* The attachment of antibodies to a foreign cell triggers the complement system.

Why are millions of different antibodies needed? Obviously, the organism manufacturing the antibodies does not know what antigens it will encounter, so apparently it synthesizes many light and heavy chains to provide enough different antibody "locks" to accept almost any antigen "key." Each lymphocyte produces only one kind of antibody molecule. When a particular antigen has been detected, an order is sent to the proper lymphocytes to produce many more copies of that antibody.

Sometimes defects occur in the combination of light and heavy chains to form complete antibodies. In 1845, English physician Henry Bence-Jones observed that the urine of people with multiple myeloma, a cancer of antibody-producing cells, was of abnormally high specific gravity. A component of the urine was later found to be a protein substance. The appearance of these proteins in the urine called for a diagnosis of multiple myeloma, but the source of Bence-Jones proteins remained unknown for over a hundred years. In 1962, Gerald M. Edelman and Joseph A. Gaily of Rockefeller University showed that **Bence-Jones proteins** *are antibody light chains made by the myeloma tumor but not incorporated into complete antibodies.*

Because each kind of lymphocyte produces only one kind of antibody, we might expect to obtain homogeneous antibodies by growing a single

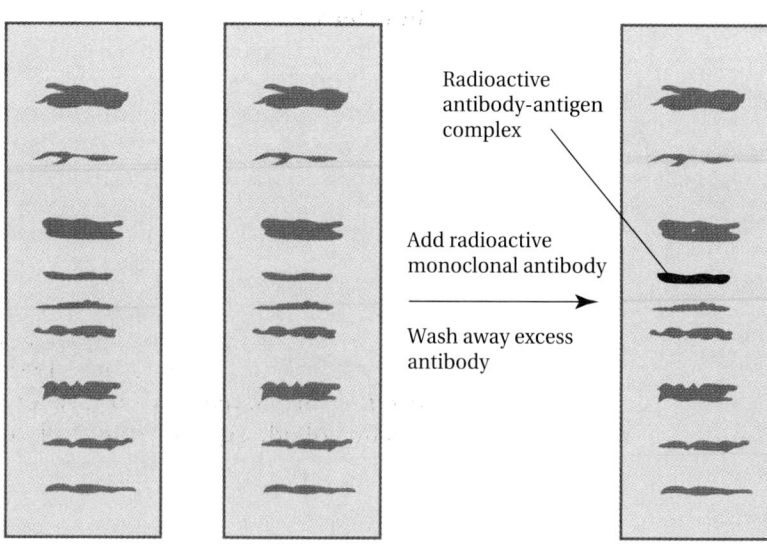

Figure 13.6
Method for detection of an antigen using a radioactive monoclonal antibody.

Electrophoresis of protein mixture containing antigen

Proteins transferred to nitrocellulose paper by blotting

Radioactive antibody-antigen complex

Add radioactive monoclonal antibody

Wash away excess antibody

kind of lymphocyte and harvesting the antibody that the cells produce. This is not possible because lymphocytes cannot be grown in cell cultures. However, crossing (cloning) lymphocytes with myeloma cells produces hybrid cells (hybridomas) that make homogeneous antibodies and can be grown in culture. **Monoclonal antibodies**—*homogeneous antibodies produced by a single kind of hybridoma*—can be obtained in large quantities. Monoclonal antibodies have become indispensable tools in the detection of infecting organisms. Consider, for example, a blood sample from a patient suspected of having a bacterial or viral infection. Blood that is infected by a foreign organism contains protein antigens characteristic of the organism. To test for these antigens, a mixture of proteins from the blood can be separated by electrophoresis and then blotted onto nitrocellulose paper (Fig.13.6). The proteins stick tightly to the nitrocellulose paper and are not removed by washing. A solution containing a radioactively labeled monoclonal antibody of the antigen is then placed on the paper, where the antibody and any antigen make a tight antibody-antigen complex. The paper is washed, which removes any excess antibody; washing does not remove the proteins or the antibody-antigen complexes. The presence of the antigen can be detected as a radioactive spot of antibody-antigen complex.

Interferons

People infected with one virus are resistant to infection by another virus at the same time. This suggests that besides producing antibodies, the body is capable of mounting special molecular defenses against viral infections. In 1957, virologists Alick Isaacs and Jean Lindenmann discovered that cells infected by viruses produce minute quantities of glycoproteins that travel to nearby cells, where they stimulate these cells to produce protective antiviral proteins. *These antiviral proteins are called* **interferons.** There are three families of interferons. Each interferon molecule contains about 150 amino acid residues. Interferons are among the most powerful known biological agents. They can be effective antiviral agents in concentrations as low as 3×10^{-14} *M*. Because of the low concentrations of interferons in the body, it is impossible to isolate from human sources the quantity needed for research and therapeutic use. Although early interferon research was hampered because so little of the substances were available, the techniques of recombinant DNA (see Sec. 11.13) now produce sufficient amounts for research and for the treatment of disease.

Recent clinical tests of interferons against several viral diseases give encouraging results. Since interferons also slow cell division, they are now used against some types of cancer, a family of diseases characterized by uncontrolled, explosive rates of cell multiplication. One type of interferon, beta-interferon (Fig. 13.7), is the only available drug for the treatment of multiple sclerosis. This neuromuscular disease affects 350,000 Americans and causes vision problems, partial paralysis, and memory loss. Multiple sclerosis is not cured by beta-interferon, but administration of the protein slows the progress of the disease for those who suffer periodic attacks.

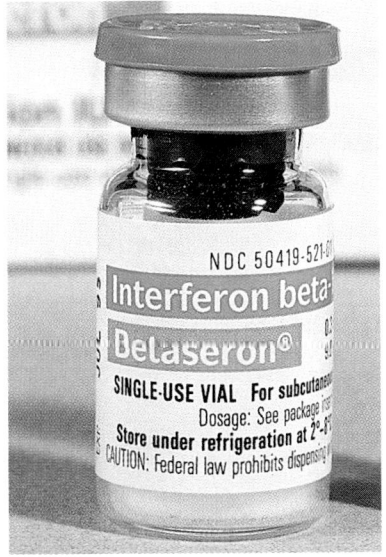

Figure 13.7
Beta-interferon, a genetically engineered human protein, is used for the treatment of multiple sclerosis.

13.4 Blood buffers

AIM: To name the major blood buffer systems and describe how each helps maintain a constant blood pH.

Our bodies function properly only when the pH of various fluids lies within certain narrow limits. To keep these fluids in a proper **acid-base balance**—*the control of pH of body fluids within a narrow range necessary for life and health*—our bodies contain buffer systems. Three independent buffer systems keep the normal pH of blood between 7.35 and 7.45, with an average normal value of 7.40. **Acidosis** *is any shift in the blood pH to below the normal value;* **alkalosis** *is a pH shift to above the normal value.* A drop in the blood pH to below 6.8 or an increase to above 7.8 is life-threatening.

The most important buffer system of the blood consists of carbonic acid (H_2CO_3) and the hydrogen carbonate (bicarbonate) ion (HCO_3^-). The HCO_3^-/H_2CO_3 ratio is normally 20:1. When protons (hydrogen ions) enter the blood, they are picked up by the bicarbonate ion to form carbonic acid.

$$HCO_3^-(aq) + H^+(aq) \longrightarrow H_2CO_3(aq)$$

As long as bicarbonate ions are present, excess protons are removed, and the pH of the blood changes very little. When hydroxide ions (bases) enter the blood, carbonic acid molecules lose a proton by reacting with hydroxide ions to form more bicarbonate ions.

$$H_2CO_3(aq) + OH^-(aq) \longrightarrow HCO_3^-(aq) + H_2O(l)$$

As long as carbonic acid molecules are present, excess hydroxide ions are removed, and the pH of the blood changes very little.

The second important buffer system in the blood consists of monohydrogen phosphate and dihydrogen phosphate ($HPO_4^{2-}/H_2PO_4^-$). This buffer system works much like bicarbonate and carbonic acid. Protons react with monohydrogen phosphate to produce the dihydrogen form, and the pH changes very little.

$$HPO_4^{2-}(aq) + H^+(aq) \longrightarrow H_2PO_4^-(aq)$$

Bases in the blood remove a proton from dihydrogen phosphate to produce the monohydrogen compound, and the pH changes very little.

$$H_2PO_4^-(aq) + OH^-(aq) \longrightarrow HPO_4^{2-}(aq) + H_2O(l)$$

The third important buffer system in the blood consists of the large protein molecules that are present in colloidal dispersion. These complex molecules contain many acidic (—COOH) and basic (—NH_2) groups that donate or accept protons.

$$
\begin{aligned}
-COOH + OH^- &\longrightarrow -COO^- + H_2O \\
-COO^- + H^+ &\longrightarrow -COOH \\
-NH_2 + H^+ &\longrightarrow -NH_3^+ \\
-NH_3^+ + OH^- &\longrightarrow -NH_2 + H_2O
\end{aligned}
$$

13.5 Oxygen and carbon dioxide transport

AIM: To describe with chemical equations the exchange of oxygen and carbon dioxide between a red blood cell and lung alveolus and between a red blood cell and a tissue cell.

Without the hemoglobin of red blood cells, 1 L of arterial blood at room temperature could dissolve and carry only about 3 mL of oxygen—not nearly enough to supply the oxygen needs of the body. With hemoglobin, the amount of oxygen that can be dissolved and carried by a liter of blood increases 70 times.

Transport of oxygen

Transport of oxygen to tissues begins when air inhaled into the lungs enters the *alveoli*—grapelike clusters of sacs responsible for the absorption of oxygen. Figure 13.8 shows how the exchange of oxygen and carbon dioxide takes place between the alveoli and red blood cells in the alveolar capillaries. The five basic steps shown in the figure are as follows:

1. Oxygen pressure is higher in the alveoli than in the red blood cell. Therefore, oxygen diffuses through the capillary wall and into the red blood cell from the alveoli. (Recall that diffusion of gases is always from higher partial pressure to lower partial pressure.)

Oxygen transported by hemoglobin is bound to the Fe^{2+} of the heme group. If Fe^{2+} is oxidized to Fe^{3+}, it loses its ability to carry oxygen molecules.

2. The concentration of protons in the red blood cell is low. This low concentration of protons means that the reaction of protonated hemoglobin (HHb) with oxygen, a reaction that releases protons, is favored. Oxyhemoglobin (HbO_2) is formed. Each molecule of oxyhemoglobin carries four molecules of oxygen.

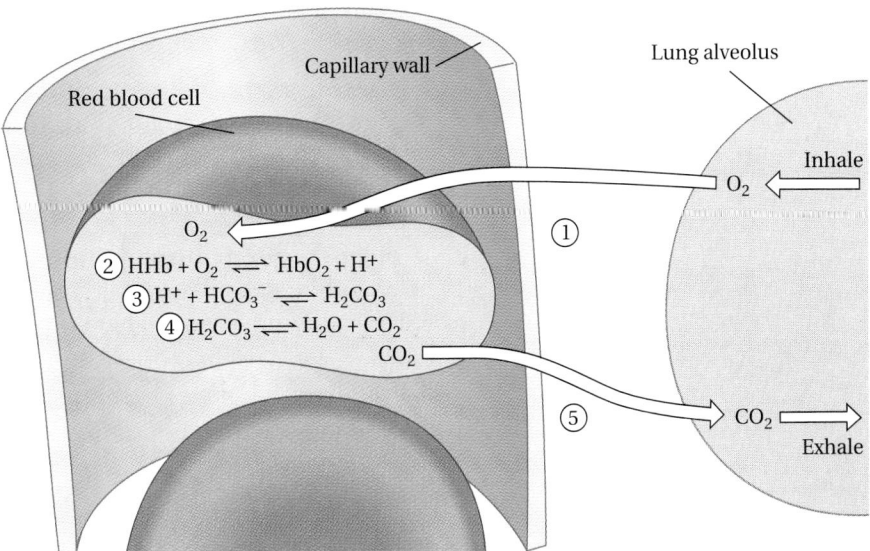

Figure 13.8
In five basic steps, red blood cells exchange carbon dioxide for oxygen at the lungs.

3. Protons released by hemoglobin as it is oxygenated react with bicarbonate ions to form carbonic acid. (The bicarbonate ions come from active tissue or the kidneys.)

4. An enzyme in the red blood cell, carbonic anhydrase, promotes the breakdown of carbonic acid to water and carbon dioxide. The reaction is reversible, but the low pressure of carbon dioxide in the lungs pulls the reaction in the direction of formation of carbon dioxide and water (Le Châtelier's principle again).

5. Since the carbon dioxide pressure is high in the red blood cell but low in the alveoli, carbon dioxide molecules diffuse out of the red blood cell and into the lung. The lungs expire the carbon dioxide and some of the water formed in step 4.

What would happen if the bicarbonate ions lost in expiration were not replaced? First of all, the body would soon deplete its store of bicarbonate ions, since a molecule of carbon dioxide is expired for each proton neutralized. Buffering of the blood by formation of carbonic acid from excess protons and bicarbonate ions would become impossible. Thus, with its load of excess protons, the blood would become acidic. Hemoglobin does not pick up oxygen in an acidic environment because the equilibrium position in the reaction shifts far to the left in the presence of a high concentration of protons.

$$HHb + O_2 \rightleftharpoons HbO_2 + H^+$$

Consequently, in severe acidosis, oxygen loses the competition with protons for hemoglobin, and body cells die for lack of oxygen. To prevent this, the necessary bicarbonate ions are partly replaced by carbon dioxide from active tissue. And as we will see later in this chapter, bicarbonate ions are also replaced by the kidneys.

Transport of carbon dioxide

Oxygenated red blood cells are carried by the bloodstream to the capillaries of active tissues. Figure 13.9 shows how oxygen is delivered to the tissues and a load of carbon dioxide is picked up by the deoxyhemoglobin of the red blood cells. We can see from the reactions in the figure that

1. Tissue cells are using oxygen to produce acids (protons) and carbon dioxide from carbon compounds. These substances are toxic to cells and must be removed. The concentration of protons and carbon dioxide is higher in the tissues than in the red blood cells, so protons and carbon dioxide diffuse into the red blood cells.

2. The high concentration of protons promotes addition of protons to oxyhemoglobin. Recall, however, that protonated hemoglobin tends not to retain its bound oxygen. Therefore, oxygen is released.

3. The released oxygen diffuses through the red blood cell membrane and the capillary wall, and oxygen is delivered to the tissue cells.

4. Protonated hemoglobin can react with carbon dioxide to form **carbaminohemoglobin** (abbreviated HHb-CO$_2$), a form of hemoglobin to which carbon dioxide is attached, and now does so.

5. The rest of the carbon dioxide that enters the red blood cell is converted to carbonic acid by the enzyme carbonic anhydrase. This reaction is the

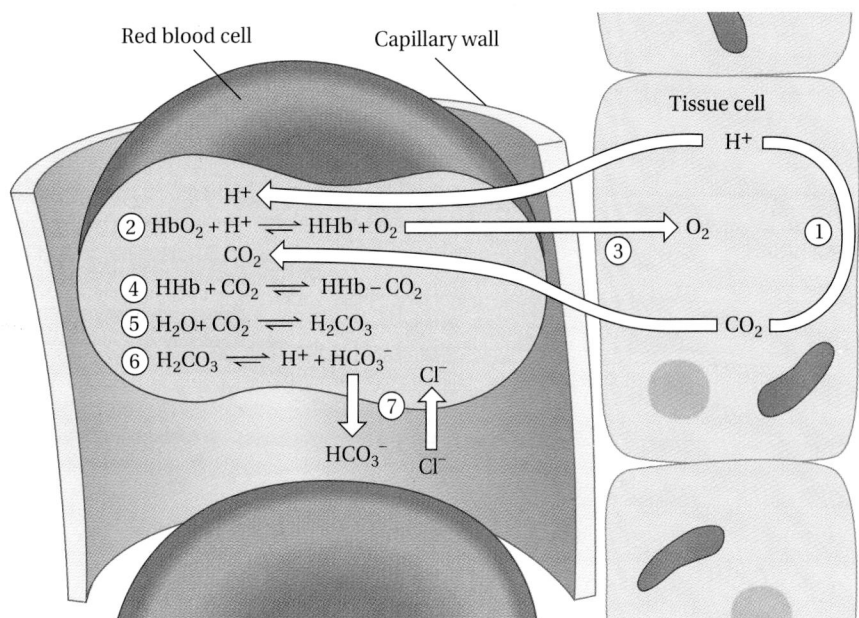

Figure 13.9
In seven steps, red blood cells exchange oxygen for carbon dioxide at the tissues.

reverse of the one carried out by carbonic anhydrase in the lungs. It is the high concentration of carbon dioxide that drives the reaction in the direction of formation of carbonic acid.

6. Some of the carbonic acid dissociates into protons and bicarbonate ions. *The released protons can be sponged up by oxyhemoglobin as in step 2 and carried to the lungs for neutralization in a step called the* **isohydric shift.**

7. Some bicarbonate ions leave the red blood cell and enter the plasma to serve as blood buffers. *The bicarbonate ions are replaced by chloride ions in a process called the* **chloride shift.** The chloride shift maintains a charge balance so that osmotic pressure relationships between the electrolytes in the blood cells and the plasma are not upset.

About 25% of the total carbon dioxide output of cells leaves tissues as carbaminohemoglobin. Some 70% is converted to bicarbonate ions. The remaining 5% is transported to the lungs as carbon dioxide dissolved in the plasma.

PRACTICE EXERCISE 13.2

Write the equation for the breakdown of carbonic acid in the lungs, and name the enzyme responsible for catalyzing the process. How would the pH of the blood be affected if this process could not occur?

PRACTICE EXERCISE 13.3

In hypochloremia, a deficiency of chloride ions in the blood caused by vomiting or upper intestinal blockage, bicarbonate ions replace the missing chloride ions. How would you expect hypochloremia to affect blood pH?

13.6 The urinary system

AIM: To state two important functions of the kidneys.

Focus

The urinary system cleanses the blood.

As we have just seen, carbon dioxide molecules produced by body cells are expelled from the blood through the lungs. A selective filtration process in the urinary system removes other waste products of cells from the blood. For example, ammonia is a by-product of amino acid breakdown in the cell. Ammonia is toxic to human cells in high concentrations and is converted to urea in the liver. The urea enters the bloodstream, circulates until it is removed by the kidneys, and is then excreted in the urine. Apart from acting as a selective filter of wastes in the bloodstream, the kidneys also play an important role in maintaining the pH of the blood by eliminating protons and replenishing bicarbonate ions lost in expiration.

Parts of the urinary system

A pair of kidneys, two ureters, a bladder, and a urethra constitute the *urinary system*, illustrated in Figure 13.10(a). Each kidney contains about a million *nephrons—narrow tubes twisted into a shape rather like the letter U* (Fig. 13.10b). Each nephron has its own miniature circulatory system. With every heartbeat, blood is forced from the renal artery into a progressively smaller system of capillaries or arterioles that enter the *renal capsule* of each nephron. Inside the capsule, the arteriole branches into even smaller vessels that are gathered into a tuft, the *glomerulus.* The vessels of the glomerulus enlarge again to form the outgoing arteriole that branches and intertwines with the nephron. The large number of very small blood vessels and their close association with the nephrons mean that all the blood that passes into the kidney can be cleansed.

Figure 13.10
(a) The human urinary system.
(b) A kidney nephron and its circulatory system.

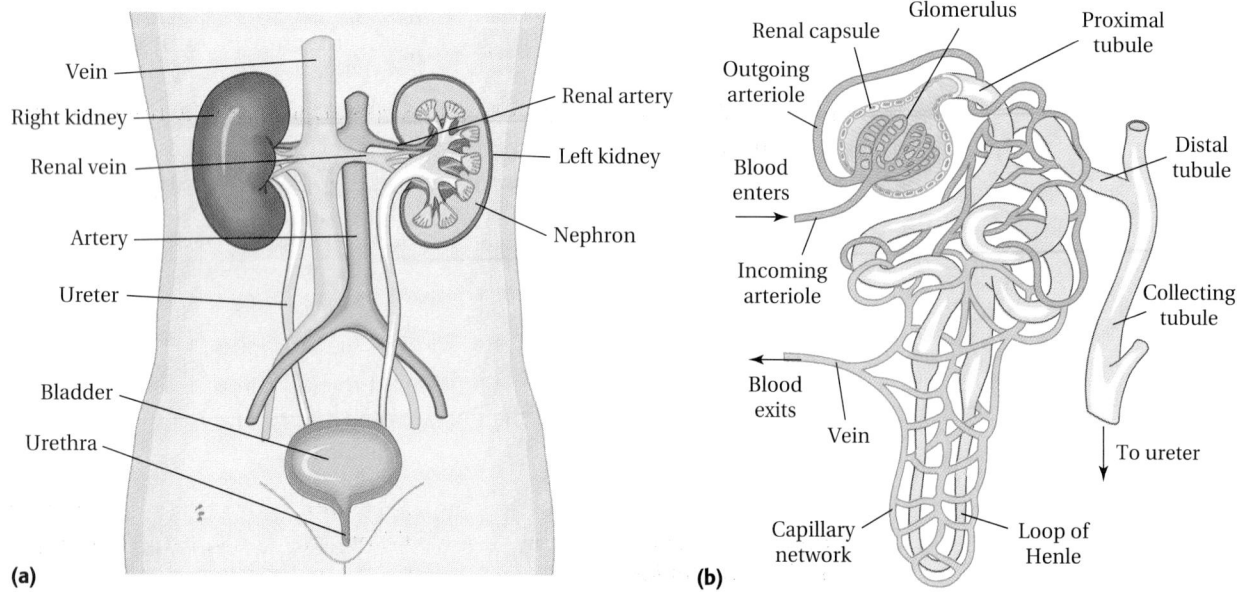

(a)

(b)

Filtration and reabsorption

The blood that enters the glomerulus is under considerable pressure. Since the walls of the capillaries of the glomerulus are very thin, water containing almost all the salts, sugars, amino acids, and other small molecules that are in the blood plasma is forced through small pores in the capillary walls. Cells and large molecules, such as proteins, do not pass through. The walls of the renal capsule are also porous, and the fluid passing through the capillaries—the *glomular filtrate*—enters the nephron at the renal capsule. The filtration of the blood by the renal capsule is rather indiscriminate, so a process of selective reabsorption must now begin.

Water and other substances needed by the body are reabsorbed into the capillaries that surround the nephron. Wastes are not reabsorbed to any great extent. As the filtered fluid passes into the *proximal tubule* of the nephron, about 80% of the water originally removed is returned to the bloodstream. All but a trace of glucose is reabsorbed in normal individuals. In untreated diabetes, where there is an excessively high concentration of blood glucose, some of the excess glucose passes into the urine. When glucose is found in the urine, the renal threshold of glucose is said to have been exceeded. In a normal adult, the renal threshold for glucose is usually exceeded when the blood glucose level rises above 140 to 170 mg of glucose per 100 mL of blood. Salt levels of blood are also controlled by the kidneys. About 75 g of sodium chloride is reabsorbed by the kidneys every day, but some is lost in the urine. About 500 g of sodium bicarbonate enters the kidneys daily, but only a few grams are excreted.

13.7 Acid-base balance

AIMS: *To relate the bicarbonate ion-producing capabilities of the kidneys to the maintenance of the acid-base balance in the blood. To describe how the pH of the urine is controlled.*

Focus

The respiratory and urinary systems help maintain the body's acid-base balance.

The control of acidity in the blood is carried out jointly by the respiratory and urinary systems. To begin this discussion, recall what happens to protons in the circulatory system. In the blood, the most important buffer system is HCO_3^-/H_2CO_3. The carbonic acid was formed by the combination of bicarbonate ions in the blood with protons expelled by active tissues.

$$H^+(aq) + HCO_3^-(aq) \longrightarrow H_2CO_3(aq)$$

We have seen that when blood arrives at the lungs, carbonic acid is degraded to water and carbon dioxide by the enzyme carbonic anhydrase.

$$H_2CO_3 \underset{\longleftarrow}{\overset{\text{Carbonic anhydrase}}{\longrightarrow}} H_2O + CO_2$$

In other words, bicarbonate ions act as "proton sponges" that protect the blood against changes in pH that would be caused by production of protons if no buffer ions were present. The water produced by the decomposi-

One aspirin (acetylsalicylic acid) tablet daily may be recommended for people with a history or risk of heart attack. Aspirin interferes with the aggregation of platelets in the blood and prevents heart attacks by inhibiting thrombosis. Could aspirin affect the blood's acid-base balance?

tion of carbonic acid is expired as vapor. The carbon dioxide formed by the breakdown of carbonic acid is also expired. For every proton produced by tissues and trapped by a bicarbonate ion, a bicarbonate ion is lost as carbon dioxide. A continuation of this process without replacement of bicarbonate ions would soon exhaust the blood's most important buffer system. This is where the kidneys make their contribution.

The kidneys are the source of new bicarbonate ions. Carbon dioxide being transported away from the tissues by the blood enters the kidney nephrons as part of the filtrate. The carbon dioxide in the fluid of the *distal tubule* enters the cells of the distal tubule walls. Figure 13.11 traces what happens to the carbon dioxide in the distal tubule cells.

1. Carbon dioxide enters the cells of the distal tubule wall.
2. Carbonic anhydrase promotes the formation of carbonic acid.
3. Carbonic acid dissociates to protons and bicarbonate ions.
4. An exchange of protons and sodium ions now occurs. For every sodium ion pumped into the tubule wall cells, a proton is pumped into the developing urine.
5. The sodium ions and bicarbonate ions enter the bloodstream through capillaries near the distal tubule.

The process that has been described results in sodium bicarbonate being added to the bloodstream and protons being added to the developing urine. The protons that enter the developing urine would make the urine very acidic if they did not encounter buffer ions in the filtrate. However, HPO_4^{2-} ions are present in the developing urine. These ions act as proton sponges.

$$H^+(aq) + HPO_4^{2-}(aq) \longrightarrow H_2PO_4^-(aq)$$

The presence of the phosphate buffer system $HPO_4^{2-}/H_2PO_4^-$ usually keeps the urine from going below a pH of 6, but too great an excess of protons can overload the system. In severe acidosis, the pH of urine may drop as low as 4.5.

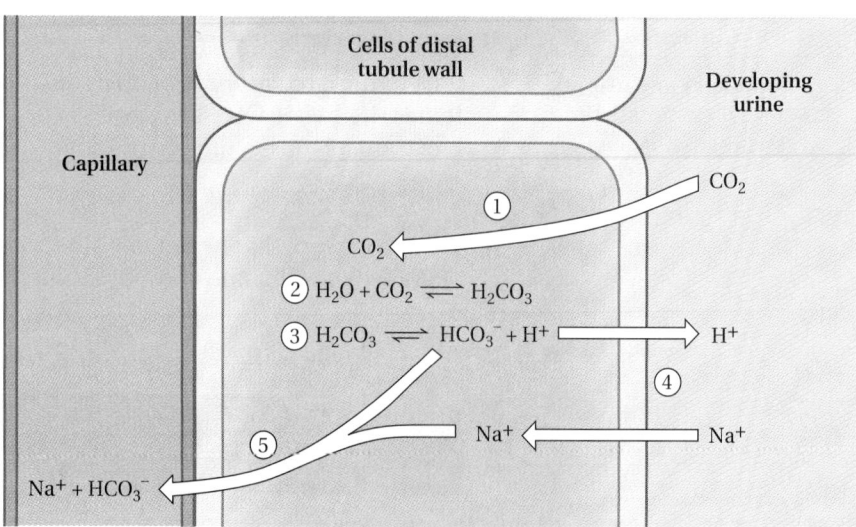

Figure 13.11
Bicarbonate ions are restored to the blood by the kidneys in five steps.

13.8 Acidosis and alkalosis

AIM: To explain alkalosis and acidosis of both of the respiratory and the metabolic types using Le Châtelier's principle.

Focus

Acidosis and alkalosis stem from a variety of causes.

As you may recall from Section 13.2, the ratio of the concentration of bicarbonate ions to the concentration of carbonic acid in blood at the normal pH of 7.4 is 20:1. The pH of blood does not, however, depend on the concentrations of these two species. Rather, as Table 13.1 shows, blood pH depends on the HCO_3^-/H_2CO_3 concentration ratio. As the ratio increases, the blood pH increases (blood becomes more basic); as the ratio decreases, the blood pH decreases. Notice that the HCO_3^-/H_2CO_3 ratio can vary only over a tenfold range—from 50:1 to 5:1—for human life to exist.

The HCO_3^-/H_2CO_3 ratio, and therefore the blood pH, may increase in two ways: by an increase in the concentration of HCO_3^- or by a decrease in the concentration of H_2CO_3. Similarly, the ratio, and therefore the blood pH, may be lowered by a decrease in the concentration of HCO_3^- or by an increase in the concentration of H_2CO_3. When the respiratory and urinary systems work together to maintain the blood at pH 7.4, they do so primarily by maintaining the HCO_3^-/H_2CO_3 ratio at 20:1. Events that tend to make the ratio larger than 20:1 can lead to alkalosis; events that make the ratio less can lead to acidosis. Alkalosis and acidosis occur only when the body's control systems cannot respond fast enough, or cannot respond at all, to maintain the 20:1 ratio.

Respiratory acidosis *and* **respiratory alkalosis** *are acid-base imbalances that result from abnormal breathing.* **Metabolic acidosis** *and* **metabolic alkalosis** *are acid-base imbalances that result from causes other than abnormal breathing.* In all types of acid-base imbalances, the major system of concern is the carbon dioxide–carbonic acid–bicarbonate equilibrium.

$$H_2O + CO_2 \rightleftharpoons H_2CO_3 \rightleftharpoons H^+ + HCO_3^-$$

Table 13.1 Relationship of Blood pH to the HCO_3^-/H_2CO_3 Concentration Ratio

HCO_3^-/H_2CO_3	pH	Remarks
50:1	7.8	highest pH compatible with life
40:1	7.7	
32:1	7.6	
25:1	7.5	slight alkalosis
20:1	7.4	pH of normal blood
16:1	7.3	slight acidosis
12.5:1	7.2	
10:1	7.1	
8:1	7.0	
6.25:1	6.9	
5:1	6.8	lowest pH compatible with life

Respiratory alkalosis

The most common way to acquire respiratory alkalosis is by **hyperventilation**—*breathing too deeply and too rapidly.* Sometimes respiratory alkalosis is called *hyperventilation alkalosis.* Hysteria, anxiety, or heavy physical exercise (such as giving birth) may result in hyperventilation, in which too much carbon dioxide is "blown off" by exhalation. The partial pressure of carbon dioxide in the lungs becomes lower than it is in the tissues. More than the normal amount of carbon dioxide therefore diffuses from the blood into the lungs. This disturbs the blood's normal carbon dioxide–carbonic acid–bicarbonate equilibrium. According to Le Châtelier's principle, the system shifts to the left—in favor of reactants—as it attempts to restore the equilibrium.

$$H_2O + CO_2 \rightleftharpoons H_2CO_3 \rightleftharpoons H^+ + HCO_3^-$$

Loss of CO_2

Direction of shift
to restore equilibrium

The result of this shift is a major loss of carbonic acid. The HCO_3^-/H_2CO_3 ratio rapidly increases to greater than 20:1, and the blood pH rises to as high as 7.7 within a few minutes. There is also a corresponding loss of bicarbonate ions, but this is not great enough to maintain the HCO_3^-/H_2CO_3 ratio at 20:1.

To understand why a similar loss of both carbonic acid and bicarbonate ion causes the HCO_3^-/H_2CO_3 ratio to *increase* instead of staying the same, we can do a simple calculation. Normal venous blood of pH 7.4 has a bicarbonate ion concentration of about 27 mEq/L and a carbonic acid concentration of about 1.35 mEq/L. The ratio is 27.0 mEq ÷ 1.35 mEq, or 20:1. If the carbonic acid concentration is reduced by half, to 0.68 mEq, the *ratio* becomes 27.0 mEq ÷ 0.68 mEq, or 40:1, corresponding to a pH of 7.7 (see Table 13.1). On the other hand, if the bicarbonate ion concentration is reduced by 0.68 mEq to 26.32 mEq, the HCO_3^-/H_2CO_3 ratio is 26.32 mEq ÷ 1.35 mEq, or 19.5:1, and the pH change is negligible. If we assume that the loss of carbonic acid equals the loss of bicarbonate ion, 0.68 mEq, we get the ratio 26.32 ÷ 0.68, or 38.7:1, corresponding to a pH very near 7.7.

Respiratory acidosis

Respiratory acidosis is sometimes the result of **hypoventilation**—*too-shallow breathing.* Lung diseases such as emphysema, an object lodged in the windpipe, or other causes of impaired breathing may result in hypoventilation. Anesthetists need to be particularly aware of respiratory acidosis because most inhalation anesthetics depress respiration. In respiratory acidosis, not enough carbon dioxide is exhaled, and its partial pressure in the lungs therefore increases. The high partial pressure of carbon dioxide in the lungs is higher than it is in the blood. Carbon dioxide therefore diffuses into the blood, pushing the carbon dioxide–carbonic acid–bicarbonate equilibrium system to the right to restore the equilibrium.

$$H_2O + CO_2 \rightleftharpoons H_2CO_3 \rightleftharpoons H^+ + HCO_3^-$$

Increase of CO_2

Direction of shift
to restore equilibrium

As a result of the shift, the concentration of carbonic acid in the blood increases. The ratio rapidly decreases to less than 20:1, and respiratory acidosis ensues.

Metabolic acidosis

Body tissues produce protons as waste products of various body processes. These processes are collectively called *metabolism*. In metabolic acidosis, the diffusion of protons from the tissues into the bloodstream pushes the carbon dioxide–carbonic acid–bicarbonate system to the left to restore the equilibrium.

$$H_2O + CO_2 \rightleftharpoons H_2CO_3 \rightleftharpoons H^+ + HCO_3^-$$

Increase of H^+

Direction of shift
to restore equilibrium

The shift decreases the concentration of bicarbonate ions but increases the concentration of carbonic acid. Since such changes affect the carbonic acid concentration much more than the bicarbonate ion concentration, the HCO_3^-/H_2CO_3 ratio decreases, and so does the blood pH. In the Follow-up to the Case in Point on the following page, we will learn more about how metabolic acidosis affects people.

Metabolic alkalosis

Metabolic alkalosis is less common than the acid-base imbalances we have seen so far. It may be caused by several conditions—including a loss of anions (primarily chloride ions), a deficiency of potassium ions, or the administration of alkaline salts in drug therapy. Excessive use of sodium bicarbonate, for example, a common home remedy for gastric acidity, increases the concentration of bicarbonate ions in the blood.

In response to a small influx of bicarbonate ions, the carbon dioxide–carbonic acid–bicarbonate equilibrium shifts to the left, decreasing the hydrogen ion concentration.

$$H_2O + CO_2 \rightleftharpoons H_2CO_3 \rightleftharpoons H^+ + HCO_3^-$$

Influx of HCO_3^-

Direction of shift
to restore equilibrium

We might expect the pH to increase, but another result of the shift is a corresponding increase in the carbonic acid concentration. The

FOLLOW-UP TO THE CASE IN POINT: A diabetic imbalance

Metabolic acidosis is often a serious problem in uncontrolled diabetes, and it also occurs on a temporary basis after heavy exercise. Both conditions result in large influxes of protons from active tissues into the bloodstream. Larry, the teacher in the Case in Point earlier in this chapter, failed to keep his schedule of insulin injections. The acid his body produced as a result of a deficiency of insulin caused the pH of Larry's blood to drop to life-threatening levels. Larry's doctor suspected that diabetes-induced metabolic acidosis might be one of Larry's immediate problems. When the acidosis was confirmed by lab tests (see figure), Larry was given an intravenous drip of a sodium bicarbonate solution to restore his blood pH to normal. The sodium bicarbonate neutralized the excess acids in his blood, and his blood was literally titrated back to normal. With a proper schedule of insulin injections, Larry was able to leave the hospital in a few days.

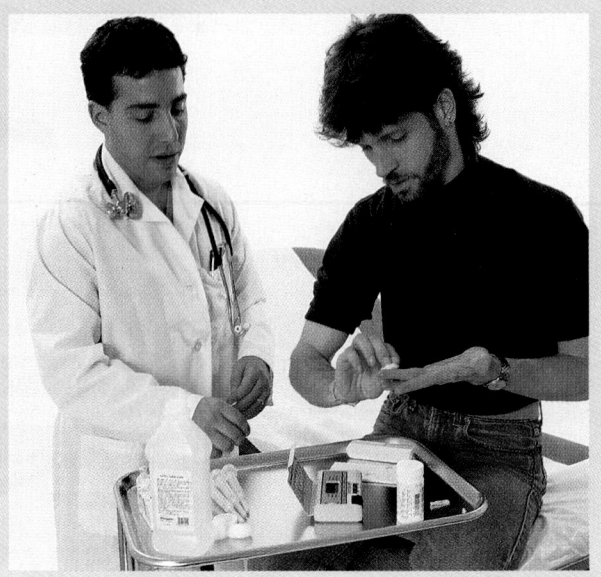

Testing for diabetes.

HCO_3^-/H_2CO_3 ratio does not change very much, even though carbonic acid and bicarbonate ions are at higher than normal concentrations. When the bicarbonate ion influx is large, however, the HCO_3^-/H_2CO_3 ratio may become greater than 20:1, and measurable alkalosis is observed.

PRACTICE EXERCISE 13.4

A first aid treatment for respiratory alkalosis is having the victim breathe into a paper bag. Using the carbon dioxide–carbonic acid–bicarbonate equilibrium and Le Châtelier's principle, explain why this treatment is effective in restoring the pH of blood to normal.

13.9 Water and salt balance

AIM: To describe the action of vasopressin and aldosterone in controlling body water and salt levels.

Focus

The endocrine system controls body water volume and salt concentration.

The volume of blood passed through the kidneys is phenomenal. In an average adult, about 100 L of blood passes through the kidneys every day. Of the water in this 100 L, only 1 or 2 L is excreted as urine. Water reabsorption and excretion by the kidneys are carefully controlled to maintain a constant volume of water in the body.

By the time the glomular filtrate reaches the loop of Henle in the kidney (Fig. 13.12), most of the solutes that are not wastes have been reabsorbed into the bloodstream. Most of the water also has been reabsorbed. However, if the dietary intake of water has been high—during a meal, for instance—a large volume of dilute urine is excreted soon after. If the intake of water has been low, a lower volume of more concentrated urine is excreted.

Vasopressin, a peptide hormone of the pituitary gland, normally controls the volume of urine. Vasopressin, also called *antidiuretic hormone* (ADH), exerts its action by affecting the permeability of the distal tubule and collecting duct to water. In the absence of vasopressin, these parts of the nephron are impermeable to water (see Fig. 13.12). Water cannot pass from the inside of the tubule or the duct to the surrounding tissues. The result is a large volume of light yellow, dilute urine. Patients with the rare disease diabetes insipidus do not produce vasopressin; their urine is extremely voluminous and dilute. In the presence of vasopressin, the distal tubule and collecting duct membranes become permeable to water. Water therefore leaves these parts of the nephron and is drawn into the surrounding tissues by osmosis, since the fluid of the tissues is saltier than the developing urine. The reabsorption of water by the tissues produces a low volume of dark yellow, concentrated urine.

The action of vasopressin is usually sufficient to maintain the proper level of body water. Sometimes, however, the water level dips dangerously low because of insufficient water intake or diarrhea. Heavy exercise or a fever also may cause dehydration due to excessive sweating. Under such circumstances, the adrenal gland secretes the steroid hormone *aldosterone.*

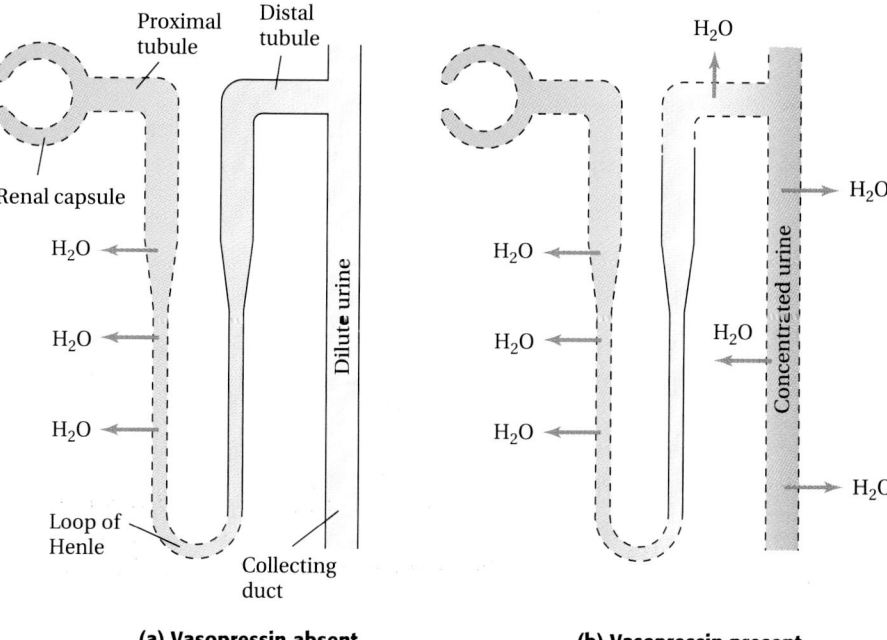

Figure 13.12
Production of dilute and concentrated urine by the kidneys. Dashed lines indicate sections of the nephron that are permeable to water in the (a) absence and (b) presence of vasopressin.

Aldosterone activates cellular pumps that transport sodium ions from the nephrons to the interstitial fluid. Chloride ions follow the sodium ions to maintain electrical neutrality, and water moves to this more concentrated salt solution. The action of aldosterone conserves both salt and water in the body. When the water level of the body returns to normal, release of aldosterone drops.

Body water can be lost by sweating. Evaporation of sweat helps to cool the body. An active person not engaged in heavy exercise loses 600 to 700 mL of water per day through perspiration. Heavy exercise, especially in a hot climate, causes more profuse sweating—up to 4 L a day. Sweat is almost pure water, but it also contains small, varying concentrations of sodium chloride. Excessive loss of water through sweating can cause dehydration unless water intake balances water loss. Excessive loss of sodium ions by sweating causes muscle cramps.

SUMMARY

The functions performed by complex living things complement one another, contributing to health and well-being. Healthy cells, tissues, and organs work in harmony to maintain homeostasis—balancing the body's needs with respect to temperature, pH of fluids, energy production, and the supply of oxygen, nutrients, water, and salts.

Body cells live in a fluid environment. The major body fluids are intracellular fluid (the fluid inside of cells), interstitial fluid (the fluid outside of cells), and plasma (the fluid portion of the blood). Cells do most of their chemical work in the intracellular fluid. The interstitial fluid provides the proper amount of salt and water for cells. Among other functions, blood carries oxygen and nutrients to cells and takes away wastes, mostly carbon dioxide and protons. Antibodies are large blood proteins produced by the body in response to a foreign substance or organism (antigen). Interferons help the body ward off viral infections.

The respiratory system and kidneys cleanse the blood of wastes and control blood pH at a value near 7.40 to avoid alkalosis or acidosis. Excess protons are neutralized by combining with bicarbonate ions (HCO_3^-) to form carbonic acid (H_2CO_3). The enzyme carbonic anhydrase breaks down carbonic acid to carbon dioxide (which is exhaled) and to water, a neutral nontoxic waste easily removed by the kidneys. The kidneys also help restore bicarbonate ions to the bloodstream by exchanging one proton of carbonic acid in the blood for one sodium ion in developing urine. If the supply of bicarbonate ions runs low either as a result of excess proton production or by retention of carbonic acid through shallow breathing, acidosis ensues. Hyperventilation causes respiratory alkalosis; too much carbon dioxide is lost through heavy breathing, the concentration of carbonic acid decreases, and the blood pH rises.

Vasopressin, a peptide hormone, controls urine concentration by affecting the permeability of parts of the kidney nephrons to water. When the hormone is present, the nephron is permeable to water, water flows from the nephron back to the saltier interstitial fluid, and the urine is scant and concentrated. When the hormone is not present, the nephron is impermeable to water, water cannot be reabsorbed, and the urine is voluminous and dilute. Aldosterone controls the body's salt retention–water volume balance. Aldosterone activates pumps that carry sodium ions from the nephron to the interstitial fluid. Chloride ions follow sodium ions, and water follows the sodium chloride.

KEY TERMS

Acid-base balance (13.4)
Acidosis (13.4)
Alkalosis (13.4)
Antibody (13.3)
Antigen (13.3)
Bence-Jones proteins (13.3)
Carbaminohemoglobin
 (13.5)

Chloride shift (13.5)
Complement system (13.3)
Erythrocyte (13.2)
Formed element (13.2)
Homeostasis (13.1)
Hyperventilation (13.8)
Hypoventilation (13.8)
Immunoglobulins (13.3)

Interferon (13.3)
Interstitial fluid (13.1)
Intracellular fluid (13.1)
Isohydric shift (13.5)
Leukocyte (13.2)
Lymph (13.2)
Metabolic acidosis (13.8)
Metabolic alkalosis (13.8)

Monoclonal antibody (13.3)
Plasma (13.2)
Respiratory acidosis (13.8)
Respiratory alkalosis (13.8)
Serum (13.2)
Thrombocyte (13.2)

EXERCISES

Body Water (Section 13.1)

13.5 Describe the distribution of body water in an average adult.

13.6 Define homeostasis.

13.7 Compare the electrolyte composition of intracellular fluid and interstitial fluid.

13.8 Compare the electrolyte composition of interstitial fluid and blood plasma.

Blood, Antibodies, and Blood Buffers (Sections 13.2, 13.3, 13.4)

13.9 Name five functions of the blood.

13.10 Distinguish among whole blood, plasma, and serum.

13.11 Name the formed elements of the blood, and describe their functions in the body.

13.12 List the plasma proteins, and describe their roles.

13.13 Describe the relationship between an antigen and an antibody.

13.14 Where are antibodies produced?

13.15 Describe a use of monoclonal antibodies.

13.16 What is the function of interferons?

13.17 What is the normal pH of the blood?

13.18 Distinguish between acidosis and alkalosis.

13.19 Name the three buffer systems of the blood.

13.20 Explain how the HCO_3^-/H_2CO_3 buffer helps to maintain a constant pH when small amounts of acid or base are added to the blood.

Oxygen and Carbon Dioxide Transport (Section 13.5)

13.21 How is oxygen transported from the lungs to active tissue?

13.22 Explain how partial-pressure differences favor uptake of oxygen by red blood cells from the lungs and release of oxygen from HbO_2 to active tissues.

13.23 Predict what would happen to the oxygen level in the tissue cells of a person who inhaled pure oxygen.

13.24 How does the pH of the blood affect the formation of oxyhemoglobin?

13.25 Describe the ways in which CO_2 is transported from the cells to the lungs.

13.26 What is the function of carbonic anhydrase?

13.27 What happens in the chloride shift?

13.28 Why does the inhalation of pure CO_2 cause asphyxiation?

Urinary System and Acid-Base Balance (Sections 13.6, 13.7, 13.8)

13.29 What essential functions do the kidneys perform?

13.30 Amino acids are broken down to ammonia in the cell. How is ammonia removed from the body?

13.31 Name three substances that are reabsorbed by the kidneys.

13.32 What has happened if glucose is present in the urine?

13.33 Explain how the kidneys supply bicarbonate ions to the bloodstream.

13.34 Why is normal urine not very acidic?

13.35 What is the cause of respiratory alkalosis?

13.36 There is a change in the blood pH when a person hyperventilates. What happens to the pH and what leads to the change?

Water and Salt Balance (Section 13.9)

13.37 What hormones control loss and retention of water by the body?

13.38 Describe the action of antidiuretic hormone (ADH).

13.39 Dehydration or low water intake causes the adrenal gland to secrete a hormone. Name this hormone, and describe its action.

13.40 What are the symptoms and cause of diabetes insipidus?

Additional Exercises

13.41 Would you expect the blood's oxygen-carrying capacity to increase or decrease in acidosis? Explain your answer.

13.42 Discuss the oxygen-carrying efficiency of hemoglobin when the pH of the blood falls below 7.35.

13.43 Explain how oxyhemoglobin releases oxygen to active tissue cells.

13.44 In what ways are the following conditions the same and in what ways are they different?
(a) respiratory acidosis and metabolic acidosis
(b) respiratory alkalosis and metabolic alkalosis

13.45 Clearly distinguish between *hypoventilation* and *hyperventilation*. Explain how each condition disturbs the HCO_3^-/H_2CO_3 ratio in the blood.

13.46 What waste products are eliminated in the urine?

13.47 How do the kidneys help to reduce acidosis?

13.48 The circulatory and urinary systems work together to maintain the fluid environment of the body. Two important reversible reactions that take place in both these systems are

Reaction A $\quad H_2CO_3(aq) \rightleftharpoons H_2O(l) + CO_2(g)$
Reaction B $\quad H_2CO_3(aq) \rightleftharpoons H^+(aq) + HCO_3^-(aq)$

For each of the three locations in the body listed below, write these two equations in the correct sequence and with the favored product on the right.
(a) at the juncture of capillary and tissue cell
(b) at the juncture of capillary and lung alveolus
(c) in the cells of the distal tubules of the kidneys

SELF-TEST (REVIEW)

True/False

1. Monoclonal antibodies are homogeneous.
2. The hormone aldosterone helps control the pH of the blood.
3. Most of the water found in the body is in the blood.
4. Removing the formed elements from unclotted blood leaves blood plasma.
5. Lymphocytes are responsible for antibody production in the body.
6. Respiratory acidosis is usually associated with high carbon dioxide concentration in the blood.
7. A high concentration of H^+ in red blood cells causes HbO_2 to dissociate.
8. Most of the carbon dioxide produced by the cells of the body leaves the tissues as $HHb\text{-}CO_2$.
9. The pH of the urine in the body is controlled by the $HPO_4^{2-}/H_2PO_4^-$ buffer system.
10. Perspiration is the major process by which sodium chloride is lost from the body.

Multiple Choice

11. The hormone that controls the volume of urine in the body under normal conditions
(a) works by causing changes in the osmotic pressure in the nephrons.
(b) is aldosterone.
(c) is vasopressin.
(d) both (a) and (b) are correct.

12. The enzyme carbonic anhydrase may
(a) speed the breakdown of carbonic acid to water and carbon dioxide.
(b) be found in the red blood cells.
(c) speed the formation of carbonic acid from carbon dioxide and water.
(d) all of the above are true.

13. The process of selective reabsorption by the kidneys
(a) applies equally to all substances in the blood.
(b) should let only a trace of glucose pass into the urine of a normal person.
(c) leads to a net increase of the concentration of CO_3^{2-} in the blood.
(d) allows more than half the water in the blood to pass into the urine.

14. The kidneys play an important part in maintaining the pH of the blood by
(a) eliminating protons in the urine.
(b) producing OH^- ions.
(c) producing bicarbonate ions.
(d) More than one are correct.

15. Hyperventilation can lead to
(a) too much carbon dioxide being retained in the blood.
(b) respiratory alkalosis.
(c) respiratory acidosis.
(d) a high blood level of carbonic acid.

16. When conditions in the body are such that body cells can operate at peak efficiency, then
(a) equilibrium has been achieved.
(b) interactions between the body's systems are minimal.
(c) homeostasis has been achieved.
(d) Both (a) and (b) are correct.

17. In the body, the majority of oxygen is carried to the cells as
(a) dissolved O_2. (b) oxyhemoglobin.
(c) H^+-O_2. (d) bicarbonate ions.

18. The kidneys are important in all the following functions *except*
(a) conversion of ammonia to urea.
(b) selective filtration of waste products from the blood.
(c) control of body water volume.
(d) regulation of blood pH.

19. The most important buffering system of the blood is the
(a) HCO_3^-/H_2CO_3 system.
(b) H_3PO_4/$H_2PO_4^-$ system.
(c) HPO_4^{2-}/$H_2PO_4^-$ system.
(d) SO_4^{2-}/HSO_4^- system.

20. At the junction between tissue cell and capillary, both CO_2 and H^+ move into the red blood cell. Then
(a) some of the CO_2 reacts with HHb to form carbaminohemoglobin (HHb-CO_2).
(b) all the carbon dioxide reacts with water to form carbonic acid.
(c) the H^+ ions react with Cl^- to form HCl.
(d) None of the above is true.

21. Which of the following is *not* a function of the blood?
(a) carrying hormones (b) maintenance of pH
(c) infection fighting (d) none of the above

22. Blood serum contains
(a) hemoglobin. (b) albumin.
(c) leukocytes. (d) clotting agents.

23. In comparing the electrolyte composition of various body fluids, we find that sodium and chloride ions predominate in interstitial fluid and that in intracellular fluid the major inorganic electrolytes are
(a) Na^+ and CO_3^{2-}. (b) K^+ and HPO_4^{2-}.
(c) Na^+ and PO_4^{3-}. (d) K^+ and SO_4^{2-}.

24. Antibodies are made of which of the following plasma proteins?
(a) lipoprotein (b) hemoglobin
(c) albumin (d) globulin

14 Energy and Life

Sources and Uses of Energy in Living Organisms

Life requires energy.

Energy and *life*—the two terms are practically synonymous. We may speak of someone who has "boundless energy" or who is "glowing with energy." And we are not far wrong, because all living organisms need ample energy to maintain their vital functions. This chapter discusses how the cells of living creatures produce and use energy. Sometimes we can better understand and appreciate what we have by seeing someone who is lacking our possessions or qualities. Our Case in Point concerns a person who lacked energy.

CASE IN POINT: A mysterious fatigue

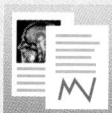

Emily had been plagued by fatigue since birth. Numerous tests eliminated the possibility of anemia or other causes of fatigue. Puzzled, her doctor referred her to a nearby medical research center for further evaluation. After extensive testing, the medical researchers pinpointed the source of Emily's fatigue as a defect in her ability to generate the central molecule in the transmission of energy. What is this molecule, and what defect interfered with its formation? We will learn the answers to these questions in Section 14.6.

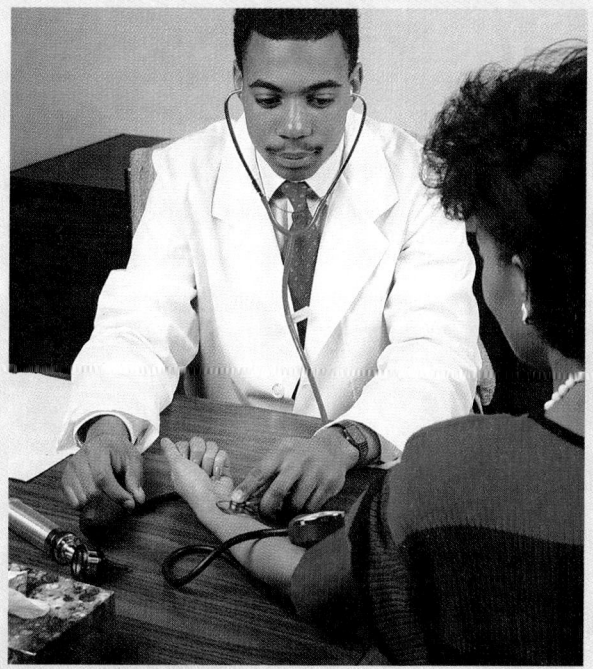

Extensive examination and testing allowed doctors to diagnose the cause of Emily's fatigue.

14.1 Metabolism

AIM: *To differentiate among metabolism, catabolism, and anabolism.*

Focus

Metabolism is all the chemical reactions that occur in living organisms.

There are three important terms that describe the chemical reactions in living organisms: *metabolism, catabolism,* and *anabolism.* **Metabolism** *is all the chemical reactions that occur in living organisms.* Virtually all metabolic reactions are catalyzed by enzymes. **Catabolism** *is a part of metabolism—the breakdown of molecules by an organism.* **Anabolism** *is another part of metabolism—the synthesis of molecules for cell growth and multiplication.* Nutrients are also converted to their storage forms by anabolic processes. Conversion of fatty acids to triglycerides for storage in fatty tissue is one example. Conversion of glucose to glycogen for storage in liver and muscle cells is another.

Anabolism and catabolism are quite distinct from each other. Cells usually employ different chemical reactions for the breakdown and synthesis of the same molecule. The reactions used to synthesize glucose, for example, are not the reverse of the reactions used to degrade it. Apart from being chemically separated, catabolic and anabolic reactions are frequently separated physically. Many important catabolic reactions occur in the mitochondria, whereas many anabolic reactions occur in the cytoplasm. The chemical and physical separation of anabolism from catabolism enables cells to regulate metabolism to make it responsive to current needs.

Metabolism
meta (Greek): beyond
ballein (Greek): to cast or throw

Catabolism
cata (Greek): down

Anabolism
ana (Greek): up

14.2 Photosynthesis

AIMS: *To write a chemical equation for photosynthesis indicating the energy-rich and energy-poor carbon compounds. To distinguish among chloroplasts, thyalkoids, and chlorophyll.*

Focus

The energy of sunlight is harnessed by plants in photosynthesis.

Photosynthesis
photos (Greek): light
synthesis (Greek): to place together

Energy production by cells involves the catabolism of carbon compounds that serve as nutrients—mainly sugars, fats, and amino acids. Oxidation reactions are generally energy-producing. Oxidation reactions that are part of cellular catabolism release the energy stored in the chemical bonds of nutrient molecules, making it available to perform the work that cells must do to stay alive.

Where do sugars, fats, and amino acids originate? Carbon dioxide in the Earth's atmosphere is the ultimate source of all carbon compounds. Carbon dioxide is an energy-poor compound because it cannot be oxidized further. Animal cells discard it as a waste product. Green plants, blue-green algae, and certain bacteria, however, conduct **photosynthesis**—*harness the energy of sunlight, convert it to useful chemical energy, and use that energy to synthesize glucose, a more reduced molecule, from carbon dioxide.* The

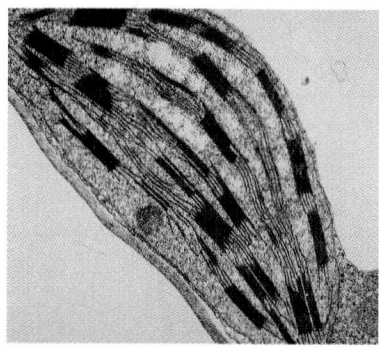

Figure 14.1
Transmission electron micrograph of a chloroplast.

processes that occur in photosynthesis are often summarized by a single equation.

$$6CO_2 \ + \ 6H_2O \ \xrightarrow{\text{Light energy}} \ C_6H_{12}O_6 \ + \ 6O_2$$

| Carbon dioxide (more oxidized) | Water | Glucose (more reduced) | Oxygen |

Reoxidation of glucose back to carbon dioxide and water releases the energy expended in forming the chemical bonds of the glucose molecule.

$$C_6H_{12}O_6 \ + \ 6O_2 \ \longrightarrow \ 6CO_2 \ + \ 6H_2O \ + \ \text{Energy}$$

| Glucose (more reduced) | Oxygen | Carbon dioxide (more oxidized) | Water |

Glucose is an energy-rich molecule and a potential source of chemical energy for any organism that obtains it.

The equation for photosynthesis is deceptively simple. Molecules of carbon dioxide and water do not simply bang together in sunlight and pop out molecules of glucose and oxygen. To carry out photosynthesis, cells must be able to absorb light energy and convert it to chemical energy. *For this purpose, photosynthetic eukaryotic cells—those of green plants are the most familiar example—contain special organelles, the* **chloroplasts,** shown in Figure 14.1. Chloroplasts can carry out this conversion. *Located within the chloroplasts are a large number of connected disks called* **thyalkoids.** *The thyalkoids contain the molecules that constitute the* **light system** *of photosynthesis.* Responsibility for trapping light energy and converting it to useful chemical energy rests with the light system. **Chlorophyll**—*the pigment that gives green plants their color*—is a molecule of chief importance to the light system. Its molecular structure is depicted in Figure 14.2. Chlorophyll initially absorbs light energy and passes it on to other molecules of the light system. The light systems of prokaryotic cells that are capable of conducting photosynthesis—certain bacteria and the blue-green algae—are located on the interior of the cell membranes.

The importance of light to life on Earth through photosynthesis in plants cannot be overemphasized. However, researchers are beginning to find that healthy human beings respond to light in ways unimaginable only a few years ago. The cure rate of certain cancers may be improved by a treatment involving light. See A Closer Look: Phototherapies, for a discussion of some connections between light and health.

answer to a) in book is porphyrin ring but hemoglobin p. 280 refers to 4 linked pyrrole ring.

PRACTICE EXERCISE 14.1

(a) Identify the structural feature that is common to vitamin B_{12} (see Figure 12.3 on page 362), hemoglobin, and chlorophyll. (b) What metals are essential to each of these molecules?

Vt B_{12}: central metal ion = cobalt, hemoglobin: iron chlorophyll: magnesium.

Figure 14.2 (left)
The molecular structure of chlorophyll. The magnesium ion is essential for chlorophyll's light-trapping function. Notice the similarity between the structure of heme (see Sec. 9.11) and the porphyrin ring of chlorophyll.

Porphyrin ring

CH_2
‖
CH H CH_3

C C C CH_3

CH_3-C C C $C-CH_2$

$C-N$ $N-C$

HC Mg^{2+} CH

CH_3 $C=N$ $N-C$

C C C $C-CH_3$

H C C C

H $HC-C=O$

CH_2 $COOCH_3$

CH_2

$O-C=O$

$H-C-H$

$H-C$

$C-CH_3$

$H-C-H$

$H-C-H$

$H-C-H$

$H-C-CH_3$

H C H

$H-C-H$

$H-C-H$

$H-C-CH_3$

$H-C-H$

$H-C-H$

$H-C-H$

$H-C-CH_3$

$H-C-H$

H

A Closer Look

Phototherapies

You may recall that the transformation of steroids in the skin by sunlight is important in the formation of vitamin D_2, but medical specialists and researchers are just beginning to recognize many of the human body's responses to light. Several conditions, including some forms of disturbed sleep, "jet lag," and *s*easonal *a*ffective *d*isorder (SAD), appear to be related to the biological cycles of wakefulness and sleep called the *circadian rhythms.* A small portion of the hypothalamus, a region at the base of the brain, supervises our bodies' clocks, telling us when it is time to eat, sleep, and wake up. This small portion of the hypothalamus, called the *suprachiasmatic nucleus,* responds to light and darkness to affect body temperature fluctuations, hormone release, blood pressure, heart rate, and the sleep-wake cycle. Light treatment is proving useful in the treatment of several human conditions related to the circadian rhythm. More than half of all Americans over age 65 suffer from disturbed sleep, waking too early in the morning and becoming sleepy early in the evening. Experts suggest that this sleep disorder is caused by a speeding up of the body's circadian rhythms in older people. To combat this troublesome shift, researchers are exposing older patients to bright light in the early evening to delay their sleep-wake cycles. Jet lag similarly affects people who fly by jet across international time zones and shift workers, such as nurses, who rotate shifts. The body's sleep-wake cycle is set to have night during the day in the new location or on the new shift. A person responds by feeling exhausted during the day and wide awake at night until the cycle is reset. This usually happens after a few days in the new location, but researchers suggest it can be speeded up by spending the first 2 days after a trip in sunlight. Exposure to bright light is also said to exert a positive effect on SAD (see figure), the name given to a form of depression experienced by some people. SAD people may feel fine in summer but become depressed during the winter months, when the days are short and sunshine may be scarce. SAD is an emotional illness more severe

Phototherapy often relieves the distressing symptoms of SAD, seasonal affective disorder.

than the tired, run-down feeling many people have during the winter months. Preliminary results indicate that light treatment also may be effective against some nonseasonal forms of depression. Light treatments are also important in the treatment of jaundice, or yellowing, in newborns, as we will see in A Closer Look: Hyperbilirubinemia, in Chapter 16.

Recent studies indicate that synchronizing cancer treatment with the body's internal rhythms gives chemotherapy drugs a stronger effect, perhaps as much as doubling their power to fight tumors. Some chemotherapy drugs work better at night; others seem more potent when given during the day. Light is also being pressed into service against certain blood cell cancers by a technology called *photopheresis.* In photopheresis, mixtures of normal and cancerous blood cells are removed from the patient and treated with a drug such as psoralen.

Psoralen

The psoralen is selectively absorbed by the cancerous cells. The interaction of light with the psoralen cross-links the DNA in the cancerous cells, making it impossible for the cells to reproduce. The treated blood is then returned to the patient.

14.3 The energy and carbon cycle

***AIM:** To describe the energy and carbon cycle.*

Life depends on the *energy and carbon cycle* (Fig. 14.3). *In the* **energy and carbon cycle,** *the photosynthetic organisms in the Earth's forests and oceans produce glucose and use it as a source of chemical energy and to build the carbon skeletons of carbohydrates, fats, amino acids, and other biological molecules.* Animals obtain these substances by eating plants, by eating animals that eat plants, or by a combination of both.

Animals and plants in need of energy unleash the energy stored in the chemical bonds of these carbon nutrients by oxidizing them back to carbon dioxide and water. The energy is used for the work that plant and animal cells must do, and the carbon dioxide and water can be recycled in photosynthesis. Plant life on Earth could probably survive without animals. However, animal life could never survive without plants, since without photosynthesis there would be no new supply of the crucial carbon compounds needed by animals for energy production.

PRACTICE EXERCISE 14.2

Explain why our existence depends on photosynthesis.

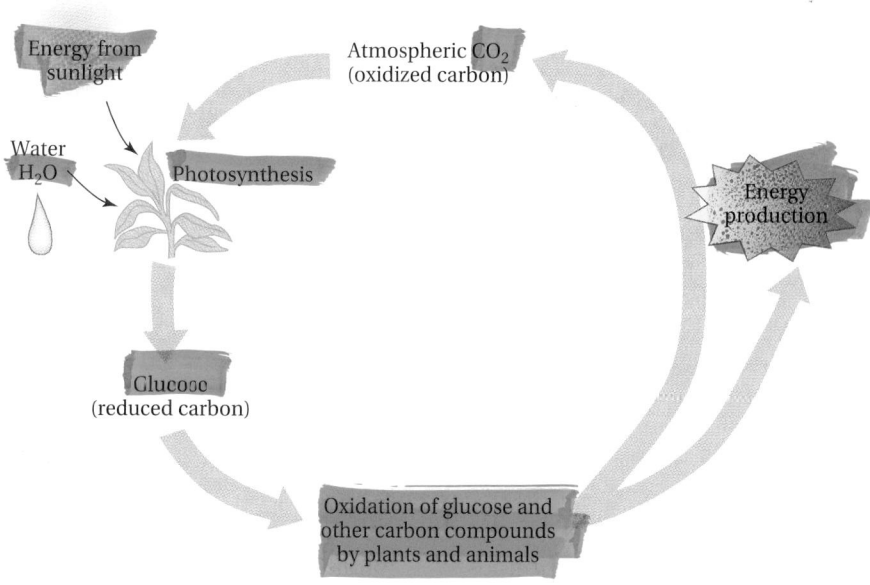

Figure 14.3
The energy and carbon cycle.

14.4 Adenosine triphosphate

AIMS: *To name the energy-transmitting molecule in nature and list its hydrolysis products. To explain why ATP could not be a very high-energy compound and still fulfill its role in cellular energetics.*

Focus

Adenosine triphosphate ties energy production to energy use.

The chemical energy released by the oxidation reactions of cellular catabolism is divisible into two parts. One part of the energy is lost as heat and is unavailable for biological use. The other part, the free energy, is available to do useful work in the cell. The free energy is transmitted to a cellular carrier molecule, which then transmits this energy to cellular processes that need it.

Adenosine triphosphate: The energy carrier

Adenosine triphosphate (ATP) *is the molecule that carries and transmits the energy needed by cells of all living things.* We could compare the function of ATP in cellular metabolism to a belt that connects a motor to a pump. The motor generates energy that is capable of operating the pump. If the motor and the pump are not connected by a belt, however, the energy produced by the motor is useless. ATP is the belt that couples the motor (catabolism) and the pump (anabolism) (Fig. 14.4). To understand how cells couple energy production and energy use, we will need to know more about the chemistry of ATP.

Structure and hydrolysis of ATP

The ATP molecule (Fig. 14.5) consists of the nucleoside adenosine covalently bonded to a triphosphate group. The triphosphate portion of the ATP molecule is an anhydride of phosphoric acid. Like other phosphoric acid anhydrides, the anhydride group of ATP is readily hydrolyzed. The

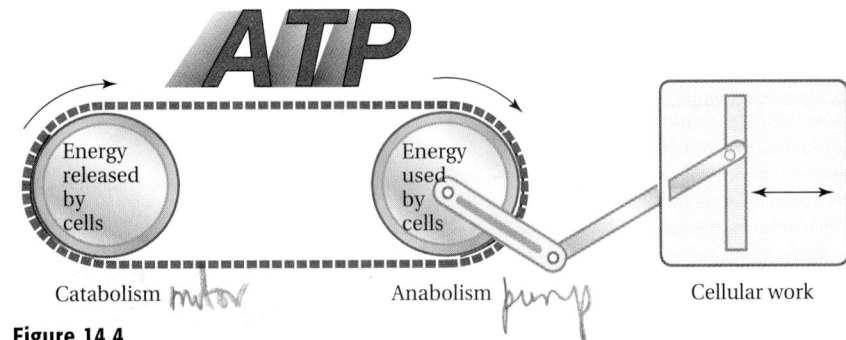

Figure 14.4
ATP is the energy carrier between the metabolic reactions that release energy and those that use energy.

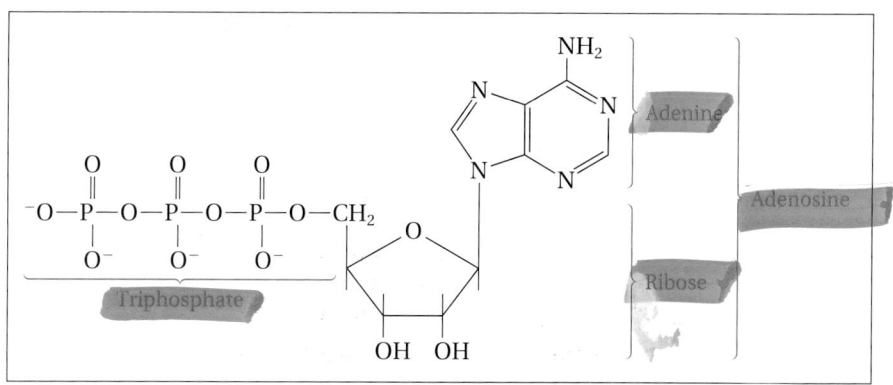

Figure 14.5
The structure of ATP.

hydrolysis products are adenosine diphosphate (ADP) and phosphate (P$_i$).

$$\text{Adenosine} - \overset{\overset{\displaystyle O}{\|}}{\underset{\underset{\displaystyle OH}{|}}{P}} - O - \overset{\overset{\displaystyle O}{\|}}{\underset{\underset{\displaystyle OH}{|}}{P}} - O - \overset{\overset{\displaystyle O}{\|}}{\underset{\underset{\displaystyle OH}{|}}{P}} - OH \; + \; H_2O \longrightarrow \text{Adenosine} - \overset{\overset{\displaystyle O}{\|}}{\underset{\underset{\displaystyle OH}{|}}{P}} - O - \overset{\overset{\displaystyle O}{\|}}{\underset{\underset{\displaystyle OH}{|}}{P}} - OH \; + \; HO - \overset{\overset{\displaystyle O}{\|}}{\underset{\underset{\displaystyle OH}{|}}{P}} - OH$$

| Adenosine triphosphate (ATP) | Water | Adenosine diphosphate (ADP) | Inorganic phosphate (P$_i$) |

Or

$$\text{ATP} + H_2O \rightarrow \text{ADP} + P_i$$

The other anhydride bond also can be broken by hydrolysis. The products are adenosine monophosphate (AMP) and pyrophosphate (PP$_i$).

$$\text{Adenosine} - \overset{\overset{\displaystyle O}{\|}}{\underset{\underset{\displaystyle OH}{|}}{P}} - O - \overset{\overset{\displaystyle O}{\|}}{\underset{\underset{\displaystyle OH}{|}}{P}} - O - \overset{\overset{\displaystyle O}{\|}}{\underset{\underset{\displaystyle OH}{|}}{P}} - OH \; + \; H_2O \longrightarrow \text{Adenosine} - \overset{\overset{\displaystyle O}{\|}}{\underset{\underset{\displaystyle OH}{|}}{P}} - OH \; + \; HO - \overset{\overset{\displaystyle O}{\|}}{\underset{\underset{\displaystyle OH}{|}}{P}} - O - \overset{\overset{\displaystyle O}{\|}}{\underset{\underset{\displaystyle OH}{|}}{P}} - OH$$

| ATP | Water | Adenosine monophosphate (AMP) | Inorganic pyrophosphate (PP$_i$) |

Or

$$\text{ATP} + H_2O \longrightarrow \text{AMP} + PP_i$$

The hydrolysis of ATP to ADP and P$_i$ or to AMP and PP$_i$ releases a good deal of free energy—7 kcal for each mole of ATP hydrolyzed.

$$\text{ATP} \longrightarrow \text{ADP} + P_i + 7 \text{ kcal free energy}$$
$$\text{ATP} \longrightarrow \text{AMP} + PP_i + 7 \text{ kcal free energy}$$

why not 14 kcal. because if 2 Pi = 14 kcal

There is no guarantee, of course, that cells will harness this free energy to do work. Indeed, if ATP were simply hydrolyzed in cells, all the energy would be lost as heat, and no useful work would be done. But cells have evolved ways of harnessing this energy. We will learn how later in this chapter, but first we will see where ATP stands in the energetics of cells and how it is made.

[handwritten note: kjoule = ? kcal]

PRACTICE EXERCISE 14.3

How much free energy would be released for each mole of ATP hydrolyzed according to the following reaction?

$$ATP + 2H_2O \longrightarrow AMP + 2P_i$$

[handwritten: ?kcal = 1 mol ATP · $\frac{7 kcal}{1 mol ATP}$ · $\frac{2 mol ATP}{1}$ = 14 kcal]

PRACTICE EXERCISE 14.4

Calculate the free energy released when 750 g of ATP is hydrolyzed according to the reaction in Practice Exercise 14.3.

[handwritten: ?kcal = 750 g ATP · $\frac{1 mol ATP}{507 g ATP}$ · $\frac{14 kcal}{1 mol ATP}$ = 21 kcal]
[handwritten: 20.7 kcal]
[handwritten: molar mass]

14.5 Cellular energetics

AIM: To state the general relationships between ATP, anabolism, and catabolism.

Focus

ATP occupies an intermediate position in the energetics of the cell.

ATP is often described as a high-energy compound. The energy released by the breakdown of ATP to ADP is not particularly high, however, compared with the energy released in some cellular oxidation reactions. To see how this works to the advantage of the cell, consider the hydrolysis of ATP, a process that releases a total of 7 kcal of free energy for each mole of ATP hydrolyzed. The reverse of hydrolysis, formation of ATP from ADP and P_i, requires at least the same amount of energy. This energy can be supplied by the energy produced by a few higher-energy oxidation reactions of catabolism. ATP is important not because it is a high-energy compound but because it occupies an intermediate position in the energetics of the cell. The energy conserved in the anhydride bonds of ATP is passed to other cellular processes.

Newborn babies and hibernating animals have brown fat tissue that contains an enzyme that prevents ATP production. The energy from brown fat tissue is released as heat used to maintain body temperature.

Cells maintain a careful balance between ATP production and use. When a cell has a good supply of nutrients to catabolize, ATP production is high, and there is enough ATP to maintain the cell with some left over. Under these conditions, the excess of ATP is used in anabolic reactions to synthesize substances needed for cell growth and multiplication. In an environment poor in nutrients, the ATP produced is used to maintain cell vitality. Under these conditions, synthesis (anabolism) is deemphasized or reversed. There is an important principle here: *Catabolism generally produces ATP; anabolism generally uses it.*

PRACTICE EXERCISE 14.5

Explain why a lack of nutrients stops cell growth.

[handwritten: For a cell to grow, it must make new subs. by anabolic reactions. Anabolic processes require ATP. If the nutrients are not available to the cell, the cell cannot make ATP. ∴ the cell stops growing.]

14.6 Oxidative phosphorylation

AIMS: *To outline the production of ATP in aerobic cells, showing the relationship between oxidation reactions, formation of reducing power, cellular respiration, and oxidative phosphorylation. To classify members of the NAD^+-NADH and FAD-$FADH_2$ pairs as oxidizing or reducing agents. To write net equations for the oxidation of NADH and $FADH_2$ in cellular respiration. To explain what happens to carriers of the electron transport chain as electrons move down the chain. To indicate the number of ATP molecules formed for every NADH and $FADH_2$ molecule that enters the electron transport chain.*

Focus

The ATP requirements of aerobic cells are met by oxidative phosphorylation.

Aerobic
aer (Greek): air
bios (Greek): life

Cellular respiration is concerned with oxygen use by aerobic cells and is *not* to be confused with pulmonary respiration, or breathing, in creatures with lungs.

Anaerobic
an (Greek): without
aer (Greek): air
bios (Greek): life

We now know in a general way that cells trap free energy released by the oxidation of sugars, fats, and amino acids. But how do cells extract this energy and use it to make ATP? This extraction of energy in **aerobic cells**—*oxygen-requiring cells*—can be described as three basic processes:

1. Some of the energy released by the oxidation of carbon compounds is conserved as **reducing power**—*the ability of the reduced forms of certain enzyme cofactors to reduce other biological molecules.*

2. Much of the energy conserved as reducing power is used for **cellular respiration**—*the energy-releasing reduction of molecular oxygen (O_2) to water.*

3. The energy released in cellular respiration is used in **oxidative phosphorylation**—*the process by which most of the cellular ATP is produced from ADP and P_i.*

These three points explain why aerobic cells need nutrients and oxygen to live. Without nutrients, there is no energy released by oxidation reactions; without energy, there is no formation of reducing power; without reducing power and oxygen, there is no cellular respiration; without cellular respiration, there is no ATP production by oxidative phosphorylation; and without ATP, there can be no life.

The need for oxygen is not universal. Certain bacteria—for example, the bacterium that causes tetanus—do not require oxygen to sustain life. *Organisms that do not need oxygen to live are called* **anaerobes.** For some strict anaerobes, oxygen is a poison, and they die in its presence.

Reducing power

Nicotinamide adenine dinucleotide (NAD^+) (Fig. 14.6a) and flavin adenine dinucleotide (FAD) (Fig. 14.6b) are compounds synthesized by cells from B-complex vitamins. NAD^+ is structurally related to nicotinic acid; FAD is structurally related to riboflavin. NAD^+ and FAD are oxidizing agents—that is, electron acceptors. Several different oxidation reactions of catabolism are catalyzed by enzymes that use either NAD^+ or FAD as their cofactor.

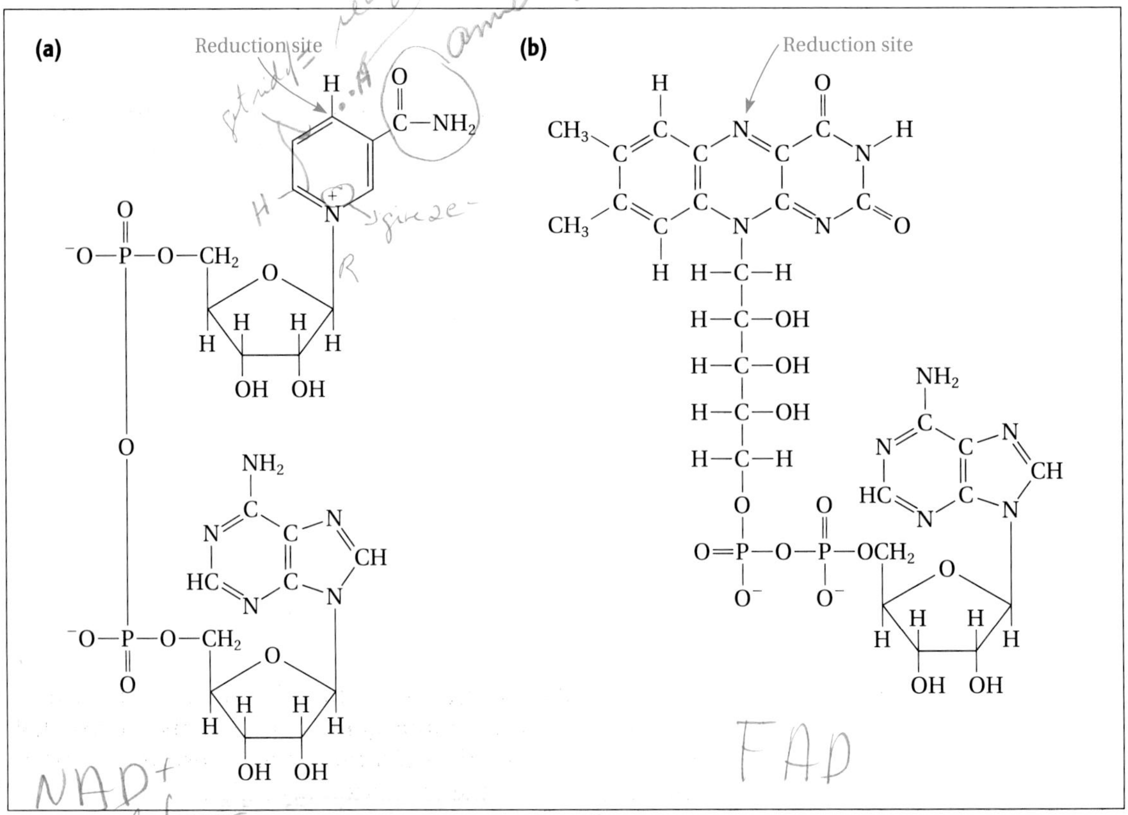

[handwritten annotations: "reduced form acid", "amide group", "H : A", "oxidized form", "H glucose", "R", "NAD⁺", "oxidized form", "FAD"]

Figure 14.6
(a) The oxidized form of nicotinamide adenine dinucleotide. Notice that the oxidized form of the coenzyme has a positive charge. Hence its abbreviation is NAD⁺. (b) The oxidized form of flavin adenine dinucleotide (FAD) does not carry a charge.

During these oxidation reactions, the substrates are oxidized, and the cofactors are reduced. NADH is the reduced form of NAD⁺, and FADH₂ is the reduced form of FAD. NADH and FADH₂ produced by the oxidation of certain substrates are then used to reduce a variety of other molecules. *This capability of NADH and FADH₂ to act as reducing agents is called their reducing power.*

Let's see how the reduction of NAD⁺ and FAD works. During oxidation reactions catalyzed by enzymes that use NAD⁺ as the coenzyme, an NAD⁺ molecule accepts a pair of electrons and a proton from a substrate molecule.

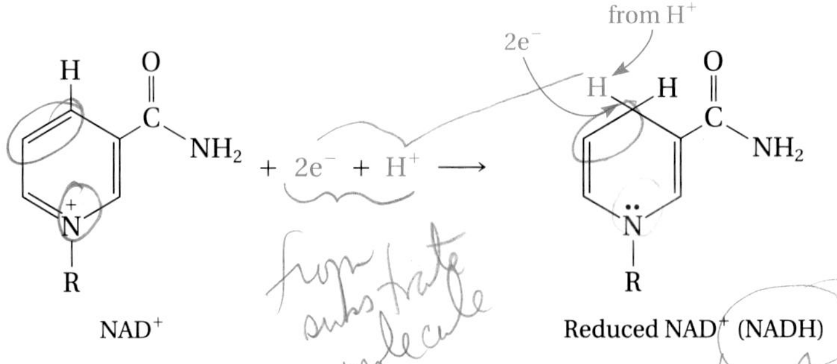

[handwritten annotations: "from H⁺", "2e⁻", "from substrate molecule", "neutral"]

NAD⁺ $+ \ 2e^- + H^+ \longrightarrow$ Reduced NAD⁺ (NADH)

Because it has gained electrons, the NAD^+ molecule is reduced to NADH. A substrate molecule has lost two electrons and a proton. The substrate has therefore been oxidized. A typical cellular reaction that results in the reduction of NAD^+ and the oxidation of a substrate is the oxidation of an alcohol to an aldehyde.

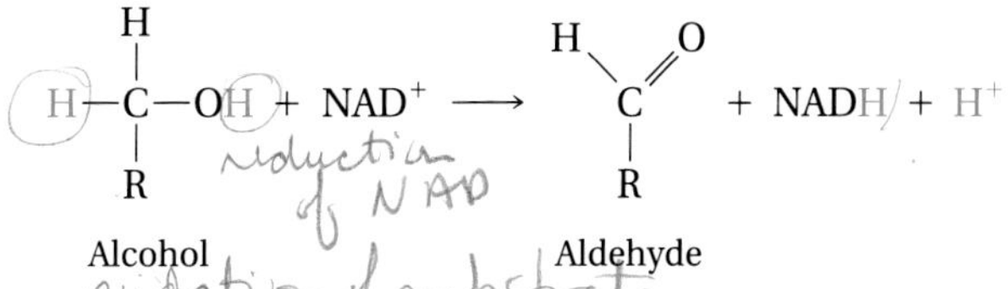

(handwritten: gain H and e⁻) *(handwritten: loss of H and e⁻)*

$$H-\underset{\overset{|}{R}}{\overset{\overset{H}{|}}{C}}-OH + NAD^+ \longrightarrow \underset{\overset{|}{R}}{C}\overset{\overset{H}{\diagdown}\quad\diagup^O}{} + NADH + H^+$$

Alcohol Aldehyde

(handwritten: reduction of NAD) *(handwritten: oxidation of substrate)*

NAD^+ is also the cofactor usually involved in the oxidation of aldehydes to carboxylic acids.

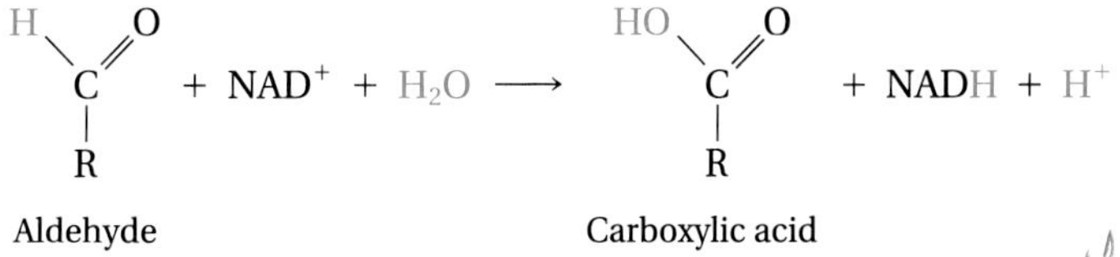

$$\underset{\overset{|}{R}}{C}\overset{\overset{H}{\diagdown}\quad\diagup^O}{} + NAD^+ + H_2O \longrightarrow \underset{\overset{|}{R}}{C}\overset{\overset{HO}{\diagdown}\quad\diagup^O}{} + NADH + H^+$$

Aldehyde Carboxylic acid

(handwritten: exothermic)

(handwritten: endothermic) In each of these reactions, the substrate is oxidized and loses energy. The NAD^+ is reduced and gains energy. This means that *some of the energy produced by the oxidation of the substrate has been conserved as reducing power by the formation of NADH.* As we will see, the reducing power of NADH will be used in two important ways: It is one of two sources of electrons that can be used to reduce oxygen to water in cellular respiration, and it can be used to reduce carbon compounds in anabolic reactions.

The reduced form of FAD ($FADH_2$) is the second source of cellular reducing power. As in the reduction of NAD^+, substrates of enzymes that use FAD as their cofactor give up two electrons to the cofactor. Upon reduction, however, FAD picks up one more proton than NAD^+. Thus, the abbreviation for the reduced form of FAD is $FADH_2$.

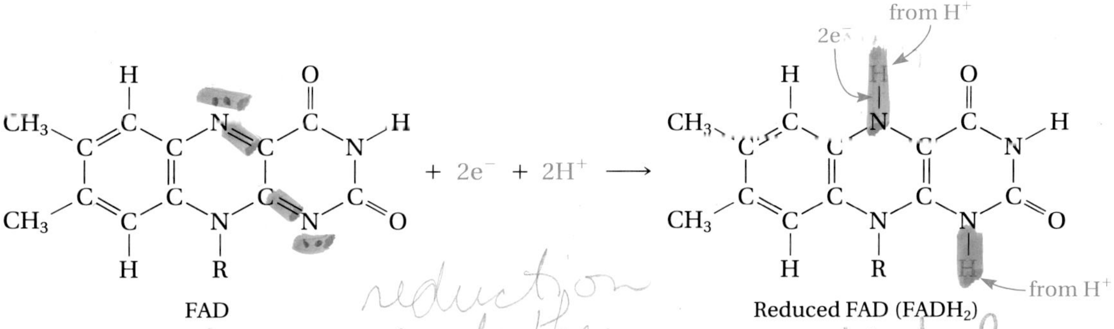

FAD $+ 2e^- + 2H^+ \longrightarrow$ Reduced FAD ($FADH_2$)

(handwritten: reduction) *(handwritten: endothermic)* *(handwritten: neutral)*

The substrates for oxidation reactions in which FAD is the cofactor are different from those involving NAD^+. FAD is involved in oxidation reactions in which a $-CH_2-CH_2-$ linkage of the substrate is oxidized to a double bond.

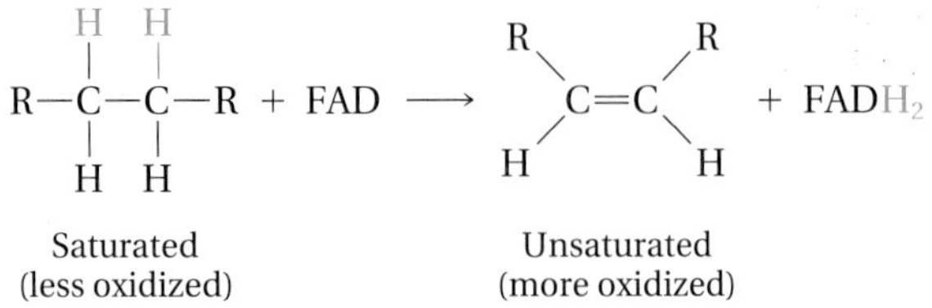

$$R-\underset{\underset{H}{|}}{\overset{\overset{H}{|}}{C}}-\underset{\underset{H}{|}}{\overset{\overset{H}{|}}{C}}-R + FAD \longrightarrow \underset{H}{\overset{R}{\diagdown}}C=C\underset{H}{\overset{R}{\diagup}} + FADH_2$$

Saturated (less oxidized) Unsaturated (more oxidized)

Handwritten left margin:

$$CH_3-\overset{\overset{O}{\|}}{C}-H + NAD^+ + H_2O \longrightarrow$$

a)
$$H-\underset{\underset{H}{|}}{\overset{\overset{H}{|}}{C}}-\overset{\overset{O}{\|}}{C}-H + NAD^+ + H_2O \longrightarrow$$

aldehyde + oxidized L

$$H-\underset{\underset{H}{|}}{\overset{\overset{H}{|}}{C}}-\overset{\overset{O}{\|}}{C}-OH + NADH + H^+$$

acid

b)
$$HO-\overset{\overset{O}{\|}}{C}-\overset{\overset{H}{|}}{\underset{\underset{H}{|}}{C}}-\overset{\overset{H}{|}}{\underset{\underset{H}{|}}{C}}-\overset{\overset{O}{\|}}{C}-OH + FAD$$

$$O-\overset{\overset{O}{\|}}{C}-C=C-\overset{\overset{O}{\|}}{C}-OH + FADH_2$$

c)
$$H-\overset{\overset{H}{|}}{\underset{\underset{H}{|}}{C}}-\overset{\overset{H}{|}}{\underset{\underset{H}{|}}{C}}-OH + NAD^+ \longrightarrow$$

oxidizing

alcohol

$$H-\overset{\overset{H}{|}}{\underset{\underset{H}{|}}{C}}-\overset{H}{C}=O + NADH + H^+$$

aldehyde

PRACTICE EXERCISE 14.6

Predict the products of the following enzyme-catalyzed oxidation reactions. The molecule on the left in the reaction is the substrate.

(a) $CH_3CHO + NAD^+ + H_2O \longrightarrow$

(b) $HOOCCH_2CH_2COOH + FAD \longrightarrow$

(c) $CH_3CH_2OH + NAD^+ \longrightarrow$

PRACTICE EXERCISE 14.7

How many moles of NADH are formed in the enzyme-catalyzed oxidation of 1 mol of ethanol (CH_3CH_2OH) to 1 mol of acetic acid (CH_3CO_2H)? Give the reaction.

Cellular respiration

Of all the metabolic processes of aerobic cells, by far the most ATP is produced by events that depend on cellular respiration. *Molecular oxygen (O_2) is reduced to water (H_2O) by NADH and FADH$_2$ in cellular respiration.* The following equation summarizes cellular respiration for NADH:

$$2NADH + 2H^+ + O_2 \longrightarrow 2NAD^+ + 2H_2O$$

Four electrons are transferred from NADH to molecular oxygen as this reduction occurs. Each molecule of NADH supplies two electrons. Two of the four protons needed come from NADH, and two come from the surroundings. FADH$_2$ is also important in cellular respiration, since it reduces oxygen in a similar way.

$$2FADH_2 + O_2 \longrightarrow 2FAD + 2H_2O$$

In the preceding equation, each FADH$_2$ molecule supplies two electrons and two protons.

Although these equations can be used to represent cellular respiration, the reduction of oxygen by NADH or FADH$_2$ does not occur in a single step. Instead, the two electrons stripped from the reduced cofactors (NADH and FADH$_2$) are passed through an *electron transport chain* (also called the *respiratory chain*). The **electron transport chain (respiratory chain)** *consists of a group of molecules that accept electrons from NADH and FADH$_2$ and transfer these electrons to molecular oxygen in a sequence of energy-releasing steps.* The components of the electron transport chain are assembled on the inner membrane of the mitochondria of eukaryotic cells (see A Closer Look:

Mitochondria). The components of the respiratory chain of aerobic prokaryotes are located on the interior of the cell membrane. The location of the respiratory assembly in the mitochondria and the importance of respiration in ATP production give mitochondria a well-deserved reputation as the "power plants" of the eukaryotic cell.

Six intermediate electron carriers separate NADH from the ultimate electron acceptor (oxygen) in the respiratory chain. The electron carriers of the respiratory chain are lined up in order of increasing affinity for elec-

A Closer Look

Mitochondria

All eukaryotic cells have mitochondria (singular: mitochondrion). Mitochondria (see figures) are often called the "power plants" of the cell, since most of the cell's energy is generated there by oxidation-reduction reactions. To understand the membrane structure of mitochondria, imagine a soft purse. A mitochondrion has a double membrane that corresponds to the outside of the purse and its lining. The interior membrane of the mitochondrion is highly folded, much like pleats in the purse lining. These folds, which extend into the interior of the mitochondrion, are *cristae.* Many of the mitochondrion's enzymes are attached to the inside surface of the interior membrane.

The chemical composition of the inner membrane of mitochondria closely resembles that of the membranes of prokaryotic cells. Moreover, mitochondria contain DNA that is different from nuclear DNA and reproduces independently of the reproduction processes of the cell. Because of these similarities, scientists believe that mitochondria are highly specialized remains of prokaryotic cells. They speculate that a mutually beneficial (symbiotic) relationship resulted when an anaerobic eukaryotic cell encapsulated an aerobic prokaryotic cell. This arrangement allowed the prokaryotic cell to use the waste products of the eukaryotic cell for ATP production; in return, the eukaryotic cell used some of the ATP produced by the prokaryotic cell. These specialized functions of the eukaryotic host and its prokaryotic guest eventually developed to the point that today neither host nor guest can survive without the other.

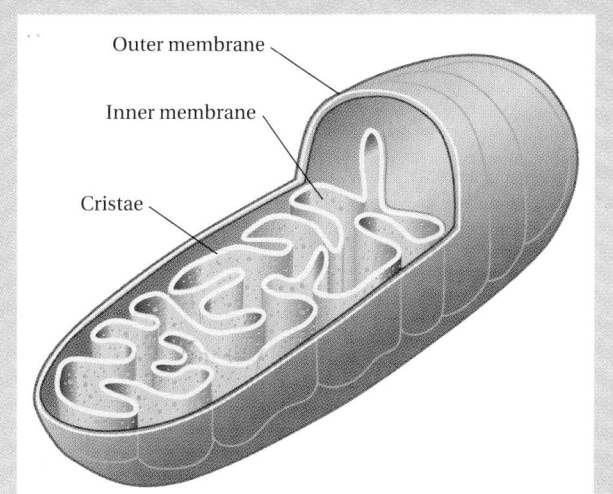

Outer membrane

Inner membrane

Cristae

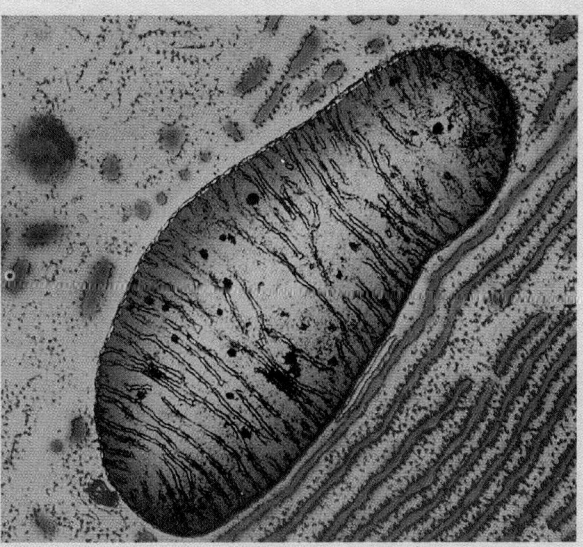

A mitochondrion.

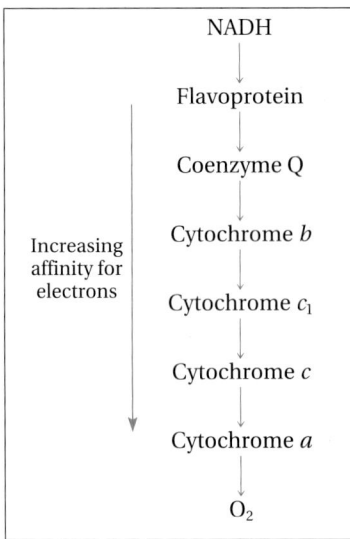

NADH
↓
Flavoprotein
↓
Coenzyme Q
↓
Cytochrome b
↓
Cytochrome c_1
↓
Cytochrome c
↓
Cytochrome a
↓
O_2

Increasing affinity for electrons

Figure 14.7
Six electron carriers move electrons from NADH to oxygen in the electron transport chain.

trons. Figure 14.7 shows these electron carriers. The first electron carrier is an enzyme that contains a tightly bound cofactor. This cofactor is similar in structure to riboflavin and FAD. Hence *the enzyme is called a* **flavoprotein.** In the first step of the respiratory chain, the cofactor portion of the flavoprotein is changed from the oxidized (ox) to the reduced (red) state by accepting an electron pair from NADH.

$$NADH + Flavoprotein_{ox} \longrightarrow NAD^+ + Flavoprotein_{red} + H^+$$

Since oxidation is a loss of electrons and reduction is a gain of electrons, NADH is oxidized and the flavoprotein is reduced. NADH has now exercised its reducing power.

Coenzyme Q (CoQ) is the second electron carrier of the respiratory chain. It accepts two electrons from the reduced flavoprotein.

$$Flavoprotein_{red} + CoQ_{ox} \longrightarrow Flavoprotein_{ox} + CoQ_{red}$$

The CoQ is reduced and the flavoprotein is reoxidized. Figure 14.8 shows the structures of the oxidized and reduced forms of CoQ.

The electron carriers beyond CoQ are the cytochromes. **Cytochromes** *are proteins that contain an iron-heme complex very similar to that of hemoglobin* (Fig. 14.9). However, there is a major chemical difference between the iron of hemoglobin and that of the cytochromes. Whereas the iron of normal hemoglobin remains in the ferrous (Fe^{2+}) state, the iron of

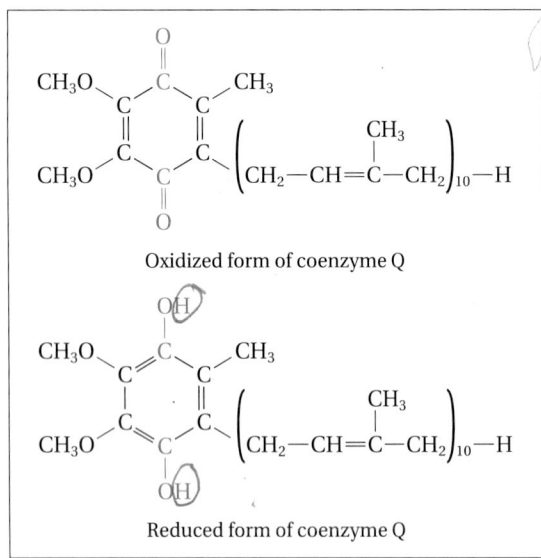

Oxidized form of coenzyme Q

Reduced form of coenzyme Q

Figure 14.8
The oxidized and reduced forms of coenzyme Q.

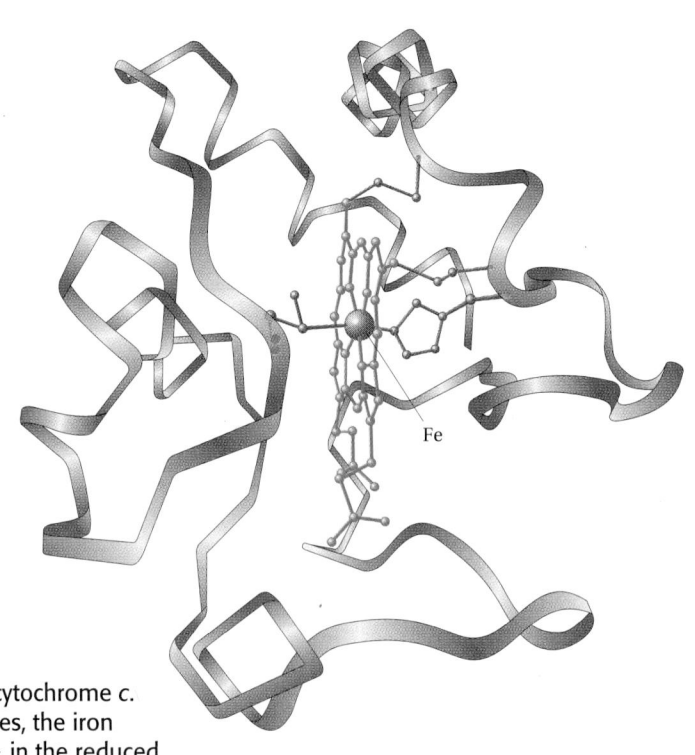

Fe

Figure 14.9
The structure of cytochrome *c*. In the cytochromes, the iron changes from 2+ in the reduced state to 3+ in the oxidized state.

the cytochromes flip-flops between the reduced ferrous (Fe^{2+}) and oxidized ferric (Fe^{3+}) states by the addition or loss of one electron.

$$Fe^{3+} + e^- \underset{\text{Oxidation}}{\overset{\text{Reduction}}{\rightleftharpoons}} Fe^{2+}$$

Cytochrome
cyto (Greek): cell
chroma (Greek): color

Thus the first cytochrome of the respiratory chain readily accepts electrons from reduced CoQ and just as readily loses them to the next cytochrome in the chain. The pattern of oxidations and reductions continues as the electrons pass through the chain single file.

$$CoQ_{red} + Cyt\ b_{ox} \longrightarrow CoQ_{ox} + Cyt\ b_{red}$$
$$Cyt\ b_{red} + Cyt\ c_{1_{ox}} \longrightarrow Cyt\ b_{ox} + Cyt\ c_{1_{red}}$$
$$Cyt\ c_{1_{red}} + Cyt\ c_{ox} \longrightarrow Cyt\ c_{1_{ox}} + Cyt\ c_{red}$$
$$Cyt\ c_{red} + Cyt\ a_{ox} \longrightarrow Cyt\ c_{ox} + Cyt\ a_{red}$$

Finally, at reduced cytochrome *a*, an oxygen molecule picks up four electrons that have traveled through the chain. With the addition of four protons from the surroundings, the reduction of oxygen to water has occurred, and the passage of an electron pair from NADH through the electron chain is complete.

$$2e^- + \overset{..}{\underset{..}{:O:}} + 2H^+ \longrightarrow H—\overset{..}{\underset{..}{O}}—H$$

From   Oxygen   Water
respiratory   atom
chain

A Closer Look: Oxygen, Disease, and Aging discusses in more detail how this water is formed and its ramifications for health.

As each carrier is reduced by the addition of electrons, the carrier from which it received the electrons becomes reoxidized and ready to receive more electrons. This feature permits electrons to flow through the respiratory chain as long as there is sufficient NADH available to donate them and sufficient oxygen available to accept them. Figure 14.10 summarizes the oxidation-reduction reactions of the respiratory chain.

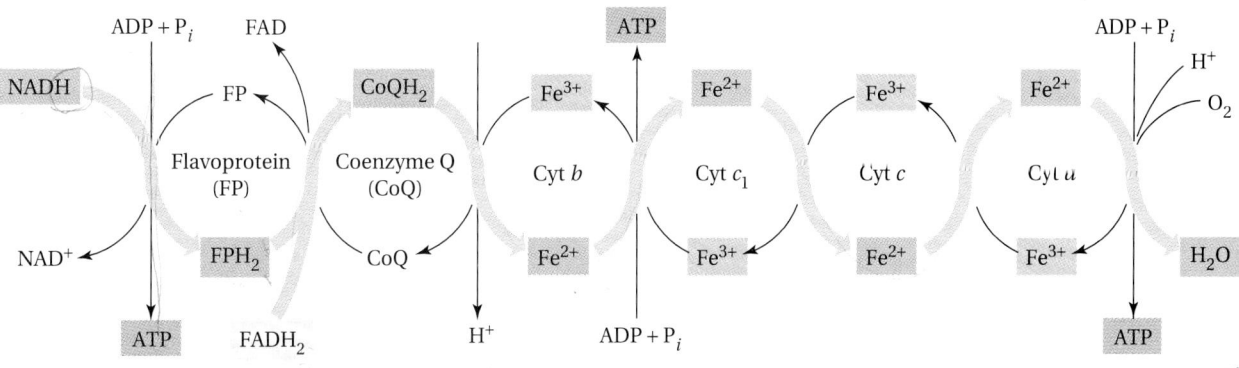

Figure 14.10   ETS
Oxidation-reduction reactions of the respiratory chain. The brown arrows show the path of the electrons as they pass from one carrier to the next. The reduced species are shown in yellow, blue, and purple. Electrons from NADH and FADH$_2$ pass through the chain in pairs until they reach the cytochromes. They pass through the cytochromes one at a time. The final electron acceptor is oxygen.

Cyanide ions are a powerful poison of the respiratory chain. These ions form an irreversible complex with the iron of cytochrome *a*, the last cytochrome in the chain. Formation of the cytochrome *a*–cyanide complex prevents electrons from being drained away by oxygen. Under these conditions, the electron carriers preceding cytochrome *a* become clogged with electrons—that is, all the carriers are soon reduced, and further electron

A Closer Look

Oxygen, Disease, and Aging

The reduction of molecular oxygen (O_2) to water (H_2O) in the electron transport chain involves the formation of several *reactive oxygen species (ROS)*. The major ROS are the *hydroxyl radical* ($\cdot OH$), the *superoxide anion* ($\cdot O_2^-$), and *hydrogen peroxide* (H_2O_2). Hydroxyl radicals and superoxide anions are free radicals (contain an unpaired electron). The rapid decomposition of ROS is vital because ROS are strong oxidizing agents that are capable of causing extensive cell damage and even death. Normal cells contain enzymes that destroy superoxide anions and hydrogen peroxide. The enzyme *superoxide dismutase* catalyzes the conversion of superoxide anions to hydrogen peroxide. Superoxide dismutase contains zinc, copper, and manganese cofactors that are important in this reaction. *Catalase*, an iron-containing enzyme, catalyzes the decomposition of the hydrogen peroxide to molecular oxygen and water. *Peroxidase*, another iron-containing enzyme, catalyzes the same reaction. Vitamin C (ascorbic acid) and vitamin E destroy ROS, especially hydroxyl radicals, by reacting with them.

The effects of faulty destruction of ROS can be seen in several human diseases and in aging. Amyotrophic lateral sclerosis (ALS), also known as Lou Gehrig's disease, is an invariably fatal degenerative disease of the nervous system (see figure). ALS results from the mutation of the single gene that codes for superoxide dismutase. The defective gene produces defective superoxide dismutase, and the damage to the nervous system by high levels of superoxide anion causes the symptoms of ALS. Oxidative damage by ROS is thought to be a factor in such diseases as arthritis, emphysema, and some cancers. A fairly extensive body of evidence indicates that normal aging is in part the result of oxidative damage by ROS and other free

Lou Gehrig, a well-known baseball player, was a victim of ALS, amyotrophic lateral sclerosis.

radicals in body cells. As we age, free radicals damage our proteins, lipids, and nucleic acids by oxidative reactions. Oxidized proteins such as enzymes may not function properly. Lipid oxidation can produce fragile cell membranes. Nucleic acid oxidation can lead to gene mutations that result in the expression of defective or nonfunctional proteins. Experimental animals fed diets high in vitamins C and E live longer and exhibit fewer of the debilitating effects of aging than control animals fed regular diets. Scientists believe that these beneficial effects of vitamins C and E are the result of their reactions with ROS and other free radicals.

transport is blocked. The blockage of electron transport has the same effect on the cell as a lack of oxygen. Respiration stops; no ATP is produced by oxidative phosphorylation; the cell, and soon the organism, dies because it cannot pass electrons to oxygen.

Oxidative phosphorylation

Oxidative phosphorylation ties cellular respiration to ATP production. Cellular respiration is an energy-yielding process. For every pair of electrons that flow through the electron transport chain—this is the number of electrons given up by one molecule of NADH—enough energy to phosphorylate a molecule of ADP is released at three points in the chain (Fig. 14.11). *For every NADH oxidized to NAD^+ in cellular respiration, three ATP molecules may be formed by oxidative phosphorylation.* The electron pairs from $FADH_2$ enter the respiratory chain further down the line than electrons from NADH. *For every $FADH_2$ oxidized to FAD, only two ATP molecules are formed by oxidative phosphorylation.* How is the energy of oxidation of NADH and $FADH_2$ used to make ATP in oxidative phosphorylation? The energy is used to activate an enzyme called an *ATP synthase.* ATP synthase catalyzes the coupling of ADP with P_i to make ATP.

Recall from the Case in Point earlier in this chapter that Emily was suffering from a mysterious fatigue. In the Follow-up to the Case in Point, on the next page, we will learn the nature of her affliction.

Figure 14.11
Energy is released as electrons pass through the electron transport chain. When two electrons from the oxidation of NADH to NAD^+ pass through the chain, enough energy is released to phosphorylate one molecule of ADP to ATP by the process of oxidative phosphorylation at each of the three sites shown. The pair of electrons from the oxidation of $FADH_2$ to FAD enters the chain at a later point. Only two molecules of ATP are formed for every two electrons from $FADH_2$ that pass through the chain.

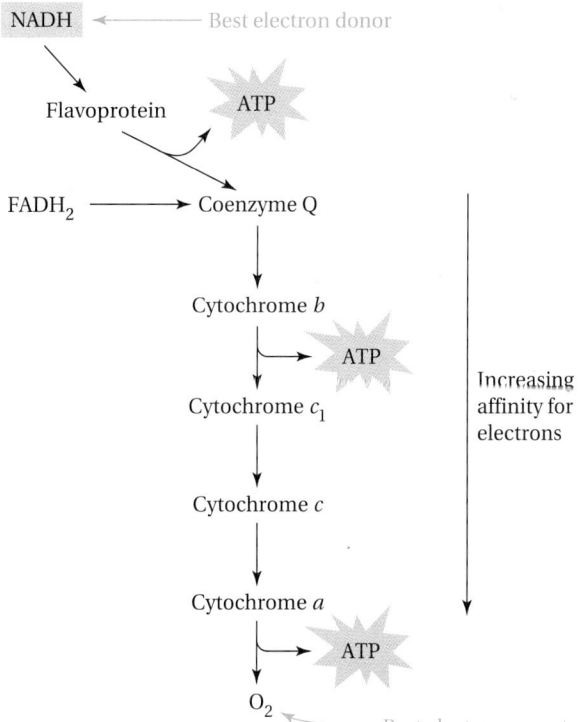

FOLLOW-UP TO THE CASE IN POINT: A mysterious fatigue

After eliminating usual causes of fatigue, the medical researchers who investigated Emily's condition sought the source of her problem in the biochemical processes of energy production: oxygen delivery, formation of NADH and $FADH_2$, the respiratory chain, and oxidative phosphorylation. They found that oxygen uptake by her tissues was normal. This eliminated the possibility of a flawed delivery of oxygen to the tissues. Her body cells' levels of NADH, NAD^+, $FADH_2$, and FAD also were normal, suggesting that enough reducing power was available to the respiratory chain and that the respiratory chain was operative. Her cellular ATP levels were considerably lower than normal.

The researchers concluded that Emily was suffering from an extremely rare defect of oxidative phosphorylation: The respiratory chain and the production of ATP were partially uncoupled. In such a situation, oxygen uptake, cellular respiration, and levels of NADH, NAD^+, $FADH_2$, and FAD would be normal. However, ATP levels would be low, leading to chronic fatigue. The molecular basis of the defect was not determined, but a mutant gene could be responsible for the defective expression of the ATP synthase that couples the respiratory chain to the production of ATP in oxidative phosphorylation. A flawed ATP synthase would result in impaired ATP production.

14.7 Cellular work

AIMS: To describe the three major types of work done by cells. To show how coupled reactions enable a cell to carry out chemical work.

Focus

Energy conserved in ATP drives cellular processes.

We have seen that cleavage of the phosphoric acid anhydride bonds in the ATP molecule releases free energy that can do cellular work. Cells do three major kinds of work: chemical, osmotic, and mechanical. In this section we will examine each kind of work and see how ATP is involved.

Chemical work

ATP often supplies energy to chemical reactions that need energy to make a product. Most of these reactions involve ATP as a phosphorylating agent. That is, because of its very reactive phosphoric acid anhydride functional groups, ATP readily transfers phosphoryl

$$\begin{array}{c} \text{O} \\ \parallel \\ -\text{P}-\text{OH} \\ \mid \\ \text{OH} \end{array}$$

or, less commonly, pyrophosphoryl

$$\begin{array}{c} \text{O} \quad\quad \text{O} \\ \parallel \quad\quad \parallel \\ -\text{P}-\text{O}-\text{P}-\text{OH} \\ \mid \quad\quad\quad \mid \\ \text{O} \quad\quad \text{O} \end{array}$$

groups to other molecules. Just as reducing power is the ability of NADH and $FADH_2$ to transfer electrons in reduction reactions, **phosphorylating power** *is the ability of ATP to transfer phosphoryl groups in phosphorylation reactions.*

Some phosphorylated enzyme substrates are activated for subsequent reactions they would not ordinarily undergo. The process of activation often involves a **coupled reaction**—*an energetically unfavorable reaction is made to occur by being linked to a reaction that is energetically very favorable (very exergonic).* In a coupled reaction, the net energy change for the linked reactions is favorable (exergonic). For example, liver cells need to use ammonia and carbon dioxide to make citrulline from the amino acid ornithine. (This reaction is important in the urea cycle, and we will return to it in Chapter 17.) A hypothetical equation for this reaction can be written:

$$
\begin{array}{c}
\text{H} \\
| \\
\text{NH} \\
| \\
\text{CH}_2 \\
| \\
\text{CH}_2 \\
| \\
\text{CH}_2 \\
| \\
\text{H}_2\text{N}\!-\!\text{CH} \\
| \\
\text{CO}_2\text{H}
\end{array}
\; + \;
\begin{array}{c}
\text{O} \\
\| \\
\text{C} \\
\| \\
\text{O}
\end{array}
\; + \; \text{NH}_3 \longrightarrow
\begin{array}{c}
\text{NH}_2 \\
/ \\
\text{O}\!=\!\text{C} \\
\backslash \\
\text{NH} \\
| \\
\text{CH}_2 \\
| \\
\text{CH}_2 \\
| \\
\text{CH}_2 \\
| \\
\text{H}_2\text{N}\!-\!\text{CH} \\
| \\
\text{CO}_2\text{H}
\end{array}
\; + \; \text{H}_2\text{O}
$$

Ornithine   Carbon dioxide   Ammonia   Citrulline   Water

This hypothetical reaction is energetically unfavorable (Fig. 14.12a). The equilibrium position greatly favors the reactants over the product. What about using an enzyme to catalyze the reaction? This would not make the reaction exergonic. As catalysts, enzymes only speed up reactions that are already exergonic; they do not alter the position of equilibrium. Cells are not stymied by this state of affairs, however, because they have ATP on their side. They carry out an exergonic reaction instead, forming a molecule of carbamoyl phosphate by using one molecule of ammonia, one of carbon

Figure 14.12
The direct conversion of ammonia to citrulline is not a favorable process (a). When the reaction is coupled with the breakdown of ATP, however, it is exergonic (favorable) (b).

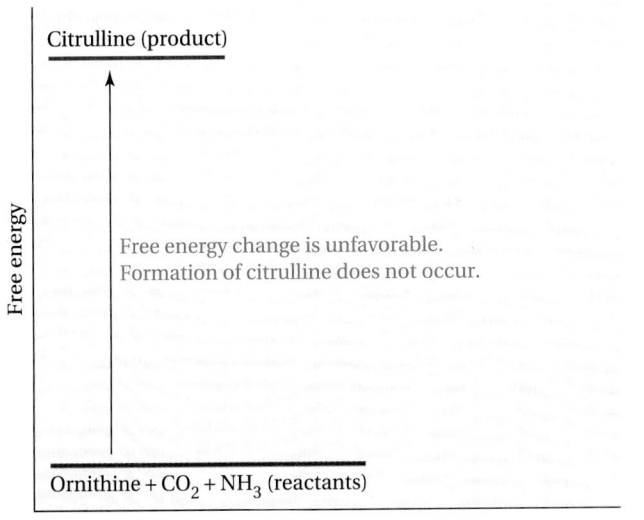

(a)

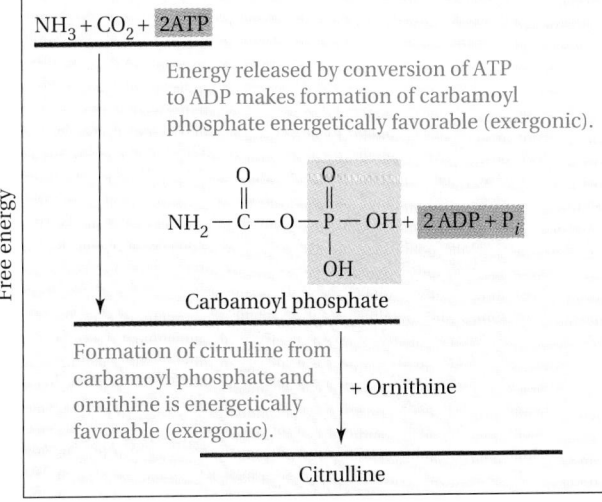

(b)

dioxide, and two of ATP (Fig. 14.12b). An enzyme speeds up the reaction.

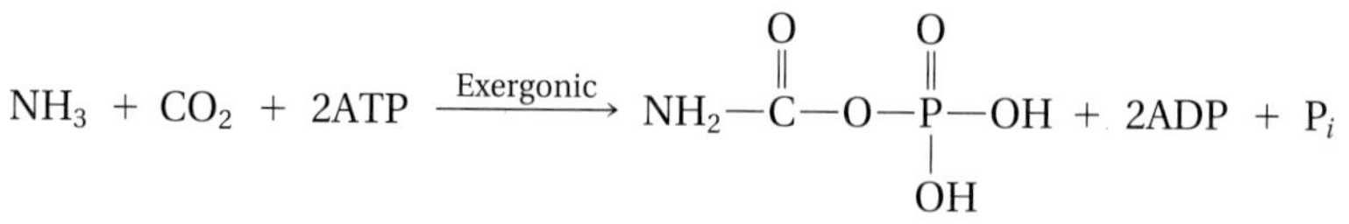

$$NH_3 + CO_2 + 2ATP \xrightarrow{\text{Exergonic}} NH_2-\overset{\overset{O}{\|}}{C}-O-\overset{\overset{O}{\|}}{\underset{\underset{OH}{|}}{P}}-OH + 2ADP + P_i$$

Carbamoyl
phosphate

The structure of carbamoyl phosphate contains a mixed anhydride linkage, one formed from a carboxylic acid and phosphoric acid. This linkage is very energetic. Carbamoyl phosphate readily reacts with ornithine to produce citrulline in another exergonic enzyme-catalyzed reaction (Fig. 14.12b).

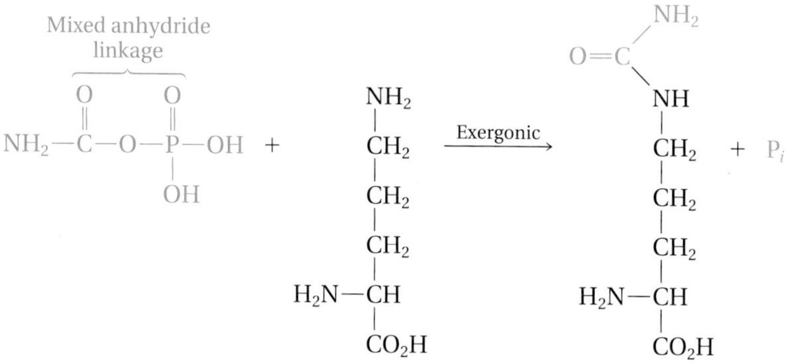

Carbamoyl   Ornithine   Citrulline
phosphate

ATP appears in the equation for the overall reaction for the formation of citrulline, and it looks as if it were used up, as it is in hydrolysis.

$$\text{Ornithine} + CO_2 + NH_3 + 2ATP \xrightarrow{\text{Exergonic}} \text{Citrulline} + 2ADP + 2P_i + H_2O$$

The reality is quite different. The free energy stored in the energetic anhydride bonds of two ATP molecules was unleashed by breaking these bonds, but some was conserved in the anhydride bond of carbamoyl phosphate. By expending the phosphorylating power of ATP to make carbamoyl phosphate, liver cells boosted the reactivity of ammonia and carbon dioxide and achieved the synthesis of citrulline. In our example, an energetically unfavorable reaction—the formation of citrulline from ammonia, carbon dioxide, and ornithine—could not occur without coupling the reaction to the favorable breakdown of ATP. Sometimes the route to energetically activated molecules requires several chemical steps, but the principle is the same. The inevitable result of substrate phosphorylation is the production of a more reactive molecule.

Like substrates, some enzymes are also phosphorylated by ATP. Phosphorylation often makes the difference between an active and inactive enzyme. A good example of enzyme activation caused by phosphorylation comes from catabolism and involves phosphorylase, the enzyme that catalyzes the breakdown of stored glycogen. The inactive form of phosphorylase (phosphorylase *b*) is a dimer consisting of identical protein subunits.

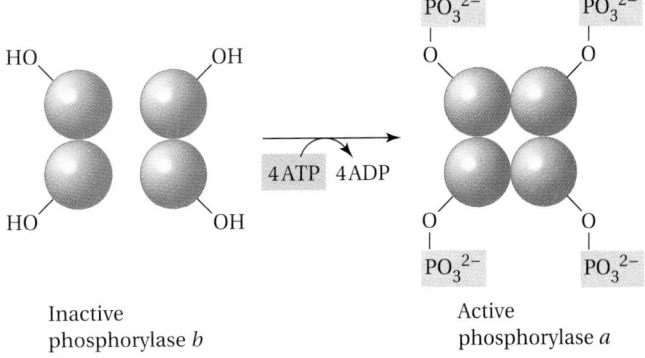

Figure 14.13
Phosphorylase is converted from an inactive to an active form upon phosphorylation.

Inactive
phosphorylase *b*
dimers

Active
phosphorylase *a*

The active form of phosphorylase (phosphorylase *a*) is a tetramer of these identical subunits.

Phosphorylation of the hydroxyl groups of a single serine residue in the amino acid sequence of each subunit causes two inactive phosphorylase *b* dimers to form the active phosphorylase *a* tetramer (Fig. 14.13). The phosphorylase *a* can now go about the work nature intended: the breakdown of glycogen. It may appear that the phosphorylation of phosphorylase, a catabolic enzyme, violates the rule that catabolism produces ATP. However, one molecule of phosphorylase *a* catalyzes the cleavage of glycogen into many molecules of glucose. And as we will see in Chapter 15, the complete oxidation of only one glucose molecule freed by the action of phosphorylase *a* on glycogen gives a return of ATP many times greater than the four molecules of ATP it takes to phosphorylate the enzyme.

Osmotic work

Cells contain molecular pumps that transport ions and molecules through cell membranes. These pumps work against the normal concentration gradient—the transport is from a lower to a higher concentration of the substance being transported—and require the expenditure of energy in the form of ATP. Such pumps often require the phosphorylation of transport proteins embedded in the cell membrane.

Mechanical work

At this point we understand that ATP is a phosphorylating agent in the chemical and osmotic work done by cells. But in mechanical work—that is, cellular movements including muscle contraction—ATP is hydrolyzed rather than acting as a phosphorylating agent. The sliding-filament model is the generally accepted depiction of muscle contraction. In this model, muscles contract when thick and thin filaments slide past each other, with ATP providing the energy of propulsion (Fig. 14.14).

Muscle contraction requires so much ATP that muscle cells cannot keep enough on hand. Nature uses ATP to *transmit* energy, not to store it. In

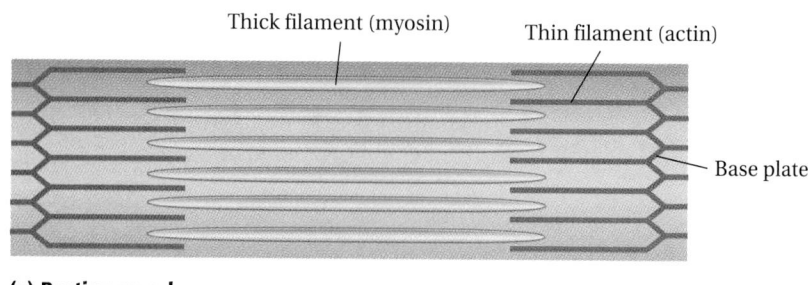

Thick filament (myosin)   Thin filament (actin)

Base plate

(a) Resting muscle

ATP   ADP + P$_i$

Figure 14.14
Muscle fibers contract when thick filaments made of myosin and thin filaments made of actin slide past each other, propelled by energy released in the hydrolysis of ATP.

(b) Contracted muscle

resting muscle, the phosphorylating power of ATP is stored in another high-energy compound called *creatine phosphate,* or *phosphocreatine.* On demand, the phosphoryl groups of creatine phosphate are transferred back to ADP.

Creatine + ATP $\underset{\text{Active muscle}}{\overset{\text{Resting muscle}}{\rightleftharpoons}}$ Creatine phosphate + ADP

The enzyme involved in this reversible reaction is *creatine phosphokinase.* (Enzymes that catalyze the transfer of a phosphoryl group to or from ATP are called *kinases.*) As soon as the demand for ATP exceeds the supply, as in heavy exercise, muscles become tired and weak.

PRACTICE EXERCISE 14.8

Outline the steps to show why the body's need for oxygen increases as physical activity increases. What happens to muscles when the demand for ATP exceeds the supply?

SUMMARY

The Sun supplies the energy needed for life through the carbon and energy cycle. Photosynthesis by plants is perhaps the most important event in the carbon and energy cycle, since it is by photosynthesis that the Sun's light energy is converted to the chemical energy stored in the bonds of sugars, fats, and proteins. Chemical energy stored in these bonds is released in catabolism—the chemical reactions that cells use to oxidize the carbons of sugars, fats, and proteins back to carbon dioxide.

Cells conserve some of the energy of oxidation as reducing power in the form of NADH and $FADH_2$. Reduction of molecular oxygen by NADH and $FADH_2$ releases enough energy to drive oxidative phosphorylation—the phosphorylation of ADP to ATP. Cells use ATP for chemical work, osmotic work, and mechanical work, as in muscle contraction.

KEY TERMS

Adenosine triphosphate (ATP) (14.4)
Anabolism (14.1)
Anaerobe (14.6)
Catabolism (14.1)
Cellular respiration (14.6)
Chlorophyll (14.2)

Chloroplast (14.2)
Coupled reaction (14.7)
Cytochrome (14.6)
Electron transport chain (14.6)
Energy and carbon cycle (14.3)

Flavoprotein (14.6)
Light system (14.2)
Metabolism (14.1)
Oxidative phosphorylation (14.6)
Phosphorylating power (14.7)

Photosynthesis (14.2)
Reducing power (14.6)
Respiratory chain (14.6)
Thyalkoids (14.2)

EXERCISES

Metabolism (Section 14.1)

14.9 Define *metabolism*.
14.10 How do *anabolism* and *catabolism* differ from each other?

The Energy and Carbon Cycle (Sections 14.2, 14.3)

14.11 Write an equation that summarizes the process of photosynthesis.
14.12 Give a balanced equation for the oxidation of glucose.
14.13 Explain what happens in photosynthesis.
14.14 Where do sugars, fats, and amino acids originate?
14.15 How do cells produce energy?
14.16 What is the ultimate source of all carbon compounds? Explain.
14.17 Interpret this statement: "Carbon dioxide is an energy-poor molecule, but glucose is an energy-rich molecule."

14.18 Arrange the following compounds starting with the most reduced (energy-rich) and ending with the most oxidized (energy-poor).

(a) 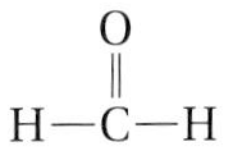 Formaldehyde

(b) 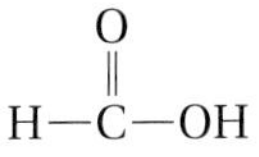 Formic acid

(c) 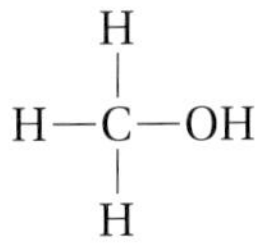 Methanol

(d)  Carbon dioxide

(e) 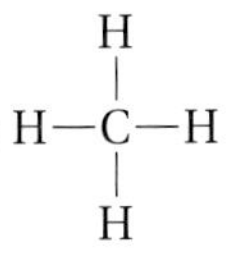 Methane

ATP (Sections 14.4, 14.5)

14.19 Draw the structure of ATP. Label the characteristic features of the molecule.

14.20 Discuss the function of ATP in cellular metabolism.

14.21 What is the charge on a fully unprotonated ATP molecule? What is the gram formula weight of the unprotonated structure?

14.22 Calculate the free energy released when (a) 2 mol of ATP is hydrolyzed to ADP and P_i and when (b) 50 g of ATP is hydrolyzed to ADP and P_i.

14.23 ATP occupies an intermediate position in the energetics of the cell. Explain what this means and why it is important.

14.24 What processes are responsible for controlling ATP concentrations in cells?

Oxidative Phosphorylation (Section 14.6)

14.25 What are aerobic cells? Outline the processes that occur when aerobic cells use energy released by the oxidation of nutrient molecules to phosphorylate ADP to ATP.

14.26 Write the abbreviations for the reduced and oxidized forms of nicotinamide adenine dinucleotide and flavin adenine dinucleotide.

14.27 Complete the following reactions.

(a) $NAD^+ + H^+ +$ _____ $\longrightarrow$ NADH

(b) $FAD +$ _____ $+$ ___ ___ $\longrightarrow FADH_2$

14.28 The coenzymes nicotinamide adenine dinucleotide (NAD^+) and flavin adenine dinucleotide (FAD) are oxidizing agents. What is their role in the enzyme-catalyzed oxidation of a substrate molecule?

14.29 Write a balanced equation for the oxidation of methanol to formaldehyde by NAD^+. What happens to NAD^+ in the course of this reaction?

14.30 NADH and $FADH_2$ are sources of cellular reducing power. What does this mean, and why is it important?

14.31 What happens during (a) cellular respiration and (b) oxidative phosphorylation?

14.32 What happens in the respiratory (or electron transport) chain?

14.33 State where cellular respiration takes place in eukaryotic cells and where ATP is produced.

14.34 What are the first and last steps in the respiratory chain?

14.35 Give the general name for the heme-containing proteins that are electron carriers in the respiratory chain. What is the major chemical difference between these molecules and hemoglobin?

14.36 Compare the effects of oxygen deficiency and cyanide poisoning on ATP production by oxidative phosphorylation.

14.37 How many molecules of ATP may be formed by oxidative phosphorylation when one molecule of NADH is oxidized to NAD^+ in cellular respiration?

14.38 If two molecules of $FADH_2$ are oxidized to FAD in cellular respiration, how many molecules of ADP may be phosphorylated?

Cellular Work (Section 14.7)

14.39 Some of the chemical energy released when nutrients are oxidized is conserved in ATP. What are the three major types of cellular work that require ATP energy?

14.40 ATP is a good phosphorylating agent. What takes place when a molecule is phosphorylated?

14.41 Why must some enzyme substrates be phosphorylated before they can undergo reaction?

14.42 Explain the importance of ATP in coupled reactions.

14.43 Why is ATP required in the transport of certain ions and molecules across cell membranes?

14.44 Creatine phosphate contains a high-energy phosphate bond. Explain its function as a store of the phosphorylating power of ATP in muscle cells.

Additional Exercises

14.45 Give the name of an important biological molecule that contains each of the following:
(a) heme ring   (b) corrin ring   (c) porphyrin ring

14.46 What do the abbreviations NAD^+ and FAD stand for?

14.47 Where do NADH and $FADH_2$ enter the electron transport chain?

14.48 Identify each of the following changes as a reduction or oxidation.
(a) $Fe^{3+} \longrightarrow Fe^{2+}$ (b) $NADH \longrightarrow NAD^+$
(c) $FAD \longrightarrow FADH_2$ (d) $RCH_2OH \longrightarrow RCHO$
(e) $RCHO \longrightarrow RCO_2H$

14.49 In the electron transport chain, how does the reduction of coenzyme Q differ from the reduction of cytochromes?

14.50 The ATP molecule has *phosphorylating power.* What does this term mean?

14.51 Draw the structure of the phosphoryl group. Why is this group important in certain metabolic reactions?

SELF-TEST (REVIEW)

True/False

1. Anabolic and catabolic reactions are reversible, the direction of the reaction depending on the needs of the cell.

2. In photosynthesis, an oxidized, energy-poor carbon-containing compound is changed into an energy-rich carbon-containing compound.

3. Photosynthesis takes place in the chloroplasts of the cells of green plants.

4. If either plant life or animal life had to exist without the other, animals would have the better chance of survival.

5. In the respiratory chain, a mole of $FADH_2$ produces twice as many molecules of ATP as does a mole of NADH.

6. At the end of a marathon race, the supply of creatine phosphate in the muscle cells of a runner would probably be low.

7. Typically, when ATP is hydrolyzed, it loses two of its phosphate groups.

8. If a cell has an excess of ATP, anabolic processes will generally take place.

9. The energy needed to drive the process of oxidative phosphorylation comes from the process of cellular respiration.

10. NADH is a reducing agent.

11. Cellular respiration takes place in the mitochondria.

12. The cytochromes in the respiratory chain are almost always in the oxidized (Fe^{3+}) state.

13. Cyanide poisons kill an organism by interfering with the transport of oxygen to the cells.

Multiple Choice

14. In an enzyme-catalyzed reaction in which the cofactor FAD is changed to $FADH_2$,
    (a) the substrate is reduced.
    (b) the FAD is oxidized.
    (c) the substrate loses two electrons.
    (d) both (a) and (b) are correct.

15. In the overall process of photosynthesis,
    (a) energy is given off.
    (b) an energy-poor compound is used to make an energy-rich compound.
    (c) water is produced.
    (d) carbon dioxide is further oxidized.

16. Cells often accomplish a desired reaction by pairing an energetically unfavorable reaction with an ener-

getically favorable one. These reactions are called
(a) paired reactions.
(b) assisting reactions.
(c) coupled reactions.
(d) enzyme-catalyzed reactions.

17. Catabolic oxidation reactions in the cell
(a) usually release energy.
(b) produce compounds with higher reducing power.
(c) consume nutrients supplied to the cell.
(d) all of the above are true.

18. ATP is used as a phosphorylating agent by a cell in doing
(a) chemical work.
(b) osmotic work.
(c) mechanical work.
(d) both (a) and (b) are true.

19. The major energy-transmitting molecule of the cell is
(a) nicotinamide adenine dinucleotide (NAD^+).
(b) adenosine triphosphate (ATP).
(c) flavin adenine dinucleotide (FAD).
(d) adenosine monophosphate (AMP).

20. A secondary alcohol is changed to a ketone in an enzymatically catalyzed reaction using nicotinamide adenine dinucleotide as a coenzyme. In this reaction,
(a) the NAD^+ acts as a reducing agent.
(b) NADH is produced.
(c) the alcohol has been reduced.
(d) the NAD^+ has been oxidized.

21. A cell converts amino acids into a protein molecule. This is an example of
(a) catabolism.
(b) oxidative phosphorylation.
(c) photosynthesis.
(d) anabolism.

22. In which of the following pairs of substances is the member with the greater reducing power listed first?
(a) $FAD/FADH_2$ (b) C_2H_5OH/C_2H_6
(c) CH_3CO_2H/CH_3CHO
(d) $NADH/NAD^+$

23. The overall equation for the cellular respiration of $FADH_2$ shows
(a) oxygen as a reactant.
(b) the $FADH_2$ is reduced to FAD.
(c) hydroxide ion as a product.
(d) that oxygen is oxidized to water.

24. The formation of ATP from ADP and P_i
(a) requires no more than 7 kcal of energy per mole.
(b) usually occurs in an anabolic process.
(c) allows energy in a cell to be conserved.
(d) is a hydrolysis reaction.

25. During cellular catabolism, the energy released
    (a) is all used to do work.
    (b) is immediately used to do work.
    (c) is immediately lost as heat.
    (d) is partially lost as heat.

26. The passage of electrons through the electron transport chain results in all the following *except*
    (a) the oxidative phosphorylation of ADP to ATP.
    (b) a series of oxidation-reduction reactions within the chain.
    (c) the formation of water molecules at the end of the chain.
    (d) the use of the oxidizing power of NADH.

27. The hydrolysis of a mole of ATP by a mole of water molecules gives
    (a) a mole of ADP.
    (b) a mole of 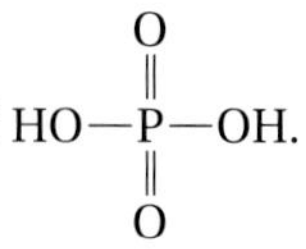
$$HO-\overset{\displaystyle O}{\underset{\displaystyle O}{\overset{\|}{\underset{\|}{P}}}}-OH.$$
    (c) about 7 kcal of free energy.
    (d) all of the above.

Carbohydrates in Living Organisms

At the Core of Metabolism

15

Bread is rich in carbohydrates; it is also a good source of vitamins and minerals.

In this chapter we will explore the relationship between carbohydrate metabolism and energy production in the cell. Why is carbohydrate metabolism important? It is because the carbohydrates occupy center stage in the metabolic drama of the cell. The star of carbohydrate metabolism is glucose. In this chapter we will discover how glucose is broken down and synthesized. Carbohydrate metabolism is important for everyone, but it can be of special concern to athletes who wish to improve their performance. This is illustrated in the following Case in Point.

CASE IN POINT: Carbohydrate loading

Barbara, the coach of a college cross-country team, has been trying to improve her runners' performance. Because Barbara studied biochemistry in her college days, she encourages her runners to eat a lot of pasta in the 2 days before a meet. What biochemical result is Barbara trying to achieve in her runners? We will see in Section 15.10.

15.1 Catabolism

AIMS: To write an equation for the catabolism of one mole of glucose. To name the three stages of aerobic catabolism of glucose.

Focus

Aerobic catabolism of glucose is accomplished in three stages.

One tomato and one potato each contain about 3.5 g of sugar; an orange contains about 12.5 g, and a banana, about 18 g.

Glycolysis
glykos (Greek): sugar or sweet
lysis (Greek): splitting

Stage I

Glucose

Glycolysis

2 Pyruvate

Stage II

2 Acetyl coenzyme A

Stage III

Citric acid cycle

The glucose produced by digestion of the carbohydrates a person eats enters the bloodstream. In a normal individual, this glucose is extracted from the bloodstream by body cells. Once inside the cells, the glucose is completely oxidized to carbon dioxide and water.

$$C_6H_{12}O_6 + 6O_2 \longrightarrow 6CO_2 + 6H_2O$$

Glucose   Oxygen   Carbon dioxide   Water

The paths that cells use to oxidize glucose completely to carbon dioxide involve many individual chemical reactions. But all aerobic cells have adopted essentially the same three-stage master plan to do the job. Briefly these three stages, shown in Figure 15.1, consist of initial breakdown in *glycolysis*, further degradation to *acetyl coenzyme A*, and, finally, complete oxidation in the *citric acid cycle*.

Glycolysis

Glycolysis *is the sequence of chemical reactions by which glucose, a six-carbon sugar, is cleaved to two molecules of pyruvate, a three-carbon acid.* Biochemists usually call organic acids produced in metabolism by the names of their dissociated forms, since these are the forms that exist at pH 7.0. Pyruvate is simply the anion of pyruvic acid.

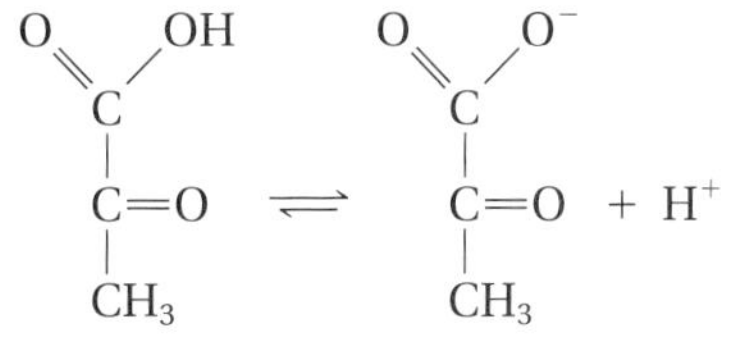

Pyruvic acid   Pyruvate

Glycolysis is sometimes called the *glycolytic pathway* or the *Embden-Meyerhof pathway* after two German biochemists, Gustav Embden and Otto Meyerhof, who proposed it in the latter part of the nineteenth century.

Acetyl coenzyme A

The second stage of glucose catabolism occurs when pyruvate ions lose carbon dioxide. The remaining two-carbon fragments end up attached to coenzyme A to form acetyl coenzyme A (Fig. 15.2). **Coenzyme A,** *usually abbreviated CoA, is a thiol;* **acetyl CoA** *is the thioester of acetic acid and CoA.* The structure of CoA contains pantothenic acid, one of the B-complex vitamins.

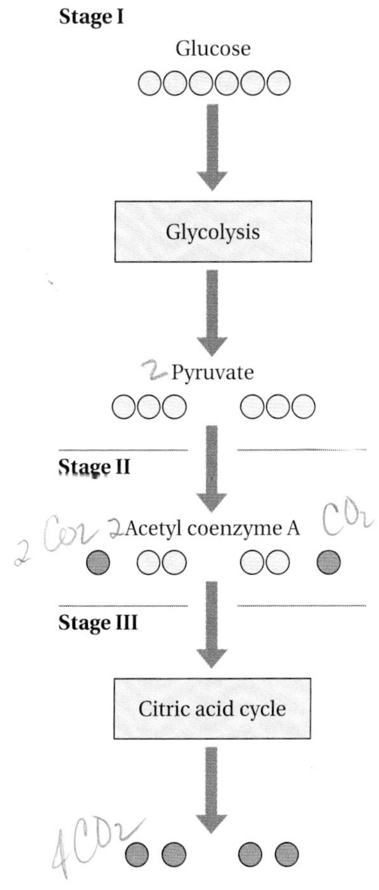

Figure 15.1 (left)
The oxidation of glucose, a six-carbon sugar (the carbon atoms are represented by yellow circles), to six molecules of carbon dioxide occurs in three stages. Two molecules of carbon dioxide (purple circles) are produced in the second stage, and four are produced in the third stage.

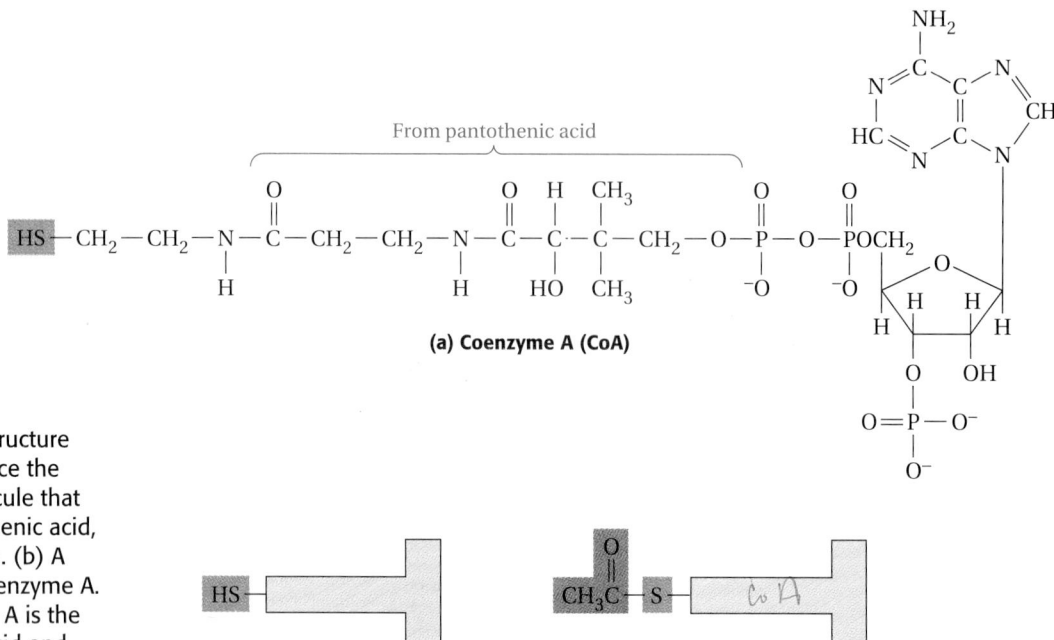

(a) Coenzyme A (CoA)

Figure 15.2
(a) The molecular structure of coenzyme A. Notice the portion of the molecule that comes from pantothenic acid, a B-complex vitamin. (b) A block diagram of coenzyme A. (c) Acetyl coenzyme A is the thioester of acetic acid and coenzyme A.

(b) Block diagram of coenzyme A   **(c) Block diagram of acetyl coenzyme A**

Like ATP, acetyl CoA is often considered an energy-rich compound. Just as the transfer of a phosphoryl group of ATP to some other molecule is made easier by its attachment to ADP, the transfer of an acetyl group is made easier by its attachment to CoA.

Citric acid cycle

The acetyl group of acetyl CoA enters the citric acid cycle. *The* **citric acid cycle** *is the pathway used by most organisms to oxidize completely to carbon dioxide the acetyl carbons of acetyl CoA formed in the breakdown of sugars, fats, and amino acids.* Production of two molecules of carbon dioxide and two of water for each molecule of acetyl CoA entering the cycle completes the third stage of glucose catabolism. The citric acid cycle is also called the *Krebs cycle* after Sir Hans Krebs, the English biochemist who proposed it in 1937.

15.2 Glucose oxidation

AIM: To state the entry points of sugars, fatty acids, and amino acids into the three stages of the aerobic catabolism of glucose.

Focus

The oxidation of glucose is the core of catabolism.

Most biochemists consider the oxidation of glucose to be the core of catabolism because the cells of virtually all organisms rely on glucose as their main source of energy. Human brain cells, for example, will accept no other nutrient to obtain energy except under dire circumstances such as starva-

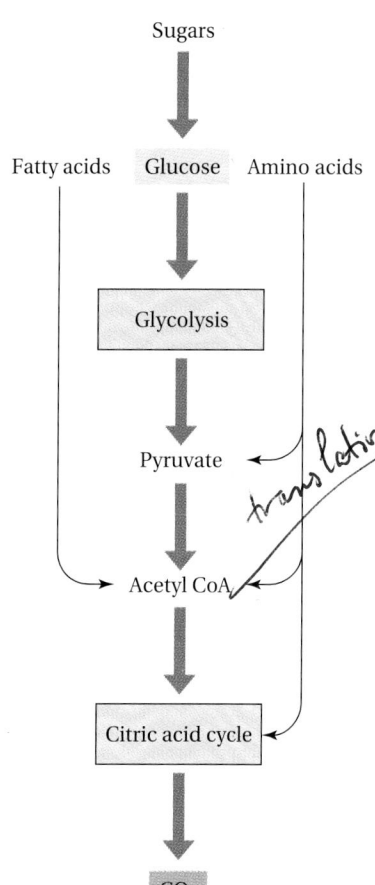

Figure 15.3
Proteins are hydrolyzed to amino acids; lipids are hydrolyzed to fatty acids and glycerol. Above, we see that the amino acids and fatty acids are converted to simpler compounds that enter the main pathways of glucose metabolism.

Focus

One glucose molecule produces two ATP and two NADH molecules in glycolysis.

The phosphorylation of glucose to form glucose 6-phosphate is so energetically favorable that essentially all the glucose that enters the cell is immediately phosphorylated.

tion. Many other cells do oxidize other sugars, fatty acids, and amino acids to obtain energy, however. Certain enzymes contained in such cells degrade these substances to compounds that eventually enter catabolism through the central core of glucose metabolism. Figure 15.3 shows the entry points.

Many organisms have the enzymes necessary to convert galactose, fructose, and other hexoses into glucose. These sugars therefore enter glycolysis as glucose. Fatty acids are oxidized and enter the central core of glucose catabolism as acetyl CoA. Because of the variety of amino acid structures, their degradation products enter the central core at several different points: at the tail end of glycolysis as pyruvate, as acetyl CoA, or as chemical intermediates of the citric acid cycle.

15.3 Glycolysis

AIM: To list the steps in the aerobic glycolysis of one molecule of glucose.

The enzymes that catalyze the steps of glycolysis are found in the cytoplasm of cells. This is where glycolysis occurs. Glycolysis begins with the phosphorylation of glucose to glucose 6-phosphate.

(The symbol ⓟ represents the phosphoryl group $-\overset{\overset{\displaystyle O}{\|}}{\underset{\underset{\displaystyle O^-}{|}}{P}}-O^-$.)

The names and structures of some of the intermediate compounds in metabolism are complex. You do not need to memorize them, but they will be used in the text to make it easier to follow what is happening. Remember also that all the steps of these reactions are catalyzed by enzymes.

The phosphoryl group of glucose 6-phosphate comes from ATP. This may seem a little surprising. Since glycolysis is a pathway of catabolism, we might expect it to *produce* ATP, not to *use* it! The important point here is that the cell is investing ATP, just as you might invest money in the stock market. Your investment—if you are lucky—will be returned with dividends of more money. The cell's investment will be repaid with dividends of more ATP. The glucose 6-phosphate is converted to fructose 6-phosphate.

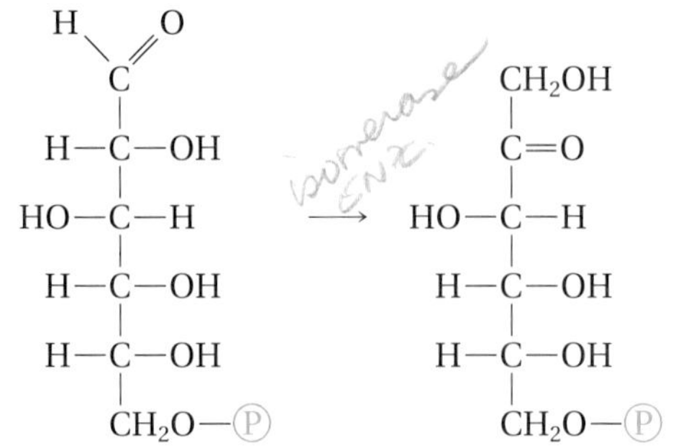

Glucose 6-phosphate   Fructose 6-phosphate

Fructose 6-phosphate undergoes phosphorylation to fructose 1,6-bisphosphate at the expense of another molecule of ATP invested.

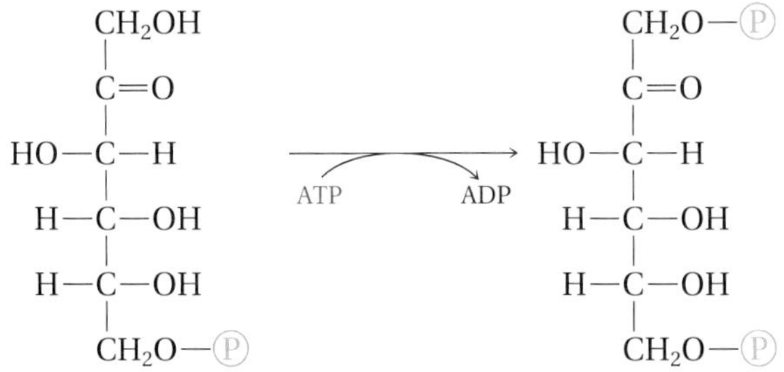

Fructose 6-phosphate   Fructose
                                        1,6-bisphosphate

The cell has now invested two molecules of ATP for every molecule of glucose to be degraded. The conversion of fructose 6-phosphate to fructose 1,6-bisphosphate is an important control step in glycolysis. Once fructose 1,6-bisphosphate is formed, it cannot escape the glycolytic pathway. The phosphorylation of fructose 6-phosphate to fructose 1,6-bisphosphate is called the *committed step* of glycolysis.

Fructose 1,6-bisphosphate is now cleaved to give a pair of three-carbon compounds, dihydroxyacetone phosphate and glyceraldehyde 3-phosphate.

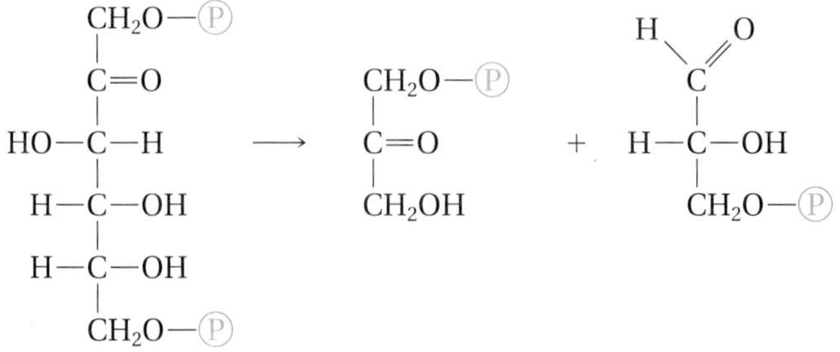

Fructose   Dihydroxyacetone   Glyceraldehyde
1,6-bisphosphate   phosphate   3-phosphate

Only glyceraldehyde 3-phosphate will be used in further steps of glycolysis.

The dihydroxyacetone is not wasted. Nature is economical, and cells have an enzyme that promotes the conversion of dihydroxyacetone phosphate to glyceraldehyde 3-phosphate.

$$
\begin{array}{ccc}
CH_2O-\textcircled{P} & & CH_2O-\textcircled{P} \\
| & & | \\
C=O & \rightleftharpoons & H-C-OH \\
| & & | \\
CH_2OH & & C \\
& & \diagup \ \diagdown \\
& & H \quad\ O
\end{array}
$$

Dihydroxyacetone phosphate   Glyceraldehyde 3-phosphate

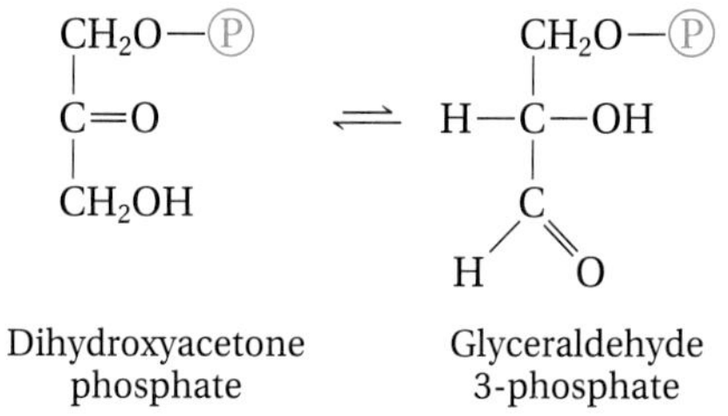

Since one molecule of glucose has provided two molecules of glyceraldehyde 3-phosphate, we will have to take this into account in our future bookkeeping. From now on, we will have to multiply the reactants and products of our reactions by 2.

An enzyme next converts glyceraldehyde 3-phosphate to 1,3-bisphosphoglycerate in the first energy-yielding oxidation reaction of glucose catabolism.

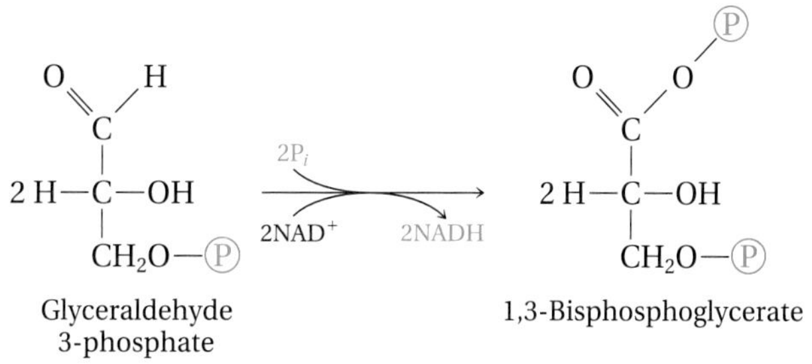

$$
\begin{array}{ccc}
O \diagdown \quad H & & O \diagdown \quad O-\textcircled{P} \\
\diagdown C \diagup & & \diagdown C \diagup \\
| & \xrightarrow{\quad 2P_i \quad} & | \\
2 \ H-C-OH & 2NAD^+ \quad 2NADH & 2 \ H-C-OH \\
| & & | \\
CH_2O-\textcircled{P} & & CH_2O-\textcircled{P}
\end{array}
$$

Glyceraldehyde 3-phosphate   1,3-Bisphosphoglycerate

The enzyme uses NAD^+ as a cofactor. The NAD^+ is reduced to NADH—it receives two electrons and a proton from the aldehyde substrate—in the course of the reaction. The new phosphoryl group of the organic product comes from inorganic phosphate ions present in the cytoplasm, so no ATP is expended here. In fact, 1,3-bisphosphoglycerate is itself a high-energy compound—a mixed anhydride of a carboxylic acid and phosphoric acid (see Sec. 14.7) that can transfer its new phosphoryl group to ADP. (The phosphorylation of ADP to ATP outside oxidative phosphorylation is called *substrate-level phosphorylation*.) This transfer occurs in the next step for glycolysis. Two ATP molecules are gained.

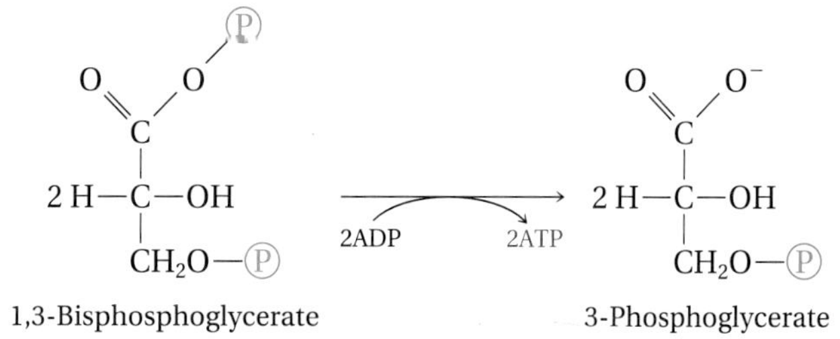

$$
\begin{array}{ccc}
O \diagdown \quad O-\textcircled{P} & & O \diagdown \quad O^- \\
\diagdown C \diagup & & \diagdown C \diagup \\
| & \xrightarrow{\quad\quad} & | \\
2 \ H-C-OH & 2ADP \quad 2ATP & 2 \ H-C-OH \\
| & & | \\
CH_2O-\textcircled{P} & & CH_2O-\textcircled{P}
\end{array}
$$

1,3-Bisphosphoglycerate   3-Phosphoglycerate

Since the cell invested two ATP molecules and now has two back, it is even in the ATP stock market. Any ATP produced from this point on is profit.

The next step in glycolysis is a shift of the phosphoryl group of 3-phosphoglycerate.

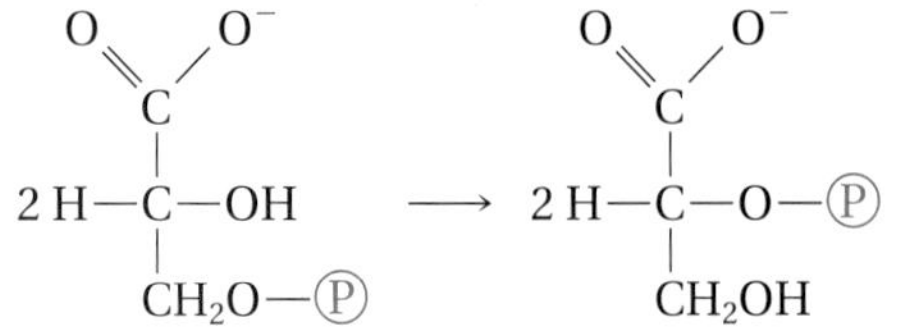

3-Phosphoglycerate 2-Phosphoglycerate

The product of this reaction, 2-phosphoglycerate, loses a molecule of water to give phosphoenolpyruvate.

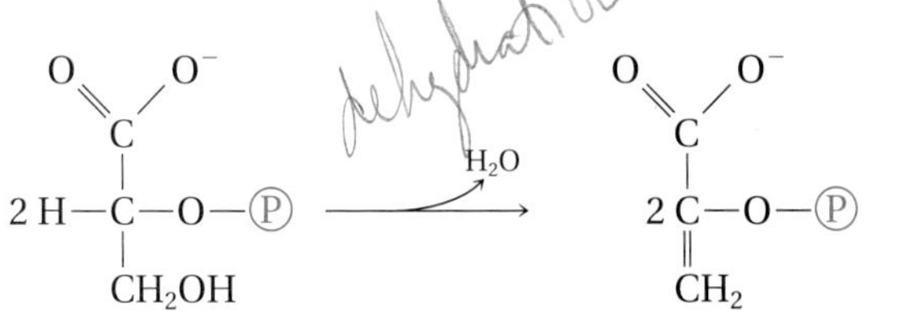

2-Phosphoglycerate 2-Phosphoenolpyruvate

Phosphoenolpyruvate is another energy-rich phosphate molecule capable of passing its phosphoryl group to ADP in another substrate-level phosphorylation. Two ATP molecules are gained.

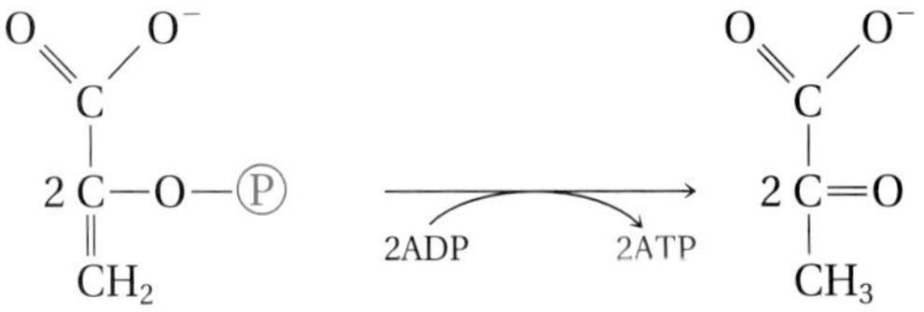

2-Phosphoenolpyruvate Pyruvate

Since the degradation of one glucose molecule eventually produces two molecules of phosphoenolpyruvate, two molecules of ADP can be phosphorylated to ATP when the phosphoenolpyruvate from one molecule of glucose is converted to pyruvate. These two molecules of ATP are the ATP dividends earned in glycolysis.

The formation of pyruvate is the final step of aerobic glycolysis. Here is what has happened in the oxidation of one molecule of glucose:

1. Two molecules of pyruvate have been formed.
2. Two molecules of NAD^+ have been reduced to NADH.
3. A net total of two ADP molecules have been phosphorylated to ATP (four ATP molecules gained minus two invested).

Table 15.1 summarizes the reactions of glycolysis.

PRACTICE EXERCISE 15.1

Write a net equation that summarizes glycolysis.

$1\ C_6H_{12}O_6 + 2\ NAD^+ + 2ADP + 2P_i \longrightarrow 2\ CH_3\overset{O}{\overset{\|}{C}}-\overset{O}{\overset{\|}{C}}-OH + 2\ NADH + 2ATP$

Table 15.1 The Reactions of Glycolysis

1. Glucose $\longrightarrow$ Glucose 6-phosphate
   ATP   ADP

2. Glucose 6-phosphate $\rightleftharpoons$ Fructose 6-phosphate

3. Fructose 6-phosphate $\longrightarrow$ Fructose 1,6-bisphosphate
   ATP   ADP

4. Fructose 1,6-bisphosphate

   Dihydroxyacetone phosphate $\rightleftharpoons$ Glyceraldehyde 3-phosphate

5. Glyceraldehyde 3-phosphate $+ P_i \longrightarrow$ 1,3-Bisphosphoglycerate
   NAD^+ NADH

6. 1,3-Bisphosphoglycerate $\longrightarrow$ 3-Phosphoglycerate
   2ADP   2ATP

7. 3-Phosphoglycerate $\rightleftharpoons$ 2-Phosphoglycerate

8. 2-Phosphoglycerate $\rightleftharpoons$ Phosphoenolpyruvate

9. Phosphoenolpyruvate $\longrightarrow$ Pyruvate
   2ADP   2ATP

PRACTICE EXERCISE 15.2

Calculate the moles of ATP produced when 90 g of glucose is broken down in glycolysis. The molar mass of glucose is 180 g.

$$? \text{ mol ATP} = 90\text{g } C_6H_{12}O_6 \cdot \frac{1 \text{ mol } C_6H_{12}O_6}{180\text{g } C_6H_{12}O_6} \cdot \frac{2 \text{ mol ATP}}{1 \text{ mol } C_6H_{12}O_6} = \boxed{1 \text{ ATP}}$$

15.4 Acetyl coenzyme A

AIM: To describe the formation and function of acetyl CoA.

Focus

Two carbons of pyruvate are incorporated into acetyl coenzyme A.

When an aerobic cell is operating with a good supply of oxygen, pyruvate molecules flow into the mitochondria. Two of the carbons of each pyruvate ion end up as acetyl CoA, and one molecule of NAD^+ is reduced to NADH. Carbon dioxide is formed as a waste product. This process can be summarized by a single equation:

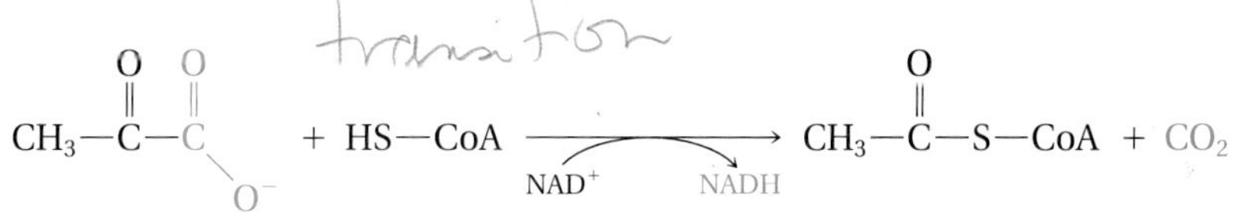

$$CH_3-\overset{O}{\underset{||}{C}}-\overset{O}{\underset{||}{C}}-O^- + HS-CoA \xrightarrow[NAD^+ \quad NADH]{} CH_3-\overset{O}{\underset{||}{C}}-S-CoA + CO_2$$

Pyruvate   Acetyl CoA

An organized assembly of three different kinds of enzyme molecules called a **multienzyme complex** *is responsible for the formation of acetyl CoA from pyruvate.* This multienzyme complex is named as if it were one enzyme—*pyruvate dehydrogenase.* Several copies of each type of enzyme are present in the pyruvate dehydrogenase multienzyme complex. Five coenzymes—

thiamine pyrophosphate, lipoic acid, FAD, NAD$^+$, and coenzyme A—are also present. Thiamine was mentioned previously (Sec. 12.8) as vitamin B$_1$. Now we see that there is a need for it, as its pyrophosphate, in converting pyruvate to acetyl CoA.

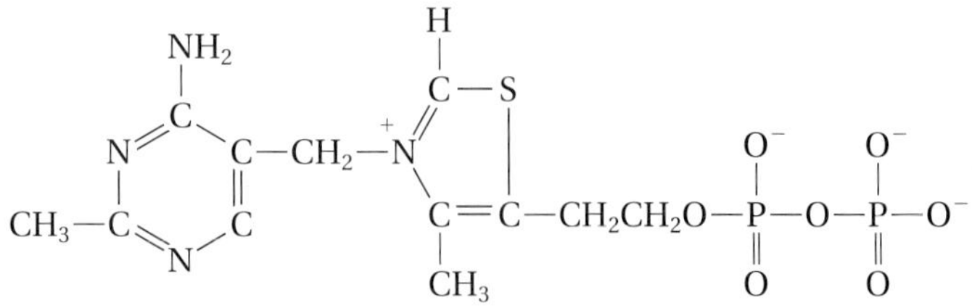

Thiamine pyrophosphate

Lipoic acid is not classified as a vitamin. Evidently humans can make their own lipoic acid. No case of lipoic acid deficiency in a human being has ever been reported.

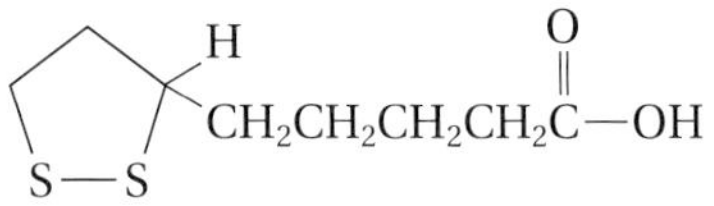

Lipoic acid

PRACTICE EXERCISE 15.3

Acetyl CoA is often considered an energy-rich compound like ATP. Explain why.

15.5 The citric acid cycle

AIM: To list the steps for the degradation of one acetyl group in the citric acid cycle.

Focus

The oxidation of glucose carbons is completed in the citric acid cycle.

The two molecules of acetyl CoA from one molecule of glucose now pass into the citric acid cycle. Figure 15.4 shows the complete cycle, which takes place in the mitochondria of eukaryotic cells. As in other metabolic pathways, all the reactions of the citric acid cycle are catalyzed by enzymes. Some of the necessary enzymes are located in the fluid contained inside the mitochondrial inner membrane; others are attached to the inner surface of the interior membrane. As we go through the steps of the cycle, be especially alert to the fates of the carbons of the reacting molecules, the various types of transformations that are occurring, and the production of NADH, FADH$_2$, and ATP.

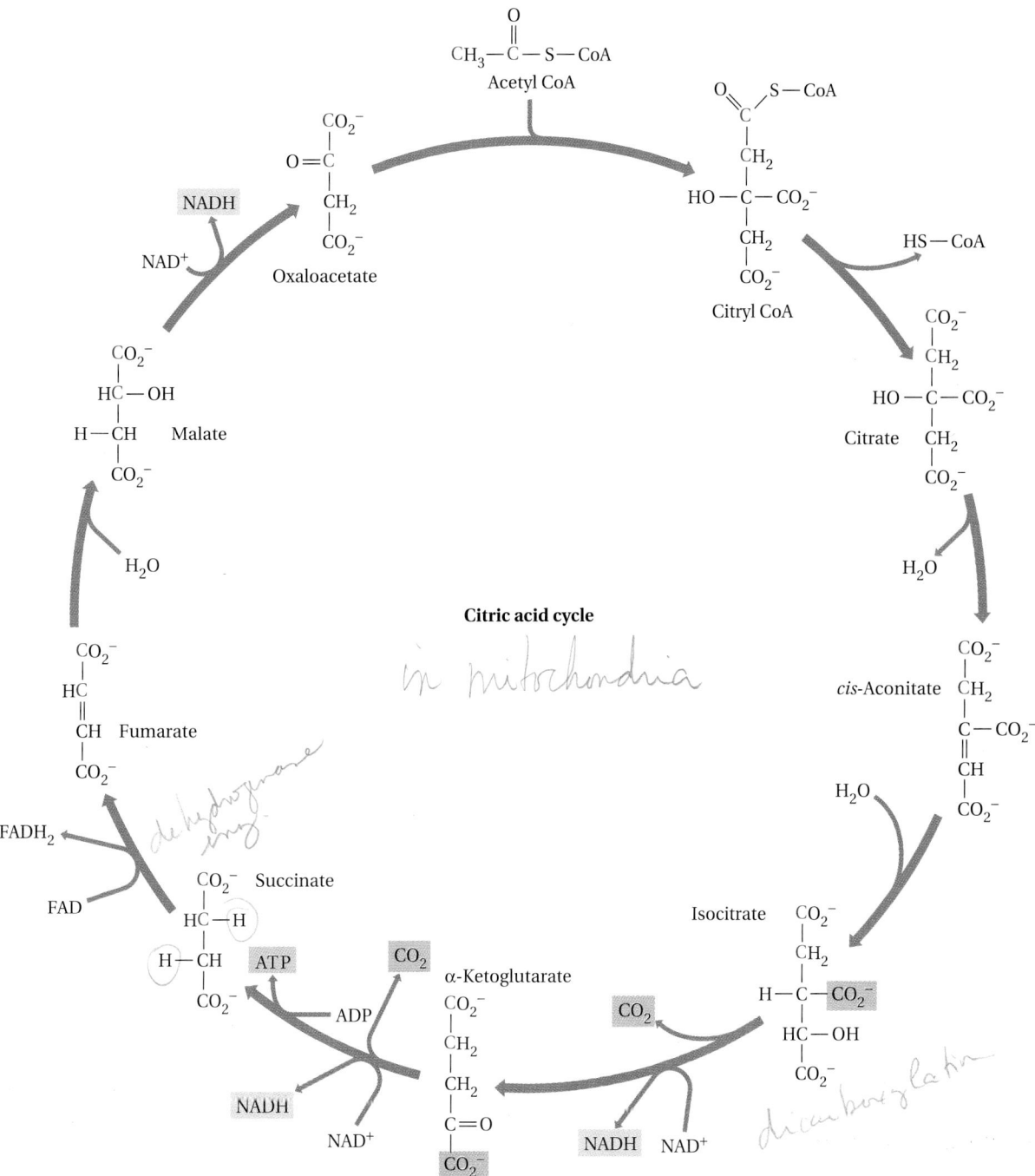

Figure 15.4
The citric acid cycle begins at the 12 o'clock position. As we follow the fate of the carbon atoms of the acetyl group of acetyl CoA, we see that they are not the ones lost as carbon dioxide in one turn of the cycle. Will either of these carbon atoms be oxidized to carbon dioxide during their second pass through the cycle?

The citric acid cycle begins when an acetyl group of acetyl CoA condenses with a molecule of oxaloacetate to give a molecule of citryl CoA.

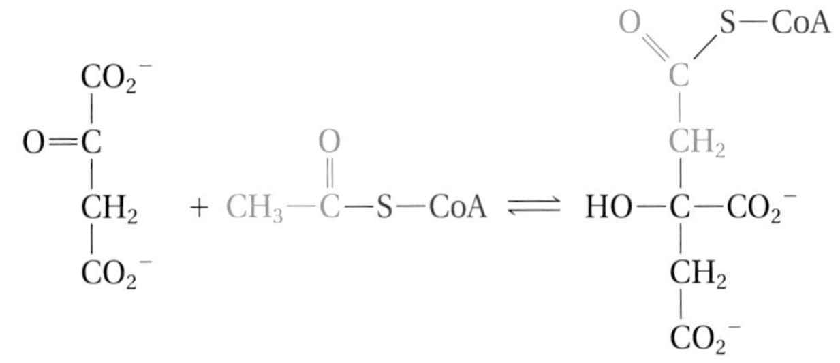

$$
\begin{array}{c}
CO_2^- \\
| \\
O=C \\
| \\
CH_2 \\
| \\
CO_2^-
\end{array}
\quad + \quad CH_3-\overset{\overset{\displaystyle O}{\|}}{C}-S-CoA
\quad \rightleftharpoons \quad
\begin{array}{c}
\overset{\displaystyle O}{\diagdown}\ \ \overset{\displaystyle S-CoA}{\diagup} \\
C \\
| \\
CH_2 \\
| \\
HO-C-CO_2^- \\
| \\
CH_2 \\
| \\
CO_2^-
\end{array}
$$

Oxaloacetate   Acetyl CoA   Citryl CoA
(CoA ester of citrate)

The thioester bond of citryl CoA is rapidly hydrolyzed to give citrate and CoA.

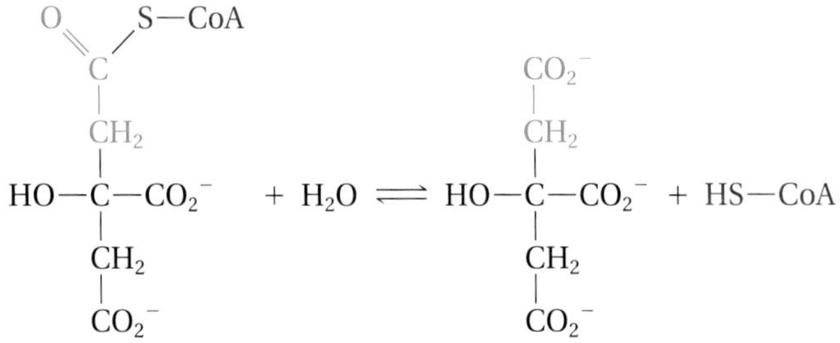

$$
\begin{array}{c}
\overset{\displaystyle O}{\diagdown}\ \ \overset{\displaystyle S-CoA}{\diagup} \\
C \\
| \\
CH_2 \\
| \\
HO-C-CO_2^- \\
| \\
CH_2 \\
| \\
CO_2^-
\end{array}
\quad + H_2O \rightleftharpoons
\begin{array}{c}
CO_2^- \\
| \\
CH_2 \\
| \\
HO-C-CO_2^- \\
| \\
CH_2 \\
| \\
CO_2^-
\end{array}
\quad + HS-CoA
$$

Citryl CoA   Citrate

The early discovery of citrate is the reason the pathway is called the *citric acid cycle.*

The next step of the cycle involves the dehydration of citrate to *cis-*aconitate.

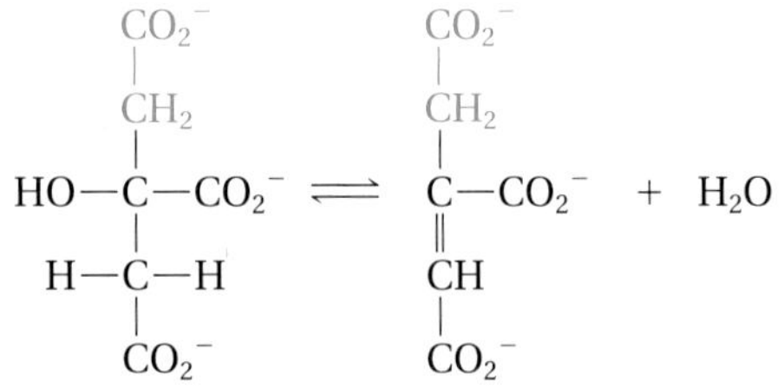

$$
\begin{array}{c}
CO_2^- \\
| \\
CH_2 \\
| \\
HO-C-CO_2^- \\
| \\
H-C-H \\
| \\
CO_2^-
\end{array}
\quad \rightleftharpoons \quad
\begin{array}{c}
CO_2^- \\
| \\
CH_2 \\
| \\
C-CO_2^- \\
\| \\
CH \\
| \\
CO_2^-
\end{array}
\quad + \quad H_2O
$$

Citrate   *cis*-Aconitate   Water

Now a hydration reaction occurs in which water is added to the double bond. Isocitrate is the product of the rehydration.

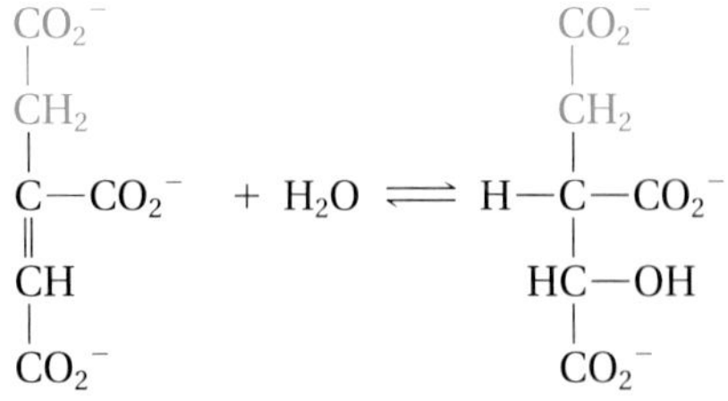

$$
\begin{array}{c}
CO_2^- \\
| \\
CH_2 \\
| \\
C-CO_2^- \\
\| \\
CH \\
| \\
CO_2^-
\end{array}
\quad + H_2O \rightleftharpoons
\begin{array}{c}
CO_2^- \\
| \\
CH_2 \\
| \\
H-C-CO_2^- \\
| \\
HC-OH \\
| \\
CO_2^-
\end{array}
$$

cis-Aconitate   Isocitrate

At this point the production of reducing power and ATP is about to begin as isocitrate is oxidized to α-ketoglutarate.

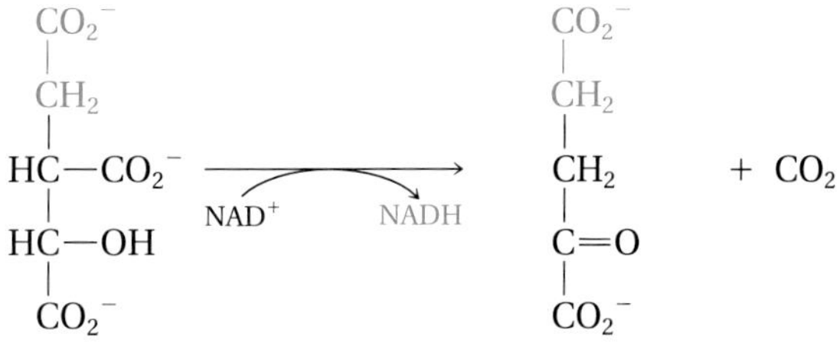

$$
\begin{array}{c}
CO_2^- \\
| \\
CH_2 \\
| \\
HC-CO_2^- \\
| \\
HC-OH \\
| \\
CO_2^-
\end{array}
\quad\xrightarrow[\text{NAD}^+ \quad \text{NADH}]{}\quad
\begin{array}{c}
CO_2^- \\
| \\
CH_2 \\
| \\
CH_2 \\
| \\
C=O \\
| \\
CO_2^-
\end{array}
\quad + \quad CO_2
$$

Isocitrate   α-Ketoglutarate

NAD^+ is the coenzyme in the reaction, and for every molecule of isocitrate oxidized, one molecule of NAD^+ is reduced to NADH. The oxidation also involves the loss of a molecule of carbon dioxide, but notice that the carbon dioxide comes from the oxaloacetate and not from the acetyl group that entered the cycle.

The next step in the cycle results in the loss of carbon dioxide, reduction of a second NAD^+, and phosphorylation of ADP to ATP (a substrate-level phosphorylation). Succinate is the product. Note that once again the carbon dioxide comes from oxaloacetate instead of acetyl CoA. A second NADH and one ATP are produced.

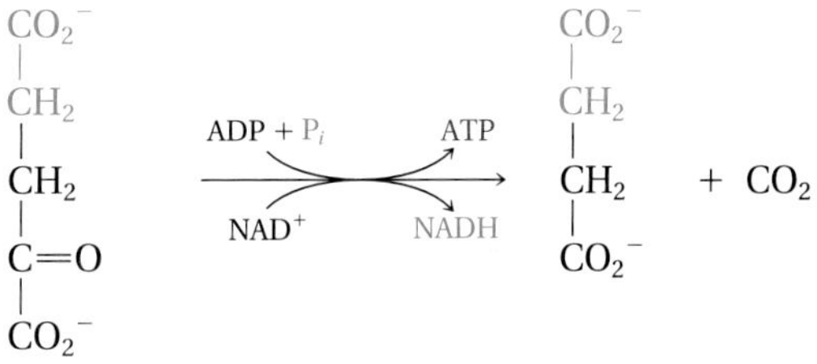

$$
\begin{array}{c}
CO_2^- \\
| \\
CH_2 \\
| \\
CH_2 \\
| \\
C=O \\
| \\
CO_2^-
\end{array}
\quad\xrightarrow[\text{NAD}^+ \quad \text{NADH}]{\text{ADP} + \text{P}_i \quad \text{ATP}}\quad
\begin{array}{c}
CO_2^- \\
| \\
CH_2 \\
| \\
CH_2 \\
| \\
CO_2^-
\end{array}
\quad + \quad CO_2
$$

α-Ketoglutarate   Succinate

Succinate is next oxidized to fumarate. Recall that oxidation-reduction enzymes that catalyze the formation of carbon-carbon double bonds usually use FAD as the coenzyme. That is the case here, where one $FADH_2$ is produced.

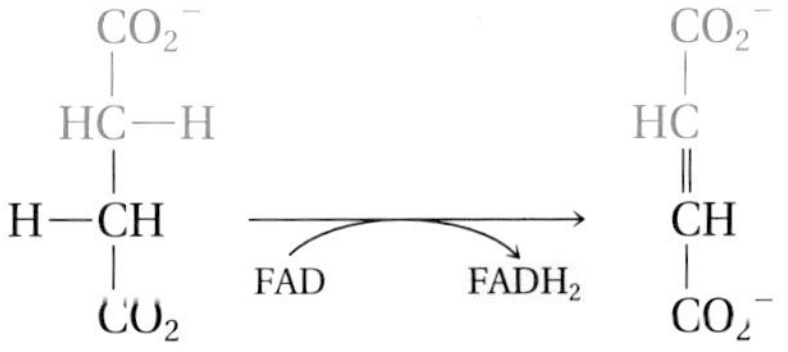

$$
\begin{array}{c}
CO_2^- \\
| \\
HC-H \\
| \\
H-CH \\
| \\
CO_2
\end{array}
\quad\xrightarrow[\text{FAD} \quad \text{FADH}_2]{}\quad
\begin{array}{c}
CO_2^- \\
| \\
HC \\
\| \\
CH \\
| \\
CO_2^-
\end{array}
$$

Succinate   Fumarate

A second hydration reaction now occurs as fumarate is converted to malate.

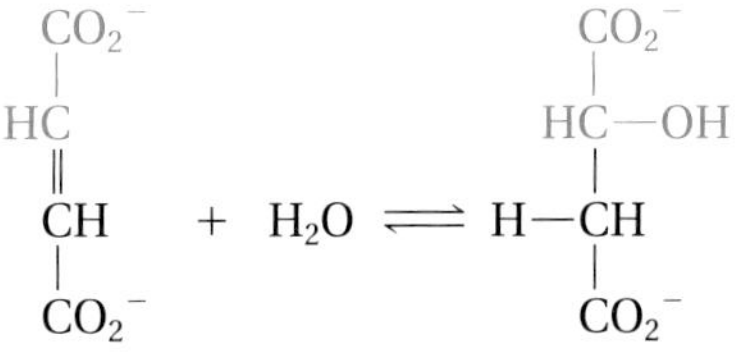

$$
\begin{array}{c}
CO_2^- \\
| \\
HC \\
\| \\
CH \\
| \\
CO_2^-
\end{array}
\quad + \quad H_2O \quad\rightleftharpoons\quad
\begin{array}{c}
CO_2^- \\
| \\
HC-OH \\
| \\
H-CH \\
| \\
CO_2^-
\end{array}
$$

Fumarate   Malate

Malate is oxidized back to oxaloacetate in the final step of the cycle, and NAD^+ is simultaneously reduced. A third NADH is produced.

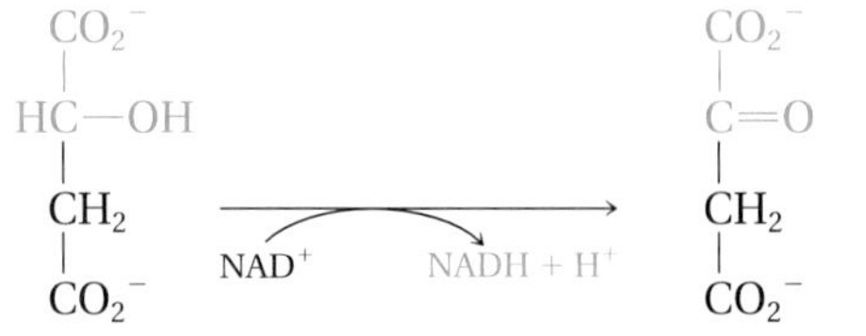

The oxaloacetate molecule is now available to start another turn of the cycle by reacting with another molecule of acetyl CoA.

This is what happens in the citric acid cycle:

1. Acetyl CoA and oxaloacetate combine to form citrate.

2. Citric acid eventually loses two carbon atoms as carbon dioxide. The carbons in the two molecules of carbon dioxide are not the same carbons that entered the citric acid cycle as acetyl groups of acetyl CoA. Nevertheless, the *net effect* of the cycle is the same as if these carbons were oxidized.

3. At the end of the pathway, a molecule of oxaloacetate remains, which is why the pathway is called a *cycle*. (The original oxaloacetate molecule could have come from several places in metabolism, but we need not worry about that here.)

4. Each turn of the citric acid cycle yields three molecules of NADH, one of $FADH_2$, and one of ATP.

PRACTICE EXERCISE 15.4

Write a net equation that summarizes one turn of the citric acid cycle.

oxaloacetate + acetyl CoA + 3 NAD⁺ + 1 ADP + 1 Pᵢ + 1 FAD ⟶ oxaloacetate + 3 NADH + 1 ATP + 1 FADH₂ + 2 CO₂

15.6 ATP yield

AIM: To give an accounting of the ATP produced by the complete oxidation of a molecule of glucose.

Focus

The complete oxidation of 1 molecule of glucose yields 38 ATP molecules.

Two turns of the citric acid cycle are necessary to oxidize completely the equivalent of two molecules of acetyl CoA (obtained from one molecule of glucose) to four molecules of carbon dioxide. Using this information, along with the yields of ATP and NADH obtained in glycolysis, we can calculate the total amount of ATP generated by the aerobic catabolism of one molecule of glucose. Remember that each NADH molecule can generate three ATP molecules and each $FADH_2$ molecule can generate two ATP molecules by oxidative phosphorylation. Table 15.2 summarizes the

Table 15.2 ATP Production from Complete Aerobic Catabolism of One Glucose Molecule

Pathway	ATP yield
glycolysis (4ATP generated minus 2 invested)	2
citric acid cycle (2 acetyl groups of acetyl CoA oxidized to carbon dioxide)	2
oxidative phosphorylation:	
2NADH from glycolysis	6
2NADH from acetyl CoA formation	6
6NADH from the citric acid cycle	18
2FADH$_2$ from the citric acid cycle	4
	38

Burning a mole of glucose to form CO_2 and H_2O produces 686 kcal of energy. When we oxidize a mole of glucose, the 38 mol of ATP represents 277 kcal of stored energy. Our bodies, then, are about 277 kcal/686 kcal $\times$ 100 = 40% efficient.

results. The data show that enough of the energy released in the complete oxidation of 1 molecule of glucose is trapped by aerobic cells to generate 38 ATP molecules.

15.7 Lactic Fermentation

AIM: To explain why cells sometimes use lactic fermentation for energy production.

Focus

Aerobic cells switch to lactic or alcoholic fermentation in the absence of oxygen.

So far our discussion of glucose metabolism has assumed that the cell doing the metabolism has plenty of oxygen available. What if it does not? We know that the mitochondria of aerobic cells need oxygen so that the electron transport chain can operate. When there is no oxygen available to drain electrons from NADH and FADH$_2$ in respiration, the electron carriers of the electron transport chain become completely reduced. More electrons cannot be passed down the chain, and oxidative phosphorylation stops. However, the levels of NADH and FADH$_2$ in the mitochondrion increase as the citric acid cycle continues to operate. Soon, not enough NAD$^+$ and FAD are regenerated by respiration to sustain the operation of the citric acid cycle, and the mitochondrial power plant shuts down.

Now the only place that ATP is being produced is in the cytoplasm. Here, two molecules of ATP are produced for every glucose molecule converted to two molecules of pyruvate in glycolysis. Cytoplasmic NAD$^+$ is also being reduced to NADH, and NAD$^+$ is needed for glycolysis to continue. (It is needed to change glyceraldehyde 3-phosphate to 1,3-bisphosphoglycerate.) If there were no way to regenerate NAD$^+$, glycolysis too would stop. With no energy production, the cell would die.

In such an emergency, the cells of many aerobic organisms regenerate NAD$^+$ from the NADH formed in glycolysis by using the NADH to reduce pyruvate to lactate.

Pyruvate → Lactate

The reduction of pyruvate to lactate is called **lactic fermentation.** (For more information, see A Closer Look: Monitoring Heart Attacks with Serum LDH Tests.) Lactic fermentation keeps glycolysis going. Since aerobic catabolism produces 38 ATP molecules from 1 molecule of glucose and lactic fermentation produces only 2, the aerobic catabolism of glucose is *19 times* more efficient than lactic fermentation. Nevertheless, considering the choices between death and life at a lower level of ATP production, lactic fermentation is not a bad bargain for the cell.

A Closer Look

Monitoring Heart Attacks with Serum LDH Tests

Many clinical laboratory tests in hospitals measure the amount of a critical enzyme in the bloodstream. One such enzyme, *lactate dehydrogenase* (LDH), is an important catalyst, promoting the reduction of pyruvate to lactate in your body. Consisting of four polypeptide chains or tetramer subunits, LDH exhibits multiple molecular forms called *isozymes*. LDH is composed of two principal kinds of subunits, H and M, that differ slightly in their primary structures. The five isozymes of LDH, H_4, H_3M, H_2M_2, HM_3, and M_4, are various combinations of these subunits. Heart and liver LDH is rich in H subunits; muscle LDH is rich in M subunits.

Because damaged tissues often release LDH into the bloodstream, certain diseases can be detected by a clinical laboratory test that measures *serum LDH* levels. In this procedure, a patient's blood sample is analyzed, and the rate at which the serum converts pyruvate to lactate is obtained. This rate, in turn, determines the level of LDH that may indicate a general abnormality. Often it is useful to track the course of disease in specific tissues. A diagnosis can be refined by using another technique called *electrophoresis* to separate the total LDH in a blood sample into its various isozymes (see figure). Abnormal amounts

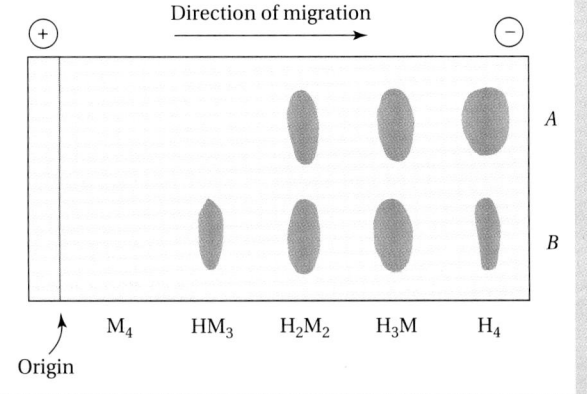

Electrophoresis of serum LDH isozymes at pH 8.6: pattern *A* belongs to a heart attack patient; pattern *B* is that of a normal individual. Leakage of cell contents of damaged heart muscle in the heart attack patient has substantially increased the serum level of the H_4 isozyme.

of a given isozyme can, in turn, narrow the search for disease to specific organs or help monitor the magnitude and course of disease in the body.

During a heart attack, for example, damaged heart muscle releases LDH, mainly the H_4 isozyme. Within 24 hours of the heart attack episode, serum LDH reaches a peak; it then returns to a normal level within 5 or 6 days. A physician often can get a good idea of the extent of a patient's heart damage by carefully monitoring the amount of H_4 isozyme over this period of time.

The product of fermentation processes is not always lactate. Some organisms—brewer's yeast is a particularly well-known example—oxidize the pyruvate formed in glycolysis to acetaldehyde.

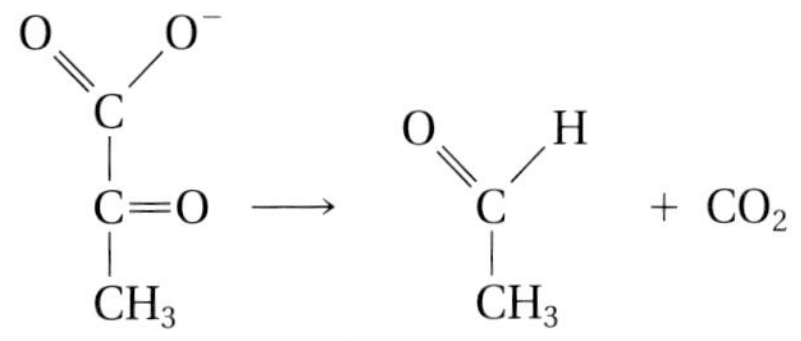

Pyruvate   Acetaldehyde

The reduction of acetaldehyde regenerates NAD^+ to keep glycolysis going. The reduced product of acetaldehyde reduction is ethanol.

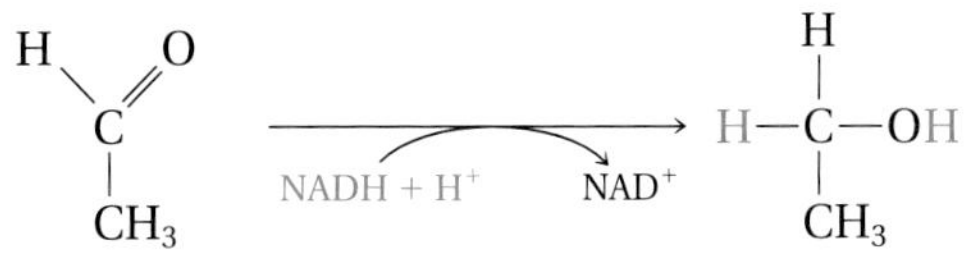

Acetaldehyde   Ethanol

The process by which glucose is degraded to ethanol is called **alcoholic fermentation.**

PRACTICE EXERCISE 15.5
Aerobic catabolism of glucose is much more efficient than fermentation. What does this statement mean?

15.8 Oxygen debt

AIMS: *To describe the conditions that could cause a state of oxygen debt. To outline two possible fates of lactate once cellular respiration is returned to normal.*

Focus

Lactate is oxidized to carbon dioxide or made into glucose.

Your respiratory system supplies the mitochondria of your body cells with enough oxygen to operate their respiratory chains efficiently at normal levels of activity. If you undertake vigorous exercise, such as sprinting or long-distance running, your cells step up respiration to make ATP to meet their increased energy needs. You breathe more rapidly and more deeply, but eventually you may develop an **oxygen debt**—*that is, not enough oxygen is available for cellular respiration.* At this point, your muscle cells begin to use the much less efficient lactic fermentation route to make ATP.

Lactate buildup

Lactate builds up as your cells regenerate NAD^+ from NADH by lactic fermentation. If you have an oxygen debt and do not stop to rest, the lactic acid concentration in your muscles continues to build. Lactic acid is toxic to muscle cells in high concentrations and results in muscle cramps and soreness. (Remember how you felt the day after a long-distance run or swim?) If you push on much further, you will soon collapse, because lactic fermentation cannot supply enough ATP to take care of your body's heavy demands.

If you do stop to rest, you will pant as your respiratory system supplies the oxygen necessary to pay the oxygen debt. Your cellular respiration is soon restored to normal, but body cells must still deal with the lactate built up during lactic fermentation.

Lactate disposal

Some of the lactate is oxidized back to pyruvate by a reversal of the lactic dehydrogenase–catalyzed reactions.

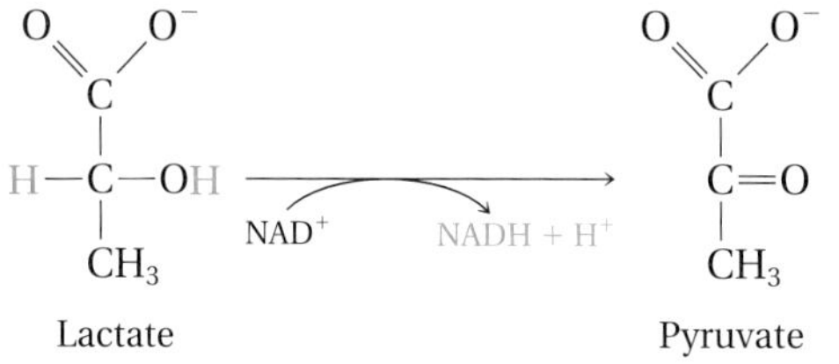

The NADH that is formed enters the respiratory chain, where ATP is produced by oxidative phosphorylation. The pyruvate is converted to acetyl CoA, which enters the citric acid cycle. There the acetyl group of acetyl CoA undergoes complete oxidation to carbon dioxide.

Some of the lactic acid formed in your muscles is not oxidized. It drains from the muscle cells and enters the bloodstream, where it could give you a severe case of acidosis. As we learned in Chapter 13, however, blood is buffered. Your blood buffers, mainly the HCO_3^-/H_2CO_3 system, will absorb the extra protons added to your bloodstream by the lactic acid. And if need be, your kidneys will generate new bicarbonate buffer ions that are lost as carbon dioxide through breathing. Your kidneys also will expel excess protons in your urine.

Gluconeogenesis

The lactic acid in your bloodstream is absorbed by your liver. Liver cells, but not muscle cells, can convert lactate back to glucose in several steps.

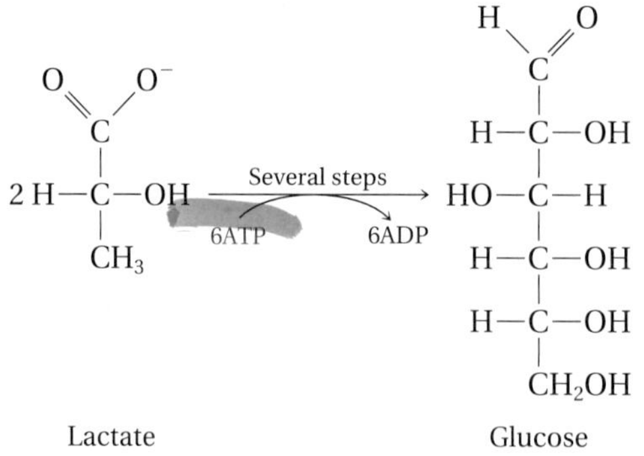

Gluconeogenesis
gluco (Greek): sweet, glucose
neo (Greek): new
genesis (Greek): creation

The pathway by which lactate is converted to glucose is called **gluconeogenesis.** Gluconeogenesis, which is the synthesis of glucose from starting materials that are not carbohydrates, is an example of an anabolic (synthetic) pathway. Like most anabolic pathways, it requires the expenditure of ATP. Six molecules of ATP are required to convert two molecules of lactate to one molecule of glucose. However, only two molecules of ATP were gained by converting one molecule of glucose to two molecules of lactate in lactic fermentation. Your liver cells must pay for this deficit of ATP. They do so with the abundant ATP produced in oxidative phosphorylation.

We will not be studying gluconeogenesis in detail. Nevertheless, the first step of the pathway is particularly interesting because it is also relevant to the citric acid cycle. It involves the chemical combination of pyruvate and carbon dioxide, a *carboxylation reaction,* to give oxaloacetate.

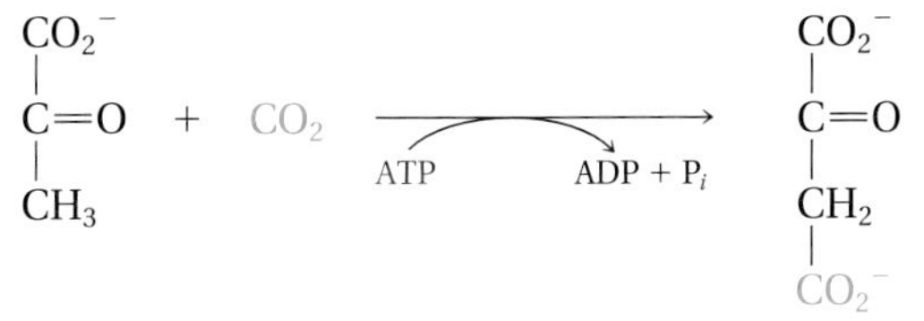

Pyruvate   Carbon dioxide   Oxaloacetate

The carboxylation of pyruvate is energetically unfavorable and requires the expenditure of one of the six ATP molecules invested in gluconeogenesis. Pyruvate carboxylase, the enzyme that catalyzes the reaction, requires the B-complex vitamin biotin as the coenzyme. The biotin serves as a carrier of carbon dioxide in this and other biological carboxylation reactions in the form of carboxybiotin. It is attached to the side-chain terminal amino group (not the alpha amino group) of a lysine residue of the protein through an amide bond.

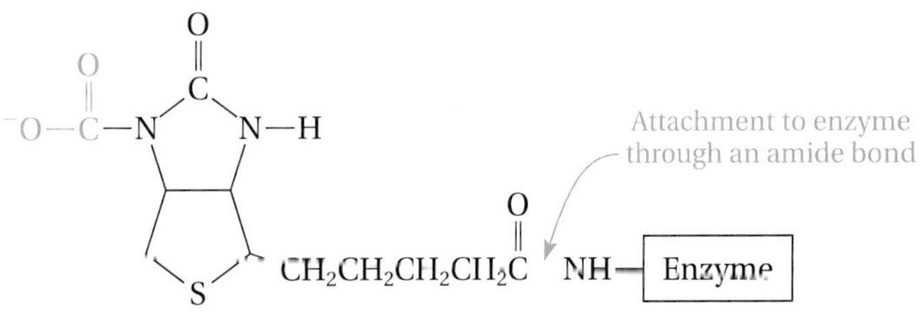

Carboxybiotin

The oxaloacetate produced in the reaction meets one of two fates: It is converted to glucose in the remainder of the reactions of gluconeogenesis, or alternatively, it may enter the citric acid cycle. Indeed, the carboxylation of pyruvate is the major source of oxaloacetate for the citric acid cycle.

15.9 Glucose storage

AIM: To describe how excess glucose is temporarily stored in the body.

Focus

Glucose is stored as glycogen.

Foods containing carbohydrates are one source of the glucose used for energy production in the cell. As you have seen, another source of glucose is gluconeogenesis. *Glucose that is not required to meet the immediate energy needs of the body is assembled into glycogen, the animal form of starch; this anabolic process is called* **glycogenesis.** Glycogenesis requires two enzymes: *glycogen synthetase* and *branching enzyme*. Glycogen synthetase promotes the formation of the $\alpha(1\longrightarrow4)$ glycosidic bonds that form the straight-chain portions of glycogen. As its name implies, the branching enzyme promotes the formation of $\alpha(1\longrightarrow6)$ glycosidic linkages at branch points of the bushy glycogen molecule.

Glucose adds to an existing glycogen chain only if it is first activated to uridine diphosphate glucose (UDP-glucose). The starting materials for the activation are glucose 1-phosphate and uridine triphosphate (UTP), one of the nucleotide triphosphates used in the synthesis of RNA. The glucose 1-phosphate comes from glucose 6-phosphate by way of the following enzyme-catalyzed reaction:

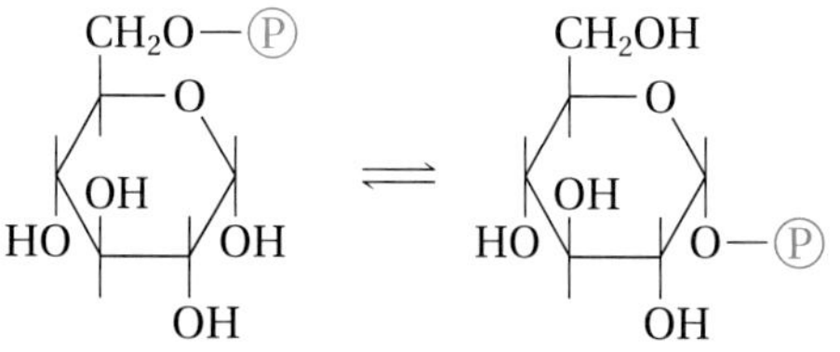

Glucose 6-phosphate   Glucose 1-phosphate

The glucose 1-phosphate now reacts with the UTP to form UDP-glucose and inorganic pyrophosphate (PP_i).

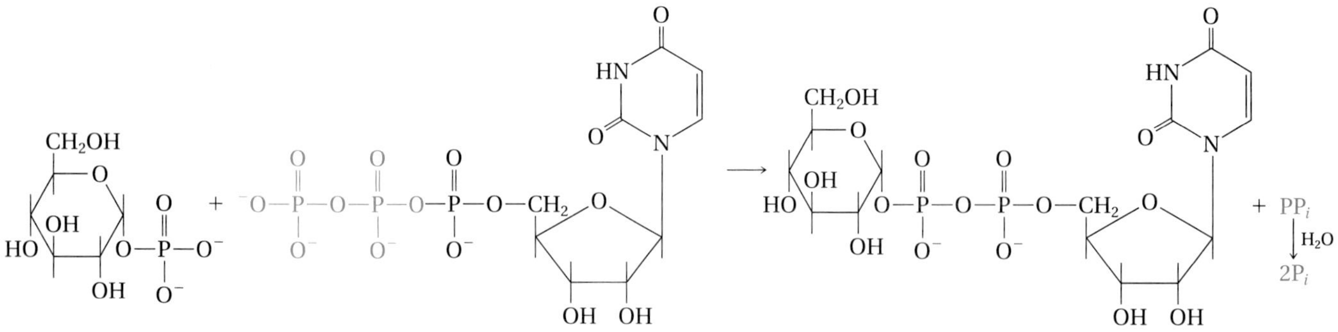

Glucose 1-phosphate   Uridine triphosphate (UTP)   Uridine diphosphate glucose (UDP-glucose)

Note that the hydrolysis of the high-energy anhydride bond of the PP_i

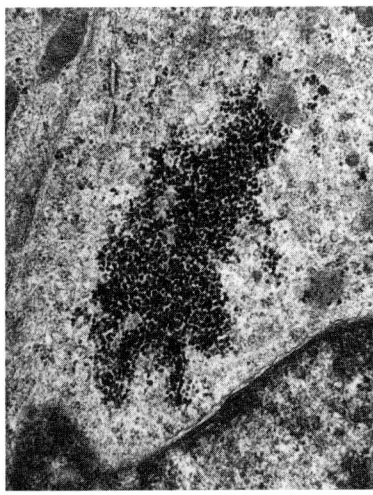

Figure 15.5
Liver cell of a rat fed a high-carbo-hydrate diet. The dark areas are masses of glycogen.

releases substantial energy and therefore helps drive the formation of the UDP-glucose. Once the UDP-glucose is formed, its glucose unit attaches to the end of an existing glycogen chain by an $\alpha(1 \longrightarrow 4)$ linkage.

UDP-glucose   Glycogen chain

Uridine diphosphate (UDP)   Glycogen chain plus 1 new glucose unit

No high-energy phosphoric acid anhydride bonds are broken in this reaction, so the cell's total investment for each molecule of glucose incorporated into glycogen is two high-energy phosphate bonds—one to make glucose 6-phosphate from glucose and one in the hydrolysis of PP_i to $2P_i$.

Glycogen is temporarily stored in muscle cells and liver cells until glucose is needed (Fig. 15.5). The amount of liver glycogen is seldom more than 5% of the weight of fresh liver tissue. Muscle glycogen seldom exceeds 1% of the weight of fresh muscle.

A typical runner can call on his or her glycogen reserves to provide glucose for energy for 2 to 3 hours. Once the glycogen reserves are gone, a runner "hits the wall" and is unable to maintain the previous level of activity.

PRACTICE EXERCISE 15.6

In glycogenesis, the glucose 1-phosphate comes from glucose 6-phosphate by way of an enzyme-catalyzed reaction. Where does the glucose 6-phosphate come from?

15.10 Glycogen breakdown

AIM: To distinguish between the body's use of liver glycogen and muscle glycogen.

Focus

Stored glycogen is degraded to meet glucose needs.

When your body's glucose supply gets low, glycogen is broken down to meet energy needs. *The process by which glycogen is broken down is called* **glycogenolysis** ("splitting of glycogen"). Glycogenolysis starts with the conversion of glycogen into glucose 1-phosphate. The enzyme *phosphorylase* catalyzes this conversion.

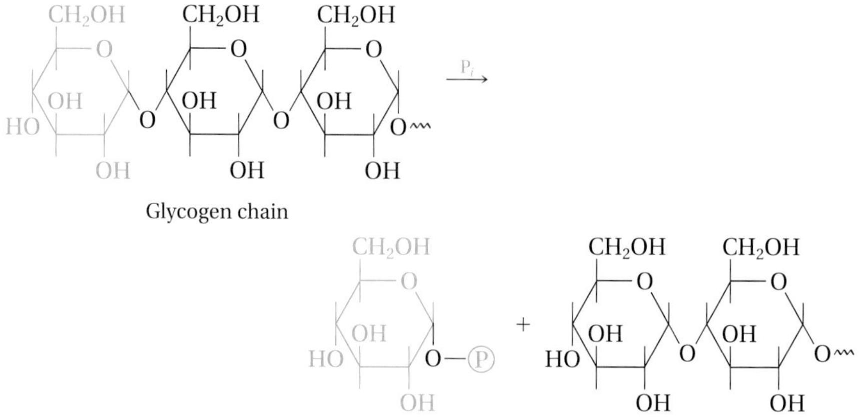

Glycogen chain

Glucose 1-phosphate   Glucose chain minus
1 glucose unit

In the process of cleaving glycogen, a phosphoryl group of inorganic phosphate (P_i) is transferred to glucose. To enter glycolysis, the glucose 1-phosphate must be converted to glucose 6-phosphate. The catalysis of this reaction is accomplished by a reversal of the reaction used to make glucose 1-phosphate.

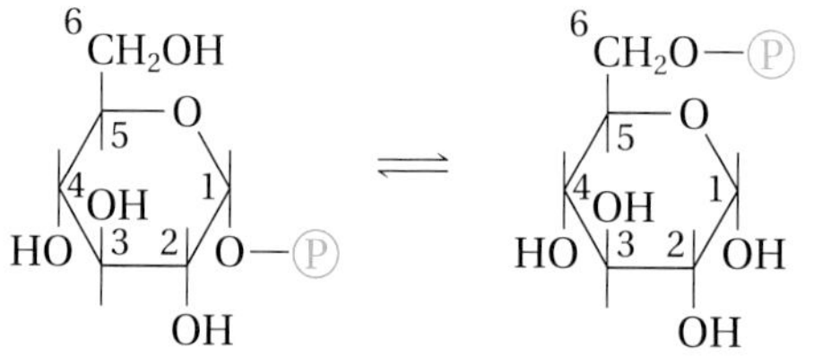

Glucose 1-phosphate   Glucose 6-phosphate
(enters glycolysis)

Negatively charged glucose 6-phosphate cannot readily pass through the membranes of cells. In muscle cells, all the glucose 6-phosphate produced by glycogenolysis is used for glycolysis. Liver cells have a broader responsibility: They must distribute glucose to other tissues via the blood. Therefore, liver cells, but not muscle cells, contain an enzyme that catalyzes the hydrolysis of glucose 6-phosphate to glucose and P_i.

$$\text{Glucose 6-phosphate} + H_2O \longrightarrow \text{Glucose} + P_i$$

The glucose leaves the liver cells and enters the blood, which delivers it to tissues where it is urgently needed. Stored glycogen does not last long. After a 24-hour fast, there is virtually no glycogen left in either the liver or the muscle.

In the preceding section we noted that the synthesis of glycogen requires the expenditure of two high-energy phosphate bonds for each molecule of glucose incorporated into glycogen (the equivalent of converting two molecules of ATP to ADP). Since the aerobic catabolism of each glucose molecule produces 38 ATP molecules, a net total of 36 ATP molecules is produced from each glucose molecule stored as muscle glycogen. Thus glycogen is nearly as good a source of cellular energy as glucose itself. Ath-

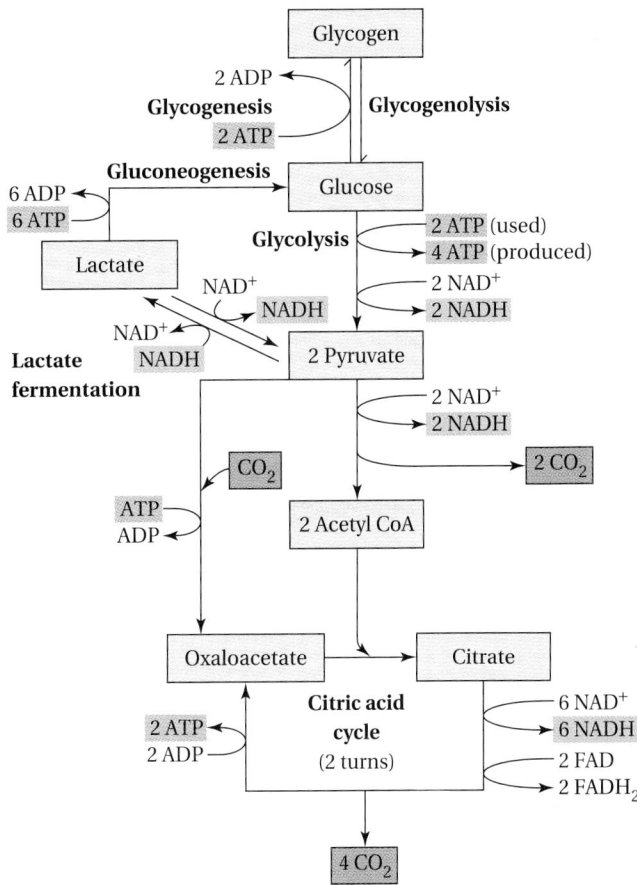

Figure 15.6
The many pathways of glucose metabolism.

letic endeavors require energy. Barbara, the biochemically savvy cross-country coach from the Case in Point earlier in this chapter, would like her athletes to store as much glycogen as possible before a meet. In the Follow-up to the Case in Point, below, we will learn more about this strategy.

An overview of the metabolic pathways we have discussed is presented in Figure 15.6.

FOLLOW-UP TO THE CASE IN POINT: Carbohydrate loading

Athletes, especially those involved in long-distance running, often try to improve their performances by maximizing the amount of stored muscle glycogen. The practice of eating large amounts of carbohydrate-rich foods such as pasta before strenuous activities is known as *carbohydrate loading*. By prescribing a pasta diet to her team, Barbara, the college cross-country coach, is attempting to maximize her runners' stores of muscle glycogen so that the glyco-

gen will be available to meet the glucose needs of their muscle cells during the meet. Because these glycogen stores are fairly small and rapidly depleted, they must be built up within a day or two of the athletic event. The maximum amount of glycogen that can be stored is different among individuals and is determined by genetics, but people can approach their personal maximum by physical conditioning.

PRACTICE EXERCISE 15.7

Explain why liver glycogen, but not muscle glycogen, is a source of glucose to other body tissues.

15.11 Metabolic regulation

AIM: To describe how hormones regulate metabolism through the use of second messengers.

One of the most fascinating aspects of metabolism is its regulation—how cellular processes are turned on when they are needed and turned off when they are not. So that you can clearly grasp the regulation of glucose metabolism, you will first need to know more about how hormones control cellular processes in general.

Examples of hormonal control

Since metabolic reactions are enzyme-catalyzed, you might reasonably expect the control of cellular metabolism to involve the regulation of enzyme activity, and it often does. In complex organisms such as human beings, many processes are controlled by hormones. In fact, over 30 effects of a single hormone, insulin, are known. You have already seen that hormones control basal metabolism (thyroxine), permeability of cellular membranes (vasopressin), and the activity of ion pumps (aldosterone). The prostaglandins also exhibit hormone-like effects. The study of prostaglandins is complicated, though, by the fact that they seem to have opposite effects on different kinds of tissue cells, whereas hormones influence all their target cells in the same way.

Second messengers

Hormones are powerful regulators of processes that occur inside cells, but few hormones are ever found inside the cells they act on. The discovery that hormones that reach target cells bind to specific protein receptors on the outer surface of the cell membrane was an important contribution to the study of hormones and cells. This discovery did not, however, explain how hormones influence processes *inside* cells. Scientists had few leads to the solution of this mystery until Earl Sutherland and his colleagues discovered *cyclic AMP* (cAMP) in 1957 (Fig. 15.7). Their work, and the work of those who followed them, has shown that formation of a hormone-receptor complex on the exterior of a cell membrane activates an enzyme, *adenylate cyclase*, that is attached to the interior of the cell membrane (Fig. 15.8).

Adenylate cyclase catalyzes the conversion of ATP to cAMP. In most instances, cAMP then binds to the allosteric sites of a group of enzymes called *protein kinases*. In other words, cAMP is a positive modulator (see Sec. 10.10) of protein kinases. Activated protein kinases catalyze the transfer of phosphoryl groups from ATP to various proteins. These proteins are

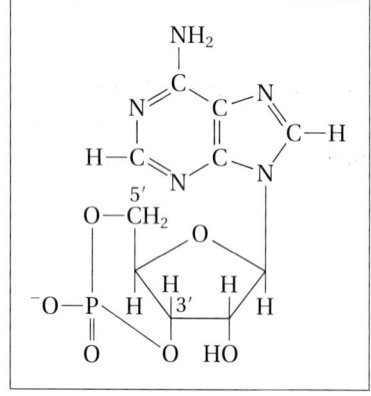

Figure 15.7
Cyclic adenosine monophosphate (cAMP). Note that in cAMP a cyclic phosphodiester linkage is formed between the hydroxyl groups on the 3' and 5' positions of the ribose ring.

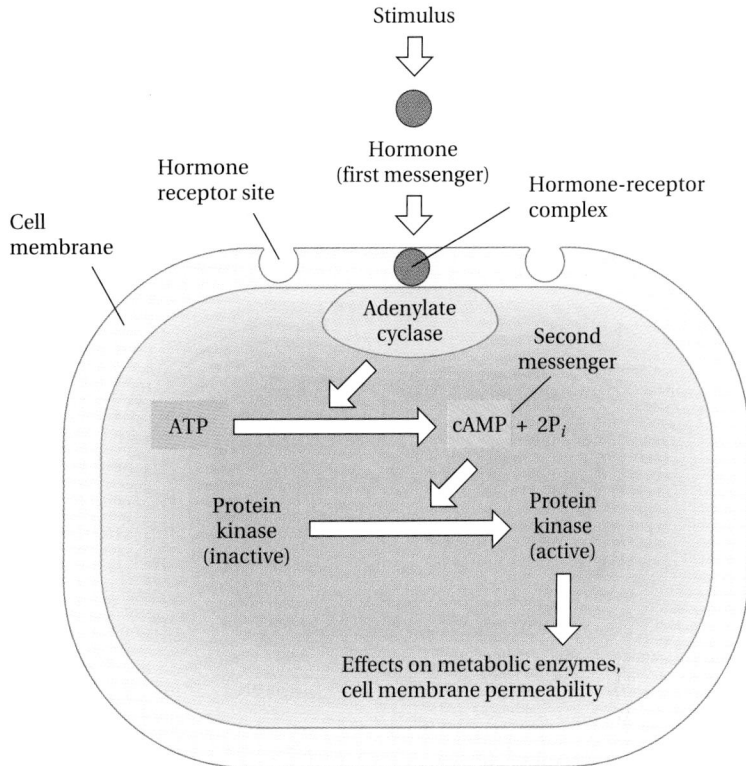

Figure 15.8
Many hormones exert their influence on cells by stimulating the production of cAMP, the second messenger.

mainly enzymes that catalyze important cellular processes. Phosphorylation activates many of these enzymes, but it deactivates some; ultimately, the presence of cAMP starts some cellular processes and stops others. Cyclic AMP also may play roles other than that of activating protein kinases. In at least one instance, cAMP activates an enzyme by releasing it from an enzyme-inhibitor complex (Fig. 15.9).

Hormone action is often compared with a messenger service in which the hormone is the **first messenger,** *delivering a regulatory message to the cell. Cyclic AMP is often the* **second messenger,** *taking the message to its final destination: the enzymes within the cell.* There are other second messengers besides cAMP. For example, scientists have found that the action of some hormones is impaired in the absence of calcium ions. They see this effect at both low and high levels of cAMP in the cellular cytoplasm.

Figure 15.9
Cyclic AMP activates some enzymes by releasing them from inactive complexes.

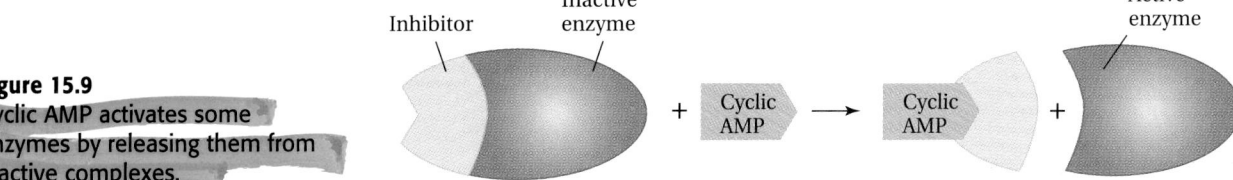

This indicates that calcium ions rather than cAMP are the second messengers. Sometimes calcium ions and cAMP work together to cause a certain effect on cellular metabolism. In addition to cAMP and calcium ions, two very simple molecules have been found to serve as second messengers in many important physiologic processes. These molecules are discussed in A Closer Look: Nitric Oxide and Carbon Monoxide as Second Messengers.

Nitric Oxide and Carbon Monoxide as Second Messengers

In a world of complex biological molecules such as cyclic AMP, simple nitric oxide (NO) seems an unlikely candidate for an important role in the body. Its simplicity aside, nitric oxide (NO) is one of the most important messengers in the body. A specific enzyme, nitric oxide synthase, makes NO from nitrogen in the amino acid arginine. Nitric oxide is a gas under atmospheric conditions. In the body it lasts only 6 to 10 seconds before it is oxidized by oxygen and water to nitrite ions (NO_2^-) and nitrate ions (NO_3^-). Despite its fleeting existence, NO is responsible for a host of effects in the body. The molecule is toxic to many kinds of cells, but certain white blood cells called *macrophages* are resistant to its toxicity. Macrophages contain NO and use it to kill tumor cells, fungi, and bacteria. NO also relaxes muscles of blood vessels, which permits the vessels to expand and reduce blood pressure. Many patients who suffer from *angina pectoralis,* chest pain that is caused by spasms of the arteries of the heart, are treated with nitroglycerin. The relaxation of these arteries relieves the symptoms of angina. Nitroglycerin is degraded to NO in the body, the spasms stop, and angina is relieved. NO simultaneously behaves much like a neurotransmitter in the brain and other parts of the body. Learning and memory are thought to involve the passage of neurotransmitter molecules between the nerve cells of the brain (neurons). NO may enable us to learn and to remember what we have learned.

Perhaps even more surprising than NO as a messenger in the body is the recent implication of carbon monoxide (CO). Most of us know that carbon monoxide is a toxic component of automobile exhaust and other incomplete combustion of hydrocarbons. Scientists have found that our bodies naturally make a small amount of carbon monoxide by the oxidation of heme, the oxygen-carrying part of the hemoglobin molecule in red blood cells. An enzyme called *heme oxygenase* catalyzes the formation of carbon monoxide from heme. The details of the action of carbon monoxide are sketchy at present, but it appears that this tiny molecule is the main regulator of the amount of the second messenger cGMP (see Sec. 15.12) in at least some brain cells.

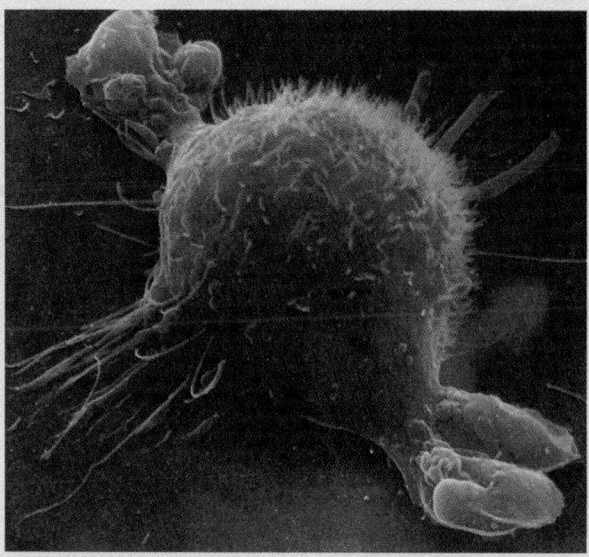

Macrophages are resistant to the toxicity of nitric oxide (NO).

15.12 Control of glycogenolysis

AIM: To describe the hormonal control of glycogenolysis.

Focus

Glucagon and epinephrine stimulate glycogenolysis.

Recall from Section 14.7 that phosphorylase, an important enzyme catalyst of glycogenolysis, is converted from an inactive form (*phosphorylase b*) to an active form (*phosphorylase a*) upon phosphorylation by ATP. This conversion is subject to hormonal control by glucagon, a peptide hormone. When levels of blood glucose are low, glucagon produced by the pancreas enters the bloodstream and is carried to its target cells—the primary target of glucagon is liver cells. Glucagon stimulates the production of cAMP (Fig. 15.10), which activates a protein kinase by binding to it. The protein kinase catalyzes the phosphorylation of phosphorylase *b*, producing phosphorylase *a*. Phosphorylase *a* then begins to catalyze the breakdown of glycogen. The end result of the action of glucagon is to raise the level of glucose in the blood.

Another hormone that stimulates the breakdown of glycogen is epinephrine (adrenaline). Epinephrine is often called the "flight or fight" hormone because it is secreted by the adrenal glands in response to a threat of bodily harm. The main effect of epinephrine is on muscle cells, which may need energy to meet the threat. The nervous system is also mobilized

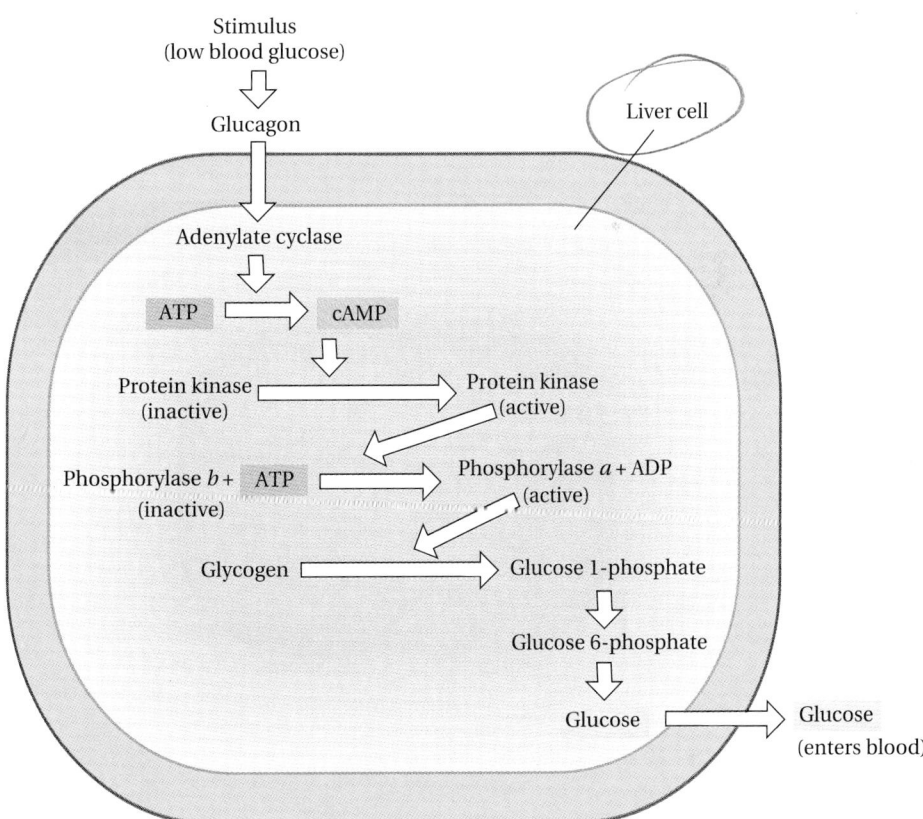

Figure 15.10
Glucagon promotes the breakdown of glycogen stored in liver cells to glucose. Some of the glucose produced by liver cells is carried by the blood to other tissues.

simultaneously. Within moments of the release of epinephrine, all systems are *go*. You may have experienced the unique feeling that results—perhaps at a crucial point in an athletic contest, in an automobile accident or near-accident, or by stepping on a snake on a woodland trail.

The path leading from epinephrine production to glycogenolysis in muscle is similar to the path used in the liver. There is, however, one major difference. Because muscle cells lack the enzyme necessary to hydrolyze glucose 6-phosphate to glucose and glucose 6-phosphate cannot easily pass through cell membranes, all the glucose 6-phosphate produced by glycogenolysis in muscle cells is used for glycolysis within the muscle.

15.13 *Glucose absorption*

AIMS: *To describe the production and function of insulin. To name the terms for low and high sugar levels in the blood. To describe the physiologic effects of low and high sugar levels in the blood. To explain the phrase "starvation in the midst of plenty" as applied to diabetes.*

Focus

Insulin stimulates the absorption of glucose by tissue cells.

Although all cells degrade glucose, most carbohydrate metabolism occurs in three places in the body: the muscle, the fatty tissue, and the liver. Liver cells present no barrier to the entry of glucose from the blood. The passage of blood glucose through the membranes of muscle and fat cells must be stimulated by the peptide hormone insulin (Fig. 15.11).

Proinsulin, an inactive form of insulin, is synthesized as a single peptide chain in the pancreas. Proinsulin is packaged in granules and converted to active insulin within the granules by the action of peptidases. The activation of proinsulin to insulin is very similar to the activation of zymogens to active enzymes.

Upon release into the bloodstream, insulin initiates a chain of events by which blood glucose is permitted to enter muscle and fat cells (Fig. 15.12).

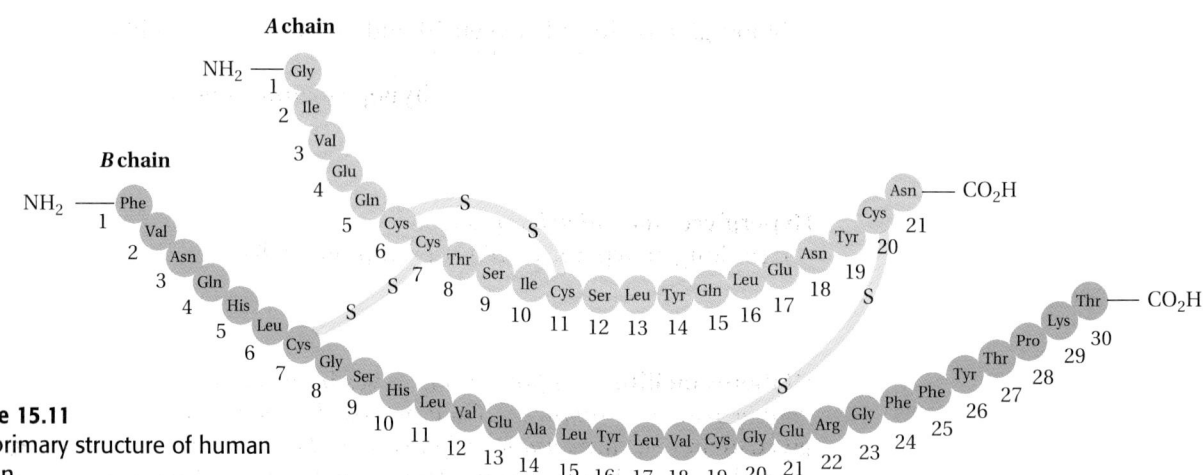

Figure 15.11
The primary structure of human insulin.

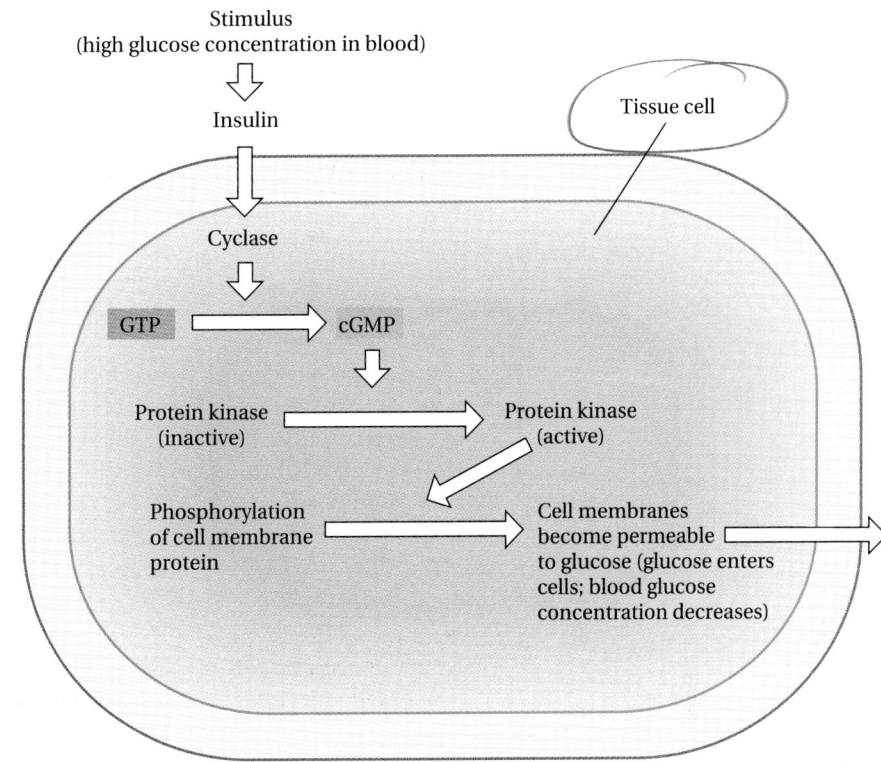

Stimulus
(high glucose concentration in blood)

Insulin

Tissue cell

Cyclase

GTP → cGMP

Protein kinase (inactive) → Protein kinase (active)

Phosphorylation of cell membrane protein → Cell membranes become permeable to glucose (glucose enters cells; blood glucose concentration decreases)

Figure 15.12
Insulin stimulates the uptake of glucose by tissue cells by starting a process that eventually renders the cell membranes permeable to glucose. The second messenger for the process is probably cGMP rather than cAMP.

Scientists believe that the second messenger for insulin is cyclic guanosine monophosphate (cGMP), a compound similar in structure to cAMP. When glucose enters tissue cells, the blood glucose concentration decreases. The drop in blood sugar is the *hypoglycemic effect* of insulin. A flaw in any link of the chain—from the production of insulin to the uptake of glucose by tissues—is a serious health problem.

Hypoglycemia and hyperglycemia

A blood glucose level between 70 and 90 mg/100 mL of blood is normal. And it is important that the blood sugar concentration be kept in this range. People who experience mild **hypoglycemia**—*low blood sugar*—may become pale, weak, and nervous. Severe hypoglycemia, a blood sugar level of less than 20 mg of glucose per 100 mL of blood, leads to shock, convulsions, coma, and finally death. There also can be too much of a good thing. **Hyperglycemia**—*high blood sugar*—for a short time is not usually harmful, but prolonged hyperglycemia is a symptom of diabetes mellitus.

Diabetes mellitus

Diabetes mellitus *is a family of diseases in which the uptake of blood glucose by tissues is impaired.* Blood glucose levels may become so high that **glycosuria**—*the appearance of glucose in the urine*—begins to occur. Loss of body water that is needed to eliminate excess glucose can severely

Hypoglycemia
hypo (Greek): below
glyc: glucose
haima (Greek): blood

Hyperglycemia
hyper (Greek): above
glyc: glucose
haima (Greek): blood

Glycosuria
glyco (Greek): sweet
uria (Greek): urine

dehydrate a patient with untreated diabetes. The name *diabetes mellitus* literally means "sweet siphon." Although there is plenty of glucose in the blood, glucose cannot reach tissues where it is urgently needed. Thus these untreated patients experience weight loss and constant hunger. This is nothing more or less than starvation in the midst of plenty.

There are many causes of diabetes, not all of which are well understood. In type I (insulin-dependent or juvenile-onset) diabetes, which usually appears in children before age 10, practically no insulin is produced by the victim's pancreas. Type I diabetes is usually treated by daily injections of insulin. The insulin dose must be carefully regulated, though; too much insulin produces *insulin shock. Caused by severe hypoglycemia,* **insulin shock** *bears the potential for coma and death.* Diabetics often carry candy bars to eat if they feel the onset of insulin shock. Type II (non-insulin-dependent or maturity-onset) diabetics, those in whom symptoms appear during young adulthood or middle age, appear to produce insulin but the release is delayed. About 20% of type II diabetics require insulin. In many others, release of their own insulin can be stimulated by drugs such as tolbutamide.

Tolbutamide

Diabetes has many more effects on metabolism and health than the few mentioned here. More of these effects are considered in the next chapter.

The insulin used to treat human diabetics formerly came from rabbits and cattle. The amino acid sequences of the insulin of these species and humans are similar enough that no severe allergic response usually occurred. Today, human insulin is being produced on a commercial scale by bacteria. A functional gene for the synthesis of human insulin was inserted into the bacteria by the gene-splicing method described in Section 11.13. The genetically engineered bacteria are a convenient source of human insulin for treatment of diabetes.

Glycosuria occasionally occurs during pregnancy because of hormonal changes. As long as blood glucose levels are normal, this is usually not a serious condition.

PRACTICE EXERCISE 15.8

Insulin is a peptide of 51 amino acid residues. Explain why this hormone cannot be administered orally to patients with diabetes but must be given intravenously.

SUMMARY

Glucose is a major source of energy for living creatures, and some of the reactions by which it is oxidized to carbon dioxide provide a common pathway for the oxidation of fats and amino acids. Glucose is oxidized in three stages: (1) oxidation to pyruvate in glycolysis, (2) formation of acetyl CoA from pyruvate, and (3) oxidation of the acetyl group of acetyl CoA to carbon dioxide in the citric acid cycle.

For each molecule of glucose oxidized, the citric acid cycle also produces six molecules of NADH, two of $FADH_2$, and two of ATP. When oxygen is available, the reducing power (NADH and FADH$_2$) formed in the oxidation of glucose is swept into cellular respiration and used to generate 36 additional ATP molecules by oxidative phosphorylation. When oxygen is not available, the citric acid cycle shuts down, and cells revert to fermentation to reoxidize NADH to NAD and thereby keep glycolysis going for a low level of ATP production.

Glycogen, the storage form of glucose, is stored mainly in the liver and muscle. Glycogen is converted on demand to glucose 1-phosphate and then to glucose 6-phosphate. Muscle cells keep glucose 6-phosphate for their own glycolysis. Liver contains an enzyme that converts glucose 6-phosphate to glucose, which passes from the liver to other tissues that need it.

Many metabolic reactions are under hormonal control. When a particular need is sensed, a hormone is sent through the bloodstream from the site of production to target cells. The hormone initiates a sequence of processes inside cells by binding to protein receptors in the target cell membrane. This triggers the formation of a second messenger, usually cyclic AMP (cAMP), which in turn triggers a series of events that ends with an effect on metabolism. Hormones also affect cell membrane permeability. The action of insulin in promoting the uptake of glucose by muscle and fat cells is one example.

Diabetes mellitus is a family of diseases that result when insulin is not produced, not released, or released at the wrong time. In all cases, the uptake of glucose by tissue cells is impaired.

KEY TERMS

Acetyl coenzyme A (15.1)
Alcoholic fermentation (15.7)
Citric acid cycle (15.1)
Coenzyme A (15.1)
Diabetes mellitus (15.13)

First messenger (15.11)
Gluconeogenesis (15.8)
Glycogenesis (15.9)
Glycogenolysis (15.10)
Glycolysis (15.1)
Glycosuria (15.13)

Hyperglycemia (15.13)
Hypoglycemia (15.13)
Insulin shock (15.13)
Lactic fermentation (15.7)
Multienzyme complex (15.4)

Oxygen debt (15.8)
Second messenger (15.11)

EXERCISES

Glycolysis (Sections 15.1, 15.2, 15.3)

15.9 Write a balanced equation for the complete oxidation of glucose.

15.10 Explain the difference between oxidation of glucose inside body cells and oxidation of glucose by burning in air.

15.11 What are the three stages by which glucose is broken down in aerobic cells?

15.12 What is the end product of glycolysis?

15.13 Where does glycolysis take place?

15.14 Write an equation for the first step in glycolysis.

15.15 Is the first step in glycolysis energetically favorable? Explain your answer.

15.16 How many molecules of ATP are required for the conversion of one molecule of glucose to two molecules of pyruvate? What is the net gain in ATP?

15.17 Write equations for the two reactions in glycolysis that lead to ATP production.

15.18 Write the equation for the reaction in glycolysis that requires NAD$^+$ as a coenzyme.

15.19 What is the fate of NAD$^+$ in Exercise 15.18?

15.20 The first oxidation reaction of glucose catabolism occurs in glycolysis. Write the equation for the reaction, and identify the oxidizing agent.

Acetyl Coenzyme A (Section 15.4)

15.21 Where does the formation of acetyl CoA take place? What is its function?

15.22 Name the B-complex vitamin that is part of the structure of CoA. What foods are a good source of this vitamin?

Citric Acid Cycle (Sections 15.5, 15.6)

15.23 Where does the citric acid cycle take place in eukaryotic cells?

15.24 How many molecules of carbon dioxide are produced by each turn of the citric acid cycle?

15.25 Write the equation for the only step in the citric acid cycle that leads directly to phosphorylation of ADP.

15.26 Draw the structure of (a) the organic molecule that ends the citric acid cycle and (b) the organic molecule that starts the cycle.

15.27 How many molecules of ATP are generated in oxidative phosphorylation by (a) $FADH_2$ and (b) NADH?

15.28 What is the total amount of ATP generated by the aerobic catabolism of one molecule of glucose?

Fermentation (Sections 15.7, 15.8)

15.29 What happens in aerobic cells when no oxygen is available to accept electrons in the electron transport chain?

15.30 Is the citric acid cycle able to operate in the absence of oxygen?

15.31 Why is lactic fermentation important?

15.32 What is produced in alcoholic fermentation?

15.33 (a) How do you get an oxygen debt? (b) What are the consequences of this condition?

15.34 What happens to the lactate that builds up in muscle cells during anaerobic glycolysis when you stop to rest and cellular respiration returns to normal?

15.35 What is gluconeogenesis?

15.36 Where does gluconeogenesis take place?

Glycogen Synthesis and Degradation (Sections 15.9, 15.10)

15.37 Describe the structure, function, and storage of glycogen.

15.38 Describe the process of glycogenesis. Why are two enzymes required?

15.39 What is glycogenolysis, and when does it occur?

15.40 Explain why muscle glycogen is not a source of glucose to other body tissues.

Hormonal Control of Glucose Metabolism (Sections 15.11, 15.12, 15.13)

15.41 Explain how hormones regulate processes that occur inside cells even though they are never found inside the cells they act on.

15.42 Where is glucagon produced, and what is its function?

15.43 (a) What stimulates the secretion of epinephrine? (b) Where is it produced, and what is its function?

15.44 Where does the majority of carbohydrate metabolism occur in the body?

15.45 Explain how insulin controls blood glucose levels.

15.46 How do blood glucose levels in hyperglycemia and hypoglycemia compare with normal blood glucose levels?

15.47 Describe the difference between juvenile-onset diabetes and maturity-onset diabetes.

15.48 What is meant by *insulin shock?*

15.49 A screening test for diabetes mellitus is done by analyzing urine for the presence of sugar. Explain why diabetes mellitus causes glycosuria.

Additional Exercises

15.50 What is the fate of the hydrogen atoms removed in oxidation reactions in glycolysis and the Krebs cycle?

15.51 Explain the role of cAMP in glycogenolysis.

15.52 Match the following:

(a) glycogenesis
(b) cAMP
(c) citric acid cycle
(d) gluconeogenesis
(e) diabetes mellitus
(f) coenzyme A
(g) lactic fermentation
(h) oxygen debt
(i) insulin
(j) glycolysis

(1) reduction of pyruvate
(2) oxidation of acetyl CoA
(3) uptake of glucose by cells
(4) insufficient oxygen for oxidative phosphorylation
(5) synthesis of glycogen
(6) oxidation of glucose to pyruvate
(7) glycosuria
(8) synthesis of glucose
(9) thiol
(10) second messenger

15.53 Why do we breathe heavily when engaged in vigorous exercise?

15.54 Describe the process by which glucose is made from noncarbohydrate substances.

15.55 Discuss the major cause of hyperglycemia, and explain how it may be treated.

15.56 Explain what happens to pyruvate in the body under (a) anaerobic conditions and under (b) aerobic conditions.

SELF-TEST (REVIEW)

True/False

1. Oral administration of insulin is an ineffective treatment of diabetes.
2. Gluconeogenesis is an anabolic process.
3. Acetyl CoA is a product of the citric acid cycle.
4. To be useful as a source of energy, all nutrients must first be changed into either glucose or pyruvate.
5. Every step in the glycolysis process is an energy-yielding one.
6. When eukaryotic cells are deprived of oxygen, oxidative phosphorylation stops, and the cell dies.
7. Lactic fermentation is a more efficient source of ATP than respiration.
8. A person's energy requirements could be met from stored glycogen for approximately 1 week.
9. Insulin decreases the blood sugar level by increasing the rate of metabolism of sugar in the liver.
10. A diabetic with a consistently high blood sugar level may feel hungry all the time and actually experience weight loss.

Multiple Choice

11. A chemical messenger produced in the endocrine glands is called
    (a) a hormone.   (b) cAMP.
    (c) a target cell.   (d) a nerve impulse.
12. The major energy product of the citric acid cycle is
    (a) ATP.   (b) cAMP.
    (c) NADH.   (d) $FADH_2$.
13. A person whose blood sugar level drops below 60 mg/100 mL of blood
    (a) will probably feel weak.
    (b) is hyperglycemic.
    (c) probably has glucose in their urine.
    (d) more than one are correct.
14. All the following occur in the mitochondria of a eukaryotic cell *except*
    (a) the citric acid cycle.
    (b) formation of acetyl CoA from pyruvate.
    (c) glycolysis.
    (d) cellular respiration.
15. cAMP is best classed as a(n)
    (a) enzyme-hormone receptor.
    (b) enzyme modulator.
    (c) first messenger.
    (d) calcium ion inhibitor.
16. The glycolysis of one molecule of glucose results in the formation of
    (a) a net total of two ATP molecules.
    (b) two molecules of NADH.
    (c) two molecules of pyruvate.
    (d) all of the above.
17. A low blood concentration of glucose
    (a) is called *hyperglycemia*.
    (b) can be countered by glycogenolysis of glycogen in the liver.
    (c) can be countered by injection of insulin.
    (d) both (a) and (c) are correct.
18. Pyruvate is the end product of
    (a) glycogenesis.
    (b) the citric acid cycle.
    (c) oxidative phosphorylation.
    (d) glycolysis.
19. Glycogenolysis can be stimulated by
    (a) epinephrine.   (b) amylase.
    (c) glucagon.   (d) Both (a) and (c) are correct.
20. Alcoholic fermentation
    (a) allows glycolysis to keep going.
    (b) results in the production of lactate.
    (c) is an anaerobic process.
    (d) answers (a) and (c) are correct.
21. Which of the following processes requires a net expenditure of ATP?
    (a) glycolysis
    (b) citric acid cycle
    (c) gluconeogenesis
    (d) lactic fermentation
22. Glucose is temporarily stored in the body as
    (a) fat.   (b) glycogen.
    (c) lipids.   (d) glucose 6-phosphate.
23. A cell in the state of oxygen debt can produce energy by
    (a) cellular respiration.
    (b) glycolysis.
    (c) glycogenesis.
    (d) gluconeogenesis.
24. You are climbing a mountain, and your partner who has diabetes is slipping into insulin shock. You should
    (a) not give her anything to eat or drink.
    (b) let her eat the last chocolate bar you own.
    (c) let her drink water but nothing else.
    (d) head for the nearest phone and call a doctor.

16

Lipid Metabolism

Fat Chemistry in Cells

Animals that live in cold climates, such as these crabeater seals, metabolize lipids for body heat. They also rely on lipids for insulation against the extremes of temperature.

CHAPTER OUTLINE

L ipids are essential to cellular life. As the material of membranes, this class of oily biological molecules is a wrapper around cells and organelles of eukaryotic cells. As sources of energy, fatty acids are preferred by liver cells and resting muscle cells, among others. Body cells also synthesize lipids and other fatty substances such as cholesterol.

This chapter is about how lipids are stored, degraded, and synthesized. Carbohydrate metabolism and lipid metabolism are intimately related, so we will examine the ways in which carbohydrate metabolism is linked to fatty acid metabolism and how diabetes affects lipid metabolism as well as glucose metabolism. Our Case in Point concerns a disorder of lipid metabolism.

CASE IN POINT: Carnitine deficiency

 Baby Clarence began to suffer from vomiting, slow heart rate, and breathing problems about 6 months after birth. One morning Clarence's mother found him, blue and not breathing, in his crib. Paramedics managed to restore Clarence's breathing, but the doctors were uncertain of the cause of his illness. Clarence's prognosis was not good. Within a few weeks, however, doctors were able to determine that the cause of Clarence's condition was a lipid metabolism defect. What was Clarence's disorder? Could it be treated? We will find out in Section 16.3.

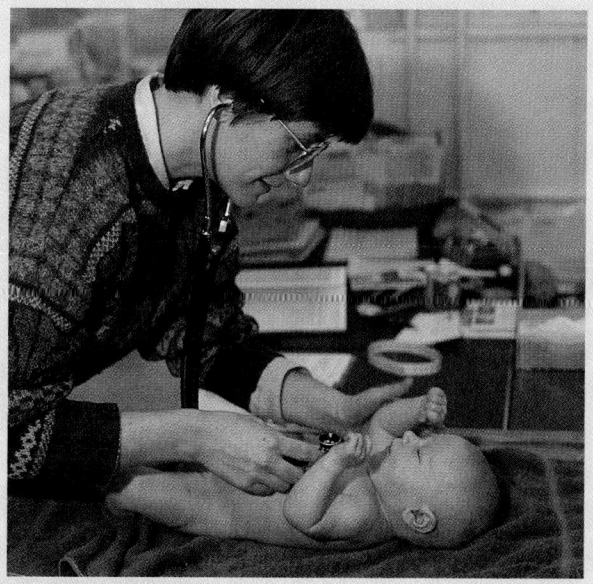

An infant suspected of having a defect in lipid metabolism is examined by a specialist.

16.1 Body lipids

AIM: To differentiate between the lipids of adipose tissue and those of working tissue on the basis of structure and function.

Focus

Body lipids either are stored or become working tissue.

About 12% of the body weight of the average male is depot fat; about 20% of the body weight of the average female is depot fat.

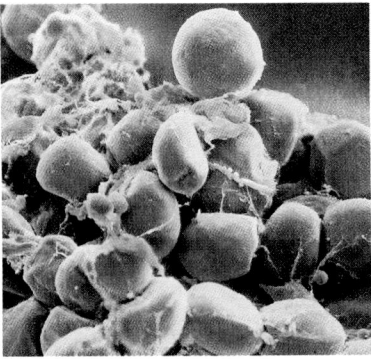

Figure 16.1
Scanning electron micrograph of fat cells (yellow) in adipose tissue (magnified 1575×).

Lipids that enter the bloodstream are distributed in the body tissues in two ways: as **storage fat (depot fat)** *or as* **working lipid (tissue lipid).** *Storage fat is collected in the cells of fat tissue, also called* **adipose tissue.** Most of the lipids stored in fat cells (Fig. 16.1) are triglycerides. Adipose tissue has a tendency to collect around the waist, hips, and buttocks.

Layers of adipose tissue also surround many body organs. These layers of body fat protect the body and its organs against bumps and blows. Fat is also a good insulator against cold. The bodies of whales and seals, mammals exposed to severe cold, are covered by thick layers of fat. At one time many biologists and physiologists thought of adipose tissue as a metabolically inactive insulating blanket. However, it now appears that metabolism in fatty tissue cells is an important source of body heat.

Complex lipids such as the phospholipids, sphingomyelins, and cerebrosides (see Chap. 8) become part of the working lipids of the body. The working lipids are mainly incorporated into cell membranes and the membranes of cell organelles such as the mitochondria, endoplasmic reticula, nuclei, and lysosomes.

Normally, both triglycerides and complex lipids are constantly being broken down and synthesized in the body. However, there are several diseases in which the enzymes required to catalyze the breakdown of complex lipids are lacking (see A Closer Look: Lipid Storage Diseases, on pages 478–479).

16.2 Fat mobilization

AIM: To describe the function of epinephrine and serum albumin in lipid metabolism.

Focus

Hormones stimulate the release of fatty acids and glycerol from adipose tissue.

Brown adipose tissue is found in newborn infants and hibernating animals. It is loaded with mitochondria and generates *heat* rather than ATP, as does the normal (*white*) adipose tissue.

Carbohydrates—through glucose catabolism—are the first choice of some tissues for energy production. The brain and active skeletal muscles are examples of such tissues. As we have seen, however, the body's stores of glycogen are used up after only a few hours of fasting. Well before the glycogen supply is used up, many body cells, such as those of resting muscle and liver, begin to break down fatty acids to produce energy. By using fatty acids for energy production, these cells help to conserve the rest of the body's glycogen store for brain cells and other cells of the nervous system. Brain and nerve cells demand a constant supply of glucose. In fact, except in emergencies such as starvation, they produce their energy only from glucose. When other body cells need fatty acids, signaled by a low blood glucose level, the endocrine system sends several hormones, including epinephrine, to the adipose tissue storehouse. In fat cells, epinephrine, through a second messenger system, stimulates **mobilization**—*the hydrolysis of triglycerides to fatty acids and glycerol, which enter the blood-*

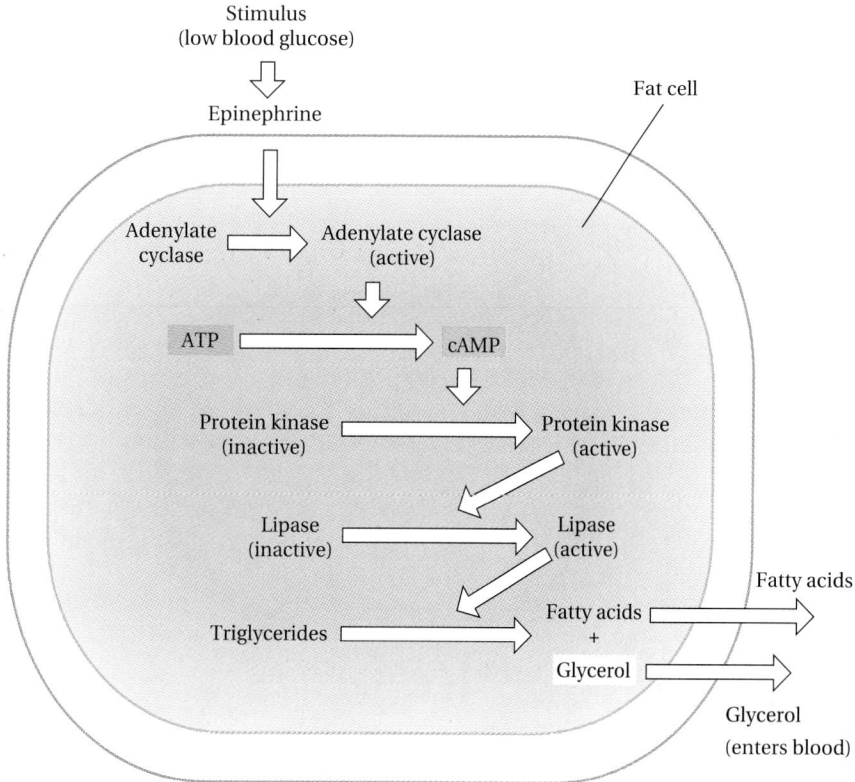

Figure 16.2
Mobilization of fatty acids from adipose tissue.

stream. Fat mobilization is shown in Figure 16.2. The mobilized fatty acids become tightly attached to a plasma protein, serum albumin. The serum albumin transports the fatty acids to tissues where they are taken up by cells that need them. The glycerol produced by the hydrolysis of the triglyceride also enters these cells.

16.3 Fatty acid oxidation

AIMS: *To name and describe the process of the catabolism of fatty acids, indicating the end products. To predict the hydrolysis products of typical lipid molecules. To determine the number of molecules of acetyl CoA, NADH, and FADH$_2$ that result from the beta oxidation of a given fatty acid.*

Focus

Fatty acids are degraded to acetyl coenzyme A.

Fatty acids enter tissue cells that need energy. Before a fatty acid molecule can be oxidized to produce energy, however, it must be converted to a **fatty acyl CoA**—*a thiol ester formed from a fatty acid and CoA.*

$$\underset{\text{Fatty acid}}{R-\overset{\overset{\text{O}}{\|}}{C}-OH} + HS-CoA \longrightarrow \underset{\text{Fatty acyl CoA}}{R-\overset{\overset{\text{O}}{\|}}{C}-S-CoA} + H_2O$$

A Closer Look

Lipid Storage Diseases

Inborn errors associated with lipid metabolism include Tay-Sachs disease, Gaucher's disease, and Niemann-Pick disease. All three conditions are lipid storage diseases. Normally, both triglycerides and complex lipids are constantly being broken down and synthesized in the body. The lipid storage diseases seem to occur because the body lacks enzymes required to catalyze the breakdown of certain complex lipids. This enzyme deficiency leads to accumulation of abnormally high levels of complex lipids in specific tissues, particularly the brain, spleen, and liver. In all lipid storage diseases, there is retardation of mental and physical development for reasons that are not well understood.

Tay-Sachs disease, in which lipids primarily accumulate in the brain, is characterized by paralysis, a deteriorated mental state, and blindness. In Tay-Sachs, the stored lipids are primarily gangliosides, the most complex members of the lipids containing sphingosine (see figure, part a). As shown in the figure, the gangliosides contain four different sugar units.

Gaucher's disease has many variants with certain clinical features, including elevated concentrations of serum acid phosphatase. The structure of a typical cerebroside that accumulates in this disease is shown in part (b) of the figure. The spleen becomes enlarged because spleen cells accumulate cerebrosides (see Sec. 16.3). Gaucher's disease may progress rapidly and be fatal in infancy, but some patients live fairly normal life spans. This suggests that Gaucher's disease is a group of inheritable diseases. The inherited mutation may be different in each form of the disease, but all these mutations affect cerebroside metabolism.

The chemical abnormality found in Niemann-Pick disease is an increased content of sphingolipids (see Sec. 16.3) in the organs, but especially the spleen (see figure, part c). As in Gaucher's disease, there are many variants of Niemann-Pick disease, indicating a variety of mutations affecting sphingolipid metabolism.

At right: The three lipids whose excessive storage causes Tay-Sachs disease, Gaucher's disease, and Niemann-Pick disease. In Tay-Sachs disease, the stored lipids are gangliosides, the most complex members of the lipids containing sphingosine (a). The sugar units of the gangliosides contain glucose (Glu), galactose (Gal), and two unusual sugars: N-acetylgalactosamine (Naga) and N-acetylneurominic acid (Nana). In Gaucher's and Niemann-Pick diseases, the lipids that cannot be broken down are the cerebrosides (b) and the sphingomyelins (c), respectively.

The conversion of fatty acids to fatty acyl CoA occurs in the cytoplasm of cells, but the subsequent oxidation of fatty acyl CoA occurs in the mitochondria. This presents a problem, since fatty acyl CoA molecules cannot pass through the mitochondrial membrane. The passage of fatty acyl CoA from the cytoplasm into the mitochondria is done indirectly. The fatty acyl CoA is first converted into an ester of the amino alcohol carnitine.

$$R-\overset{\overset{\displaystyle O}{\|}}{C}-SCoA + (H_3C)_3\overset{+}{N}-CH_2-\overset{\overset{\displaystyle OH}{|}}{CH}-CO_2H \longrightarrow (H_3C)_3\overset{+}{N}-CH_2-\overset{\overset{\displaystyle O}{\underset{\displaystyle \|}{R-C-O}}|}{CH}-CO_2H + H-SCoA$$

Fatty acyl CoA   Carnitine   Fatty acyl carnitine

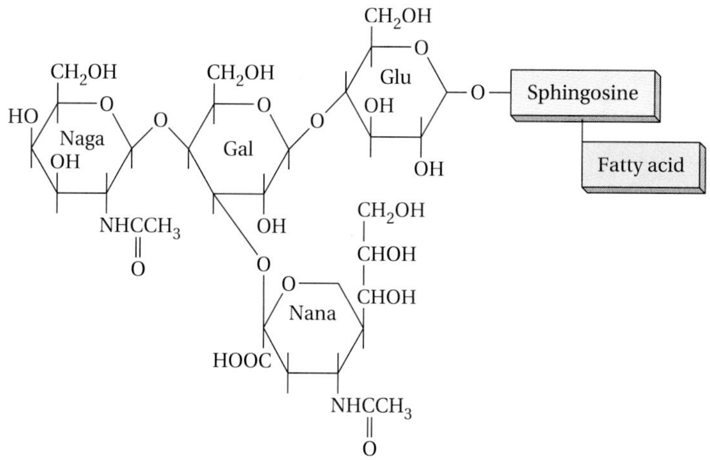

(a) A ganglioside (accumulates in Tay–Sachs disease)

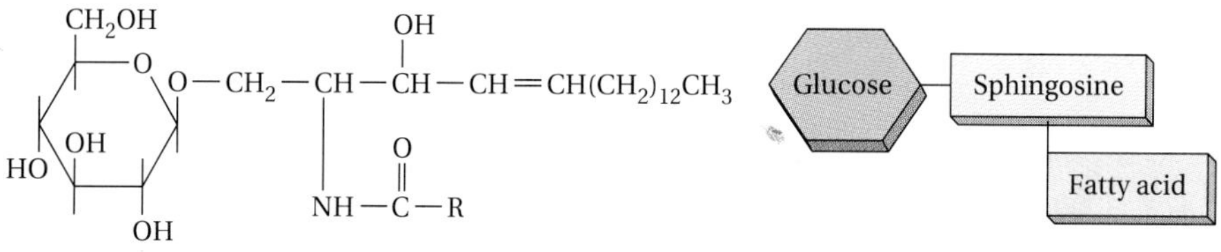

(b) A cerebroside (accumulates in Gaucher's disease)

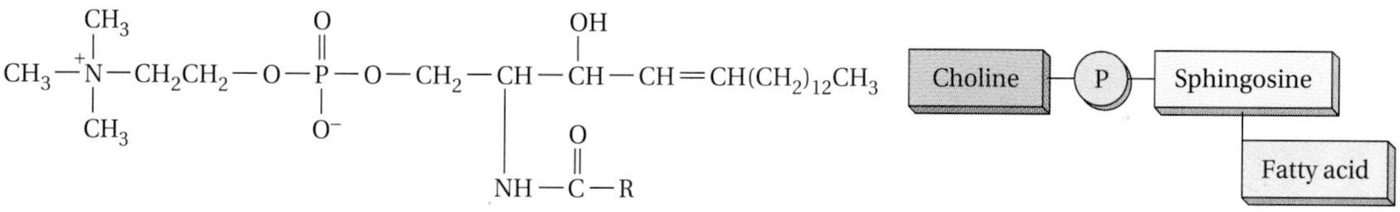

(c) A sphingomyelin (accumulates in Niemann–Pick disease)

The fatty acyl carnitine passes through the mitochondrial membrane, where reaction with CoA converts it back to fatty acyl CoA.

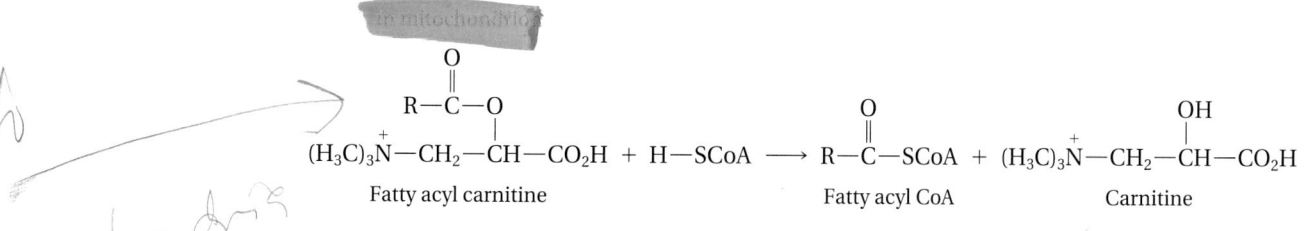

Fatty acyl carnitine   Fatty acyl CoA   Carnitine

The fatty acyl carnitine is so important in the metabolism of fatty acids that a carnitine deficiency can be life-threatening. Clarence, the sick baby in the Case in Point earlier in this chapter, is an example.

FOLLOW-UP TO THE CASE IN POINT: Carnitine deficiency

Clarence was found to be suffering from a potentially fatal carnitine deficiency. This deficiency was the result of a defect in the genes responsible for fatty acid metabolism. Clarence's symptoms—vomiting, slow heart rate, and breathing problems—were general, and the condition is rare. He was fortunate that his carnitine deficiency was diagnosed before it took its fatal course. Most of the symptoms of his disease disappeared when carnitine was added to his formula three times a day. Since his carnitine deficiency was accompanied by other defects in his fatty acid metabolism, Clarence also was placed on a low-fat diet. Clarence, whose prognosis was poor, seems to be thriving on his carnitine supplement and low-fat diet.

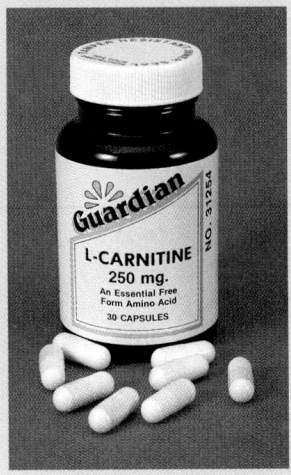

Carnitine supplements help in the oxidation of long-chain fatty acids.

Fatty acid activation

Formation of fatty acyl CoA molecules shows how the energy stored in the phosphate bonds of ATP can be used to drive chemical processes that normally could not occur. Neither the fatty acid nor the CoA is sufficiently energetic to form the fatty acyl CoA directly, so the cell boosts the chemical reactivity of the fatty acid carboxyl group by investing ATP. The triphosphate group of the ATP molecule is cleaved to give pyrophosphate (PP_i), and the remaining adenosine monophosphate (AMP) forms a high-energy anhydride bond with the fatty acid.

$$\underset{\text{Fatty acid}}{R-\overset{O}{\overset{\|}{C}}-OH} + \underset{\text{ATP}}{\begin{array}{c}\text{Adenosine}\\ ^{-}O-\overset{O}{\overset{\|}{P}}-O-\overset{O}{\overset{\|}{P}}-O-\overset{O}{\overset{\|}{P}}-O^{-}\\ \underset{O^-}{}\quad \underset{O^-}{}\quad \underset{O^-}{}\end{array}} \longrightarrow \underset{\text{Mixed anhydride}}{R-\overset{O}{\overset{\|}{C}}-O-\overset{\overbrace{O}^{\text{High-energy anhydride linkage}}}{\underset{O^-}{\overset{\|}{P}}}-O\;\text{Adenosine}} + PP_i$$

The anhydride linkage formed between the fatty acid and AMP in this reaction is energetic enough to be broken by reaction with the thiol group of CoA.

$$\underset{\text{Mixed anhydride}}{R-\overset{O}{\overset{\|}{C}}-O-\overset{O}{\underset{O^-}{\overset{\|}{P}}}-O\;\text{Adenosine}} + \underset{\text{CoA}}{HS-CoA} \longrightarrow \underset{\substack{\text{Fatty}\\\text{acyl CoA}}}{R-\overset{O}{\overset{\|}{C}}-S-CoA} + \underset{\substack{\text{Adenosine}\\\text{monophosphate}\\\text{(AMP)}}}{^{-}O-\overset{O}{\underset{O^-}{\overset{\|}{P}}}-O\;\text{Adenosine}}$$

The thioester bond of the fatty acyl CoA is a high-energy bond, just as it is in acetyl CoA. Therefore, some of the energy invested in the breakdown of ATP to AMP has been conserved in the thioester bond of the fatty acyl CoA. The cost to the cell of activating one molecule of fatty acid is the loss of two high-energy phosphoric acid anhydride bonds: one in the formation of the anhydride between the fatty acid and AMP and one when PP_i is hydrolyzed to $2P_i$.

Beta oxidation

The formation of fatty acyl CoA molecules prepares fatty acids for entry into the mitochondria. *In the mitochondria, fatty acyl CoA molecules are degraded to acetyl CoA in the catabolic process called* **beta oxidation.** During beta oxidation, the third (beta) carbon of the saturated fatty acid chain of the fatty acyl CoA molecule is oxidized to a ketone.

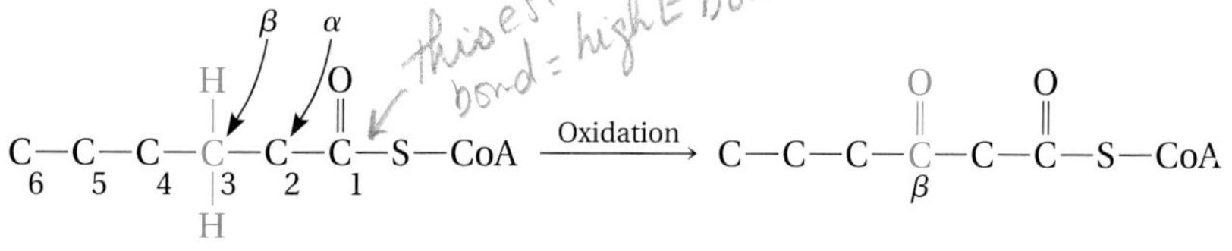

Fatty acyl CoA

There are four reactions involved in beta oxidation. An example is the fatty acyl CoA formed from a 16-carbon acid, palmitic acid. Any NADH or $FADH_2$ produced in these reactions must be accounted for, since the reducing power of these molecules may be used to produce ATP in respiration. As in all the metabolic reactions we have discussed, every step of beta oxidation is catalyzed by enzymes.

Step 1: Dehydrogenation. A carbon-carbon double bond is formed between the second and third carbons of the palmitic acid chain. One molecule of $FADH_2$ is produced.

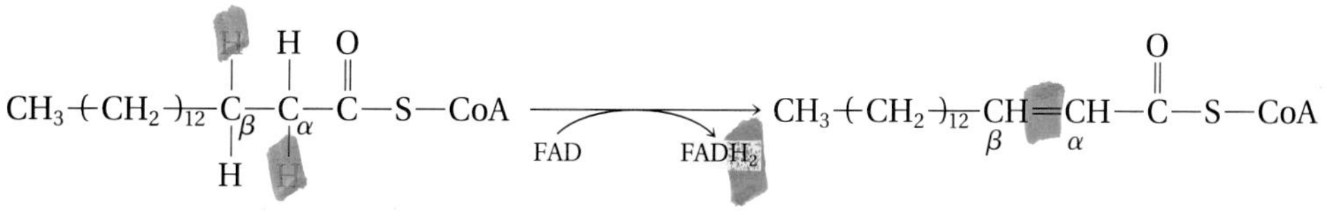

As we have seen, FAD is often the coenzyme for enzymes that catalyze the removal of hydrogen from saturated carbon compounds. This is the case here, and one $FADH_2$ molecule is produced.

Step 2: Hydration. A molecule of water is added to the double bond of the fatty acyl CoA. This step is the hydration reaction of beta oxidation.

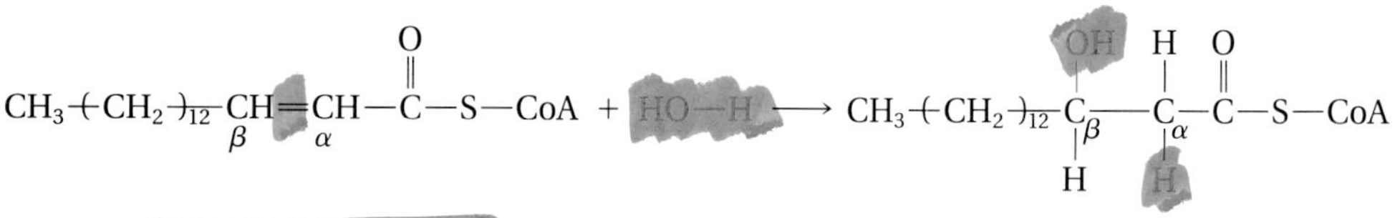

Step 3: Oxidation. The hydroxyl group is now oxidized to a ketone. NAD^+ is the coenzyme that acts as the oxidizing agent in this second oxidation

reaction of beta oxidation. Thus beta oxidation has now produced one molecule of $FADH_2$ and one molecule of NADH.

$$CH_3 + CH_2 \rightarrow_{12} C_\beta \overset{\underset{|}{H}}{C_\alpha} \overset{O}{C} - S - CoA \xrightarrow[NAD^+ \quad NADH + H]{} CH_3 + CH_2 \rightarrow_{12} \overset{O}{C} - \overset{\underset{|}{H}}{C} - \overset{O}{C} - S - CoA$$

Step 4: Carbon-carbon bond cleavage. The carbon-carbon bond between the second and third carbons is cleaved to give acetyl CoA and a new fatty acyl CoA that is two carbons shorter than the starting thioester. This reaction involves another molecule of CoA, but no ATP is expended.

$$CH_3 + CH_2 \rightarrow_{12} \overset{O}{C} - \overset{\underset{|}{H}}{C} - \overset{O}{C} - S - CoA + S - CoA \longrightarrow CH_3 + CH_2 \rightarrow_{12} \overset{O}{C} - S - CoA + CH_3 \overset{O}{C} - S - CoA$$

16 carbons   14 carbons   2 carbons

The fragmentation of fatty acyl CoA molecules into molecules of acetyl CoA is similar to other metabolic cycles. The steps of beta oxidation repeat over and over again until the fatty acyl CoA is completely degraded to acetyl CoA. Look at Figure 16.3. The fatty acyl CoA that starts each round of the

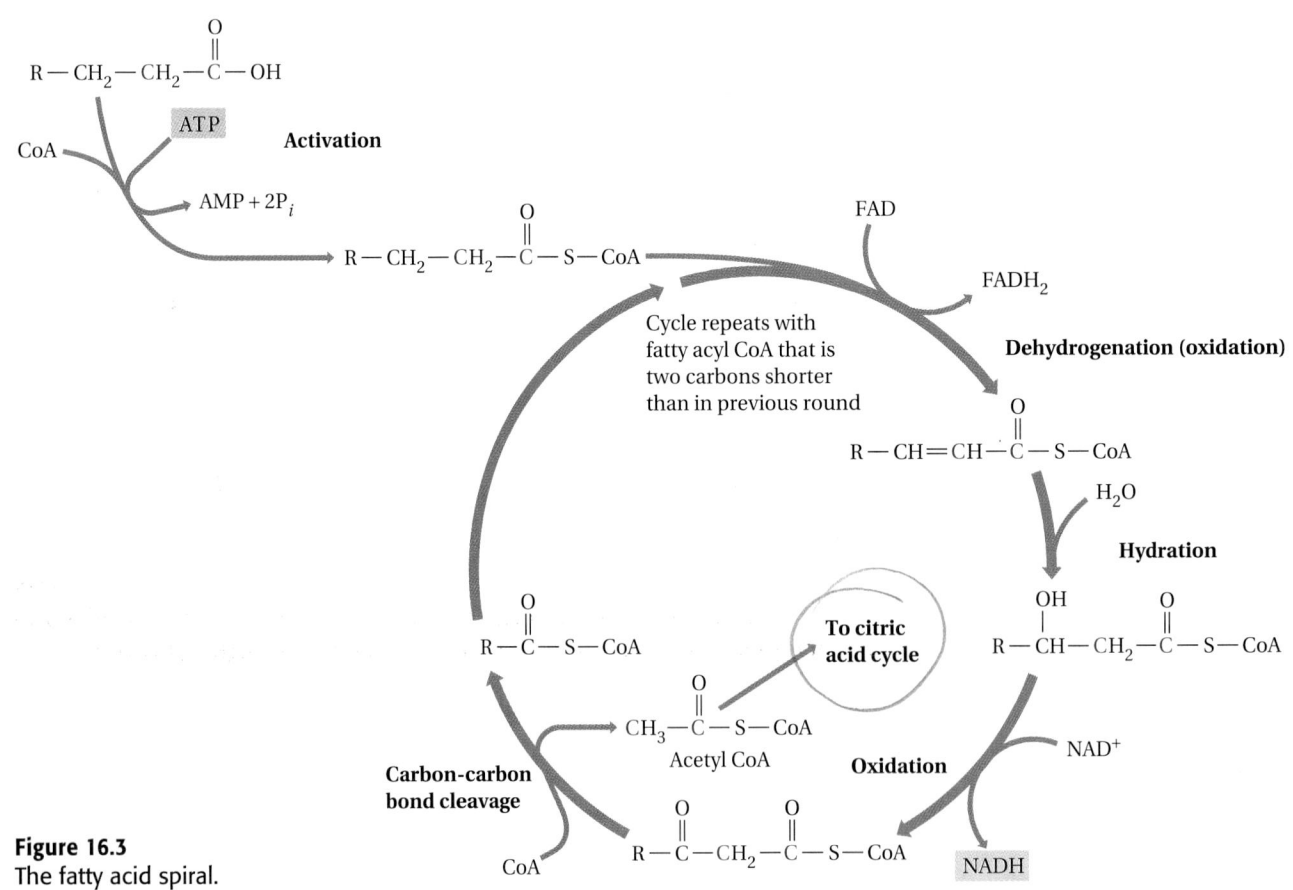

Figure 16.3
The fatty acid spiral.

beta-oxidation cycle is two carbons shorter than in the previous round. *For this reason, the pathway for the degradation of fatty acids to acetyl CoA is often called the* **fatty acid spiral.**

Every round of the spiral produces one molecule each of acetyl CoA, NADH, and FADH$_2$ until the fatty acyl CoA molecule is only four carbons long. At this point, the first three steps of the final round of beta oxidation produce the compound acetoacetyl CoA. The fourth step, the reaction of acetoacetyl CoA with CoA, produces an extra molecule of acetyl CoA from the tail end of the fatty acid without formation of NADH and FADH$_2$.

$$CH_3-\overset{\overset{\displaystyle O}{\|}}{C}-CH_2-\overset{\overset{\displaystyle O}{\|}}{C}-S-CoA \ + \ HS-CoA \ \longrightarrow$$

Acetoacetyl CoA

$$CH_3-\overset{\overset{\displaystyle O}{\|}}{C}-S-CoA \ + \ CH_3-\overset{\overset{\displaystyle O}{\|}}{C}-S-CoA$$

In other words, the complete conversion of a fatty acyl CoA to two-carbon fragments of acetyl CoA always produces one more molecule of acetyl CoA than of NADH or FADH$_2$. To summarize, the breakdown of palmitic acid gives eight molecules of acetyl CoA, but only seven molecules of NADH and seven molecules of FADH$_2$ are produced.

PRACTICE EXERCISE 16.1

Lauric acid is converted to acetyl CoA in beta oxidation. Determine the yields of (a) acetyl CoA, (b) NADH, and (c) FADH$_2$.

16.4 ATP yield

***AIM:** To calculate the number of ATP molecules formed by the oxidation of a fatty acid molecule.*

In Chapter 15 we saw that the carbons of the acetyl CoA produced by the catabolism of glucose can be completely oxidized to carbon dioxide in the citric acid cycle. Each molecule of acetyl CoA oxidized in this fashion yields enough energy to make one molecule of ATP, one molecule of FADH$_2$, and three molecules of NADH. The reducing power of each molecule of NADH can make three molecules of ATP by cellular respiration; FADH$_2$ produces two molecules of ATP in the same way. It was shown that 38 ATP molecules is the total useful energy yield of aerobic glucose catabolism.

Molecules of acetyl CoA are the same, regardless of their source. Like acetyl CoA molecules produced from glucose, the acetyl CoA molecules formed in the fatty acid spiral can be oxidized in the citric acid cycle. Since we can find the yield of NADH, FADH$_2$, and ATP from the beta-oxidation reactions of the fatty acid spiral and from the citric acid cycle, we can calculate how many molecules of ATP are produced by the total oxidation of one molecule of any fatty acid to carbon dioxide and water. Table 16.1 shows a calculation of this kind for palmitic acid.

In calculating the total ATP yield obtained from the complete oxidation of the fatty acid, we can count the investment of two high-energy phos-

Table 16.1 ATP Production from Complete Aerobic Catabolism of One Molecule of Palmitic Acid

Pathway	ATP yield
fatty acid spiral (2 ATP invested to make palmitoyl CoA to start spiral)	−2
citric acid cycle (8 molecules of acetyl CoA degraded)	8
oxidative phosphorylation	
7NADH from fatty acid spiral	21
7FADH$_2$ from fatty acid spiral	14
24NADH from citric acid cycle	72
8FADH$_2$ from citric acid cycle	16
	129

The oxidation of 1 g of a fatty acid yields about 0.5 mol ATP. The oxidation of 1 g of glucose yields about 0.2 mol ATP. This ratio, 0.5:0.2, is approximately the same as the ratio of the energy yield for the complete oxidation of fats and carbohydrates, 9 kcal:4 kcal.

[handwritten: LAURIC ACID = 12 C]
[handwritten: Activation. −2 ATP]
[handwritten: 6 acetyl CoA 6]
[handwritten: degraded in Krebs]
[handwritten: x3 ATP 5 NADH from spiral 15]
[handwritten: x2 ATP 5 FADH$_2$ from spiral 10]
[handwritten: x3 ATP 6x3 NADH from Krebs 54]
[handwritten: x2 6x1 FADH$_2$ from Krebs 12]
[handwritten: 95 ATP]

phate bonds required to activate the fatty acid as two ATP molecules. We can do this because hydrolysis of one molecule of ATP to AMP and 2P$_i$ is equivalent to the hydrolysis of 2ATP to 2ADP and 2P$_i$. Table 16.1 shows that for every molecule of palmitic acid completely oxidized to carbon dioxide and water, 129 molecules of ATP are formed. No wonder fats are an important source of energy for cellular work.

Energy production is not the only useful function of beta oxidation. Cells using NADH and FADH$_2$ to reduce oxygen also produce a good deal of water as a by-product of cellular respiration. Certain animals have become physiologically adapted to take advantage of this fact. A camel's hump, for example, consists of fat that has been stored in times of plenty. Aerobic catabolism of this fat supplies enough energy and water to permit camels to survive during long periods of famine and drought.

PRACTICE EXERCISE 16.2

Calculate how many molecules of ATP are produced by the total oxidation of lauric acid to carbon dioxide and water. *[handwritten: 95 ATP]*

16.5 Glycerol metabolism

AIM: To describe how glycerol is used as an energy source.

The hydrolysis of triglycerides produces glycerol as well as fatty acids. Glycerol is converted by cells to dihydroxyacetone phosphate in two steps.

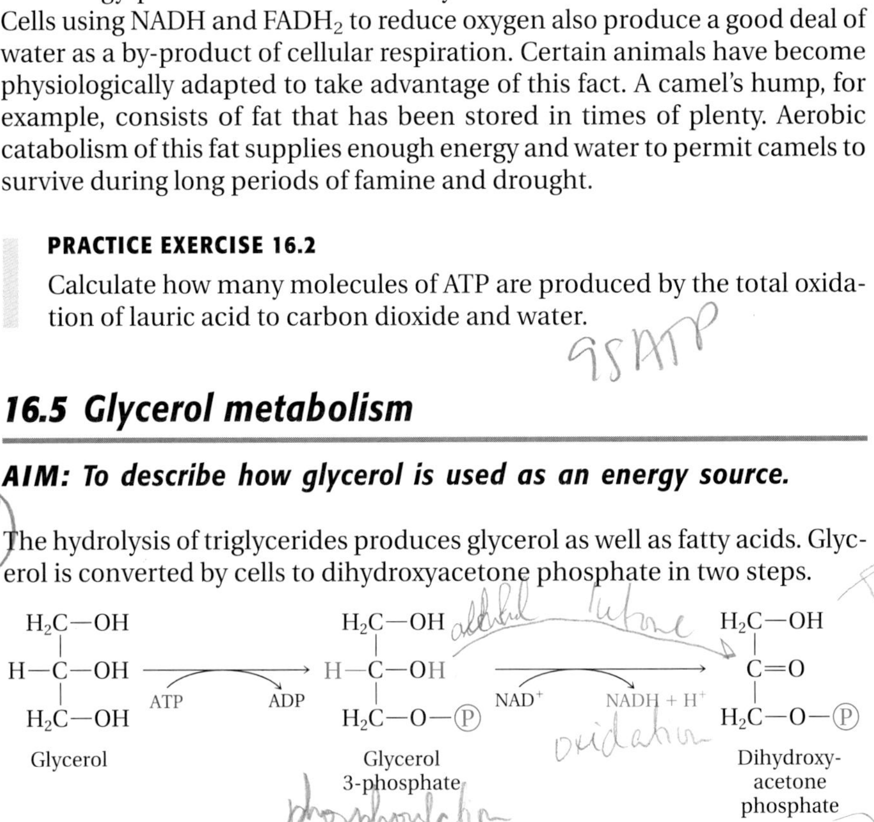

Dihydroxyacetone phosphate is one of the chemical intermediates of glycolysis. Thus the glycerol produced by hydrolysis of triglycerides contributes to energy production by entering glycolysis as dihydroxyacetone phosphate.

16.6 The key intermediate–acetyl CoA

AIMS: To name the shared intermediate of both carbohydrate and fatty acid metabolism. To list four fates of acetyl CoA in the liver.

Now that we have seen how the body oxidizes fatty acids, we can form an overall picture of the various parts of fatty acid metabolism. We can examine the relationships between carbohydrate metabolism and fatty acid metabolism at the same time. Since the liver conducts more carbohydrate metabolism and fatty acid metabolism than any other organ, this discussion will focus on it.

Figure 16.4 shows the relationships we will be examining in the remainder of this chapter. Refer to it often as you read on. The figure shows that

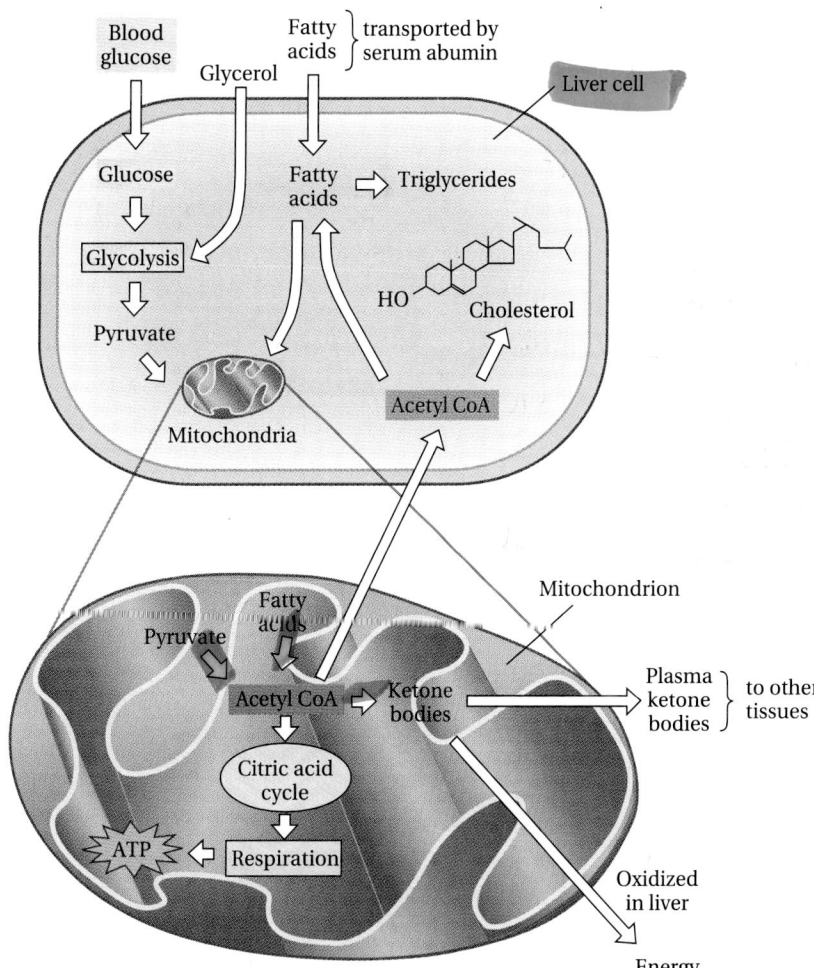

Figure 16.4
The major pathways of lipid metabolism in the liver and their relationship to carbohydrate metabolism.

fatty acids entering the liver from the blood may be resynthesized into triglycerides and stored in the adipose tissue there. Alternatively, fatty acids may be broken down to acetyl CoA. Glucose is also broken down to acetyl CoA. If you are beginning to suspect that acetyl CoA must be a key compound in the metabolic interplay between carbohydrate and fatty acid metabolism, you are certainly correct. Four possible fates await the acetyl CoA produced from fatty acids or glucose in the liver:

1. The acetyl CoA in the mitochondria may be oxidized to carbon dioxide and water in the citric acid cycle and respiration. This pathway, which is used if the liver cells need to generate energy through respiration, makes it clear that the citric acid cycle is shared by both glucose metabolism and fatty acid metabolism.

2. The acetyl CoA in mitochondria may be used to synthesize the substances called *ketone bodies*. Some ketone bodies are oxidized for energy production by the liver. The remainder are transported to other tissues that can use them for energy production. Extensive production of ketone bodies from acetyl CoA occurs only when no glucose is available, as in starvation or diabetes. When no glucose is available to starving brain cells, for example, they adapt to the use of ketone bodies as an energy source in about 48 hours.

3. The cytoplasmic acetyl CoA may be used to resynthesize fatty acids. Since some of the acetyl CoA may have come from glucose, this means that carbohydrates may be converted to fatty acids.

4. The acetyl CoA may be used to synthesize cholesterol in the cellular cytoplasm. As with other synthetic processes, the synthesis of fatty acids and cholesterol occurs when the body is well supplied with energy.

Cells other than liver cells can carry out most of these four functions. However, only the liver adds significant quantities of ketone bodies to the bloodstream for transport to other tissues.

PRACTICE EXERCISE 16.3

What is the fate of acetyl CoA when the body needs energy? What is its fate when energy needs are met?

16.7 Fatty acid synthesis

AIM: To explain the advantages of lipid energy storage over carbohydrate energy storage.

Carbohydrates can be converted to fats. The key compound that links carbohydrate metabolism to fatty acid synthesis is acetyl CoA. When body cells have more glucose than they require to meet their energy needs, they can divert some of the acetyl CoA produced by glucose catabolism to the synthesis of fatty acids.

Fatty acid synthesis and fatty acid oxidation are not the reverse of one another, though both processes do occur in cycles. In fatty acid synthesis, a

Focus

Converting carbohydrates to fatty acids is an efficient way to store energy.

two-carbon fragment is added to the growing fatty acid chain in each turn of the cycle rather than being removed, as it is in fatty acid oxidation. Moreover, the synthesis and degradation of fatty acids involve different sets of enzymes. The cellular locations of the two processes are different as well: Fatty acid synthesis occurs in the cytoplasm; fatty acid oxidation takes place in the mitochondria. Like most synthetic or anabolic pathways, the synthesis of fatty acids requires the expenditure of cellular reducing power and ATP. This requirement is clearly shown by the overall equation for the synthesis of palmitic acid.

$$8CH_3\overset{\overset{O}{\|}}{C}-S-CoA + 7ATP + 14NADH \longrightarrow CH_3(CH_2)_{14}\overset{\overset{O}{\|}}{C}-OH + 7ADP + 14NAD^+ + 8CoA + 6H_2O + 7P_i$$

Acetyl CoA $\qquad\qquad\qquad\qquad\qquad$ Palmitic acid

The newly synthesized fatty acids are incorporated into triglycerides and are stored as depot fat in adipose tissues.

Lipids are a concentrated storage form of energy compared with carbohydrates. There are two reasons for this. First, fatty acids are in a more reduced state than carbohydrates; the carbon chains of fatty acids are generally saturated hydrocarbons, whereas the carbon chains of most carbohydrates are already partially oxidized to alcohols. Thus the bonds of fatty acids have more energy to release upon complete oxidation to carbon dioxide and water than the bonds of carbohydrates. This is why 1 g of fat releases 9 kcal of energy but 1 g of carbohydrate releases only 4 kcal. Second, fats are stored in a nearly anhydrous, or dry, state. In contrast, for each gram of carbohydrate stored as glycogen, nearly 2 g of water is carried along.

A reasonable amount of body fat is necessary and beneficial, but maintaining proper body weight is a problem for many people. From the metabolic point of view, dieting consists of reducing caloric intake so that stored fats must be degraded to meet energy needs. Some people try to reduce by eliminating all fatty foods from their diets. By eliminating foods that contain fats, however, one runs the risk of also eliminating essential vitamins and minerals contained in these foods. Moreover, if the decreased fat intake is more than compensated by an increased intake of carbohydrates—in starchy foods, sweets, soda, and so forth—a person not only fails to lose weight but actually gains.

Human beings convert carbohydrates to fats, but people do not convert fats to carbohydrates. Human cells have no enzyme that can catalyze the conversion of acetyl CoA to pyruvate, a compound required for gluconeogenesis. Some bacteria, however, do have such an enzyme; they convert fats to carbohydrates as part of their normal metabolism.

The amount of energy that is released upon complete oxidation of a fuel can be measured in a **bomb calorimeter.** *A weighed sample of a fuel is burned in a closed chamber situated in a water-filled, insulated container. The heat given off is calculated from the measured increase in the temperature of the known mass of water.*

PRACTICE EXERCISE 16.4

A person has a basal metabolism of 1800 Calories per day. Calculate (a) the weight of fat that must be oxidized to CO_2 and H_2O to provide this amount of energy and (b) the weight of glucose that must be metabolized to release this amount of energy.

16.8 Ketone bodies

AIM: To list the three ketone bodies and the conditions that cause their production.

Focus

Liver cells produce ketone bodies when glucose is in short supply.

Under certain circumstances, body cells do not have enough glucose even for brain cells to use as an energy source. This happens most often in starvation or in untreated diabetes. In starvation, no supply of glucose is available; in diabetes, glucose is present in the blood, but it cannot penetrate cell membranes.

A lack of glucose causes the cells of many organs to step up the beta oxidation of fatty acids. However, when glucose levels are low, there is not enough oxaloacetate available to condense with acetyl CoA in the first step of the citric acid cycle. This is so because oxaloacetate comes from the carboxylation of pyruvate, and pyruvate comes from the breakdown of glucose in glycolysis. At low glucose levels, therefore, the concentration of acetyl CoA produced by the beta oxidation of fatty acids builds up. Under these conditions, the liver manufactures three special compounds from the excess acetyl CoA—the ketone bodies. Ketone bodies may be oxidized by many tissues to meet energy needs.

The **ketone bodies** *are acetoacetic acid, β-hydroxybutyric acid, and acetone.* We can see from their structural formulas that one of these compounds, β-hydroxybutyric acid, is inaccurately named as a ketone body, since it does not contain a ketone group.

$$CH_3-\overset{\overset{\displaystyle O}{\|}}{C}-CH_2-\overset{\overset{\displaystyle O}{\|}}{C}-OH \qquad CH_3-\overset{\overset{\displaystyle O}{\|}}{C}-CH_3 \qquad CH_3-\overset{\overset{\displaystyle OH}{|}}{\underset{\underset{\displaystyle H}{|}}{C}}-CH_2-\overset{\overset{\displaystyle O}{\|}}{C}-OH$$

Acetoacetic acid   Acetone   β-Hydroxybutyric acid
(not a ketone)

The liver does not use ketone bodies for energy production but releases them into the bloodstream. From the bloodstream, the ketone bodies reach other tissues—mainly the brain, the heart, and skeletal muscle.

The only ketone body that is in a form that can be used directly to produce energy is acetoacetic acid. The acetoacetic acid is converted to its thioester with CoA.

$$CH_3-\overset{\overset{\displaystyle O}{\|}}{C}-CH_2-\overset{\overset{\displaystyle O}{\|}}{C}-OH + HS-CoA \longrightarrow CH_3-\overset{\overset{\displaystyle O}{\|}}{C}-CH_2-\overset{\overset{\displaystyle O}{\|}}{C}-S-CoA + H_2O$$

Acetoacetic acid   Acetoacetyl CoA

The thioester that is formed, acetoacetyl CoA, may look familiar. If you recall our discussion of beta oxidation, you will see that acetoacetyl CoA is the same compound that is formed at the end of the fatty acid spiral. Tissue cells can cleave the acetoacetyl CoA back to two molecules of acetyl CoA.

$$CH_3-\overset{\overset{\displaystyle O}{\|}}{C}-CH_2-\overset{\overset{\displaystyle O}{\|}}{C}-S-CoA + HS-CoA \longrightarrow 2CH_3-\overset{\overset{\displaystyle O}{\|}}{C}-S-CoA$$

Acetoacetyl CoA

The acetyl CoA is then oxidized to carbon dioxide in the citric acid cycle, thereby providing NADH and $FADH_2$ for ATP production by cellular respiration. The thioester of the ketone body β-hydroxybutyric acid is also formed in cells, but the hydroxyl group of the acid portion of the ester must be oxidized to a ketone. Acetoacetyl CoA, useful for energy production, is formed as a result of this oxidation.

$$CH_3-\underset{\underset{H}{|}}{\overset{\overset{OH}{|}}{C}}-CH_2-\overset{\overset{O}{\|}}{C}-S-CoA \xrightarrow[\text{NAD}^+ \qquad \text{NADH}+\text{H}^+]{} CH_3-\overset{\overset{O}{\|}}{C}-CH_2-\overset{\overset{O}{\|}}{C}-S-CoA$$

β-Hydroxybutyryl CoA   Acetoacetyl CoA

Acetone, the third ketone body, is not used as an energy source.

16.9 Ketosis

AIMS: *To characterize the following aspects of ketosis: ketonemia, ketonuria, acetone breath, and ketoacidosis. To describe how the effects of ketosis are counteracted by mechanisms within the body and by the administration of external agents.*

Focus

Prolonged ketosis starves cells for oxygen.

In normal metabolism, some ketone bodies are continuously produced and broken down in energy production. The normal blood level of ketone bodies seldom exceeds 3 mg/100 mL of blood. In diabetes, however, the liver produces large quantities of ketone bodies, releasing them into the bloodstream for delivery to other tissues. This causes a substantial increase in the level of ketone bodies in the blood of untreated diabetics. *A level of ketone bodies greater than about 20 mg/100 mL of blood is called* **ketonemia** *("ketones in the blood").*

Tissue cells cannot use all the ketone bodies produced. But the liver does not stop production, and eventually, a surplus builds up. *At a level of about 70 mg/100 mL of blood, ketone bodies are excreted in the urine. This condition is* **ketonuria** *("ketones in the urine").* At high levels of ketone bodies in the blood, acetone is excreted by the lungs. The sweet, minty smell of *acetone breath* becomes apparent.

The conditions of ketonuria, ketonemia, and acetone breath together are symptoms of **ketosis** *(also called* **ketoacidosis***)—blood acidosis caused by an excess of the ketone body acids, acetoacetic acid, and β-hydroxybutyric acid.* Diabetic ketosis involves the same problem as respiratory acidosis and lactic acidosis. This is the problem: The ketone body acids in the blood will lower the blood pH unless enough bicarbonate ions are present to act as buffers (proton sponges).

$$\underset{\substack{\text{from ketone} \\ \text{body acid}}}{} H^+ + \underset{\substack{\text{Bicarbonate} \\ \text{ion}}}{HCO_3^-} \rightleftharpoons \underset{\substack{\text{Carbonic} \\ \text{acid}}}{H_2CO_3}$$

$$\quad\;\; \underset{\text{Proton}}{}$$

The pH of the blood is maintained at 7.40 as long as the kidneys can regenerate new bicarbonate ions. In severe cases of diabetic ketosis, however, the kidneys cannot supply enough bicarbonate ions to keep up with the production of ketone bodies. More ketone bodies are produced, insufficient bicarbonate ions are available, and the blood pH drops. This sequence has a disastrous effect. Hemoglobin can pick up oxygen only in an environment with a low concentration of protons. The lower the pH of the blood, the higher is the concentration of protons, and the less oxygen can be transported by hemoglobin. Brain cells become starved for oxygen. If this lack of oxygen continues, coma and death will follow.

The first step in the treatment of patients with diabetes who are exhibiting ketosis is usually the administration of insulin. This should restore normal glucose metabolism and reduce the formation of ketone bodies. Like glycosuria, ketonuria results in the loss of a large volume of body water, often causing dehydration. In diabetic patients with severe dehydration and ketosis, fluids and buffering power are restored by intravenous administration of solutions containing sodium bicarbonate.

16.10 Cholesterol synthesis

AIM: To show how the synthesis of cholesterol and ketone bodies illustrates the compartmentalization of cellular processes.

Focus

The carbon skeleton of cholesterol is formed from acetyl CoA.

Cholesterol is formed in the cytoplasm of liver cells. Biochemists have shown that the 27 carbons of the carbon skeleton of this important steroid come entirely from acetyl CoA (Fig. 16.5). The total biosynthesis of 1 molecule of cholesterol uses up 15 molecules of acetyl CoA and involves at least 13 separate chemical reactions. A few of the main steps in cholesterol synthesis are shown in Figure 16.6.

The synthesis of cholesterol in the cytoplasm and the synthesis of ketone bodies in mitochondria are excellent examples of the compartmentalization of cellular processes. The cytoplasm lacks the enzymes necessary to synthesize ketone bodies from acetyl CoA; the mitochondrion lacks the enzymes necessary to synthesize cholesterol from acetyl CoA. Compartmentalization is often important in balancing the synthetic and energy needs of cells. When a cell needs energy, most of the acetyl CoA produced in its mitochondria is oxidized in the citric acid cycle. When large amounts of acetyl CoA are being oxidized, lesser amounts can be furnished to the cytoplasm. The decreased level of acetyl CoA in the cytoplasm slows down or

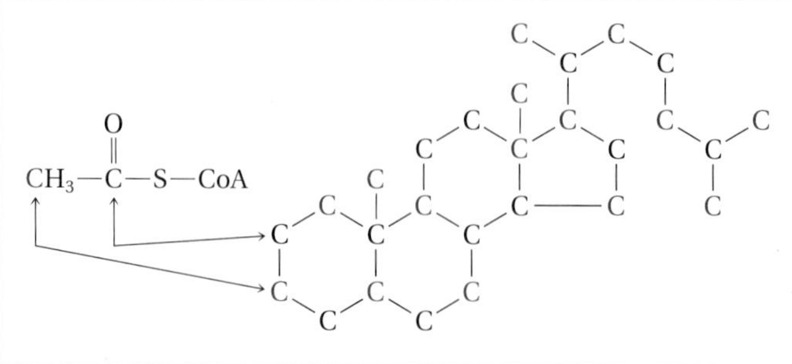

Figure 16.5
The incorporation of the acetyl carbons of acetyl CoA into the carbon skeleton of cholesterol. The carbons of cholesterol shown in color come from the methyl group of acetyl CoA. The remainder of the carbons come from the carbonyl group (—C=O).

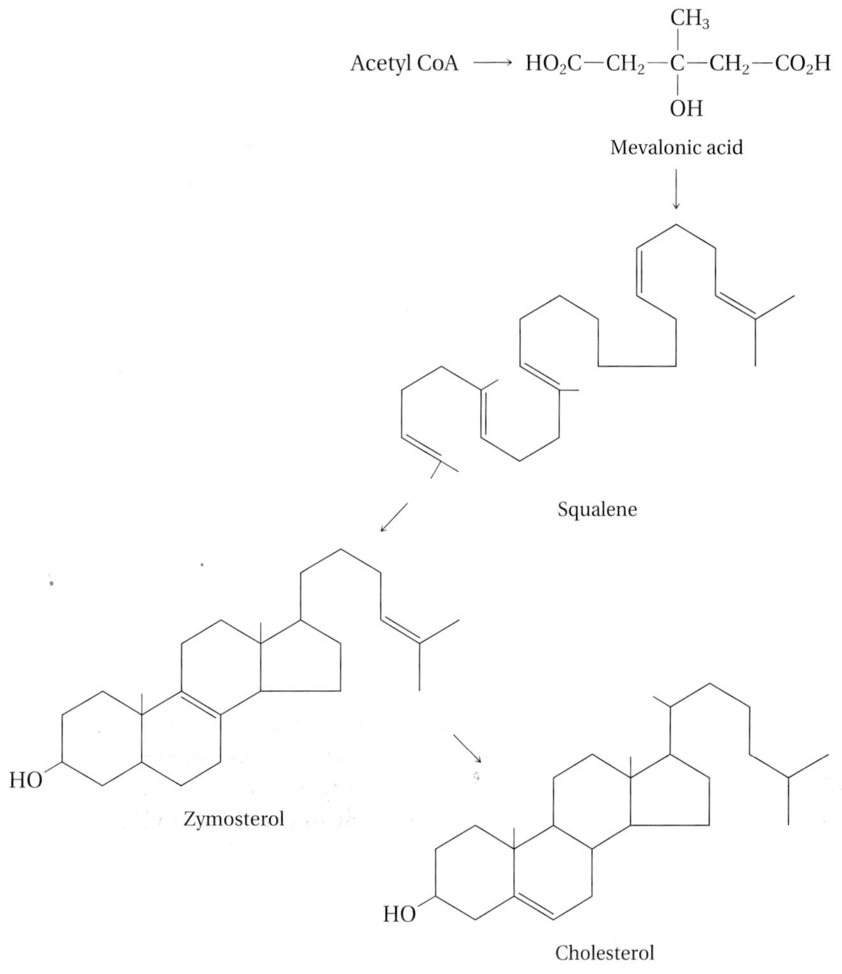

Figure 16.6
Some intermediates in the synthesis of cholesterol.

The level of cholesterol in the blood is influenced by a combination of heredity, diet, and metabolic diseases.

stops cholesterol synthesis. The synthesis of cholesterol is an anabolic (synthetic) pathway. As such, it requires the expenditure of the reducing power of NADH and the energy of ATP. The slowdown of cholesterol synthesis has the effect of conserving reducing power and energy in the cell.

16.11 Blood cholesterol

AIM: To explain how an increase in the percentage of unsaturated fatty acids in the diet could lead to lower plasma cholesterol levels.

Focus

Blood cholesterol levels are related to the incidence of atherosclerosis.

Cholesterol is the most abundant steroid in the animal body. About 90% of the cholesterol synthesized by mammals is formed in the liver and distributed to blood serum and other tissues. The serum cholesterol is bound to lipoproteins (see A Closer Look: Blood Lipoproteins and Heart Disease, page 363). The amount of cholesterol bound to a lipoprotein increases in proportion to the density of the lipoprotein. *Chylomicrons,*

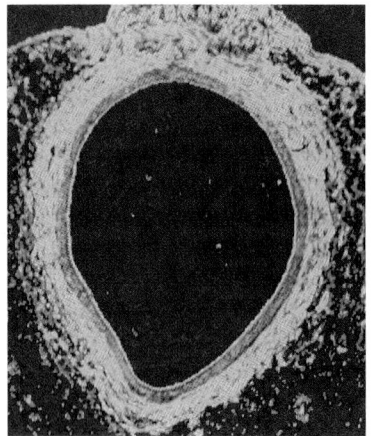

(a)

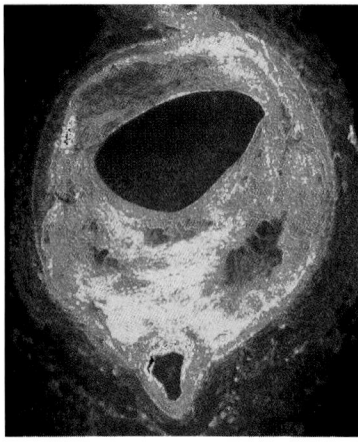

(b)

Figure 16.7
(a) A cross-section of a normal artery. (b) A cross-section of an artery constricted by the formation of atherosclerotic plaque.

Atherosclerosis
athera (Greek):   gruel
skleros (Greek):   hard

According to a study by the Center for Science in the Public Interest, the American Cancer Society, and the American Public Health Association, if Americans reduced their daily intake of saturated fats by just 8 grams, health care costs would go down as much as $17 billion per year.

which have the lowest density of all lipoproteins, mainly contain triacyl glycerols and a relatively small amount of cholesterol. *Very-low-density lipoproteins* (VLDLs) contain more cholesterol than is present in chylomicrons. *Low-density lipoproteins* (LDLs) contain more cholesterol than VLDLs, and *high-density lipoproteins* (HDLs) are very rich in cholesterol. Normal blood contains 120 to 200 mg of cholesterol per 100 mL of serum. Elevated levels of 200 to 300 mg of cholesterol per 100 mL of serum are usually considered a threat to health, having been associated with *atherosclerosis*.

Atherosclerosis *is a disease in which cholesterol and arterial tissue cells build up in layers, called* **plaques** (Fig. 16.7). The buildup of arterial plaque can block arteries and cause impaired blood circulation. Heart failure from blocked coronary (heart) arteries is the leading cause of death in industrialized nations. Doctors advise many people to take steps to reduce their serum cholesterol because high levels stimulate the onset of atherosclerosis.

Three major strategies for the reduction of serum cholesterol are increasing the level of HDLs, a low-saturated-fat diet, and cholesterol-lowering medications. HDLs can extract cholesterol from the blood, which tends to lower serum cholesterol levels. This explains the protection from heart attack conferred by high levels of serum HDLs mentioned in A Closer Look: Blood Lipoproteins and Heart Disease (page 363). Because of this protection, the cholesterol in HDLs is sometimes called "good cholesterol." Physically fit people tend to have higher levels of HDLs than unfit people, so exercise is often an important component of a program to reduce serum cholesterol.

A low-saturated-fat diet—one in which fat intake is mainly restricted to lipids containing unsaturated fatty acids—is sometimes helpful in reducing serum cholesterol. The biochemical basis for the low-saturated-fat diet is the relationship between the lipid composition of cell membranes and the cholesterol concentration of plasma. Cell membranes need just the right degree of rigidity to function normally, and body cells control membrane rigidity by continuously adjusting the mix of lipids they use to construct their membranes. As we learned in Chapter 8, saturated lipids and cholesterol tend to make membranes rigid; unsaturated lipids tend to make them flexible. A person on a low-saturated-fat diet consumes enough unsaturated lipids to make very flexible membranes, but the dietary supply of saturated lipids and cholesterol needed to rigidize the membranes is limited, so the dieter's body cells use plasma cholesterol instead. Consequently, the low-saturated-fat diet results in increased tissue cholesterol and a corresponding decrease in plasma cholesterol. For information on the control of serum cholesterol by medications, see A Closer Look: Drug Strategies for Reducing Serum Cholesterol.

A Closer Look

Drug Strategies for Reducing Serum Cholesterol

Exercise and a low-saturated-fat diet may not bring blood cholesterol under control for two reasons: (1) As dietary cholesterol is reduced, the body increases its synthesis to compensate for the decrease and there is no net change, and (2) the human body lacks a way to excrete cholesterol. When exercise and a low-saturated-fat diet fail to reduce serum cholesterol, a physician may prescribe medication. Several drugs, called *hypolipidemic* drugs, are available for the reduction of serum cholesterol.

The action of hypolipidemic drugs is based on two different strategies. The first strategy is to promote the excretion of cholesterol. The patient takes oral doses of a cholestyramine or similar resin. The resin is an edible but indigestible polymer that contains hydrophobic benzene rings and hydrophilic quaternary amine groups.

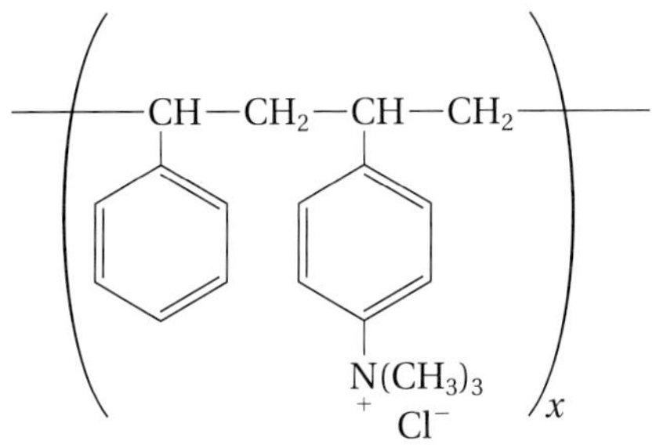

Cholestyramine resin
(Questran)

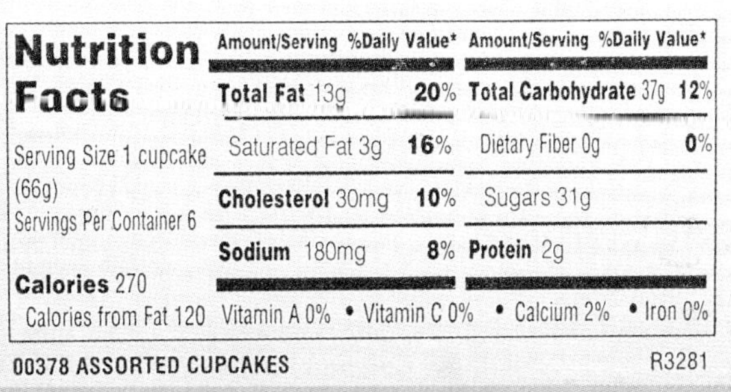

Cholesterol content and fat content are given on the labels of most consumable items. A reduced intake of cholesterol does not necessarily result in lowered cholesterol levels in the blood.

Ordinarily, cholesterol in the intestines is reabsorbed into the bloodstream and is not excreted. However, cholesterol binds tightly to the resin present in the intestines. The resin and its bound cholesterol are indigestible, and the resin-cholesterol complex is excreted in the feces. This loss results in lowered serum cholesterol. The second strategy aims to inhibit cholesterol synthesis. An oral drug called *lovastatin* exemplifies this strategy.

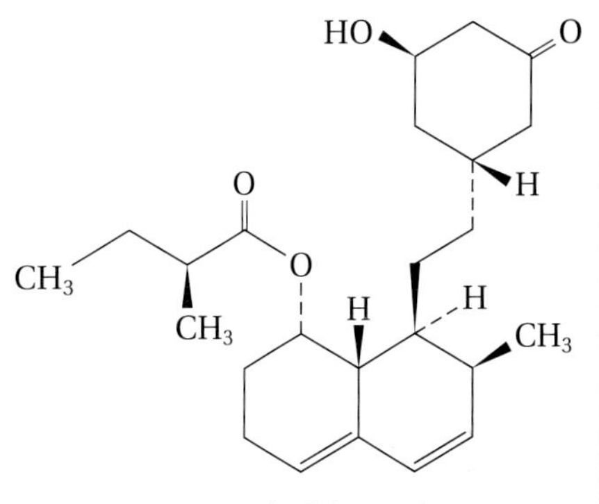

Lovastatin (Mevacor)

Lovastatin is an inhibitor of the enzyme *hydroxymethylglutaryl-CoA reductase* (HMG-CoA reductase). HMG-CoA reductase catalyzes the reduction of its substrate, hydroxymethylglutaryl-CoA (HMG-CoA), to the product mevalonate. This reaction is a necessary early step in the synthesis of cholesterol from acetyl CoA.

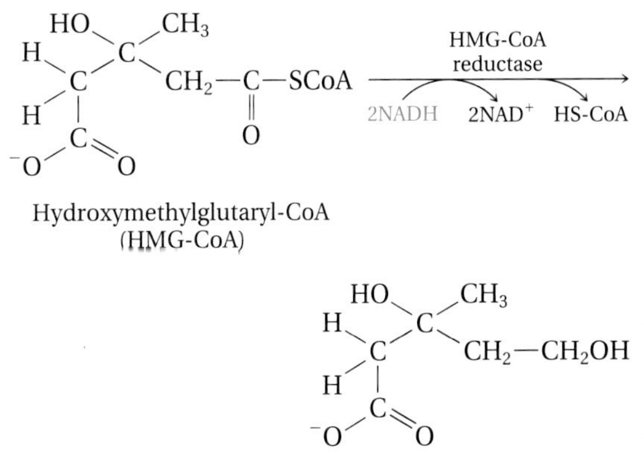

Hydroxymethylglutaryl-CoA
(HMG-CoA)

Mevalonate

The inhibition of the HMG-CoA reductase reduces the amount of cholesterol synthesized, so the level of serum cholesterol decreases.

SUMMARY

Lipids are extremely important as nutrients and in cell membranes. Reserves of stored triglycerides are mobilized as needed for energy production. Fat mobilization is stimulated by epinephrine (adrenalin). The triglycerides are hydrolyzed into fatty acids and glycerol, which enter the bloodstream. In tissues, free fatty acids are converted to fatty acyl CoA molecules, which are broken down to acetyl CoA by beta oxidation in the mitochondrion. The acetyl CoA may be used for energy production by way of the citric acid cycle and the electron transport chain.

Beta oxidation of fatty acids is a good source of energy. One molecule of palmitic acid produces enough energy to phosphorylate 129 molecules of ADP to ATP. When ATP is in abundant supply, acetyl CoA is resynthesized back to fatty acids or to cholesterol. When ATP is in short supply and glucose is not available, as in starvation or diabetes, the liver uses acetyl CoA to make the ketone bodies acetone, β-hydroxybutyric acid, and acetoacetic acid; the latter

two compounds are used for energy production in the citric acid cycle.

The brain usually needs glucose, but it adapts to ketone body oxidation in about 2 days. Excess production of ketone body acids leads to ketonemia (ketone bodies in the blood). The ketone bodies eventually appear in the urine (ketonuria). Chronic overproduction of ketone body acids leads to ketone body acidosis (ketosis) of the blood because the kidneys cannot supply enough bicarbonate ions to neutralize the acids.

The liver synthesizes cholesterol from acetyl CoA. The cellular cytoplasm is the site of cholesterol synthesis. Too much cholesterol in the blood may cause the formation of atherosclerotic plaque—a buildup of layers of cholesterol and cells on arterial walls. The resulting constriction of the blood vessels causes harmful effects on circulation; a blockage in the arteries of the heart can cause heart failure. Blood cholesterol is sometimes helped by exercise, a low-saturated-fat diet, or medication.

KEY TERMS

Adipose tissue (16.1)
Atherosclerosis (16.11)
Beta oxidation (16.3)
Depot fat (16.1)

Fatty acid spiral (16.3)
Fatty acyl CoA (16.3)
Ketoacidosis (16.9)
Ketone bodies (16.8)

Ketonemia (16.9)
Ketonuria (16.9)
Ketosis (16.9)
Mobilization (16.2)

Plaque (16.11)
Storage fat (16.1)
Tissue lipid (16.1)
Working lipid (16.1)

EXERCISES

Body Lipids (Sections 16.1, 16.2)

16.5 Draw the structure of a triglyceride that contains stearic, oleic, and linoleic acids.

16.6 What are the products of hydrolysis of the triglyceride in Exercise 16.5? Name the enzyme that catalyzes the process.

16.7 Describe how lipids are distributed in body tissues.

16.8 Discuss the functions of adipose tissue.

16.9 What stimulates the hydrolysis of triglycerides?

16.10 Explain the function of serum albumin in fatty acid transport.

16.11 Which tissues prefer glucose to fatty acids for energy production?

16.12 What is the cause of lipid storage diseases?

Fatty Acid Metabolism and Storage (Sections 16.3, 16.4, 16.5, 16.6, 16.7)

16.13 Where in the cell does fatty acid catabolism occur?

16.14 How is a fatty acid prepared for entry into a mitochondrion?

16.15 What is meant by the term *beta oxidation?*

16.16 Why is the degradation pathway of fatty acids called the fatty acid *spiral* rather than the fatty acid *cycle?*

16.17 The fatty acyl CoA formed from the 10-carbon acid capric acid is degraded by beta oxidation to acetyl CoA. What is the yield of (a) acetyl CoA, (b) NADH, and (c) $FADH_2$?

16.18 What is the maximum number of ATP molecules that can be produced in the complete oxidation of capric acid to carbon dioxide and water? (See Table 16.1.)

16.19 Why is acetyl CoA a key compound in the interplay between fatty acid and carbohydrate metabolism?

16.20 Describe the four fates that await acetyl CoA in the liver.

16.21 How does the camel's hump permit it to survive famine and drought?

16.22 Explain why fatty acids are more efficient energy stores than glycogen.

Ketone Bodies and Diabetes (Sections 16.8, 16.9)

16.23 What is the difference between starvation and diabetes?

16.24 What are the ketone bodies?

16.25 How do ketone bodies form?

16.26 Explain why untreated diabetics accumulate large amounts of ketone bodies.

16.27 Define (a) *ketonemia*, (b) *ketonuria*, (c) *acetone breath*, and (d) *ketosis*.

16.28 Describe how untreated diabetes mellitus leads to coma and death.

Cholesterol (Sections 16.10, 16.11)

16.29 Where does cholesterol synthesis take place? What is used as "starting material"?

16.30 Explain why cholesterol synthesis slows down when a cell needs energy.

16.31 What is atherosclerosis?

16.32 Explain why an increased intake of unsaturated fats is more helpful in reducing plasma cholesterol than decreasing dietary cholesterol.

Additional Exercises

16.33 Match the following:
(a) beta oxidation
(b) ketonemia
(c) fatty acid synthesis
(d) ketonuria
(e) mitochondria
(f) cholesterol
(g) epinephrine
(h) glycerol
(i) acetone
(j) ketosis

(1) ketone body
(2) blood acidosis caused by ketone bodies
(3) synthesized from acetyl CoA
(4) degradation of fatty acids
(5) ketone bodies in blood
(6) enters glycolysis
(7) site of beta oxidation
(8) fatty acid mobilization
(9) utilizes ATP
(10) ketone bodies in urine

16.34 What is the total ATP yield from the complete oxidation of 1 mol of myristic acid, $CH_3(CH_2)_{12}CO_2H$, to carbon dioxide and water?

16.35 How are fatty acid molecules prepared for transport across the mitochondrial membrane?

16.36 Name and describe the several conditions that are symptoms of ketosis.

16.37 Acetoacetic acid and β-hydroxybutyric acid, two of the ketone bodies, are used by the brain, the heart, and skeletal muscle as a source of energy. Why do these compounds present a serious problem in diabetes or starvation?

16.38 Where in the cell does (a) fatty acid catabolism take place? (b) fatty acid activation occur?

16.39 How do low blood glucose levels trigger the mobilization of fatty acids from adipose tissue?

16.40 Explain why the mobilization of fatty acids from adipose tissue stops when blood glucose level rises.

16.41 For each statement, write the number of the correct name and the number for the correct structure.

(a) A ketone body that is not used as an energy source
(b) Produced by reaction of a fatty acid with CoA
(c) A product of the hydrolysis in adipose tissue
(d) A ketone body that is not really a ketone
(e) Reacts with CoA to produce two molecules of substance (14)
(f) One of the ketone bodies
(g) Intermediate of both carbohydrate and lipid metabolism

(1) acetoacetyl CoA
(2) fatty acyl CoA
(3) β-hydroxybutyric acid
(4) acetyl CoA
(5) acetone
(6) acetoacetic acid
(7) a fatty acid

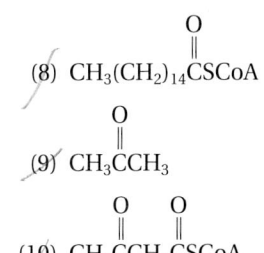

(8) $CH_3(CH_2)_{14}\overset{\overset{\displaystyle O}{\|}}{C}SCoA$

(9) $CH_3\overset{\overset{\displaystyle O}{\|}}{C}CH_3$

(10) $CH_3\overset{\overset{\displaystyle O}{\|}}{C}CH_2\overset{\overset{\displaystyle O}{\|}}{C}SCoA$

(11) $CH_3\overset{\overset{\displaystyle OH}{|}}{C}HCH_2CO_2H$

(12) $CH_3\overset{\overset{\displaystyle O}{\|}}{C}CH_2CO_2H$

(13) $CH_3(CH_2)_{14}CO_2H$

(14) $CH_3\overset{\overset{\displaystyle O}{\|}}{C}SCoA$

SELF-TEST (REVIEW)

True/False

1. Working lipids are generally localized in muscle tissue.
2. For equal weights, fats are better sources of energy than carbohydrates.
3. Epinephrine stimulates the hydrolysis of glycogen and triglycerides.
4. Once acetyl CoA is in the mitochondria, it is always used for energy production in the citric acid cycle.
5. Fatty acids can be converted to carbohydrates through an acetyl CoA intermediate.
6. The principal site of fatty acid metabolism is the liver.
7. A body with low ATP demand and excess glucose responds by making and storing fatty acid molecules.
8. The beta oxidation of a molecule of

$$CH_3(CH_2)_{16}CO_2H$$

would yield nine molecules each of acetyl CoA, NADH, and $FADH_2$.
9. Glucose metabolism has no effect on the formation of ketone bodies.
10. The rate of synthesis of cholesterol increases as the rate of acetyl CoA oxidation in the citric acid cycle increases.

Multiple Choice

11. An untreated severe diabetic would probably experience
    (a) acetone breath. (b) glucosuria.
    (c) ketonemia. (d) all of these.
12. Which of the following does not occur in the mitochondria of liver cells?
    (a) synthesis of ketones
    (b) beta oxidation
    (c) citric acid cycle
    (d) synthesis of cholesterol
13. The glycerol produced by the hydrolysis of triglycerides enters the metabolic pathway in
    (a) the citric acid cycle.
    (b) oxidative phosphorylation.
    (c) glycolysis.
    (d) the fatty acid spiral.
14. Fat tissue in the body is
    (a) an important source of body heat.
    (b) also called *adipose tissue.*
    (c) a good insulator.
    (d) all of the above.

15. In the aerobic catabolism of a molecule of a fatty acid, a majority of the energy is produced
    (a) directly as ATP in the citric acid cycle.
    (b) from NADH from the fatty acid spiral.
    (c) from NADH from the citric acid cycle.
    (d) from $FADH_2$ from the fatty acid cycle.
16. In the actual oxidation step of beta oxidation,
    (a) an alcohol is oxidized to an acid.
    (b) an alcohol is oxidized to a ketone.
    (c) an alcohol is oxidized to an aldehyde.
    (d) an aldehyde is oxidized to an acid.
17. Acetyl CoA in the liver may be converted to all the following *except*
    (a) cholesterol. (b) glycogen.
    (c) fatty acid. (d) ketone bodies.
18. Efficient energy storage is achieved by
    (a) converting fats to glycogen.
    (b) eliminating all fatty foods from the diet.
    (c) converting excess acetyl CoA into stored triglycerides.
    (d) avoiding the use of ATP in forming storage compounds.
19. Tay-Sachs disease is associated with
    (a) a defect in lipid anabolism.
    (b) the unavailability of complex lipids for use in the brain and liver.
    (c) a defect in lipid catabolism.
    (d) a defect in the process by which cholesterol is synthesized.
20. Suppose that you haven't eaten for 5 hours and you begin jogging. Which of the following statements about your system is *not* true?
    (a) Triglycerides are being hydrolyzed.
    (b) Glucose is being changed to glycogen.
    (c) Fatty acids are being broken down to acetyl CoA.
    (d) Cholesterol synthesis is minimal.
21. A deficiency of glucose in cells
    (a) increases the rate of beta oxidation of fatty acid.
    (b) can be a result of severe diabetes.
    (c) results in the formation of ketone bodies.
    (d) all of the above are true.
22. In severe diabetic ketosis,
    (a) the concentration of H^+ eventually drops.
    (b) oxygen transport by hemoglobin is impaired.
    (c) the pH of the blood remains unchanged.
    (d) the bicarbonate ion concentration increases.
23. Ketone bodies are not utilized as an energy source in the
    (a) brain. (b) liver.
    (c) heart. (d) skeletal muscles.

Metabolism of Nitrogen Compounds

17

Nitrogen and Life

Because of nitrogen fixation, legumes, such as these soybeans, can incorporate nitrogen into amino acids and eventually into important dietary proteins.

CHAPTER OUTLINE

This chapter is about the metabolism of biological molecules containing nitrogen. Nitrogen compounds are so diverse—amino acids, proteins, vitamins, and nucleic acids, to name a few—that we will cover only the principal aspects of nitrogen metabolism. In this chapter we will learn how nitrogen compounds are degraded and synthesized and how the metabolism of nitrogen compounds is related to the metabolism of carbohydrates and lipids. We also will learn about the diseases that can result from defects in the metabolism of nitrogen compounds. There are many such defects because there are so many nitrogen compounds and their metabolism is so complex. The following Case in Point describes the beginning of symptoms caused by one of these defects.

CASE IN POINT: A painful episode

 Ralph, a 45-year-old accountant for a large soda bottler, was not very concerned about maintaining a healthy diet. One recent night Ralph awoke from a sound sleep with an excruciating pain in the big toe of his right foot. In the morning his toe was still hurting and had become inflamed at the joint. What disease has caused Ralph's symptoms? What does his condition have to do with nitrogen metabolism, and how can it be treated? These questions will be answered in Section 17.9.

When uric acid crystals deposit in joints they cause pain, inflammation, and swelling of the affected region (magnified 150×).

17.1 Nitrogen fixation

AIM: To name and describe the process by which atmospheric nitrogen is made available to plants and animals.

Few plants can form nitrogen-containing compounds from nitrogen in the air, no animals can, but certain bacteria can. It is through nitrogen-fixing bacteria that atmospheric nitrogen enters the *biosphere*—the domain of living things. **Nitrogen-fixing bacteria** *are organisms that reduce atmospheric nitrogen to ammonia, a water-soluble form of nitrogen that can be used by plants and animals.*

$$N_2 + 3H_2 \longrightarrow 2NH_3$$

In soil and biological fluids, most ammonia is present as ammonium ions.

$$NH_3 \;+\; H^+ \;\rightleftharpoons\; NH_4^+$$
Ammonia   Proton   Ammonium
ion

Plants and animals incorporate ammonia into nitrogen compounds such as proteins and nucleic acids. The plants and animals die and decay, aided by other bacteria. Decaying matter returns nitrogen to the soil as ammonia, nitrite ions (NO_2^-), or nitrate ions (NO_3^-). Moreover, some nitrogen gas is returned to the atmosphere. *The flow of nitrogen between the atmosphere and the Earth and its living creatures is the* **nitrogen cycle,** shown in Figure 17.1.

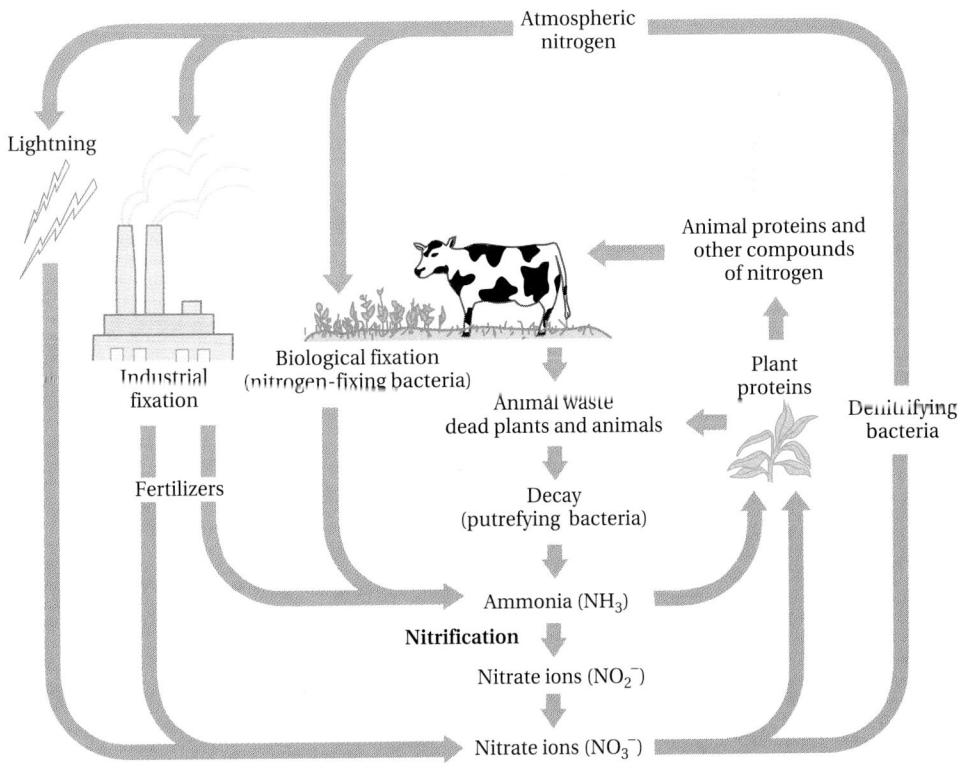

Figure 17.1
The nitrogen cycle.

It has been estimated that each year upwards of 10 million tons of nitrogen is fixed by natural and industrial processes. Some of this fixed nitrogen accumulates in soil and lakes and promotes algae growth. There are bacteria that convert fixed nitrogen back to N_2.

Modern agriculture intervenes to a great extent in the nitrogen cycle. For the past several years, the daily amount of atmospheric nitrogen fixed by industrial processes for the production of fertilizers has actually exceeded the amount fixed by living organisms in the Earth's forests and oceans. Besides bacterial and industrial nitrogen fixation, a smaller amount of atmospheric nitrogen is fixed by lightning discharges, which produce the soluble nitrogen oxides (NO, NO_2, N_2O_4, N_2O_5).

17.2 Protein turnover

AIM: To explain the function of cathepsins and state their cellular location.

Focus

Proteins are continuously hydrolyzed and synthesized by the body.

The rate of turnover varies by protein function. Plasma proteins have a half-life of about 10 days; 50% are hydrolyzed in a 10-day period. Muscle protein has a half-life of 180 days, some connective tissue proteins, as high as 1000 days.

In the human body, dietary proteins are hydrolyzed to their constituent amino acids in the small intestine as part of digestion. Many of the body's proteins are continuously hydrolyzed and synthesized within body cells. *The continuous hydrolysis and synthesis of proteins in the body is called* **protein turnover.** The enzymes responsible for the intracellular hydrolysis of proteins are a class of proteases called the *cathepsins*. In cells, the cathepsins are confined to the lysosomes, where protein degradation occurs. We discussed translation, the process of protein synthesis from genetic information, in Section 11.9.

PRACTICE EXERCISE 17.1
Why are cathepsins confined to the lysosomes in cells?

17.3 Transamination reactions

AIMS: To describe some uses of the amino acids in the amino acid pool. To use words and equations to describe transamination reactions, indicating the function of pyridoxal phosphate. To interpret the significance of increased levels of transaminases in a person's blood serum.

Focus

Amino acids part with their amino groups by transamination reactions.

Amino acids produced by digestion of dietary protein and during protein turnover in body cells become part of the body's amino acid pool. *The* **amino acid pool** *is the total quantity of free amino acids present in tissue cells, plasma, and other body fluids.* The amino acids of the amino acid pool are available for cellular needs, as shown in Figure 17.2.

Many of the reactions of amino acid metabolism require that amino acids first lose their alpha amino groups. The most common way for this to occur is by **transamination**—*the transfer of an amino group from one molecule to another.* An intermediate of the citric acid cycle, α-ketoglutaric

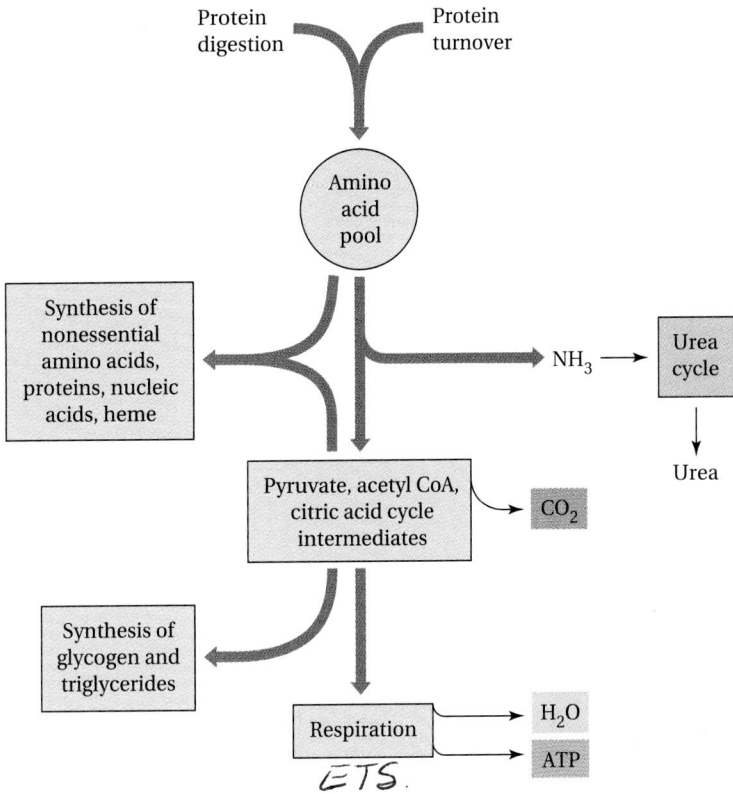

Figure 17.2
The major pathways of amino acid metabolism.

acid, is the usual acceptor of the amino group. The products of the reaction are an α-ketoacid and glutamic acid.

An α-amino acid α-Ketoglutaric acid   An α-keto acid   Glutamic acid

Transamination reactions are catalyzed by enzymes called *transaminases*. The transaminases were mentioned in Table 10.1 as indicators of disease or trauma that affects tissues. The principal transaminase of the liver is *glutamic-pyruvic transaminase* (GPT), an enzyme that catalyzes the formation of pyruvate from alanine

Alanine   Pyruvic acid

Liver disease causes an increase of blood serum GPT (SGPT). Elevated levels of the enzyme in serum can be used as a diagnostic test for liver disease. The principal transaminase of the heart muscle is *glutamic-oxaloacetic transaminase* (GOT), which catalyzes the formation of oxaloacetate, one of the intermediates of the citric acid cycle, from aspartic acid.

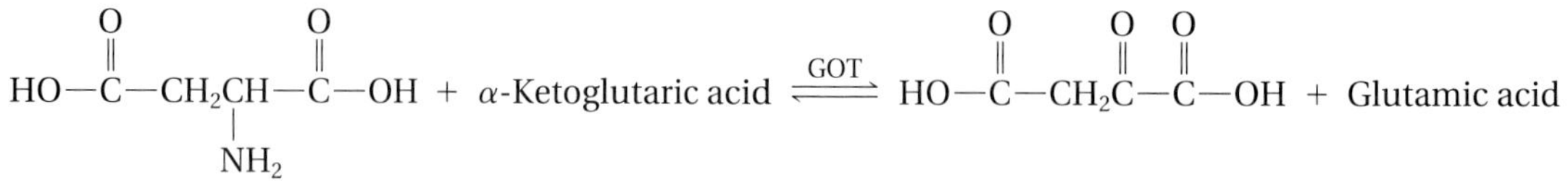

$$\underset{\text{Aspartic acid}}{HO-\overset{\overset{\displaystyle O}{\|}}{C}-CH_2\underset{\underset{\displaystyle NH_2}{|}}{CH}-\overset{\overset{\displaystyle O}{\|}}{C}-OH} + \alpha\text{-Ketoglutaric acid} \overset{GOT}{\rightleftharpoons} \underset{\text{Oxaloacetic acid}}{HO-\overset{\overset{\displaystyle O}{\|}}{C}-CH_2\overset{\overset{\displaystyle O}{\|}}{C}-\overset{\overset{\displaystyle O}{\|}}{C}-OH} + \text{Glutamic acid}$$

Damaged heart cells die and split open, but some of the GOT molecules that spill into the blood are still active. The levels of serum GOT (SGOT) activity are a measure of the extent of damage to heart muscle caused by a heart attack.

All known transaminases require pyridoxal phosphate, a derivative of pyridoxol (vitamin B_6), as the cofactor.

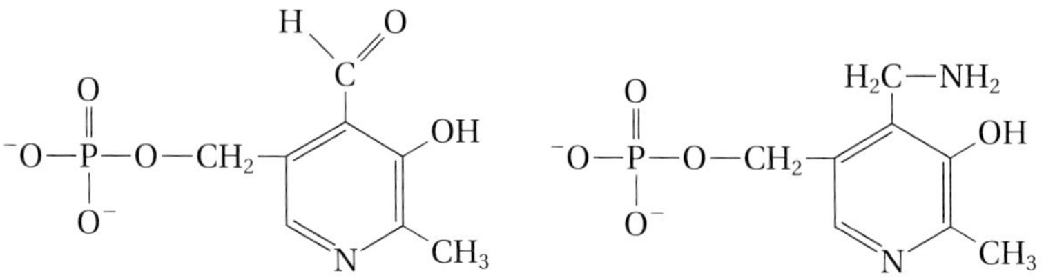

Pyridoxal phosphate   Pyridoxamine phosphate

The aldehyde carbon of pyridoxal phosphate is the acceptor of the amino group of amino acids in the first stage of reactions catalyzed by transaminases. An α-ketoacid and pyridoxamine phosphate are the products of the reaction.

Amino acid + Pyridoxal phosphate $\longrightarrow$ α-Ketoacid + Pyridoxamine phosphate

The second stage of transamination consists of the transfer of the amino group of the pyridoxamine phosphate to α-ketoglutarate with regeneration of the pyridoxal phosphate.

α-Ketoglutaric acid + Pyridoxamine phosphate $\longrightarrow$ Glutamic acid + Pyridoxal phosphate

PRACTICE EXERCISE 17.2

Write an equation for the transamination of tyrosine with α-ketoglutarate as the acceptor.

PRACTICE EXERCISE 17.3

How does an amino acid lose its amino group in transamination?

17.4 The urea cycle

***AIMS:** To explain the net results of the urea cycle on a molecular and a physiologic scale. To list two causes of a negative nitrogen balance.*

Focus

Ammonia produced from the oxidation of glutamic acid is excreted as urea.

Glutamic acid serves as the depot for receiving amino groups removed from amino acids by transamination reactions. Since there is a limited quantity of α-ketoglutarate in cells, it must be regenerated so that transamination reactions and the citric acid cycle can continue. The α-

ketoglutarate is regenerated from glutamic acid by an oxidation reaction catalyzed by glutamate dehydrogenase. This enzyme uses NAD^+ as the coenzyme. *The removal of the amino group by oxidation is called* **oxidative deamination.**

$$HO-\underset{\overset{\|}{O}}{C}-CH_2CH_2\underset{\underset{NH_2}{|}}{\overset{\overset{H}{|}}{C}}-\underset{\overset{\|}{O}}{C}-OH + H_2O \xrightarrow[\underset{NAD^+ \quad NADH}{}]{\text{Glutamate dehydrogenase}} NH_4^+ + HO-\underset{\overset{\|}{O}}{C}-CH_2CH_2\underset{\overset{\|}{O}}{C}-\underset{\overset{\|}{O}}{C}-OH$$

Glutamic acid   Ammonium   α-Ketoglutarate
                                        ion

With the regeneration of α-ketoglutarate, the goal of the cell has been met—the cellular concentration of α-ketoglutarate is restored. Oxidative deamination also has produced ammonia, however. And ammonia, even in low concentrations, is especially toxic to brain cells and can result in coma and death. A Closer Look: Hyperammonemia, examines some reasons for this toxicity. It is important that ammonia be removed from the body. This is accomplished by the reactions of the *urea cycle,* shown in Figure 17.3.

The **urea cycle** *is a metabolic pathway by which land-dwelling animals prepare waste ammonia and ammonium ions for excretion.* The excretion of nitrogen as urea helps to conserve body water and prevent dehydration. If ammonia were excreted directly in the urine, the ammonia would have to be kept very dilute. The resulting loss of large volumes of body water would make dehydration an ever-present danger. Urea, on the other hand, is very soluble in water and is nontoxic except at high concentrations. Urea formation permits land-dwellers to eliminate large quantities of nitrogen while losing relatively little water as urine. Other animals, primarily the fishes,

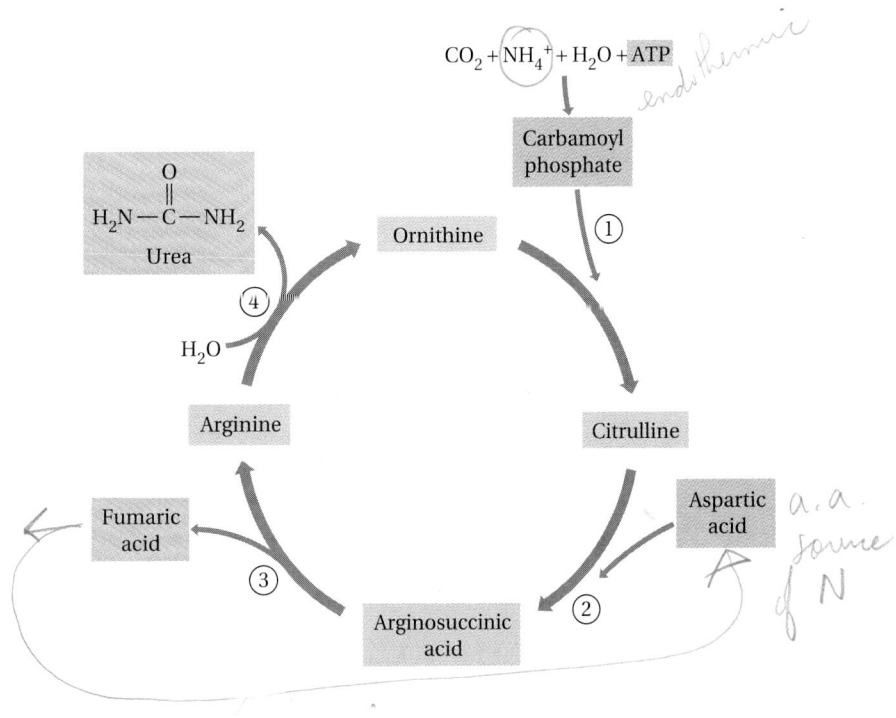

Figure 17.3
The urea cycle.

excrete ammonia as a waste product. Because they live in a water environment, fishes are not in danger of dehydration.

PRACTICE EXERCISE 17.4

What toxic substance is produced when α-ketoglutarate is regenerated from glutamic acid? How is this substance removed from the body?

Steps of the urea cycle

Let's examine the chemical reactions of the urea cycle in more detail. Before ammonia can enter the urea cycle, it is converted to a more energetic form: the compound carbamoyl phosphate. The formation of carbamoyl phos-

---

A Closer Look

Hyperammonemia

Hyperammonemia is an increase in ammonia in the blood to higher concentrations than the normal 30 to 60 mM. This condition is usually caused by the inability of patients to form urea fast enough to keep ammonia in the blood at normal levels. Several conditions can lead to hyperammonemia. Some newborns exhibit a delay in the development of urea cycle enzymes. The hyperammonemia in these infants clears up when the enzymes begin to function. In adults, causes of hyperammonemia include hereditary defects in the enzymes of the urea cycle and blockage of the blood flow through the liver, as might occur in cirrhosis (see figure).

Coma (an unconscious state) may occur in cases of hyperammonemia that are caused by liver (hepatic) disease. The reason for hepatic coma is uncertain, but it may be due to the depletion of ATP in the brain. To help us understand how ammonia may contribute to ATP depletion, let's review some aspects of metabolism. You may recall that the citric acid cycle is an important source of reducing power in the form of NADH and FADH$_2$ for oxidative phosphorylation and that oxidative phosphorylation is responsible for the production of most of the ATP needed to provide the energy to keep cells alive and healthy. In hyperammonemia, ammonia may cross the blood-brain barrier and enter brain cells. In the brain cells, the excess ammonia could drive the reaction between glutamate and α-ketoglutarate

in the direction of the formation of glutamate. The shift to glutamate would reduce the amount of α-ketoglutarate available for the citric acid cycle. As a result, the concentrations of NADH and FADH$_2$ also would decrease, less ATP would be formed in oxidative phosphorylation, and hepatic coma could result from the depletion of ATP. Some cases of hyperammonemia can be treated by the addition of α-ketoacids to the diet of the hyperammonemic patient. This diverts some of the excess ammonia into amino acids through transamination of the α-ketoacids. The amino acids are used by body cells to make proteins instead of having the ammonia contribute to the patient's hyperammonemia.

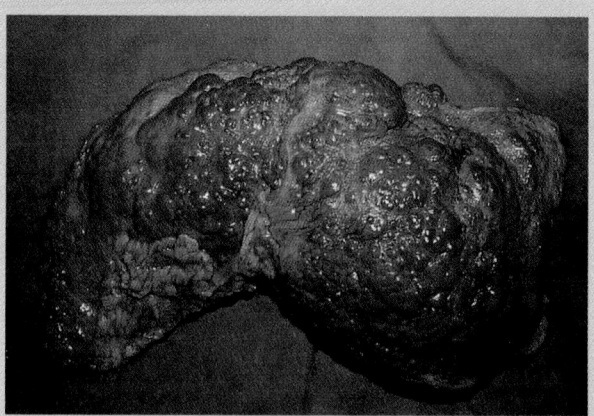

Cirrhosis of the liver impairs the urea cycle and can cause hyperammonemia.

phate requires, in addition to ammonia, one bicarbonate ion and the expenditure of two molecules of ATP.

$$NH_3 + HCO_3^- + 2ATP \longrightarrow H_2N-\overset{\overset{\displaystyle O}{\|}}{C}-O-\overset{\overset{\displaystyle O}{\|}}{\underset{\underset{\displaystyle O^-}{|}}{P}}-O^- + 2ADP + P_i$$

Carbamoyl
phosphate

With the formation of carbamoyl phosphate, ammonia has been energetically primed for entry into the urea cycle. The cycle can be summarized in four steps:

Step 1: Carbamoyl phosphate reacts with ornithine (see Sec. 14.7). Ornithine, an amino acid not found in proteins, is already present in the cell. The product of this reaction is citrulline.

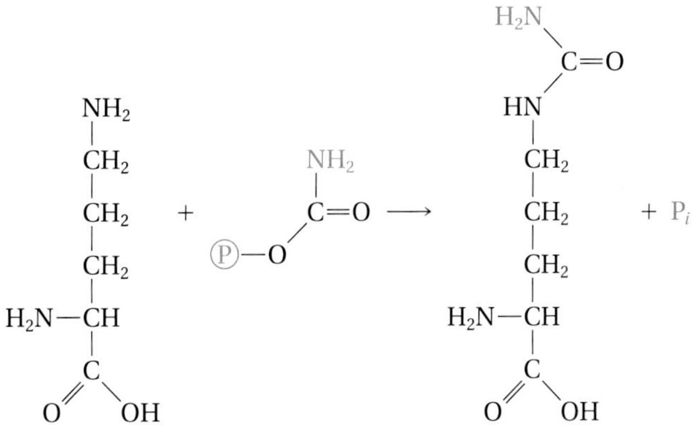

Ornithine   Carbamoyl   Citrulline
                   phosphate

Step 2: Citrulline reacts with aspartic acid to give a complex molecule called *arginosuccinic acid*. This reaction requires the expenditure of two more high-energy phosphate bonds of ATP. You need not memorize such complex molecular structures, but you should be aware of the path that nitrogen follows through the cycle.

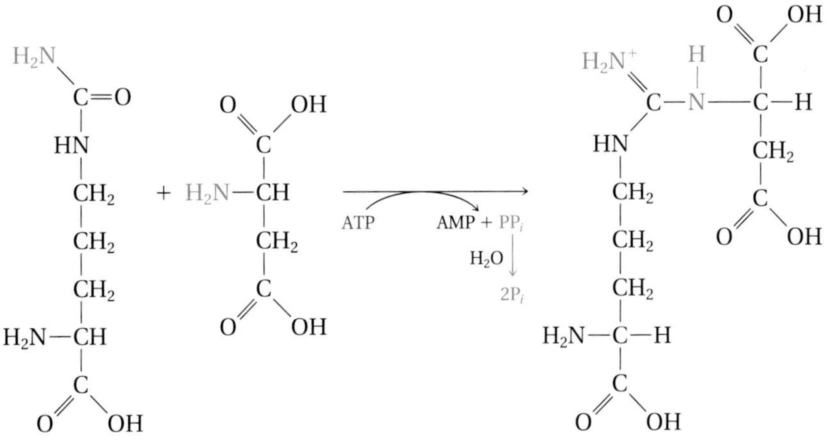

Citrulline   Aspartic   Arginosuccinic acid
                    acid

Keep in mind that both the new nitrogens that have been added to ornithine can originate as amino groups of amino acids: The first nitrogen can come from ammonia produced by the deamination of an amino acid and used to form carbamoyl phosphate; the second comes from succinic acid.

Step 3: An enzyme now cleaves arginosuccinic acid into arginine and fumaric acid. Fumaric acid is one of the intermediates of the citric acid cycle.

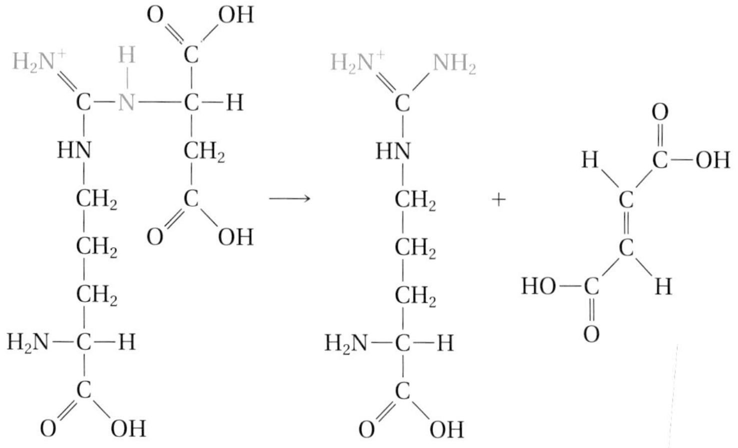

Arginosuccinic acid → Arginine + Fumaric acid

We have followed the synthesis of the nonessential amino acid arginine.

Step 4: The production of urea occurs when arginine is hydrolyzed to ornithine and urea.

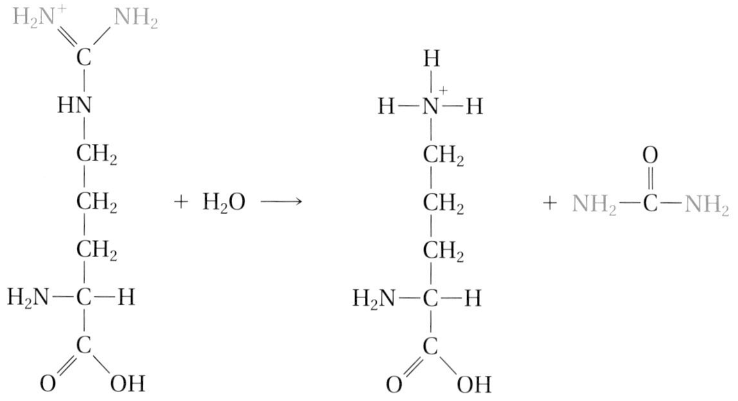

Arginine + H_2O → Ornithine (as its ammonium ion) + Urea

The ornithine reenters the urea cycle by reacting with another molecule of carbamoyl phosphate. After entering the bloodstream, urea is filtered by the kidneys and disposed of in the urine.

Nitrogen balance

Normal adults generally maintain a **nitrogen balance**—*the quantity of nitrogen excreted daily equals the intake.* About 80% of the nitrogen excreted in the urine is in the form of urea, and almost all the body's waste nitrogen of metabolism is excreted in the urine; the nitrogen compounds in the feces come mainly from indigestible materials. Children have a **positive nitrogen balance**—*the excretion of less nitrogen than is consumed.* The nitrogen balance is positive because children are growing and their cells are making new proteins and other nitrogen compounds. Several conditions result in a **negative nitrogen balance**—*the excretion of more nitrogen than is consumed.* During starvation and certain diseases, the carbon skeleton of amino acids derived from the breakdown of muscle proteins must be catabolized as an energy source. Since no new protein is available to eat, starving people excrete more nitrogen than they consume. The lack of even one essential amino acid in the diet results in a negative nitrogen balance. With an essential amino acid missing, the other amino acids cannot be used to make complete proteins. These other amino acids are deaminated, and the nitrogen is excreted as urea.

17.5 Catabolism of amino acids

***AIMS:** To distinguish between glucogenic and ketogenic amino acids. To discuss how amino acids can be used for energy production, gluconeogenesis, and the synthesis of fats.*

Focus

The carbon skeletons of amino acids are converted to intermediates of glucose metabolism and fatty acid metabolism.

Once α-ketoacids have been formed from amino acid by transamination reactions, their carbon skeletons are subjected to further chemical changes. One set of amino acids is converted to pyruvate, oxaloacetate, or α-ketoglutarate (Fig. 17.4). *Amino acids that are converted to these intermediates are called* **glucogenic,** *since these compounds are also important to glucose metabolism.* Pyruvate, formed at the end of glycolysis, and oxaloacetate are intermediates of the citric acid cycle. The remainder of the amino acids are converted to acetyl CoA, which is also a product of fatty acid metabolism. *The amino acids that are converted to acetyl CoA are called* **ketogenic.**

The conversion of all the amino acids to intermediates of glucose or fatty acid metabolism demonstrates the highly organized character of metabolism and the economy of nature. By using a single, central pathway for the metabolism of sugars, fats, and amino acids, the cell greatly decreases the number of enzymes and chemical steps that otherwise might be required to accomplish the same task.

Cells have priorities for the utilization of amino acids present in the amino acid pool. In a normal individual, well nourished with carbohydrates, fats, and proteins, the synthesis of nonessential amino acids, proteins, and other nitrogen-containing compounds is at the top of the prior-

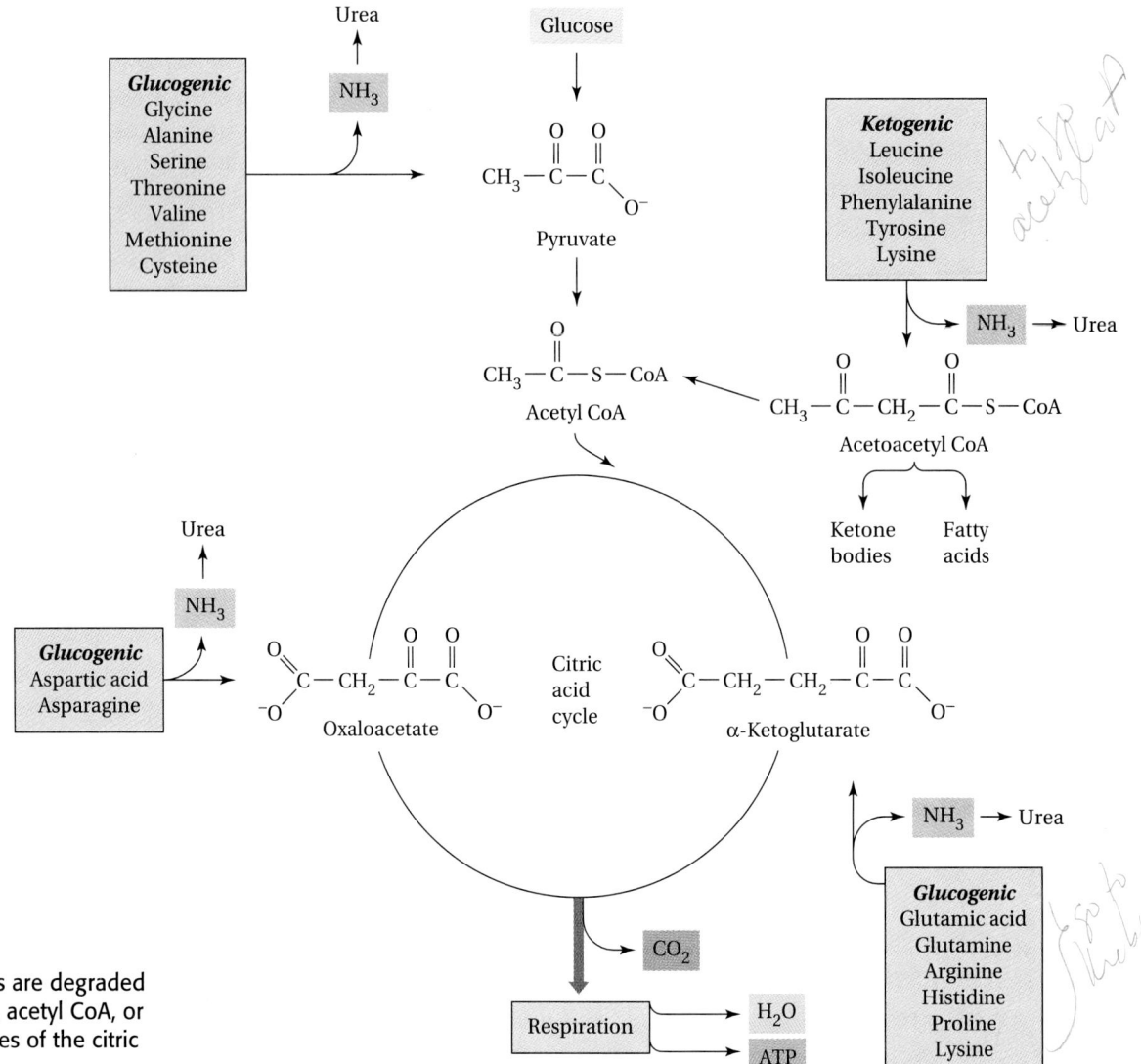

Figure 17.4
Amino acids are degraded to pyruvate, acetyl CoA, or intermediates of the citric acid cycle.

ity list. Energy production comes second. The body does not store amino acids as such. Any amino acids that remain after synthetic and energy needs are met are converted to glucose and fatty acids.

PRACTICE EXERCISE 17.5

Explain how transamination provides the link between amino acid metabolism and glucose metabolism or fat metabolism.

Energy production

The carbons of amino acids are used for energy production when needed. Upon demand, pyruvate and acetyl CoA derived from amino acids are oxidized to carbon dioxide in the citric acid cycle. Moreover, by forming oxaloacetate and α-ketoglutarate from glucogenic amino acids, cells can replenish or increase the concentrations of intermediates of the citric acid cycle. An increase in these intermediates enables cells to step up energy production. You may recall that oxaloacetate for the citric acid cycle can come from several places in metabolism. We have seen that the carboxylation of pyruvate in gluconeogenesis is one of those places. Now we see that amino acid metabolism is another.

Certain emergencies such as diabetes or starvation result in a reduction in the amount of acetyl CoA in the liver. Liver cells respond by using acetyl CoA produced in amino acid metabolism to make ketone bodies. The ketone bodies are transported to other tissues, where they are oxidized for energy production.

Synthesis of glycogen and triglycerides

From our study of glucose metabolism, we know that glucose is formed from pyruvate by gluconeogenesis, in which oxaloacetate is an intermediate (see Sec. 15.8). The cell does not recognize whether the pyruvate has come from glucose or from amino acid metabolism. Once glucose has been synthesized, it can be assembled into glycogen and stored in muscle or liver cells. Oxaloacetate from amino acid metabolism also can be converted to glucose in gluconeogenesis.

Fatty acids, as we have seen, are synthesized from acetyl CoA. The acetyl CoA can come from glucose metabolism, from fatty acid metabolism, or from amino acid metabolism. Newly synthesized fatty acids are either used immediately for energy production or converted to triglycerides or membrane lipids. The triglycerides are stored in adipose tissue as an energy reserve. Humans cannot synthesize glucose from acetyl CoA, since people lack the enzyme that converts acetyl CoA to pyruvate.

17.6 Synthesis of amino acids

AIM: To show the relationship between the citric acid cycle and the synthesis of nonessential amino acids.

Our bodies need to synthesize nonessential amino acids (see Sec. 12.6) because their proportions in our diet seldom match our bodies' needs. The main starting materials for this synthesis are pyruvate and two intermediates of the citric acid cycle: α-ketoglutarate and oxaloacetate. As we have seen, α-ketoglutarate accepts amino groups from other amino acids in transamination to give *glutamic acid,* and *arginine* is formed in the urea cycle. Two other nonessential amino acids—*aspartic acid* and *alanine*—may be synthesized directly from α-ketoacids because the reactions catalyzed by the transaminases are reversible. Reversals of transamination reactions form alanine from pyruvic acid and aspartic acid from oxaloacetic acid.

$$CH_3-\overset{\overset{\displaystyle O}{\|}}{C}-\overset{\overset{\displaystyle O}{\|}}{C}-OH \;+\; \text{Glutamic acid} \;\rightleftharpoons\; CH_3\underset{\underset{\displaystyle NH_2}{|}}{CH}-\overset{\overset{\displaystyle O}{\|}}{C}-OH \;+\; \alpha\text{-Ketoglutarate}$$

Pyruvic acid   Alanine

$$HO-\overset{\overset{\displaystyle O}{\|}}{C}-CH_2-\overset{\overset{\displaystyle O}{\|}}{C}-\overset{\overset{\displaystyle O}{\|}}{C}-OH \;+\; \text{Glutamic acid} \;\rightleftharpoons\; HO-\overset{\overset{\displaystyle O}{\|}}{C}-CH_2\underset{\underset{\displaystyle NH_2}{|}}{CH}-\overset{\overset{\displaystyle O}{\|}}{C}-OH \;+\; \alpha\text{-Ketoglutarate}$$

Oxaloacetic acid   Aspartic acid

Glutamine and *asparagine* are formed from glutamic acid and aspartic acid by reaction of the side-chain carboxyl groups with ammonia.

$$HO-\overset{\overset{\displaystyle O}{\|}}{C}-CH_2\underset{\underset{\displaystyle NH_2}{|}}{CH}-\overset{\overset{\displaystyle O}{\|}}{C}-OH \;+\; NH_3 \;\longrightarrow\; H_2N-\overset{\overset{\displaystyle O}{\|}}{C}-CH_2\underset{\underset{\displaystyle NH_2}{|}}{CH}-\overset{\overset{\displaystyle O}{\|}}{C}-OH \;+\; H_2O$$

Aspartic acid   Asparagine

$$HO-\overset{\overset{\displaystyle O}{\|}}{C}-CH_2CH_2\underset{\underset{\displaystyle NH_2}{|}}{CH}-\overset{\overset{\displaystyle O}{\|}}{C}-OH \;+\; NH_3 \;\longrightarrow\; H_2N-\overset{\overset{\displaystyle O}{\|}}{C}-CH_2CH_2\underset{\underset{\displaystyle NH_2}{|}}{CH}-\overset{\overset{\displaystyle O}{\|}}{C}-OH \;+\; H_2O$$

Glutamic acid   Glutamine

Tyrosine, the only nonessential amino acid with an aromatic side chain, is produced from the essential amino acid phenylalanine. The conversion requires a single oxidation step catalyzed by the enzyme phenylalanine hydroxylase.

Phenylalanine   Tyrosine

So far we have seen how 7 of the 12 nonessential amino acids are synthesized. The syntheses of the remaining 5—cysteine, histidine, glycine, proline, and serine—are more complex. Their synthesis will not be considered here. The syntheses of 4 of the nonessential amino acids from oxaloacetate and α-ketoglutarate, two intermediates in the citric acid cycle, demonstrate an important metabolic principle: Besides being a pathway of catabolism, the citric acid cycle is a pathway of anabolism—a metabolic switch-hitter.

17.7 Defects of amino acid metabolism

AIM: To describe the cause, effects, and treatment of phenylketonuria.

Focus

There are many known defects of amino acid degradation.

Since there are a large number of amino acids, the possibilities for diseases related to amino acid metabolism are also great. Many of the diseases caused by defects of amino acid metabolism are rare. The most common are the **amino acidurias**—*conditions in which amino acids or related compounds are excreted in large quantities in the urine.* One example of an amino aciduria is phenylketonuria, which occurs about once for every 10,000 births.

Phenylketonuria (PKU) *is the result of an inborn error of metabolism in which phenylalanine hydroxylase, the enzyme responsible for the conversion of phenylalanine to tyrosine, is inactive.* This conversion is necessary for the complete catabolism of the benzene ring of phenylalanine. Since phenylalanine cannot be degraded further without being converted to tyrosine, the levels of phenylalanine and its deamination product, phenylpyruvic acid, build up until they are excreted in the urine in large quantities.

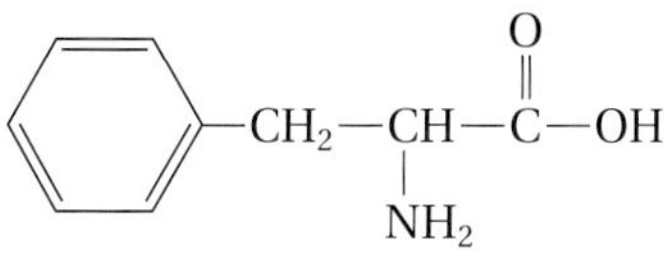

Phenylalanine

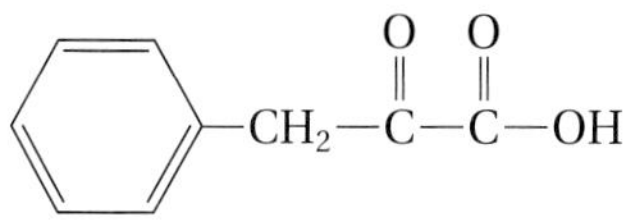

Phenylpyruvic acid

In the inherited disease iminoglycinuria, the urine contains excessive amounts of proline, hydroxyproline, and glycine.

Left untreated, phenylketonuria results in severe mental retardation by age 6 months. Why retardation occurs is not known, but it can be prevented or greatly alleviated by feeding the newborn infant a diet low in phenylalanine. (Phenylalanine cannot be eliminated from the diet entirely because it is an essential amino acid.) Since the retardation due to phenylketonuria can be prevented but not reversed, many states require the routine screening of urine samples of all newborn infants for signs of the disease. One screening test is very simple. Addition of a few drops of a dilute solution of ferric chloride to a small urine sample gives an olive-green color if the baby has phenylketonuria. The color results from a complex formed by ferric ions and phenylpyruvic acid. A Closer Look: The Amino Acidurias, examines other defects of amino acid metabolism.

17.8 Hemoglobin and bile pigments

AIMS: To trace the degradation of hemoglobin, indicating two important results of the process. To list some possible causes of jaundice.

Focus

Heme groups of hemoglobin are degraded to the bile pigment bilirubin.

Hemoglobin is synthesized by means of many complex reactions that take place in immature red blood cells. In the mature red blood cells, the hemoglobin does its work by transporting oxygen to body tissues. The life span of

A Closer Look

The Amino Acidurias

Phenylketonuria is only one of many amino acidurias that are recorded in the medical literature (see figure). All are the result of inborn errors of metabolism. *Alkaptonuria* is a disease related to the inability to break down phenylalanine, although the error in the catabolism of the phenylalanine is different from that in phenylketonuria. In phenylketonuria, as we have mentioned, the body lacks the enzyme necessary to put a phenolic hydroxyl group on the benzene ring of phenylalanine in order to make tyrosine. Alkaptonuria is a disease of tyrosine metabolism. In alkaptonuria, the conversion of phenylalanine to tyrosine occurs, but an enzyme necessary to degrade the benzene ring of tyrosine is lacking or defective. Ingested tyrosine in the body is incompletely converted into its normal degradation products, and the partially degraded products are excreted in the urine. These products are colorless when excreted but soon form a dark red pigment when exposed to air. The urine of people who have alkaptonuria may be colored from wine red to black depending on the concentration of the pigment. The pigment also forms in the bones, connective tissue, and organs of alkaptonuric patients. This deposition is thought to be the cause of arthritis that develops in many individuals with alkaptonuria. Except for the discomfort of arthritis, many alkaptonuric individuals have lived long and reasonably healthy lives.

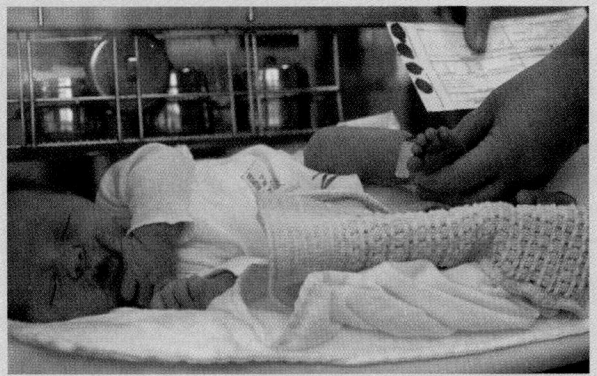

The blood of newborn infants is routinely screened for phenylketonuria (PKU).

An interesting group of amino acidurias is the family of diseases grouped as the *maple syrup urine diseases.* Many genetic diseases are found mainly in families and small groups that are highly intermarried. Maple syrup urine disease is a thousand times more common among the Old Order Mennonites of Pennsylvania than in the general population. The genetic flaws in the maple syrup urine diseases are in the catabolism of the branched-chain amino acids such as leucine and valine. The excretion of the products of incomplete breakdown of branched-chain amino acids imparts the odor of maple syrup to the urine. The product or products that give the urine this characteristic odor are unknown. Many individuals with maple syrup urine disease are mentally retarded and generally have short life spans.

the average red blood cell is about 120 days. The spleen—an organ of the lymphatic system—filters out cells that have reached the end of their useful lives. In the spleen, red blood cell membranes are broken down and hemoglobin spills out. Globin, the protein portion of hemoglobin, is hydrolyzed to its individual amino acids.

Degradation of heme begins with the removal of a single carbon from the heme ring by an oxidation reaction (Fig. 17.5). *The product of oxidation of heme is* **biliverdin,** *a green pigment. The iron released when heme is oxidized to biliverdin is retained in the body as a complex with the protein* **ferritin.** *Reduction converts biliverdin to* **bilirubin,** *an orange-red pigment.* Bilirubin enters the circulation and is transported to the liver as a complex with serum albumin. From the liver it moves to the gallbladder, where it is

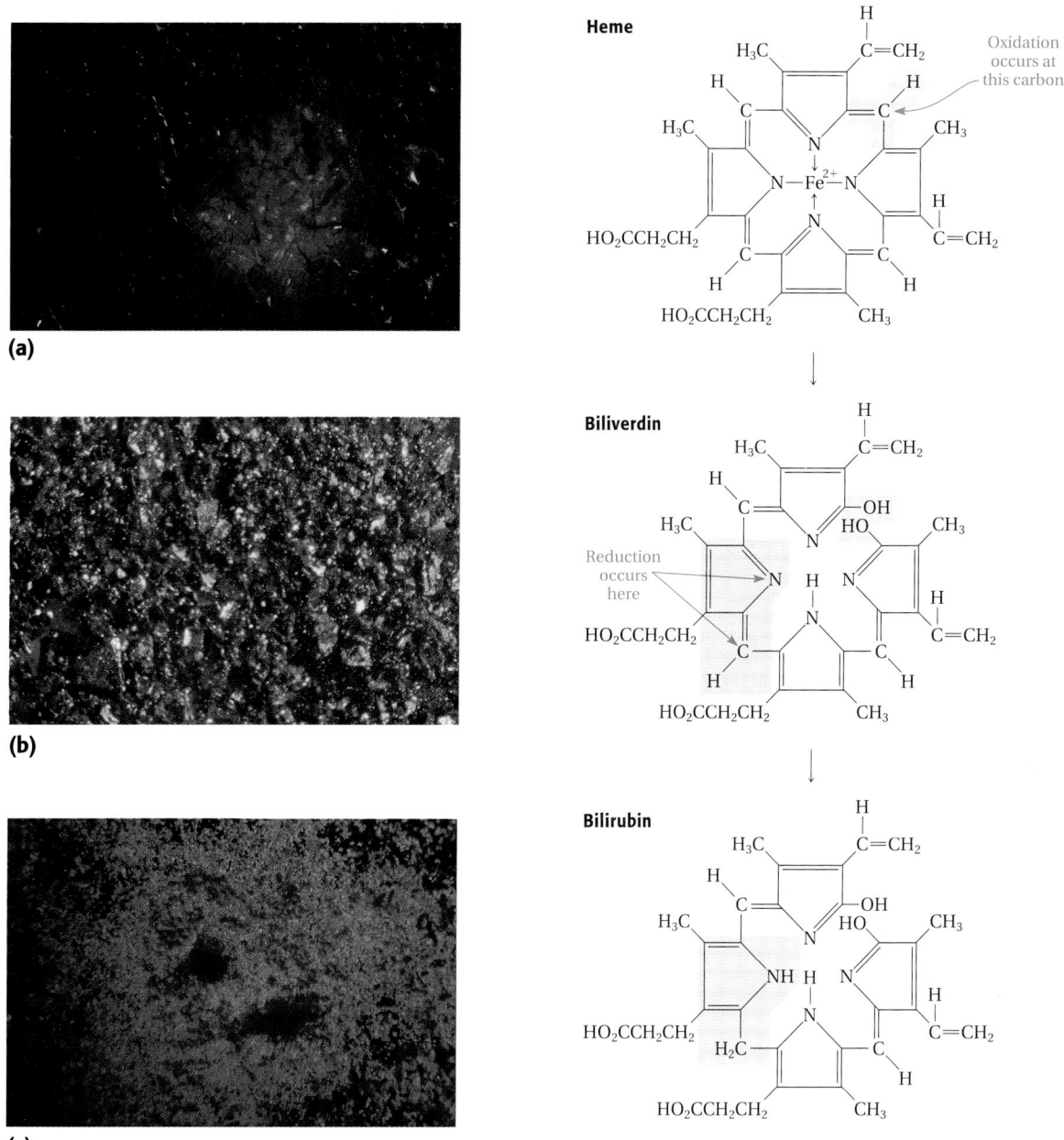

Figure 17.5
Hemoglobin (a) releases heme which is degraded to the bile pigments biliverdin (b) and bilirubin (c).

stored as part of the bile and eventually excreted into the small intestine along with the bile salts.

An excess of bilirubin in circulating blood is responsible for **jaundice**— *a yellow color of the skin and the whites of the eyes.* Jaundice is caused by any of a number of malfunctions of the bile production and storage system (see A Closer Look: Hyperbilirubinemia). If the bile duct is obstructed, the bile enters the circulation rather than the small intestine. The obstruction may be due to gallstones, which often consist of nearly pure cholesterol. Why the cholesterol forms these hard, insoluble lumps in the gallbladders of some people and not in others is still not understood. *In certain diseases such as* **infectious hepatitis,** *the liver cannot remove bile pigments as they are formed, and they enter the circulation.* **Hemolytic jaundice** *occurs when breakdown of heme groups by the spleen is faster than the liver can remove the bile that is produced.*

Bilirubin that passes from the small intestine into the large intestine is oxidized to colorless **urobilinogen** *by bacteria residing there. The urobilinogen oxidizes in air to orange-yellow* **urobilin.** The excreted urobilin gives feces their color. The yellow color of urine is the result of a small amount of urobilin filtered from the bloodstream by the kidneys.

A Closer Look

Hyperbilirubinemia

Before birth, biliverdin (from the breakdown of fetal hemoglobin) is converted to bilirubin and crosses the placenta into the mother's liver. It is then secreted in her bile. At birth an infant's liver must take over this vital function; to do this, the liver cells must be mature. Usually the liver becomes fully functional within the first week after birth. But if it does not, high serum bilirubin levels accumulated in the blood (hyperbilirubinemia) and skin (jaundice) cause the infant's skin to turn yellow. If the maturation period of the liver is prolonged, bilirubin may start to accumulate in brain tissue. Left untreated, this condition can lead to cerebral palsy, brain damage, and death.

Phototherapy (see A Closer Look: Phototherapies, on page 418) is used to treat hyperbilirubinemia. When bilirubin is exposed to white fluorescent light or sunlight, it is converted to *photobilirubin.* Therefore, if an infant suffering from hyperbilirubinemia is exposed to fluorescent light, some of the bilirubin in blood flowing near the skin is converted to photobilirubin. Photobilirubin is more water-soluble than bilirubin because of a slight difference in molecular structure. Because this photoproduct is more soluble in water than in fatty tissues, it leaves the skin and enters the blood circulation. From the blood it passes to the liver and is readily secreted in bile and excreted in the urine and feces.

During phototherapy, infants may be exposed to florescent light (see figure) for periods of 8 to 10 hours daily for a week or until the liver cells reach maturity. The infant's eyes are covered to prevent possible damage. He or she is fed intravenously to minimize dehydration and turned frequently; only the skin exposed to light loses its yellow color as the bilirubin is changed to photobilirubin and excreted.

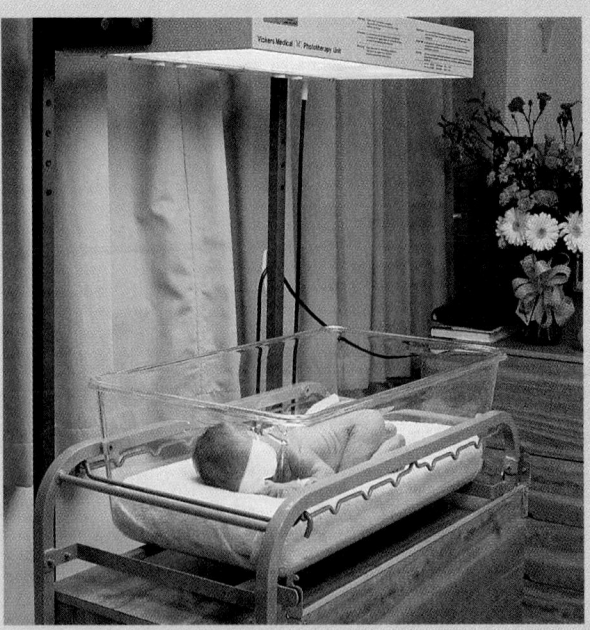

Infant undergoing phototherapy for hyperbilirubinemia.

17.9 Purines and pyrimidines

AIMS: To show the link between amino acids and nucleic acids in the body. To name the product of purine catabolism and name the disease that results from excessive concentrations of this substance in the blood.

Focus

Many of the ring atoms of pyrimidines and purines come from amino acids.

Organisms need to synthesize purine and pyrimidine bases for incorporation into the nucleic acids RNA and DNA. Moreover, nucleosides such as adenosine are found as part of ATP, cyclic AMP, CoA (coenzyme A), NAD^+ (nicotinamide adenine dinucleotide), and FAD (flavin adenine dinucleotide). The atoms that constitute both pyrimidine and purine ring systems come from amino acids, ammonia, and carbon dioxide, as shown for uracil and adenine in Figure 17.6.

The pathway of purine degradation is shown in Figure 17.7. The end product of this pathway is uric acid. Uric acid does not have a carboxyl group, but it is an acidic compound because one of the hydrogens on its five-membered ring readily dissociates. Human beings normally excrete uric acid as a minor waste product in the urine. For reptiles and birds, uric acid is the major form of excreted waste nitrogen. The white part of the deposits that pigeons leave on statues in the park is almost pure uric acid.

Patients with high serum uric acid concentrations may have to take drugs for life to prevent development of hypertension or kidney disease. The drugs act either by preventing the formation of uric acid or by increasing its rate of excretion by the kidneys.

Some people, almost always male, make too much uric acid or fail to excrete it. Uric acid is quite insoluble, and it also readily forms an insoluble sodium salt, sodium urate. *In* **gout,** *uric acid or sodium urate exceeds its solubility in plasma and forms needle-like crystals that are deposited in joints, especially in the big toe.* Gout attacks, which result in swelling and inflammation of the affected joint, are agonizingly painful. Ralph, the accountant in the Case in Point earlier in this chapter, has gout.

Lesch-Nyhan syndrome, *a genetic disease, is caused by the lack of an enzyme needed to catalyze the synthesis of purines.* At about 2 or 3 years of

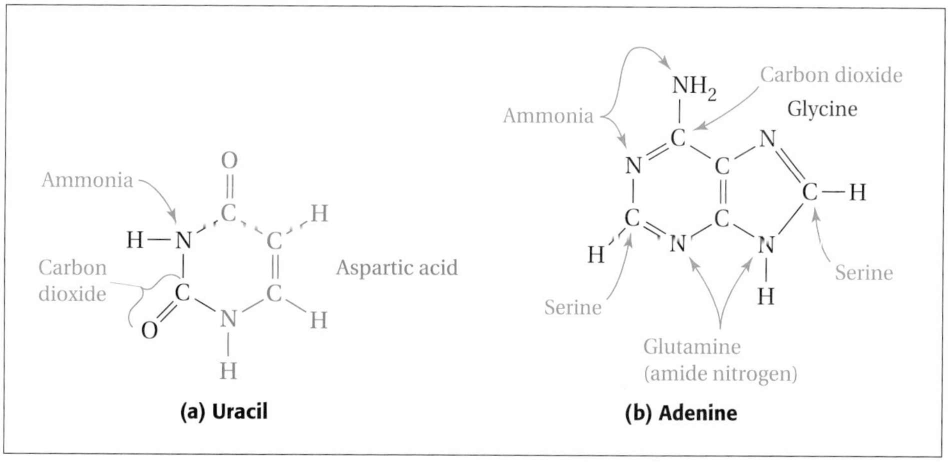

Figure 17.6
The atoms that comprise pyrimidines and purines come from amino acids. The figure shows the pyrimidine uracil (a) and the purine adenine (b).

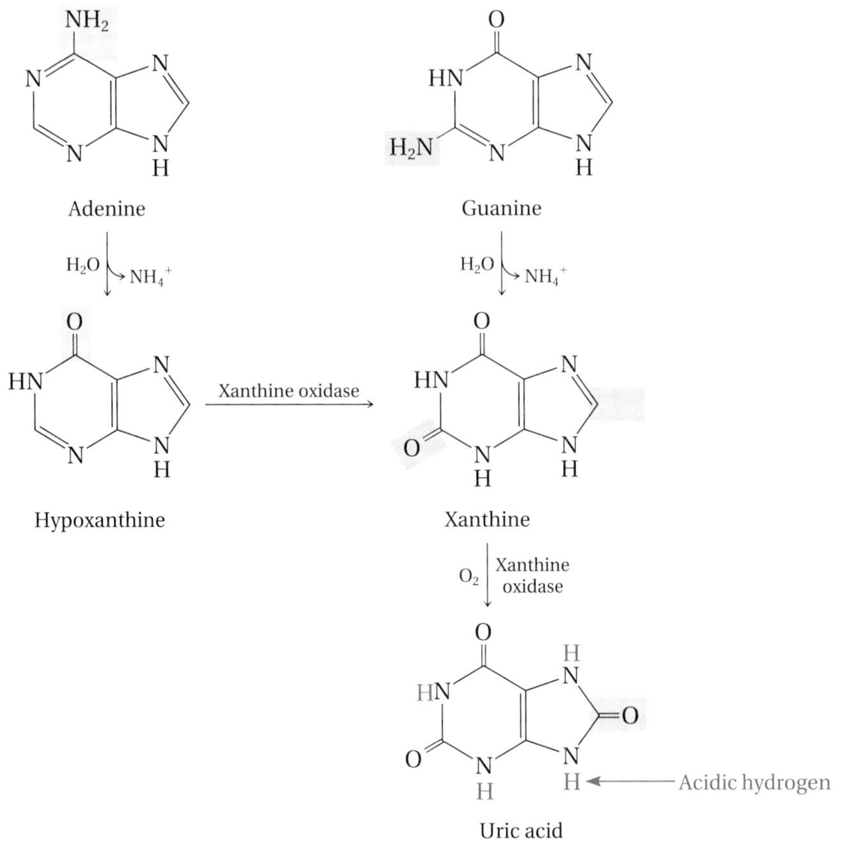

Figure 17.7
Purines are degraded to uric acid.

age, Lesch-Nyhan victims begin to show an uncontrollable urge to bite themselves. Unless they are forcibly restrained, some Lesch-Nyhan patients will literally bite off their fingers and lips. This disease is rare; about 60 cases are reported in the medical literature. Victims of the Lesch-Nyhan syndrome often show symptoms of gout. These symptoms are alleviated by allopurinol, but there is no alleviation of the tendency for self-destruction.

FOLLOW-UP TO THE CASE IN POINT: A painful episode

Ralph's painful symptoms indicated gout, and a blood test showed that he had elevated levels of uric acid in his blood. The symptoms of gout may often be relieved by a diet that restricts the intake of foods high in purines, such as shellfish, bacon, beef, and turkey. In severe cases, the drug allopurinol is sometimes effective.

Allopurinol prevents the synthesis of uric acid by inhibiting *xanthine oxidase*, the enzyme that converts xanthine to uric acid. A low-purine diet and allopurinol have helped put Ralph back on his feet.

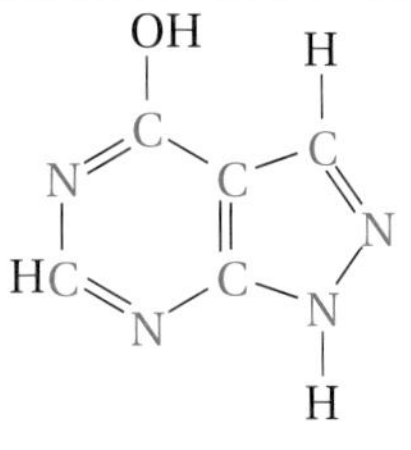

Allopurinol

PRACTICE EXERCISE 17.6

Match the following:

Medical condition   *Cause or result of condition*

(a) bile duct obstruction *jaundice*   (1) gout

(b) high serum GOT *damage to heart muscle* (2) absence of methionine
                                                   from diet
(c) high serum GPT *liver disease*
                                              (3) jaundice
(d) phenylpyruvic acid in
     urine *PKU*   (4) liver disease

(e) negative nitrogen balance   (5) damage to heart muscle

(f) deposit of uric acid   (6) PKU
     in joints *gout*

SUMMARY

Long before nitrogen was fixed (reduced to ammonia) by industrial processes, nitrogen-fixing bacteria were the link between atmospheric nitrogen and living organisms. Fixed nitrogen is used by living creatures to synthesize important classes of biological molecules: amino acids, nucleic acids, and heme, to name but three. Upon dying, organisms decay and return nitrogen compounds to the soil, from which molecular nitrogen eventually returns to the atmosphere to complete the nitrogen cycle.

The chief use of amino acids in the body is for making proteins. Proteins are continuously synthesized and hydrolyzed in the body. The lack of even one essential amino acid causes death because the amino acid is not available for protein synthesis. Amino groups are liberated from amino acids by transamination reactions. The amino groups are transferred to α-ketoglutarate to form glutamic acid, which serves as a nitrogen storage pool. The release of ammonia from glutamic acid by oxidative deamination restores the supply of α-ketoglutarate. Ammonia is very toxic to humans. It is converted to relatively nontoxic urea by the reactions of the urea cycle and excreted in the urine.

The carbon skeletons of the lipogenic amino acids are degraded to acetyl CoA. Those of the glucogenic amino acids are degraded to pyruvate or intermediates of the citric acid cycle. When ATP is abundant, acetyl CoA is used to make fatty acids for storage as triglycerides. Pyruvate and intermediates of the citric acid cycle are used to make glycogen. When ATP supplies are low, acetyl CoA, pyruvate, and intermediates of the citric acid cycle are degraded by the cycle to produce ATP by way of respiration and oxidative phosphorylation. Many malfunctions of amino acid metabolism are known.

Hemoglobin is synthesized in immature red blood cells and degraded in the spleen. Globin, the protein part of hemoglobin, is hydrolyzed to its individual amino acids. The heme is degraded to green biliverdin and orange-red bilirubin. Bilirubin is the coloring agent in jaundice.

The carbons and amine nitrogens of the purines and pyrimidines come from amino acids. The ultimate breakdown product of purine metabolism is uric acid.

KEY TERMS

Amino acid pool (17.3)
Amino aciduria (17.7)
Bilirubin (17.8)
Biliverdin (17.8)
Ferritin (17.8)
Glucogenic amino acid
  (17.5)
Gout (17.9)

Hemolytic jaundice (17.8)
Infectious hepatitis (17.8)
Jaundice (17.8)
Ketogenic amino acid (17.5)
Lesch-Nyhan syndrome
  (17.9)
Negative nitrogen balance
  (17.4)

Nitrogen balance (17.4)
Nitrogen cycle (17.1)
Nitrogen-fixing bacteria
  (17.1)
Oxidative deamination
  (17.4)
Phenylketonuria (PKU)
  (17.7)

Positive nitrogen balance
  (17.4)
Protein turnover (17.2)
Transamination (17.3)
Urobilin (17.8)
Urobilinogen (17.8)

EXERCISES

Nitrogen Fixation (Section 17.1)

17.7 How does atmospheric nitrogen enter the biosphere?

17.8 Describe the nitrogen cycle in your own words.

Protein and Amino Acid Catabolism (Sections 17.2, 17.3, 17.4, 17.5)

17.9 What is meant by the term *protein turnover*?

17.10 What are cathepsins?

17.11 What is the amino acid pool, and where is it in the body?

17.12 Discuss two metabolic fates of amino acids in the body.

17.13 What is a transamination reaction?

17.14 Explain the function of pyridoxal phosphate in transamination reactions.

17.15 Write an equation for the transamination of valine with α-ketoglutarate as the amino group acceptor.

17.16 Describe how an assay for GPT and GOT in blood serum can give information about liver or heart damage.

17.17 What is the function of the urea cycle?

17.18 How does the excretion of nitrogen as urea conserve body water in land-dwelling animals?

17.19 Define nitrogen balance. Why do growing children have a positive nitrogen balance?

17.20 Describe one medical condition that produces a negative nitrogen balance.

17.21 Amino acids are glucogenic or ketogenic. What do these terms mean?

17.22 Identify the locations at which the carbon skeletons derived from amino acids enter the citric acid cycle.

Amino Acid Synthesis (Section 17.6)

17.23 Describe the difference between an essential amino acid and a nonessential amino acid.

17.24 Write a reaction for the synthesis of a nonessential amino acid.

Defects of Amino Acid Metabolism (Section 17.7)

17.25 Draw the structural formula of phenylalanine. What compound is phenylalanine converted to early in its catabolism?

17.26 Explain why newborn infants are often given PKU tests.

Degradation of Hemoglobin (Section 17.8)

17.27 What is the average life span of a red blood cell?

17.28 Outline the degradation of a hemoglobin molecule.

17.29 What are the bile pigments?

17.30 Name two malfunctions of bile production and storage that give rise to a jaundiced appearance.

Degradation of Purines (Section 17.9)

17.31 Draw the structural formula of uric acid. Why is this compound acidic?

17.32 Relate the cause of gout to its symptoms. How does the drug allopurinol relieve the symptoms of gout?

Additional Exercises

17.33 How many high-energy phosphate bonds are expended in the urea cycle? In which steps?

17.34 What α-ketoacid is the transamination product of alanine? Of glutamic acid?

17.35 In what form is most of the nitrogen in amino acids and proteins excreted from the body? Give the name of the compound and draw its structure.

17.36 What are the two kinds of compounds that react in transamination reactions?

17.37 Describe the structure of the compounds that are catabolized in the synthesis of uric acid.

17.38 The purine and pyrimidine ring systems are synthesized in the body. Where do the atoms that make up these ring systems come from?

17.39 The body synthesizes urea. What is the source of each amino group and the carbonyl group in this compound?

17.40 Write equations to show how a molecule of glycine is deaminated in the body.

SELF-TEST (REVIEW)

True/False

1. The following compounds are both α-ketoacids:

$$CH_3-\overset{\overset{O}{\|}}{C}-\overset{\overset{O}{\|}}{C}-OH$$

pyruvate

$$HO-\overset{\overset{O}{\|}}{C}-CH_2C-\overset{\overset{O}{\|}}{C}-\overset{\overset{O}{\|}}{C}-OH$$

oxaloacetate

2. In periods of starvation, amino acids are used for energy production.

3. Symbiotic nitrogen-fixing bacteria always work in conjunction with plants.

4. A protein that contains 14 of the 20 common amino acids would have to be an incomplete protein.

5. The only dietary source of nitrogen is protein.

6. Damage to liver tissue is usually indicated by an increased level of serum glutamic-pyruvic transaminase (SGPT).

7. All animals eliminate nitrogen from their system primarily in the form of urea.

8. Overall, the urea cycle is energy producing.

9. A normal, healthy adult should, on average, take in more nitrogen than he or she excretes.

10. The supply of iron in the body is conserved by complexing with the protein ferritin.

Multiple Choice

11. The carbon skeletons of glucogenic amino acids are used
(a) for energy production.
(b) to synthesize triglycerides.
(c) for gluconeogenesis.
(d) all of the above are true.

12. A nitrogen atom from an ingested protein could eventually be found in which of the following types of molecules?
(a) heme   (b) urea
(c) nonessential amino acid   (d) all of the above

13. Inorganic nitrogen is usually found in the body in the form of
(a) N_2 (b) NO_3^-
(c) NH_3 (d) NH_4^+

14. An elevation of serum glutamic-oxaloacetic transaminase (SGOT) activity is indicative of
(a) heart muscle damage.   (b) starvation.
(c) liver damage.   (d) Lesch-Nyhan syndrome.

15. The orange-red bile pigment produced by the degradation of heme is
(a) ferritin.   (b) bilirubin.
(c) citrulline.   (d) biliverdin.

16. Energy produced from the metabolism of ketogenic amino acids would be stored as
(a) triglycerides.   (b) glycogen.
(c) uric acid.   (d) More than one are correct.

17. Phenylketonuria is the result of
(a) the faulty metabolism of phenylalanine.
(b) the lack of the essential amino acid phenylalanine in the diet.
(c) an inborn error of metabolism.
(d) more than one are correct.

18. A transamination reaction results in
(a) the formation of ammonium ion.
(b) an intermolecular amino group transfer.
(c) the formation of urea.
(d) amino aciduria.

19. The process by which bacteria change atmospheric nitrogen to ammonia is called
(a) ammoniation.   (b) transamination.
(c) nitrogen cycle.   (d) nitrogen fixation.

20. Which of the following pairs of compounds is used both as intermediates in the citric acid cycle and as starting materials for the synthesis of nonessential amino acids?
(a) pyruvate and oxaloacetate
(b) α-ketoglutarate and glutamic acid
(c) oxaloacetate and α-ketoglutarate
(d) aspartic acid and pyruvate

21. The chemical equation below represents a reaction called

$$HO-\overset{\overset{O}{\|}}{C}-CH_2CH_2\overset{\overset{H}{|}}{\underset{\underset{NH_2}{|}}{C}}-\overset{\overset{O}{\|}}{C}-OH + H_2O \xrightarrow[\text{NAD} \quad \text{NADH}]{}$$

$$NH_4^+ + HO-\overset{\overset{O}{\|}}{C}-CH_2\overset{\overset{O}{\|}}{C}-\overset{\overset{O}{\|}}{C}-OH$$

(a) hydrolysis.   (b) transamination.
(c) oxidative deamination.   (d) α-ketogenesis.

22. Gout is characterized by high blood levels of
 (a) uric acid.
 (b) ornithine.
 (c) urea.
 (d) phenylalanine.

23. The enzymes found in the lysosomes of cells that hydrolyze proteins are called
 (a) cathepsins.
 (b) peptidases.
 (c) transaminases.
 (d) hydrolases.

24. On a daily basis, a child consumes more nitrogen than he excretes. This child has
 (a) a positive nitrogen balance.
 (b) a serious medical problem.
 (c) a negative nitrogen balance.
 (d) more than one are correct.

25. The majority of the nitrogen eliminated from the body is in the form of
 (a) urea.   (b) ammonium ions.
 (c) uric acid.   (d) bilirubin.

Glossary

Acetals Compounds produced by the reaction of two molecules of an alcohol with an aldehyde. The general structure is

$$R-CH-OR$$
$$|$$
$$OR \qquad (4.6)$$

Acetyl coenzyme A (acetyl CoA) A thioester of acetic acid and coenzyme A, which is important for the transfer of acetyl groups in many processes of metabolism. (15.1)

Acid anhydride Substances formed when a total of one molecule of water is removed from two acid molecules. (5.5)

Acid-base balance A state in which the pH of body fluids is controlled so that no dramatic changes in acidity or basicity occur; blood has a normal pH of 7.40. (13.4)

Acid mucopolysaccharide A class of carbohydrates in which some carbohydrate units contain acidic ($-CO_2H$ or $-SO_3H$) groups; the fluid that lubricates joints contains these substances. (7.10)

Acidosis A condition in which excessive acid production or faulty acid elimination lowers the body's alkali reserves. (13.4)

Activated transfer RNA (activated tRNA) A tRNA molecule attached to an amino acid through an ester bond; the amino acid is transferred to a growing peptide chain in protein synthesis. (11.8)

Active site A groove or pocket in an enzyme molecule into which the substrate (reactant molecule) fits; the substrate is converted to products in a subsequent reaction. (10.4)

Acyl group A group with the structure $R-\overset{\overset{\displaystyle O}{\|}}{C}-$; the acetyl group has the structure

$$CH_3-\overset{\overset{\displaystyle O}{\|}}{C}- \qquad (5.5)$$

Acylation reaction A reaction in which an acyl group ($R-\overset{\overset{\displaystyle O}{\|}}{C}-$) is transferred from one molecule to another; the accepting group is usually an alcohol (to form an ester) or an amine (to form an amide). (5.5)

Addition reaction A reaction in which one reactant divides into two parts, each part joining to a carbon of a multiple carbon-carbon bond to form the product. (3.2)

Adenosine triphosphate (ATP) The major source of usable energy in all cells. When an ATP molecule loses a phosphate group to form ADP, free energy is released. (14.4)

Adipose tissue Tissue that stores large amounts of fats, mainly as triglycerides. (16.1)

Aerobic cell Cell that requires the presence of free oxygen to carry out oxidation reactions vital to cellular life. (12.1)

Alcohol A compound with the general structure $R-OH$. (3.3)

Alcoholic fermentation The process by which yeast converts glucose to ethyl alcohol (CH_3CH_2OH) in absence of oxygen. (15.7)

Aldehyde Compound containing an aldehyde functional group, $-\overset{\overset{\displaystyle O}{\|}}{C}-H$; the general formula RCHO. (4.1)

Aldose A carbohydrate containing an aldehyde group; glucose, galactose, and ribose are all simple aldoses. (7.1)

Aliphatic compound An open-chain compound of carbon that contains no aromatic rings. (2.9)

Alkaloids A class of nitrogen-containing organic compounds obtained from plants. Many have striking biological effects on animals and humans. The alkaloids are weak bases (hence their name). (6.8)

Alkalosis A condition in which the pH of body fluids such as blood increases when the buffering capacity of the body is exceeded. (13.4)

Alkane A compound composed only of hydrogen and carbon; all the bonds are single covalent bonds. Alkanes are saturated hydrocarbons. (2.2)

Alkene A hydrocarbon containing one or more double bonds, such as 1-butene ($CH_3CH_2CH=CH_2$). Alkenes are unsaturated hydrocarbons. (2.8)

Alkoxide A salt produced by the reaction of a reactive metal with an alcohol. Sodium methoxide (CH_3O-Na^+) is the alkoxide produced by the reaction of methanol and sodium. (3.6)

Alkyl group A hydrocarbon group such as methyl (CH_3-), ethyl (CH_3CH_2-), or n-propyl ($CH_3CH_2CH_2-$). (2.5)

Alkyl halide A derivative of a hydrocarbon in which one or more hydrogens are replaced by halogens. Alkyl halides are also called halocarbons. (3.1)

Alkylammonium ion The cation produced by protonation of an aliphatic amine. (6.2)

Alkyne A hydrocarbon containing a triple bond, such as ethyne ($CH\equiv CH$). Alkynes are unsaturated hydrocarbons. (2.8)

Allosteric enzyme   An enzyme whose activity is changed by the binding of a small molecule (a modulator) to a site other than the active site. (10.10)

Alpha carbon   The carbon to which the carboxyl and amino groups are attached in the twenty common amino acids. (9.1)

Alpha helix   A winding chain conformation in proteins that is formed and maintained by hydrogen bonds. (9.6)

Alpha-hydroxy ketone   A compound of the general structure

$$\underset{\qquad\qquad\quad OH}{R-\overset{\overset{O}{\|}}{C}-\overset{\overset{H}{|}}{C}-R}\qquad(4.5)$$

Alpha keratin   A type of protein found in hair, wool, nails, and hooves. The protein chain conformation is mostly alpha helix. (9.6)

Amide   An organic compound in which nitrogen is linked to a carbonyl carbon; general structures are

$$R-\overset{\overset{O}{\|}}{C}-NH_2 \qquad R-\overset{\overset{O}{\|}}{C}-\overset{\overset{H}{|}}{N}-R \quad \text{and} \quad R-\overset{\overset{O}{\|}}{C}-\overset{\overset{R}{|}}{N}-R$$
$$(6.4)$$

Amine   An organic compound of general structure

$$R-\overset{\overset{H}{|}}{N}-H \qquad R-\overset{\overset{R}{|}}{N}-H \quad \text{and} \quad R-\overset{\overset{R}{|}}{N}-R \quad(6.1)$$

Amino acid   Any carboxylic acid that contains an amino ($-NH_2$) group; the name is usually reserved for the twenty naturally occurring alpha amino acids. (9.1)

Amino acid pool   The total supply of amino acids present in the body. (17.3)

Amino acid residue   The portion of an amino acid that remains after it has been incorporated into a protein molecule; the general formula is

$$-\overset{\overset{H}{|}}{N}-\overset{\overset{R}{|}}{\underset{\underset{H}{|}}{C}}-\overset{\overset{O}{\|}}{C}-\qquad(9.4)$$

Amino acid side chain   The substituent group attached to the alpha carbon of an alpha amino acid in addition to the carboxyl and amino groups. (9.1)

Amino aciduria   A condition in which amino acids are excreted in excessive amounts in the urine. (17.7)

Ammonium salt   Salt formed by the neutralization of aqueous ammonia or an amine with an acid. (6.2)

Anabolism   The synthetic processes in the metabolism of cells; it usually requires the expenditure of ATP and reducing power in the form of NADH and $FADH_2$. (14.1)

Anaerobe   A cell that can live in the absence of oxygen. (14.6)

Analgesic   Any painkilling drug. (5.6)

Anilide   An amide in which the amine portion comes from the parent amine aniline or a derivative of aniline. (6.4)

Anion   A negatively charged ion. (1.2)

Anomers   Two stereoisomers of the closed-chain form of a sugar that differ only in the position of the hydroxyl group at carbon 1. (7.4)

Antibiotic   A drug that kills bacteria, usually by interfering with some vital process. (10.13)

Antibody   Protein synthesized in response to a foreign substance (antigen) in the body; the antigen is destroyed after the formation of an antibody-antigen complex. (13.3)

Anticodon   A three-base group on a tRNA molecule that pairs with a three-base codon of mRNA, leading to the correct incorporation of an amino acid into a growing protein chain in protein synthesis. (11.8)

Antidiuretic   A substance that blocks the loss of body water through urination. (9.4)

Antigen   A substance that stimulates production of antibodies and reacts with them. (13.3)

Antihistamine   Any drug that relieves the symptoms of colds and allergies by blocking the release of histamine. (6.7)

Antiparallel strands   In nucleic acids, the structural feature in which the two strands of a double helix run in opposite directions. (11.2)

Antipyretic   Any drug used to reduce a fever. (5.6)

Apoenzyme   The protein portion of an enzyme that requires a cofactor for activity. (10.6)

Arene   Any compound that contains a benzene ring. (2.9)

Aromatic compound   An organic compound that contains a benzene ring or closely related ring. (2.9)

Aromatic substitution   A chemical reaction in which a substituent replaces hydrogen on an aromatic ring. (3.2)

Aryl halide   A benzene ring with one or more hydrogens replaced by halogen. (3.1)

Arylammonium ion   The cation produced by protonation of an aromatic amine. (6.2)

Asymmetric carbon atom   A carbon atom covalently bonded to four different groups; two different tetrahedral arrangements of these groups in space are possible, leading to stereoisomerism. (7.1)

Atherosclerosis   Hardening of the arteries; caused by the buildup of a high-cholesterol substance called plaque; also called arteriosclerosis. (16.11)

Atomic orbital   A region of space around the nucleus of an atom in which an electron is most likely to be found. (1.1)

Aufbau principle   The arrangement of electrons in an atom is the one requiring the least amount of energy to attain. (1.1)

Autoactivation The process by which active proteases (protein-cutting enzymes) produce more of the same protease by catalyzing cleavage of peptide bonds in the zymogen. (10.11)

Barbiturates Salts of barbituric acid and its derivatives used (and abused) as sedatives and hypnotics. (6.9)

Basal metabolism The minimum metabolic activity required to sustain life in a resting but awake human being. (12.5)

Base triplet A sequence of three nitrogen bases that codes for an amino acid in DNA or mRNA. (11.8)

Bence-Jones protein Light chains of single antibody species produced by myeloma cells; detected in the urine of multiple myeloma patients. (13.3)

Benedict's reagent An alkaline solution of copper(II) sulfate ($CuSO_4$) used to test for the presence of reducing sugars. (4.5)

Beriberi A disease affecting the nervous system, caused by a thiamine deficiency. (12.8)

Beta keratin The major protein of spiderwebs and silk; the protein chain conformation is primarily the beta pleated sheet. (9.6)

Beta oxidation The metabolic pathway in which fatty acids are broken down in steps into two-carbon fragments of acetyl coenzyme A. (16.3)

Beta pleated sheet A protein chain conformation in which two protein strands running in opposite directions are held together by hydrogen bonds. (9.6)

Bile A soap-like substance produced by the gallbladder. Primarily consisting of carboxylic acid derivatives of cholesterol, bile is used in the dissolution of fats in digestion. (12.4)

Bile pigment A group of substances produced by the breakdown of hemoglobin; many, such as biliverdin and bilirubin, are highly colored. (12.4)

Bile salt Salts of one of several carboxylic acids derived from cholesterol. (12.4)

Bilirubin An orange-red bile pigment. (17.8)

Biliverdin A green bile pigment. (17.8)

Biodegradable Substances capable of being broken down to simple nontoxic components, usually by the action of microorganisms, but sometimes by water, sunlight, and other means. (5.3)

Bone matrix The molecular framework of collagen upon which calcium salts precipitate to form hard bone. (12.9)

Branched-chain alkane An alkane containing carbon groups attached to the longest continuous carbon chain. (2.5)

Caloric value The quantity of heat evolved (in calories) per gram of food burned. (12.5)

Carbaminohemoglobin The product of the reaction of hemoglobin with carbon dioxide; it is a carrier of carbon dioxide away from tissue cells that produce the carbon dioxide as a waste product. (13.5)

Carbohydrates Polyhydroxyaldehydes, polyhydroxyketones, or substances that yield these substances upon hydrolysis. (7.1)

Carbonyl group The functional group $-\overset{\overset{\displaystyle O}{\|}}{C}-$, found in aldehydes, ketones, esters, and amides. (4.1)

Carboxyhemoglobin A hemoglobin molecule in which carbon monoxide has reacted with the heme iron; it does not react with oxygen. (9.9)

Carboxyl group The functional group $-\overset{\overset{\displaystyle O}{\|}}{C}-OH$ found in carboxylic acids. (5.1)

Carboxylic acid One of a class of organic molecules containing the carboxyl group ($-\overset{\overset{\displaystyle O}{\|}}{C}-OH$); the general formula is $R-CO_2H$. (5.1)

Catabolism The reactions in living cells in which substances are broken down and energy is produced. (14.1)

Cation A positively charged ion. (1.2)

Cell membrane The partition made of phospholipid and protein that forms the outermost edge of the cell; it is the barrier between the cell and its environment. (8.3)

Cellular respiration The process in living cells in which oxygen is reduced to water and NADH and $FADH_2$ are oxidized to NAD^+ and FAD, respectively. It is an energy-releasing process that is tightly coupled to the production of ATP by oxidative phosphorylation. (14.6)

Cellulose A polysaccharide produced by plants as a structural material. Humans and most other animals cannot digest it, since they lack the enzyme to hydrolyze it. A typical cellulose molecule contains from 1000 to 3000 glucose units. (7.6)

Central dogma Hypothesis that hereditary information flows from DNA to RNA to protein. (11.3)

Chloride shift A process in which chloride ions (Cl^-) replace bicarbonate ions (HCO_3^-); it is important in maintaining a constant osmotic pressure in red blood cells. (13.5)

Chlorophyll The green pigment found in plants that is a major light-trapping molecule of photosynthesis. (14.2)

Chloroplast An organelle found in photosynthetic eukaryotic cells; chloroplasts contain the chlorophyll necessary for the conversion of light energy into chemical energy in photosynthesis. (14.2)

Chylomicron A large particle consisting of lipoproteins and their associated lipids found in the bloodstream after a fatty meal. (12.4)

***Cis* configuration**   Alkenes in which groups attached to each carbon of the double bond are on the same side of the bond. (2.8)

Citric acid cycle   The central metabolic pathway, in which the acetyl group of acetyl coenzyme A is oxidized to two molecules of carbon dioxide. (15.1)

Codon   A sequence of three adjacent nucleotides in mRNA that codes for a specific amino acid. (11.8)

Coenzyme A   Coenzyme that functions as a carrier in the acetyl group and other acyl groups. (15.1)

Cofactor   A small organic molecule or metal ion necessary for an enzyme's biological activity; one of the B-complex vitamins is often part of coenzyme molecules. (10.6)

Collagen   The structural protein of tendons, ligaments, bones, and teeth. (9.6)

Competitive inhibitor   A molecule resembling the normal substrate molecule of an enzyme; it blocks the catalytic site of the enzyme by binding to it so that the normal enzymatic reaction cannot occur. (10.10)

Complement system   A complex series of interacting plasma proteins that "complements" the function of antibodies in destroying antigens. (13.3)

Complementarity   The fit between substrate and enzyme; also refers to the pairing of purine and pyrimidine bases in DNA. (10.4)

Complementary base pair   Interacting pair of nitrogen bases on opposing strands of DNA; adenine (A) pairs with thymine (T), and cytosine (C) pairs with guanine (G). (11.2)

Complete protein   A protein containing the essential amino acids. (12.6)

Complex lipid   A lipid molecule that is not a triglyceride; the major complex lipids are the phospholipids, the sphingomyelins, and the glycolipids. (8.3)

Condensed structural formula   Formulas of organic molecules that leave out part of the structure of a molecule. (2.4)

Conjugated protein   A protein molecule that requires tight binding of a nonprotein ion or small organic molecule for biological activity. (9.7)

Coordinate covalent bond   A covalent bond between two atoms in which the shared electron pair comes from only one of the atoms; one of the bonds in the ammonium ion (NH_4^+) is coordinate covalent since both electrons come from nitrogen. (1.3)

Corticoids   A group of steroid hormones produced by the adrenal cortex. (8.6)

Coupled reactions   Two chemical reactions, one spontaneous and the other nonspontaneous, that have an intermediate in common; because of the common intermediate, the free energy released from the spontaneous reaction may be used to drive the nonspontaneous reaction. (14.7)

Covalent bond   A chemical bond in which two atoms share a pair of electrons. (1.3)

Cracking   Decomposing large hydrocarbons to smaller ones by heat or pressure with or without a catalyst; this process is used to increase the supply of low-formula mass hydrocarbons from petroleum for gasoline. (2.10)

Cyanohemoglobin   The binding of carbon monoxide to the heme iron of hemoglobin. (9.9)

Cycloalkane   A saturated hydrocarbon in which the ends are joined to form a ring. (2.7)

Cytochromes   A group of heme proteins that transport electrons in the respiratory chain. (14.6)

Decongestant   Medical drug that causes shrinkage of membranes lining the nasal passages, relieving stuffed-up feeling. (6.7)

Degenerate code   Property of the genetic code; more than one three-letter code word exists for each amino acid. (11.9)

Degradation   Hydrolysis of enzymes to their constituent amino acids as part of the protein turnover in cells. (10.9)

Dehalogenation   Elimination of a halogen (Cl_2, Br_2) from an organic molecule containing halogen constituents to produce a carbon-carbon double bond. (3.5)

Dehydration   Any loss of water; in organic chemistry, loss of water from a primary or secondary alcohol to form an alkene. (3.5)

Dehydrogenation reaction   A reaction in which two hydrogens are lost from a compound; all dehydrogenations are oxidations. (4.3)

Dehydrohalogenation   Loss of a hydrohalogen (HF, HCl, HBr, HI) from a halocarbon to produce an alkene. (3.5)

Denaturation   The partial or complete unfolding of the biologically active chain conformation of a protein; the process may be reversible or irreversible. (9.12)

Denatured alcohol   Ethanol (CH_3CH_2OH) to which a poisonous substance has been added to make it unfit to drink. (3.3)

Deoxyhemoglobin   A hemoglobin molecule to which no oxygen is bound. (9.8)

Deoxyribonucleic acid (DNA)   A giant polymer of repeating deoxyribonucleotide units joined by phosphodiester linkages; the storage form of genetic information in living organisms. (11.1)

Depot fat   Stored fat, mainly in the form of triglycerides. (16.1)

Detergent   A cleaning agent, often the sodium salt of a long-chain sulfonic acid. (5.3)

Diabetes mellitus   A disease characterized by the inability of glucose to enter tissue cells; caused by a deficiency of the peptide hormone insulin, which is needed to make cell membranes permeable to the sugar. (15.13)

Dicarboxylic acid   Any carboxylic acid molecule containing two carboxylic acid groups ($-CO_2H$). (5.1)

Dimer   A hydrogen-bonded carboxylic acid pair. (5.2)

Dipole interaction   Attractions between the positive and negative poles of polar molecules. (1.6)

Disaccharide   Carbohydrate made from two monosaccharide units joined by a glycosidic linkage. (7.7)

Dispersion forces   Weak attractive forces between molecules associated with the number of electrons in the molecule; the greater the number of electrons, the stronger are the dispersion forces. (1.6)

Displacement reaction   Chemical reaction in which a substituent on carbon is replaced by a different substituent. (3.4)

Disulfide   An organic compound containing the $-S-S-$ functional group; the general formula is $R-S-S-R$. (3.9)

Disulfide bridge   A covalent bond between sulfurs of two cysteine residues in a protein; important in holding proteins in their native chain conformations. (9.4)

DNA double helix   A form of DNA in which two antiparallel single polynucleotide strands are twisted into a helix in which opposing nitrogen bases are paired (A = T and G = C). (11.2)

DNA polymerase   An enzyme that catalyzes the synthesis of DNA from its deoxyribonucleoside 5′-triphosphate precursors. (11.4)

Double covalent bond   A covalent bond involving two pairs of electrons; each atom donates one pair of electrons. (1.3)

Electron configuration   The arrangement of electrons around the nucleus of an atom. (1.1)

Electron dot structure   Notation for writing the valence-shell electron configurations of atoms; valence electrons are depicted as dots surrounding the inner electrons, and the nucleus is represented by the symbol of the atom being considered. Also called a Lewis dot structure. (1.1)

Electron transport chain   A sequence of molecules that pass electrons from one molecule to the next in line; the original electron donors are NADH and $FADH_2$, and the final receptor is oxygen. Energy released in the process may be used to phosphorylate ADP in oxidative phosphorylation. Also called the respiratory chain. (14.6)

Electronegativity   A measure of the ability of an atomic nucleus in a covalent bond to attract electrons. (1.5)

Electrophoresis   A method of separating substances of different charges by placing them in an electrical field; positively charged ions (cations) migrate to the negatively charged electrode (cathode), and negatively charged ions (anions) migrate to the positively charged electrode (anode). (9.10)

Elimination reaction   Reaction of organic molecule in which substituents on adjacent carbons leave and an alkene is formed. (3.5)

Elongation   Steps in protein synthesis in which amino acid residues are added to a growing peptide chain. (11.8)

Endopeptidase   A protein-cutting enzyme that specifically cleaves peptide bonds in a protein chain. (10.11)

Energy and carbon cycle   The vast and complex process in which atmospheric carbon dioxide is reduced to organic molecules in photosynthesis, animals eat the organic molecules, and then they oxidize these molecules to carbon dioxide that reenters the atmosphere. (14.3)

Enkephalins   A group of hormonelike peptides that transmit sensations of pain and pleasure in the brain; opiate drugs appear to act by binding to brain receptors normally reserved for enkephalins. (9.4)

Enzyme   A biological catalyst, always a protein; enzymes are very specific with respect to the reactions they catalyze. (10.1)

Enzyme activity   The speed at which an enzyme catalyzes a biological reaction; a high activity may be due to a high concentration of enzyme or a favorable set of reaction conditions. (10.7)

Enzyme assay   An experiment done to find the concentration of an enzyme, usually in a biological fluid such as blood; important in medical diagnosis. (10.7)

Enzyme-substrate complex   The weak association of an enzyme and its substrate; necessary for the enzyme to exert its catalytic activity. (10.4)

Erythrocyte   A red blood cell. (13.2)

Essential amino acids   The 8 common amino acids, out of a total of 20, that must be included in the diet because they cannot be synthesized by humans. (12.6)

Ester   The product of an acid and an alcohol. Organic esters contain the functional group

$$-\overset{\overset{\displaystyle O}{\|}}{C}-O-$$

and have the general structure

$$R-\overset{\overset{\displaystyle O}{\|}}{C}-OR;$$

esters of phosphoric acid are important in biological systems. (5.6)

Esterification (reaction)   A chemical reaction in which an alcohol reacts with an acid to produce an ester. (5.6)

Estrogen   A female sex (steroid) hormone that produces secondary sex characteristics; it also prepares the lining of the uterus for pregnancy. (8.8)

Ether   An organic compound that contains the functional group

$$-\overset{\overset{\displaystyle |}{}}{C}-O-\overset{\overset{\displaystyle |}{}}{C}-;$$

the general formula is $R-O-R$. (3.7)

Exon   Portion of the DNA of eukaryotic cells that contains genetic information; about 5% of the DNA of eukaryotes consists of exons. (11.5)

Exopeptidase   A protein-cutting enzyme that cleaves consecutive amino acid residues from a protein, starting from the N-terminal or C-terminal end. (10.11)

Fats   Lipids (especially triglycerides) that exist as solids at room temperature. (8.2)

Fatty acid   Any continuous-chain carboxylic acid, but especially those with chains containing 12 or more carbon atoms; the natural fatty acids contain an even number of carbons. (5.1)

Fatty acid spiral   Consecutive rounds of beta oxidation in which a long-chain fatty acid molecule is degraded to acetyl coenzyme A with a loss of one molecule of acetyl coenzyme A per round. (16.3)

Fatty acyl coenzyme A   The thioester between a fatty acid and coenzyme A. (16.3)

Feedback inhibition   Process in which the end product of a sequence of enzyme-catalyzed reactions "feeds back" to inhibit an earlier step in the sequence; many feedback inhibitors are allosteric modulators; also called end-product inhibition. (10.10)

Fehling's reagent   An alkaline solution of copper(II) sulfate used to detect the presence of reducing sugars. (4.5)

Ferritin   The protein used to store iron in the body. (17.8)

Fibrous proteins   A class of long, rod-shaped protein molecules that play mainly structural roles in cells; usually insoluble in water and dilute salt solutions. (9.6)

First messenger   Hormones; act to carry signals for cellular activity to receptors on the external surfaces of eukaryotic cells. (15.11)

Fischer projection   A method for depicting the three-dimensional shapes of molecules, especially carbohydrates. (7.1)

Flavoprotein   An oxidation-reduction protein having a molecule related to riboflavin as its prosthetic group. (14.6)

Fluid mosaic model   A theory describing cell membranes as dynamic, with lipids and some proteins moving about within the lipid bilayer. (8.4)

Fluorosis   A condition in which excessive fluoride ion (F^-) is deposited in bones and teeth. (12.10)

Formed element   The insoluble substances of blood, consisting mainly of erythrocytes, leukocytes, and thrombocytes. (13.2)

Free amine   An amine in which the nitrogen is not protonated to an ammonium ion; free amines are electrically neutral. (6.2)

Functional group   A specific arrangement of atoms in an organic compound that is capable of characteristic chemical reactions; the chemistry of an organic compound is determined by its functional groups. (3.2)

Furanose ring system   The fundamental closed-chain form of several sugars; contains four carbons and one oxygen. (7.3)

Fused-ring aromatic compound   An organic compound containing two or more benzene rings; each ring shares two ring carbons with one or more other rings. (2.9)

Gastric juice   The fluid of the stomach; it consists of 0.01 M hydrochloric acid and enzymes that aid digestion. (12.3)

Gene   The segment of a DNA molecule coding for a single polypeptide chain. (11.5)

Gene mutation   Any change in the base sequence of DNA; a protein synthesized from a mutated gene may be fully functional, partially functional, or nonfunctional. (11.10)

Gene therapy   Cure or alleviation of a hereditary disease by replacement of a missing or defective gene. (11.11)

Genetic code   The set of 61 triplet code words used to specify amino acids in protein synthesis plus three triplet code words that signal "stop" when a peptide chain has been synthesized. (11.9)

Geometric isomer   An organic structure that differs from another only in the geometry of the molecule and not in the order of linkage of atoms. (2.8)

Globin   The protein portion of hemoglobin; that is, hemoglobin from which the heme prosthetic group has been removed. (9.8)

Globular protein   Any protein that has its peptide chain folded into a roughly spherical shape; most globular proteins are soluble in water and dilute salt solutions. (9.7)

Glucocorticoids   A group of steroid hormones produced in the adrenal cortex; they regulate several body functions, including glucose breakdown. (8.6)

Glucogenic amino acids   Amino acids that are broken down to pyruvic acid in cellular metabolism. (17.5)

Gluconeogenesis   The metabolic pathway leading to the synthesis of glucose. (15.8)

Glycogen   Animal starch consisting of polymers of D-glucose; storage form of glucose in animals. (7.6)

Glycogenesis   The series of chemical reactions by which living organisms convert glucose to glycogen. (15.9)

Glycogenolysis   The cleavage of stored glycogen to glucose-1-phosphate in living organisms. (15.10)

Glycolipids   Lipids that have polar, hydrophilic carbohydrate "heads" attached to nonpolar, hydrophobic "tails." (8.3)

Glycolysis   The sequence of chemical reactions that living organisms use to oxidize glucose to pyruvic acid; also called the glycolytic pathway or Embden-Meyerhof pathway. (15.1)

Glycoprotein   Protein molecules that have carbohydrate units attached to side chains of certain amino acid residues; many membrane proteins are glycoproteins, as are the protein blood-clotting factors and collagen. (9.11)

Glycoside   A carbohydrate in which two or more sugar units are connected by glycosidic bonds; also any acetal or ketal of a saccharide. (7.5)

Glycosidic bond   The ether linkage produced by the reaction between the anomeric hydroxyl group of one sugar unit and a hydroxy group of another sugar or an alcohol. (7.5)

Glycosuria   A condition in which reducing sugars, mainly glucose, are found in the urine; a symptom of diabetes mellitus. (15.13)

Goiter   A swelling of the neck due to the enlargement of the thyroid gland; the enlargement is sometimes caused by an iodine deficiency. (12.9)

Gout   A painful hereditary disorder of purine metabolism, mainly in males, in which crystals of uric acid precipitate in the joints. (17.9)

Hallucinogen   A substance that produces hallucinations and other bizarre mental effects. (6.7)

Halocarbon   A hydrocarbon in which one or more halogens replace hydrogen. (3.1)

Halogenation   Addition of a halogen (Cl_2, Br_2) to the carbon-carbon double bond of an alkene. (3.2)

Hard water   Water that contains dissolved ions of calcium, magnesium, or iron. (5.3)

Haworth projection   A method of depicting the structures of carbohydrates in three dimensions. (7.4)

Hemiacetal   The product of the reaction between an aldehyde molecule and one molecule of an alcohol; the general formula is

$$R-\underset{\underset{H}{|}}{\overset{\overset{OH}{|}}{C}}-OR \quad (4.6)$$

Hemiketal   The product of the reaction between a ketone molecule and one molecule of an alcohol; the general formula is

$$R-\underset{\underset{R}{|}}{\overset{\overset{OR}{|}}{C}}-OR \quad (4.6)$$

Hemoglobin   The oxygen-carrying heme protein of blood; it consists of two pairs of identical polypeptide chains and an iron-heme prosthetic group. (9.8)

Hemolytic jaundice   A yellow appearance of the skin due to the excessive breakdown of red blood cells, with the resultant formation and circulation of excessive amounts of bile pigments. (17.8)

Heterocyclic ring   A ring in which one or more atoms is an element other than carbon. (3.7)

Holoenzyme   An enzyme that contains both the protein portion and a cofactor required for activity. (10.6)

Homeostasis   The maintenance of constant conditions within the body, even when external conditions change. (13.1)

Hormone   A substance produced by the body in minute amounts that causes profound physiologic effects; many, but not all, are either peptides or steroids. (8.6)

Hund's rule   In orbitals of equal energy, each orbital gets an electron until all of the orbitals contain one electron; additional electrons add to the orbitals until they all contain two electrons. (1.1)

Hydrate   Any substance that contains water in a tightly bound form, such as sodium sulfate decahydrate ($Na_2SO_4 \cdot 10H_2O$). (4.6)

Hydration reaction   Addition of water to an alkene to form an alcohol. (3.4)

Hydrocarbon   An organic compound containing only carbon and hydrogen. (2.2)

Hydrogen bond   A weak attractive force; very important in determining the properties of water, proteins, and other biological substances. (1.6)

Hydrogenation   The addition of hydrogen to a multiple covalent bond; the hydrogenation product of an alkene is an alkane. (3.2)

Hydrohalogenation   Addition of a hydrohalogen (HCl, HBr, HI) to the carbon-carbon double bond of an alkene to produce a halocarbon. (3.2)

Hydrolysis   Reaction that involves the splitting of a covalent O—H bond of water. (5.5)

Hydrophilic   Water-loving; polar and ionic species are hydrophilic owing to the favorability of their interactions with water. (3.8)

Hydrophobic   Water-hating; nonpolar molecules are hydrophobic owing to the lack of favorable interactions with water. (3.8)

Hydroxy acid   A carboxylic acid (RCO_2H) that contains a hydroxyl group (HO—). (5.1)

Hydroxy function   The —OH functional group in alcohols; also called a hydroxyl group. (3.3)

Hydroxyl group   See *Hydroxy function*. (3.3)

Hyperglycemia   A condition in which the level of blood sugar is abnormally high; a symptom of diabetes mellitus. (15.13)

Hyperventilation   Respiratory condition characterized by deep, rapid breathing that can lead to respiratory alkalosis. (13.8)

Hypervitaminosis A condition resulting from excessive intake and storage of a vitamin; a fat-soluble vitamin is usually involved. (12.7)

Hypnotic A sleep-producing drug. (6.9)

Hypoglycemia A condition in which the level of blood sugar is abnormally low. (15.13)

Hypoventilation Respiratory condition caused by too-shallow breathing that can lead to respiratory acidosis. (13.8)

Immunoglobulin Y-shaped antibody protein molecule that binds to antigens and thereby neutralizes them. (13.3)

Induced-fit model A description of the binding of substrates to enzymes: The enzyme active site continuously adjusts to fit the substrate as it approaches so that the fit is perfectly complementary at contact. (10.4)

Inducible enzyme An enzyme that is synthesized in response to a cellular need. (10.9)

Induction The synthesis of enzymes by cells as the enzymes are needed. (10.9)

Infectious hepatitis A disease of the liver in which the liver cannot remove bile pigments from the body, allowing the pigments to circulate. (17.8)

Initiation The first step in protein synthesis; in eukaryotic cells initiation involves the binding of tRNA for methionine to mRNA. (11.8)

Insulin shock A serious condition of hypoglycemia caused by an overdose of insulin; insulin lowers blood sugar by increasing the permeability of cell membranes to glucose. (15.13)

Integral protein A protein molecule that is embedded in the lipid bilayer of a cell or organelle membrane. (8.4)

Interferon Protein of the immune system that has evolved specifically for defense against viral infections. (13.3)

International unit A measure of enzyme activity; 1 international unit (1 IU) is the amount of enzyme required to convert 1 micromole of substrate to product in 1 minute at specified conditions. (10.7)

Interstitial fluid The water and dissolved and suspended substances that surround tissue cells. (13.1)

Intracellular fluid The water and dissolved and suspended substances contained inside cells. (13.1)

Intron The noncoding portion of the DNA of eukaryotic cells; about 95% of human DNA consists of introns. (11.5)

Invert sugar The syrupy mixture of equal amounts of D-glucose and D-fructose produced by the hydrolysis of sucrose. (7.8)

Ion Atoms or group of atoms bonded together that have acquired a charge by gaining or losing electrons. (1.2)

Ionic bond The force of attraction between positively and negatively charged ions. (1.2)

Ionic compounds Electrically neutral substances consisting of particles of positive and negative charge (ions). (1.2)

Iron-deficient anemia A defect in red blood cell production owing to a lack of dietary iron with which to make hemoglobin. (12.9)

Isoelectric pH The pH at which there is no net charge on a protein or amino acid; a substance at its isoelectric point will not migrate in an electric field. (9.3)

Isoelectric point The condition of having no net charge; a protein is at its isoelectric point when the positive and negative charges on the side chains of its amino acid residues balance. (9.3)

Isohydric shift Transport of protons from red blood cells to the lungs. (13.5)

IUPAC system A method for naming organic compounds proposed by the International Union of Pure and Applied Chemistry. (2.4)

Jaundice Yellow coloration of skin and, sometimes, whites of eyes due to accumulation of bilirubin or other bile pigments in the bloodstream. Caused by a number of diseases involving the liver and gall-bladder. (17.8)

Keratinization The hardening of soft tissues, especially those of the eyes, due to hypervitaminosis A. (12.7)

Ketal The organic product of the reaction between one molecule of a ketone and two molecules of an alcohol; the general formula is

$$R{-}\overset{\displaystyle OR}{\underset{\displaystyle R}{\overset{|}{\underset{|}{C}}}}{-}OR \qquad (4.6)$$

Ketoacidosis Lowering of the blood pH due to excessive production of ketone body acids. (16.9)

Ketogenic amino acids Amino acids that are broken down to acetyl coenzyme A in metabolism. (17.5)

Ketone An organic compound with the general formula $R{-}\overset{\displaystyle O}{\overset{\|}{C}}{-}R$. (4.1)

Ketone bodies Substances produced by the liver as an alternative energy source in a carbohydrate-deficient diet: acetoacetic acid, β-hydroxybutyric acid, and acetone. (16.8)

Ketonemia A condition characterized by an excess of ketone bodies in the blood; often found in untreated diabetics. (16.9)

Ketonuria A condition in which ketone bodies are excreted in the urine; often found in diabetics. (16.9)

Ketose A carbohydrate that contains a ketone functional group. (7.2)

Ketosis Another name for ketoacidosis. (16.9)

Kwashiorkor A protein deficiency disease, especially prevalent in certain parts of Africa. (12.6)

Lactic fermentation The process by which glucose is degraded to lactic acid in the absence of oxygen. (15.7)

Lecithin Another name for phosphatidyl choline. (8.3)

Lesch-Nyhan syndrome A hereditary disorder of purine metabolism characterized by the self-destructive tendencies of its victims. (17.11)

Leukocyte A white blood cell. (13.2)

Leukotriene A group of C_{20} compounds related to the prostaglandins; leukotrienes are involved in triggering the physiological effects of such processes as inflammation and allergic reactions. (8.11)

Light system A highly organized group of molecules in the chloroplasts of photosynthetic cells; responsible for trapping light energy and converting it to chemical energy. (14.2)

Lipid bilayer A model of the structure of cell membranes; polar head groups of the constituent lipid molecules face the solvent, and nonpolar tail groups cluster toward the interior. (8.4)

Lipids A broad class of organic compounds from natural sources; soluble in organic solvents and insoluble in water. (8.1)

Lipoprotein A protein that associates with and transports lipids; interaction with a lipoprotein is necessary to make most lipids soluble in body fluids. (12.4)

Liposome A spherical body consisting of a lipid bilayer that surrounds a droplet of aqueous solvent. (8.4)

Lock-and-key model A description of the binding of a substrate to an enzyme active site; the fit between the substrate and enzyme active site is analogous to the fit of a key into a lock. (10.4)

Lymph A clear fluid generated by filtration of blood into tissues from the capillaries; it contains no blood cells but is otherwise similar in composition to blood plasma. (13.2)

Macronutrient Essential mineral element present in relatively large amounts in the body. (12.9)

Marasmus Another name for starvation. (12.6)

Markovnikov's rule In additions of hydrohalogens (H—Cl, H—Br) or water (H—OH) to a double bond of an alkene, the hydrogen of the adding reagent ends up on the hydrogen of the double bond that already has the most hydrogens. (3.2)

Mercaptan An older name for a thiol or thioalcohol (general formula R—SH). (3.9)

Messenger RNA (mRNA) A class of RNA molecules, complementary to one strand of cellular DNA, that carry hereditary information from the nucleus to the ribosomes for protein synthesis. (11.3)

Metabolic acidosis A decrease of blood pH to lower than normal because of excessive influx of protons into the bloodstream; metabolic acidosis is a common complication of diabetes mellitus. (13.8)

Metabolic alkalosis An increase in blood pH to higher than normal values, usually because of excessive influx of alkaline substances, such as the bicarbonate ions of antacids, into the bloodstream. (13.8)

Metabolism The sum of the enzyme-catalyzed chemical and energy changes that occur in cells. (14.1)

Methemoglobin A form of hemoglobin in which the heme iron is Fe^{3+} rather than Fe^{2+}; methemoglobin does not bind oxygen. (9.8)

Methemoglobinemia A hereditary disease in which the heme iron of either the alpha or beta chains of hemoglobin is Fe^{3+} rather than Fe^{2+}. (9.8)

Micelle An association in water of molecules possessing polar heads and nonpolar tails; the polar heads face the solvent, and the nonpolar tails cluster together in the interior. (5.3)

Micronutrient Essential mineral element present in trace amounts in the body. (12.9)

Mineral element In a living system, any element other than carbon, oxygen, hydrogen, nitrogen, phosphorus, or sulfur. (12.9)

Mineralocorticoids A class of steroid hormones produced in the adrenal cortex; they regulate mineral and water metabolism in the body. (8.6)

Mixed triglycerides A triglyceride in which more than one kind of long-chain fatty acid is esterified to the glycerol hydroxyl groups. (8.2)

Mobilization Hydrolysis of stored fats in preparation for the degradation of the product fatty acids in energy production within cells. (16.2)

Modulator binding site A location in an allosteric enzyme, not the active site, to which a regulatory substance (modulator) binds. (10.10)

Molecular disease Any disease caused by an absent or defective enzyme or other protein as a result of heredity; also known as inborn errors of metabolism. (11.10)

Molecule Two or more atoms of the same or different elements joined by chemical bonds. (1.3)

Monoclonal antibody Homogeneous antibody; antibody made by one kind of lymphocyte. (13.3)

Monomer The individual building block unit from which a polymer is constructed. (3.2)

Monomolecular layer One-molecule-thick layer of a fatty acid or detergent on the surface of water. (5.2)

Monosaccharide A carbohydrate consisting of one sugar unit; also called a simple sugar. (7.2)

Multienzyme complex   A highly organized assembly of enzyme molecules that together catalyze one or more steps in a metabolic pathway. (15.4)

Myoglobin   A protein used to store oxygen in muscle; it consists of a heme-iron prosthetic group associated with a single polypeptide chain. (9.7)

Native state   The normal, physiologically active chain conformation of a protein. (9.12)

Negative modulator   A molecule that can bind at the modulator binding site of an allosteric enzyme and inhibit enzyme action by distorting the peptide chain conformation of the enzyme active site. (10.10)

Negative nitrogen balance   The excretion of more nitrogen than is consumed, a situation that occurs in starvation. (17.4)

Neurotransmitter   A chemical substance released from an activated neuron that diffuses across a synapse to an adjacent neuron or muscle cell; acts as a chemical bridge in nerve impulse transmission. (6.7)

Nitrogen balance   Equilibrium reached when the quantity of nitrogen a person excretes daily equals the intake. (17.4)

Nitrogen cycle   A chemical cycle in which nitrogen, as an element or in compounds, circulates between the atmosphere and the earth and its living creatures. (17.1)

Nitrogen-fixing bacteria   Bacteria that convert atmospheric nitrogen to ammonia, a water-soluble compound of nitrogen that can be used by growing plants. (17.1)

Nonessential amino acid   An amino acid synthesized by the body. (12.6)

Nonionic detergent   A detergent whose molecules have a nonpolar hydrocarbon tail and a polar, but uncharged, head. (5.3)

Nonpolar covalent bond   The equal sharing of bonding electrons by two atoms. (1.5)

Nonreducing sugar   A sugar that fails to give a positive test with Benedict's or Tollens' reagents. (7.9)

Nonsuperimposable mirror image   A molecule or other object that exists in right-handed and left-handed forms. (7.1)

Nucleic acid   A polymer of ribonucleotides (RNA) or deoxyribonucleotides (DNA); found primarily in the nucleus of eukaryotic cells, they play a role in the transmission of hereditary characteristics, the control of cellular activities, and protein synthesis. (11.1)

Nucleoside   A compound that on hydrolysis yields a purine or pyrimidine base and a pentose. (11.1)

Nucleotidase   An enzyme that catalyzes the hydrolysis of nucleotides to give nucleosides and phosphate. (12.4)

Nucleotide   A compound consisting of a nitrogen-containing base (a purine or pyrimidine), a sugar (ribose or deoxyribose), and a phosphate; a monomer of a nucleic acid. (11.1)

Oil   In biochemistry, a triglyceride that is a liquid at room temperature. (8.2)

Okazaki fragments   Pieces of DNA that are connected to form a new DNA strand, by action of an enzyme, during replication. (11.4)

Opiate   A substance that acts as an analgesic and narcotic; for example, morphine. (6.8)

Optical isomers   Mirror-image isomers that contain at least one asymmetric carbon. They are identical in their chemical and physical properties with one exception—they rotate plane-polarized light in different directions; also called stereoisomers. (7.1)

Orbital hybridization   The mixing of atomic orbitals to produce the same number of orbitals of lower energy than the original orbitals. (2.2)

Osteoporosis   A condition of bone fragility caused by the desorption of calcium from the bones. (12.9)

Oxidative deamination   Loss of the amino group of glutamic acid to produce ammonium ions and α-ketoglutarate; oxidative deamination is catalyzed by the enzyme glutamate dehydrogenase. (17.4)

Oxidative phosphorylation   The phosphorylation of ADP to ATP in cellular respiration. (14.6)

Oxygen debt   The condition that exists when not enough oxygen is available for cellular respiration; it occurs during vigorous exercise. (15.8)

Oxyhemoglobin   The compound formed when oxygen unites with hemoglobin; it is the red pigment of red blood cells. (9.8)

Pacemaker enzyme   An enzyme that helps control the rate of cellular processes; often an allosteric enzyme. (10.10)

Pancreatic juice   A digestive secretion that flows from the pancreas into the upper region of the small intestine; enzymes in pancreatic juice help hydrolyze proteins, carbohydrates, and fats. (12.4)

Pauli exclusion principle   An atomic orbital can contain at most two electrons and two electrons must have opposite spins to occupy the same orbital. (1.1)

Pellagra   A disease caused by a deficiency of niacin (nicotinic acid); it is characterized by diarrhea, dementia, and dermatitis. (12.8)

Peptidase   An enzyme that catalyzes the hydrolysis of peptide bonds. (10.2)

Peptide   A compound formed when two or more amino acids are joined by peptide bonds. (9.4)

Peptide bond The amide bond joining amino acid residues in a peptide; it has the structure

$$\underset{\substack{\| \\ -C-N-}}{\overset{\substack{O\quad H \\ \| \quad |}}{}} \quad (9.4)$$

Peripheral protein A cell membrane protein that perches on either side of the lipid bilayer. (8.4)

Pernicious anemia A disease caused by an inability to absorb vitamin B_{12} (cobalamin); it is characterized by general fatigue. (12.8)

Petroleum refining The distillation of crude oil to divide it into fractions according to boiling point. (2.10)

pH activity profile A measure showing how the activity of an enzyme varies with changes in pH. (10.8)

pH optimum The pH at which an enzyme has maximum catalytic activity. (10.8)

Phenol A class of organic compounds having the hydroxyl group (—OH) attached directly to a carbon of a benzene ring. (3.3)

Phenylketonuria (PKU) A disease caused by an inborn error of metabolism in which the enzyme responsible for the conversion of phenylalanine to tyrosine is inactive. If untreated, PKU results in severe mental retardation by the age of 6 months. (17.7)

Phosphodiester Any molecule that has the general formula

$$R-O-\underset{\substack{| \\ OH}}{\overset{\substack{O \\ \|}}{P}}-O-R \quad (11.1)$$

Phosphoglyceride A type of phospholipid molecule that is built from long-chain fatty acids, glycerol, and phosphoric acid. (8.3)

Phospholipid A complex lipid that contains a phosphate group and is a major component of most cell membranes. There are two main types of phospholipids: phosphoglycerides and sphingomyelins. (8.3)

Phosphoryl group A functional group with the structure

$$-\underset{\substack{| \\ OH}}{\overset{\substack{O \\ \|}}{P}}-OH \quad (5.9)$$

Phosphorylating power The ability of ATP to transfer phosphoryl groups to other molecules. (14.7)

Phosphorylation The transfer of a phosphoryl group to a molecule. (5.9)

Photosynthesis The process in which green plants and algae convert radiant energy from the sun into useful chemical energy in order to synthesize glucose from carbon dioxide and water. (14.2)

Pi bond Chemical bond formed by the side-by-side overlap of two atomic p orbitals, each containing one electron. (2.8)

Plane-polarized light Light that has only one plane of vibration. (7.1)

Plaque The layers of cholesterol and arterial tissue cells that build up in the arteries. The disease in which plaque formation occurs—arteriosclerosis—often leads to a heart attack due to impaired blood circulation. (14.11)

Plasma The amber fluid that remains after red blood cells, white blood cells, and platelets are removed from whole blood. (13.2)

Polar bond A covalent bond in which the bonding electrons are shared unequally by the bonding atoms. (2.5)

Polar molecule A molecule that has a dipole; molecules of water and ammonia are polar. (1.6)

Polyamide A polymer in which the constituent units are joined by amide bonds; nylon and proteins are polyamides. (6.5)

Polycyclic aromatic compound A derivative of benzene in which carbons are shared between benzene rings; also called a fused-ring compound. (2.9)

Polyester A polymer that consists of many repeating units of dicarboxylic acids and dihydroxy alcohols joined by ester bonds. (5.6)

Polyfunctional molecule An organic compound that contains two or more functional groups. (3.10)

Polymer A very large molecule formed when large numbers of small molecules, known as monomers, are joined by covalent bonds. (3.2)

Polynucleotide A polymer in which the repeating units are nucleotides; DNA and RNA molecules are polynucleotides. (11.1)

Polypeptide Any peptide with more than ten amino acid residues. (9.4)

Polysaccharide A polymer of monosaccharide units; examples include starch, cellulose, and glycogen. (7.6)

Positive modulator A molecule that binds to the modulator site of an allosteric enzyme and induces formation of an active site. (10.10)

Positive nitrogen balance The excretion of less nitrogen than is consumed. Children have a positive nitrogen balance; since they are growing, their cells are making new proteins and other nitrogen compounds. (17.4)

Posttranslational processing The modification of proteins after the proteins are synthesized at the ribosomes. (11.8)

Primary structure The sequence of amino acids in a protein molecule. (9.5)

Proenzyme A physiologically inactive form of an enzyme; also called a zymogen. (10.11)

Progesterone   A female sex hormone that helps ready the uterus for pregnancy, prevents the release of further ova (eggs), and prepares the breasts for lactation. (8.8)

Prostaglandin   A potent hormonelike substance; prostaglandins are present in nearly all tissues and organs of the body and are derived from unsaturated fatty acids. (8.11)

Prosthetic group   The nonprotein coenzyme attached to a conjugated protein. Prosthetic groups are often metal ions or small organic molecules. (9.7)

Protease   Any enzyme that catalyzes the hydrolysis of one or more peptide bonds in a protein. (10.2)

Protein   Any peptide with more than 100 amino acid residues. (9.5)

Protein turnover   The dynamic process by which many of the body's proteins are continuously hydrolyzed and synthesized within body cells. (17.2)

Protonated amine   A cation produced by the combination of an amine with a hydrogen ion. (6.2)

Purine base   A component of nucleic acids related to purine. Adenine and guanine are purine bases. (11.1)

Pyranose ring system   A six-membered sugar ring system that is considered a derivative of pyran. (7.3)

Pyrimidine base   A component of nucleic acids related to pyrimidine. Cytosine, thymine, and uracil are pyrimidine bases. (11.1)

Quantum   A small package of energy. (1.1)

Quaternary ammonium salt   An ammonium salt with four organic groups (the same or different); the general formula is $R_4N^+Cl^-$. (6.3)

Quaternary structure   The organization of protein subunits into a biologically active assembly. (9.8)

Rancid   Having an unpleasant smell or taste because of chemical decomposition. (8.2)

Recombinant DNA   New DNA chains produced in the laboratory by breaking apart and recombining the DNA chains of different organisms. (11.11)

Reducing power   The ability of NADH to supply electrons that can be used to reduce oxygen to water in cellular respiration and to reduce carbon compounds in synthetic or anabolic reactions. (14.6)

Reducing sugar   A sugar with an aldehyde group or potential aldehyde group that reduces cupric ions (Benedict's reagent) or silver ions (Tollens' reagent) in alkaline solution. (7.9)

Replication   The process by which a single DNA molecule produces two exact copies of itself; replication occurs during mitosis (cell division). (11.3)

Resonance hybrid   An average of two or more resonance structures of a molecule or ion; the true molecule is a resonance hybrid of all of these structures. (2.9)

Resonance structure   Structural formulas for a molecule in which the nuclei have the same arrangement, but the arrangement of the electrons is different. (2.9)

Respiratory acidosis   A decrease in blood pH caused by hypoventilation (very shallow breathing). (13.8)

Respiratory alkalosis   A rise in blood pH caused by hyperventilation (rapid, deep breathing). (13.8)

Respiratory chain   See *Electron transport chain*. (14.6)

Retrovirus   A virus that contains RNA as its genetic material. (11.3)

Ribonucleic acid (RNA)   A nucleic acid important in the synthesis of proteins; a polymer of ribonucleotides. See also *Messenger RNA (mRNA), Ribosomal RNA (rRNA),* and *Transfer RNA (tRNA)*. (11.1)

Ribosomal RNA (rRNA)   The nucleic acid component of ribosomes. (11.6)

Ribosome   A small spherical body or organelle found in cells; composed largely of ribosomal RNA and protein. Ribosomes are the site of protein synthesis in a cell. (11.6)

Rickets   A disease of children caused by a vitamin D deficiency; the characteristic symptoms of rickets are bowed legs and malformed ribs. (12.7)

RNA polymerase   An enzyme that catalyzes the formation of RNA from ribonucleotides, using a strand of DNA as a template. (11.7)

Saliva   The digestive juice secreted in the mouth by the salivary glands; it contains the enzyme amylase (ptyalin) that catalyzes starch hydrolysis. (12.2)

Salt bridge   An ionic bond that contributes to peptide chain conformations in proteins. (9.7)

Saponification   The alkaline hydrolysis of fats or oils (triglycerides) yields glycerol and salts of fatty acids; soaps are the alkali metal salts of fatty acids. (8.2)

Saturated compound   An organic compound in which the carbons are joined to each other by single covalent bonds. (2.8)

Saytzeff's rule   The major product in an elimination reaction is the alkene with the largest number of carbon groups on the double bond. (3.5)

Scurvy   A disease caused by lack of vitamin C (ascorbic acid) in the diet. (12.8)

Second messenger   A molecule used to transmit signals within cells; cyclic AMP (cAMP), calcium ions, and nitric oxide (NO) are second messengers. (15.11)

Secondary structure   Certain regular arrangements of protein chains, typically the beta pleated sheet and the alpha helix. (9.6)

Sedative   A substance that tends to tranquilize the human mind. (6.9)

Serum   The straw-colored liquid that separates from blood that has been allowed to clot; serum lacks the formed elements and clotting agents. (13.2)

Sex hormones Steroid hormones that produce the primary and secondary sex characteristics in males and females. (8.5)

Sigma bond Chemical bond formed by overlap of two spherical half-filled *s* orbitals, overlap of half-filled *s* and *p* orbitals, or end-to-end overlap of two half-filled *p* orbitals. (2.2)

Simple sugar Another name for a monosaccharide. (7.2)

Simple triglyceride A triester made from glycerol and three molecules of one kind of fatty acid. (8.2)

Single covalent bond A covalent bond in which only one pair of electrons is shared by two bonded atoms. (1.3)

Soap The alkali metal salts of long-chain fatty acids; sodium stearate ($C_{17}H_{35}COO^-Na^+$) is a typical soap. (5.3)

Specificity The ability of many enzymes to catalyze only one chemical reaction using only one substrate. (10.3)

Sphingomyelin A phospholipid that contains sphingosine, a long-chain unsaturated amino alcohol; large amounts of sphingomyelins are found in brain and nerve tissue. (8.3)

Starch A polysaccharide found in plants that yields only glucose upon complete hydrolysis; amylose starch is a linear chain, whereas amylopectin is a branched molecule. (7.6)

Stereoisomers Isomers that are nonsuperimposable mirror images of each other. (7.1)

Steroid A class of organic compounds that contain four fused rings; many biologically important compounds are steroids (cholesterol, the sex hormones, vitamin D). (8.5)

Steroid hormone An organic compound that acts as a chemical messenger and has a steroid nucleus; progesterone and testosterone are important steroid hormones. (8.6)

Sterol A steroid molecule containing the hydroxy function; cholesterol is a sterol. (8.5)

Storage fat Fat collected in the cells of adipose tissue (fat tissue); also called depot fat. Triglycerides are the major component of storage fat. (16.1)

Straight-chain alkane A saturated open-chain hydrocarbon in which all carbons are arranged consecutively; that is, there is no branching. (2.4)

Structural formula A formula that shows the arrangement of atoms in a molecule; each covalent bond is drawn as a dash. (1.3)

Structural isomers Compounds that have the same molecular formula but different molecular structures. (2.6)

Substituent Any group attached to a parent hydrocarbon chain. (2.5)

Substrate The molecule on which an enzyme acts. (10.3)

Subunit One of the individual protein molecules in the organization of a protein's quaternary structure. (9.8)

Sugar A general name for any carbohydrate, but most often applied to monosaccharides and disaccharides. (7.1)

Template In DNA replication: A DNA strand used as a pattern for the formation of a complementary strand of DNA. In transcription: A DNA strand used as a pattern for the formation of a complementary strand of mRNA. (11.4)

Termination The end of peptide chain growth in protein synthesis (translation). (11.8)

Termination codons Three noncoding base triplets of mRNA that block further peptide chain growth in translation: UAA, UAG, and UGA. (11.8)

Tertiary structure The overall folding in space of the peptide chain of a protein; the tertiary structure includes the secondary structure of the protein and is determined by the primary structure. (9.7)

Testosterone The major male sex hormone. (8.7)

Thioalcohol A group of organic compounds with the general formula R—SH; also called a thiol. (3.9)

Thioester An ester that contains sulfur and has the general structure

$$R-\overset{\overset{\textstyle O}{\|}}{C}-S-R \quad (5.7)$$

Thioether An organic derivative of hydrogen sulfide; a thioether has the general formula R—S—R. (3.9)

Thiol Another name for a thioalcohol. (3.9)

Thrombocyte A formed element of the blood that functions in blood-clotting; also known as a platelet. (13.2)

Thyalkoid Structure located in the chloroplasts of green plants; thyalkoids contain the molecules that comprise the light system. (14.2)

Tissue lipid Those lipids that comprise the membrane structures of cells; also called working lipids. (16.1)

Tollens' reagent A mild oxidizing agent used to test for the presence of an aldehyde; the reagent is an alkaline solution of silver nitrate. (4.5)

Trans **configuration** Geometric isomer of an alkene in which groups attached to each carbon of the carbon-carbon double bond are on different sides of the bond. (2.8)

Transamination The enzyme-catalyzed transfer of an amino group from an amino acid to a keto acid; some nonessential amino acids are synthesized by this reaction. (17.3)

Transcription A process involving base-pairing in which genetic information contained in a DNA molecule is used to assemble a complementary sequence of bases in an RNA chain. (11.3)

Transfer RNA (tRNA)   A class of short-stranded RNA molecules (molecular weight about 25,000) that bind with amino acids and carry them to the ribosomes in protein synthesis; the nucleotide triplet on tRNA is known as an anticodon. (11.6)

Translation   The process by which the genetic information in an mRNA strand directs the sequence of amino acids during protein synthesis. (11.3)

Triglyceride   An ester in which all three hydroxyl groups on glycerol have been esterified by saturated or unsaturated fatty acids; natural oils and fats are triglycerides. (8.2)

Triple covalent bond   A covalent bond in which three pairs of electrons are shared by two bonded atoms. (1.3)

tRNA binding site   Positions on ribosomes at which activated tRNA molecules bind to complementary mRNA codons during translation; there are two tRNA binding sites on a ribosome. (11.8)

Unsaturated compound   An organic compound with one or more double or triple bonds. (2.8)

Unshared pair   Pair of valence electrons not involved in covalent bonding in molecules. (1.3)

Unwinding protein   A polypeptide that tends to unwind the double helix because it can bind to and stabilize single-stranded DNA. (11.4)

Urobilin   A colorless product of the degradation of heme, the prosthetic group of hemoglobin. (17.8)

Urobilinogen   An orange-yellow product of the degradation of heme, the prosthetic group of hemoglobin, by oxidation of bilirubin. (17.8)

Valence electron   An electron in the highest occupied energy level of an element's atoms; valence electrons largely determine the chemistry of an element. (1.1)

van der Waals forces   Weakest of the attractive forces between molecules, they include dispersion forces and dipole interactions. (1.6)

Vitamin   An organic compound essential in the diet in small amounts; vitamins are fat-soluble or water-soluble. (12.1)

Wax   An ester of a long-chain fatty acid and a long-chain monohydric alcohol. (8.1)

Wilson's disease   An inherited disease in which large amounts of copper accumulate in the liver and brain. (12.9)

Working lipid   The lipid in cell membranes and the membranes of cell organelles such as mitochondria. (16.1)

Zwitterion   The internal salt of an amino acid; a dipolar ion. (9.3)

Zymogen   The inactive precursor of an enzyme; for example, pepsinogen is the zymogen of pepsin. (10.11)

Zymogen granule   Package of enzyme precursors (zymogens) encased in coats of lipids and proteins. (10.11)

Answers to Selected Exercises

Chapter 1

Practice Exercises

1.1 (a) 6   (b) 8

1.2 (a) three   (b) one   (c) seven   (d) three
(e) five   (f) nine

1.3 (a) one   (b) one

1.4 (a) one   (b) four   (c) two   (d) six

1.5 (a) 2+   (b) 1+   (c) 3+

1.6 (a) 2−   (b) 3−   (c) 1−

1.7 (a) H· ·S̈·   H:S̈:H   or   H—S̈—H
         H·

(b) H· ·P̈·   H:P̈:H   or   H—P̈—H
     H·   ·   H   H
     H·

(c) :C̈l· ·F̈:   :C̈l:F̈:   or   :C̈l—F̈:

(d) H· ·Ö: H:Ö: or   H—Ö—Ö—H
                  ·
     H· ·Ö: H:Ö:

1.8
H⁺   H—Ö—H   H:Ö⁺:H   or   H—Ö⁺—H
                                 |
                                 H   H

1.9 The molecule would have a tetrahedral shape.

        :C̈l:
     :C̈l:C:C̈l:
        :C̈l:

1.10 decreasing order of polarity: (c), (d), (a), (f), (b), (e)

1.11 38 kcal

1.12

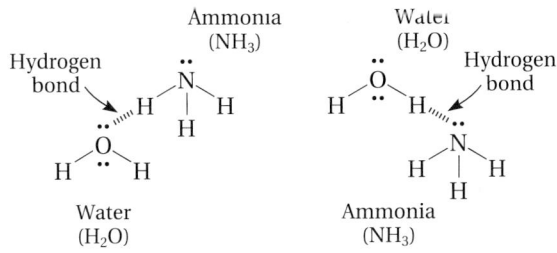

Water (H₂O) | Ammonia (NH₃) | Water (H₂O) | Ammonia (NH₃) — Hydrogen bond diagrams

Exercises

1.13 seven

1.15 (a) O   (b) Mg   (c) B   (d) S

1.17 *Valence electrons* are the electrons in the highest occupied principal energy level of an element's atoms; (a) 5   (b) 6   (c) 7   (d) 1

1.19 An anion has more electrons than protons; a cation has more protons than electrons.

1.21 (a) shared pair of electrons   (b) pair of valence electrons that are not involved in a bond. The hydrogen sulfide molecule, H₂S, is an example of a molecule that contains two single covalent bonds and two unshared pairs of electrons.

H:S̈:H   or   H—S̈—H

1.23 (a) both shared electrons come from one atom
(b) two shared pairs electrons

1.25 Valence-electron pairs arrange themselves about the nucleus of an atom to be as far apart from one another as possible. The molecules (a) methane (b) ammonia (c) water are described in detail in Section 1.4.

1.27 (a) O   (b) C and S the same   (c) C   (d) N

1.29 *Bond dissociation energy* is the energy required to break a single covalent bond.

1.31 Van der Waals forces are very weak attractions between molecules.

1.33 See Figure 1.3.

1.35 Electrons do not pair up until they have to. Silicon $1s^2 2s^2 2p^6 3s^2 3p^2$; phosphorus $1s^2 2s^2 2p^6 3s^2 3p^3$; sulfur $1s^2 2s^2 2p^6 3s^2 3p^4$. Silicon has 2 unpaired electrons, phosphorus has 3 unpaired electrons, sulfur has 2 paired and 2 unpaired electrons.

1.37 (a) 2   (b) 6   (c) 10

1.39 Repels the bonding electrons; H—N—H bond angles are less than 109.5°.

1.41 (a) BrCl   :B̈r:C̈l: Br—Cl

(b) HCN   H:C::N: H—C≡N

(c) SO₂   :Ö::S̈:Ö: O=S—O

(d) HCCH   H:C::C:H   H—C≡C—H

Self-Test

1. False   2. True   3. False   4. True   5. True
6. False   7. True   8. True   9. True   10. True
11. c   12. a   13. c   14. b   15. d   16. a
17. c   18. b   19. d   20. d   21. b   22. c
23. b   24. d   25. c   26. a   27. d   28. d
29. a   30. b

A15

Chapter 2

Practice Exercises

2.1

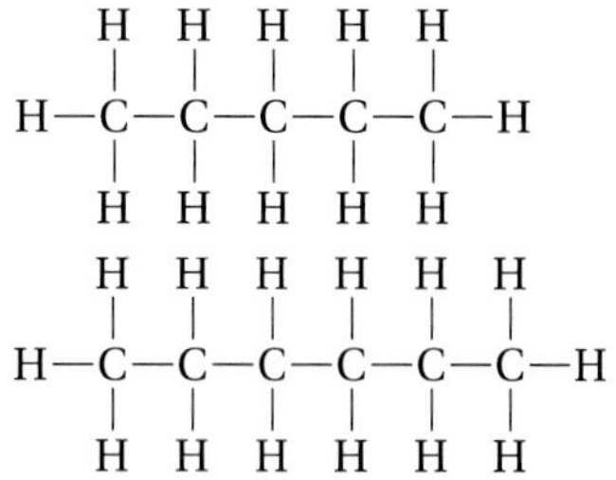

2.2 $CH_3-CH_2-CH_2-CH_2-CH_3$

$CH_3CH_2CH_2CH_2CH_3$

$CH_3(CH_2)_3CH_3$

$C-C-C-C-C$

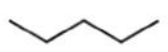

$CH_3-CH_2-CH_2-CH_2-CH_2-CH_3$

$CH_3CH_2CH_2CH_2CH_2CH_3$

$CH_3(CH_2)_4CH_3$

$C-C-C-C-C-C$

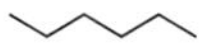

2.3   (a) propane   (b) hexane   (c) pentane

2.4   (a) 2-methylbutane   (b) 3-methylpentane
(c) 3-ethylhexane

2.5

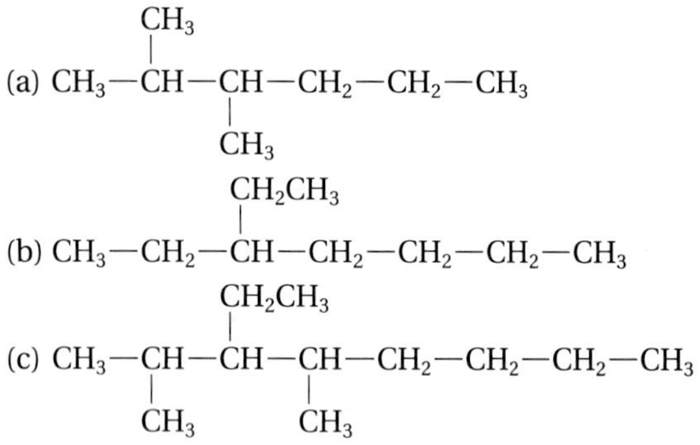

(a)

(b)

(c)

2.6

(a) $CH_3CH_2CH_2CH_2CH_2CH_3$

Hexane

(b) $CH_3CHCH_2CH_2CH_3$ with CH_3

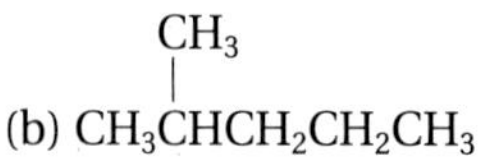

2-Methylpentane

(c)

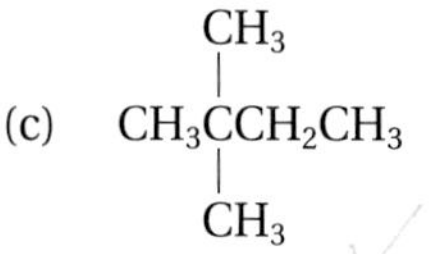

2,2-Dimethylbutane

(d)

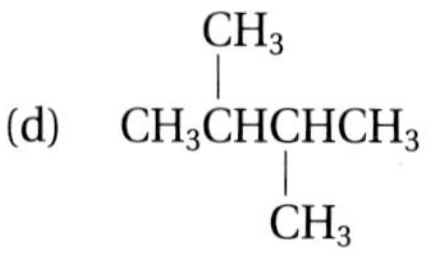

2,3-Dimethylbutane

(e)

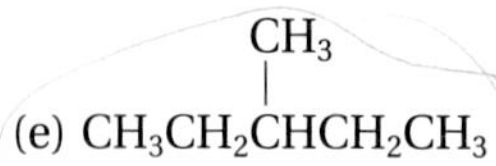

3-Methylpentane

2.7
(a) 1,2,3-trimethylcyclobutane   (b) methylcyclohexane

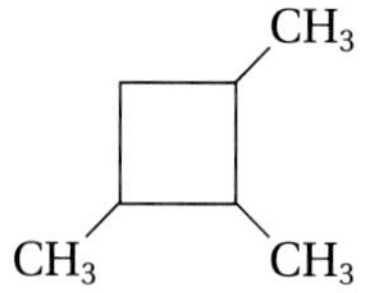

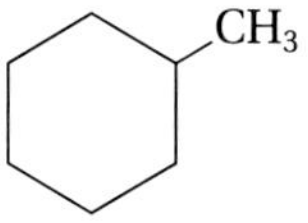

(c) 1,3-dimethylcyclopentane

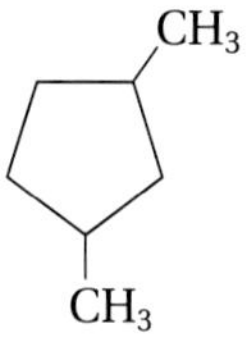

2.8
(a)

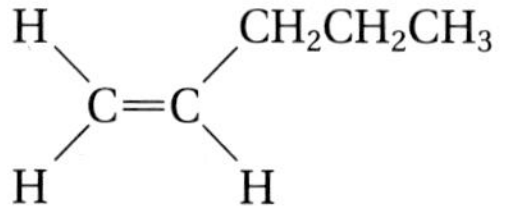

(b)

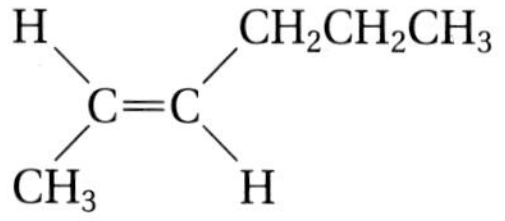

Trans 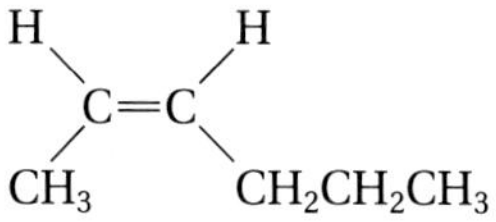   *Cis*

(c)

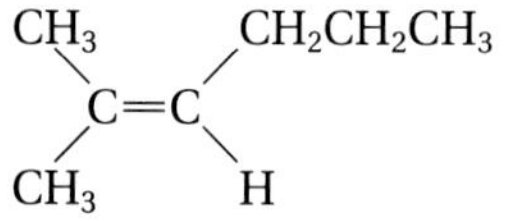

(d)

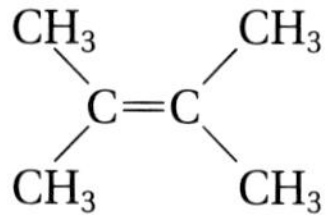

2.9   (a) 1-phenylpropane (propylbenzene)
(b) 2-phenylpropane (isopropylbenzene)
(c) 2,5-dimethyl-3-phenylhexane

Exercises

2.11   (a) ethyl   (b) octane   (c) propane
(d) pentane   (e) pentane   (f) propyl

2.13   A group that replaces a hydrogen on a parent alkane; the isopropyl group, $(CH_3)_2CH—$.

2.15   (a) methylpropane   (b) 3-ethyl-2,5-dimethyl-hexane   (c) 2-methylhexane   (d) 2-methyl-hexane   (e) 4-ethyl-2,5-dimethylheptane
(f) 4-ethylheptane

2.17   (a) $CH_3—CH_2—CH_2—CH_2—CH_3$
(b) 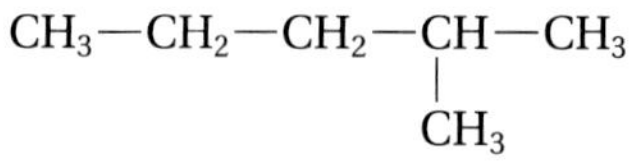
(c) 
(d) $CH_3—CH_2—CH_2—CH_3$

2.19 CH₃—CH₂—CH₂—CH₂—CH₃

Pentane

2.21 (a) 1,2-dimethylcyclohexane (b) methylcy-
clopentane (c) 1,1-dimethylcyclopropane

2.23 Number the longest carbon chain containing the
carbon-carbon double bond so that the double
bond is given the lowest number. The name of this
longest chain is the name of the corresponding
alkane, but with an *-ene* rather than an *-ane* ending. Substituents on the longest chain are numbered and named according to the rules for
naming alkanes.

2.25 No

2.27

(a)

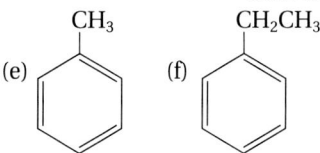

Cis-2-pentene   *Trans*-2-pentene

(b) CH₃CH₂   CH₂CH₃   CH₃CH₂   H
          C=C   C=C
     H   H   H   CH₂CH₃

Cis-3-hexene   *Trans*-3-hexene

2.29 (a) 1-butene (b) 2-methyl-2-nonene
(c) 2-methyl-2-butene (d) 1-hexene

2.31 The term *aromatic character* refers to hybrid bonding in benzene and related arenes.

2.33 Each structure depicts a different resonance form
of 1,2-diethylbenzene.

2.35 (a) cyclohexene (b) 1-3-ethylmethylbenzene
(c) naphthalene (d) methylcyclohexane; b, c, e,
and f are aromatic.

(e)   CH₃   (f)   CH₂CH₃

Toluene   Ethylbenzene

2.37 The hydrocarbons in crude oil are mostly alkanes
and cycloalkanes; the hydrocarbons in coal are
mostly aromatic.

2.39 (a) hydrocarbons contain only carbon and hydrogen (b) alkane is a saturated hydrocarbon
(c) alkene contains the carbon-carbon double
bond (d) alkyne contains the carbon-carbon
triple bond (e) cycloalkane is an alkane with
ends of the carbon chain joined together to form a
ring (f) arene molecules contain the benzene
ring

2.41 (a) 3 (b) 10 (c) 6 (d) 7
(e) 8 (f) 1 (g) 9 (h) 2
(i) 5 (j) 4

2.43 (a) cyclopentane (b) 2-methyl-2-butene
(c) 1,2,4-trimethylcyclohexane
(d) 1,3-dimethylcyclobutane

2.45

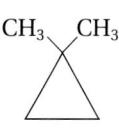

1,1-Dimethylcyclopropane   1,2-Dimethylcyclopropane

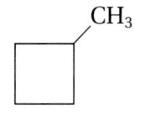   

Methylcyclobutane   Cyclopentane

Self-Test

1. False 2. False 3. False 4. True 5. False
6. True 7. False 8. True 9. True 10. False
11. c 12. c 13. c 14. c 15. b 16. a
17. b 18. b 19. b 20. a 21. b 22. c
23. a 24. d 25. a 26. b 27. d

Chapter 3

Practice Exercises

3.1 (a) bromobenzene (b) chloroethane
(c) 3-chloro-1-butene

3.2

(a)   CH₃
        H—C—Cl
             CH₃

                              CH₃
(b) CH₃—CH₂—CH₂—C—CH₂—I
                              CH₃

(c)

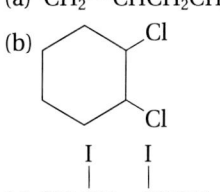

3.3
          Br   Br
(a) CH₂—CHCH₂CH₃
(b)   Cl

          Cl

          I   I
(c) CH₃CH—CHCH₃

3.4

(a)
$$\underset{\text{(major product)}}{Br-\underset{\underset{CH_3}{|}}{\overset{\overset{CH_3}{|}}{C}}-\underset{\underset{H}{|}}{\overset{\overset{H}{|}}{C}}-H} \quad + \quad \underset{\text{(minor product)}}{H-\underset{\underset{CH_3}{|}}{\overset{\overset{CH_3}{|}}{C}}-\underset{\underset{H}{|}}{\overset{\overset{H}{|}}{C}}-Br}$$

(b) [benzene ring with Cl] + HCl

(c) [cyclohexane ring with Cl and H]

3.5 (a) 1-butanol   (b) 2-propanol (isopropyl alcohol)   (c) 2-methyl-1-butanol   (d) cyclohexanol (cyclohexyl alcohol)   (e) 2-methyl-2-propanol (*tert*-butyl alcohol)

3.6 (a) primary alcohol   (b) secondary alcohol   (c) primary alcohol   (d) secondary alcohol   (e) tertiary alcohol

3.7 (a) $CH_2{=}CH-CH_3$
(b) $CH_3CH{=}CHCH_3$
(c) $CH_3CH{=}CHCH_3$ $\quad$ $CH_2{=}CHCH_2CH_3$
$\quad\quad$ (major product) $\quad\quad$ (minor product)

3.8 (a) $2Na + 2CH_3OH \longrightarrow 2CH_3O^-Na^+ + H_2$

(b) $2Na + 2$ [benzene ring]$-OH \longrightarrow$

2 [benzene ring]$-O^-Na^+ + H_2$

3.9 (a) ether   (b) halocarbon   (c) alcohol   (d) aliphatic hydrocarbon (alkane); (c) is the most polar compound.

3.10 (d), (b), (a), and (c)

3.11 Urushiol contains two phenolic hydroxyl groups, two carbon-carbon double bonds, and an aromatic ring.

3.12 Estradiol contains the hydroxyl group of an aliphatic alcohol, a phenolic hydroxyl group, and an aromatic ring.

Exercises

3.13 (a) [benzene ring with two Cl]
(b) 3-chloropropene

(c) 1,2-dichloro-4-methylpentane
(d) [cyclohexane ring with two Cl]

3.15 (a) $\underset{}{CH_3CH_2\underset{\underset{}{|}}{\overset{\overset{Cl}{|}}{C}}H-\overset{\overset{Cl}{|}}{C}H_2}$

(b) $\underset{\text{(major product)}}{CH_3CH_2\overset{\overset{Br}{|}}{C}H-\overset{\overset{H}{|}}{C}H_2}$ $\quad$ $\underset{\text{(minor product)}}{CH_3CH_2\overset{\overset{H}{|}}{C}H-\overset{\overset{Br}{|}}{C}H_2}$

(c) $CH_3CH_2\underset{\underset{H}{|}}{\overset{\overset{CH_3}{|}}{C}}{-}\overset{\overset{H}{|}}{C}HCH_3$

3.17 $Cl-\underset{\underset{}{}}{\overset{\overset{Cl}{|}}{C}}H-CH_2-CH_3$
1,1-Dichloropropane

$Cl-CH_2-\overset{\overset{Cl}{|}}{C}H-CH_3$
1,2-Dichloropropane

$CH_3-\underset{\underset{Cl}{|}}{\overset{\overset{Cl}{|}}{C}}-CH_3$
2,2-Dichloropropane

$Cl-CH_2-CH_2-CH_2-Cl$
1,3-Dichloropropane

3.19 (a)
$XCH_3CH_2CH{=}CH_2 \longrightarrow -(-\underset{\underset{}{}}{\overset{\overset{CH_3}{\overset{|}{CH_2}}}{C}}HCH_2-)_x-$
1-Butene $\quad\quad$ Poly-1-butene

(b)
$X\overset{\overset{Cl}{|}}{C}H{=}\underset{\underset{Cl}{|}}{C}H \longrightarrow -(-\overset{\overset{Cl}{|}}{C}H\underset{\underset{Cl}{|}}{C}H-)_x-$
1,2-Dichloroethene $\quad$ Poly-1,2-dichloroethene

3.21 (a) 2-butanol
(b) 2-methyl-1-propanol
(c) $HOCH_2CH_2OH$

(d)

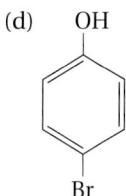

3.23 (a) secondary   (b) secondary   (c) tertiary
(d) primary
3.25 Addition of water to an alkene; displacement of halide ion from an aliphatic halocarbon.

$$CH_2=CH_2 + H_2O \xrightarrow[100\,°C]{H^+} CH_3CH_2OH$$

Ethene   Water   Ethanol

$$CH_3CH_2Br + OH^- \longrightarrow CH_3CH_2OH + Br^-$$

Bromoethane   Hydroxide   Ethanol   Bromide
ion   ion

3.27 (a) CH_3CH_2OH

Ethanol

OH
|
(b) $CH_3CHCH_2CH_3$

2-Butanol

CH₃   CH₃
| |
(c) $CH_3CH_2CCH_3$ $CH_3CH_2CCH_2OH$
| |
OH   H

2-Methyl-2-butanol   2-Methyl-1-butanol
(major product)   (minor product)

3.29 (a)

(b) $CH_3CH_2=CH_2$

Propene

Cyclohexene

CH₃
|
(c) $CH_3CH=CH_2$ (d) $CH_3CH_2CH=CHCH_3$

2-Pentene

2-Methylpropene

3.31 (a) 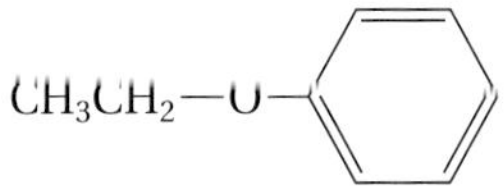 CH_3CH_2-O-⟨⟩

(b)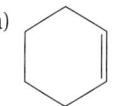

(c) butylethyl ether
(d) dipropyl ether

3.33 (a) 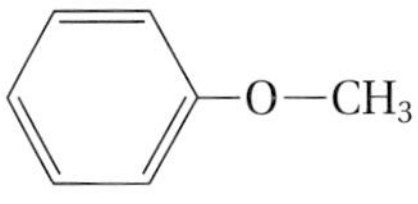 ⟨⟩$-O-CH_3$

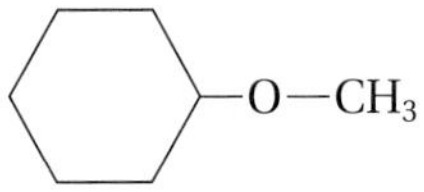

(b) ⟨⟩$-O-CH_3$

(c) $CH_3CHOCH_2CH_3$
            |
           CH_3

3.35 Polarity of the hydroxyl group of ethanol overcomes hydrophobicity of the short carbon chain; ethanol is very soluble. Hydrophobicity of the long carbon chain of decanol dominates polarity of the hydroxyl group; the decanol is almost insoluble.

3.37 (a) 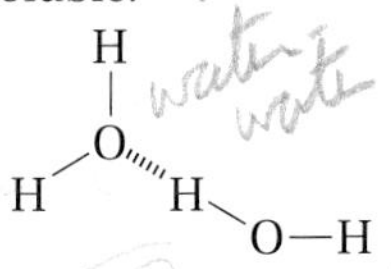   (b)

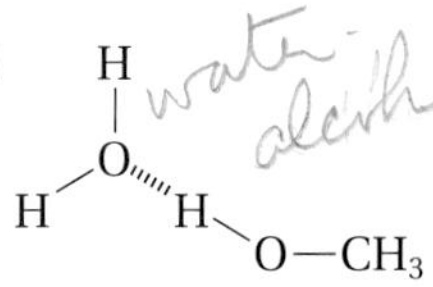

(c)

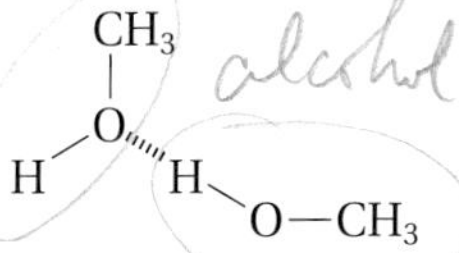

3.39 phenolic hydroxyl group, a primary alcohol, an ether, an alkene, and a chloro group
3.41 (a) phenol, more hydrogen bonding
(b) propanol, more hydrogen bonding
(c) 1-butanol, branched molecules tend to have lower boiling points   (d) methanol, more hydrogen bonding   (e) chlorocyclohexane, higher molar mass and polar molecule
3.43 (a)

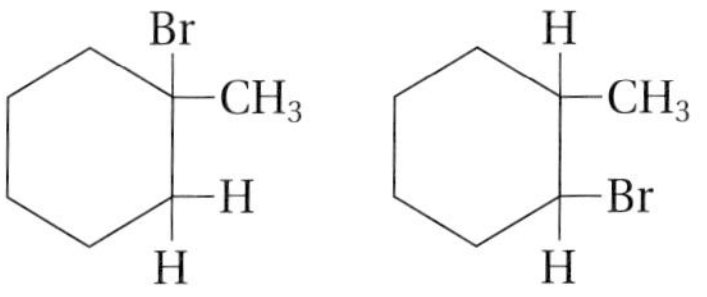

(major product)   (minor product)

(b)

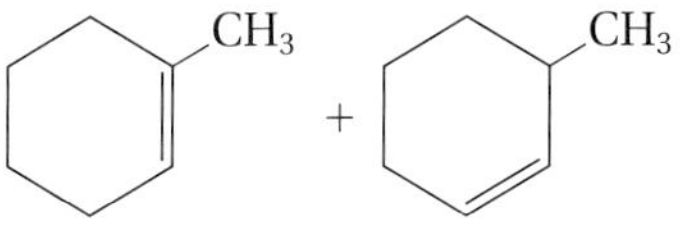

+

(major product)   (minor product)

(c)

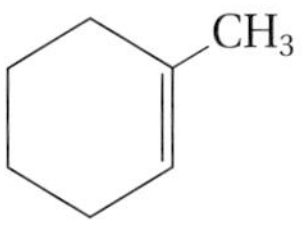

(d)

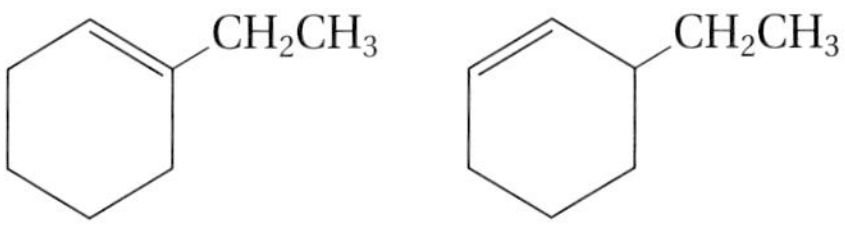

(major product)   (minor product)

3.45 (d), (c), (a), (b)

Self-Test

1. True   2. True   3. True   4. False   5. True
6. True   7. True   8. False   9. True   10. True
11. a   12. b   13. b   14. b   15. d   16. a
17. b   18. d   19. b   20. d   21. b   22. b
23. c   24. d   25. a   26. c

Chapter 4

Practice Exercises

4.1   propanal   (b) 3-methylpentanal
(c) pentanal   (d) 3-chlorobutanal

4.2   (a) 3-hexanone   (b) butanone   (c) 4-methyl-
2-pentanone

4.3   (a) CH₃CHO   (b) CH₃CCH₃ (with O double bond)

$$\text{(a) } CH_3CHO \qquad \text{(b) } CH_3\overset{\displaystyle O}{\overset{\|}{C}}CH_3$$

$$\text{(c) } CH_3CH_2\overset{\displaystyle O}{\overset{\|}{C}}CH_2CH_3 \qquad \text{(d) } CH_3\overset{\displaystyle CH_3}{\overset{|}{C}H}CH_2CHO$$

4.4   (d), (e), (c), (b), (a)

4.5   (a) 1-butyne   (b) propanal   (c) cyclohexanol
(d) 3-pentanone

4.6   (a) CH₃CH₂CH₂CHO   (b) CH₃CH₂CCH₃ (with O)

$$\text{(a) } CH_3CH_2CH_2CHO \qquad \text{(b) } CH_3CH_2\overset{\displaystyle O}{\overset{\|}{C}}CH_3$$

Butanal   Butanone

(c) No reaction   (d)

Cyclobutanone

4.7   (a) CH₃CH₂CH₂OH   (b) CH₃CH₂CHCH₃ (with OH)

$$\text{(a) } CH_3CH_2CH_2OH \qquad \text{(b) } CH_3CH_2\overset{\displaystyle OH}{\overset{|}{C}H}CH_3$$

1-Propanol   2-Butanol

(c) CH₃CH₂CHCH₂OH (with CH₃)

$$\text{(c) } CH_3CH_2\overset{\displaystyle }{\underset{\underset{\displaystyle CH_3}{|}}{C}H}CH_2OH$$

2-Methyl-1-butanol

4.8   Positive Tollens' and Benedict's tests for pentanal;
same tests negative for 2-pentanone.

4.9   (b), (c), (e)

4.10   Three ketone carbonyl groups, one carbon-carbon
double bond, two aliphatic —OH groups.

Exercises

4.11   The carbonyl group is —C— (with O double bond); bonding electrons
pulled closer to electronegative oxygen, making
oxygen partially negative and carbon partially

positive. The electronegativities of carbons in the
carbon-carbon double bond are equal; therefore,
the carbon-carbon double bond is nonpolar.

4.13   (a) ethanal (acetaldehyde)   (b) benzaldehyde
(c) 3-methylbutanal   (d) diphenyl ketone
(benzophenone)   (e) 5-methyl-3-hexanone

4.15   Acetaldehyde molecules attract one another
through polar-polar interactions; polar-polar
interactions are not possible between propane
molecules.

4.17   (a) oxidation   (b) oxidation   (c) reduction
(d) oxidation

4.19   (a) CH₃CH₂CHO   (b) No reaction; the alcohol
is tertiary.

Propanal

$$\text{(c) } CH_3CH_2\overset{\displaystyle O}{\overset{\|}{C}}CH_2CH_3 \qquad \text{(d)}$$

3-Pentanone

Cyclohexanone

4.21   (a)   CH₂O   (b) CH₃CCH₃ (with O)

$$\text{(b) } CH_3\overset{\displaystyle O}{\overset{\|}{C}}CH_3$$

Methanal   Propanone
(formaldehyde)   (acetone)

(c)   CH₃CHCHO (with CH₃)

$$\text{(c) } CH_3\overset{\displaystyle CH_3}{\overset{|}{C}H}CHO$$

Methylpropanal

4.23   Of oxidation products from Exercise 4.20,
(b) octanal and (c) 2-methylbutanal will give a
positive Benedict's test.

4.25

$$CH_3CH_2OH + CH_3CH_2\overset{\displaystyle H}{\overset{|}{C}}{=}O \longrightarrow CH_3CH_2\overset{\displaystyle H}{\underset{\underset{\displaystyle OH}{|}}{\overset{|}{C}}}{-}OCH_3$$

4.27   (a) neither   (b) neither   (c) acetal
(d) acetal

4.29

$$\text{(a) } CH_3\overset{\displaystyle O}{\overset{\|}{C}}CH(CH_3)_2 \qquad \text{(b) } CH_3CH_2\overset{\underset{\underset{\displaystyle CH_3}{|}}{C}H}{}\overset{\displaystyle O}{\overset{\|}{C}}{-}OH$$

(c) no reaction   (d) CH₃CHCH₂C—OH (with CH₃ and O)

$$\text{(d) } CH_3\overset{\underset{\underset{\displaystyle CH_3}{|}}{C}H}{}CH_2\overset{\displaystyle O}{\overset{\|}{C}}{-}OH$$

(e) no reaction   (f) CH₃CH₂OH

(g) $CH_3\overset{\displaystyle H}{\underset{\displaystyle OH}{C}}{-}OCH_3$ (h) $CH_3{-}\overset{\displaystyle OH}{\underset{\displaystyle CH_3}{C}}{-}OCH_2CH_3$

4.31 The structures given are (a) ketal, (b) hemiacetal, (c) acetal, (d) hemiketal, (e) acetal, and (f) acetal.

4.33 Ketone molecules attract each another through polar-polar interactions; propanone boils at a higher temperature than ethyl methyl ether.

4.35 $CH_3CH_2CH_2CHO$ gives positive tests with Benedict's reagent or Tollens' reagent; $CH_3CH_2COCH_3$ gives negative tests.

4.37 (a) methanal (formaldehyde)   (b) propanone (dimethyl ketone, acetone)   (c) butanal (butyraldehyde)   (d) benzaldehyde   (e) cyclohexanone   (f) methylphenylmethanone (methyl phenyl ketone)   (g) 3-pentanone

4.39 (a) butanol   (b) ethane   (c) formaldehyde   (d) isopropyl alcohol

Self-Test
1. False   2. False   3. False   4. False   5. False
6. True   7. False   8. True   9. False   10. False
11. b   12. c   13. d   14. c   15. b   16. d
17. d   18. c   19. d   20. c   21. b

Chapter 5

Practice Exercises
5.1 (a) propanoic acid (propionic acid)
(b) succinic acid   (c) 4-hydroxypentanoic acid
(d) phthalic acid

5.2 (a), (c), (d), (b)

5,3 (a) $CH_3CH_2COO^-Na^+$ + H_2O

Sodium propanoate   Water

(b) $(CH_3COO^-)_2Ca^{2+}$ + $2H_2O$

Calcium acetate   Water

5.4 $(CH_3(CH_2)_{16}COO^-)_3Fe^{3+}$

5.5 (a) CH_3COOH (b) $CH_3\overset{\displaystyle O}{\overset{\displaystyle \|}{C}}CH_3$
(c) CH_3CH_2COOH (d) no reaction

5.6 Butanoic anhydride; hydrolysis gives butanoic acid.

$CH_3(CH_2)_2\overset{\displaystyle O}{\overset{\displaystyle \|}{C}}{-}O{-}\overset{\displaystyle O}{\overset{\displaystyle \|}{C}}(CH_2)_2CH_3$ + H_2O $\longrightarrow$

Butanoic anhydride

$2CH_3(CH_2)_2\overset{\displaystyle O}{\overset{\displaystyle \|}{C}}{-}OH$

Butanoic acid

5.7 (a) phenyl propanoate (b) ethyl benzoate

5.8 (a) $CH_3CH_2CH_2\overset{\displaystyle O}{\overset{\displaystyle \|}{C}}OCH_3$

Methyl propanoate (methyl propionate)

(b) ⬡$-\overset{\displaystyle O}{\overset{\displaystyle \|}{C}}OCH_2CH_2CH_3$

Propyl benzoate

(c) $CH_3\overset{\displaystyle O}{\overset{\displaystyle \|}{C}}OCH_2CH_3$ + CH_3COOH

Ethyl ethanoate (ethyl acetate)   Ethanoic acid (acetic acid)

5.9 (a) $CH_3CH_2COO^-Na^+$ + $HOCH_2CH_3$

Sodium propanoate   Ethanol

(b) $CH_3COO^-K^+$ + $HO{-}$⬡

Potassium acetate   Phenol

5.10 Transfer of a phosphoryl group ($-PO_3H_2$) to an alcohol or other functional group.

Exercises
5.11 (a) methanoic acid (formic acid)   (b) lactic acid
(c) salicylic acid   (d) pyruvic acid
(e) $CH_3(CH_2)_{16}COOH$
(f) $CH_3\overset{\displaystyle OH}{\overset{\displaystyle |}{C}}HCOOH$

(g) ⬡ with COOH at top and Cl at bottom

(h) $HOOCCH_2COOH$

5.13 (h) dicarboxylic acid; (b), (c), (d), (f), (g) polyfunctional acids; (e) fatty acid.

5.15 Because propanoic acid has the strongest intermolecular hydrogen bonding, it has the highest boiling point.
(a) $CH_3CH_2CH_2CH_2CH_3$ (b) $CH_3CH_2CH_2CHO$
Pentane   Butanal
(c) $CH_3CH_2CH_2CH_2OH$ (d) CH_3CH_2COOH
Butanol   Propanoic acid

5.17 Carboxylic acids dissociate to $RCOO^-$ and H^+ to a small degree in aqueous solution.

5.19 (a) $HCOO^-K^+ + H_2O$

      Potassium   Water
      formate

$$\begin{array}{c} COO^- \\ | \\ COO^- \end{array} Ca^{2+} + H_2O$$

(b)
                Water
     Calcium
     oxalate

(c) $CH_3(CH_2)_{12}COO^-Na^+ + H_2O$

      Sodium   Water
      myristate

5.21 A soap is an alkali-metal salt of a long-chain carboxylic acid. The hydrophobic hydrocarbon chains of soap molecules emulsify grease droplets, enabling them to be rinsed away with water.

5.23 (a) CH_3COOH (b) CH_3CH_2COOH
     (c) No reaction

5.25 (a) butanal   (b) methanal (formaldehyde)

5.27 (a) $R-\overset{\overset{\displaystyle O}{\|}}{C}-O-\overset{\overset{\displaystyle O}{\|}}{C}-R$

(b) $R-\overset{\overset{\displaystyle O}{\|}}{C}-OR$

5.29 (a) methyl ethanoate (methyl acetate)   (b) ethyl ethanoate (ethyl acetate)   (c) ethyl benzoate

(d) $H\overset{\overset{\displaystyle O}{\|}}{C}OCH_3$

     Methyl formate
     (methyl methanoate)

(e) $CH_3CH_2\overset{\overset{\displaystyle O}{\|}}{C}OCH_2CH_2CH_3$

     Propyl propanoate
     (propyl propionate)

5.31 $R-\overset{\overset{\displaystyle O}{\|}}{C}-O-R$ $R-\overset{\overset{\displaystyle O}{\|}}{C}-S-R$

   Oxyester   Thioester

5.33
(a) Methyl acetate + Water $\xrightarrow{HCl}$ $CH_3COOH + CH_3OH$

                    Acetic acid   Methanol

(b) Phenyl propanoate + Sodium hydroxide $\longrightarrow$

$CH_3CH_2COO^-Na^+$ +

     Sodium propanoate   Phenol

(c) Ethyl benzoate + Sodium hydroxide $\longrightarrow$

       $-COO^-Na^+ + CH_3CH_2OH$

    Sodium benzoate   Ethanol

(d) Propyl butanoate + Sodium hydroxide $\longrightarrow$
   $CH_3CH_2CH_2COO^-Na^+ + CH_3CH_2CH_2OH$

    Sodium butanoate   1-Propanol

(e) Ethyl formate + Potassium hydroxide $\longrightarrow$
         $HCOO^-K^+ + CH_3CH_2OH$

    Potassium   Ethanol
    formate

(f) Phenyl benzoate + Water $\xrightarrow{HCl}$

      $-COOH$ + $-OH$

    Benzoic acid   Phenol

5.35 The monoanion and the dianion dominate at pH 7.

$$HO-\overset{\overset{\displaystyle O}{\|}}{\underset{\underset{\displaystyle OH}{|}}{P}}-OH \quad \text{Completely undissociated}$$

$$HO-\overset{\overset{\displaystyle O}{\|}}{\underset{\underset{\displaystyle OH}{|}}{P}}-O^- \quad \text{Monoanion}$$

$$HO-\overset{\overset{\displaystyle O}{\|}}{\underset{\underset{\displaystyle O^-}{|}}{P}}-O^- \quad \text{Dianion}$$

$$^-O-\overset{\overset{\displaystyle O}{\|}}{\underset{\underset{\displaystyle O^-}{|}}{P}}-O^- \quad \text{Trianion}$$

5.37 (a) $HO-\overset{\overset{\displaystyle O}{\|}}{\underset{\underset{\displaystyle OH}{|}}{P}}-O-\overset{\overset{\displaystyle O}{\|}}{\underset{\underset{\displaystyle OH}{|}}{P}}-OH + CH_3CH_2OH \longrightarrow$

    Pyrophosphoric acid

$$HO-\overset{\overset{\displaystyle O}{\|}}{\underset{\underset{\displaystyle OH}{|}}{P}}-O-OCH_2CH_3 + HO-\overset{\overset{\displaystyle O}{\|}}{\underset{\underset{\displaystyle OH}{|}}{P}}-OH$$

    Ethyl phosphate   Phosphoric acid

(b) phosphorylation

5.39 (d), (c), (b), (a)

5.41 Acetic acid (K_a 1.8×10^{-5}) dissociates only to a small extent in aqueous solution. Water abstracts a hydrogen ion from only a small fraction of acetic acid molecules, so acetate ion must be a stronger base than water.

5.43 (a) $CH_3CH_2CH_2CH_2CH_2CH_2CH_2CH_2CH_2CH_3$
(b) $CH_3CH_2CH_2CH_2CH_2COOH$
(c) CH_3OCH_3
(d) $CH_3CH_2CH_2CH_2CH_2CH_2CH_2CHO$
(e) $CH_3CH_2CH_2\overset{O}{\overset{\|}{C}}OCH_2CH_3$
(f) $CH_3CH_2CH_2CH_2CH_2OH$
(g) $CH_3CH_2\overset{Cl}{\overset{|}{C}}HCOOH$
(h) 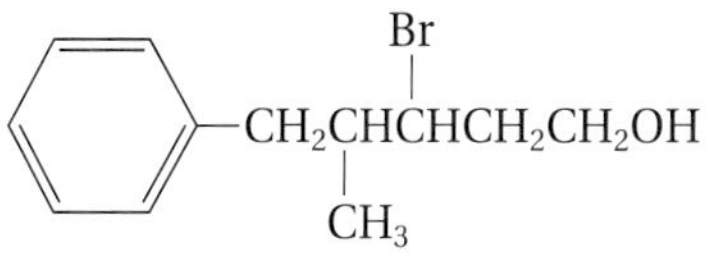

Self-Test

1. False 2. True 3. True 4. True 5. True
6. True 7. True 8. True 9. True 10. True
11. c 12. b 13. a 14. d 15. d 16. d
17. a 18. e 19. c 20. a 21. e 22. d
23. d

Chapter 6

Practice Exercises

6.1 (a) ethylmethylamine (secondary) (b) propylamine (primary) (c) ethyldimethylamine (tertiary) (d) triethylamine (tertiary)

6.2 (a) *N*-ethylaniline (secondary) (b) *N*,*N*-diethylaniline (tertiary)

6.3 (a) $CH_3CH_2NH_2$ (b) 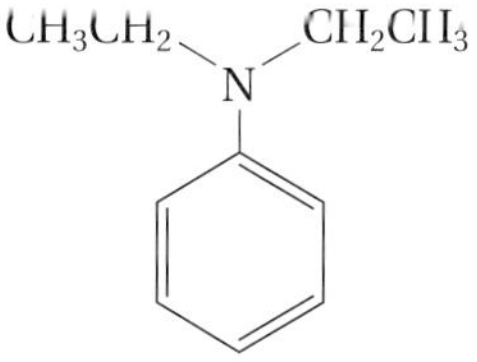

(c) $(CH_3CH_2CH_2)_3N$ (d) 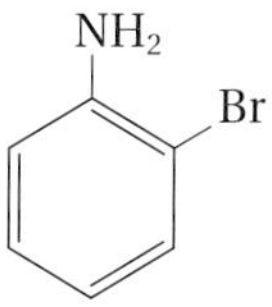

6.4 (d) pyridine

6.5

(a) $(CH_3)_2NH + HCl \longrightarrow (CH_3)_2NH_2{}^+Cl^-$
Dimethylammonium chloride

(b) $CH_3NH_2 + HNO_3 \longrightarrow CH_3NH_3{}^+NO_3{}^-$
Methylammonium nitrate

(c) 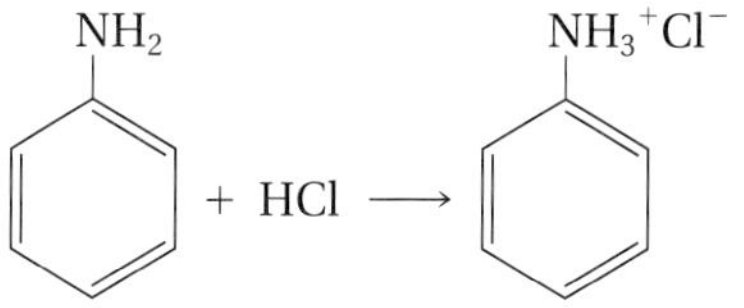
Anilinium chloride

(d) $CH_3CH_2NH_2 + H_3PO_4 \longrightarrow CH_3CH_2NH_3{}^+H_2PO_4{}^-$
Ethylammonium dihydrogen phosphate

6.6

(a) $CH_3CH_2CH_2CH_2Cl + NH_3 \longrightarrow$
1-Chlorobutane, Ammonia

$CH_3CH_2CH_2CH_2NH_3{}^+Cl^-$
Butylammonium chloride

(b) $CH_3CH_2CH_2CH_2NH_3{}^+Cl^- + NaOH \longrightarrow$
Butylammonium chloride, Sodium hydroxide

$CH_3CH_2CH_2CH_2NH_2 + NaCl + H_2O$
Butylamine, Sodium chloride, Water

6.7 (a) propanamide (b) 3-methylbutanamide (c) *N*-methylpropanamide

6.8

$$CH_3CH_2\overset{O}{\overset{\|}{C}}-O^-\ H-\overset{H}{\overset{|}{\underset{H}{\underset{|}{N^+}}}}-CH_2CH_3 \xrightarrow{\text{Heat}}$$

$$CH_3CH_2\overset{O}{\overset{\|}{C}}-NHCH_2CH_3 + H_2O$$
N-Ethylpropanamide, Water

6.9

$$CH_3\overset{O}{\overset{\|}{C}}-O-CH_2CH_3 + H-\overset{H}{\overset{|}{N}}-CH_3 \longrightarrow$$
Ethyl acetate, Methylamine

$$CH_3\overset{O}{\overset{\|}{C}}-NHCH_3 + CH_3CH_2OH$$
N-Methylacetamide, Ethanol

6.10

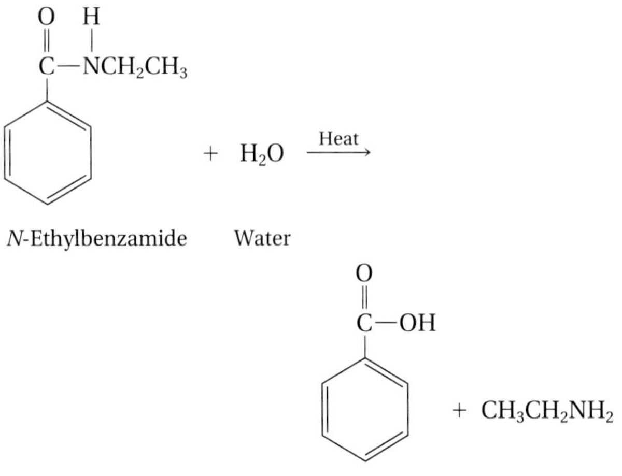

N-Ethylbenzamide   Water

Benzoic   Ethylamine
acid

6.11 phenolic hydroxyl group, secondary hydroxyl group, secondary amine group; decongestant

6.12 (a) painkiller   (b) nervous system stimulant (c) dilates pupils of eyes   (d) antimalarial agent

6.13

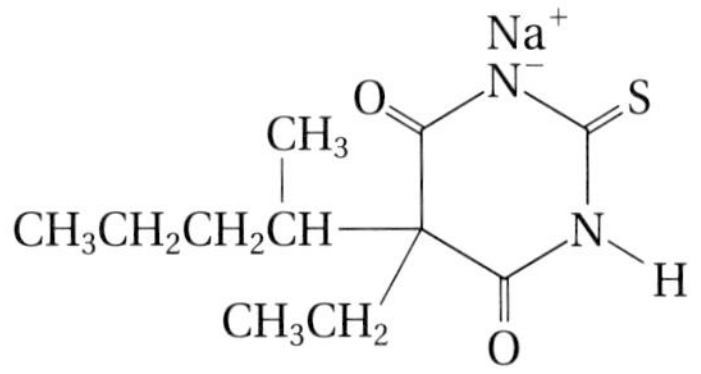

Sodium pentothal

6.14 Electron-withdrawing power of the adjacent carbonyl groups weakens the N—H bond.

Exercises

6.15 (a) Dimethylamine (secondary)

(b)   (c) 3-Chloroaniline (primary)

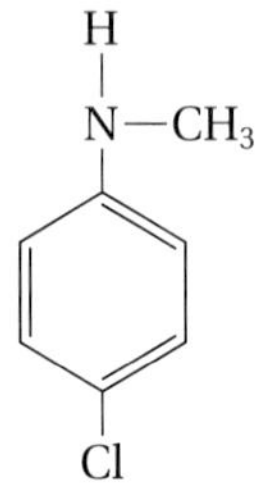

(secondary)

(d) CH_3CH_2—N—CH_2CH_3 (tertiary)
         |
         CH_3

6.17 The amine nitrogen is contained within a ring in a heterocyclic amine.

6.19

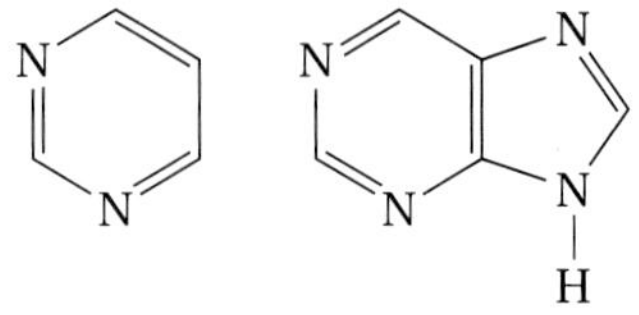

Pyrimidine   Purine

6.21 Amines are weak bases because they abstract a hydrogen ion (proton) from water to a slight extent.

6.23 (a) $CH_3CH_2NH_3{}^+Cl^-$ (b) $(CH_3)_2NH_2{}^+NO_3{}^-$

Ethylammonium   Dimethylammonium
chloride   nitrate

(c) $CH_3NH_3{}^+HSO_4{}^-$ (d) $(CH_3)_3CNH_3{}^+Cl^-$

Methylammonium   tert-Butylammonium
hydrogen sulfate   chloride

6.25 The solution tests acidic because a small fraction of the alkylammonium ions dissociate to produce hydrogen ions.

6.27 (a) ethanamide (acetamide)   (b) N-methyl-propanamide   (c) N-methylbenzamide

(d) CH_3CH_2C—N—CH_2CH_3
         ‖ |
         O CH_3

(e) CH_3—C—N—
         ‖ |
         O   H

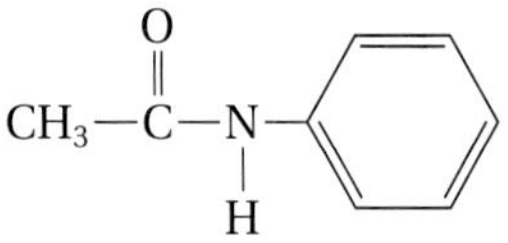

6.29

(a) $CH_3CH_2CNH_2$ (b) CH_3CNH_2
         ‖   ‖
         O   O

Propanamide   Ethanamide
                     (acetamide)

(c) $HCNHCH_2CH_3$
       ‖
       O

N-Ethylmethanamide
(N-ethylformamide)

6.31 Dopamine cannot cross the blood-brain barrier. Dopa penetrates the barrier and is converted to dopamine in the brain.

6.33 Class of amines obtained from plants; gives alkaline (basic) aqueous solutions

6.35 Morphine and codeine

6.37 Barbituric acid is not physiologically active. Barbiturates can be hypnotics (sleep-inducers) and sedatives (tranquilizers).

6.39 (a) pyrimidine; (b) pyridine; (c) pyrrole; (d) purine; (e) imidazole; (f) indole.

6.41

(a) $(CH_3CH_2)_3N^+Br^-$

(b)
$$\underset{\displaystyle \bigcirc}{\overset{\displaystyle H}{N}}-CH_2CH_2CH_3$$

(c)
$$CH_3\overset{O}{\overset{\|}{C}}-\overset{H}{\underset{|}{N}}-CH_3$$

(d)
$$\overset{O}{\overset{\|}{C}}-N\overset{CH_3}{\underset{CH_3}{}}$$

6.43 (a) $(CH_3CH_2)_3N$

(b) $CH_3CH_2\overset{O}{\overset{\|}{C}}NH_2$

(c)
$$\overset{H}{N}-CH_2CH_3$$

(d)
$$\overset{O}{\overset{\|}{C}}-\overset{H}{\underset{|}{N}}-$$

(e)
$$NH_2 \quad CH_2CH_3$$

(f) $CH_3\overset{O}{\overset{\|}{C}}N(CH_2CH_3)_2$

6.45

$$RCOOH + NH_3 \xrightarrow{Heat} R\overset{O}{\overset{\|}{C}}NH_2 + H_2O$$

$$RCOOH + RNH_2 \xrightarrow{Heat} R\overset{O}{\overset{\|}{C}}NHR + H_2O$$

$$RCOOH + R_2NH \xrightarrow{Heat} R\overset{O}{\overset{\|}{C}}NR_2 + H_2O$$

6.47

(a) Preparation of propylamine

$$CH_3CH_2CH_2OH \xrightarrow[Heat]{H_2SO_4} CH_3CH=CH_2 + H_2O$$

$$CH_3CH=CH_2 + HCl \longrightarrow$$
$$\underset{(major\ product)}{CH_3\overset{Cl}{\underset{|}{CH}}-\overset{H}{\underset{|}{CH_2}}} + \underset{(minor\ product)}{CH_3\overset{H}{\underset{|}{CH}}-\overset{Cl}{\underset{|}{CH_2}}}$$

$$CH_3CH_2CH_2Cl + NH_3 \longrightarrow CH_3CH_2CH_2NH_3{}^+Cl^-$$
$$CH_3CH_2CH_2NH_3{}^+Cl^- + NaOH \longrightarrow$$
$$CH_3CH_2CH_2NH_2 + NaCl + H_2O$$

(b) Preparation of *N*-ethylpropanamide

$$CH_3CH_2CH_2OH \xrightarrow{K_2Cr_2O_7} CH_3CH_2COOH$$
Propanoic acid

$$CH_2=CH_2 + HCl \longrightarrow CH_3CH_2Cl$$
$$CH_3CH_2Cl + NH_3 \longrightarrow CH_3CH_2NH_3{}^+Cl^-$$
$$CH_3CH_2NH_3{}^+Cl^- + NaOH \longrightarrow$$
$$CH_3CH_2NH_2 + NaCl + H_2O$$
Ethylamine

$$CH_3CH_2COOH + CH_3CH_2NH_2 \xrightarrow{Heat}$$
$$CH_3CH_2\overset{O}{\overset{\|}{C}}NHCH_2CH_3 + H_2O$$
N-Ethylpropanamide

Self-Test

1. True   2. True   3. False   4. True   5. True
6. True   7. True   8. True   9. True   10. True
11. a   12. b   13. d   14. d   15. a   16. c
17. c   18. d   19. c   20. b   21. b   22. a
23. c   24. a

Chapter 7

Practice Exercises

7.1   (a), (d), (e), (f)
7.2   (asymmetric carbons marked with asterisks)

(a) $CH_3CH_2-\overset{O}{\overset{\|}{C}}-H$ (b) $CH_3-\overset{H}{\underset{|}{\overset{O}{\overset{\|}{C}}}}\!\!-\overset{O}{\overset{\|}{C}}-H \atop CH_3$

(c) $CH_3-\overset{H}{\underset{Cl}{\overset{|}{C}^*}}-OH$ (d) $CH_3-\overset{H}{\underset{CH_2CH_3}{\overset{|}{C}^*}}-\overset{O}{\overset{\|}{C}}-H$

7.3

(a)
$$\begin{array}{c} CH_3 \\ H-\!\!\!\blacktriangleright\!\!|\!\!\blacktriangleleft\!\!-OH \\ Cl \end{array} \quad \begin{array}{c} CH_3 \\ HO-|\!\!-H \\ Cl \end{array}$$

(b)
$$\begin{array}{c} CHO \\ H-\!\!\!\blacktriangleright\!\!|\!\!\blacktriangleleft\!\!-OH \\ CH_3 \\ CHO \end{array} \quad \begin{array}{c} CHO \\ HO-\!\!\!\blacktriangleright\!\!|\!\!\blacktriangleleft\!\!-H \\ CH_3 \end{array}$$

(c)
$$\begin{array}{c} CHO \\ H-\!\!\!\blacktriangleright\!\!|\!\!-CH_3 \\ CH_3 \end{array}$$

7.4

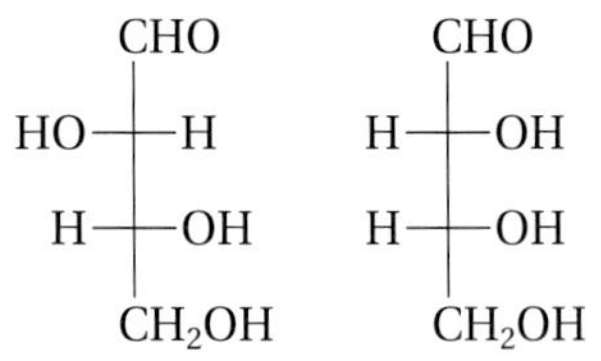

CHO   CHO
(Fischer projections)

HO—H   H—OH
H—OH   H—OH
CH₂OH   CH₂OH

7.5 (a) L, (b) D, (c) D, (d) L isomer.

7.6

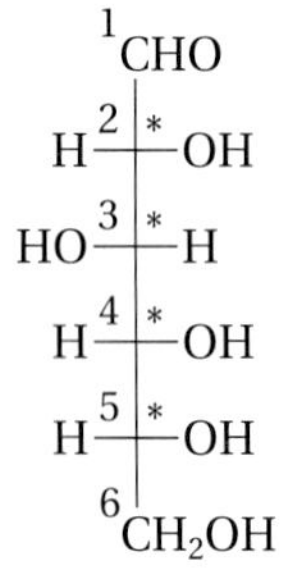

^{1}CHO
H—^{2}OH *
HO—^{3}H *
H—^{4}OH *
H—^{5}OH *
^{6}CH₂OH

7.7

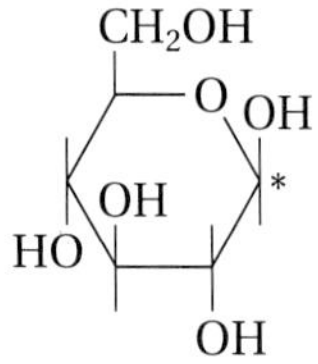

CH₂OH
O OH
OH *
HO
OH
(pyranose ring)

7.8

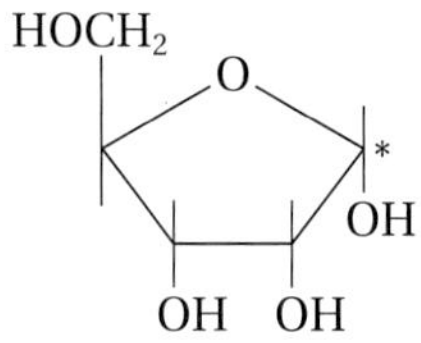

HOCH₂
O
*
OH
OH   OH
(furanose ring)

7.9

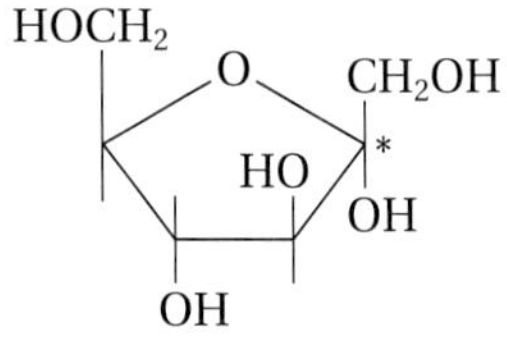

HOCH₂
O   CH₂OH
HO *
OH
OH
(furanose ring)

7.10

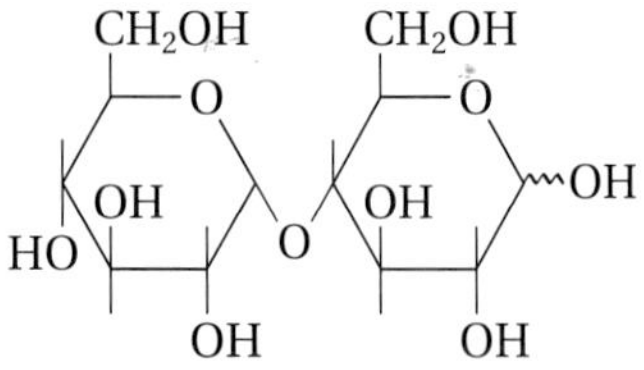

CH₂OH   CH₂OH
O   O
OH   OH   ·····OH
HO   O
OH   OH
(disaccharide)

7.11

CH₂OH   HOCH₂
O   O   OH
OH   OH
HO   OH   CH₂OH
OH   OH

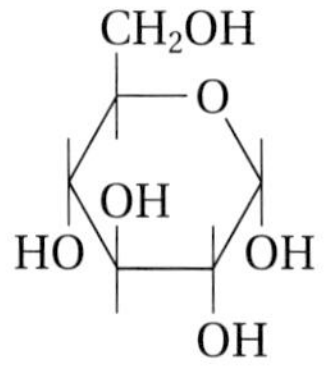

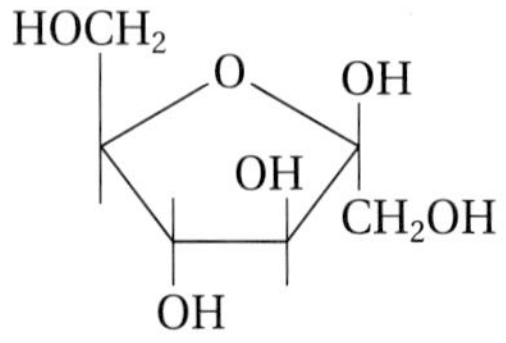

D-Glucopyranose   D-Fructofuranose

7.12 Amylose: long polymer chains of D-glucose linked
α(1 ⟶ 4). Amylopectin: polymeric chains with
α(1 ⟶ 4) and α(1 ⟶ 6) cross-links.

7.13 All four polymers yield D-glucose upon hydrolysis.

7.14 Both are disaccharides composed of D-glucose, but
the linkage between the sugars is α(1 ⟶ 4) in
maltose and β(1 ⟶ 4) in cellobiose.

7.15

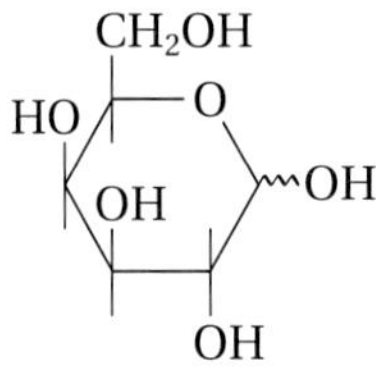

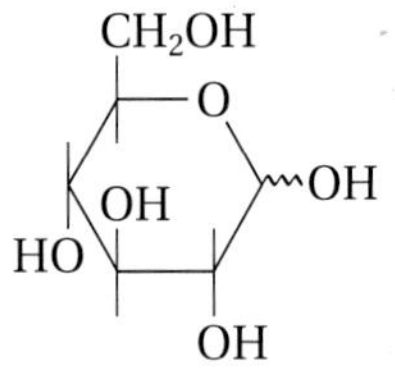

CH₂OH   CH₂OH
O   O
HO   OH   ·····OH
OH   ·····OH   HO
OH   OH

D-Galactose   D-Glucose

7.16 1 primary hydroxyl group, 6 secondary hydroxyl
groups, 2 acetal groups, and 1 phenyl group

Exercises

7.17

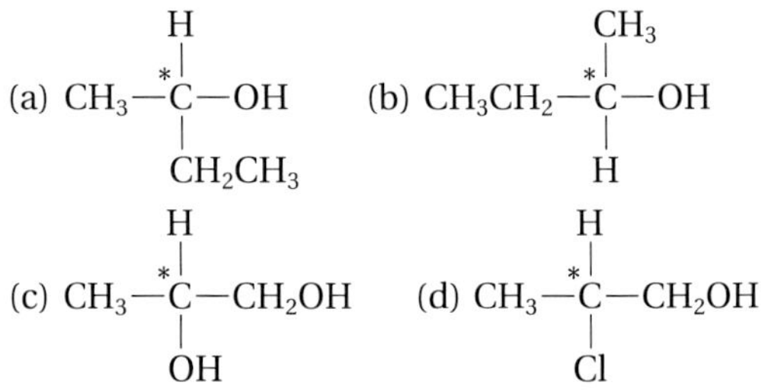

(a) CH₃—$\overset{H}{\underset{CH_2CH_3}{\overset{*}{C}}}$—OH
(b) CH₃CH₂—$\overset{CH_3}{\underset{H}{\overset{*}{C}}}$—OH

(c) CH₃—$\overset{H}{\underset{OH}{\overset{*}{C}}}$—CH₂OH
(d) CH₃—$\overset{H}{\underset{Cl}{\overset{*}{C}}}$—CH₂OH

7.19

Fischer
projection

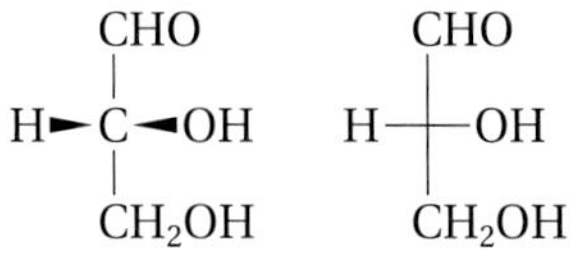

CHO   CHO
H►C◄OH   H—OH
CH₂OH   CH₂OH

D-Glyceraldehyde

Fischer
projection

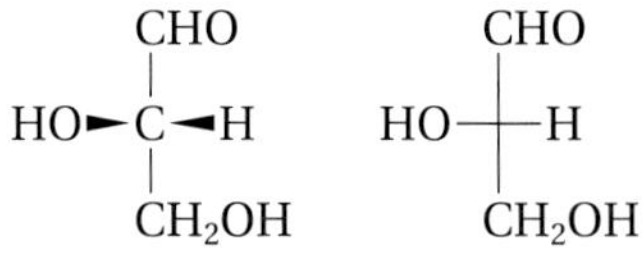

CHO   CHO
HO►C◄H   HO—H
CH₂OH   CH₂OH

L-Glyceraldehyde

7.21 (a) D-threose (an aldotetrose)   (b) D-glucose (an
aldohexose)   (c) D-fructose (a ketohexose)

7.23 (a) L   (b) D   (c) D   (d) D

7.25

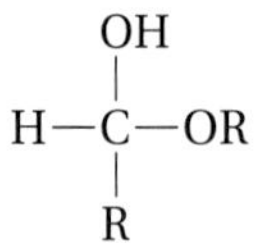

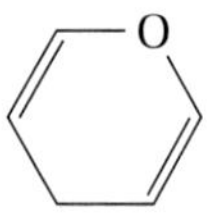

OH
H—C—OR
R

Hemiacetal

(six-membered oxygen ring)

Pyran

7.27

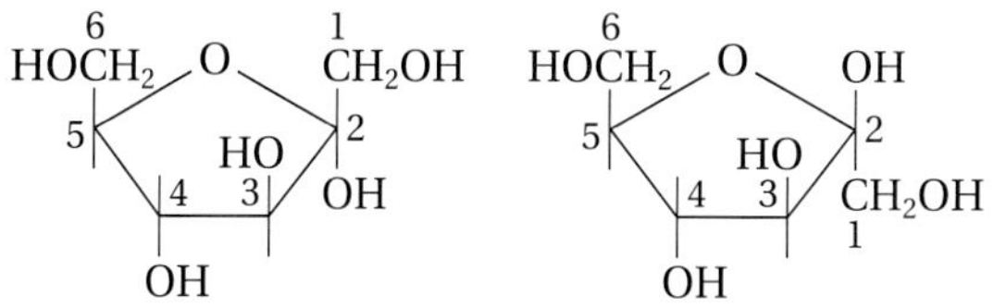

α-anomer β-anomer

7.29

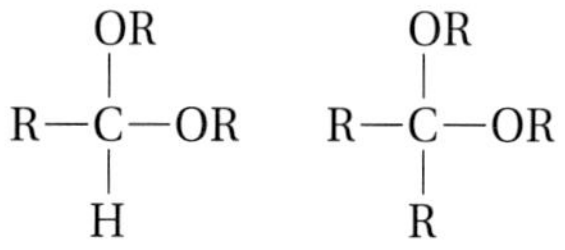

Acetal Ketal

7.31 (a) edible (b) wood and cotton (c) storage form of glucose

7.33 Starch (amylose): linear chain of glucose linked $\alpha(1 \longrightarrow 4)$. Cellulose: linear chain of glucose linked $\beta(1 \longrightarrow 4)$. Humans can digest (hydrolyze) amylose, but not cellulose.

7.35 (a) Sucrose: from juice or sap of sugar cane, sugar beets, and maple trees; sweetener and constitutes ordinary table sugar; a disaccharide, hydrolysis gives D-glucose and D-fructose. (b) Maltose (malt sugar): from hydrolysis of starch; a disaccharide; hydrolysis D-glucose.

7.37 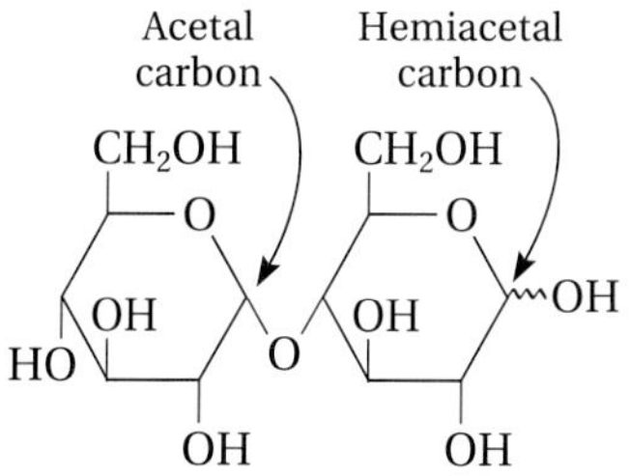
Acetal carbon, Hemiacetal carbon,
Maltose

7.39 Open-chain form of lactose is an easily reduced aldehyde group; sucrose has neither a cyclic hemiacetal group that could ring-open to form an easily oxidized aldehyde group nor a cyclic hemiketal group that could ring-open to form an easily reduced alpha-hydroxy ketone.

7.41 (a) glucose positive, starch negative (b) fructose positive, sucrose negative (c) fructose and glucose positive

7.43 (a) and (b) no assymetric carbon and not optically active (c) assymetric carbon and optically active (d) asymmetric carbon and optically active

7.45 α anomer β anomer

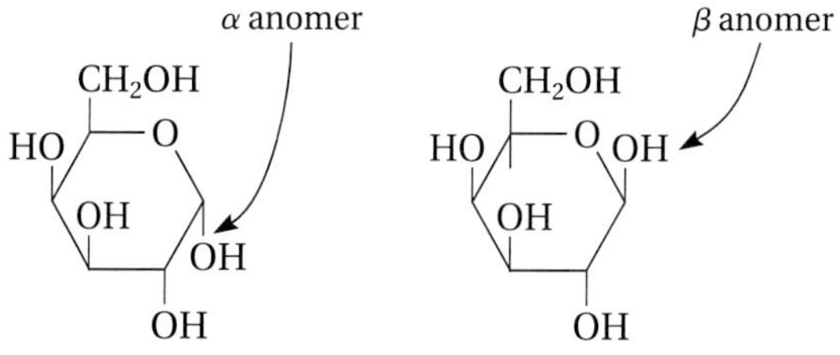

7.47 (a) 3 (b) 2 (c) 4 (d) 3 (e) 4 (f) 4
7.49 (a) L (b) D

Self-Test
1. True 2. True 3. True 4. False 5. True
6. True 7. True 8. True 9. False 10. True
11. b 12. c 13. b 14. c 15. a 16. d
17. d 18. a 19. d 20. a 21. a 22. d
23. b 24. c 25. b 26. d

Chapter 8

Practice Exercises
8.1 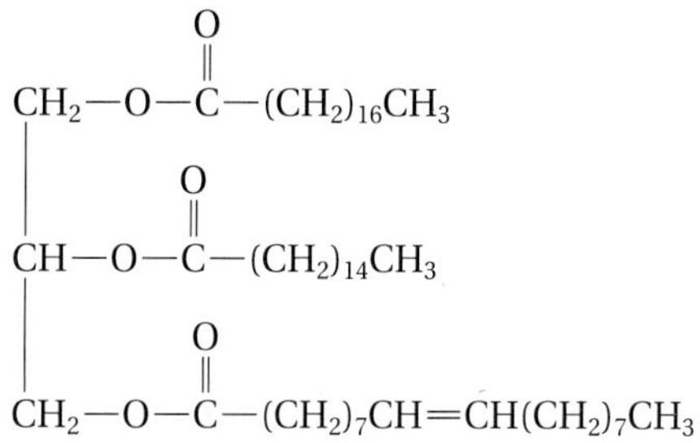
8.2 Tripalmitin
8.3 Tristearin (glyceryl tristearate)
8.4 Sodium oleate (2 mols) Sodium stearate (1 mol) Glycerol (1 mol)
8.5 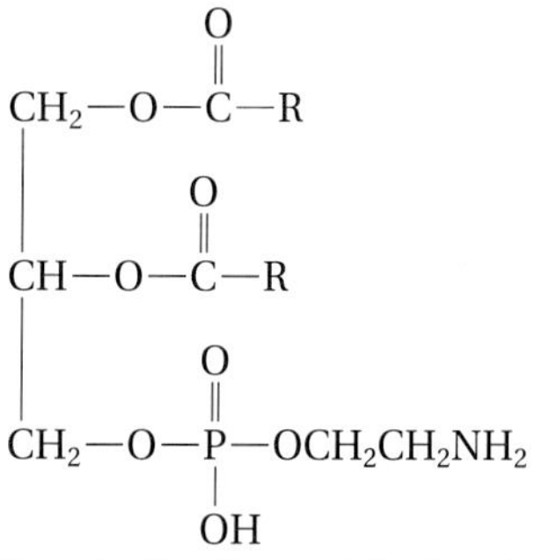
8.6 No micelles formed; hydrophobic tails of phosphatidyl choline molecules would not aggregate to exclude water.
8.7 Polar —OH group on one end and the remainder of the molecule is hydrophobic.
8.8 (a) Aldosterone: 2 ketone carbonyl groups, 1 aldehyde group, 1 C—C double bond, 1 secondary hydroxyl group, and 1 primary hydroxyl group. (b) Cortisone: 3 ketone carbonyl groups, 1 primary hydroxyl group, 1 tertiary hydroxyl group, and 1 C—C double bond.

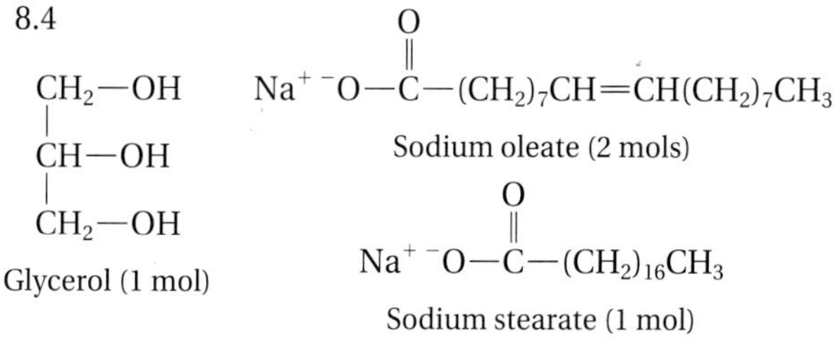

Exercises

8.9

$$\underset{\text{RCOR}}{\overset{\overset{\textstyle O}{\|}}{}}$$

8.11 (a) solid (mp 44 °C); (b) liquid (mp −5 °C); (c) solid (mp 63 °C); (d) liquid (mp 14 °C).

8.13 Triester made from 1 molecule of glycerol and 3 molecules of fatty acids.

8.15 (a) Carbon-carbon double bonds of fatty acids become saturated; (b) used to make butter and lard substitutes.

8.17

$$\underset{\text{CH}_3(\text{CH}_2)_{14}\text{CO}^-\text{Na}^+}{\overset{\overset{\textstyle O}{\|}}{}}$$

8.19 Phospholipids, glycolipids

8.21 Sphingomyelins: 1 fatty acid, sphingosine, phosphoric acid, and choline. Phosphoglycerides: 2 fatty acids, glycerol, and phosphoric acid.

8.23 Micelle: hydrophobic tails of the lipids point to interior of the aggregate, away from water; polar or ionic heads face water. Liposome: hydrophobic tails of the lipids form a bilayer; polar or ionic heads face solvent water both inside and outside of the liposome.

8.25 Membrane flexibility increases in proportion to the amount of unsaturated fatty acids in membrane lipids.

8.27 Facilitate the transport of ions and molecules into and out of cells.

8.29

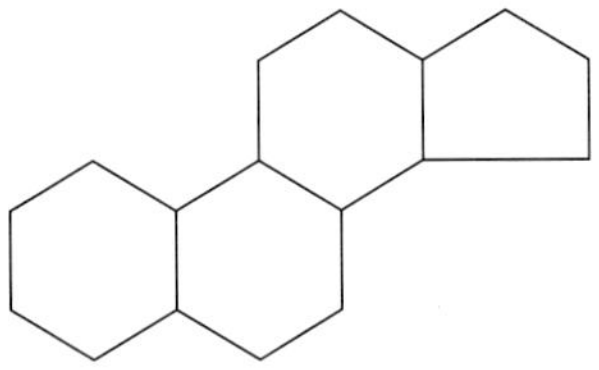

8.31 Chemical messengers produced by endocrine glands; carried through blood to sites where they produce dramatic physiological effects.

8.33 To reduce body's immunity to the introduction of an organ.

8.35 (a) testosterone   (b) estrone, estradiol, and progesterone

8.37 Increase muscle mass; can cause testicular atrophy and cancer.

8.39 Degraded before it reaches the ovaries.

8.41 Contains an attached trisaccharide.

8.43

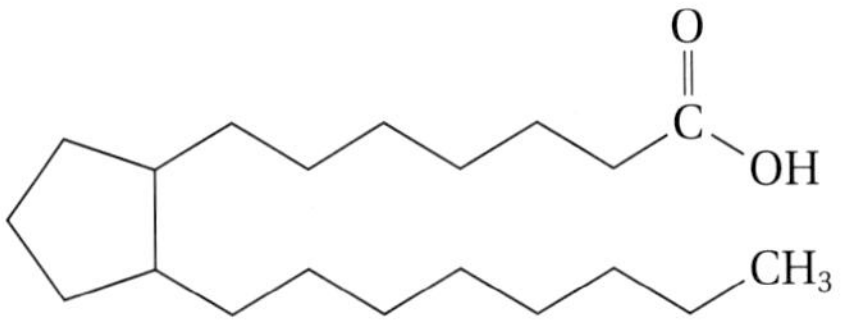

8.45 Control of acid secretions in stomach, relaxation and contraction of smooth muscle, vascular permeability, inflammation, and body temperature.

8.47 Phospholipids and glycolipids

8.49

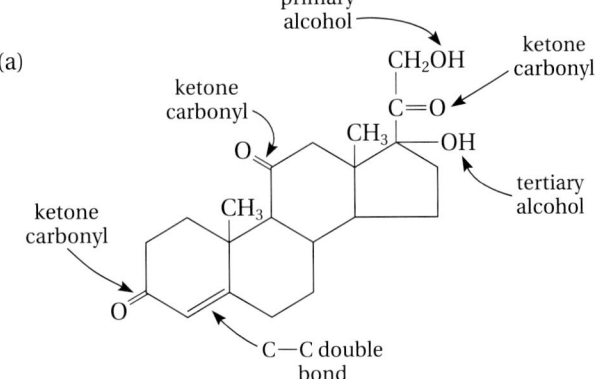

(a)

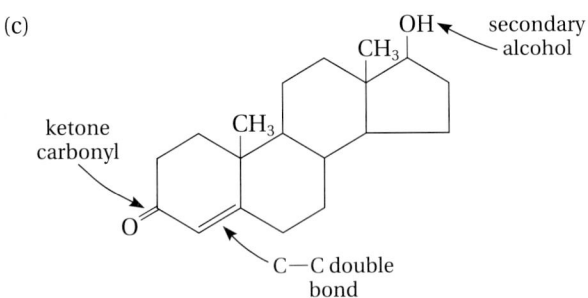

(b)

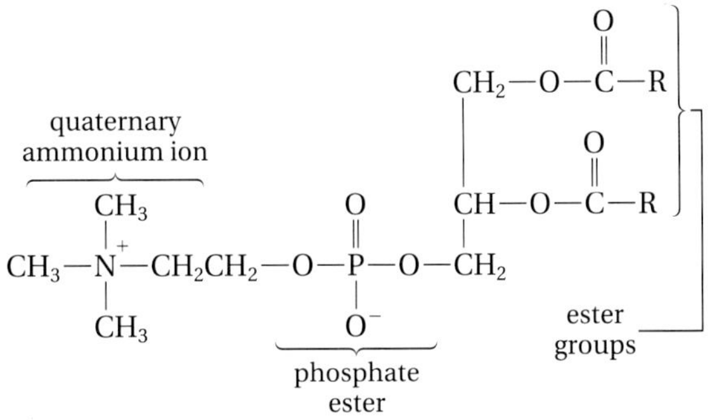

(c)

(d)

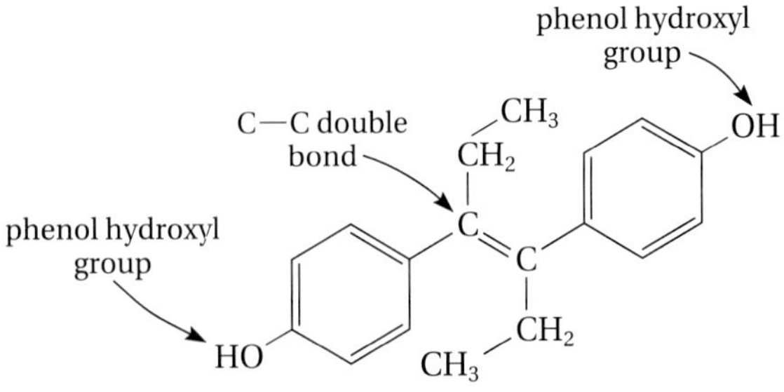

(e)

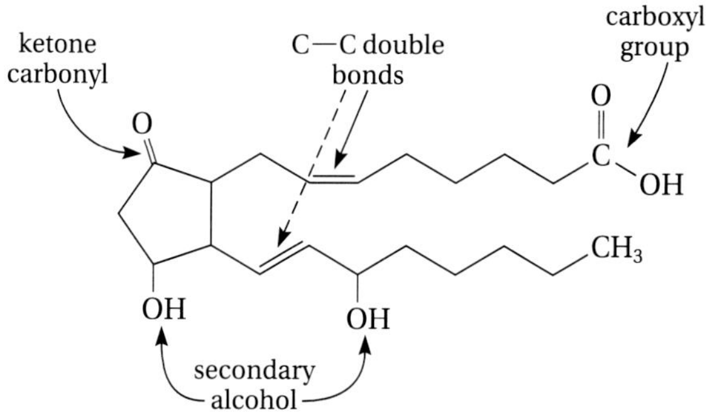

ketone carbonyl

C—C double bonds

carboxyl group

ketone carbonyl, C—C double bonds, O, C, OH, CH$_3$, OH, OH, secondary alcohol

(f)

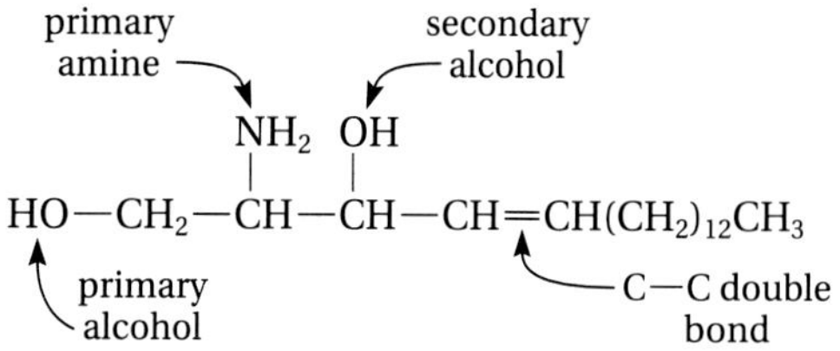

primary amine → NH$_2$ secondary alcohol → OH

HO—CH$_2$—CH—CH—CH=CH(CH$_2$)$_{12}$CH$_3$

primary alcohol

C—C double bond

(g)

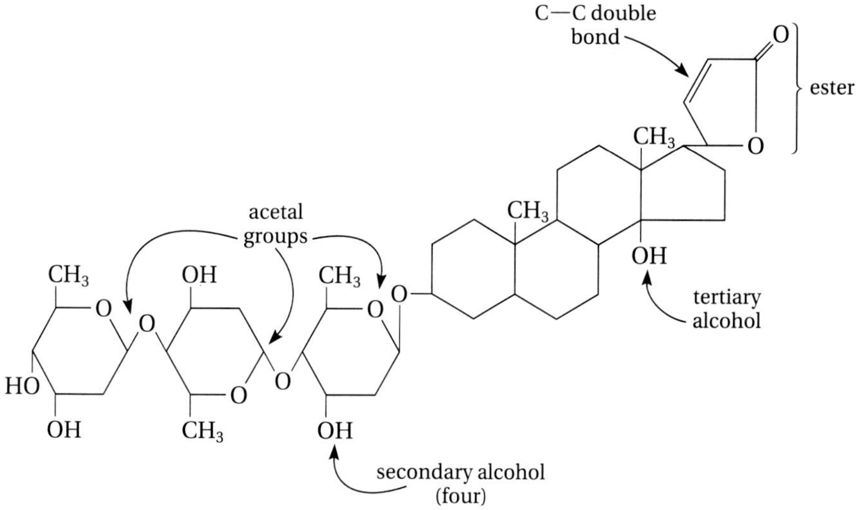

C—C double bond

ester

CH$_3$

CH$_3$ OH   CH$_3$ O

tertiary alcohol

acetal groups

CH$_3$ OH   CH$_3$

HO

OH   CH$_3$ OH

secondary alcohol (four)

(h)

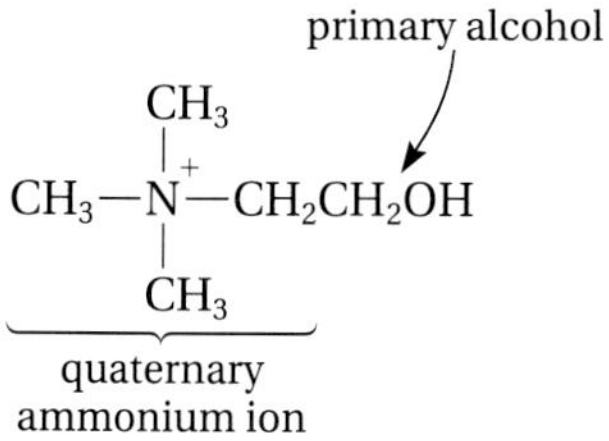

primary alcohol

CH$_3$

CH$_3$—N$^+$—CH$_2$CH$_2$OH

CH$_3$

quaternary ammonium ion

(i)

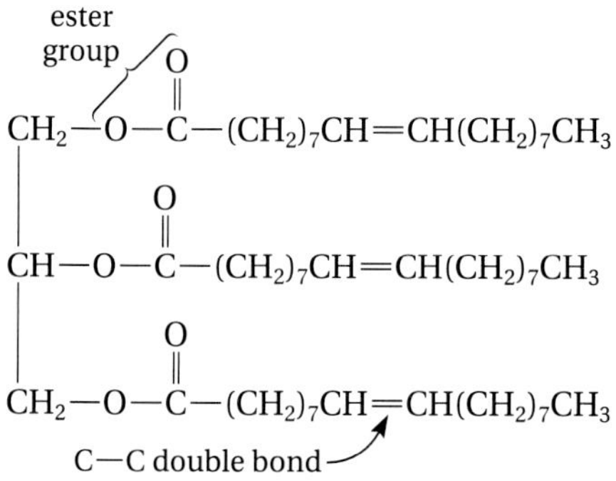

ester group   O

CH$_2$—O—C—(CH$_2$)$_7$CH=CH(CH$_2$)$_7$CH$_3$

O

CH—O—C—(CH$_2$)$_7$CH=CH(CH$_2$)$_7$CH$_3$

O

CH$_2$—O—C—(CH$_2$)$_7$CH=CH(CH$_2$)$_7$CH$_3$

C—C double bond

8.51 Hydrocarbon chains of saturated fatty acids fit better into crystals than hydrocarbon chains of unsaturated fatty acids.

Self-Test

1. True   2. True   3. False   4. False   5. True
6. False   7. False   8. False   9. True   10. True
11. c   12. d   13. c   14. b   15. a   16. c
17. d   18. b   19. a   20. d   21. c   22. d
23. b   24. b   25. c   26. a

Chapter 9

Practice Exercises

9.1   (a) aliphatic   (b) basic   (c) hydroxylic
(d) aromatic

9.2   (a) L-Cysteine   (b) D-Cysteine

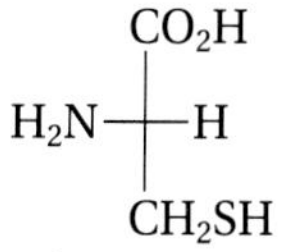

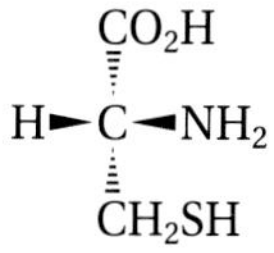

CO$_2$H

H$_2$N—H

CH$_2$SH

CO$_2$H

H—C—NH$_2$

CH$_2$SH

(c) L-isomer

9.3 (a)

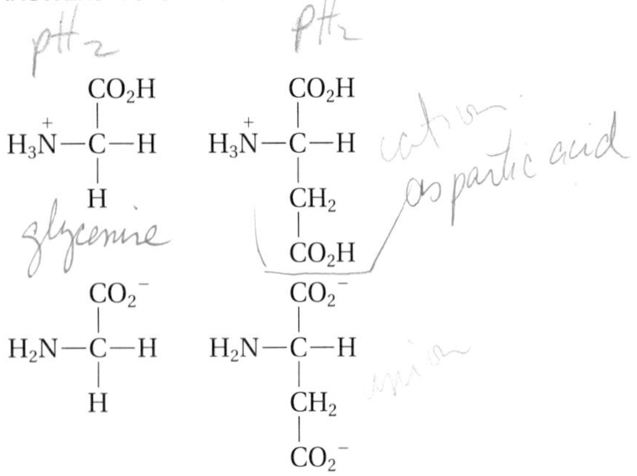

pH 2 *pH 2* (handwritten)

CO_2H CO_2H

$H_3\overset{+}{N}-C-H$ $H_3\overset{+}{N}-C-H$

H CH_2

glycine (handwritten) CO_2H

cation *aspartic acid* (handwritten)

(b)

CO_2^- CO_2^-

H_2N-C-H H_2N-C-H

H CH_2

    CO_2^-

anion (handwritten)

9.4 (a)

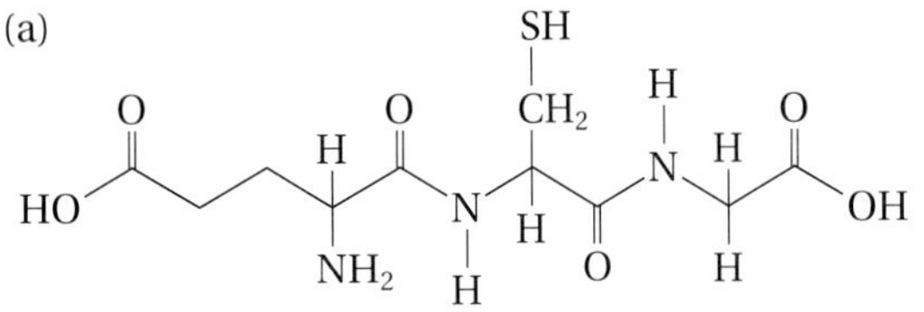

(b) Glu-Cys-Gly

9.5 Glu-Cys-Gly; Glu-Gly-Cys; Cys-Gly-Glu; Cys-Glu-Gly; Gly-Glu-Cys; Gly-Cys-Glu.

9.6

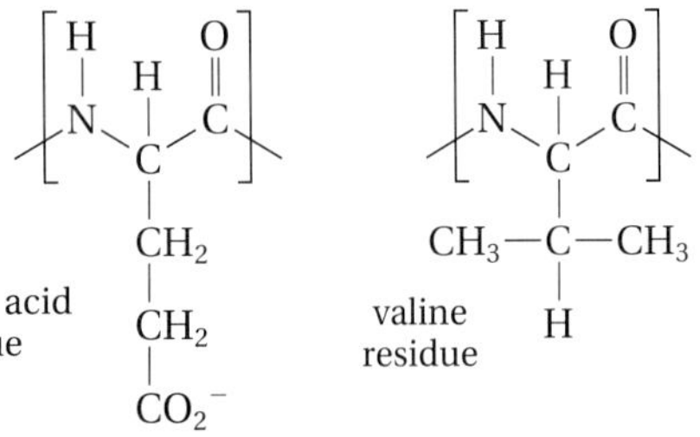

glutamic acid residue   valine residue

Side chain of glutamic acid residue has a carboxyl group dissociated at pH 7. Side chain of valine residue is uncharged at all values of pH and is hydrophobic.

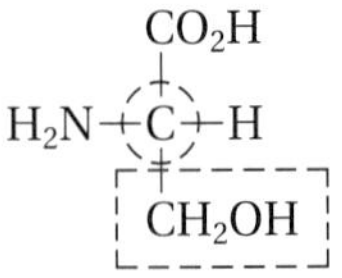

Exercises

9.7 (Alpha carbons circled and side chain R groups boxed)

(a)
CO_2H
H_2N-C-H
CH_3

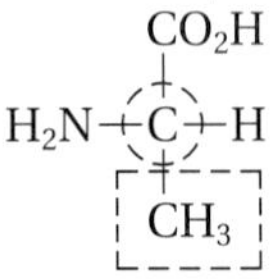

(b)
CO_2H
H_2N-C-H
CH_2OH

(c)
CO_2H
H_2N-C-H
CH_2
CH_2
CO_2H

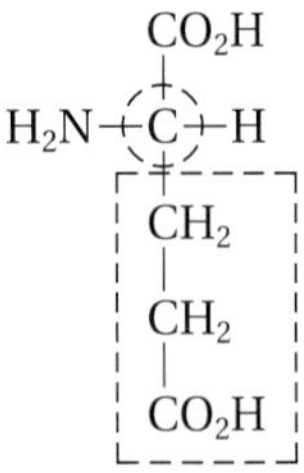

9.9 (a) D-Serine 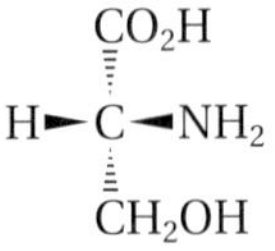 (b) L-Alanine

CO_2H CO_2H

$H-C-NH_2$ H_2N-C-H

CH_2OH CH_3

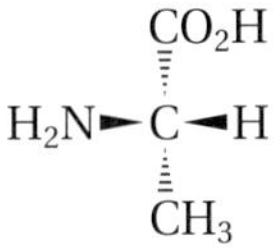

9.11 The internal salt of an amino acid.

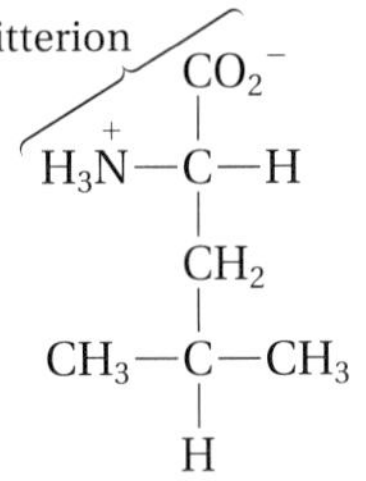

zwitterion

CO_2^-

$H_3\overset{+}{N}-C-H$

CH_2

CH_3-C-CH_3

H

9.13 The pH at which the positive and negative charges on an amino acid balance each other, giving a net charge of zero.

9.15 Portion of an amino acid that is incorporated into a peptide chain.

9.17 (a)

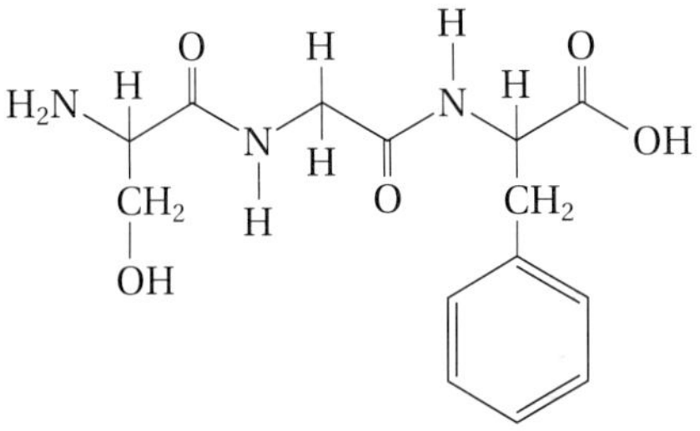

(b) Ser-Gly-Phe   (c) Two peptide bonds

9.19 (a) Stimulates milk ejection in females, contraction of the uterus in labor, and feelings of satisfaction. (b) Antidiuretic in both sexes.

9.21 The *primary structure* of a protein is the sequence of amino acid residues in a protein.

9.23 Alpha helix, beta pleated sheet, and collagen helix.

9.25 Major structural protein of the body, found in skin, bone, teeth, cartilage, and tendon.

9.27 A *conjugated protein* contains an organic non-protein portion called a prosthetic group.

9.29 Salt bridges, hydrogen bonds, hydrophobic aggregation, and disulfide bridges.

9.31 (a) Iron(II) (Fe^{2+}) (b) Iron(III) ion (Fe^{3+})

9.33 They bind to the heme iron of hemoglobin, preventing oxygen from binding.

9.35 Sickle cell hemoglobin (HbS) in its deoxygenated form is less soluble than normal adult hemoglobin (HbA). Precipitation of HbS causes red blood cells to sickle and burst.

9.37 High oxygen pressure used in hyperbaric oxygenation keeps sickle cell hemoglobin (HbS) oxygenated.

9.39 Subunits (quaternary structure) dissociate; secondary and tertiary structure unfolds.

9.41 The correct matches are as follows.
(a) (3) (b) (2) (c) (5) (d) (8) (e) (7)
(f) (1) (g) (6) (h) (4)

9.43 His-Met-Glu; His-Glu-Met; Met-His-Glu; Met-Glu-His; Glu-Met-His; Glu-His-Met.

9.45 (a)

CO_2H
$H_2N-\overset{|}{\underset{|}{C}}-H$
CH_2
SH
Cysteine

CO_2H
$H_2N-\overset{|}{\underset{|}{C}}-H$
H
Glycine

CO_2H
$H_2N-\overset{|}{\underset{|}{C}}-H$
CH_3
Alanine

(b)

CO_2H
$H_2N-\overset{|}{\underset{|}{C}}-H$
CH_2
CH_2
CO_2H
Glu

CO_2H
$H_2N-\overset{|}{\underset{|}{C}}-H$
CH_2
CH
CH_3 CH_3
Leu

CO_2H
$H_2N-\overset{|}{\underset{|}{C}}-H$
$CH_3-\overset{|}{\underset{|}{C}}-CH_3$
H
Val

Pro

CO_2H

(c)

CO_2H
$H_2N-\overset{|}{\underset{|}{C}}-H$
CH_2
OH
Ser

CO_2H
$H_2N-\overset{|}{\underset{|}{C}}-H$
CH_2
His

CO_2H
$H_2N-\overset{|}{\underset{|}{C}}-H$
CH_2
Phe

CO_2H
$H_2N-\overset{|}{\underset{|}{C}}-H$
CH_2
OH
Tyr

CO_2H
$H_2N-\overset{|}{\underset{|}{C}}-H$
CH_2
Trp

9.47 (a) Ionic bond formed between the negatively charged carboxylate ion of the side chains of aspartate or glutamate and the positively charged amino side chains of lysine or arginine. (b) Nonprotein group attached to a protein by covalent bonds. (c) Folding of a polypeptide chain into a relatively stable three-dimensional shape. (d) A protein with a more or less spherical shape. (e) Peptide chains arranged side by side to form a structure that resembles pleats. (f) Specific, repeating patterns of folding of the peptide backbone of a protein. (g) Coiling of the peptide backbone of a protein into a spiral shape resembling a corkscrew. (h) Order in which the amino acids of a peptide or protein are linked by peptide bonds.

9.49 (a) (6) (b) (4) (c) (7) (d) (9) (e) (1)
(f) (2) (g) (3) (h) (5) (i) (8)

Self-Test
1. True 2. False 3. True 4. True 5. True
6. False 7. True 8. True 9. True 10. False
11. c 12. c 13. b 14. a 15. d 16. d
17. a 18. a 19. a 20. b 21. c 22. b
23. a 24. c 25. c 26. b 27. b 28. b

Chapter 10

Practice Exercises
10.1 (a) lipase (breakdown of lipids) (b) cellulase (breakdown of cellulose)
10.2 (a) oxidase (b) hydrolase
10.3 When a substrate is bound at the active site of an enzyme, (a) there is a high concentration of substrate and enzyme, and (b) the substrate is "locked" in the ideal position for bond-breaking or bond-making processes to occur.
10.4 Pancreatic disorder
10.5 Enzyme assays measure the activity of an enzyme. The activity of an enzyme depends upon the pH and the temperature.
10.6

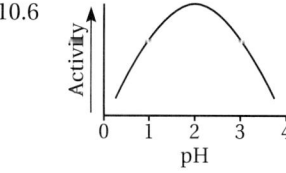

10.7 A positive modulator increases the activity of the enzyme; a negative modulator decreases the activity of an enzyme.
10.8 An enzyme cofactor takes part in the chemical reaction that the enzyme catalyzes; a positive modulator activates the enzyme but does not take part in the reaction.

10.9 (a) bonds 2, 3, 4   (b) bonds 1, 5
10.10 The conversion of fibrinogen to fibrin is catalyzed by thrombin. Active thrombin must be made from its inactive zymogen, prothrombin.

Exercises

10.11 A biological catalyst that speeds up chemical reactions.
10.13 (a) hydrolysis of esters   (b) hydrolysis of lactose (c) removal of hydrogen   (d) hydrolysis of amylose
10.15 The active site of the enzyme is the "lock" that the substrate "key" must fit.
10.17 The location where the action takes place.
10.19 To form an active enzyme.
10.21 Enzyme activity measures the speed at which an enzyme changes substrate to product(s).
10.23 An enzyme lowers the energy of activation.
10.25 (a) activity decreases   (b) denatured irreversibly, all activity lost.
10.27 An enzyme produced in response to a temporary need of the cell.
10.29 Competes with the substrate for the active site of the enzyme.
10.31 allosteric enzyme
10.33 inactive form of an enzyme
10.35 Refer to the drawing in Section 10.12 of your text.
10.37 Penicillin inhibits an enzyme and as a consequence the bacterial cell walls are defective.
10.39 enzyme inhibition
10.41 Ability for an enzyme to catalyze one chemical reaction with only one substrate.
10.43 A zymogen is an inactive precursor of an enzyme (Table 10.2).
10.45 See Section 10.10 of the text.

Self-Test

1. False   2. False   3. True   4. False   5. True
6. True   7. True   8. False   9. True   10. False
11. d   12. b   13. b   14. a   15. d   16. c
17. c   18. a   19. b   20. c   21. c   22. d

Chapter 11

Practice Exercises

11.1

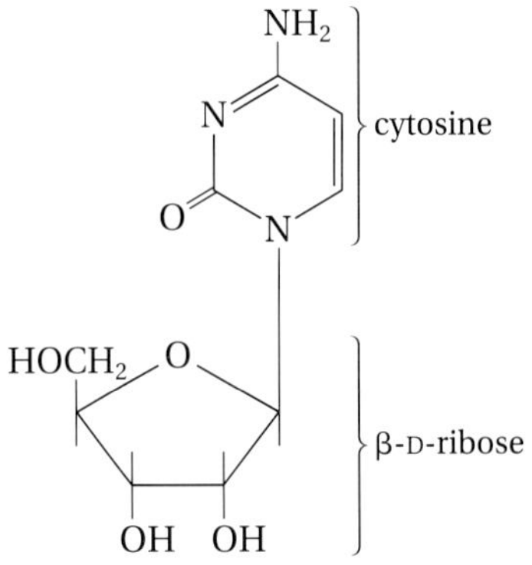

Cytidine

11.2

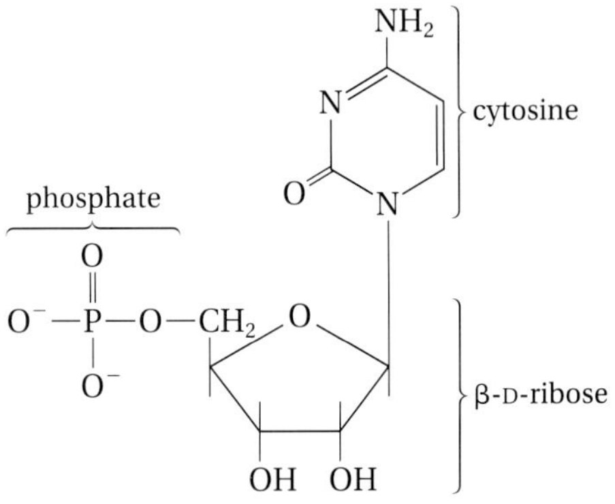

Cytidine 5′-monophosphate (CMP)

11.3 (a) adenine-cytosine-guanine-uracil (from RNA)
(b) thymine-adenine-cytosine-guanine (from DNA)
11.4 A phosphodiester bridge links the 3′-hydroxyl group of one nucleotide to the 5′-hydroxyl group of another nucleotide in DNA and RNA.
11.5 You should include the following ideas: (1) hydrogen bonding, (2) complementary base pairing, and (3) antiparallel strands.
11.6 d T=A
          C≡G
          C≡G
          C≡G
          A=T
          A=T
          G≡C d
11.7   3′   5′
      RNA: U—A—G—C—U—U
11.8 5′   3′
      G—C—U—G—G—U—U—C—U
11.9 The amino acid sequence of the tetrapeptide is Arg-Arg-Asn-Thr.

11.10 When the sequence of bases on DNA is altered through the deletion, addition, or substitution of a single nucleotide, the DNA is faulty. Faulty DNA leads to gene mutations. Faulty DNA gives faulty mRNA that translates into proteins with the wrong amino acid sequence or no protein at all.

11.11 (a) The pentapeptide is Thr-Val-Ser-Glu-Pro. (b) The mutation leads to the pentapeptide Thr-Asp-Ser-Glu-Pro.   (c) Deletion of uracil gives the tetrapeptide Thr-Ala-Ala-Ser.

11.12 Answers will depend on personal opinions.

Exercises

11.13 Deoxyribonucleic acid, DNA, and ribonucleic acid, RNA.

11.15 A nucleoside is composed of a nitrogen base and a sugar unit. Adding a phosphate group to a nucleoside forms a nucleotide.

11.17 (a) adenine, guanine, thymine, and cytosine
(b) Adenine and guanine are derivatives of purine thymine and cytosine are derivatives of pyrimidine

11.19 (a)

Deoxyadenosine

(b) deoxyadenosine.

11.21 Nucleic acids can lose a proton from the phosphate group of their phosphodiester bridges, making them highly acidic.

11.23 Hydrogen bonds between complementary pairs of bases hold the two DNA strands together in the double helix. Adenine and thymine form one base pair and cytosine and guanine form the other base pair.

11.25 Antiparallel strands run in different directions. One DNA strand runs in the 3′ to 5′ direction and the complementary DNA strand runs in the 5′ to 3′ direction.

11.27 (a) *replication*   (b) *transcription*
(c) *translation*

11.29 Replication involves the following steps: (1) unwinding the original double helix, (2) synthesis of complementary strands in the 5′ to 3′ direction,

(3) joining together the segments of 3′ to 5′ strand, and (4) assembly of two identical DNA double helixes.

11.31 DNA polymerase catalyzes the joining of nucleotides to form complementary strands of DNA.

11.33 A gene is a segment of a DNA molecule that codes for the synthesis of one kind of protein molecule; a polypeptide is a chain of amino acids that is formed when the genetic message is translated in protein synthesis.

11.35 (a) rRNA is a structural component of a ribosome. (b) mRNA carries the genetic information from DNA to the ribosome, where the message is translated in protein synthesis.

11.37 (1) DNA template, (2) RNA polymerase, (3) nucleoside triphosphates containing the nitrogen bases adenine, guanine, cytosine, and uracil

11.39 (a) A codon is a base triplet on mRNA.   (b) An anticodon is a base triplet on tRNA.

11.41 Written from 5′ to 3′ the anticodon for methionine is CAU.

11.43 (a) 3′ U—C—G—A—C—C—C—U—G 5′
(b) Val-Pro-Ala

11.45 Substitution, addition, or deletion of one or more nucleotides in the DNA molecule.

11.47 Increased ability to manipulate genes and genetic material. Answers will vary.

11.49 (a) 10   (b) 6   (c) 5   (d) 7   (e) 8   (f) 2
(g) 4   (h) 9   (i) 3   (j) 1

11.51 (a) One correct answer is 5′ UUU—UCU—GCU—CAU 3′   (b) 5′ *d*A—T—G—A—G—C—A—G—A—A—A—A 3′

11.53 rRNA: A structural component of ribosomes.
mRNA: Carries information from DNA to the ribosome to direct protein synthesis and contains the codons.
tRNA: Carries amino acids to the ribosome. The anticodon on tRNA base pairs with a codon of mRNA to ensure that the correct amino acid is incorporated into the peptide chain.

11.55 ATC

11.57 Answers will vary. A change of a single base could result in premature termination of protein synthesis. Insertion or deletion of a base will cause a change in the codons and result in a nonsense protein or no protein.

Self-Test
1. False   2. True   3. False   4. False   5. True
6. False   7. False   8. False   9. True
10. False   11. c   12. d   13. a   14. d
15. b   16. b   17. b   18. c   19. a   20. d
21. a

Chapter 12

Practice Exercises

12.1   Protein breakdown in the stomach is catalyzed by the enzyme pepsin, which works best at very low pH. Antacids neutralize stomach acid, raising the pH and lowering the catalytic activity of pepsin.

12.2   The enzyme amylase in saliva catalyzes the breakdown of carbohydrates to maltose. Maltose, sucrose, and lactose are hydrolyzed to give monosaccharides in the small intestines, where they pass into cells of the intestinal wall and then into the bloodstream.

12.3   Lipase catalyzes the hydrolysis of lipids in the small intestines. The products of lipid hydrolysis pass into the intestinal cells, where they are reassembled into lipid molecules. Insoluble lipids interact with lipoproteins to form chylomicrons, which then pass into the bloodstream.

12.4 8.9×10^{19} molecules niacin

12.5   (a) 4   (b) 6   (c) 1   (d) 8   (e) 2   (f) 10
       (g) 3   (h) 9   (i) 11   (j) 5   (k) 7

Exercises

12.7   amylase; catalyzes the hydrolysis of starch and glycogen. Mucin is a glycoprotein.

12.9   pepsinogen

12.11 It catalyzes the hydrolysis of peptide bonds.

12.13 chymotrypsin, trypsin, carboxypeptidase, maltase, sucrase, lactase, lipases, nucleotidases

12.15 (a) a liquid consisting of cholesterol, bile salts, and bile pigments   (b) in the liver   (c) to emulsify fats so they can be hyrolyzed by lipase action

12.17 fats

12.19 Essential amino acids cannot be synthesized by the body; nonessential amino acids can be synthesized by the body.

12.21 (a) eggs, dairy products, kidney, and liver
       (b) corn, wheat, and gelatin

12.23 vitamin A, vision; vitamin D, promotion of calcium and phosphorus uptake; vitamin E, antioxidant; vitamin K, blood clotting.

12.25 They are not stored in the body and are readily excreted.

12.27 calcium (Ca), phosphorus (P), potassium (K), sulfur (S), chlorine (Cl), sodium (Na), magnesium (Mg)

12.29 sodium ion (Na^+), potassium ion (K^+), and chloride ion (Cl^-)

12.31 If phosphorus is too high, the excess is eliminated from the body as calcium phosphate. This leads to a calcium deficiency.

12.33 A deficiency in children causes rickets; in adults a deficiency causes osteoporosis.

12.35 bowed legs, pigeon breast, and poor tooth development

12.37 iron (Fe), zinc (Zn), copper (Cu), manganese (Mn), cobalt (Co), chromium (Cr), selenium (Se), iodine (I), and molybdenum (Mo)

12.39 animal organs, such as liver and kidney, and green vegetables

12.41 for the formation of hemoglobin; deficiency can lead to anemia

12.43 copper; Wilson's disease

12.45 An iodine deficiency causes the thyroid gland to grow large—a goiter—as it tries to meet the body's demand for thyroid hormones.

12.47 Fluoride ions make teeth and bones stronger. Excess of fluoride makes teeth mottled.

12.49 leafy green vegetables; produced by intestinal bacteria

12.51 (a) amylase   (b) pepsinogen or pepsin
       (c) trypsinogen or trypsin, chymotrypsinogen or chymotrypsin, procarboxypeptidase or carboxypeptidase, amylase, maltase, sucrase, lactase, lipase, nucleotidases

12.53 See Section 12.4.

12.55 vitamin C and the B vitamins

12.57 (a) part of vitamin B_{12} (b) trace element for growth   (c) hemoglobin formation   (d) enzyme cofactor   (e) functioning of the CNS   (f) nerve impulse transmission, muscle contraction, and bond formation

Self-Test

1. False   2. True   3. True   4. False   5. False
6. False   7. True   8. False   9. True   10. True
11. a   12. c   13. a   14. c   15. b   16. c
17. c   18. d   19. b   20. a   21. c   22. a
23. d

Chapter 13

Practice Exercises

13.1   Blood: carries food, oxygen, hormones, and antibodies and removes wastes. The lymphatic system traps bacteria and other foreign substances, produces antibodies, and transports lipids to the bloodstream.

13.2   Carbonic anhydrase, $H_2CO_3 \rightleftharpoons CO_2 + H_2O$; blood would be more acidic.

13.3   The pH would increase.

13.4   Carbon dioxide/carbonic acid/bicarbonate ion equilibrium shifts to the right, the HCO_3^-/H_2CO_3 ratio is restored to 20:1, and blood pH decreases.

Exercises

13.5   42 L water in 70-kg person: 28 L intracellular, 11 L interstitial, and 3 L in blood plasma.

13.7   Intracellular fluids are high in K^+ and HPO_4^{2-}; interstitial fluids are high in Na^+ and Cl^-.

13.9   carrying food, oxygen, wastes, hormones, antibodies; distributing body heat; maintaining pH

13.11   erythrocytes, carry oxygen; leukocytes, fight bacteria; thrombocytes, role in blood clotting

13.13   An antigen is a foreign particle; an antibody is formed by the immune response to antigen.

13.15   identification of infecting organisms

13.17   7.40

13.19 H_2CO_3/HCO_3^-; $HPO_4^{2-}/H_2PO_4^-$; —COOH/—COO⁻ and —NH_2/—NH_3^+

13.21   as oxyhemoglobin, HbO_2

13.23   It would increase.

13.25   carbaminohemoglobin, dissolved in blood, bicarbonate ion

13.27   Chloride ions move from plasma into red blood cells and bicarbonate ions move from red blood cells into the plasma.

13.29   filtration of wastes from bloodstream and maintenance of blood pH

13.31   water, glucose, and electrolytes

13.33   Carbonic acid dissociates into bicarbonate ions and hydrogen ions; hydrogen ions are exchanged for sodium ions in developing urine; sodium ions and bicarbonate ions diffuse into the bloodstream.

13.35   hyperventilation

13.37   vasopressin and aldosterone

13.39   aldosterone, conserves salt and water in the body

13.41   decrease

13.43   Addition of H^+ to oxyhemoglobin releases the oxygen.

13.45   hypoventilation, decreases HCO_3^-/H_2CO_3 ratio; hyperventilation, increases HCO_3^-/H_2CO_3 ratio

13.47   The kidneys are the source of bicarbonate ions.

Self-Test

1. True   2. False   3. False   4. True   5. True
6. True   7. True   8. False   9. True   10. False
11. c   12. d   13. b   14. d   15. b   16. c
17. b   18. a   19. a   20. a   21. d   22. b
23. b   24. d

Chapter 14

Practice Exercises

14.1   (a) porphyrin ring   (b) vitamin B_{12}, cobalt; hemoglobin, iron; and chlorophyll, magnesium

14.2   Plants in photosynthesis produce complex molecules that all animals need for survival. Animals cannot make these essential energy-rich compounds.

14.3   14 kcal per mole of ATP

14.4   21 kcal free energy

14.5   A cell needs ATP to grow. Without nutrients the cell cannot make ATP and dies.

14.6   (a) $CH_3CH_2COOH + NADH + H^+$
        (b) $HOOCCH=CHCOOH + FADH_2$
        (c) $CH_3CHO + NADH + H^+$

14.7 $CH_3CH_2OH + 2NAD^+ + H_2O \longrightarrow$
        $CH_3COOH + 2NADH + H^+$

14.8   Energy for muscle contraction comes from the hydrolysis of ATP. With heavy exercise, the demand for ATP exceeds the supply and muscles become tired and weak.

Exercises

14.9   all chemical reactions in living organisms

14.11 $$6CO_2 + 6H_2O \xrightarrow[\text{energy}]{\text{Light}} C_6H_{12}O_6 + 6O_2$$
        carbon dioxide   water   glucose   oxygen

14.13   Energy-poor carbon dioxide is converted into energy-rich glucose.

14.15   catabolism

14.17   Carbon dioxide is an oxidized compound of carbon; it is energy-poor. Glucose is a reduced compound of carbon; it is energy-rich.

14.19   Structure of ATP (adenosine triphosphate)

14.21   Unprotonated ATP has a charge of 4−; its molar mass is 504.

14.23   To form ATP requires 7 kcal/mol of free energy. Since some energy is always lost as heat, any reaction used to make ATP must release more than 7 kcal/mol.

14.25   Aerobic cells need oxygen to live. Energy is conserved as "reducing power" by formation of NADH and $FADH_2$; NADH and $FADH_2$ reduce oxygen to water and produce energy; energy is used to convert ADP to ATP.

14.27 (a) $NAD^+ + H^+ + 2e^- \longrightarrow NADH$
(b) $FAD + 2H^+ + 2e^- \longrightarrow FADH_2$

14.29 $CH_3OH + NAD^+ \longrightarrow$

methanol

$HCHO + NADH + H^+$

methanal
(formaldehyde)

NAD^+ is reduced to NADH.

14.31 (a) oxygen is reduced to water and energy is released (b) energy released is used to convert ADP to ATP.

14.33 inside mitochondria

14.35 cytochromes; the iron in cytochromes changes back and forth between Fe^{2+} and Fe^{3+}, the iron in hemoglobin is Fe^{2+}.

14.37 three

14.39 chemical, osmotic, mechanical

14.41 to make them active

14.43 to supply energy to transport them against the osmotic gradient

14.45 (a) hemoglobin, cytochromes (b) vitamin B_{12} (c) chlorophyll

14.47 NADH enters at the top; $FADH_2$ enters at the second step.

14.49 Reduction of coenzyme Q requires *two* electrons; reduction of a cytochrome requires *one* electron.

14.51

$$\begin{array}{c} O \\ \| \\ HO{-}P{-} \\ | \\ OH \end{array} \quad \text{for energy-transfer}$$

Self-Test

1. False   2. True   3. True   4. False   5. False
6. True   7. False   8. True   9. True   10. True
11. True   12. False   13. False   14. c   15. b
16. c   17. d   18. d   19. b   20. b   21. d
22. d   23. a   24. c   25. d   26. d   27. d

Chapter 15

Practice Exercises

15.1

$C_6H_{12}O_6 + 2NAD^+ + 2ADP + 2P_i \longrightarrow$

$$\begin{array}{cc} O & O \\ \| & \| \\ 2CH_3C{-}C{-}O^- & + 2NADH + 2ATP \end{array}$$

15.2 1 mol ATP

15.3 An acetyl group attached to CoA is easily transferred to some other molecule, such as water, with the release of energy.

15.4 oxaloacetate + acetyl CoA + $3NAD^+$ + FAD + ADP + P_i $\longrightarrow$ oxaloacetate + $2CO_2$ + 3NADH + $FADH_2$ + ATP

15.5 Aerobic glycolysis produces 38 ATP molecules per molecule of glucose; anaerobic glycolysis produces 2 ATP per molecule of glucose.

15.6 phosphorylation of glucose

15.7 Muscle glycogen is converted to glucose 6-phosphate but not to glucose. Glucose 6-phosphate cannot escape from muscle cells. In liver cells glucose 6-phosphate is hydrolyzed to glucose that can leave liver cells to be distributed throughout the body.

15.8 Insulin taken orally would be hydrolyzed in the stomach and small intestine.

Exercises

15.9 $C_6H_{12}O_6 + 6O_2 \longrightarrow 6CO_2 + 6H_2O$

15.11 (1) glycolysis, (2) acetyl CoA, (3) citric acid cycle

15.13 the cytoplasm

15.15 Yes, it involves ATP.

15.17

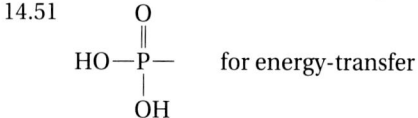

2(1,3-bisphosphoglycerate) →
2ADP   2ATP
2(3-phosphoglycerate)
2(2-phosphoenolpyruvate) → 2(pyruvate)
2ADP   2ATP

15.19 reduced to NADH

15.21 in mitochondria; to transfer an acetyl group

15.23 in mitochondria

15.25 alpha-ketoglutarate + ADP + P_i + NAD^+ $\longrightarrow$ succinate + CO_2 + NADH + ATP

15.27 (a) two molecules of ATP (b) three molecules of ATP

15.29 oxidative phosphorylation stops

15.31 Lactic fermentation produces NAD^+ from NADH when cellular respiration has slowed down or stopped.

15.33 (a) vigorous exercise results in insufficient oxygen for cellular respiration (b) concentration of lactate increases, causing muscle cramps

15.35 synthesis of glucose from noncarbohydrate materials

15.37 a branched polysaccharide; source of glucose formed and stored in muscle and liver cells

15.39 Glycogenolysis is the release of glucose from glycogen. It occurs when glucose is low in the body.

15.41 Hormones act through "second messengers" like cyclic AMP.

15.43 (a) a scare (b) adrenal gland; stimulates breakdown of glycogen in muscle cells

15.45 controls the movement of glucose into cells

15.47 Juvenile-onset: pancreas does not produce insulin. Maturity-onset: delayed insulin release.

15.49 In diabetes mellitus the renal threshold for glucose is exceeded; glucose appears in the urine (glycosuria).

15.51 Cyclic AMP catalyzes the production of phosphorylase *a*.

15.53 To increase the supply of oxygen for oxidative phosphorylation.

15.55 Chronic hyperglycemia is a symptom of diabetes mellitus. It may be treated by injections of insulin.

Self-Test
1. True   2. True   3. False   4. False   5. False
6. False   7. False   8. False   9. False
10. True   11. a   12. c   13. a   14. c   15. b
16. d   17. b   18. d   19. d   20. d   21. c
22. b   23. b   24. b

Chapter 16

Practice Exercises
16.1 (a) six acetyl CoA molecules   (b) five NADH molecules   (c) five $FADH_2$ molecules
16.2 95 ATP
16.3 Body needs energy; used in citric acid cycle and for production of ketone bodies. Body does not need energy; used to synthesize fatty acids or cholesterol.
16.4 (a) 200 g   (b) 450 g

Exercises
16.5

$$CH_2-O-\overset{\overset{\displaystyle O}{\|}}{C}-(CH_2)_{16}CH_3 \quad \text{(stearic)}$$

$$CH-O-\overset{\overset{\displaystyle O}{\|}}{C}-(CH_2)_7CH=CH(CH_2)_7CH_3 \quad \text{(oleic)}$$

$$CH_2-O-\overset{\overset{\displaystyle O}{\|}}{C}-(CH_2)_7CH=CHCH_2CH=CH(CH_2)_4CH_3$$
(linoleic)

16.7 as adipose tissue and working lipids

16.9 the hormone epinephrine
16.11 brain, nervous system, and active skeletal muscle
16.13 mitochondria
16.15 oxidation of the *beta* carbon of a fatty acid to a ketone
16.17 (a) five   (b) four   (c) four
16.19 Both fatty acids and carbohydrates are degraded to acetyl CoA.
16.21 It is a store of triglycerides that produces energy and water when oxidized under aerobic conditions.
16.23 starvation, no glucose in the blood; diabetes, glucose in the blood but none in cells
16.25 When glucose in cells is low the citric acid cycle ceases to function; acetyl CoA therefore produces ketone bodies in the liver.
16.27 (a) ketone bodies in blood greater than 20 mg/100 mL   (b) ketone bodies in the urine (c) acetone expired by the lungs   (d) a form of blood acidosis
16.29 liver cells; acetyl CoA
16.31 buildup of cholesterol in the arteries
16.33 (a) 4   (b) 5   (c) 9   (d) 10   (e) 7   (f) 3
(g) 8   (h) 6   (i) 1   (j) 2
16.35 converted to fatty acyl CoA
16.37 They contribute to ketosis.
16.39 Epinephrine stimulates hydrolysis of triglycerides to fatty acids and glycerol for use by cells in energy production.
16.41 (a) 5, 9   (b) 2, 8   (c) 7, 13   (d) 3, 11
(e) 1, 10   (f) 3, 5 or 6; 9, 11   (g) 4, 14

Self-Test
1. False   2. True   3. True   4. False   5. False
6. True   7. True   8. False   9. False   10. False
11. d   12. d   13. c   14. d   15. c   16. b
17. b   18. c   19. c   20. b   21. d   22. b
23. b

Chapter 17

Practice Exercises
17.1 Cathepsins are peptidases, and would destroy all proteins found in the cytoplasm if not confined.
17.2

$$HO-\bigcirc-CH_2\overset{\overset{\displaystyle H}{|}}{\underset{\underset{\displaystyle NH_2}{|}}{C}}-CO_2H + HO-\overset{\overset{\displaystyle O}{\|}}{C}-CH_2CH_2-\overset{\overset{\displaystyle O}{\|}}{C}-\overset{\overset{\displaystyle O}{\|}}{C}-OH \xrightarrow[\text{Phosphate}]{\text{Pyridoxal}}$$

Tyrosine α-Ketoglutaric acid

$$HO-\bigcirc-CH_2\overset{\overset{\displaystyle O}{\|}}{C}-CO_2H + HO-\overset{\overset{\displaystyle O}{\|}}{C}-CH_2CH_2-\overset{\underset{\underset{\displaystyle NH_2}{|}}{}}{CH}-\overset{\overset{\displaystyle O}{\|}}{C}-OH$$

Glutamic acid

17.3   initially transferred to the cofactor pyridoxal phosphate

17.4   ammonia; converted to urea, excreted in urine

17.5   Loss of amino group from amino acid gives alpha-keto acid, which enters glucose metabolism (glucogenic) or is converted to acetyl CoA, as are fatty acids (ketogenic).

17.6   (a) 3   (b) 5   (c) 4   (d) 6   (e) 2   (f) 1

Exercises

17.7   nitrogen-fixing bacteria, lightning discharge, industrial processes

17.9   Protein molecules are continually being degraded and synthesized.

17.11 all free amino acids in cells and tissues of the body

17.13 transfer of an amino group from one molecule to another

17.15

$$(CH_3)_2CH{-}\overset{\overset{\displaystyle H}{|}}{\underset{\underset{\displaystyle NH_2}{|}}{C}}{-}CO_2H \; + \; \text{alpha-ketoglutarate} \longrightarrow$$

valine

$$(CH_3)_2CH{-}\overset{\overset{\displaystyle O}{\|}}{C}{-}CO_2H \; + \; \text{glutamic acid}$$

an alpha-keto acid

17.17 converts ammonia to urea

17.19 difference between nitrogen intake and nitrogen excretion; a positive nitrogen balance in children means amino acids are being used for growth

17.21 Glucogenic amino acids form intermediates of glucose metabolism. Ketogenic amino acids are degraded to acetyl CoA.

17.23 Essential amino acids are essential in the diet. Nonessential amino acids are made by the body.

17.25 converted to tyrosine

$$\overset{\overset{\displaystyle H}{|}}{\underset{\underset{\displaystyle NH_2}{|}}{C}}$$
—CH$_2$C—CO$_2$H ⟶

Phenylalanine

HO—⟨benzene ring⟩—CH$_2$$\overset{\overset{\displaystyle H}{|}}{\underset{\underset{\displaystyle NH_2}{|}}{C}}$—CO$_2$H

Tyrosine

17.27 120 days

17.29 partially degraded heme

17.31 All of the hydrogens attached to nitrogen in uric acid can be lost as hydrogen ions.

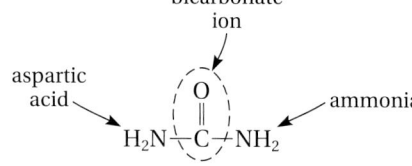

Uric acid

17.33 four; two to make carbamoyl phosphate, two to make arginosuccinic acid

17.35 urea, NH$_2$CONH$_2$

17.37 purine ring system

17.39

aspartic acid ⟶ bicarbonate ion

$$H_2N{\dashv}\overset{\overset{\displaystyle O}{\|}}{C}{\vdash}NH_2$$ ⟵ ammonia

Self-Test

1. True   2. True   3. True   4. False   5. False
6. True   7. False   8. False   9. False   10. True
11. d   12. d   13. d   14. a   15. b   16. a
17. d   18. b   19. d   20. c   21. c   22. a
23. a   24. a   25. a

Photograph Credits

Page 1 ©CNRI/Science Photo Library/Photo Researchers, Page 2 ©Kathy Merrifield/Photo Researchers, Takahara/ Photo Researchers, Page 15 ©Tino Hammid/ Gemological Institute, Page 16 Courtesy, Southern Illinois University Photographic Services, Page 18 ©Scott Camazine/Photo Researchers, Page 27 ©Richard Megna/ Fundamental Photographs, Page 28 Courtesy, BRK Electronics, Page 55 ©Richard Megna/Fundamental Photographs, Page 57 ©Jeffrey Mark Dunn/Stock Boston, Page 58 (a) ©Ted Clutter/Photo Researchers, Page 58 (b) ©Charlie Ott/Photo Researchers, Page 58 (c) ©Sean Brady/24th Street Group, A Division of American Color, Page 59 ©Rob Crandell/Stock Boston, Page 66 ©SIU/ Photo Researchers, Page 67 ©Sean Brady/24th Street Group, A Division of American Color, Page 73 (a) ©Custom Medical Stock Photo, Page 73 (b) Goddard Space Center, Page 75 ©Sean Brady/24th Street Group, A Division of American Color, Page 77 Courtesy, Exxon Company, Page 79 ©Dr. E. R. Degginger, Page 83 ©Phil Degginger, Page 89 ©Bettmann, Page 103 ©Richard Megna/Fundamental Photographs, Page 104 ©Phil Degginger, Page 111 ©Cyril C. Laubscher/Bruce Coleman, Page 116 ©Biophoto Associates/Science Source/Photo Researchers, Page 119 ©Bill Beatty/Visuals Unlimited, Page 124 ©Omikron/Photo Researchers, Page 132 ©Sal Maimone/FPG, Page 133 ©Kenneth Greer/Visuals Unlimited, Page 137 ©Custom Medical Stock Photo, Page 146 ©Nalco Chemical Company, Page 155 ©Biophoto Associates/Science Source/Photo Researchers, Page 156 ©Phil Degginger, Page 157 Photograph courtesy of Meadox Medicals, Inc., Page 171 ©Richard Megna/ Fundamental Photographs, Page 172 ©Jim Pickerell/ The Image Works, Page 183 Reproduced with permission of Astra USA, Inc., 50 Otis Street, Westborough, MA 01581-4500, Page 186 ©Bob Daemmrich/Stock Boston, Page 187 ©George Mars Cassidy/The Picture Cube, Page 194(a) ©V. I. Thaytl/Dinodia Picture Agency, Page 194 (b) ©Kjell B. Sandved/Photo Researchers, Page 203 ©Scott Camazine/Photo Researchers, Page 204 ©Charles D. Winters/Photo Researchers, Page 222 ©John Curtis/The Stock Market, Page 225 ©Toni Michaels, Page 227 ©Larry Mulvehill/Photo Researchers, Page 234 ©Funk/ Schoenberger/Grant Heilman Photography, Page 235 ©Eric Neurath/Stock Boston, Page 238 ©Mark Antman/ The Image Works, Page 263 ©Ron Goor/Bruce Coleman, Page 264 ©Brian Yarvin/The Image Works, Page 273 ©Phil Degginger, Page 278 ©S. J. Krasemann/Photo Researchers, Page 281 Courtesy, Dr. John Kendrew, Page 293 ©John Bennett Dobbins/Photo Researchers, Page 294 ©C. C. Duncan/Medical Images, Inc., Page 591 ©Eric Neurath/Stock Boston, Page 601 ©Joyce Photographics/ Science Source/Photo Researchers, Page 315 ©Richard R. Hansen/Photo Researchers, Page 319 ©Mark Burnett/ Stock Boston, Page 320 Courtesy, National Institutes of Health, Page 349 March of Dimes Birth Defects Foundation, Page 335 ©SIU/Visuals Unlimited, Page 346 (a) and (b) ©Astrid & Hanns-Frieder Michler/Science Photo Library/Photo Researchers, Page 355 ©Michael Keller/FPG, Page 356 ©Steven Hansen/Stock Boston, Page 364 ©Joseph Nettis/Photo Researchers, Page 368 ©Nancy D. McKenna/Photo Researchers, Page 370 (a) ©Biophoto Associates/Science Source/Photo Researchers, Page 370 (b) Courtesy, Dr. Mark Moldawer, Page 378 (a) and (b) ©SPL/Custom Medical Stock Photo, Page 387 ©William R Sallaz/Duomo, Page 388 ©Martin/Custom Medical Stock Photo, Page 390 (a), (b), and (c) ©Organogenesis, Inc., Page 390 (bottom right) ©SIU/Peter Arnold, Page 397 ©Phil Degginger, Page 408 ©Custom Medical Stock Photo, Page 414 ©Michael Kevin Daly/The Stock Market, Page 415 ©Stacy Pick/Stock Boston, Page 417 ©Dr. J. Burgess/ Science Photo Library/Photo Researchers, Page 418 ©John Griffin/Medichrome, Page 427 ©Photo Researchers, Page 430 ©UPI/Bettmann Newsphotos, Page 441 ©Mark Antman/The Image Works, Page 442 ©Bob Daemmrich/ Stock Boston, Page 461 ©David M. Phillips, The Population Council/Photo Researchers, Page 466 ©David M. Phillips/Visuals Unlimited, Page 474 ©M. P. Kahl/Bruce Coleman, Page 475 ©Owen Franken/Stock Boston, Page 476 ©Prof. P. Motta/Dept. of Anatomy/University "La Sapienza", Rome/Science Photo Library/Photo/Researchers, Page 480 ©Phil Degginger, Page 492 ©Custom Medical Stock Photo, Page 492 ©Custom Medical Stock Photo, Page 493 ©Phil Degginger, Page 497 ©Norm Thomas/Photo Researchers, Page 498 ©Richard J. Green/Photo Researchers, Page 504 ©Biophoto Associates/Science Source/Photo Researchers, Page 512 ©Grace Moore/Medichrome, Page 513 ©Thomas E. Tottleben, Page 514 ©Ron Sutherland/Science Photo Library/Photo Researchers. Note: Photographs by Sean Brady were arranged with the help of Professor Doug Sawyer.

Index

The page on which a term is defined is in **boldface** type. The letters t, f, c, and p following page numbers indicate table, figure, A Closer Look, and Case in Point respectively.

Some Functional Groups Encountered in Organic and Biological Chemistry

Functional group	Compound type	Functional group	Compound type
R — X (X = F, Cl, Br, I)	halocarbon	R — S — R	sulfide
R — OH	alcohol	R — S — S — R	disulfide
R — O — R	ether	R — C(=O) — S — R	thioester
R — C(=O) — H	aldehyde		
R — C(=O) — R	ketone	R — O — P(=O)(OH) — OH	phosphate ester
R — C(=O) — OH	carboxylic acid		
R — C(=O) — O — R	ester	R — O — P(=O)(OH) — O — P(=O)(OH) — OH	diphosphate ester
R — C(=O) — O — C(=O) — R	anhydride	R — O — P(=O)(OH) — O — P(=O)(OH) — O — P(=O)(OH) — OH	triphosphate ester
R — NH$_2$, R — N(H) — R, R — N(R) — R	amine		
R — C(=O) — N(H) — R	amide	R — O — P(=O)(OH) — O — R	phosphodiester
R — SH	thiol		
R — S(=O)(=O) — OH	sulfonic acid		

SI Units and Conversion Factors*

Length
SI unit: meter (m)

1 meter	= 1.0936 yards
1 centimeter	= 0.39370 inch
1 inch	= 2.54 centimeters (exactly)
1 kilometer	= 0.62137 mile
1 mile	= 5280 feet
	= 1.6093 kilometers

Volume
SI unit: cubic meter (m³)

1 liter	= 10^{-3} m^3
	= 1 dm^3
	= 1.0567 quarts
1 gallon	= 4 quarts
	= 8 pints
	= 3.7854 liters
1 quart	= 32 fluid ounces
	= 0.94633 liter

Mass
SI unit: kilogram (kg)

1 kilogram	= 1000 grams
	= 2.2046 pounds
1 pound	= 453.59 grams
	= 0.45359 kilogram
	= 16 ounces
1 atomic mass unit	= 1.66056×10^{-27} kilograms

Pressure
SI unit: pascal (Pa)

1 atmosphere	= 101.325 kilopascals
	= 760 torr (mm Hg)
	= 14.70 pounds per square inch

Energy
SI unit: joule (J)

1 joule	= 0.23901 calorie
1 calorie	= 4.184 joules

*Note: Conversion factors are given with more significant figures than those in the text.